Electric Motor Handbook

Electric Motor Handbook

Editor: E. H. Werninck

McGRAW-HILL Book Company (UK) Limited

London · New York · St Louis · San Francisco · Auckland · Beirut
Bogotá · Düsseldorf · Johannesburg · Lisbon · Lucerne · Madrid
Mexico · Montreal · New Delhi · Panama · Paris · San Juan
São Paulo · Singapore · Sydney · Tokyo · Toronto

Published by
McGRAW-HILL Book Company (UK) Limited
MAIDENHEAD · BERKSHIRE · ENGLAND

British Library Cataloguing in Publication Data

Electric motor handbook.
1. Electric motors – Design and construction
2. Electric motors – Maintenance and repair
I. Werninck, E H
621.46'2 TK2511 77–30474

ISBN 0–07–084488–7

2000/379
HERTFORDSHIRE
LIBRARY SERVICE
621·462
9089622

Copyright © 1978 McGraw-Hill Book Company (UK) Limited. All rights reserved. No part of this publication may be reproduced, stored in a retrieval system, or transmitted, in any form or by any means, electronic, mechanical, photocopying, recording, or otherwise, without the prior permission of McGraw-Hill Book Company (UK) Limited.

12345 JWA 80798

PRINTED AND BOUND IN GREAT BRITAIN

Contents

Preface

Men have said that they climb mountains, or cross seas and continents because they are there, presenting a challenge. The motivation to edit a Handbook of the nature of this one is that it is *not* there. However, editing a handbook cannot be said to require the same personal attributes and the only courage required is to publish and face the consequences.

In the very early days of my career, the then current edition of Knowlton's *Standard Handbook for Electrical Engineers* was of considerable help to me. General reference books, though uniquely useful particularly to the young and still inexperienced electrical engineer, cannot treat specific subjects in depth, or from all aspects, and so later it was a more specialized Handbook, *Truxall's Control Engineer's Handbook*, which gave me an insight into control engineering and the special requirements for electric motors in this field. It also helped me to understand better the designer's need to express performance mathematically, and the usefulness of transfer functions. These two examples from my personal experience illustrate the evolution of technical handbooks in the last three decades and the development of the engineering sciences which resulted in an information explosion and successively more specialized generations of books.

Thus, the second generation of handbooks dealt with specific subjects, such as electronics, process-control and instrumentation, radio, maintenance engineering, and so on. Another specialization which developed was aimed at engineers working in specific branches of their profession. This third generation was more selective and catered for industrial electronic control, industrial power systems, radio engineering, as well as many other subjects within the more general generic classification.

This *Handbook* I suggest is a member of a fourth generation, in that it deals with a specific type of electrical machine and is directed at those engineers who have to apply, use, or maintain it. Aspects of design are only covered to the extent that they may give a better understanding of characteristics, particularly under abnormal conditions.

Optimization of electric motor applications requires not only the matching of load, motor, and control gear, but such considerations as capital and operational costs, the need to increase productivity as well as the interaction with the environment. Thus it was considered essential to include gearing, couplings, brakes, and clutches, as well as a short section on noise. Many features of electrical equipment are easily changed in theory to meet particular system requirements exactly. In practice, economic production of such equipment demands considerable standardization which, however, still leaves many options. The special design approach can be thought of as 'synthesis', whereas the examination of the considerable number of systems which can be made up from standard motors, transmission, and control gear require analysis and decision-making skills.

Many engineers still underestimate the wealth of information and expert assistance which can be theirs for the asking, and consider analysis and evaluation of the alternatives offered a far inferior task to designing and developing at great cost their own special solution. Let me hasten to add that this generalization is, like many others, confirmed by many excellent exceptions.

In the early days of handbooks they were often the only sources of technical information for engineers remote from libraries or immediate advice. Today, it is often more the wealth of information and conflicting advice which can make decisions difficult. So many developments and solutions are due to particular circumstances or even industries with special conditions. It would be hopeless to try to cover the millions of permutations and combinations of electric motor types, sizes, and applications, but it is at this stage that I wish to thank all my contributors for their efforts. If they feel that my editing has at times been heavy, I ask their indulgence on the grounds that I have made an earnest attempt to bring together and summarize as much relevant information as possible without duplication.

The complexities of practical engineering problems are such that relatively few decisions can be supported by calculations based on fundamental principles. Similar applications are generally studied, scaled, modified and generally used as criteria for such performance requirements as reliability, quiet running and transient overloads. An attempt to define a 'good engineer' could easily take up the next few hundred pages and unleash fierce arguments. It can be stated with conviction and little fear of contradiction that an innovatory approach tempered by experience is the most likely to produce a satisfactory solution.

Every effort has been made to acknowledge the sources from which I have drawn and where diagrams have been modified, or scales changed, the object has been to unify and standardize the information. Here it is appropriate also to thank my wife for so patiently retyping much of the manuscript time and time again.

Acknowledgements

The editor, the contributors and the publishers are indebted to a number of individuals and establishments for help in the preparation of this handbook. Certain illustrations are acknowledged individually, and valuable information has been made available by Demag of Hamburg, Eberhard Bauer of Esslingen-Neckar, Elektra-Faurndau, and *Philips Technical Review*. The tables from British Standards are included by permission of BSI, 2 Park Street, London W1A 2BS from whom complete copies of the Standards can be obtained.

Contributors

N. Binks. Chief Engineer, Newman Electric Motors (Manufacturing) Ltd.
Co-author of Chapter 20.
G. Biscoe. Rotating Machines Division, Electrical Research Association.
Author of Chapters 7, 18.
J. C. H. Bone. Laurence, Scott & Electromotors Ltd.
Author of Chapter 11.
M. Bradford. Head of Electrical Machines Department, Electrical Research Association.
Author of Chapter 5.
F. E. Butcher. Deceased.
Co-author of Chapter 19.
P. B. Greenwood. Director, Brook Motors Ltd.
Author of Chapter 9.
H. Greiner. Eberhard Bauer, Esslingen–Neckar.
Author of Chapter 16.
E. C. Haverley. Maintenance Engineer, Battersea Power Station.
Co-author of Chapter 19.
D. Knights. Power Engineering Division, Electrical Research Association.
Author of Chapters 8, 10.
Professor E. R. Laithwaite. Imperial College of Science and Technology.
Author of Chapter 6.
D. Legg. Chief Design Engineer, D. C. Machines, Brush Electrical Machines Ltd.
Author of Chapter 3.
R. F. Mathieson. Consultant.
Author of Chapter 14.
R. G. Oakes. Technical writer.
Author of Chapter 17.
Professor G. H. Rawcliffe.
Co-author of Chapter 4.
D. D. Stephen. G.E.C. Electrical Projects Ltd.
Author of Chapter 15. Co-author of Chapters 2, 4.
K. H. Williamson. Head of Product Development and Circuit Design, Electrical Research Association.
Author of Chapter 12.
E. H. Werninck. Walter Jones and Co. (Engineers) Ltd.
Author of Chapters 1, 13. Co-author of Chapters 2, 4, 14, 20, translator of Chapter 16. Editor.

1 Units, Mathematics, and Formulae

1.1 Systems of units

The introduction of a metric system for practical measurements is said to have been sponsored by the French politician, Talleyrand, at the beginning of the nineteenth century on the advice of the scientists of the day. The system took its name from the proposed unit of length, the 'metre', the magnitude of which was to be one ten millionth of the distance between the North Pole and the equator measured along a meridian passing through Paris.

In 1873 the British Association for the Advancement of Science adopted the centimetre and the gramme as the basic units of measurement. Engineers used and continued to use units based on the pound and the foot, that is the FPS system, while scientists worked in the CGS system. In 1875 the Conference Générale des Poids et Mesures (CGPM) was established as the International Authority on the metric system, and with the Bureaux International des Poids et Mesures (BIPM) set up the basic standards at Sèvres. The standard metre was defined by the distance between two marks on a standard bar and the kilogramme in terms of a prototype mass. These units were of a more convenient practical magnitude than the centimetre and the gramme, and were adopted as two of the fundamental units of yet another system, namely the MKS system of units. Other national laboratories such as the National Physical Laboratory in London (NPL) and the Physikalische Technische Bundesanstalt in Berlin which had collaborated in the setting up of standards became custodians of their own secondary standards.

From the basic dimensions of mass, length, and time other units were derived and some of these required the use of the 'constants' such as gravitational acceleration (g), permeability (μ_0), and permittivity of free space (ε_0) to link them into a compatible, practical system. To take account of these complications and 'rationalize' units the International Electrotechnical Commission (IEC) in 1950 accepted a suggestion made some 50 years earlier by the Italian Professor Giorgi. The MKSA system designated the ampere as a fourth basic unit and was named in his honour the 'Giorgi' system. Units for this system were defined as 'international' units. To simplify the system of practical units even more the SI Units now include absolute temperature (kelvin), luminous intensity (candela), amount of substance (mole).

Before these now universally accepted units were adopted, science and engineering students had to familiarize themselves with electrostatic, electromagnetic, and fundamental practical units. Some of these units, such as the slug and the poundal, which were introduced for mechanical calculations in the FPS system, were occasionally used for mass and force respectively. This confusion has now disappeared for at its tenth meeting in 1954 the CGPM adopted a rationalized and coherent system of metric units which, in 1960, was given the title 'Système International d'Unités' and is now universally designated as 'SI'.

Advances in science have led to the metre being defined more fundamentally, accurately, and conveniently as 1 650 763·73 times the wavelength in vacuum of the orange light of krypton 86 on an interference comparator. The convenience here is one reserved for scientists in the national laboratories who, with the necessary equipment can calibrate and certify sub- or secondary standards and measuring instruments. When linear dimensions must be maintained with tolerances of only a few hundredths of a millimetre then temperature and surface finish become significant and must be specified to ensure agreement between supplier and buyer.

The number of basic quantities on which the SI system has been based has recently been increased and though not all are directly applicable to electrical machines they have been included here.

Table 1.1 Basic SI units

length	metre	m
mass	kilogramme	kg
time	second	s
current	ampere	A
light	candela	cd
temperature	kelvin	K
substance	mole	mol

The *metre* as a unit of length has been defined above.
The *kilogramme* will continue to be defined by the mass of the platinum-iridium prototype in Sevres until scientists find a satisfactory more fundamental and readily reproducible measure of mass.

Table 1.2 Derived SI units

Physical quantity	Symbol	Unit	Definition	Definition in terms of basic units
Frequency	f	hertz Hz	s^{-1}	s^{-1}
Force	F	newton N	$kgms^{-2}$	$kgms^{-2}$
Work or Energy	W	joule J	Nm	kgm^2s^{-2}
Power	P	watt W	Js^{-1}	$kgm^2 s^{-3}$
Pressure	P	pascal Pa	Nm^{-2}	$kgm^{-1} s^{-2}$
Electric Charge	Q	coulomb C	As	As
Potential difference	V	volt V	JC^{-1}	$kgm^2 s^{-3} A^{-1}$
Capacitance	C	farad F	F	$A^2 s^4 kg^{-1} m^{-2}$
Resistance	R	ohm Ω	VA^{-1}	$kgm^2 s^{-3} A^{-2}$
Conductance	G	siemens S		$kg^{-1} m^{-2} s^3 A^2$
Flux Density	B	tesla T	N m	$kgs^{-2} A^{-1}$
Magnetic flux	Φ	weber Wb	Tm^2	kgm^2
Inductance	L or M	henry H	Wb/A = Vs/A	kgm

The *second* is the duration of 9 192 631 770 periods of a specified radiation from an atom of caesium-133.
The *ampere* is that constant electric current which, maintained in two infinitely long conductors 1 metre apart causes each to exert a force of 2×10^{-7} N per metre length on the other.
The *kelvin* is the 1/273·16 part of the thermodynamic temperature of the triple point of water.
The *candela* is the unit of luminous intensity.
The *mole* is the amount of substance of a system which contains as many elementary units as there are carbon atoms in 12×10^{-3} kg of carbon-12. Thus a mole contains $6{\cdot}022169 \times 10^{23}$ specified elementary units, such as atoms, molecules, ions, electrons, etc. From this follows that a mole of electrons has a charge of $-9{\cdot}6 \times 10^{4}$ C (coulomb), a quantity also known as the Faraday constant.

From these basic units the other SI units are derived as shown in table 1.2.

1.2 Metric Multiples and Submultiples

Table 1.3

T	Tera	10^{12}	p	Pico	10^{-12}
G	Giga	10^{9}	n	Nano	10^{-9}
M	Mega	10^{6}	μ	Micro	10^{-6}
k	Kilo	10^{3}	m	Milli	10^{-3}
h	Hecto	10^{2}	c	Centi	10^{-2}
da	Deca	10	d	Deci	10^{-1}

NOTE: 10^{18} = Trillion, 10^{12} = Billion (except in USA, France, Spain, and Italy where 1 billion is 10^{9} and 10^{12} is then called a trillion).

Double prefixes, even if more easily understood, may not be used. For example:

the tonne (t) = 1000 kg = 10^{6} g or 1 Mg and *not* 1 kkg.
Millimicro is to be replaced by *nano*-, and
Micro-micro is *pico*-.

When using decimal multiples of units in formulae care must be taken to ensure that they remain coherent; thus Ohm's law for example is coherent for:

a) volts, amperes, and ohms
b) milli-volts (mV), milli-amperes (mA), and ohms (Ω)
c) kilo-volts (kV), milli-amperes (mA), and mega-ohms (MΩ)

In practice the only safe method of ensuring the correct order of magnitude is to always work in the basic units for which the SI formulae apply.

1.3 Units of Length

Table 1.4

	m	cm	mm	µm	yd	ft	in	µ in
1 m	1	100	1000	1×10^{6}	1·0936	3·2808	39·37	—
1 cm	0·01	1	10	$1{\cdot}10^{4}$	—	—	0·3937	—
1 mm	0·001	0·1	1	1000	—	—	0·0394	$3{\cdot}937\times10^{-4}$
1 µm	1×10^{-6}	1×10^{-4}	0·001	1	—	—	—	39·37
1 yd	0·9144	91·44	914·4	—	1	3	36	—
1 ft	0·3048	30·48	304·8	—	$\frac{1}{3}$	1	12	—
1 in	0·0254	2·54	25·4	—	$\frac{1}{36}$	$\frac{1}{12}$	1	1×10^{6}
1 µin	$2{\cdot}54\times10^{-6}$	$2{\cdot}54\times10^{-4}$	$2{\cdot}54\times10^{-3}$	0·0254	—	—	1×10^{-6}	1

1 dm = 10 cm = 0·1 m
1 mil (USA) = 1 thou (UK) = 0·001 in = 0·0254 mm
1 statute mile = 1760 yards = 5280 ft = 1·6093 km
1 (nautical) mile = 6080 ft = 1·8532 km
1 fathom = 6 ft = 1·828 m
1 µm 'micron' is strictly 1 micrometre

1.4 Units of Area

Table 1.5

	m^2	cm^2	mm^2	sq. yd	sq. ft	sq. in
1 m^2	1	1×10^{4}	1×10^{6}	1·196	10·764	—
1 cm^2	1×10^{-4}	1	100	—	—	0·1550
1 mm^2	1×10^{-6}	0·01	1	—	—	0·0016
1 sq. yd	0·8136	8361	$8{\cdot}36\times10^{5}$	1	9	—
1 sq. ft	0·0929	929·03	$9{\cdot}29\times10^{4}$	$\frac{1}{9}$	1	144
1 sq. in	$6{\cdot}45\times10^{-4}$	6·4516	645·16		$\frac{1}{144}$	1

1 circular mil = area of circle 0·001 in diameter $= \pi/4$ sq. mil $= 5{\cdot}076.10^{-4}$ mm^2
1 ha (hectare) = 100 ar = 10 000 m^2 = 2·471 acres
1 acre = 0·4047 ha
1 sq. mile = 640 acres = 2·590 km^2 = 259 ha

1.5 Units of Volume

Table 1.6

	m^3	dm^3 (l)	cm^3 (ml)	imp. gal	imp. pt	cu. yd	cu. ft	cu. in
1 m^3	1	1000	$1{\cdot}10^{6}$	264·2	—	1·3079	35·31	$6{\cdot}1\times10^{4}$
1 dm^3 (l)	0·001	1	0·2642	0·2642	2·1136	—	—	61·02
1 cm^3 (ml)	1×10^{-6}	0·001	1	—	—	—	—	0·061
1 gal imp	$4{\cdot}55\times10^{-3}$	4·546	4546	1	8	$5{\cdot}91.10^{-3}$	0·165	—
1 pint imp	—	0·5682	—	—	1	—	—	34·67
1 cu. yd	0·7645	764·55	$7{\cdot}65\times10^{5}$	168·2	—	1	27	46656
1 cu. ft	0·0283	28·317	$2{\cdot}832\times10^{4}$	6·232	49·856	$\frac{1}{27}$	1	1728
1 cu. in	—	0·0164	16·3871	—	—	—	—	1

1 US gallon = 4 liquid quarts = 8 liquid pints = 3·7856 litres
1 US fluid oz = 8 US fl. drams = 29·57 ml, 1 minim = 0·0616 ml (cm^3)
1 imp. fluid oz = 8 imp. fl. drachms = 28·41 ml, 1 imp. minim = 0·0592 cm^3
1 barrel (petrochem./oil) = 42 gallons = 158·9952 litres
1 bushel (UK) = 4 pecks = 8 gallons = 36·37 litres

1.6 Units of Mass, Weight and Force

The unit of mass in the SI system is the kilogramme (kg) and Newton's law connects this with the unit of force as follows:

> a force of 1 newton acting on a mass of 1 kilogramme produces an acceleration of 1 m/s^2

$$F(\text{N}) = \text{mass (kg)} \times \text{acceleration } (\text{m/s}^2)$$

Weight is the special force a body exerts on its support due to gravity. The earth's gravitational field varies from about 9·832 m/s^2 at the poles to 9·780 m/s^2 at the equator and for all but the most exact calculations it is sufficient to substitute 9·81 m/s^2 for gravitational acceleration (g). The confusion between mass and weight which has been mentioned in section 1.1 is recognized, and laws are being passed to make such technical units as the kilopond disappear, at least from legal documents, before the 1980s. The kilogramme force (kgf), or as it is known in some European countries the kilopond (kp), is defined as the force which gives a mass of one kilogramme an acceleration of 9·8066 metres per second per second, but as it does not fit into the coherent SI system must eventually disappear.

The measurement of weight must eventually be only in newtons and may either be made by a spring balance, or by comparison on some form of beam balance. The spring balance, calibrated in newtons, will at balance register the 'apparent' weight not corrected for any variation of the earth's gravitational field, wherever it is used. Beam balances compare masses by comparing their weight, and the true weight in SI units must be calculated from Newton's equation. The ballistic balance which is sometimes used for the measurement of impact compares inertial masses using the earth's gravitational pull.

Table 1.7

	N	kgf (kp)	dynes	(UK) tonf	lbf	poundal
1 N	1	0·1020	1×10^{5}	$1{\cdot}004\times10^{-4}$	0·2248	7·2464
1 kgf (kp)	9·807	1	$9{\cdot}81\times10^{5}$	$9{\cdot}84\times10^{-4}$	2·205	71·0672
1 dyne	1×10^{-5}	$1{\cdot}02\times10^{-6}$	1	$1{\cdot}10^{-9}$	$2{\cdot}25\times10^{-6}$	—
1 (UK) tonf	9964	1016	9964×10^{5}	1	2240	—
1 lbf	4·448	0·4536	4448×10^{2}	$4{\cdot}46\times10^{-4}$	1	32·23
1 poundal	0·138	0·0141	$1{\cdot}38\times10^{4}$	—	—	1

Imperial system

1 ozf = 28·35 gf (p), 16 ozf = 1 lbf = 453·6 gf (p),
112 lbf = 1 cwt = 50·8024 kgf (kp), 20 cwt = 1 (UK) ton
1 stone = 14 lbf = 6·35 kgf (kp)
Newton's law: Force (lbf) = mass (slugs) × acceleration (ft/s^2) therefore 1 slug = 32·2 lbf = 14·594 kgf (kp)
gravitational force (average value) 32·2 ft/s^2
1 long (UK) ton = 1·016 t
1 short (US) ton = 0·907 t = 2000 lb

1.7 Units of Torque

1 Nm = 0·10197 kgf m = 0·73756 lbf ft = 8·8507 lbf in
1 kgf (kp) m = $9{\cdot}81 \times 10^{-7}$ Nm = 0·0139 ozf in
1 ozf in = 72·01 cpm (gf cm) = $7{\cdot}062 \times 10^{-3}$ Nm
1 lbf in = 0·1129 Nm = 1152 gf cm (cpm)
1 lbf ft = 1·356 Nm = 0·1383 kgf (kpm)

1.8 Units of Power, Work and Energy

Power, which in the SI system is Joules/s or watts can be thought of as the capability of transferring energy.

In rotating electrical machines Power (W) = 2π(rev/s)T(N/m) as long as the torque T and the speed n remain constant.

Similarly for linear motion Power = force × speed

Table 1.8

	kW	H.P.	PS	kpm/s	ftlbf/s	kcal/s	Btu/s
1 kW =		1·3410	1·3596	102·0	737·6	0·2388	0·9478
1 HP =	0·7457	1	1·014	76·04	550·0	0·1781	0·7068
1 PS =	0·7355	0·9863	1	75·00	542·5	0·1757	0·6971
1 kp/s =	$9{\cdot}81 \times 10^{-3}$	$1{\cdot}315 \times 10^{-2}$	$1{\cdot}333 \times 10^{-2}$	1	7·233	$2{\cdot}342 \times 10^{-3}$	$9{\cdot}295 \times 10^{-3}$
1 ftlbf/s =	$1{\cdot}36 \times 10^{-3}$	$1{\cdot}360 \times 10^{-3}$	$1{\cdot}84 \times 10^{-3}$	0·1383	1	$0{\cdot}324 \times 10^{-3}$	$1{\cdot}185 \times 10^{-3}$
1 kcal/s =	4·1868	5·615	5·692	426·9	3088	1	3·968
1 Btu/s =	1·055	1·415	1·435	107·6	778·2	0·2520	1

Work done, or the total power over a specific distance moved in the direction of the applied force, or interval of time is a useful concept for the solution of certain drive problems. The SI unit of work is the joule, i.e., a force of 1 N having moved its point of application through 1 m.

Energy, or the capacity of doing work has the same units as work and not only appears in many forms but is subject to transformation from one form to another. Engineers considering electric motors are particularly concerned with the transformation of electrical into mechanical energy, and in this process most losses appear in the form of heat energy. In some applications the stored mechanical-kinetic energy will be converted back to electrical energy (regenerative braking).

1.9 Units of Temperature

The SI unit of thermodynamic temperature is the kelvin, which, as a measure of temperature differences, is equal in magnitude to the degree Celsius (Centigrade). The difference is in the zero point, which, for the kelvin, is based on the absolute or lowest theoretically obtainable temperature.

Some typical calibration points are as follows:

Table 1.9 Calibration points for thermometers

Calibration points	Kelvin	Celsius	Fahrenheit
Absolute zero	0	−273·15	−459·67
Liquid oxygen boils	90·18	−182·97	
Ice point	273·15	0	32
Steam point	373·15	100	212
Freezing point of zinc	692·7	419·55	
Freezing point of silver	1234·0	960·85	
Freezing point of gold	1336·2	1063·05	

Conversion: $t_C = \frac{5}{9}(t_F - 32)$
$t_F = 1{\cdot}8 t_C + 32$

1 calorie = heat energy required to raise 1 gramme of water from 14·5 to 15·5°C = 4·187 J

Thermal constants

$$1\ \text{Btu/sq in} = 391\ \text{kcal/m}^2 = 1{\cdot}64\ \text{MJ/m}^2$$

$$1\ \text{Btu/sq ft} = 2{\cdot}71\ \text{kcal/m}^2 = 11{\cdot}35\ \text{kJ/m}^2$$

$$1\ \text{Btu/cu ft} = 8{\cdot}90\ \text{kcal/m}^3 = 37{\cdot}26\ \text{kJ/m}^2$$

$$1\ \text{Btu/lb} = 0{\cdot}556\ \text{kcal/kg} = 2{\cdot}2328\ \text{kJ/kg}$$

1.10 Units of Pressure

When specifying pressure it must be made clear if it is absolute, or above or below atmospheric, and if the latter the value of atmospheric pressure at the time of measurement must also be specified.

Table 1.10

	N/m^2	m bar	kgf/m^2 = mm WG	kp/cm^2 = 1 at	Torr = mm Hg	atm	lbf/sq.ft	lbf/sq.in (p.s.i.)
1 N/m^2 (Pascal)	1	0·01	1·02	$1{\cdot}02 \times 10^{-5}$	0·0075	9×10^{-5}	0·0289	—
1 m bar	100	1	10·2	—	0·7501	—	2·089	0·0145
1 kgf/m^2 = mm WG	9·807	—	1	0·0001	—	—	0·2048	—
1 kgf/cm^2 = 1 at.	98·067	980·07	10 000	1	735·6	0·9678	2048	14·22
1 Torr = mm Hg	133·32	1·333	13·6	0·00136	1	—	2·785	0·01934
1 atm	101 325	1013	10 332	1·033	760	1	2116	14·7
1 lb/sq. ft (p.s.f.)	47·88	0·4788	4·882	—	0·3591	—	1	—
1 lb/sq. in (p.s.i.)	6894·8	68·95	703·1	0·07031	51·71	0·068	144	—

1.11 Units of Viscosity

Viscosity is defined as the internal friction between different layers of a fluid moving with different velocities.

Dynamic viscosity is the force in dynes necessary to move a layer of liquid 1 cm high and area 1 cm^2 with a speed of 1 cm/s.

Unit 1 P (Poise) = 100 cP (centipoises).

Kinematic viscosity = Dynamic viscosity/density (g/cm^2)

Unit: Stokes = 100 cST (centistokes)

Most common liquids

Frictional force, $F = nA$ newton

A = area in m^2

n = coefficient of viscosity (frictional force/unit area of a liquid when it is in a region of unit velocity gradient (i.e. 1 m/s),

i.e. Ns/m^2 dekapoise

1.12 Trigonometric Functions and Equations

Figure 1.1 treats the trigonometric ratios from a geometrical aspect. Circular functions can also be expressed as series by using Maclaurin's theorem: i.e.

$$f(x) = f(0) + xf'(0) + \frac{x^2}{2!}f''(0) \ldots \frac{x^n}{n!}f^{n'}(0) \ldots$$

$$\sin x = x - \frac{x^3}{3!} + \frac{x^5}{5!} - \frac{x^7}{7!} \cdots = \frac{e^{jx} + e^{-jx}}{2j}$$

$$\cos x = 1 - \frac{x^2}{2!} + \frac{x^4}{4!} - \frac{x^6}{6!} \cdots = \frac{e^{jx} + e^{-jx}}{2}$$

Hyberbolic functions $\sinh x$ and $\cosh x$ relate to a hyperbola in the same way as $\sin x$ and $\cos x$ relate to a circle. Thus

$$\sinh x = \frac{e^x - e^{-x}}{2} = -j \sin jx = x + \frac{x^3}{3!} + \frac{x^5}{5!} + \cdots$$

and

$$\cosh x = \frac{e^x + e^{-x}}{2} = \cos jx = 1 + \frac{x^2}{2!} + \frac{x^4}{4!} + \cdots$$

1.12.1 *Equations for triangles*

In any triangle such as those shown in Fig. 1.2 the sum of the internal angles is 180 degrees. Thus $\alpha + \beta + \gamma = 180$.

The most suitable equation for solving triangles depends on the known parameters.

Sine rule

$$\frac{a}{\sin \alpha} = \frac{b}{\sin \beta} = \frac{c}{\sin \gamma}$$

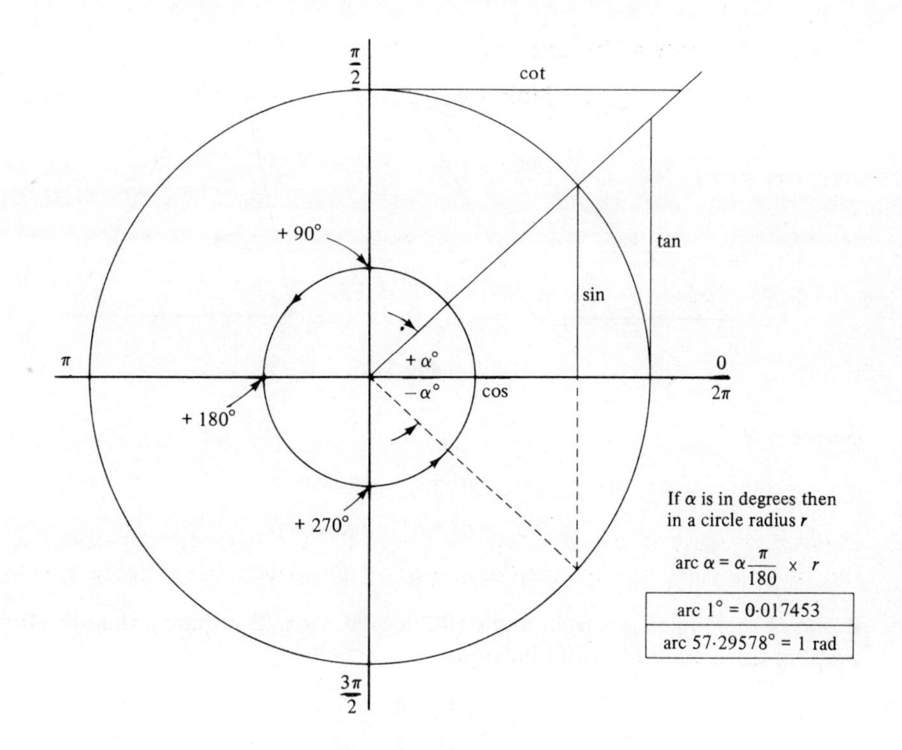

$\theta =$	$\pm \alpha$	$90 \pm \alpha$	$180 \pm \alpha$	$270 \pm \alpha$	0° / 2π	30° / $\frac{\pi}{6}$	45° / $\frac{\pi}{4}$	60° / $\frac{\pi}{3}$	90° / $\frac{\pi}{2}$
$\sin \theta =$	$\pm \sin \alpha$	$+ \cos \alpha$	$\mp \sin \alpha$	$- \cos \alpha$	0	$\frac{1}{2}$	$\frac{\sqrt{2}}{2}$	$\frac{\sqrt{3}}{2}$	1
$\cos \theta =$	$+ \cos \alpha$	$\mp \sin \alpha$	$- \cos \alpha$	$\pm \sin \alpha$	1	$\frac{\sqrt{3}}{2}$	$\frac{\sqrt{2}}{2}$	$\frac{1}{2}$	0
$\tan \theta =$	$\pm \tan \alpha$	$\mp \cot \alpha$	$\pm \tan \alpha$	$\mp \cot \alpha$	0	$\frac{1}{\sqrt{3}}$	1	$\sqrt{3}$	∞
$\cot \theta =$	$\pm \cot \alpha$	$\mp \tan \alpha$	$\pm \cot \alpha$	$\mp \tan \alpha$	∞	$\sqrt{3}$	1	$\frac{\sqrt{3}}{2}$	0

$\sec \alpha = \frac{1}{\cos \alpha}$	$\text{cosec } \alpha = \frac{1}{\sin \alpha}$	$\tan \alpha = \frac{\sin \alpha}{\cos \alpha} = \frac{1}{\cot \alpha}$
$\cos^2 \alpha + \sin^2 \alpha = 1$	$\sin 2\alpha = 2 \sin \alpha \cos \alpha$	$\sin 3\alpha = 3 \sin \alpha - 4 \sin^2 \alpha$
	$\cos 2\alpha = \cos^2 \alpha - \sin^2 \alpha$	$\cos 3\alpha = 4 \cos^3 \alpha - 3 \cos \alpha$

$\sin(\alpha \pm \beta) = \sin \alpha \cos \beta \pm \cos \alpha \sin \beta$	$\sin \alpha \sin \beta = 2 \sin \frac{\alpha \pm \beta}{2} \cos \frac{\alpha \mp \beta}{2}$
$\cos(\alpha \pm \beta) = \cos \alpha \cos \beta \mp \cos \alpha \sin \beta$	$\cos \alpha + \cos \beta = 2 \cos \frac{\alpha + \beta}{2} \cos \frac{\alpha - \beta}{2}$
	$\cos \alpha - \cos \beta = -2 \sin \frac{\alpha + \beta}{2} \sin \frac{\alpha - \beta}{2}$
$\tan(\alpha \pm \beta) = \frac{\tan \alpha \pm \tan \beta}{1 \mp \tan \alpha \tan \beta}$	$\tan \alpha \pm \tan \beta = \frac{\sin(\alpha \pm \beta)}{\cos \alpha \cos \beta}$
$\cot(\alpha \pm \beta) = \frac{\cot \alpha \cot \beta \mp 1}{\cot \pm \cot \alpha}$	$\cot \alpha \cot \beta = \frac{\sin(\beta \pm \alpha)}{\sin \alpha \sin \beta}$

More accurate values: $\pi = 3{\cdot}141\ 592\ 65$ 1 rad = 57·295 7·79 513 degrees
e = 2·718 281 828 5

Fig. 1.1 Trigonometric functions (on circle of unit radius)

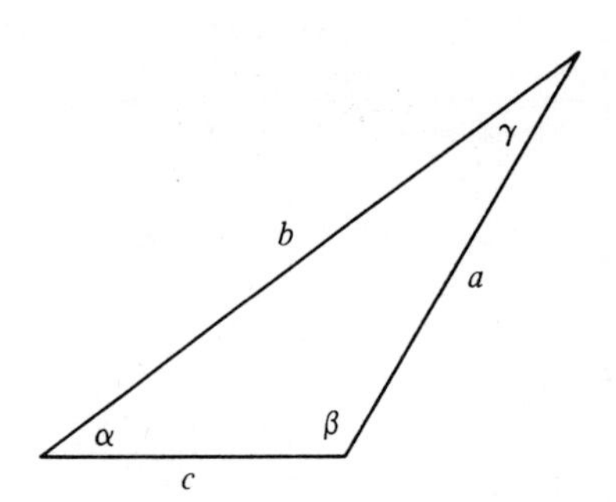

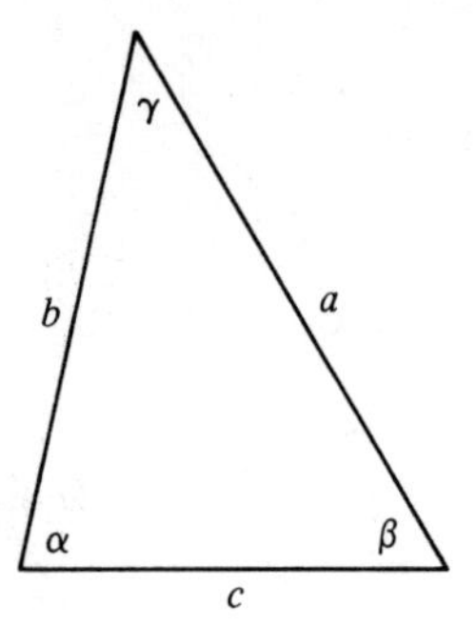

Fig. 1.2

Cosine rule

$$a^2 = b^2 + c^2 - 2bc \cos \alpha$$

$$b^2 = c^2 + a^2 - 2ca \cos \beta$$

$$c^2 = a^2 + b^2 - 2ab \cos \gamma$$

If one of the angles is a right angle (90 degrees or $\pi/2$ radians), then the formulae simplify since $\cos 90 = 0$ and become

$$a^2 = b^2 + c^2$$

$$b^2 = c^2 + a^2$$

$$c^2 = a^2 + b^2$$

usually known as Pythagoras' theorem.

1.12.2 *Equations*

Straight line (Fig. 1.3)

$$y = mx + c$$

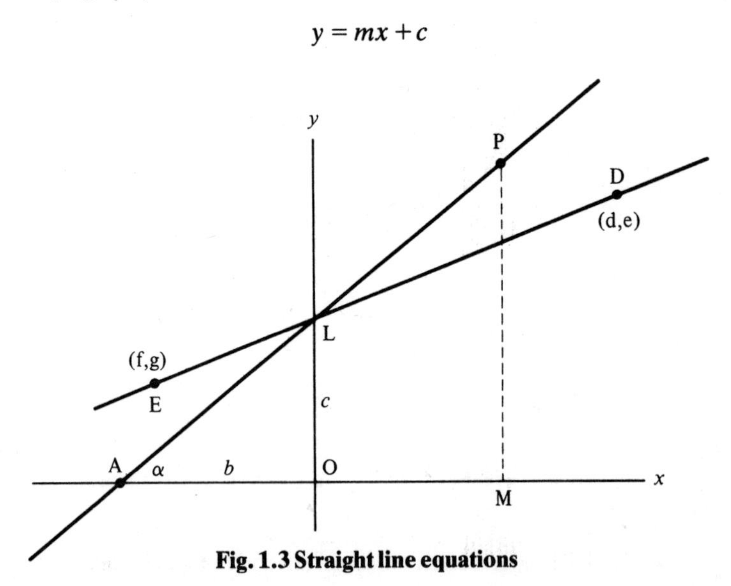

Fig. 1.3 Straight line equations

gradient

$$m = \tan \alpha = \frac{PM}{AM} = \frac{y-c}{x}$$

intercept c = value of y when $x = 0$. Given intercepts A(b, 0) and L(0, c) then:

$$\frac{x}{b}\frac{y}{c} = 1$$

Equation of straight line passing through points D(d, e) and E(f, g)

$$\frac{y-e}{e-g} = \frac{x-d}{d-f}$$

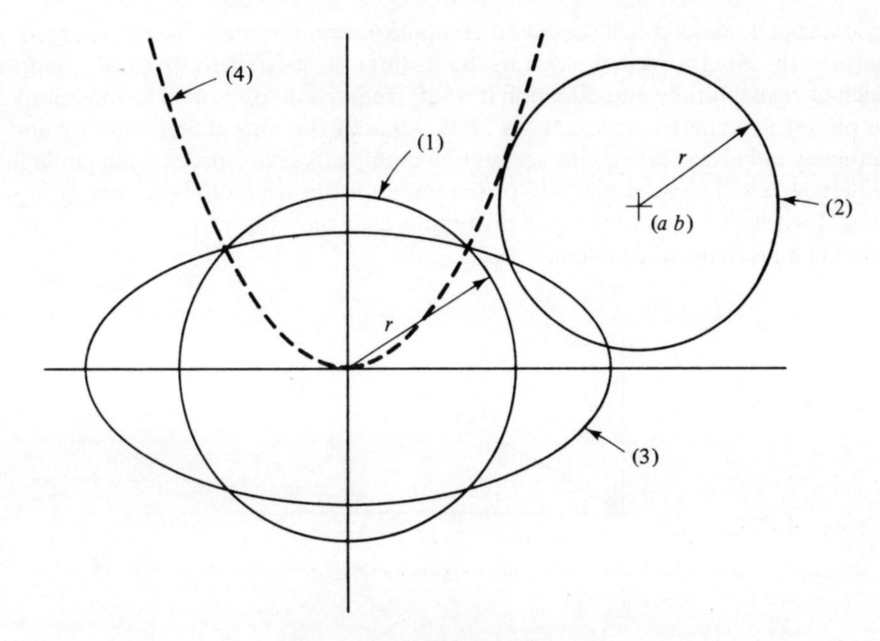

Fig. 1.4 Conic section equations

Conic sections (Fig. 1.4)
General equation: $ax^2 + 2hxy + by^2 + 2gx + 2fy + c = 0$
(1) Circle radius r, centre at the origin $x^2 + y^2 = r^2$
(2) Circle radius r, centre at $(a, b)(x-a)^2 + (y-b)^2 = r^2$
(3) Ellipse semi-axes c and d centre at the origin

$$\frac{x^2}{c^2} + \frac{y^2}{d^2} = 1$$

(4) Ellipse semi-axes c and d and centre at (a, b) with axes parallel to coordinate axes

$$\frac{(x-a)^2}{c^2} + \frac{(y-b)^2}{d^2} = 1$$

Parabola, vertex at origin and axis along x coordinate $y^2 = 4ax$
Hyperbola with foci on x axis and centre at the origin

$$\frac{x^2}{a^2} - \frac{y^2}{b^2} = 1$$

Solution of quadratic $ax^2 + bx + c = 0$

$$x_1 = \frac{-b + \sqrt{(b^2 - 4ac)}}{2a}, \qquad x_2 = \frac{-b - \sqrt{(b^2 - 4ac)}}{2a}$$

1.13 Vectors and Phasors

Periodic phenomena can, if they are of a simple harmonic nature, be represented and analysed by using a rotating vector. To distinguish it from true vector quantities which have magnitude and direction it is referred to as a 'phasor'. The magnitude of the phasor is made to represent the peak value of the alternating quantity and its frequency in hertz is taken into account by making its coordinates rotate at angular frequency $2\pi f$. The manipulation of the equations in which the phasors represent r.m.s. quantities follows the same procedure as if they were for example coplanar forces in a mechanical problem.

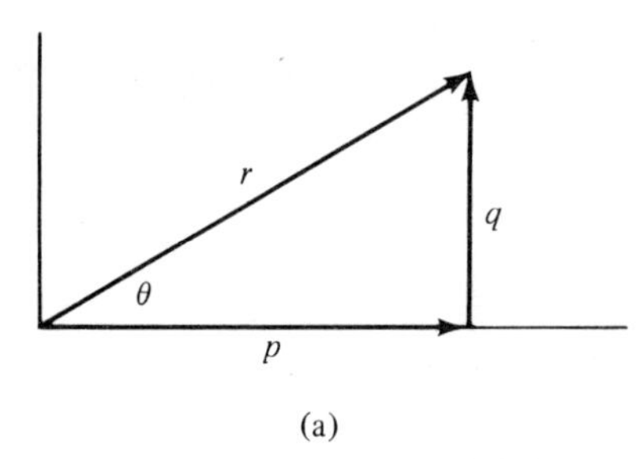

(a)

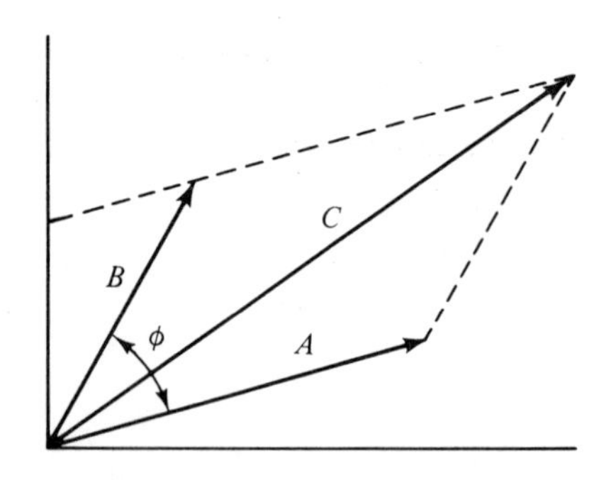

(b) Addition of two phasors representing sinusoidally varying quantities e.g.
$a = A \sin \omega t \quad b = B \sin \omega t + \phi$

Fig. 1.5 Vectors and phasors

The vector or phasor shown in Fig. 1.5 can be defined:

(a) by rectangular coordinates

$$r = p + jq$$

(b) by polar coordinates

$$r = \sqrt{(p^2 + q^2)}$$

$$\theta = \tan^{-1} q/p$$

(c) exponentially

$$r\,e^{j\theta}$$

(d) trigonometrically

$$r(\cos\theta + j\sin\theta)$$

where as above

$$\theta = \tan^{-1} q/p$$

1.14 Fourier Analysis

Where the quantities involved are non-sinusoidal they may be broken down into their harmonics and each of the most significant dealt with in the normal way. This analysis may be made by using Fourier's series.

Any single valued periodic function may be expressed in the form

$$f(t) = f(t + kT) = \frac{a_0}{2} + a_1 \cos \omega t + \ldots a_n \cos n\omega t + b_1 \sin \omega t$$

$$+ b_2 \sin 2\omega t \ldots + b_n \sin n\omega t$$

The Fourier coefficients are given by

$$a_0 = \frac{2}{T}\int_0^T f(t)\,dt; \qquad a_n = \frac{2}{T}\int_0^T f(t) \cos n\omega t\,dt$$

$$b_n = \frac{2}{T}\int_0^T f(t) \sin n\omega t\,dt$$

Characteristic equations of periodic functions

	General form	*purely sinusoidal*
	$f(t) = \frac{c_0}{2} + \Sigma c_n \sin(n\omega t + \phi)$	$f(t) = H \sin(\omega t + \phi)$
Periodic time	$T = \frac{2\pi}{\omega}$	$T = \frac{2\pi}{\omega}$
Circular frequency	$\omega = \frac{2\pi}{T}$	$\omega = \frac{2\pi}{T}$

	General form	*purely sinusoidal*
Frequency of fundamental	$f = \frac{\omega}{2\pi} = \frac{1}{-T}$	$f = \frac{\omega}{2\pi} = \frac{1}{T}$
Frequency of nth harmonic	$f_n = \frac{n\omega}{2\pi} = \frac{n}{T}$	$f_n = 0$
Linear mean value of $f(t)$ (over 1 cycle)	$f_m = \frac{1}{T} f(t)\,\mathrm{d}t = \frac{a_0}{2}$	$f_m = 0$
Mean value per half cycle	$f_a = \frac{1}{T}\int_0^T (f)\,\mathrm{d}t$	$f_a = \frac{2}{n} H = 0{\cdot}63662H$ where H = peak value
Root mean square (r.m.s.) value (effective or virtual value)	$f_e = \frac{1}{T}\int_0^{\pi} f(t)\,\mathrm{d}t$	$f_e = \frac{1}{\sqrt{2}} H = 0{\cdot}7071H$
Peak factor	$k_p = \frac{\text{peak}}{\text{r.m.s.}} = \frac{H}{f_e}$	$k_p = \sqrt{2} = 1{\cdot}4142$
Form factor	$k_f = \frac{\text{r.m.s.}}{\text{mean}} = \frac{f_c}{f_e}$	$k_f = \frac{\pi}{2\sqrt{2}} = 1{\cdot}1107$

In machines the non-sinusoidal distribution of flux in the airgap is the main cause of harmonics.

1.15 Circuit and Network Calculations

The following circuit theorems apply to d.c. and sinusoidal a.c. circuits provided the complex values of impedances are used:

Ohm's law

$V = IZ$ where for a series circuit Fig. 6:

$$Z = \sqrt{\left[R^2 + \left(\omega L - \frac{1}{\omega C}\right)^2\right]} \qquad \phi = \tan^{-1}\frac{\omega L - 1/\omega C}{R}$$

$\omega = 2\pi f.$

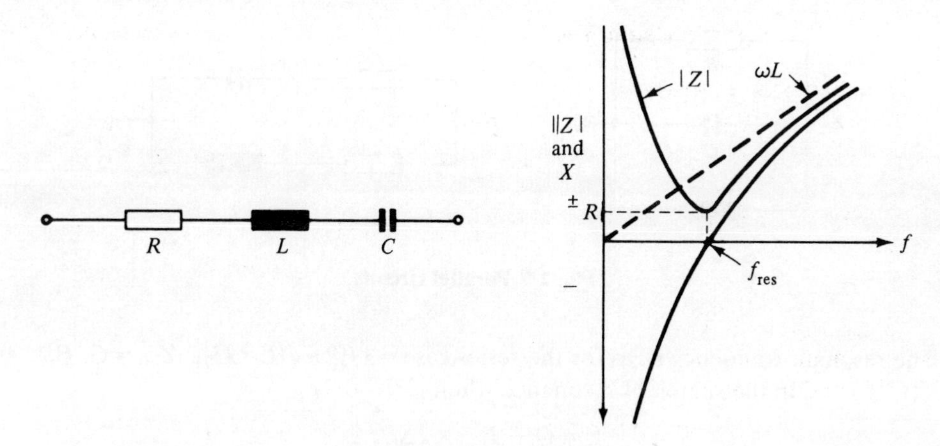

Fig. 1.6 Series circuit

At resonance

$$\omega L = \frac{1}{\omega C}, \qquad f_{res} = \frac{1}{2\pi\sqrt{(LC)}}$$

Voltage across inductance

$$E_L = E\frac{\omega_0 L}{R} = QE$$

Voltage across capacitance

$$E_C = I\omega_0 L = QE$$

At resonance impedance is minimum ($=R$) 'acceptor' circuit.
If loss angle of inductance is small

$$\delta \approx \tan\delta = \frac{R}{\omega L}$$

and Z becomes

$$Z = \omega L\delta + \mathrm{j}\omega L\theta = \tan^{-1}\frac{\omega L}{R}$$

$$= \omega L(\delta + \mathrm{j})$$

when $\delta = 4°$ error is about 0·1 per cent.

Parallel resonance (Fig. 1.7)

Parallel circuits are easiest dealt with by considering the admittance $Y(=1/Z)= G+\mathrm{j}B$. G is the conductance in S (siemens), B is the susceptance.
For any number of parallel impedances the resultant total is

$$Y_T = Y_1 + Y_2 + \cdots = (G_1 + G_2 + \cdots) + \mathrm{j}(B_1 + B_2 + \cdots)$$

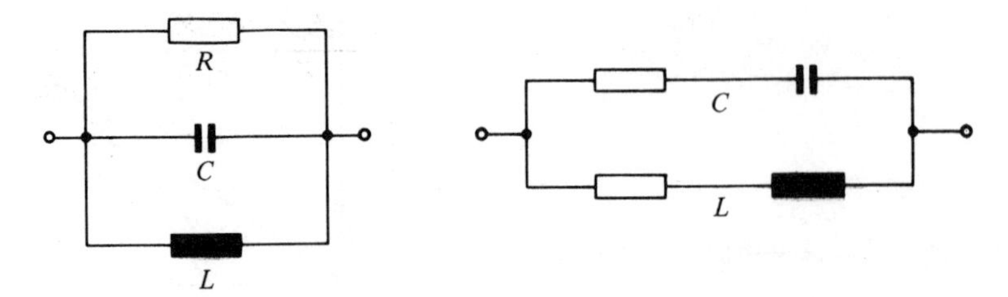

Fig. 1.7 Parallel circuits

The resonant frequency is, as for the series case $= 1/[2\pi\sqrt{(L \times C)}]$, $Y_{res} = G$, $B_{res} = \sqrt{(C/L)}$ and in the region of resonance when $|\Delta f| < 0{\cdot}1 f_{res}$

$$Y = G + jB_{res} \times 2\Delta f/f_{res}$$

Note: when $f < f_{res}$ Δ is negative!

If V is the voltage applied to a parallel resonant circuit with a

$$Q = \left(\frac{2\pi f_{res} L_{res}}{R}\right)$$

I_C or $I_L = QI$ where I is the total current through the circuit, there exists therefore 'current resonance'.

Parallel equivalent of $Z = R_s + jX_s$ is

$$R_p = \frac{R_s^2 + X_s^2}{R_s} \quad \text{and} \quad X_p = \frac{R_s^2 + X_s^2}{X_s}$$

i.e.,

$$Y = \frac{1}{Z} = \frac{1}{R_p} - j\frac{1}{X_p}$$

Series equivalent of parallel circuit $Y = G_p + jB_p$ is

$$Z = \frac{G_p}{G_p^2 + B_p^2} - j\frac{B_p}{G_p^2 + B_p^2}$$

1.16 Kirchhoff's Rules

1. The sum of the currents flowing into a circuit junction is zero. For a.c. this is true at every instant and for r.m.s. values provided magnitude and phase are taken into account.
2. The e.m.f. round a closed loop in a network is equal to the sum of the voltage drops. This is again applicable to alternating quantities considered as instantaneous or vector values.

To simplify complex networks the following *star-delta* conversion may be used:

In Fig. 1.8 for the two networks to be equivalent between any pair of terminals the following equations must be satisfied

$$Z_a = \frac{Z_3 Z_1}{Z_1 + Z_2 + Z_3} \qquad Z_1 = \frac{Z_a Z_b + Z_b Z_c + Z_c Z_a}{Z_c}$$

$$Z_b = \frac{Z_1 Z_2}{Z_1 + Z_2 + Z_3} \qquad Z_2 = \frac{Z_a Z_b + Z_b Z_c + Z_c Z_a}{Z_b}$$

$$Z_c = \frac{Z_2 Z_3}{Z_1 + Z_2 + Z_3} \qquad Z_3 = \frac{Z_a Z_b + Z_b Z_c + Z_c Z_a}{Z_a}$$

The general rule is that at any given single frequency an n-branch star network may be replaced by a mesh of $\frac{1}{2}n(n-1)$ branches, but not conversely.

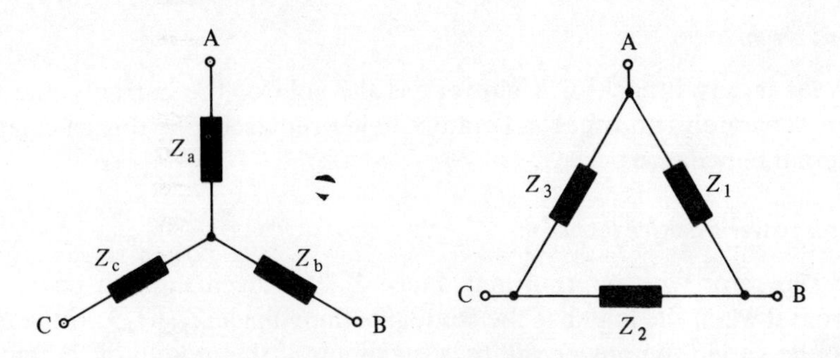

Fig. 1.8 Star-delta equivalent circuits

Thevenin's (Helmholtz's) theorem

The current in any branch of a network is also given if it is connected to a generator with an e.m.f. E_1 and an internal impedance Z_1, if E_1 is the potential difference appearing across it when open circuited and Z_1 its total impedance when all sources of e.m.f. are replaced only by their internal impedance.

Example: A network consisting of generators and impedances has two output terminals at which the following measurements were made:

a) open circuit voltage 100 V
b) Short-circuit current 2,00 A
c) Current with 12 Ω resistance load 1,66 A

Find output impedance.

Using Thévenin's theorem and the values of $E_1 = 100$ V and $I_{sc} = 2$A allows the magnitude of the circuit impedance to be calculated:

$$|Z| = \frac{V_1}{I_{sc}} = \frac{100}{2} = 50\ \Omega \qquad \text{(i)}$$

With the 12 Ω load the circuit impedance is given by:

$$Z' = (R + 12) + jX \qquad \text{which has a magnitude of}$$

$$|Z'| = \sqrt{(R^2 + X^2 + 24R + 144)} \qquad \text{(ii)}$$

Substituting (ii) in the equation $100 = 1{,}66\,Z'$, squaring and simplifying gives

$$R^2 + X^2 + 24R = 3484{,}97 \qquad \text{(iii)}$$

Since $Z^2 = R^2 + X^2$ (i) can also be written in the form $R^2 + X^2 = 2500$ and when this is subtracted from (iii) the value of R is obtained:

$$24R = 984{,}97 \quad \text{hence} \quad \underline{R = 41{,}04\ \Omega}.$$

Now $X = \sqrt{(2500 - 41{,}04^2)} = 28{,}56$ so that the output impedance

$$\underline{\underline{Z = 41{,}0 + j28{,}6}}$$

Other useful rules for network calculations are:

Superposition theorem

The current in any branch of a network is the sum of the currents due to each generator separately, all other generators being replaced, for this calculation, by their internal impedances.

Maximum power transfer theorem

Given a generator with internal impedance $Z_G/\varnothing$ the maximum power will be drawn from it when the load has the conjugate impedance $Z_G/-\varnothing$. If the modulus alone can be varied the power will be a maximum if the moduli of the motor and generator impedances are equal.

Reciprocity theorem (for passive element networks only)

If an e.m.f. in a circuit A produces a current in circuit B then the same e.m.f. in circuit B produces the same current in circuit A. The ratio the e.m.f. to the current is termed the transfer impedance.

2 Application Considerations

2.1 Introduction

The extensive utilization of many types and constructional variants of electric motors in industry, commerce, and the home would make a comprehensive guide for their selection, application, and maintenance run into many volumes. There are, fortunately, many similarities between applications for which existing standard motors (see chapter 5) provide a wide choice of characteristics, mountings, enclosures, and shaft configurations. For certain applications, for example conveyors, steel rolling mills, lifts, etc, manufacturers have evolved ranges of special purpose motors designed to meet special regulations and performance requirements. Whether the equipment or machine is designed around the motor, as for example in portable tools, or requires special considerations for mounting and coupling, the quality and cost of the final design benefits greatly from the earliest possible consideration of the prime mover. In many cases each completed design stage introduces constraints, which finally may not limit the performance, but make the use of an existing standard motor impossible.

The complex selection process involves analysis and synthesis of not only the load and the proposed motor, but the complete drive assembly and the control equipment, which may include rectification or frequency changing. If the characteristics of the load and the motor can be expressed mathematically these and the criteria for optimum results can be analysed with a computer program. An alternative is the construction of a suitable model and the use of an analogue method. Pumps, fans, and similar devices lend themselves particularly to the construction of scale models from which the horsepower requirements for a range may be determined. The scale factors of electromagnetic machines are very complex, and computer-aided design optimization of the motor should only be attempted by those experienced in their design and manufacture.

The determination of motor requirements for machine tools involves the assessment of loading torques resulting from speeds and feeds determined over a wide range. For such applications careful consideration must be given to protection against short- and long-term overloads to prevent damage to the motor, or machine parts being over-stressed. In such application the mechanical control and protection by torque limiting will have high priority in meeting design objectives.

Since machine tools and specially designed production machines such as are used by the automotive industry are required to carry out complex duty cycles involving frequent stops and reversals they helped to create a market for brake- and low-inertia motors. The brake motors are worthy of special mention as they incorporate another element of the 'system', particularly those in which the rotor moves also axially to overcome the brake spring. Such motors are also useful in lifting applications where the fact that the brake operates when the supply fails is an automatic 'fail-safe' feature.

Other motors of special interest to machine tool or control equipment designers are the special low-inertia and printed circuit motors described in chapter 18.

2.2 Problem Statements

Before the optimum selection of type of motor can be made it is essential that all relevant factors are fully considered for any particular application. There is a large number of possible factors, but not all are necessarily relevant for any particular installation although they should be considered, and they can be grouped:

(a) System behaviour
(b) Motor characteristics
(c) Load requirements.

2.2.1 *System behaviour*

The effect of the electrical system on the motor must be considered as well as the effect of the motor on the system, and relevant factors are:

System voltage under normal conditions, including variation.
System frequency under normal conditions, including variations.
Voltage disturbance caused by starting current taken by motor.
Effect on system resulting from reactive kVA taken by motor.
Effect on motor due to transient voltage disturbance in system.
Contribution by motor to fault level of system.

2.2.2 *Motor characteristics*

The most important of these are:

Speed.
Torque: starting breakaway, sychronizing, normal, pull out, transient.
Efficiency.
Power factor.
Reactance: steady state and transient, and stored energy constant, and the interaction of these with system parameters and with each other.

2.2.3 *Load requirements*

The essential data consist of:

Speed required.
Inertia to be accelerated and/or synchronized.
Torque requirements including breakaway, dynamic acceleration torque as a function of speed, normal full load torque. Also abnormal torque conditions likely to be encountered.

Type of load, i.e., constant, irregular, cyclic, and, if the last, details of typical cycles.
Behaviour of load to transient irregularities likely to be produced by motor; including any control systems, valves, etc, which could interact to cause extension of any such irregularity.

In addition to all the above there are many other factors which affect the best choice of motor drive but they are not always significant with regard to the choice between synchronous and induction motor.

2.3 Small Power Motors

In small power ratings cage induction motors are the most commonly produced type of a.c. motor and are, consequently, relatively inexpensive and easy to procure compared with other types of motors. In fact it is often cheaper to use this type of motor together with additional equipment to offset any deficiencies in its behaviour than to try and use another type of motor with more suitable inherent characteristics. Thus, for example, the power factor at rated load could be in the region of 0·8 to 0·9 lagging for small induction motors and this implies that the distribution system including transformers, cables, switches, etc, have to be rated for 10 to 20 per cent more current than the actual kilowatt power assumed would require.

By fitting capacitors at the motor terminals the amount of lagging kVAR drawn from the system, and consequently the amount of current, can be reduced resulting in a cost saving, but this operation also presents a possible hazard in that if the capacitor rating is too great a phenomenon known as self-excitation can occur under some conditions. This occurs when the charging current taken by the capacitor exceeds the magnetizing current required by the induction motor, because if the motor plus capacitor is disconnected from the system the motor self-excites as an induction generator and its terminal voltage will rise until the magnetizing current equals the capacitor current. This rise in voltage can damage the capacitor and, in some cases, the motor itself. Because of the voltage and frequency range over which a motor may have to operate the capacitor should be so rated as to avoid this trouble at the worst combination producing the lowest flux in the motor. A value of 85 per cent of the rated voltage magnetizing kVAR of the motor is usually a safe maximum rating for capacitors connected to the terminals of a normally rated induction motor.

This obviously reduces the amount of power factor improvement which can be used with induction motors. Alternative arrangements can be made to prevent any induction motor being left connected to an oversize capacitor when disconnected from the supply but this requires that capacitors be permanently connected to the system, which can result in over-improvement of power factor, or alternatively necessitates switching of the capacitors as required. This latter alternative involves either interlocking with motor switches or an independent monitoring and control system. When compared with the behaviour of synchronous motors these arrangements can prove expensive both initially and during maintenance, service, etc, and are normally only adopted for installations of small motors. When the installation includes mixed sizes of motors it is usually economic to use induction motors for all small powers and synchronous machines for some of the large, continuously running drives, to give overall power factor improvement.

The only economic forms of small power synchronous motors are the reluctance or permanent magnet types which are usually built in standard induction motor frames, utilizing a suitable design of rotor. These motors are normally used solely because of their synchronous speed characteristic in that they provide, in effect, an 'electric gear box' system which can maintain constant speed ratio between the generator speed (or equivalent system frequency) and each motor shaft. Such drives are useful for speed signals for governors, where a permanent magnet generator and such a synchronous motor can be given a synchronous drive independent of all other sources of electrical power. They are also used for multi-shaft drives such as tissue machines where all motors can be started, run, and stopped in absolute synchronism.

2.4 Medium and Large Power Motors

These tend to be selected individually and it is usual to consider more of the factors listed previously in more detail. In addition, the choice of type of motor is increased and this in itself needs further consideration. It is preferable to start the analysis from the Load Requirements.

2.4.1 *Speed*

Depending on the system frequency and possible integral numbers of poles, only a limited number of actual synchronous motor speeds are available, e.g., at 50 Hz, 3000, 1500, 1000, 750, 600 rev/min, etc, and if the drive requires some other value then some speed-changing device such as a gearbox, belt drive, etc, would be required. This is true for both synchronous motors and cage induction motors the speed of which at full load is a few per cent below synchronous speed, and there is little to choose between these two types. In a few special cases the slight variation in speed that occurs with load in an induction motor is prohibitive, and the synchronous motor, which maintains its speed with the frequency variation of the power system, offers a more economic form of drive.

When the load requires a single constant speed the above applies but if it requires to be driven at two or more definite speeds, or at any of a range of speeds neither of these simple types of motors can be applied directly. It is possible to use either type of motor with a mechanical or hydraulic device to produce the special speed requirements and this is often the best solution. Change-pole motors can provide alternative definite speeds, and although both synchronous and induction motors have been used as change-pole motors the latter are usually preferred since no rotor circuits need switching. This is required with the synchronous motor, which produces extra cost and complication, and if this is the factor of prime importance the choice is obvious.

The wound rotor induction motor can provide speed control by using an adjustable rotor resistor, and although considerable losses are involved with such an arrangement, it may be preferable to the above mentioned alternatives. A variable speed synchronous motor fed directly from a fixed frequency system is not economically available.

2.5 Inertia

The most important aspect of the load inertia is the energy required to accelerate it to full speed. In addition to the rotor loss associated with the load torque during acceleration, an amount of heat equal to the total kinetic energy stored in the rotating parts of the load plus driving motor must be dissipated in the motor starting winding circuit. In the case of wound rotor starting winding motors, both synchronous and induction, this circuit includes the external starting resistor and in practice much of this 'acceleration' heat is dissipated outside the motor. These types of motor are thus equally to be preferred for very high inertia drives rather than the cage or solid pole types of synchronous or induction motors.

Synchronous motors must accelerate this load inertia from full induction motor speed attained by means of the starting winding to full synchronous speed by means of the synchronous torque produced by the excitation winding. This requirement may involve extra cost due to increased excitation requirements for very high inertias and all other factors being equal would favour the use of the induction motor.

Another aspect of load inertia is associated with transient system disturbances since it tends to maintain motor speed when the motor torque is reduced. In conjunction with the motor torque capability during reduced supply voltage conditions it determines the capability of the drive to withstand such disturbances, related to a Transient Stability Factor, as is discussed in more detail later.

2.6 Starting Torque

Starting consists of separate stages; breakaway, acceleration, and synchronizing in the case of synchronous motors, and the torque developed by the motor must exceed the load torque during starting if normal full load full speed operation is to be obtained. All synchronous motors start as induction motors and utilize similar types of starting windings and have very similar starting characteristics. Because of the requirements of synchronous characteristics synchronous motors have longer radial airgaps than induction motors and have, consequently, a higher value of magnetizing current during starting. The current drawn from the supply during starting consists basically of this current in, approximately, quadrature with the current associated with useful starting torque and as a consequence on low torque conditions synchronous motors take a relatively higher value of starting current, but on high torque drives the difference is not great.

Because an induction motor operates continuously on the same winding that is used for starting there has to be some measure of compromise in its design since any change in breakaway torque affects the normal full speed operation characteristic. This limitation does not apply to a synchronous motor as different windings are used for starting and running. However, the torque provided by the starting winding near full speed is important with regard to the synchronizing operation so that there are limitations on the breakaway torque which can be obtained depending on the torque and slip values required for synchronizing.

Salient pole synchronous motors do not have the same uniformity of starting winding distribution as round rotor type motors and the two-axis method of analysis used for determining synchronous characteristics must also be used to evaluate the starting behaviour. This shows that the starting torque has a superimposed pulsating torque produced by the saliency effect in addition to a torque essentially similar to the non-salient pole type of motor. In general this has no significant effect on the starting procedure, but because the frequency of the pulsating torque varies as the motor accelerates it is possible for this torque to excite torsional oscillations in the shaft system at some point during run-up. When acceleration is rapid the time spent at the speed producing the critical frequency is very short and resonance is unlikely. However, if the shaft critical speed is close to an exciting torque frequency which can be sustained, such resonance could occur with consequent risk of damage to some component in the shaft system. It is usual on all high-power shaft systems to do a torsional study at the design stage and when synchronous motors are involved the effect of the pulsating torque should be included at this time to ensure freedom from trouble.

In general, therefore, the starting torque characteristics do not, in themselves, become a deciding factor between synchronous and induction motors, only between the types of starting winding available with either.

2.7 Pull-Out Torque

The torque–speed characteristic of the induction motor exhibits a peak value of torque, the magnitude of which is approximately proportional to the supply voltage squared, and which occurs at a slip two or three times the full-load torque slip. Thus if the load torque increases above the full-load value the slip increases to this pull-out value, or if the torque remains constant at full load and the voltage decreases the slip again increases to this value. At this value of slip the torque starts to decrease rapidly and the speed will drop further and the motor will stall. If an induction motor has to run through a voltage reduction to 80 per cent of the rated value, the peak torque must be at least 1·6 times rated torque, and for a voltage of 70 per cent the peak torque must be twice rated torque.

The pull-out torque of a synchronous motor is proportional to the product of supply voltage and the motor flux (which depends primarily on the value of excitation) and for a motor with fixed excitation the pull-out is approximately proportional to the supply voltage. Thus for a voltage reduction to 70 per cent the peak torque need only be 1·5 times rated torque. Synchronous torque varies with load angle, that is displacement between rotating flux field and rotor pole axis, but the speed remains synchronous. When the load torque exceeds the pull-out torque or the voltage decreases low enough, the load angle exceeds the pull-out torque value, the motor starts to 'slip poles', and loses synchronism.

Synchronous motors can be provided with automatic excitation which can increase the pull-out torque when required and enable a smaller and cheaper motor to be used than one designed to provide the same pull-out torque inherently.

Pull-out torque is a steady-state characteristic and relates to slowly applied low torque or changes in system voltage. Where the magnitudes of these are great the synchronous motor has distinct advantages if it is desirable to maintain full-speed operation of the drive.

When either, or both, of these changes occur rapidly the effect of mechanical inertia must be considered and the 'transient' characteristics of the motor must be used to determine response.

2.8 Transient Response

The analysis of transient response of a.c. motors is complicated in that it involves a large number of machine parameters and also depends on the magnitude, duration, and type of system disturbance.

During a transient disturbance an induction motor loses speed and to avoid instability the disturbance must have disappeared or become less severe before the 'critical speed' is reached. This speed is that from which the motor can re-accelerate to full speed under the conditions existing at the end of the disturbance, and these are often still subnormal. Procedures for analysing this are available, see references (below).

The synchronous motor transient behaviour is more complicated and it is usual to examine the problem using the generalized 'transient parameter' which is a ratio indicating the capability of a machine to withstand a disturbance and includes both electrical and mechanical effects. This approach enables the limiting machines on a multi-machine system to be selected and analysed in detail. A similar approach is available for induction motors.

In general, if it is not critical to maintain full speed on the drive an induction motor has advantages as it will slow down during the disturbance, and possibly stall, and start on restoration of the supply if it recovers to normal. This assumes that protection devices such as overcurrent relays, starting winding protection, overtemperature relays, etc, do not trip the motor supply switch during the disturbance.

If it is desirable to maintain the drive at full speed the synchronous machine, if provided with a high enough value of transient parameter, is to be preferred. This may involve increased size and cost, however, and is found to be economical only on very essential drives or where the duration of the disturbance is short, for example, fault clearance time on feeder breakers.

In each particular installation where this problem is an important factor in selecting the type of motor a careful analysis must be carried out.

2.9 Inherent Motor Characteristics

Two factors which do not really affect either the system or the load are the motor winding temperature rises and the operating efficiencies. The permitted standard temperature rises are the same for both types of motor for similar winding and classes of insulation, but the physical nature of the machine results in quite different ventilating systems being used. Also, because of the different nature of torque production the individual losses in the machines are quite different in their nature and their relationship to load. For these reasons and the fact that their operating characteristics cannot always be directly compared, it is not easy to evaluate relative efficiencies directly. In general, however, if efficiency is of prime importance a unity power factor synchronous motor will be used and some useful comparisons can be made. Since the induction motor operates on a lagging power factor at rated load its kVA and current must be slightly higher and its primary winding copper losses would be of the same order or higher for an equivalent design of winding. Likewise the friction and windage loss would be of the same order as the synchronous motor, although as the unity power factor synchronous motor could have a much larger radial airgap than the induction motor it would tend to have a much lower surface loss. The only other significant losses are the load losses and secondary winding losses. The latter are determined by the types of winding used, which cannot be regarded as equivalent. In the induction motor there is a definite relationship between the secondary and primary losses but in the synchronous motor there is no such relationship between the excitation loss and the primary loss, and, as a result, the excitation loss can be reduced appreciably by appropriate design. As a general rule, therefore, a large synchronous motor can always be designed with a higher full-load efficiency than the equivalent induction motor. The amount of difference possible depends on the actual value of efficiency, being less the nearer the efficiency approaches 100 per cent.

The relationship between full-load and other load efficiencies is determined by the ratio of constant load to variable loss and the latter is virtually current-related for an induction motor and hence the efficiency/load characteristic is predetermined by the loss ratio to a first order of accuracy. For the synchronous motor, however, the type of excitation determines how the excitation loss varies with load and it can be either constant, adjusted to give unity power factor thus producing minimum primary and stray load loss, or adjusted in some other way. Thus for maximum efficiency over a range of loads a synchronous motor with excitation control is to be preferred.

Motors required to have very high efficiencies are often supplied with stated guarantees on the total, or component losses. Testing procedures for synchronous motors have been developed and agreed internationally by the IEC so that these losses can be measured accurately at the manufacturer's works and compliance with the specification can be proved absolutely. Induction motors, however, have no such procedures available and methods used in attempting to evaluate the total loss on rated load are subject to considerable difficulties and consequent inaccuracies.

As a result where high efficiency is the most important aspect of a particular drive synchronous motors are used.

2.9.1 *Combination of factors*

The choice between induction and synchronous motor for any particular drive is usually quite definite when all the parameters of load and system are known accurately and the significant factors are few and specific.

Difficulties arise, however, when several conflicting factors are involved since it is not easy to give relative weight to these several aspects. The probability of particular situations arising often has to be assessed, for example, the likelihood of transient disturbances of a serious nature.

In situations where the number of conflicting requirements renders the choice difficult it is wise to analyse the behaviour of existing motors in similar situations both as regards trouble-free operation and optimum performance, before making the final choice. The advice of users of such machines should be sought and their opinion weighed against the purely technical advantages described since the benefit of continuous trouble-free operation may be of over-riding importance since it could result in saving due to reduced maintenance, reduced shut-down or other outage time and, possibly, reduced supervision.

2.10 Summary Table of Major Differences in Motor Characteristics

Factor	*Induction Motors*	*Synchronous Motors*
Speed	Nearly synchronous, varies with load.	Synchronous with supply frequency.
	Variable speed using wound rotor.	—

Factor	*Induction Motors*	*Synchronous Motors*
Power factor	Always lagging.	Depends on excitation, can be leading, or unity power factor. Varies with load but can be adjusted as required by excitation control. Excitation control can be used to give constant correction to system, or voltage control over small ranges.
Efficiency	Losses consist of constant loss plus load dependent loss. Efficiency variation with load.	Losses consist of constant loss, load dependent loss plus excitation loss. Excitation control can be used to give optimum efficiency over any load range.
Pull-out torque	Related to breakaway torque and full-load slip. Proportional to square of voltage.	Related to rated power factor, and excitation. Proportional to voltage with fixed excitation. Can be adjusted using excitation control.
Starting	Involves breakaway and acceleration.	Involves breakaway, acceleration, and pulling into synchronism.

2.11 Rated Output

In most electric motor drives the total power may be regarded as consisting of the following main components:

International system (SI)	MKS system	FPS system
Lifting power		
$P_H = \frac{mgv_{vert}}{\eta 1000}$	$P_H = \frac{Gv_{vert}}{\eta 75}$	$P = \frac{Wv_{vert}}{\eta 550}$
Friction Power		
$P_R = \frac{F_R v}{1000}$	$N_R = \frac{Rv}{75}$	$P_F = \frac{Rv}{550}$
$F_R = \mu mg$	$R = \mu G$	$R = \mu W$

International system (SI)	MKS system	FPS system
Friction Power—continued		
$F_R = F_{R_1} + F_{R_2} + F_{R_3}$	$R = R_1 + R_2 + R_3$	$R = R_1 + R_2 + R_3$
Accelerating Power		
$P_a = \dfrac{M_a n}{9550}$	$N_b = \dfrac{M_a n}{716}$	$P_a = \dfrac{T_a n}{5250}$
P—power in kW	*P*—power in PS (NB 1 PS = 735 W)	*P*—power in h.p. (NB 1 h.p. = 746 W)
F—frictional resistance in N	*R*—frictional resistance in kp	*R*—frictional resistance in lb
m—mass (weight in kg)	*G*—weight in kp	*W*—weight in lb
g—acceleration due to gravity 9·81 m/s^2		
v—velocity in m/s	*v*—velocity in m/s	*v*—velocity in ft/s
η—efficiency of external transmission (fractional)	η—efficiency of external transmission (fractional)	η—fractional efficiency of external transmission
μ—coefficient of friction	μ— coefficient of friction	μ—coefficient of friction
M—torque in Nm	*M*—torque in kpm	*T*—torque in lbf.ft
n—rotational speed in 1/min	*n*—rotational speed in rev/min	*n*—rotational speed in rev/min
Suffixes:		
H—lift	H—lift	H—lift
R—friction	R—friction	F—friction
a—acceleration	a—acceleration	a—acceleration

2.12. Criteria for Selecting Motors

Table 2.1 Generalized equations for basic types of electric motors

		Alternating current	
Direct current		Induction	Synchronous
Torque, T	$K_1 \Phi I_a$	$K_2 I_2 R_2 / S$	$K_3 I^2 R_2$
Speed, n	$E - I_a R_a / K_4 \Phi$	$(120f/P)-S$	$120f/P$
Power,			
Watts, W	EI Eff	KEI Eff PF	KI Eff PF
elctr	above: 746	KEI Eff PF	KI Eff PF
mech		$Tn/5252$	

Losses and efficiency

(a) Electric; comprising copper losses; I^2R + brush contact.
iron losses

(b) Mechanical; Friction and windage.

(c) Stray e.g. eddy current loss in primary winding.

Excitation and power factor

Direction of rotation

Conditioned by load requirements

Acceleration and deceleration

Factors involved are:

Motor accelerating torque at operating input conditions.
Total motor and load inertia.
Motor thermal capacity.
Frequency and number of accelerations and brakings with respect to time.
Ability to dissipate the losses for the entire duty cycle:

(a) Number of starts and stops may be limited (consult manufacturer).
(b) The time of acceleration or deceleration, e.g.

$$t = \frac{Wk^2 \Delta n}{308T} \text{ seconds}$$

Wk^2 total rotating system inertia referred to motor shaft at rated speed.
Δn speed change rev/min.
T net accelerating or decelerating torque.
For acceleration $T_{acc} = T_M - T_L - T_F$.

2.13 Calculation of Load Power Requirements

2.13.1 *Introduction*

Ideally the motor selected for a particular drive should:

(a) Start under all specified conditions.
(b) Require minimum starting current.
(c) Be able to accelerate to rated full load speed within the limit set by operational requirements or by its own thermal capacity determined by the frequency of starting, and the possible application of reverse-current or d.c. injection braking.
(d) Have maximum efficiency at rated load.
(e) Have adequate overload capacity.

In practice motor applications are complicated by random variations in the magnitude and duration of the rated load, demands for various or varying speeds and a wide variety of operating conditions. The accuracy with which it is possible to determine the optimum power, rating, speed, and other motor characteristics depends in the first instance on the precision with which the load parameters can be determined and defined. In some cases the load will vary according to the way in which a machine or device is used and the drive must be designed to allow maximum performance without detriment to either motor or machine. The ideal motor, like a good design, is a concept which depends on the judge, the circumstances and, in fact, so many factors as to be esoteric. It is, however, generally agreed that cost-effectiveness is the most essential attribute. The process of selecting the most suitable motor must, therefore, begin by examining standard motors normally stocked by suppliers, then investigate what manufacturers can assemble from standard elements, and only as a last resort consider special design. This consideration would, of course, not apply to equipment such as, for example, a portable tool or domestic appliance which is to be manufactured in such large quantities that the achievement of particular cost to performance ratio makes a special design economically desirable.

The power requirements of a particular drive may be met by various types of a.c. and d.c. motors having the appropriate output at different speeds. The alternative between a multipole motor and a geared unit, for example, is considered in chapter 16. In some applications such as fans in ducted air-conditioning quiet running is of paramount importance so that a larger and more costly slow-running motor is preferable, if not essential. The various requirements of each application must be considered in their order of importance with due regard to their effect on prime and operational costs, availability, and overall performance. In some cases, for example, the cost of providing a very large speed-range cannot be justified economically and the designer of the drive must consult with the user to arrive at an acceptable compromise.

Often there is the dilemma of the 'safety (or ignorance) factor' and the implications of a larger motor which can, at higher capital and running cost, provide a greater torque and thermal capacity must be weighed against the operational limitations of the smaller, just adequate, motor. In the case of the larger, more powerful motor the load may have to be protected against high accelerating and maximum torques, whereas the smaller machine should be protected against sustained overload. Modern insulating materials and bearing lubricants, though permitting considerably greater latitude on the specified temperature limits, must still be considered to have their life shortened in proportion to the frequency and magnitude of excess temperatures. To ensure continued reliability of operation motors subject to periodic overloading must, therefore, be inspected more frequently and replaced or rewound as soon as the probability of failure exceeds operational reliability requirements.

The approximate motor power and system performance for a particular application can be calculated and optimized by using the fundamental equations of Newtonian mechanics, and the speed/torque curves of motor and load. The motor characteristic cannot, in general, be expressed as a simple, easy-to-integrate mathematical function, so that a number of usually quite justified simplifying assumptions are made. More accurate calculations of complex load-, motor-, and control equipment systems are obviously necessary where large powers or critical performance criteria are involved. In such cases transfer functions for the various elements must be determined and the well-developed methods of system stability analysis applied. Inertia, compliance and, in particular, friction are parameters whose exact magnitude can usually only be obtained from tests on the actual system or the analysis of similar systems. Practical experience is thus an invaluable ingredient of good design as long as there is also an awareness that it may also inhibit innovatory solutions.

The most commonly used standard motors are continuously rated with limits on the permissible number of starts per hour. Where drives are subject to frequent starting it may, therefore, be necessary to choose a special design in a larger frame size. Conversely, where truly intermittent duty operation obtains it may be possible to select a much smaller unit. Automation of manufacturing processes causes electric drives to be either subjected to cyclic loading or, where positioning is involved, frequent starting and braking duty. In order to define thermal rating more accurately the basic definitions of maximum continuous- and short time rating have been supplemented by equivalent continuous- and duty type rating and further duty types within those categories defined. (Section 5.2.1). These definitions are contained in IEC 34-1, and appear in many national standards such as BS 4999, Part 30, 1972 and VDE 0530. Where the load cycles are more complex than those defined in these standards the root mean square of the load cycles is taken to represent the continuous power requirement. This applies, for example, to presses and reciprocating pumps using totally enclosed fan cooled motors (IP 44 enclosure and P 41 cooling, see section 5.2.9 and 10). Where large inertia such as a flywheel is added to smooth out load peaks and the driving motor is a cage induction motor, a high resistance rotor version has been found to be more economical. Where, as is generally recommended, the torque required is used to determine motor size it is regarded as inadvisable to let the maximum load torque exceed 75 per cent of the motor's pull-out torque at rated voltage.

In cases where the motor is supplied with power through rectifiers, thyristors, static inverters, or other control devices introducing ripple or non-sinusoidal voltages, respectively, account must be taken of any reduction in torque and increase in heating.

2.13.2 *Fundamental equations of motion*

The following equations are valid for constant acceleration and velocity:

Linear motion	*Rotary motion*
1. $v = u + at$	1. $\omega_2 = \omega_1 + \alpha t$
2. $s = ut + \frac{1}{2}at^2$	2. $\theta = \omega_1 + \frac{1}{2}\alpha t^2$
3. $v^2 = u^2 + 2as$	3. $\omega_1^2 = \omega_2^2 + 2\alpha\theta$
u = initial velocity (m/s)	ω_1 = initial velocity rad/s
v = final velocity (m/s)	ω_2 = final velocity rad/s
a = acceleration (m/s^2)	α = acceleration rad/s^2
s = distance travelled (m)	θ = angular travel in radians
	$\omega = \frac{2\pi n}{60} = 0{\cdot}10472n$
	where n is the number of revolutions per minute
$P = Fv$	$P = T\omega$
$F = ma$	$T = Fr = J\alpha$
$W = Fs = \frac{1}{2}mv^2$	$W = T\theta = \frac{1}{2}J\omega^2$
P = power in W	J = moment of inertia kgm^2

Linear motion	*Rotary motion*
F = force in N	M = torque in Nm
m = mass in kg	r = radius arm in m
W = work in J	F = force in N
	W = work in J

Since the watt is only convenient for small motors the unit for power used hereafter will be the kilowatt unless otherwise stated.

Symbol	Quantity	SI unit	MKS unit	FPS unit
P	Power	kW	PS (= 0·735 kW)	h.p. (= 0·746 kW)
M, T	Torque	Nm	kgfm	lbf.ft
n	Rotation in revolutions	1/min	rev/min	rev/min
ω	Rotational speed	rad/s	rad/s	rad/s
α	Rotational acceleration	rad/s^2	rad/s^2	rad/s^2
v	Linear speed	m/s	m/s	ft/s
a	Linear acceleration	m/s^2	m/s^2	ft/s^2
g	Acceleration due to gravity	9·81 m/s^2 (average values)	9·81 m/s^2	32·2 ft/s^2
m	Mass	kg	kg	(slug)
G, W	Weight	—	kgf	lbf
F_f	Frictional force (μW)	N	kgf	lbf
J, GD^2, Wk^2	Moment of inertia and flywheel effect	kgm^2 $\left[=\frac{GD^2}{4}\right]$	kgf m^2 GD^2	lbf. ft^2 Wk^2
t_a	Acceleration or run-up time	s	s	s
η	Fractional efficiency			
μ	Coefficient of friction			

For drives with relatively low rotational speed or linear velocities the nominal torque gives a more realistic indication of the size of motor required than the power rating, which as the following example shows may be surprisingly small:

> A 215 mm (8 in) diameter drum motor with a peripheral speed of 3·8 cm/s (7·5 ft/min), and producing a torque of 215 Nm (158 lbf. ft) has a rated output of 0·075 kW (0·1 h.p.). The tension in the conveyor band would be 2000 N (204 kgf or 450 lbf).

Figure 2.1 shows an inclined conveyor, where a linear speed of v causes the load to be raised with a vertical velocity of $v \sin \alpha$. The diagram also shows where friction has to be overcome.

The equations in the three systems of units are:

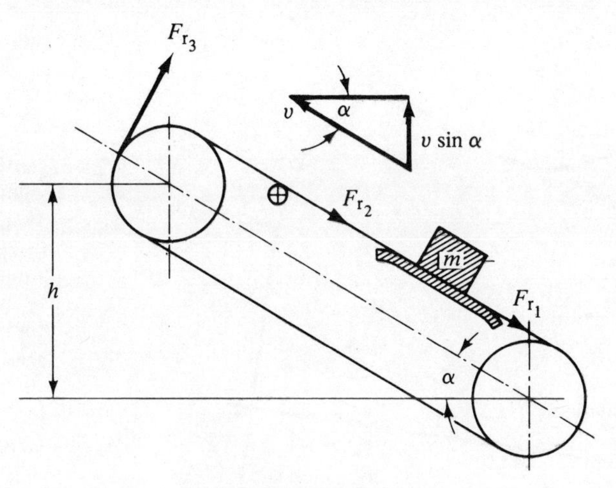

Fig. 2.1 Conveyor drive

	SI	MKS	FPS	
Power to raise load against gravity				
	$P_h = \dfrac{mgv \sin \alpha}{\eta 1000}$	$\dfrac{Gv \sin \alpha}{\eta 75}$	$\dfrac{Wv \sin \alpha}{\eta 550}$	(2.1)
Power to overcome friction				
	$P_f = \dfrac{F_f v}{1000}$	$\dfrac{F_f v}{75}$	$\dfrac{F_f v}{550}$	(2.2)
	$F_{f_1} = \mu mg$	μG	μW	
	$F_f = F_{f_1} + F_{f_2} + \cdots$			
Power to accelerate				
	$P_a = \dfrac{M_a n}{9550}$	$\dfrac{M_a n}{716}$	$\dfrac{T_a n}{5250}$	(2.3)

If the pull on the belt can be measured the power required can be calculated from the values of force on the belt F_b and the motor speed

SI	MKS	FPS	
$P = \dfrac{F_b v}{1000}$	$P = \dfrac{F_b v}{75}$	$P = \dfrac{F_b v}{550}$	(2.4)

n from the drive drum diameter D and the gear ratio i

SI	MKS	FPS
$n = \dfrac{19{,}lvi}{D}$	$\dfrac{19{,}lvi}{D}$	$\dfrac{19{,}lvi}{D}$

To calculate the time the motor requires to reach its rated speed the net torque available to accelerate the motor, transmission, and load must be determined.

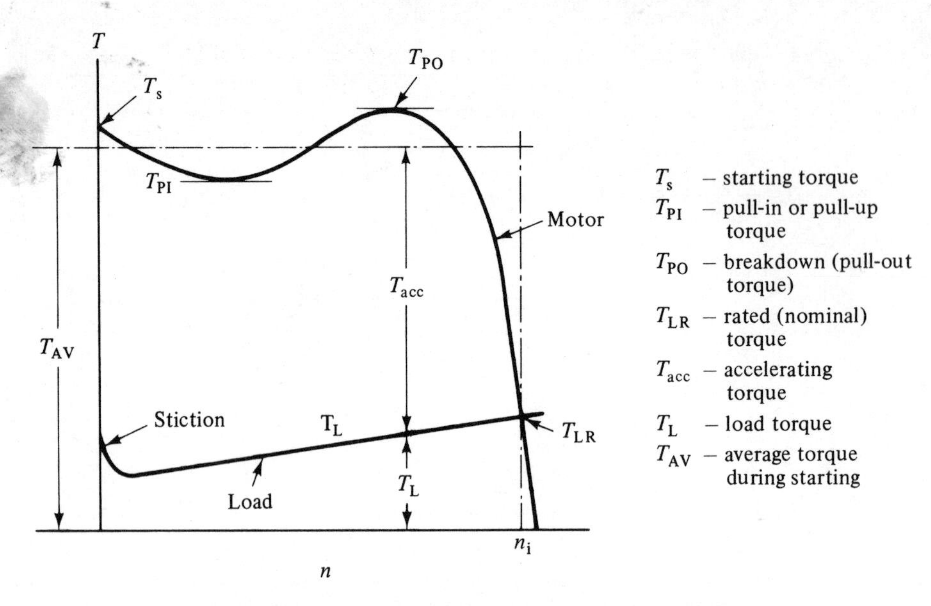

Fig. 2.2 Induction motor torque/ speed characteristic

Figure 2.2 shows a typical induction motor speed/torque curve and a nearly constant torque load. Where the accelerating torque varies with speed an expression must be found for the curve, e.g.

$$T = F(\omega) \quad \text{and from } T = J\alpha = J\frac{d\omega}{dt}$$

it follows that

$$dt = J\frac{d\omega}{F(\omega)} \tag{2.5}$$

and the time for any change of speed from

$$\omega_1 \quad \text{to} \quad \omega_2 \quad \text{will be } t = J\int_{\omega_1}^{\omega_2} \frac{d\omega}{F(\omega)} \tag{2.6}$$

(This will be dealt with in more detail in section 2.13.3).

For constant values of accelerating torque, T_a, the time to reach the speed n will be

	SI	MKS	FPS	
$t_a =$	$\dfrac{Jn}{9{\cdot}55T_a}$	$\dfrac{GD^2n}{375T_a}$	$\dfrac{Wk^2n}{308T_a}$	(2.7)

The moments of inertia in the above equation must be the sum of all inertias of the system. Where there is a change in speed due to the gearbox or other form of transmission all inertias must be referred to either the input or output shaft.

$$J = \frac{J_e}{i^2} + J_r \qquad GD^2 = \frac{GD_e^2}{i^2} + GD_r^2 \qquad Wk^2 = \frac{Wk_e^2}{i^2} + Wk_r^2 \tag{2.8}$$

suffix e = external $\quad r$ = rotor or armature

In slow-speed drives the motor inertia is as a rule appreciably larger than the inertia of the load when referred to the motor shaft through the gear reduction ratio i.

Figure 2.3 shows a load moving at a linear speed v which when referred to the rotor shaft gives the following total inertia:

SI	MKS	FPS	
$J = 91{\cdot}2m\dfrac{v^2}{n^2}$	$GD^2 = 364G\dfrac{v^2}{n^2}$	$Wk^2 = \dfrac{W}{39{\cdot}5} \cdot \dfrac{v^2}{n^2}$	(2.9)

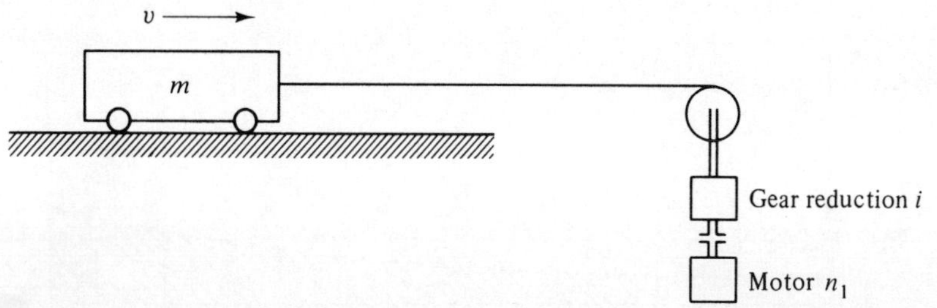

Fig. 2.3 Diagram for referring the inertia of a load moving in a straight line to the motor shaft

In roller table drives the tangentially driven load (Fig 2.4) will have the following values of inertia:

$$J = mr^2 \qquad GD^2 = Gd^2 \qquad Wk^2 = \frac{Wd^2}{4} \tag{2.10}$$

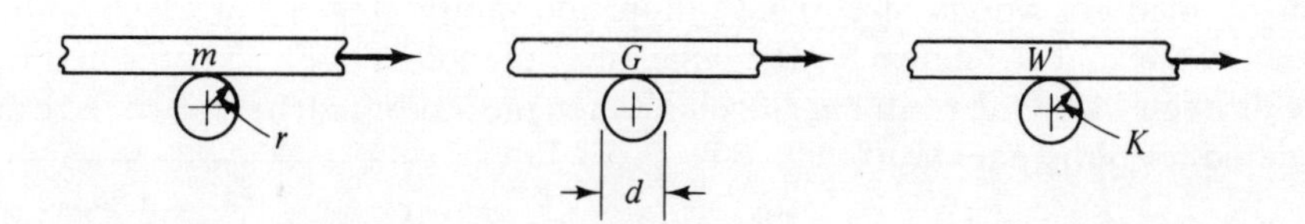

Fig. 2.4 Inertia of load moving in a straight line to drum type motor or driven roller

Inertias of a simple cylinder and hollow cylinder are:

Inertia of hollow cylinder (for solid cylinder d_i and $r_i = 0$)

SI	MKS	FPS	
$J = \frac{1}{2}m(r_e^2 + r_i^2)$	$GD^2 = \frac{1}{2}G(d_e^2 + d_i^2)$	$Wk^2 = \frac{1}{2}W(r_e^2 + r_i^2)$	
$= 98 \times \rho \times l \times (d_e^4 - d_i^4)$	$= 393 \times \rho \times l \times (d_e^4 - d_i^4)$	$= 0{\cdot}1\rho \times l \times (d_e^4 - d_i^4)$	(2.11)
	Suffix e = external, i = inside		
J kgm^2	GD^2kgfm^2	lb.ft^2	
d, r, l in m	m	ft	
ρ in kg/dm^3	ρ in kgf/dm^3	ρ in lb/ft^3	(2.12)

For standard geared three-phase induction motors it is possible to give guide values for the mean accelerating torque as defined in Fig. 2.2 and express it as a multiple of rated torque. Figure 2.5 shows the values of accelerating torque for 2 to 12 pole cage motors up to 50 kW rating. When starting with full voltage applied

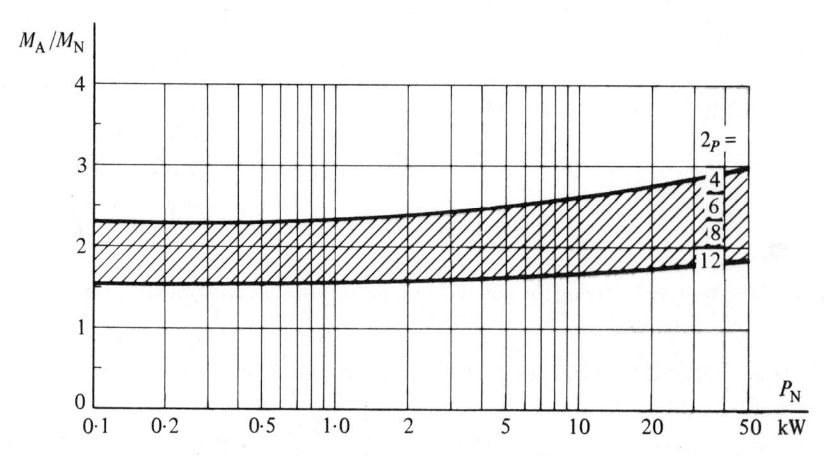

Fig. 2.5 Guide values for per unit starting torque M_A/M_N for standard geared motors with normal rating P_N (Courtesy Eberhard Bauer)

under no load conditions three-phase induction motors reach full speed within a fraction of a second, as shown for the same range of motors in Fig. 2.6. From this the angle through which the rotor has travelled when the motor reaches full speed can be calculated from the equation

$$\varphi_{a01} = 3n_1 t_{a0} \tag{2.13}$$

where φ_{a01} will be in degrees if n_1 is the no-load rotor speed in rev/min, and t_{a0} the time to reach that speed, ascertained from Fig. 2.6.

In many applications, particularly in hoisting loads, braking is required and the following gives the formulae for braking times for lifting and lowering of loads.

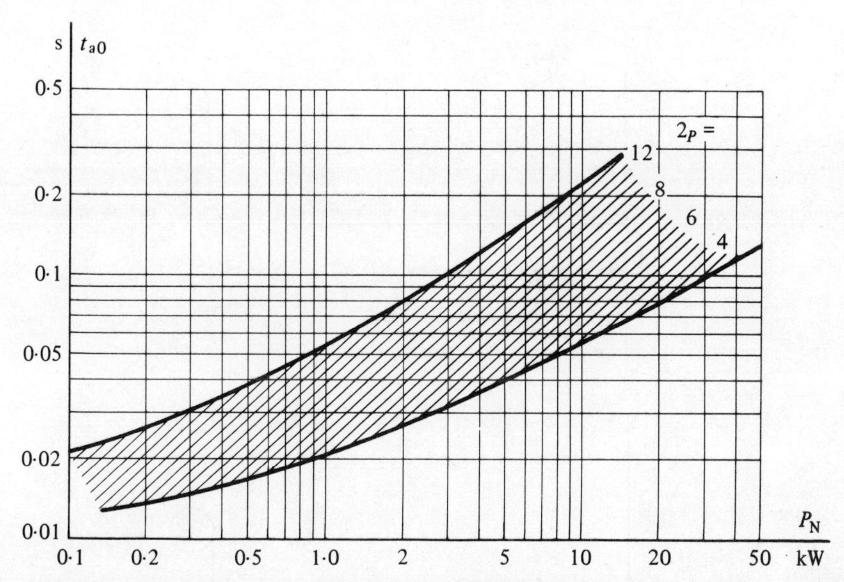

Fig. 2.6 Guide values for the running-up time ta_0 of standard geared motors with rated output P_N (Courtesy Eberhard Bauer)

	SI	MKS	FPS	
Lifting				
	$t_{a1} = \dfrac{Jn}{9{\cdot}55(M_{BR}+M_L)}$	$t_{a1} = \dfrac{GD^2n}{375(M_{BR}+M_L)}$	$t_{a1} = \dfrac{Wk^2n}{308(M_{BR}+M_L)}$	(2.14)
Lowering				
	$t_{a2} = \dfrac{Jn}{9{\cdot}55(M_{BR}-M_L}$	$t_{a2} = \dfrac{GD^2n}{375(M_{BR}-M_L)}$	$t_{a2}\dfrac{Wk^2n}{308(M_{BR}-M_L)}$	(2.15)
t_a	s (seconds)	s	s	
J	kgm^2	GD^2 in $kgfm^2$	$lb.ft^2$	
n	rotor speed rev/min	rev/min	rev/min	

A graphical illustration is given in Fig. 2.7(a) and (b).

2.13.3 *General equations for rotating systems with variable torques*

The typical system shown in Fig. 2.8(a) contains inertia and load-torque components which can be classified as follows:

Constant loads—independent of speed, T_3.
Loads whose magnitude is directly proportional to speed, T_1.
Loads varying with some power (a) of speed, T_2 a is, for example, $=2$ for certain types of fans.

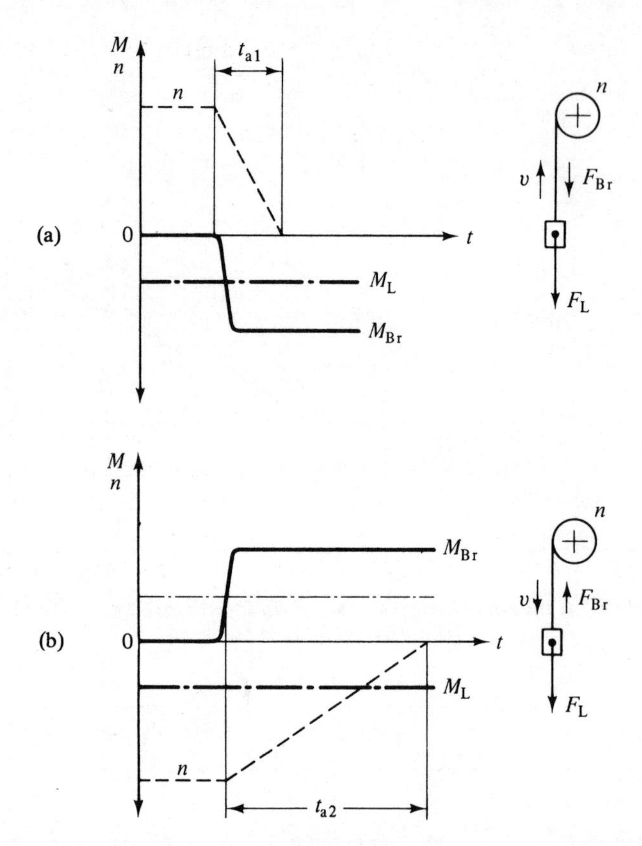

Fig. 2.7(a) Load torque M_L and braking torque M_{BR} act against motion and have a decelerating effect when the load is being lifted. (b) The load torque M_L has an accelerating effect whereas the braking torque causes decelerating so that the net-decelerating torque when lowering is $M_{a2} = M_{BR} - M_L$

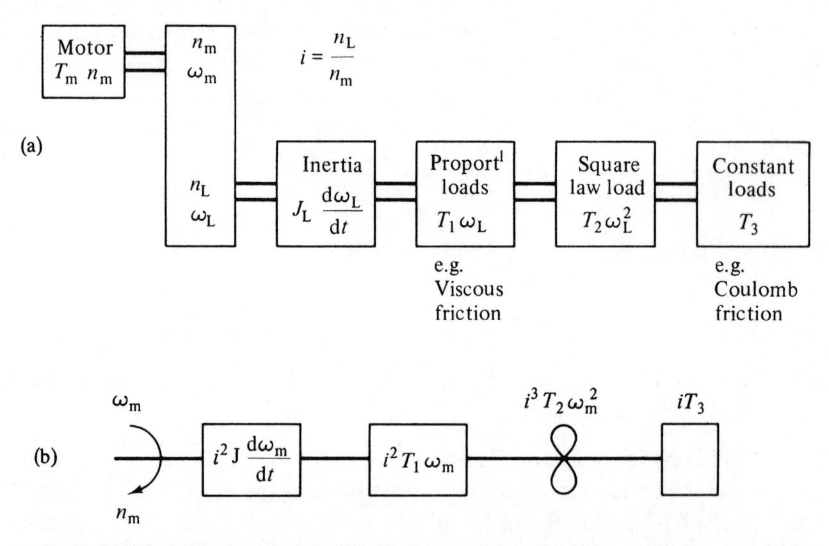

Fig. 2.8(a) Rotating system. (b) Equivalent system, load referred to motor shaft

The equivalent load torque referred to the motor shaft for the system equivalent to Fig. 2.8(a) (Fig. 2.8(b)) is given by

$$T_L = i^2 J \frac{d\omega_n}{dt} + i^2 T_1 \omega_M + i^3 T_2 \omega_M + iT_3 \tag{2.16}$$

Unless subject to special control the motor torque also varies with speed and generally the speed/torque curve cannot be expressed in the form of a simple equation for the whole of the speed range from standstill to no-load speed. In section 2.13.2, Eq. (2.6) for the time an induction motor takes to run up to speed when started on full voltage was obtained by taking an arithmetic average for the running-up torque and assuming it to remain constant. The load torque which has to be subtracted to obtain the accelerating torque was also assumed to remain constant. If more accurate calculations are demanded the equations of motor and load characteristics must be determined. If integration and evaluation of the resulting equations proves too difficult, graphical methods of integration may be used.

The speed/torque characteristic of a typical d.c. series motor is shown in Fig. 2.9. The portion of the characteristic from starting to point A corresponding to a speed of 4500 rev/min is considered to be near enough linear and can, therefore, be represented by an equation of the form

$$T = T_s - \omega \tan \alpha \tag{2.17}$$

where

$$\tan \alpha = \frac{T_s}{\omega_0}$$

T_s is the starting torque of the motor and ω_0 the no load speed which is the intercept on the speed axis when the linear portion T_s to n_1 is produced.

The section of the speed/torque characteristic from about 4500 rev/min deviates only slightly from a curve of the form $T\omega^3 = C$.

Dealing firstly with the straight line portion: the starting torque of the motor is generally given for the full value of rated voltage. Starting resistances or voltage drops in the main supply to the motor considerably reduce the value of T_s and must be duly taken into account. Where the starting resistance is reduced in steps the complex speed/torque characteristic is better dealt with graphically.

Substituting $T = T_s - T_s/\omega_0\, \omega$ into the formula gives

$$t = J \int_0^{\omega} \frac{d\omega}{T_s - (T_s/\omega_0)\omega} \tag{2.18}$$

which after integration becomes

$$t = \frac{J\omega_0}{T_s} \times lg_e \frac{\omega_0}{\omega_0 - \omega} \tag{2.19}$$

(Where the straight line actually represents the motor characteristic $\omega_0 = \omega_{NL}$.)

The evaluation of this equation can be facilitated if the particular value of ω for which t is required is expressed as a fraction of ω_0.

$$\omega = K\omega_0 \qquad (K < 1)$$

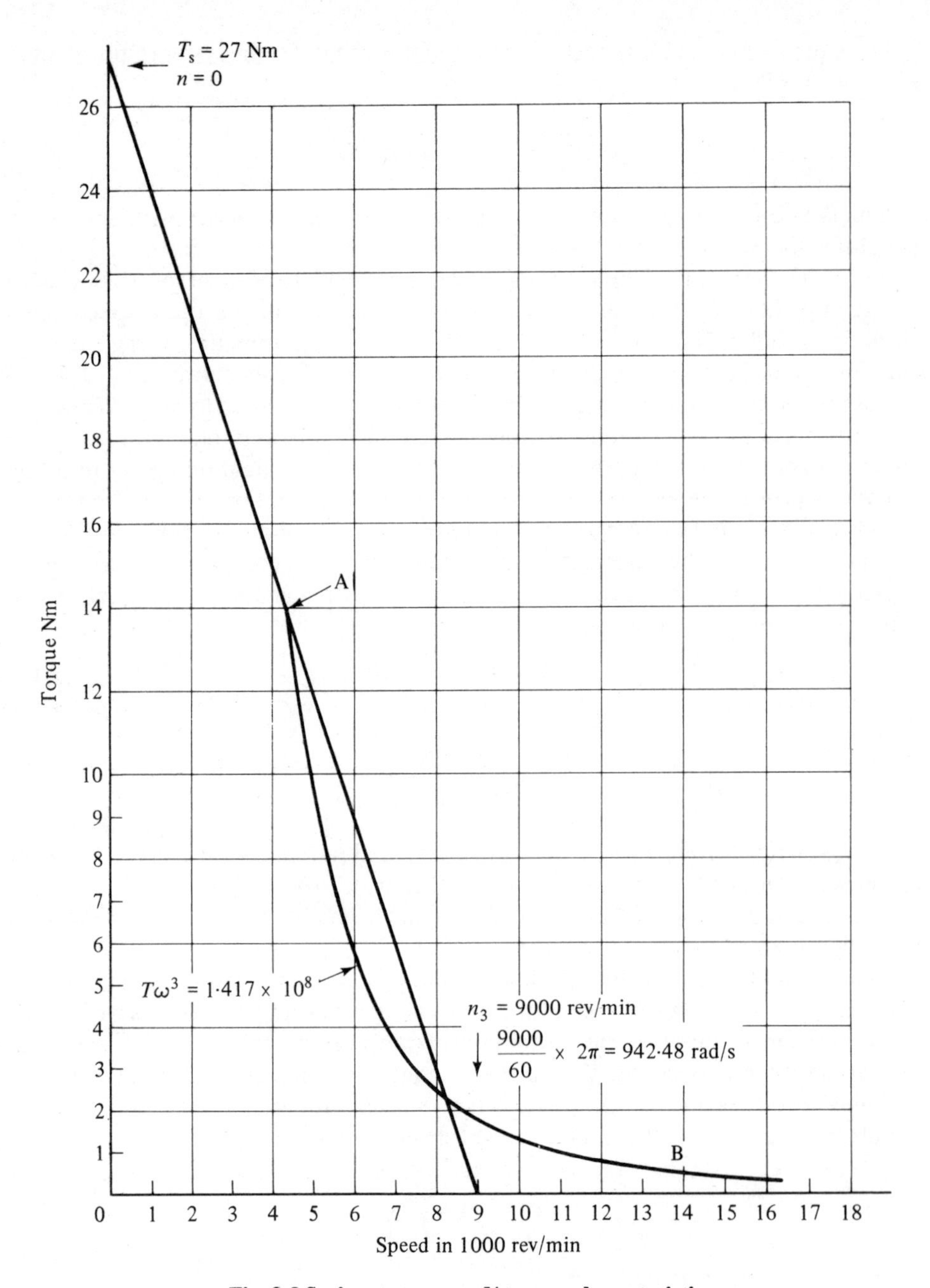

Fig. 2.9 Series motor speed/ torque characteristic

The equation (2.19) can be written in the form

$$t = \frac{J\omega_0}{T_s} \quad lg_e \frac{1}{1-K} \tag{2.20}$$

Plotting $lg_e \cdot 1/(1-K)$ for values of K from 0 to about 0·95 as shown in Fig. 2.10

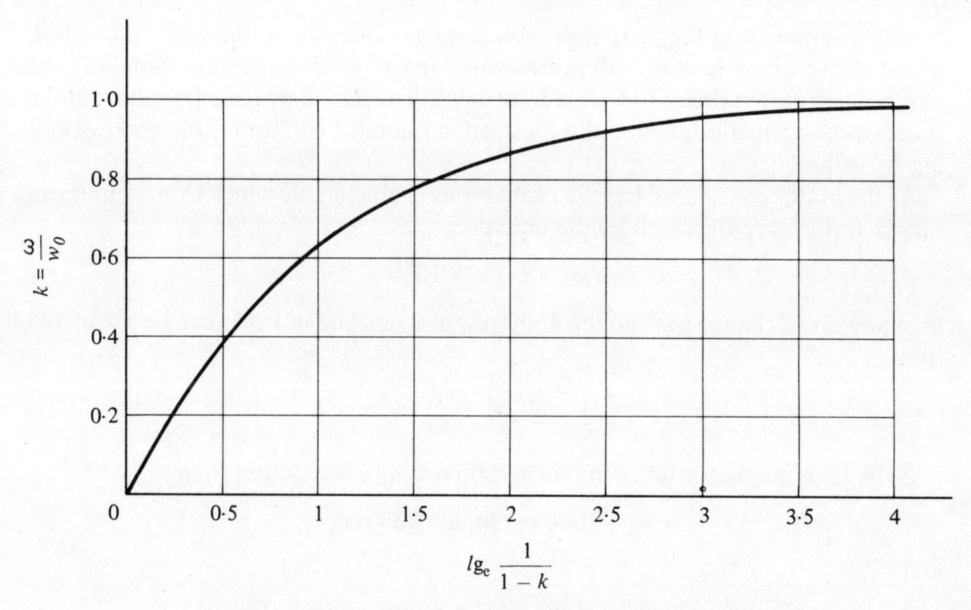

Fig. 2.10 Values of $lg_e(1/1-K)$ for per unit speeds $K(\omega/\omega_0)$

then makes it possible to write the equation in a form readily evaluated as the following example shows:

A motor with linear speed/torque characteristic has a starting torque of 27 Nm and a no-load speed of 9000 rev/min. The inertia of the rotating parts is 0·0547 kgm^2. Find the time it takes to reach 4500 rev/min.

The equation for the speed/torque curve will be

$$T = T_s - \frac{T_s}{\omega_0}\omega$$

Now

$$\omega_0 = \frac{9000\ 2\pi}{60} = 942{\cdot}48 \text{ rad/s}$$

Therefore

$$T = 27 - \frac{27}{942{\cdot}48}\omega$$

$$K = \frac{\omega}{\omega_0} = \frac{4500}{9000} = 0{\cdot}5$$

and from the graph, Fig. 2.10,

$$lg_e \frac{1}{1-K} = 0{\cdot}693$$

Therefore

$$T = \frac{0{\cdot}0547 \times 942{\cdot}48 \times 0{\cdot}693}{27} = 1{\cdot}32 \text{ s}$$

The ordinates K of Fig. 2.10 represent angular velocity ω to the scale $K\omega_0$. Thus, in the above example $K=0{\cdot}50$ represents a speed of 4500 rev/min. Similarly the abscissae represent time t to a scale factor also determined by the parameters of the speed-torque equation, and in the illustration quoted 1·905 times the corresponding lg_e value.

As the motor considered has the usual series characteristic which from A to B can satisfactorily be represented by the equation

$$T\omega^3 = 1{\cdot}417 \times 10^9 = k$$

then any speed change beyond the 4500 rev/min reached in 1·32 s can be calculated from

$$t = \frac{J}{k}\int_{\omega_1}^{\omega_2} \omega^3 \, d\omega = \frac{J}{4k}(\omega_2^4 - \omega_1^4) \tag{2.21}$$

If the time the motor takes to reach 7000 rev/min is required then

$$\omega_2 = 7000 \text{ rev/min} = 733 \text{ rad/s}$$

and

$$\omega_1 = 4500 \text{ rev/min} = 471 \text{ rad/s}$$

therefore

$$t = \frac{0{\cdot}0547\,(733^4 - 471^4)}{4 \times 1{\cdot}417 \times 10^9} = 2{\cdot}31 \text{ s}$$

The total time the motor would require to reach a speed of 7000 rev/min when started from rest would, therefore, be 3·63 s.

The slight discontinuity at point A, though affecting the result, will still give a value which is of practical use. In practice many of the factors such as friction, values of torque, and corresponding speeds are subject to manufacturing tolerances and these, combined with environmental and supply conditions will give a range of possible acceleration times. At least two calculations will, therefore, be necessary to ascertain the times the motor requires to reach a certain speed under worst and most favourable conditions.

When the shape of experimentally determined or calculated speed/torque curves makes reasonable mathematical approximations with simple and easy-to-integrate equations difficult, graphic integration will not only be found simpler but quicker. The procedure is to choose increments of speeds which take account of the nature of the curve. In the example quoted the interval from 0 to 4000 rev/min is divided into four equal intervals, and that from 4000 to 7000 into six as shown in table 2.2. The value of torque at the mid-point of each interval is read from the graph and entered in column 3, whereupon the time taken to make the speed changes $\Delta\omega_i$ is calculated from $\Delta t_i = IJ \cdot \Delta\omega_i / T_i$. The cumulative total will then give the time to reach a certain speed, and, as can be seen in this case for 7000 rev/min, table 2.2 agrees with the results obtained from the calculations based on the two equations made to represent the motor characteristic.

The time a rotating mass takes to change speed is given by Eq. (2.6) and if the net accelerating torque varies in a complex, discontinuous manner due, for example, to switching steps during starting it is useful to plot a graph representing this time more

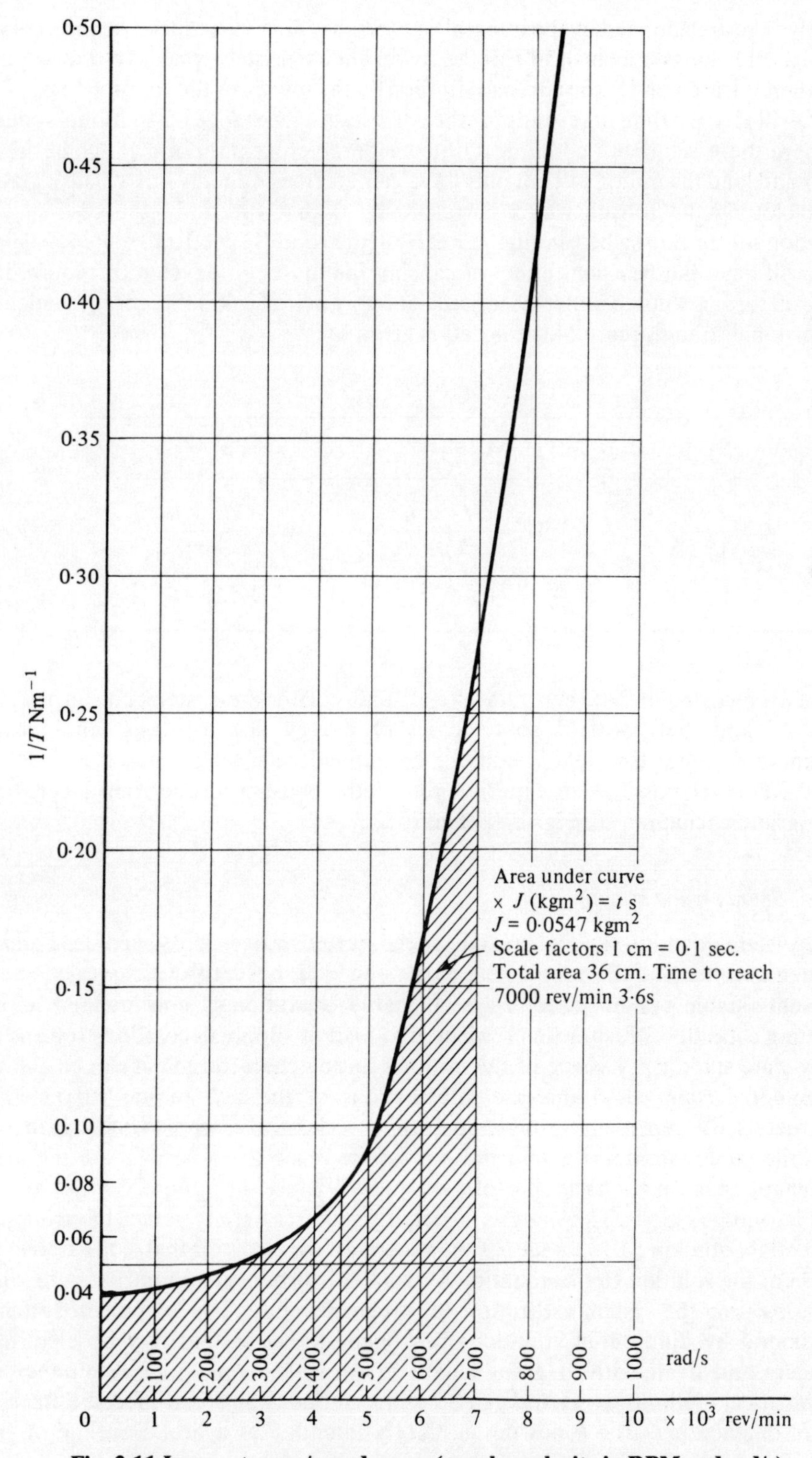

Fig. 2.11 Inverse torque/speed curve (angular velocity in RPM and rad/s)

directly. The area defined by the integral is that under an inverse speed/torque curve, and Fig. 2.11 shows ω against $1/T$. If the correct units, namely Nm and radians/s, are used then the area under the curve multiplied by the inertia of the rotating mass, J in kg m^2, will give the time in seconds for the corresponding change of rotational speed.

Where there is either a coupling with considerable backlash or a clutch between motor and load the motor will not only have gained kinetic energy but will be capable of developing the torque which corresponds to that speed. In the case of the induction motor it may be running at nearly synchronous speed and clutch engagement will have the transient effect of causing the motor to develop, if required, a value of torque limited only by its pull-out torque. The kinetic energy which is proportional to the square of the speed is given by:

SI	MKS	FPS	
$W=\dfrac{J\times n^2}{182{\cdot}5}$	$W=\dfrac{GD^2\times n^2}{7160}$	$W=\dfrac{Wk^2\times n^2}{5880}$	(2.22)
Joules	kgfm	lbf.ft	

When engagement between motor and load occurs the system inertia will be increased and there will be some losses of energy due to elastic and plastic deformation. These transient conditions are difficult to calculate and it is usual to resort, wherever possible, to experimental methods using torque transducers and oscillographs. (chapter 16, Fig. 14(a) and (b).)

2.13.4 *Steady-state stability*

In the system equation the steady-state speed/torque curves have been used and it has been assumed that the point where motor and load speed/torque curves intersect represents stable system equilibrium. If stable operation is now defined as an operating condition which, when subjected to load or supply fluctuations returns to steady-state stability by virtue of the motor's torque characteristic it can be shown that under certain conditions the intersections of the two torque curves may nevertheless also represent unstable operation. A classical example (Fig. 2.12(a)) of this is the induction motor whose pull-up torque is too low to accelerate the load through the spectrum of harmonic torques, and will cause the motor to 'crawl' at one of the harmonic speeds. The analysis of stability requires a study of the characteristic differential equation (2.16), page 39. In general, it may be said that if the transient portion of the solution to the equation for small disturbances approaches zero with increasing time the system will return to its original state. The analysis of systems represented by differential equations has been studied extensively by electrical engineers and dealt with at great length in control engineering. A number of mathematical methods exists and can be found in books on servo-mechanisms and control engineering. The following is merely intended as a broad outline of the principles involved and a possible appraisal of stability.

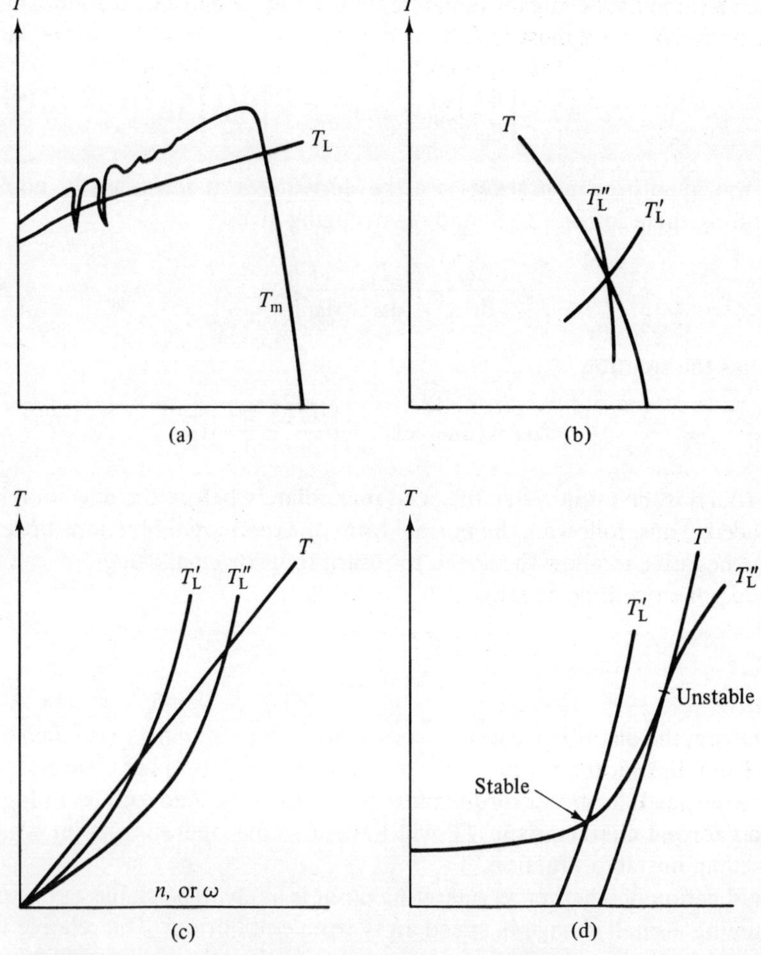

Fig. 2.12 (a) Induction motor with excessive torque load. (b) Motor torque characteristic with load lines showing stable, T'_L, conditions. (c) d.c. motor driving compressor. (d) reel motor with stable, T', and unstable, T'', load characteristics

Taking the simplified form of Eq. (2.16):

$$T(\omega) = T_L(\omega) + J\frac{d\omega}{dt} \tag{2.23}$$

where T is the net torque available for accelerating or decelerating the inertia and dealing with other torques, T_L is a function of speed whose small but finite deviations are AT, AT_L, and ACO, so that after a small disturbance Eq. (2.23) becomes

$$J\frac{d\omega}{dt} + J\frac{d(\Delta\omega)}{dt} + T_L + \Delta T_2 - T - \Delta T = 0 \tag{2.24}$$

and subtracting Eq (2.23) yields

$$J\frac{d(\Delta\omega)}{dt} + \Delta T_L - \Delta T = 0 \tag{2.25}$$

If the increments are made so small that they may be expressed as linear functions of the change in speed then

$$\Delta T = \left[\frac{dT}{d\omega}\right]\Delta\omega \quad \text{and} \quad T_L = \left[\frac{dT_L}{d\omega}\right]\Delta\omega$$

where the quantities in brackets are the derivatives at the point of equilibrium. Substituting them in Eq. (2.25) and re-arranging gives

$$J\frac{d(\Delta\omega)}{dt} + \left[\frac{dT_L}{d\omega} - \frac{dT}{d\omega}\right]\Delta\omega = 0$$

which has the solution

$$\Delta\omega = (\Delta\omega)_0 \exp\left\{\frac{1}{J}\left[\frac{dT_L}{d\omega} - \frac{dT}{d\omega}\right]t\right\} \tag{2.26}$$

where $(\Delta\omega)_0$ is the initial value of speed immediately before the deviation has been introduced. Thus, following the normal form of exponential functions the exponent must be negative to allow the speed to return to its original value.

The exponent will be negative if

$$\frac{dT_L}{d\omega} - \frac{dt}{d\omega} > 0 \tag{2.27}$$

If, therefore, the disturbance has caused the motor speed to decrease then to restore equilibrium the motor torque must exceed the load torque. Conversely for an increase in speed the motor torque must be less than the load torque. In Fig. 2.12(b) the load torque characteristic T'_L will have a stable operating point whereas T''_L results in an unstable situation.

An indication of whether an operating point is likely to be stable can be obtained by assuming a small change in speed away from equilibrium. The relative values of motor and load torque will then determine whether the speed will return to its previous value. If the increase in load torque due to an increment in speed is greater than the corresponding increase in motor torque then the speed will tend to be reduced and return to its original value, indicating stable operation. Conversely if the speed continues to increase the system will be unstable. In practice transient speed/torque characteristics often deviate from those of the steady-state curve and the complex interaction of load, motor, and control gear characteristics must be considered. In particular it must be borne in mind that in addition to mechanical time constants the electrical ones may become significant. The energy stored in the magnetic field, and the kinetic parameters can be considered analogous to a spring and disturbances in a motor drive can cause damped, undamped, or even resonant conditions.

Figure 2.12(c) shows the speed/torque curve T of a specially designed variable voltage d.c. motor driving the air compressor for cabin pressurization in an aircraft. The motor speed/torque curve is almost linear and that of the compressor approximately parabolic. The load curve T'_L will obtain for a lower altitude than the curve T''_L and though the motor torque increases rapidly with speed both operating points will be perfectly stable.

The curves in Fig. 2.12(d) relate to a reeling motor in a strip-mill. Since the speed of the strip is constant through the rollers the speed of the reel motor must be decreased as the diameter of the coil increases. Furthermore, the reel motor must develop sufficient torque to maintain tension which is its principal load. The curve T_L in Fig. 2.12(d) indicates the load characteristic. Should the speed of the reel motor fall below that of the final roll, the strip tension disappears and load torque drops to a very low value. If the reel motor has the drooping characteristic of a normal shunt motor, T', the operating point, will be perfectly stable. Should there, however, be a tendency, due to overcompensation, for the speed to rise at high loads, T'', the operation becomes unstable.

Table 2.1

1	2	3	4	5
$\Delta\omega$	rev/min	T_i Nm	$\Delta t_i \dfrac{J\,\Delta\omega_i}{T_i}$	$\sum \Delta t_i$
1	1000 =	25·63	0·224	0·224
2	104·72	22·60	0·254	0·478
3	rad/s	19·59	0·293	0·771
4		16·58	0·346	1·117
5		14·35	0·200	1·317
6	500 =	500 =	11·39	0·252
7	52·36	8·54	0·336	1·905
8	rad/s	6·50	0·441	2·346
9		5·06	0·568	2·914
10		4·03	0·713	3·627

$J = 0{\cdot}05473$ kgm^2

Bibliography

1. D. D. Stephen, 'Effect of system voltage depressions on large a.c. motors', *Proc. IEE*, **113** No. 3 (March 1966).
2. D. D. Stephen, 'Motor Stability during system disturbances', *Electrical Times* (17 July 1969).
3. D. D. Stephen, 'Generalized approach to motor stability', *Electrical Times* (19 February 1970).
4. D. D. Stephen, 'Simulating multiple machines for stability studies', *Electrical Times* (3 September 1971).
5. R. W. Jones, *Electrical Control Systems*, Chapman & Hall, London (1953).
6. H. Greiner, *Electrical Drives with Geared Motors*, Eberhard Bauer Publication SD 475 E.
7. J. D. Burby, 'Acceleration of Inertia Loads with a Varying Torque', *Product Engineering* (December 1947).
8. V. H. Seliger, 'Speed-torque Analysis for Rotating Mechanical Systems', *Product Engineering* (December 1956).
9. V. H. Seliger, 'Speed-torque Analysis for Complex Mechanical Systems', *Product Engineering* (February 1957).

3 Direct Current Motors

The d.c. motor has maintained its role as a power drive for industry due to its inherent advantages in flexibility and speed control compared to the a.c. induction or synchronous motor. The d.c. motor is essentially a variable speed motor and with forced ventilation it is capable of operating continuously over a very wide range. By variation of field windings its characteristics can be matched to most industrial applications. Designs have been improved to give better reliability and reduced maintenance to suit modern industry. This has been helped by the widespread use of thyristor control which demands a motor of good basic design to cope with the unusual waveform of the supply.

3.0.1 *List of symbols*

V = Applied voltage (volts)
E = back e.m.f. (volts)
I = armature current (amps)
R = armature circuit resistance (ohms)
ϕ = magnetic flux (webers)
p = number of main poles
a = number of armature circuits
n = speed (revs per second)
Z = armature conductors
C = number of commutator segments
L = inductance (henrys)
l = conductor length or core length (metres)
D = armature diameter (metres)
B_{av} = specific magnetic loading (tesla)
ac = specific electric loading (ampere conductors/metre)

3.1 Motor Equations

The d.c. motor consists of an armature with conductors which rotates in a fixed magnetic field. There are two basic types, homopolar and heteropolar. We will consider mainly the normal heteropolar type in which the main poles are of alternate polarity. Figure 3.1 shows the simplest two-pole motor and how the conductors move alternately under a north pole and then a south pole. The voltage generated in each conductor is thus an alternating one. The conductors are, however, connected by a commutator so that the voltage at the brushes is direct and opposes the d.c. supply

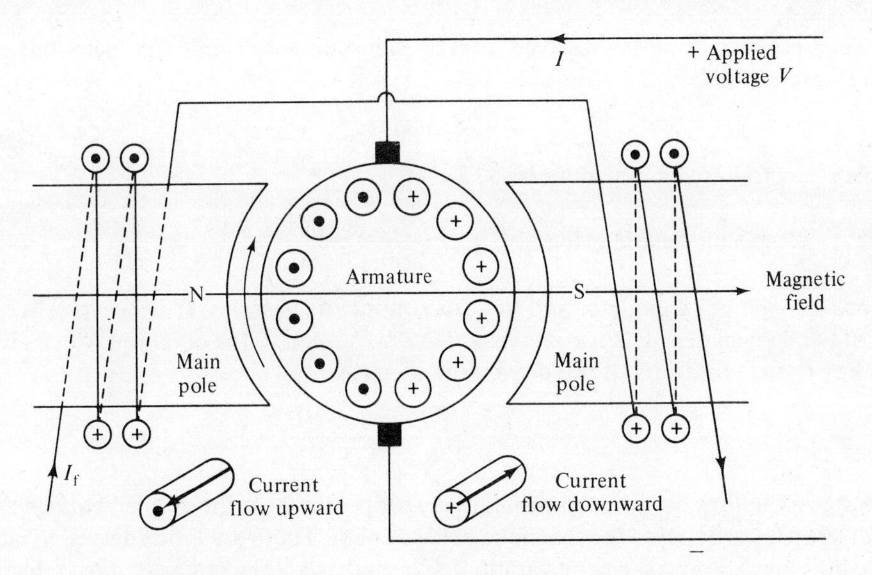

Fig. 3.1 Direction of field, motion, and current in a d.c. motor

voltage. The brushes are shown for simplicity between the poles. For many practical motors they are under the poles. See Fig. 3.14.

3.1.1 *Voltage, flux, and speed*

The applied voltage V is opposed by electromotive force E giving the steady state equation

$$V = E + IR \tag{3.1}$$

where IR is the voltage drop in the armature circuit due to current I and includes the drop of the brushes.

The e.m.f. is that resulting from electromagnetic induction as discovered by Faraday and for a coil of N turns and flux ϕ is given by

$$E = -N\frac{d\phi}{dt} \text{ volts.} \tag{3.2}$$

E is given the negative sign or alternatively is described as a back e.m.f. because the direction of the induced e.m.f. in a closed circuit would produce a current with magnetic effect opposing the change of flux (Lenz's law).

If a motor has p poles then the flux changes from $+\phi$ to $-\phi$ in $1/p$ of a revolution or $1/pn$ seconds.

Thus $d\phi/dt = 2\phi\, pn$ and back e.m.f. $E = 2N\phi pn$.

If Z is the total conductors on the armature and a is the number of parallel circuits then $N = Z/2d$,

$$\text{Hence } E = \frac{Z}{a}\phi pn \tag{3.3}$$

This expression may also be derived from consideration of a single conductor moving in a magnetic field.[1]

$$e = B_{av} l v \text{ volts} \tag{3.4}$$

Where $v = \pi D n$ = velocity in metres/second

$$\text{specific magnetic loading } B_{av} = \frac{\text{flux per pole} \times \text{poles}}{\text{armature surface area}} = \frac{\phi p}{\pi D l} \tag{3.5}$$

Hence $e = \phi p n$ per conductor and for Z/a conductors in series. $E = Z\phi pn/a$, as Eq. (3.3). For a given design Zp/a is constant and so we derive the equation which gives the key to the versatility of the d.c. motor, $E = K\phi n$

$$\text{or speed } n = \frac{E}{K\phi} = \frac{V - IR}{K\phi} \tag{3.6}$$

The speed can thus be varied approximately proportional to the applied voltage and inversely proportional to the magnetic field strength. There are limits discussed later as to how much V and ϕ can be varied. If V is increased or ϕ reduced, thus reducing E, the current I is temporarily increased giving an accelerating torque which increases the speed until the steady-state condition at a higher speed is reached.

3.1.2 *Torque*

The conductors in the magnetic field move because there are forces acting on them. The force is produced at right angles to both the field and the conductor. This is illustrated in Fig. 3.15 which shows Fleming's left-hand rule for motors. If i = conductor current,

$$\text{Force } F = B_{av} i l \text{ newtons} \tag{3.7}$$

$$\text{Total force } F_T = \text{sum of forces on all the conductors}$$

$$= il \sum B_1 \cdots B_Z = Z B_{av} i l$$

From Eq. (3.5) for B_{av},

$$F_T = \frac{Z\phi p}{\pi D} \times \frac{I}{a}$$

and torque = force × radius.

$$\text{Torque} = \frac{1}{2\pi} \times \frac{Z\phi I p}{a} \text{ newton metres} \tag{3.8}$$

The torque is proportional to the flux and current on a given motor design. Note that this is the gross torque. The net shaft output torque is less than this by a small percentage which is absorbed by the windage and friction losses, the iron losses and some of the stray losses. (See section 3.1.9.)

Since in the normal type of d.c. motor with the conductors in slots the force is not produced directly on the conductors but mainly on the teeth it is preferable to derive the torque by an alternative method.[2]

$$\text{Gross shaft power} = VI - I^2R = I(V - IR) = EI \tag{3.9}$$

$$\text{Gross torque} = \frac{EI}{2\pi n} \text{ (and using Eq. (3.3))} = \frac{I}{2\pi} \times \frac{Z\phi Ip}{a} \qquad \text{(as 3.8)}$$

The specific electric loading is equal to the ampere conductors per metre of armature periphery.

$$ac = \frac{ZI}{a\pi D} \tag{3.10}$$

3.1.3 *Power*

$$\text{Gross shaft power (watts)} = \text{Gross torque} \times 2\pi n \text{ (newton-metre/second)} \tag{3.11}$$

$$\text{From eq. (3.9) gross power} = EI \text{ (watts)}$$

The output shaft power is slightly less than this and is expressed usually in kilowatts. 1 kW = 1000 watts = 1·34 h.p.

From Eq. (3.11), Power $= Z\phi Ipn/a$, Eq. (3.5), $B_{av} = \phi p/\pi Dl$, and Eq. (3.10), $ac = ZI/a\pi D$, we get the expression

$$\text{Power} = K_0 B_{av} ac D^2 l n \tag{3.12}$$

where

$$K_0 B_{av} ac = \text{output coefficient.}$$

Power is proportional to $D^2 l$ and speed. Note that l is the motor core length.

3.1.4 *Output with constant field*

From Eq. (3.8) for a given design where Z, p, and a are fixed,

$$\text{Torque} = K_T \phi I. \tag{3.13}$$

If ϕ is kept constant the torque is proportional to I and if I is also kept constant then the torque is constant. To vary the speed n means that the e.m.f. and hence the armature voltage must vary. (Equations (3.1) and (3.3)).

Figure 3.2(a) shows the power output capability of a d.c. motor with constant flux and armature current up to speed n_0. Figure 3.2(b) shows the torque output. The speed n_0 is the speed attained with maximum armature voltage and maximum field current or flux. It is known as the base speed.

3.1.5 *Output with variable field*

With torque $= K_T \phi I$, if I is kept constant then as the field is weakened the torque will reduce. But from Eq. (3.3) the speed will increase. If V and I are kept constant then the power will be constant (neglecting losses). Figure 3.2(a) shows the power output capability of a d.c. motor with constant armature voltage and current and variable field between speeds n_0 and n_t. Figure 3.2(b) shows the torque output. n_t is the maximum speed and the range n_t/n_0 is the speed range with field weakening. There are limitations on this range due to commutation and stability.

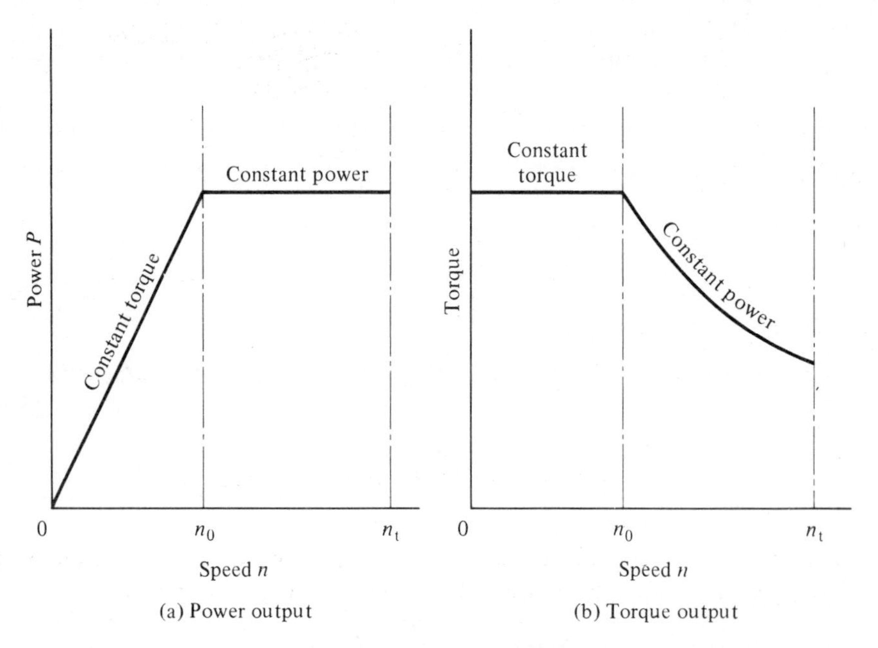

Fig. 3.2 Output power and torque; $0-n_0$, variable armature volts, constant field, n_0-n_t, constant armature volts, variable field

3.1.6 *Field strength and saturation*

The magnetic field strength or flux ϕ is measured in webers (1 weber $= 10^8$ lines of flux) and the density of the field B is given in tesla (1 tesla $=$ 1 weber/sq. metre). It results from the current flowing through the field coils which gives a value of ampere turns or it can result from the use of permanent magnets. The field strength is not a linear function of the ampere turns but depends on the permeability and saturation of the magnetic materials used in the construction. Each material has a curve relating flux density B and magnetizing force H which is stated in ampere turns per metre.

The curve is expressed in the form

$$B = \mu H \tag{3.14}$$

where μ is the permeability of the material and is a variable except for non-magnetic materials including air when $\mu = \mu_0$ and is equal to $4\pi \times 10^{-7}$. The magnetic materials used in d.c. motors are steel in various forms and μ varies over a wide range. Working from a value of flux required, and knowing the magnetic circuit areas, the densities B are calculated. From the standard B–H curves for the materials and the lengths of the magnetic paths the required ampere turns are calculated.

Figure 3.3 shows the normal parts of the magnetic circuit and these are as follows:

Air gap, armature teeth, and armature core, main poles and yoke, pole teeth if compensated.

Figure 3.4 shows a typical magnetization curve. The working flux ϕ requires AT ampere turns. It is economically necessary to work in the saturated region. The curve does not pass through the origin but with zero ampere turns there is a flux ϕ_R known

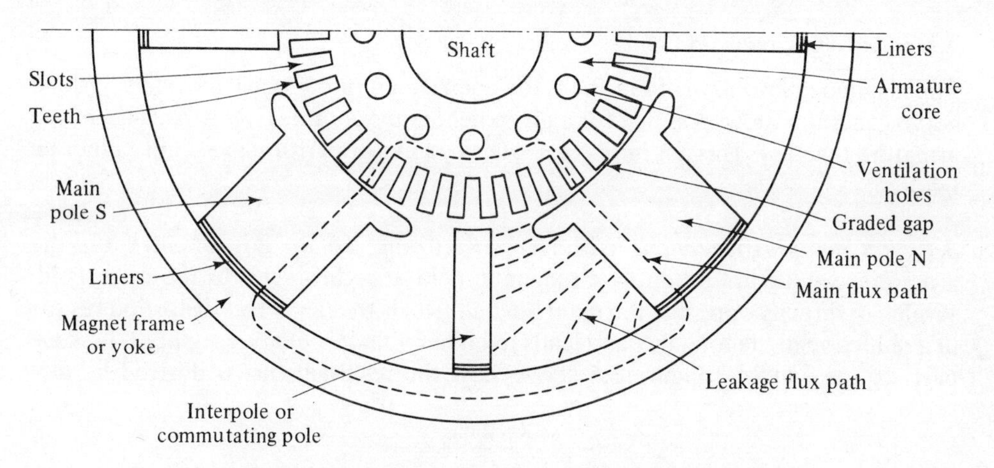

Fig. 3.3 Parts of the magnetic circuit and no-load flux paths

as the residual flux. This is due to the hysteresis effect in magnetic materials whereby as H is decreased and reversed the same values of B are not obtained.

Another factor in the estimation of the ampere turns is the leakage flux since there are no effective barriers to magnetism.

Figure 3.3 shows lines of leakage flux from main poles to interpoles. The useful flux is that in the air gap. Due to this leakage the flux in the poles and yoke on a d.c. motor may be 1·1 to 1·2 times the working flux. More vital still is interpole leakage flux and here the total flux at the back of the pole can be several times the working flux, thus making it difficult to avoid interpole saturation which can upset commutation.

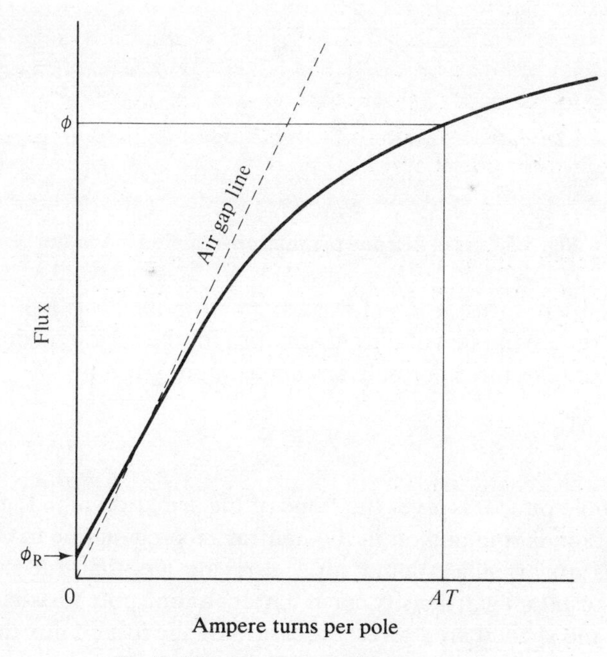

Fig. 3.4 No-load magnetization curve

3.1.7 *Armature reaction*

The magnetization curve of Fig. 3.4 is the open circuit or no-load curve. The on-load curve generally shows less flux for a given field ampere turns largely because of the armature reaction. This has important effects on motor performance, and cannot be ignored.

3.1.7.1 Cross magnetization. With reference to Fig. 3.1 the flow of current in the armature conductors produces a magnetic field at right angles to the main field. While not directly opposing the main field it distorts the field. This distortion results in a reduction in the total flux and leads to higher voltage between segments in some parts of the winding. Figure 3.5 shows how the resultant flux is derived by first

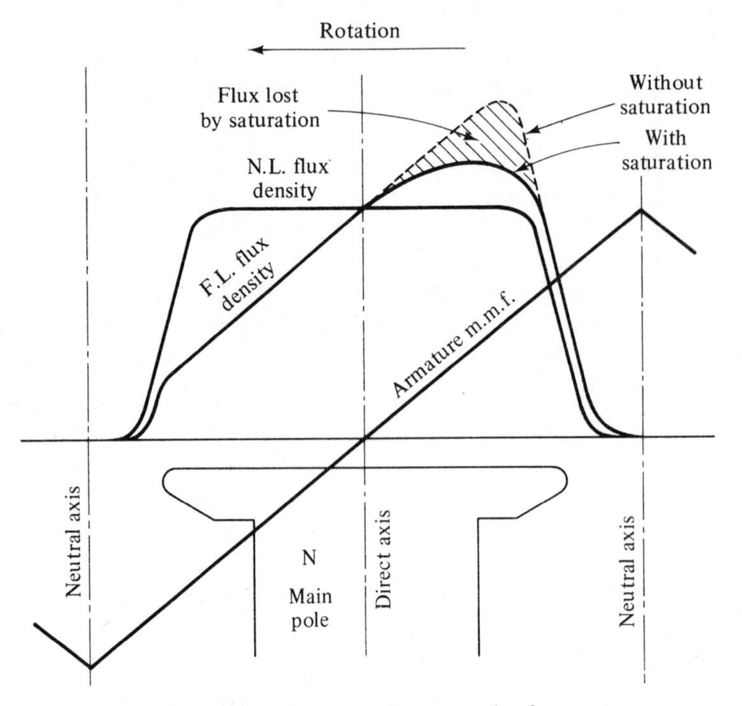

Fig. 3.5 Loss of flux due to armature reaction in a motor

summing the ampere turns or m.m.f.s at any point in the air gap of the motor. The field ampere turns are the sum of the ampere turns of the various main pole windings. The armature ampere turns per pole are a maximum given by

$$AT_a = \frac{ZI}{2ap} \tag{3.15}$$

Taken over a pole pitch this gives the slope of the armature m.m.f. graph. In actual practice, the graph is rounded off in the neutral axis region due to chording effects and some coils undergoing commutation, carrying less than full current. Due to saturation the resultant gap density curve flattens at one pole tip where the m.m.f. is increased and the shaded area gives a measure of the loss of flux due to armature reaction.

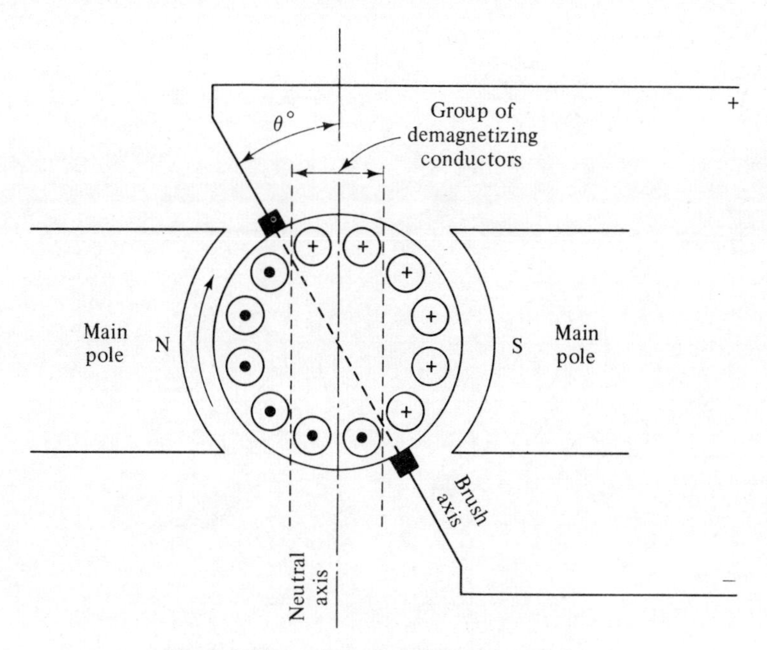

Fig. 3.6 Effect of brush shift against rotation

3.1.7.2 Brush shift. The above applies with the brushes in the neutral position or in the quadrature axis. If the brushes are rocked from this axis then further armature reaction effects come into play. Figure 3.6 shows the brushes rocked backwards from neutral by $\theta°$. The armature m.m.f. is now split into two components and part now becomes direct demagnetizing. Hence, in a motor, if the brushes are shifted against rotation, it increases the armature reaction and more excitation is required on the main field. The opposite effect is attained if the brushes are rocked forward.

$$\text{Ampere turns per pole for brush shift} = \pm\frac{ZI}{2ap}\times\frac{H\theta}{180} \tag{3.16}$$

or

in terms of commutation segments

$$\pm\,\text{segment shift}\times\frac{ZI}{a\times C} \tag{3.17}$$

3.1.7.3 Compensating winding. On certain types of motor the effects of armature reaction are undesirable and if they cannot be accepted it is necessary to use a compensated motor. This has a pole face compensating winding carrying the armature current and producing an m.m.f. which cancels or partially cancels the armature m.m.f. in the main pole region. The latter case is sometimes known as a semi-compensated motor. Figure 3.7 shows the basic principle. With full compensation there is virtually no reduction in flux. The compensating winding can also be displaced from the quadrature axis to give a compounding effect.

3.1.7.4 Effect of interpoles. As detailed in section 3.4 it is normal practice on most d.c. motors to provide interpoles or commutating poles to achieve satisfactory

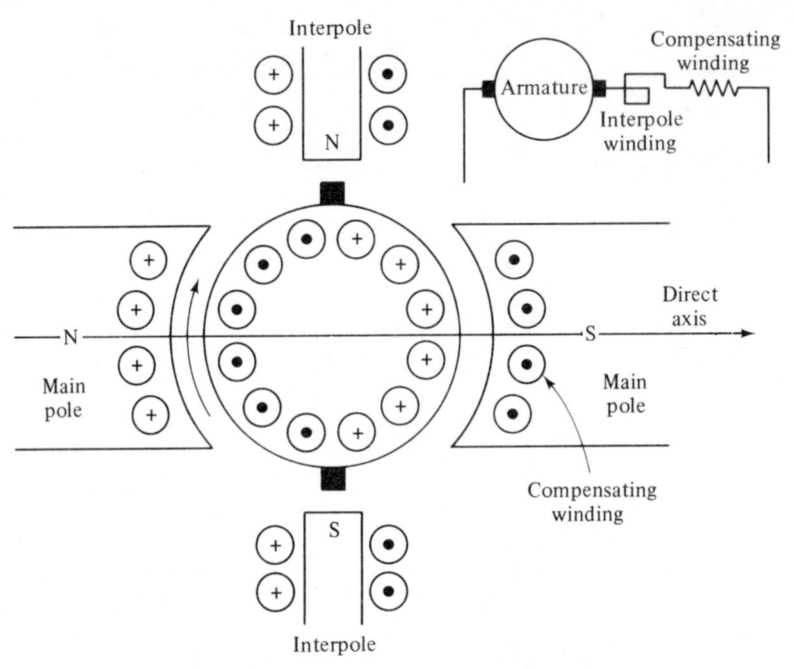

Fig. 3.7 Interpole and compensating windings. (Brushes in neutral)

commutation. These poles with their windings again carrying the armature current come between the main poles. (See Figs. 3.3 and 3.7.) In this region they neutralize the armature m.m.f. and in addition have extra ampere turns to provide a field which produces a commutating voltage. If the interpoles are not exactly of the right strength, and due to various factors this does apply, then there is a compounding effect which alters the effective armature reaction. In a d.c. motor if the interpoles are strong then the excitation is increased and this gives a larger value for armature reaction. If the interpoles are too weak then the opposite effect applies.

3.1.7.5 Gap grading. To reduce the effects of armature reaction it is normal practice now to use graded main pole air gaps. The gap at the tip may be 2 to 4 times the centre gap. (See Fig. 3.21.) The normal flux density at the tip is low and hence any change due to the armature m.m.f. is low. This particularly keeps the peak voltage per segment down.

3.1.8 *Resistance and inductance*

In the circuits for d.c. motors, apart from the applied voltage and the back e.m.f. we have to consider winding resistances, winding inductances, and the effect of brush voltage drop.

3.1.8.1 Winding resistances. These are derived from the basic formula. Resistance = resistivity × length/area.

$$\text{Armature resistance} = \frac{\rho \times Z \times LMT_{\text{a}}}{2A_{\text{a}} \times a^2} \text{ ohms} \tag{3.18}$$

$$\text{Field or interpole resistance} = \frac{\rho \times \text{poles} \times \text{turns/pole} \times LMT_f}{A_f \times (\text{no of parallel circuits})^2} \text{ ohms} \quad (3.19)$$

where

ρ = resistivity of the wire (normally copper)

$= 0{\cdot}173 \times 10^{-7}$ ohm per metre cube at 20°C.

A_a, A_f = conductor cross section (sq. metres)

LMT_a, LMT_f = mean length of turn (metres)

The materials normally used for machine windings increase in resistance with temperature. For copper this amounts to 0·4 per cent increase per °C. This is very significant particularly on class F motors working to 100°C rise by resistance and means that the hot resistance could be 40 per cent more than the cold value.

3.1.8.2 Brush voltage drop. The voltage drop at the brushes has to be included in the armature circuit IR drop. It does not vary linearly with current and can be expressed as $V_b \approx kI_d^m$ where $m < 1$ and I_d is the current density. It is usual practice to either take a nominal value of 1 to 3 volts or the value read from the brush manufacturer's graph. Typical current densities are up to 10 A/cm^2 for electrographitic brushes.

$$\text{Current density} = \frac{2I}{\text{No. brush arms} \times \text{brushes/arm} \times \text{brush area}} \quad (3.20)$$

3.1.8.3 Winding inductances. A winding which produces a magnetic field, and therefore stores energy has an inductance L (henry). The units are volt seconds per amp. If i is the winding current (amps)

$$\text{Stored energy} = Li^2 \text{ joules or watt seconds} \quad (3.21)$$

With an inductive circuit it is not possible to change the current instantaneously and there is a time delay which depends on the time constant of the circuit. If R is the circuit resistance (ohms)

$$\text{Time constant} = \frac{L}{R} \text{ seconds} \quad (3.22)$$

This is the time to reach 63·2 per cent of the final value.

Inductance depends on a number of factors including saturation. The field winding inductance for main pole windings with all coils in series is

$$L_f = \text{No. of coils} \times \text{Flux per amp} \times \text{Turns per coil} \times K_L \text{ (henrys)} \quad (3.23)$$

where K_L is the pole leakage factor.

L_f is proportional to the flux per amp which is the slope of the magnetization curve. From Fig. 3.4 the working range of a machine using field weakening can be 3 to 1 or even higher. The saturated inductance may therefore be only a fraction of the unsaturated inductance. Windings which are closely coupled magnetically such as shunt and series windings on the same pole also have mutual inductance M which tends to delay flux changes due to current changes. On switching on a main pole winding the flux is not established immediately. Also there is a problem involved in switching off such a winding. If the circuit is opened without a discharge path then a very high voltage is induced resulting from the stored energy. A discharge path is essential for switching.

The transient equation for the field circuit is

$$V_f = R_f I_f + L_f \frac{dI_f}{dt} \tag{3.24}$$

The armature circuit inductance can involve a detailed calculation.[3] For practical purposes a formula is available requiring no more than the basic output data, the number of poles and whether compensated or not. This formula has now been adopted by NEMA. See MG1–23.81, April, 1973. With the base speed n_0 in revs. per second,

$$\text{Approximate inductance } L_a = \frac{V}{\pi p \times n_0 \times I} \times C_x \text{ henrys} \tag{3.25}$$

where $C_x \geqslant 0{\cdot}4$ for uncompensated machine, $0{\cdot}1$ for compensated machine.

The value of I in this formula must be taken as the full load rated current as a ventilated machine.

The transient equation for the armature circuit is now given by

$$V = E + IR + L_a \frac{dI}{dt} \tag{3.26}$$

3.1.9 *Losses and efficiency*

The efficiency of a d.c. motor may be tested in different ways but for an initial estimation this is calculated from a knowledge of the individual losses and a summation of these generally in accordance with BS 269 or BS 4999 part 33. IEC 34-2 is on similar lines.

3.1.9.1 Armature iron losses. The two main components are eddy loss and hysteresis loss. This latter is due to the energy absorbed in making excursions round the hysteresis loop for the magnetic material due to the flux reversals. The eddy loss is due to the voltages induced in the conducting steel of the armature and is the reason the core has to be laminated. Insulated stampings reasonably free from burrs are used. Iron loss varies as B^2 and $n^{1{\cdot}5}$. The a.c. frequency $= np/2$. If the ratio of slot opening to gap length is large then a further loss caused by flux pulsation due to the slotting can be significant. The iron losses vary with load and increase with an ungraded gap due to the distortion discussed in section 3.1.7.1. With fully graded gaps the iron loss tends to decrease on load and BS 269 allows a reduction to be made.

3.1.9.2 Armature copper losses. These are the normal I^2R loss of any winding. Additional losses due to eddy current effects can be calculated but will normally be grouped under stray losses. This I^2R loss is computed at the hot value of resistance which is taken as 75°C temperature for class A, E, and B insulated motors and 115°C for class F and H.

3.1.9.3 Interpole, compensating, and series field copper losses. These are the normal I^2R losses.

3.1.9.4. Brush electrical loss. This is taken as the product of the current and a nominal voltage drop at the contact surface per pair of brushes (positive + negative). Other losses in the brush material, flexibles, and connections are neglected. The usual value for the nominal voltage drop is taken as 2 volts.

3.1.9.5 Excitation loss. This is the power required to provide the excitation and normally includes the I^2R loss in the shunt or separate field plus the loss in any regulator.

3.1.9.6 Light running loss. This includes the windage loss, bearing loss, and brush friction loss. It can be easily determined by test but in the design stage has to be partly calculated and partly estimated from tests on similar motors. This loss varies as speed to speed cubed.

3.1.9.7 Stray loss. This is the most difficult to calculate or to measure. Because of this BS 269 allows the use of empirical estimates depending on the motor output, degree of compensation, degree of field weakening, and whether the gap is graded or not. In actual practice the stray losses derive mainly from the following:

Eddy losses in the armature conductors due to skin effects on load.
Stray losses in the magnetic circuit or other metal parts, e.g., steel banding wire.
Commutation losses due to currents in the short-circuited coils.
The basic stray loss for an uncompensated motor on full field and without special gap grading is 1 per cent of the output.

3.1.9.8. Additional losses. Where a separate blower is used then the input to the blower motor would have to be included to give an overall assessment of the losses. Where a motor is supplied from rectified a.c. there are additional losses due to the ripple in the waveform.

$$\text{Copper loss} = I^2R \times (\text{form factor})^2 \tag{3.27}$$

$$\text{where form factor} = \frac{\text{r.m.s. current}}{\text{average current}} \tag{3.28}$$

This type of supply may also affect the iron loss and the stray loss.

3.1.9.9 Efficiency. By the summation of losses method.

$$\text{Output} = \text{Input} - \text{losses} \tag{3.29}$$

$$\text{Efficiency } \% = \frac{\text{output}}{\text{input}} \times 100\% = \left(1 - \frac{\text{losses}}{\text{input}}\right) 100\% \tag{3.30}$$

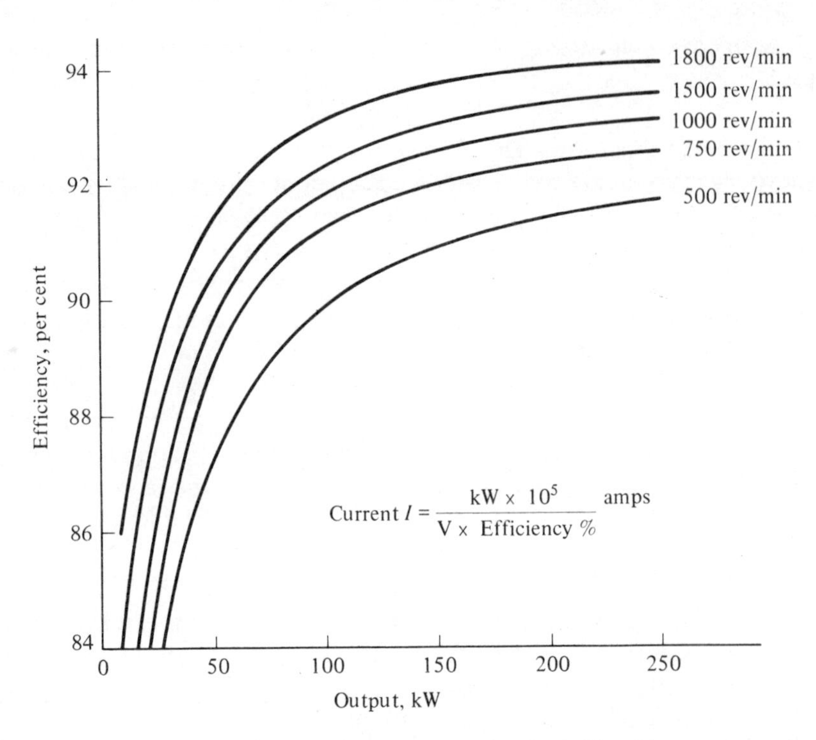

Fig. 3.8 Full load motor efficiency. Force ventilated. Excluding blower motor input or excitation loss

Most specifications allow a tolerance on declared efficiency. (See BS 4999 part 69.) Typical efficiency values are given in Fig. 3.8.

3.2 Circuits and Characteristics

To establish the type of windings and speed/torque characteristics required by a d.c. motor it is necessary to study the types of load which may be met and the criterion for steady state stability. Figure 3.9 shows the differing types of load.

Graph A—Constant power as in reeling drives.

Graph B—Constant torque as in extruder drives.

Graph C—Torque increasing slightly with speed.

Graph D—Torque varies as speed squared as in fan drives.

Graph E—An extreme case where torque varies without speed change as in driving an alternator supplying power to infinite busbars.

A drive is stable if the rate of change of the accelerating torque with respect to speed is negative. If T_M = motor torque and T_L = load torque then

$$\text{Accelerating torque} = T_A = T_M - T_L \tag{3.31}$$

If dT_A/dn is negative then the drive is stable.

$$\text{For stability} \frac{dT_M}{dn} - \frac{dT_L}{dn} < 0 \qquad (3.32)$$

In general it will be found that a drive will be stable if the motor has a drooping characteristic where the speed reduces as the torque increases. For only a small drop in speed with a large increase in load dT_M/dn has a large negative value. However, motor characteristics are rarely strictly linear and the effect of curvature of the speed/torque curve becomes important. Factors aiding stability are armature circuit resistance, cumulative series field, forward lead of brushes, interpoles too weak, and compensating winding particularly if displaced against rotation. Factors hindering stability are armature reaction, differential series field, backward lead of brushes, interpoles too strong, and compensating winding if displaced with rotation.

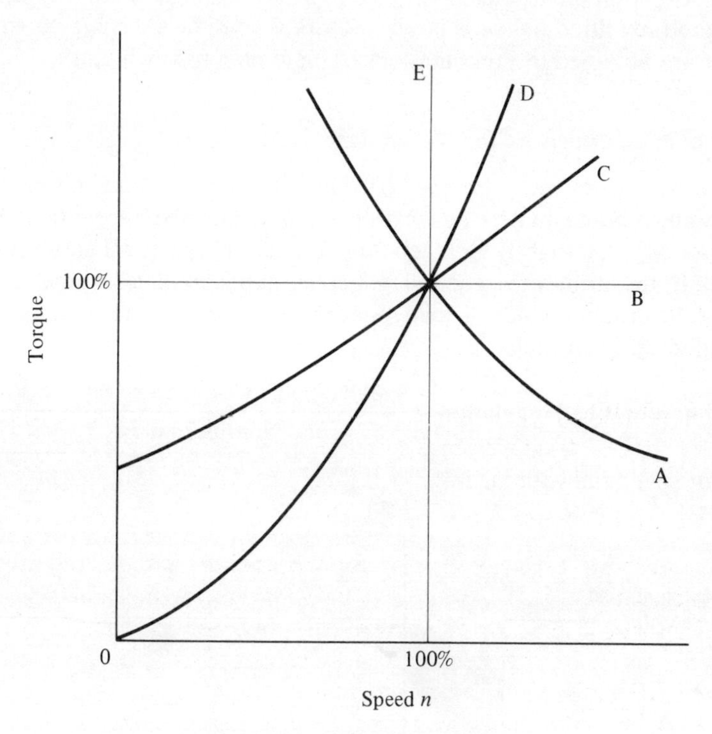

Fig. 3.9 Load characteristics

3.2.1 *Basic windings and terminal markings*

The windings of a d.c. motor can be divided into two main groups:

(a) Armature circuit windings comprising armature, interpole, compensating, and series field windings.
(b) Separately excited or self-excited shunt field windings.

Terminal markings are normally as BS 822; part 6:1964 or BS 4999 part 3. The equivalent IEC specification is No. 34 part 8.

Table 3.1 Main letter markings

	BS 822	BS 4999 IEC 34
Armature	A, AA	A1, A2
Series field, whether for main excitation or stabilizing	Y, YY	D1, D2
Shunt field, whether for main excitation or stabilizing	Z, ZZ	E1, E2
Separately excited field windings on main poles, whether having shunt or series characteristics	X, XX	F1, F2
Commutating windings (compole or interpole) whether series or shunt, and windings comprising interconnected commutating and compensating windings	H, HH	B1, B2 or C1, C2
Compensating pole-face windings	K, KK	C1, C2

Interpoles are fitted on all except very small or low voltage motors. Compensating windings are only fitted on some larger motors. The differing characteristics of the d.c. motor are obtained by combinations of main pole field windings.

3.2.2 *Shunt or separately excited d.c. motors*

Without armature reaction the speed/torque curve of a shunt motor would fall linearly by an amount equal to the *IR* drop as a percentage of the voltage. However, the curve tends to fall slightly and then due to armature reaction bends upwards, see Fig. 3.10. If the armature reaction effect as a percentage is greater than the percentage *IR* drop then the full-load speed is higher than the no-load speed and the motor is inherently unstable.

$$\text{Inherent speed regulation} = \frac{\text{Speed change with load} \times 100 \text{ per cent}}{\text{Nominal speed}} \qquad (3.33)$$

Positive for speed fall with load.

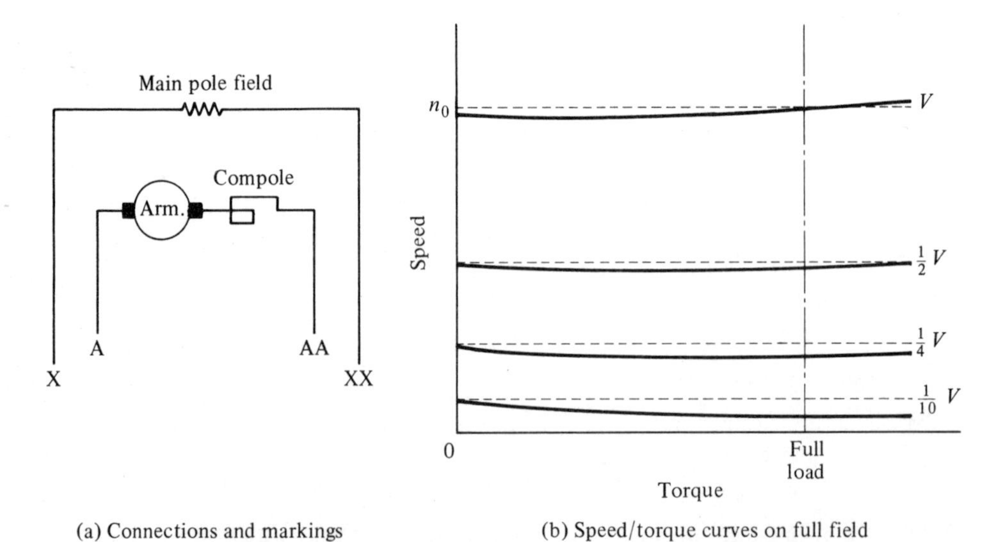

(a) Connections and markings (b) Speed/torque curves on full field

Fig. 3.10 Circuit and characteristics for separately excited motor

Most d.c. motors operating off a fixed supply are therefore fitted with a cumulative series stabilizing winding, see section 3.2.3. When operating on a thyristor supply system incorporating automatic speed control it is possible to dispense with the stabilizing winding provided the inherent regulation is not more than about -2 to -3 per cent. The thyristor equipment has some voltage regulation which aids stability and the speed control system also helps. As the applied voltage is reduced the *IR* drop becomes a larger percentage and the motor may change from negative to positive regulation or from positive to larger positive regulation. As the field is weakened the armature reaction tends to make the speed curve rise so again a plain shunt motor is not widely used.

3.2.3 *Compound d.c. motors*

The stablilized shunt motor is really a light compound motor. The series winding is described as cumulative since it adds to the strength of the shunt field. The speed regulation with light compounding is about 5 to 10 per cent. With heavy compounding it may be 25 to 50 per cent. Compound motors are used where high starting torques apply, see Fig. 3.11. Compound motors are also used where field weakening is required as a shunt motor is more unstable on weak field than full field. The armature reaction is less on weak field than full field but so is the field itself and the net result is a tendency to a rising characteristic. It is possible to have a differential compound motor where the series opposes the shunt but this is very unlikely as it will almost certainly be very unstable. To achieve load sharing between two similar

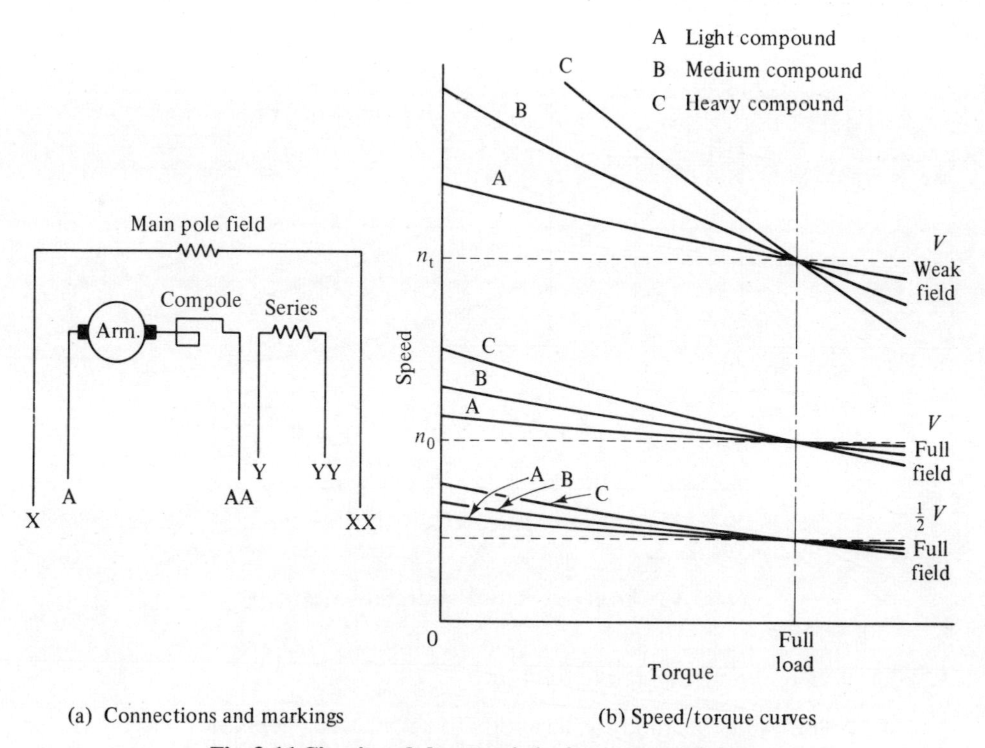

(a) Connections and markings (b) Speed/torque curves

Fig. 3.11 Circuit and characteristics for compound motor

motors coupled to the same load it is usual to have a drooping characteristic. The speed/torque curves can be made similar by adjustment of the shunt field. Motors cannot be equalized in the same manner as d.c. generators unless two series fields are used, one differential and one cumulative.

3.2.4 *Series d.c. motors*

The series motor is used in particular applications demanding a high starting torque, a good acceleration torque, and a high speed at light loads. It finds its main use in crane, steelworks, and traction duty. The light load speed is limited only by windage and friction. For this reason a plain series motor cannot be used where there is a danger of the load going to zero, e.g., in the case of a belt driven load if the belt breaks. In such cases it would be possible to use a series motor with a shunt limiting winding. Alternatively this could be described as a very heavily compounded motor. Series motors are often operated with diverted fields or series-parallel control of the fields and Fig. 3.12 shows the effect on the speed/torque curves. Because of its drooping speed/torque curve the series motor is suited to drives where more than one motor is connected to the same load as if one motor tends to take more than its share of the load its speed will fall and tend to reduce its load. As explained in section 3.2.5 the series motor cannot regenerate without reconnection. This is done by separately exciting the field.

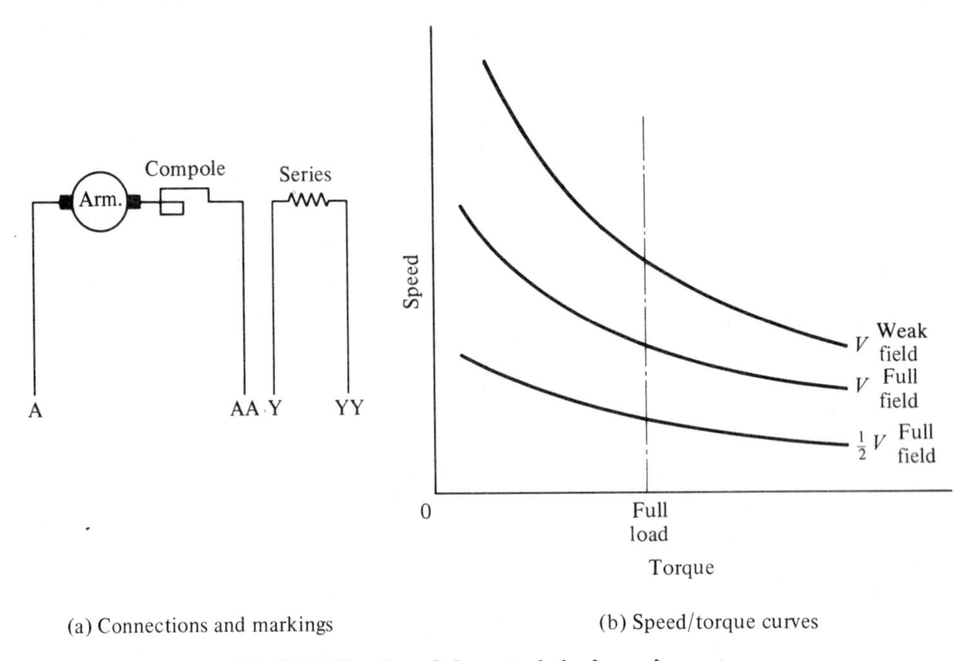

(a) Connections and markings (b) Speed/torque curves

Fig. 3.12 Circuit and characteristics for series motor

3.2.5 *Motoring, regeneration, and braking*

A useful feature of a d.c. motor is its ability to act also as a generator and provide a braking torque. This is subject to certain conditions and limitations. From Eq. (3.1),

$I = (V - E)/R$. Thus if the back e.m.f. exceeds the voltage at any time, I will become negative and the d.c. machine will regenerate and provide a braking torque. This will occur for example on a d.c. compound motor operating on a constant voltage in the field weakening range. If when running at top speed the field is strengthened then E will exceed V. A braking torque is produced and the motor decelerates until the new steady-state condition is reached. The flux must remain of the same polarity otherwise E would change sign and I would be positive. For this reason a series motor cannot regenerate without reconnection of the field winding, as if the current reversed so would the field polarity. Apart from regeneration on field weakening it is possible in the full field condition by reduction of the armature voltage. A prime condition for regeneration is that the supply must be capable of accepting power back. This is inherent in supply systems involving d.c. generators driven by a.c. motors. Engine-driven generators can only accept limited power. Modern thyristor control schemes in general cannot accept power back unless designed for this purpose with a regenerative circuit.

Motoring and generating in one rotation are two modes of operating. Two modes in opposite rotation are also possible and can occur in certain applications. The four-quadrant operation of the d.c. machine is shown in Fig. 3.13.[4] This shows

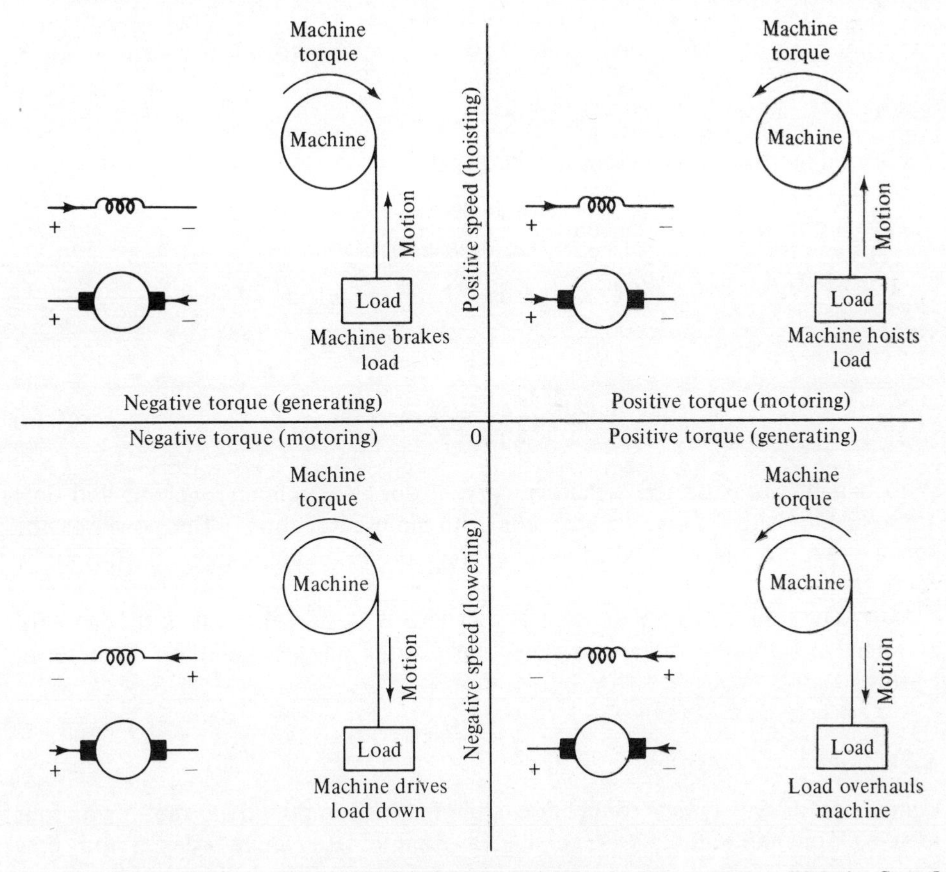

Fig. 3.13 Four-quadrant operation of d.c. machine. As illustrated by a crane or lift hoist. Speed reversing by field reversal

reversal by field reversal but this can be achieved alternatively with constant polarity field by armature polarity reversal. Where a motor is compound to achieve field reversal then both shunt and series fields must be reversed otherwise the series winding changes from cumulative to differential. Sometimes this is acceptable if the torque required in reverse is small and the motor is only going to operate at low speeds for inching.

Where regenerative braking is not possible it is still practical to use dynamic or resistance braking. The motor is disconnected from the supply and the armature is connected to a dynamic braking resistance. The field circuit is kept energized either from its original source or by reconnection. The peak braking current is limited by commutation and mechanical limits. The braking torque falls to zero at zero speed so for high inertia loads a mechanical brake may have to be added at low speed. However, this form of braking is often used for rapid emergency stopping with loads of low inertia and a good friction content. With a compound motor it is useful practice to connect the braking resistance across the armature plus interpole windings. If the series winding is left in circuit it would oppose the shunt winding as the armature circuit current is reversed.

The time to decelerate with regenerative braking with full-load armature current is given by two equations. If P = full load motor power output in kW,

$$\text{With constant field, Time} = \frac{n^2 \times J}{25{\cdot}4 \times P} \text{ seconds} \qquad (3.34)$$

where

J = total inertia referred to motor shaft (kgm^2)

$$J = \text{motor inertia} + \text{load inertia} \times \frac{(\text{load speed})^2}{(\text{motor speed})^2} \qquad (3.35)$$

$$\text{Factor of inertia} = J/\text{motor inertia}$$

With field weakening

$$\text{Time} = \frac{(n_t^2 - n_0^2)J}{50{\cdot}8 \times P} \text{ seconds} \qquad (3.36)$$

With dynamic braking the armature current does not remain constant and only approximate calculations can be made with simple equations. The power in the above equations has to be taken as an average of the braking power. Speeds are in rev/s.

Note: The value of J used above is Wk^2 where W is the weight and k the radius of gyration. In Europe a value four times this is often quoted, based on diameter of gyration.

3.2.6 *Connection diagrams*

Figure 3.14 shows a typical connection diagram for a compound d.c. motor with four main ends out and suitable for reversing by armature reversal. In order to study how this diagram is evolved we need certain basic facts and rules which are set out in Fig. 3.15 and cover both motors and generators. Part 1 gives the electrical polarity of the

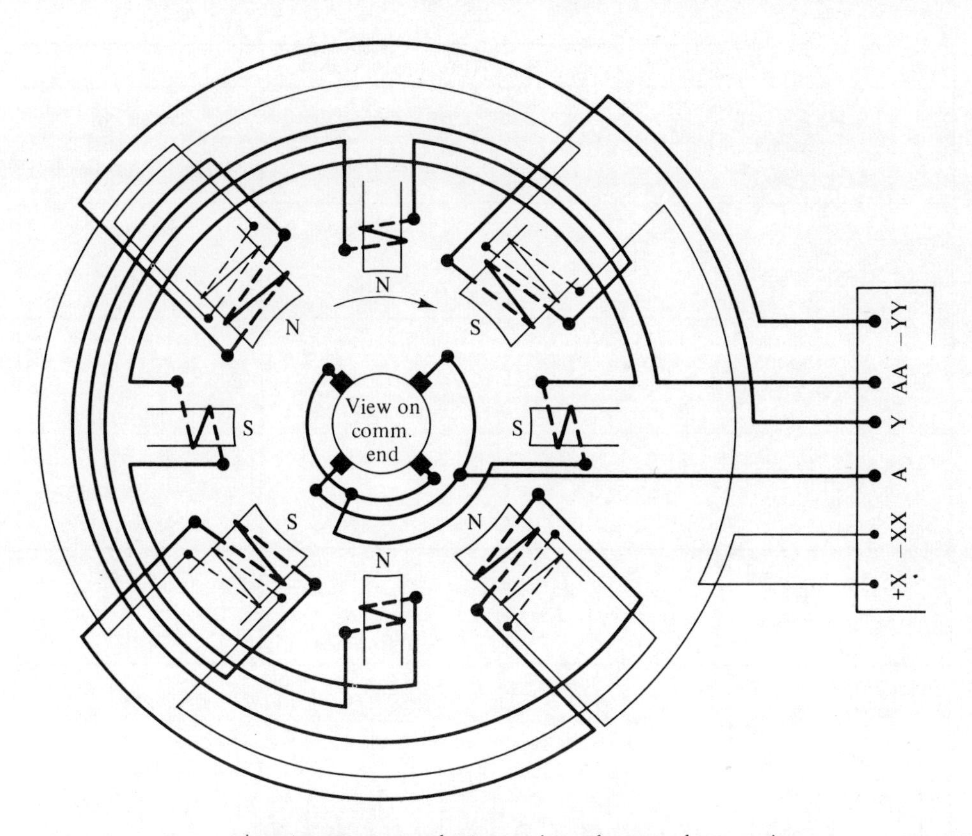

Armature wave wound retrogressive or lap wound progressive
All connections and rotation at commutator end.
Clockwise rotation. Link Y and AA. A is positive.
Anticlockwise rotation. Link Y and A. AA is positive.
Magnetic polarities shown for clockwise rotation.

Fig. 3.14 Connection diagram for compound motor

brushes according to the type of armature winding, the magnetic polarity and the direction of rotation. Part 2 shows the direction of current flow in armature and shunt field circuits. Part 3 shows the relationship between interpole and main pole polarities. Part 4 shows how to determine magnetic polarity when the direction of current flow in a coil is known and the way in which the coil is wound. Part 5 depicts Fleming's left-hand rule for motors (see section 3.1.2) and his right-hand rule for generators. It will be noted that the coil connections for interpoles and series are not the shortest possible. The reason is to balance out the connections so that the resultant ampere turns around the shaft is reasonably low. This is to prevent shaft and bearing currents. The coils are connected 1, 3, 4, 2 instead of 1, 2, 3, 4. Figure 3.14 shows the magnetic polarities for the particular case of clockwise rotation. The motor is cumulative compound wound and the series coil and shunt coil on one pole have the same polarity. In BS 4999 the direction of rotation is specified at the drive end. The commutator is normally at the non-drive end and hence direction of rotation applying to connection diagrams is opposite to the BS rotation. If a compass needle is used to check polarity from outside the magnet then the N pole of the needle will

Part 1. Brush polarity for motors and generators		
Armature winding	Rotation looking at commutator end	Polarity of brush under N main pole
Lap progressive	Clockwise Anti-clockwise	Negative Positive
Lap retrogressive	Clockwise Anti-clockwise	Positive Negative
Wave progressive	Clockwise Anti-clockwise	Positive Negative
Wave retrogressive	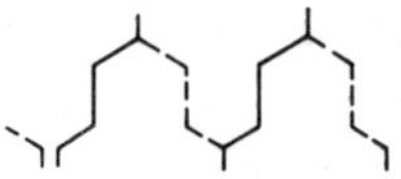 Clockwise Anti-clockwise	Negative Positive

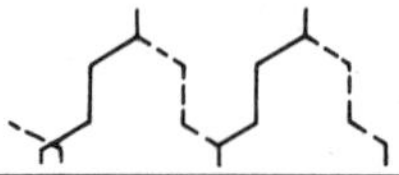

Part 2. Direction of current flow.

–VE Gen. +VE

To shunt coils

To shunt coils

–VE Motor +VE

Part 3. Polarity of interpoles + compensating winding

MOTOR

S N N N N S

The polarity of the interpole must be of the same polarity as the main pole which it follows in direction of rotation.

GENERATOR

N N S S N N

The polarity of the interpole must be of the same polarity as the main pole ahead in direction of rotation.

Part 4. Pole polarity motor or generator

N

i

S

Flux lines

N S

Part 5. FLEMING'S RULE

Left hand for motor

Direction of mag. lines

Direction of current

Direction of motion

Right hand for generator

Direction of induced E.M.F.

Direction of mag. lines

Direction of motion

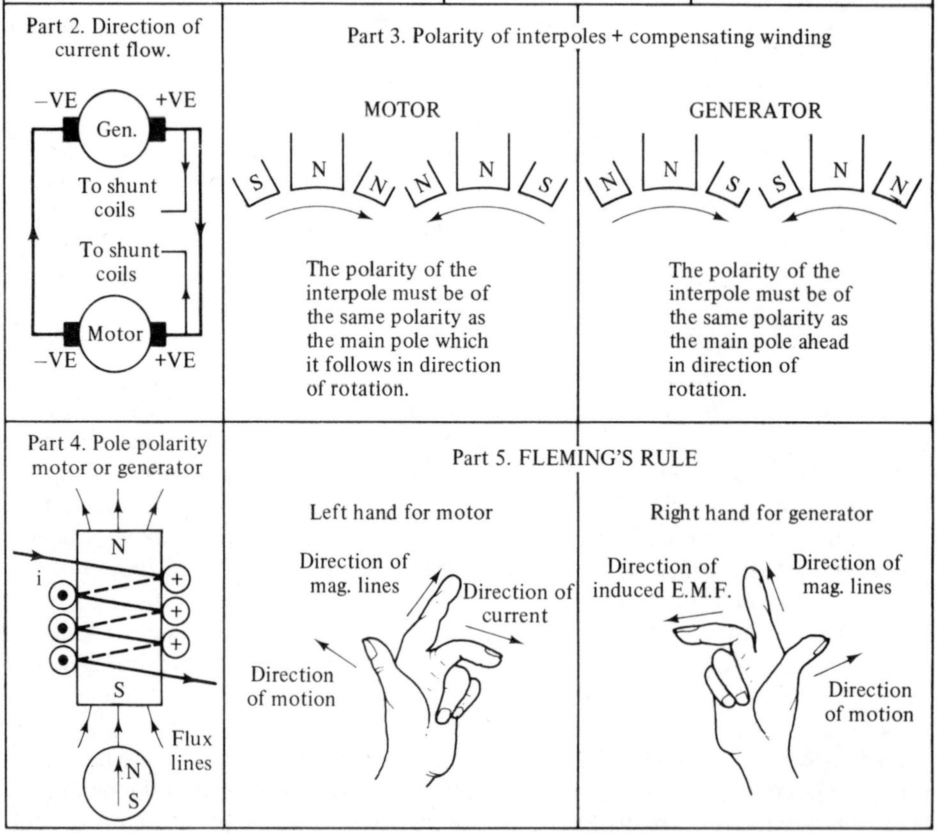

Fig. 3.15 Electrical and magnetic polarity rules

point to a N pole as this is really a S pole on the outside. If checking inside the motor then the N pole of the needle will point to a S pole. Care must be taken with this test not to remagnetize the compass needle as the motor fields are extremely powerful.

3.2.7 *Starting*

For motors operating from a fixed voltage supply there is a starting problem. From Eq. (3.1), $I=(V-E)/R$. At start $E=0$ and therefore neglecting inductance effects $I=V/R$. The armature circuit resistance R is very low on most d.c. motors. On large motors the IR drop is only 2 to 4 per cent of V. Even on 1 kW motors it may be as low as 10 to 15 per cent. The armature circuit inductance would prevent the starting current reaching V/R but even so it could reach ten or more times full load current and the motor would probably flash over. Some form of reduced voltage starting is necessary. For full details of d.c. starters see chapter 12 and BS 587. The usual method is one or more steps of starting resistance which are cut out in turn either manually or automatically. The current is limited to a value which gives sufficient starting and acceleration torque for the application. The usual values are between one and two times full load though higher values do occur. For values above twice full load or for repeated starting it is necessary to specify clearly the starting performance required so that the motor can be suitably designed. Energy is wasted during resistance starting and roughly 50 per cent of the starting energy is lost, the other 50 per cent goes towards accelerating the motor and load. For infrequent starting this loss is not important. For frequent starting other methods may have to be considered such as series parallel control of twin motors or booster control.

For motors operating from a variable voltage supply such as a thyristor control panel or a Ward–Leonard control system there is no basic problem. The voltage is increased from zero to circulate sufficient current to overcome the starting torque and then the voltage and speed are brought up together maintaining the current at a reasonable level to provide the accelerating and load torques. Table 3.2 gives the approximate starting currents for the different types of motor to give certain starting torques. It is assumed that the motor is starting on full shunt or separately excited field as this is normal practice.

Table 3.2 Percentage starting current (× full load)

	FL Torque	1·5×FLT	2×FLT	2·5×FLT
Shunt	100	160	220	—
Light compound	100	150	200	250
Heavy compound	100	140	180	220
Series	100	130	160	190

3.2.8 *Speed control*

From Eq. (3.6), $n=(V-IR)/K\phi$ and the speed is varied by controlling the armature voltage and/or the field strength. Where a fixed voltage d.c. supply is being used then field control only applies. The motor is brought up to the base speed n_0 using a starter and then the speed is varied by controlling the excitation. In its simplest form this means a shunt field regulator. The field current must not be changed too rapidly otherwise the motor may flash over due to the transient armature current.

On uncompensated motors the usual limits for field weakening are about 4 : 1 speed range on 220 volts and 3 : 1 range on 440 volts. Compensated motors may work to wider ranges up to 5 : 1. Motors must, however, be designed for these wide ranges. If it is required to add field weakening to an existing motor then the manufacturer should be consulted. Apart from the electrical limitations due to commutation there are mechanical limitations due mainly to banding, commutator and fan construction, also type and size of bearings. Other methods of field control include variable voltage excitation supply and use of diverters, or tapped field. Field rheostats are covered by BS 280. The most common form of speed control now is armature voltage control. On small motors this can be achieved by inserting resistance but this is inefficient and leads to stalling as the speed characteristic is much more drooping. Armature divert can also be used but again is limited to small motors. A combination of resistance and divert control is useful in crane control. For many years the main method of armature control was the Ward–Leonard system in which the motor was supplied by a d.c. generator.[5] By varying the generator excitation, full control of the voltage from zero to maximum could be achieved. The generator was driven by an engine or an a.c. motor. The system was expensive and fairly inefficient but was widely used and many equipments are still in operation. It had the advantage of ease of control and gave automatic regenerative braking.

Most methods of speed control have now been superseded by thyristor control schemes.[6] These are dealt with in chapter 11. They offer wider speed ranges and built-in automatic speed control. There were important repercussions on d.c. motor design which had to be adapted to cater for the unusual waveforms of the supply voltage and current. See section 3.4.4.

3.2.9 *Variable characteristic d.c. machines*

In an attempt to combine the more desirable characteristics of d.c. machines into one type of motor it has been suggested that a separately excited field be used and its excitation controlled in relation to the armature current to give the required speed/torque characteristic.[7] To give a series characteristic the voltage applied across the separately excited field would have to be proportional to the armature current. The control would be done electronically by thyristor or transistor units. The motor might have to be slightly larger than normal. The extent to which this system may be used is not clear but it is an indication of the versatility of the d.c. motor.

3.3 Constructional Features

The basic d.c. motor comprises a rotating armature system, a static field system, and associated endframes and brushgear. There are many variations in detail construction particularly to meet special applications. Details are given of a typical industrial drive d.c. motor and most of the main features are described. The new ranges of d.c. motors are generally smaller in diameter and shaft centre height due to the use of class B and class F insulation with improved space factor and better cooling. They may be broadly classified as fractional up to 1 kW, small from 1 to 30 kW, medium from 30 to 500 kW, and large above 500 kW. The main variants and their applications are covered in chapters 7, 8, and 10.

3.3.1 *Basic construction*

Figure 3.16 shows a sectional arrangement of a medium-size motor rated at 56 kW, 1800 rev/min, 440 volts.

3.3.2 *Frame, bearings, terminals, and brushgear*

The magnet or yoke forms the main frame of the motor but is also the magnetic circuit linking the poles. It consists of a rolled steel fabricated ring of low carbon content to give good permeability. A homogeneous weld is essential and it is usually near the centre line of a main pole. Feet are welded on or bolted for small sizes. Lifting facilities are provided. The complete fabrication may be annealed. See Fig. 3.17. Larger motors may use split construction for ease of assembly or maintenance. In this case the yoke is in two parts, jointed near the centre line. Severe duty motors may have bracing rings or ribs. On small motors steel tube is often used. Cast steel and cast iron are not common now but may be found on older motors. A laminated yoke is now being used by some manufacturers for thyristor-fed d.c. motors to improve the electrical performance. In this the yoke is built up from specially blanked stampings which may be square or circular.

The main poles are built up of high permeability steel laminations securely riveted together under pressure. Each pole is bolted to the frame yoke by bolts fitted from the outside. Older motors with cast magnets may have cast-in poles and bolted-on laminated tips to reduce pole face losses. One manufacturer uses split poles to reduce armature reaction. Interpoles, when fitted, come between the main poles. The same number as main poles are usually fitted but some small motors may have only half the number. The pole bricks may be solid of mild steel plate or of laminated construction. They are bolted to the yoke from the outside. Interpoles are sometimes shorter than the main poles but the tendency now is towards full length interpoles to avoid saturation on peak currents. Non-magnetic winding or pole supports may be fitted.

Circumferential spacing of the poles is important and the accuracy required depends on difficulty of commutation. The correct air gap is set by means of liners behind the poles and if they are removed they should be put back in the same manner. Main pole liners are magnetic sheet steel of full area. On some special machines reduced area saturation liners may be used. Interpole liners are often slotted for ease of adjustment at the test stage and may consist of magnetic and non-magnetic material. Non-magnetic liners provide an effective air gap at the back of the interpole which assists in reducing leakage and helps commutation by maintaining the interpole flux proportional to armature current. On large motors the interpole bolts may be non-magnetic.

The shape of the pole face can take different forms. Older type main poles are concentric with the armature but with chamfers towards the tips. Modern designs generally use the fully graded gap. The compensated motor uses a deep pole tip with slots. The interpole tip can take many forms depending on individual designers. The width and shape become important on motors where commutation is difficult. The frame is completed by two endframes of cast iron or fabricated steel which house the bearings. The drive-end endframe is short and accommodates the internal fan. The

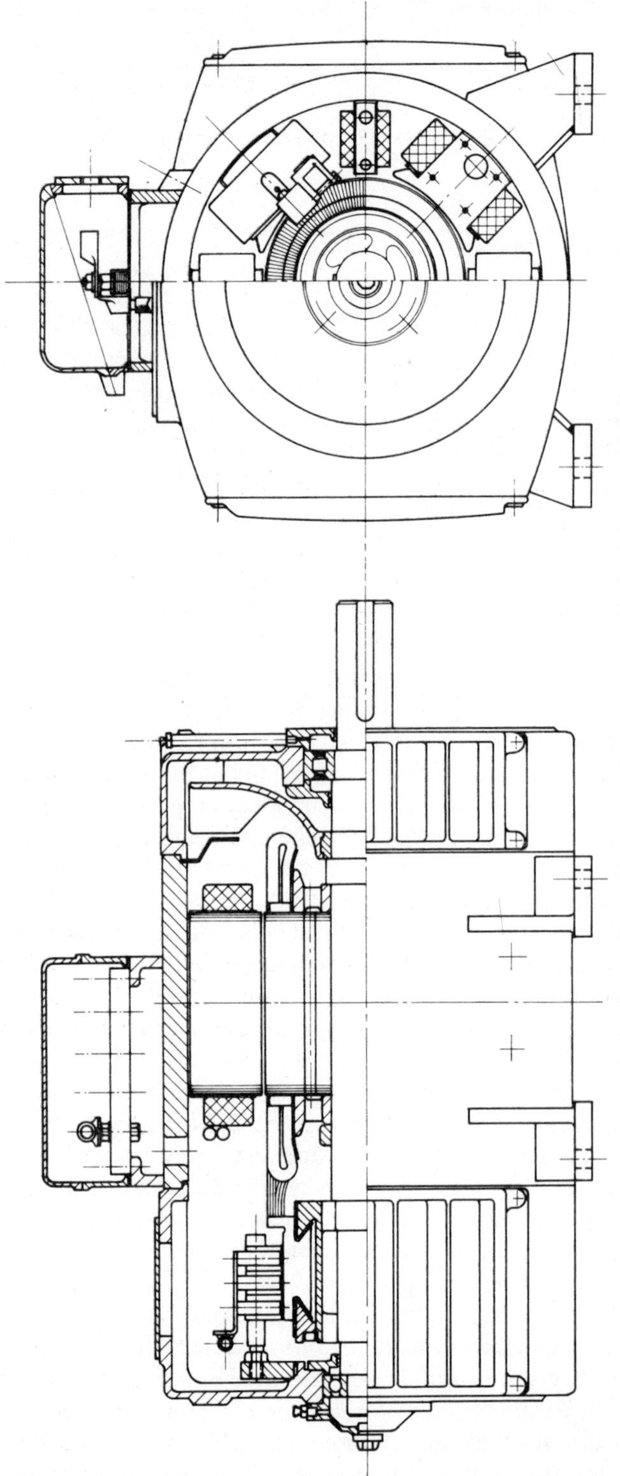

Fig. 3.16 Sectional arrangement of a thyristor drive motor, 56 kW, 1800 rev/ min., 440 volts

Fig. 3.17 Yoke complete with poles and field windings. Resin bonded coils. View at drive end

commutator endframe is the longer of the two and its length depends on the number of brushes. Both endframes, while being strong enough for the framework of the motor, must have ventilation openings to give a good air flow and openings large enough and in the right place to give good access to the brushgear. They should be adaptable for different ventilation schemes. For this reason one manufacturer has adopted a square box type of construction.

Ventilation openings are provided with removable covers to suit the enclosure, e.g., screen-protected, drip-proof, splash-proof, etc. Where the brush gear access covers restrict the view of commutation, some users prefer inspection windows. A more robust frame is provided by the 'through' barrel construction where a long yoke barrel is fitted with flat endplates.

3.3.2.1. Bearings. Standard motors are fitted with ball and/or roller journal bearings. Smaller sizes use ball type and larger sizes one of each. The roller bearing is fitted at the drive end to withstand belt pull or gear shock. The ball journal bearing at the non-drive end takes any axial thrust from the driven load. Figure 3.16 shows normal construction with the outer races fitting directly into the endframes. The bearings are an interference fit on the armature shaft and may be retained if necessary by a circlip or locking nut. The method will depend on the degree of endthrust which may come from the load, expansion with temperature, or skewing. The bearings may be mounted in separate cartridge housings which permit removal of the armature without dismantling the bearings. This is an advantage where there is a danger of contamination from dirt or moisture. All bearing housings should be provided with some form of seal to keep out dirt. In its simplest form this consists of an extended close running surface with grooves (see Fig. 3.16). For special environments such as in the cement industry or iron ore plants or for weather-proofing, an improved seal may be necessary. Flange-mounted motors coupled to gearboxes use a rubbing shaft seal to prevent oil getting into the motor.

Limit type lubricators allow adequate grease supply without danger of overgreasing. On some larger motors or for high speed requiring more frequent or positive lubrication, grease escape valves are fitted. On small motors it is common practice to fit sealed bearings which are pre-packed with grease and do not require any attention between major overhauls when new bearings will be fitted. The lithium-based grease supplied with standard industrial motors is able to operate at temperatures well in excess of normal working. For more details of motor bearings see chapter 13.

3.3.2.2 Terminals. A standard reversible compound motor will have four heavy and two light current terminals. The terminal box may be on the yoke or on the commutator endframe depending on manufacturer's preference. Magnet mounted boxes can be in alternative positions on top or either side. Figure 3.18 shows a typical

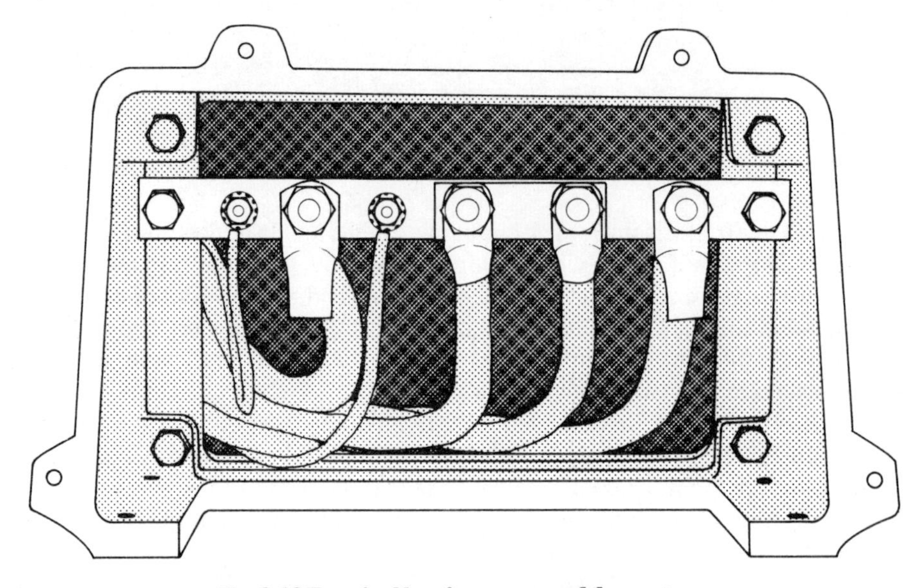

Fig. 3.18 Terminal box for compound d.c. motor

box for a 50 kW, 440 volt motor with cover and gland plate removed. The box should give good access to the terminals and be capable of being arranged for either top or bottom entry. A steel adaptor or gland plate is usually provided and may be tapped for conduit or bushes. If PILC cable is to be used then a deep compound sealing chamber is fitted. On large current motors with heavy cables a limit may be set by the bending radius of the cables and a spreader box may be required. Special precautions may be necessary on large thyristor drives to prevent overheating due to the current ripple. This may mean non-magnetic gland plates or arranging cables so that the magnetic effects cancel out. The terminals should preferably be of the non-turning type and connections arranged so that the current does not go through the studs.

Creepage and clearance distances must be adequate for the voltage. Typical values for medium size 440 volt motors are creepage 20 mm and clearance 10 mm. Segregated boxes are not necessary on d.c. motors compared to some a.c. motors as the maximum fault power at a d.c. motor box is comparatively small. An earthing terminal is provided on the frame.

3.3.2.3 Brushgear. This must be robustly constructed, accurately assembled, and well maintained. On most small and medium sized motors the brushgear is carried from the commutator endframe, see Fig. 3.16. On large motors it may be carried from the yoke. Most motors have adjustable brushgear and it can be rocked circumferentially to alter the brush position. This is done to simplify construction and the brushes can be rocked from neutral to improve the commutation or alter the speed characteristic. Motors can be made with fixed brushgear but accurate construction is required involving lining up of the commutator, core, and winding. On some small motors it is possible to rock the endframe but generally the brushgear is mounted on a rocker ring which can be adjusted inside the fixed endframe. Either the rocker or the brush arms must be insulated. Accurate circumferential spacing of the brushes is essential for good collection and commutation. On small motors ±0.8 mm is satisfactory but on larger motors ± 0.4 mm is better. This entails accurate drilling of holes for brush arms, good brushholders, and accurate assembly. Since carbon dust can accumulate on brushgear a good creepage distance to earth is essential. This will vary with the size of motor and voltage. Electrical connections between brushes and brush arms must be designed to ensure good current sharing. Brushes must be staggered axially to ensure proper wear of the commutator and the golden rule is that all parts of the commutator must be swept by an equal number of positive and negative brushes, see Fig. 3.19. On larger motors use is made also of circumferential

Fig. 3.19 Brushgear axial stagger pattern

brush stagger to increase the arc of contact. There are rules governing this type of stagger.[8] There should be provision for radial adjustment of the holders to allow for commutator wear. Brushholders should be set within 1·5 to 3 mm from the commutator to support the brushes properly. Most British constructors use radial brushgear on small to medium reversible motors. American practice favours an angle of $22\frac{1}{2}°$. For single rotation 8° trail is probably best. With brushholders there has been a big swing towards constant pressure type brushgear typified by the use of the 'Tensator' type spring, see Fig. 3.20. It is not necessary to adjust or set brush pressure

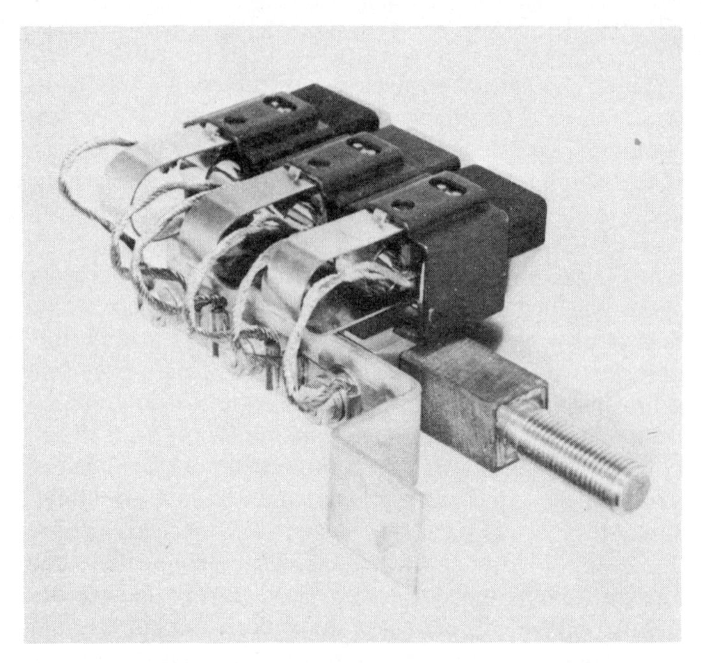

Fig. 3.20 Brush arm with constant pressure type holders

and the special spring gives virtually constant pressure throughout the life of the brush. Some small motors use fixed pressure type clock springs. Normal spring pressures are about 17·5 mN/mm^2 ($2·5 lb/in^2$) on industrial motors to about twice this value on traction type duty. A wide range of brush grades and types are offered by the brush manufacturers. Thyristor drive motors normally use the open texture electrographitic type which give good commutation and reasonable wear. It is impossible to be dogmatic about life of brushes as so much depends on individual operating conditions and environment. Under reasonable conditions it should be possible to obtain 2000 to 4000 hours running time from a set of brushes. To improve commutation, split or sandwich type brushes are fairly common now. Other features to consider on brushes are type of flexible connection, tinning and insulation of flexibles, insulation of top of brush, angle of top of brush, etc. Brushes must be a good fit in holders and tolerances are given in BS 4362 which covers metric brushes and holders. See also BS 4999 part 70 and IEC 136 and 276. For further details refer to chapter 13.

3.3.3 *Field windings*

Due to the use of modern insulation materials there have been major changes in field winding design. Coils are often wound direct on the insulated pole brick and held in place by resin bonding. The heat dissipation is much improved and smaller coils are used. If a rewind is undertaken it must be done in a similar manner to the original otherwise overheating may result. A spare coil must include the pole brick.

3.3.3.1 Main pole winding. Modern class F construction is shown in Figs. 3.17 and 3.21. The pole brick is insulated with glass tape treated with a special resin mix during taping. Resin is applied between all layers during winding of the coil. The shunt coil is usually insulated round wire. A series winding is added on top of the shunt winding with full insulation between. The complete coil is well coated in resin which is cured by baking. Other forms of winding usually involve an insulation spool or an insulated sheet steel spool. The coil is wound on the spool and is separate from the pole brick. This type involves packing to prevent movement of the spool on the brick.

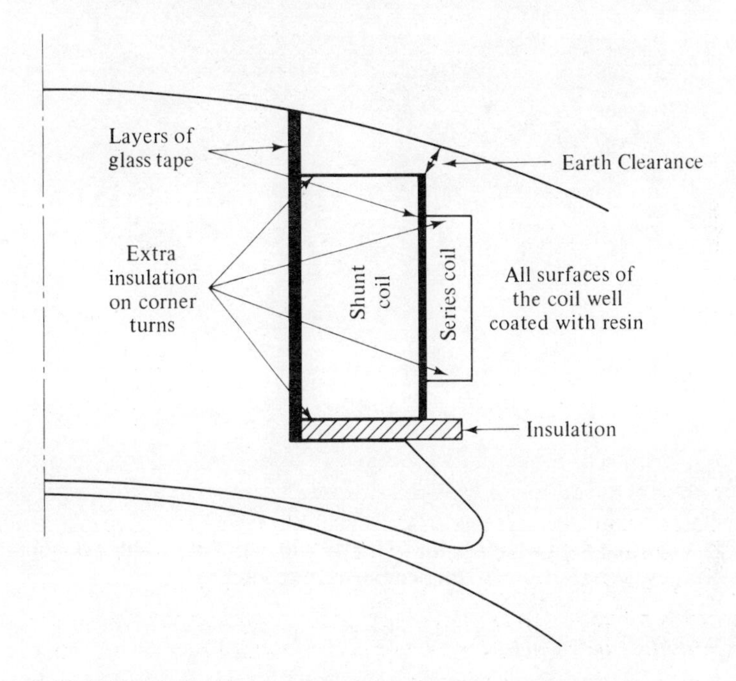

Fig. 3.21 Insulation for resin bonded main pole compound coil

3.3.3.2 Interpole winding. This follows a similar pattern to the main pole windings except that rectangular strip is used instead of round wire. The interpole turns must be exact in number to give the correct interpole field strength. On large motors the coils may be wound strip on edge or fabricated and may be only partially insulated. Auxiliary interpole windings may be found on motors with wide field weakening ranges.

3.3.3.3 Compensating winding. This is wound in slots in the main pole face. Insulation is very similar to armature slot insulation, see Fig. 3.26. Large motors have only one bar per slot with separate end connections at each end of the machine, see Fig. 3.22. Alternatively a U-shaped coil may be formed with connections only at the commutator end.

Fig. 3.22 Yoke and field windings for 850 kW, 800 volt, 900/1200 rev/min motor with pole face compensating winding

3.3.4 *Armature construction*

See Fig. 3.23 for a typical 50 kW, 440 volts, 1500 rev/min armature. The shaft is machined from steel bar and is accurately ground on all fitting and bearing surfaces. The core is built of insulated laminations of high permeability steel with low loss factor and with notching burrs removed. The core may have axial or radial ventilation ducts or both. The stampings are keyed to the shaft and clamped in position under pressure. Small motors may dispense with the key and use an interference fit. Endwinding supports are fitted depending on size and speed. Larger armatures are built with a hub or on ribs welded to the shaft. Above 1·1 metre diameter it is necessary to use segmental type stampings. Skewing of cores is often used for noise reduction, prevention of cogging at low speed, and to improve commutation. Mill type motors may have withdrawable shafts for easy replacement. A ventilating fan may be fitted. The complete armature is dynamically balanced.

3.3.4.1 Commutator. This is made from hard drawn copper bars (to BS 1434). These are insulated with micanite. The ends are machined to take the micanite and

Fig. 3.23 Armature for force ventilated motor. 50 kW, 440 volts, 1500 rev/ min

steel V rings which are clamped together under pressure by riveting, ring nut, or through bolts. The completed commutator is seasoned to ensure a stable condition. This may include both static and rotational seasoning. Normal construction uses an arch bound type commutator where the pressure is applied in the inner flank of the V ring. The Pollock type commutator provides support for the copper bars along the whole of the length which is useful for high speed or long commutators.[9] Moulded commutators are used on small sizes and offer cheaper construction. Wearing depth on commutators is usually sufficient to last the lifetime of the motor. There are two main types of commutator riser. Figure 3.23 shows solid risers. The open type where the riser is riveted and sweated to the segment is used on the larger industrial motors and gives better cooling but is not so robust and also allows carbon dust to penetrate more easily on to the armature winding. The open type is soldered to the armature winding whereas the solid type permits the use of soldering, brazing or tungsten inert gas (TIG) welding. The commutator mica is undercut and the edges of the segments bevelled. For further details see chapter 13.

3.3.5 *Armature winding*[10]

The conductors which are in two layers in slots close to the surface are connected in series and in parallel circuits. The two main windings are simplex wave where $a = 2$ and simplex lap where $a = p$. Figure 3.24 shows the basic arrangement. The coil span is close to the pole pitch to give the optimum e.m.f. Small amounts of chording are used to improve commutation. If t_c = turns per coil

$$\text{Number of conductors } Z = 2Ct_c \tag{3.37}$$

The progression of the winding affects the direction of rotation, see section 3.2.6. Wave windings are used wherever possible as they are simpler and require no equalizers. On larger heavy current motors it is necessary to use lap windings to keep

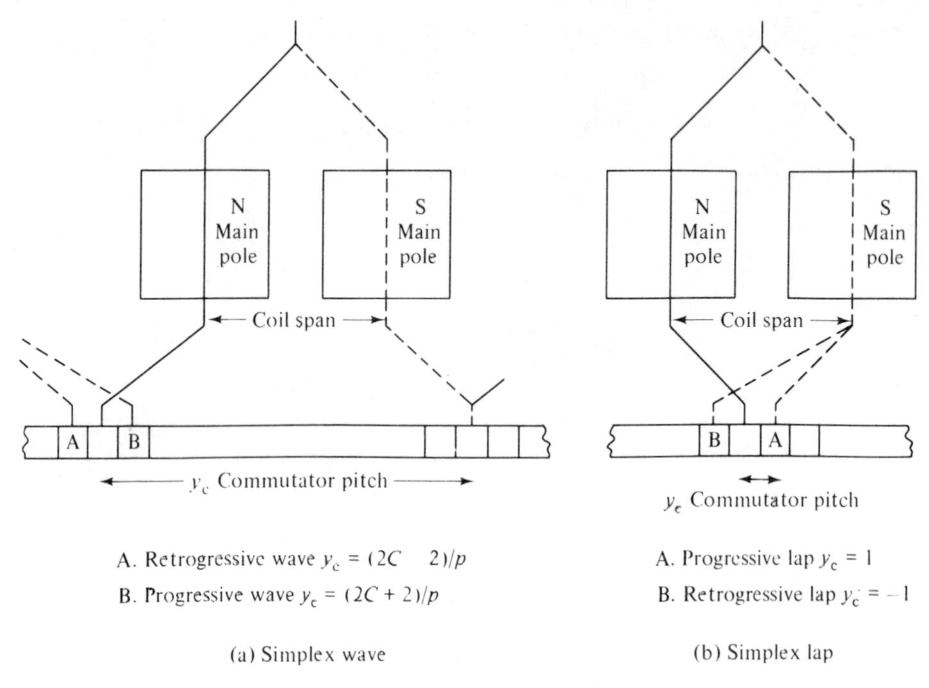

Fig. 3.24 Basic armature windings

down the voltage per segment and the current per circuit. The number of turns per coil is kept to a minimum to help commutation, see section 3.4.1.1. With four-pole wave windings the number of commutator segments must be odd. On a four-pole motor with $C = 2S - 1$, $4S - 1$, or $6S - 1$, the result is a dummy coil which is wound into the armature but not connected, e.g., 45 slots with 89, 179, or 269 segments are all cases involving a dummy coil. These are avoided where possible as they introduce a degree of dissymmetry. The lap winding requires as many brush arms as poles. The wave winding could operate with only two brush arms but in order to keep the commutator length down it is usual to use p brush arms.

3.3.5.1 Equalizers and self-equalizing windings. One disadvantage of lap windings is that to maintain equal current distribution through the parallel circuits it is necessary on most machines to fit an equalizing winding. These are separate windings at either the commutator end or back end of the armature connecting together points of equal potential. The number of equalizers depends on the size of motor and difficulty of commutation and there may be a tapping point every slot. There are also special self-equalizing windings which are equivalent to a fully equalized lap winding. They consist of a simplex lap coil together with a multiplex wave coil which acts as a normal conductor generating an e.m.f. and producing torque but also acts as an equalizer. Figure 3.25 shows the basic arrangement. The points of equal potential are marked E.

3.3.5.2 Multiplex windings. On large motors it may be necessary for commutation reasons to use more segments and circuits in the armature winding. The duplex lap winding with $a = 2p$ and $y_c = 2$ can be used for this purpose. It also enables motors of

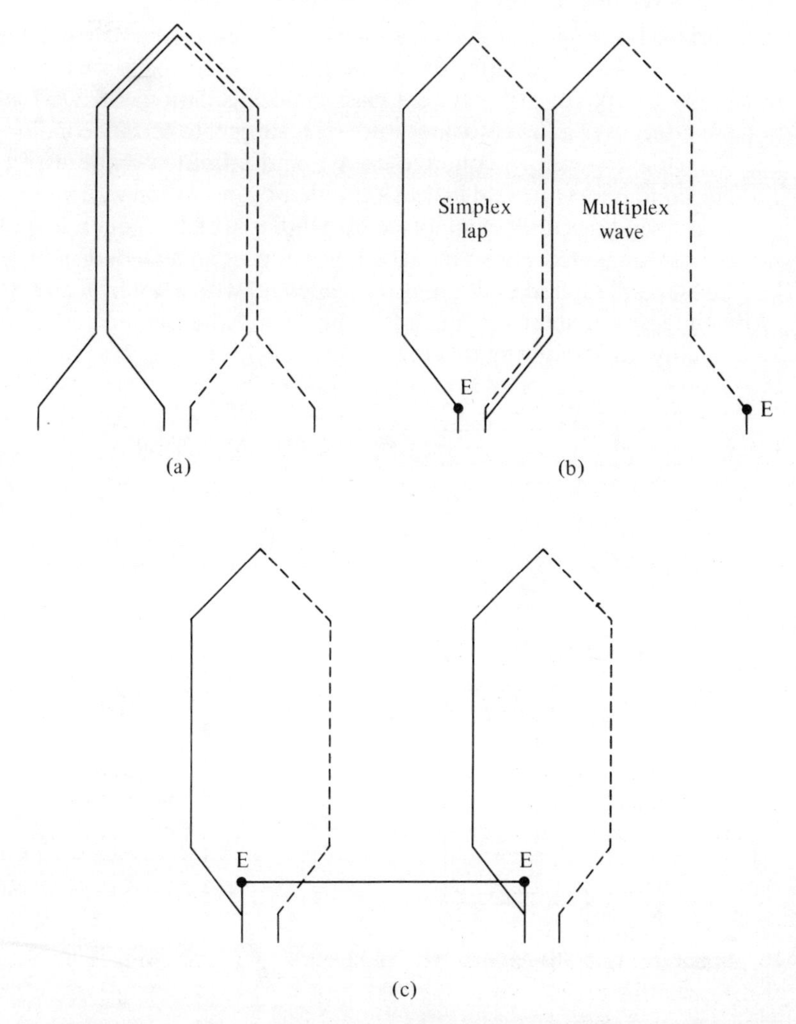

Fig. 3.25 (a) Double coil of self-equalizing winding as wound. (b) The equalizing action of the self-equalizing winding. (c) The equalizing action of an equalizer in a lap winding

longer core length to be built giving low inertia. Special winding rules apply and good equalization is required with more than one set of equalizers in some cases.[11] The self-equalizing winding in section 3.3.5.1 uses a multiplex wave coil. In a four-pole motor this would be duplex and on a six-pole triplex.

3.3.5.3 Coils and insulation. Most integral power d.c. motor armatures have former-wound coils placed in open or semi-closed slots. On the smaller motors multi-turn wire wound coils are made on special coil forming machines which pull them to the shape required. The coils are then insulated by taping. On the larger motors the coils are bar wound from rectangular strip with one turn per coil. Multi-turn bar windings are also used. The strip may be insulated with glass braid or polyester enamel or taped. With single-turn coils there is a choice between a complete coil with an evolute at the rear end or two half coils with clips at the rear end. Single-turn coils are generally former wound after cutting the strip to length and forming the evolute. After insulating the winding supports the coils are inserted. See Fig. 3.26 for an example of 100 per cent class F slot insulation. BS 2757 allows a proportion of materials of a lower temperature class subject to certain conditions and some class F systems therefore combine class F and a limited amount of class B insulation. The conductors are held in the slots either by insulation wedges or by core banding. The endwindings are held in place by banding which is generally polyester or acrylic resin impregnated glass tape which has almost superseded steel banding wire. This glass tape is applied under tension, sealed off with a hot iron and cured by baking. Advantages over steel wire include reduced endwinding reactance, no stray losses, and no clips and soldering to come adrift.

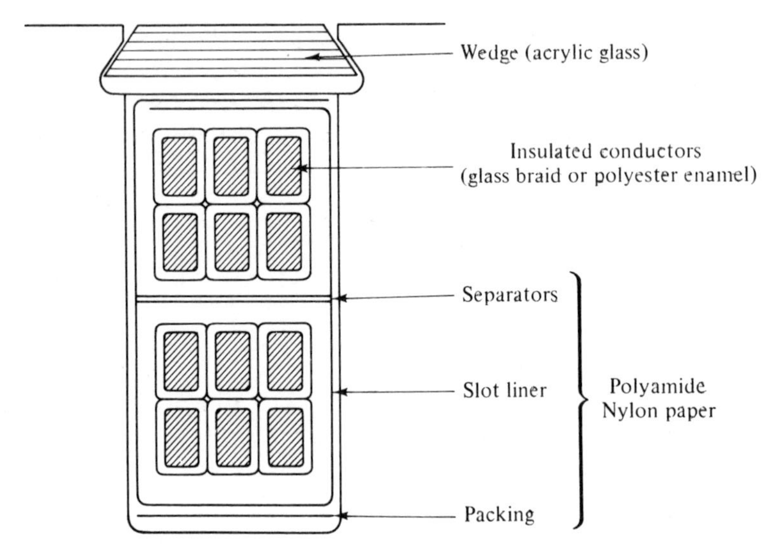

Fig. 3.26 Armature slot insulation. Six conductors per slot with split conductor. Class F

3.3.6 *Ventilation and enclosures*

The various motor enclosures were defined in BS 4727: part 2: Group 03. These have been transferred to BS 4999 part 20, and are dealt with in chapter 5.

3.3.7 *Auxiliary equipment*

Direct-current motors may be fitted with auxiliary equipment, some essential for operation and some for protection.

3.3.7.1 Tachometer generators. For automatic speed control systems a d.c. tacho-generator may be fitted. This must be accurately lined up and well mounted. The coupling is carefully chosen to avoid injecting any unwanted ripple into the system. The tacho itself must be accurate enough to meet the demands of the speed control system allowing for temperature drift. On some drives a digital device of the toothed wheel type with transducer may be used.

3.3.7.2 Blowers. For forced ventilated motors. They are driven by two- or four-pole a.c. motors and are usually of the centrifugal type though some axial ones are fitted. They should be solidly built and well balanced to prevent build up of vibration. Noise is important and the blower will probably contribute most of this.

3.3.7.3 Heaters. These are fitted to prevent condensation when motors operating under humid conditions are switched off. The wattage is fixed to keep the temperature of the motor 5°C above ambient. They are of the strip or tubular type mounted either between the poles or in the endframe. They are derated so that the surface temperature is not excessive. The shunt field is sometimes used instead of fitting heaters and run at a reduced current.

3.3.7.4 Brakes. Magnetic disc type brakes can be fitted at the non-drive end and form a compact system. The tacho-generator may have to be mounted on the end. The main types are electrically energized or electrically released which fails to safety. The size of brake has to be carefully selected to match the particular application.

3.3.7.5 Overspeed switches. If the tacho-generator cannot be used for overspeed protection then a separate mechanical device is fitted often mounted on the end of the tacho. The trip speed will usually be at least 10 per cent above top running speed.

3.3.7.6 Pressure switches. To protect against failure of the ventilating air it is preferable to have a pressure switch. This operates on air failure. A motor on full load with no ventilation will soon burn out. The water-gauge drop across a filter can also be monitored by a pressure switch arranged to give a signal when filter maintenance is required.

3.3.7.7 Over temperature protection. Various devices including thermostat, thermistor, and thermocouples can be fitted on windings to give warning of overheating. The degree of protection is limited on a d.c. motor. It is not easy to use such devices on the armature which is the main winding so they are usually installed on the interpole winding. Here they can only give limited protection. The interpole winding temperature is not always in a fixed relationship to the armature temperature and responds differently to overload. Indicating thermometers with alarm contacts can be fitted to bearings and in air circuits.

3.3.7.8 Radio interference suppressors. These usually consist of small capacitors from positive and negative to earth, fitted on the brush-gear or in the terminal box.

3.3.8 *Other types of d.c. motors*

Low voltage motors for battery powered vehicles are dealt with in chapter 8, section 8.2.[12] Traction motors used for trains, rail cars, and buses are specially designed to provide the desired electrical and robust mechanical characteristics.[15] The special requirements of crane, lift, and conveyor motors are mentioned in chapter 8, section 8.3,[13] and those of marine motors in section 8.6.[14] Brushless d.c. motors are discussed in chapter 10, section 10.2.

In certain applications timing motors require to be run on d.c. supplies and their accuracy, which can be made to be better than 1 per mille is elaborated on in section 10.4 of chapter 10.

Direct-current motors are inherently particularly suitable for automatic control and servo-systems (see chapter 18, section 18.3) but special forms of construction such as ironless or disc armatures improve the performance further (see chapter 10, sections 10.4, Moving coil motors and 10.5, Printed circuit motors.)

Interest in homopolar motors is mainly centred in super-conducting motors of very large power.[16] The main limitation on homopolar machines of low voltage has been overcome by the development of new methods to connect conductors in series. The use of super-conductivity whereby at very low temperatures near absolute zero many metals and alloys lose their resistance enables very powerful magnetic fields to be employed with low power input. The principles of the axial and radial types are shown in Fig. 3.27.

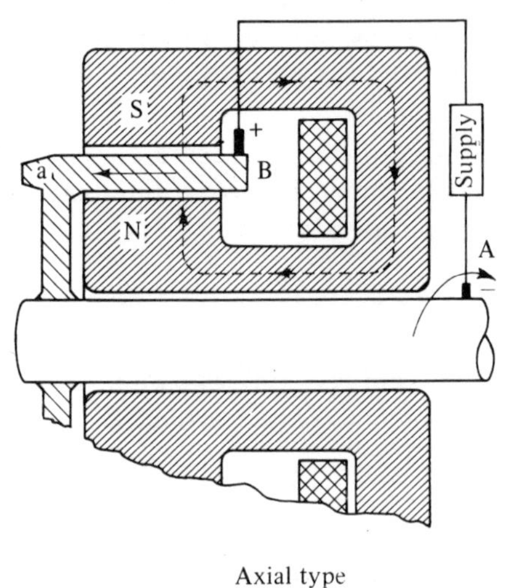

Axial type

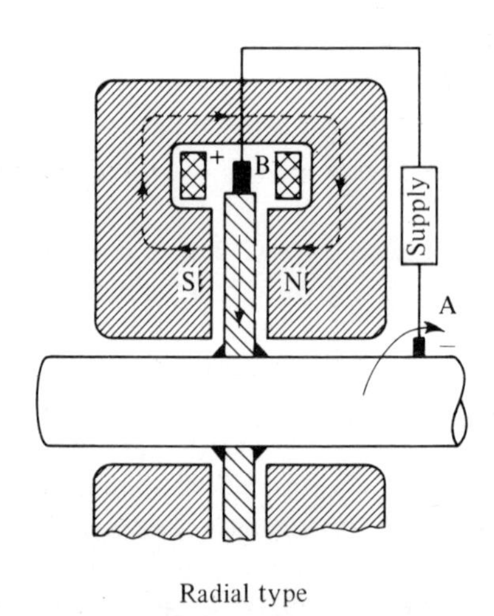

Radial type

Fig. 3.27 Principles of homopolar motors

3.4 Commutation

The main function of the commutator is to convert the generated a.c. voltage into a d.c. voltage. It will be noted from Fig. 3.1 that as the armature rotates and a conductor passes a brush the direction of current has to reverse. This is called commutation or switching. Since the armature coils have inductance and stored energy there is a commutation problem since this change of current and energy can produce sparking at the brushes.[17] The commutator also has to carry the current from the supply to the armature. The two duties of the commutator then are current collection and commutation.

3.4.1 *Basic problem*

Figure 3.28 shows the three positions of the commutator relative to a brush when (a) coil 1 is about to start commutation, (b) coil 1 is short circuited and in the middle of commutation, and (c) coil 1 has just finished commutation. The time of commutation

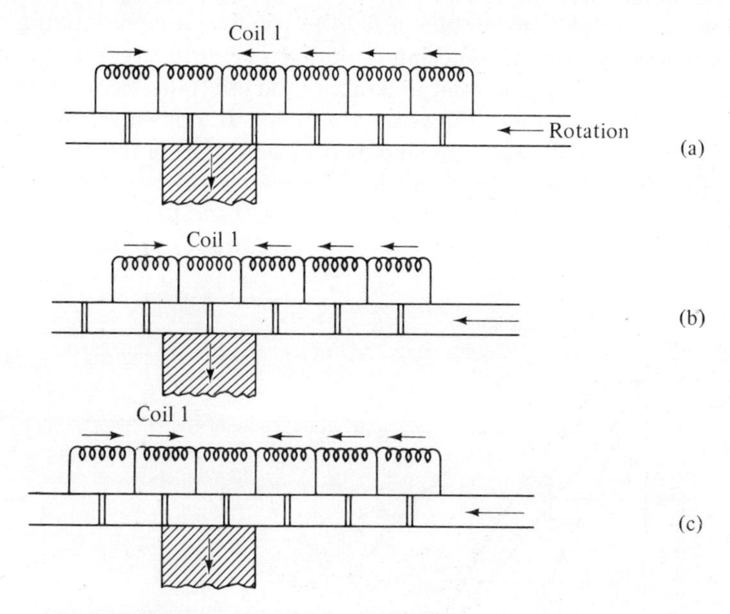

Fig. 3.28 Current reversal in a coil during the commutation process

for one coil is given by $T=(t-m)/1000v_c$ where t = brush thickness (mm), m = mica thickness (mm) and v_c = commutator peripheral speed in m/s. The commutation zone is defined as the distance moved by the armature during the time that all the conductors in one slot are undergoing commutation. There are two layers, several conductors per layer and it depends also on the amount of chording and the type of winding.

$$\text{Commutation zone C.Z.} = \frac{p_s}{C/S}\left(\frac{C}{S}+\frac{C}{S}\times \text{chord}+Q-\frac{a}{p}\right) \tag{3.38}$$

where p_s = slot pitch; S = slots; Q = brush span in segments; chord = amount in slot pitches by which the armature winding differs from full pitch (positive).

Equation (3.38) applies for number of brush arms = number of poles. If the current density remained constant in the brush during the time of commutation the current would change linearly in the coil giving straight-line commutation. This ideal condition is not achieved because of the coil inductance which prevents the current changing quickly enough. If at the end of the commutation period the current is not correct then it has to suddenly change causing sparking unless the brush resistance can prevent this. This is known as resistance commutation.

3.4.1.1 Reactance voltage. The voltage required to reverse the current in the coil is given by $L\,\mathrm{d}i/\mathrm{d}t$, where L is the coil inductance, and is known as the reactance voltage e_r. Since $\mathrm{d}i/\mathrm{d}t = 2I/aT$,

$$e_r = \frac{2IL}{aT} \qquad (3.39)$$

L is a complex function of inductance and mutual inductance and there is no simple or accurate formula. On quite small motors, except for low voltage, it is necessary to fit compoles or interpoles, to provide a field to produce a commutating e.m.f. to oppose this reactance voltage. The interpole has first to overcome the armature m.m.f. and then to produce the interpole flux so the interpole has a strong winding with a ratio of 1·2 to 1·4 times the armature m.m.f. In non-interpole motors it is possible to provide a small reversing field by backward shift of the brushes (see Fig. 3.6). Since the commutating e.m.f. never quite matches the reactance voltage all motors rely to some extent on resistance commutation and hence the importance of choice of brush grade. Figure 3.29 shows how the current reverses with interpoles

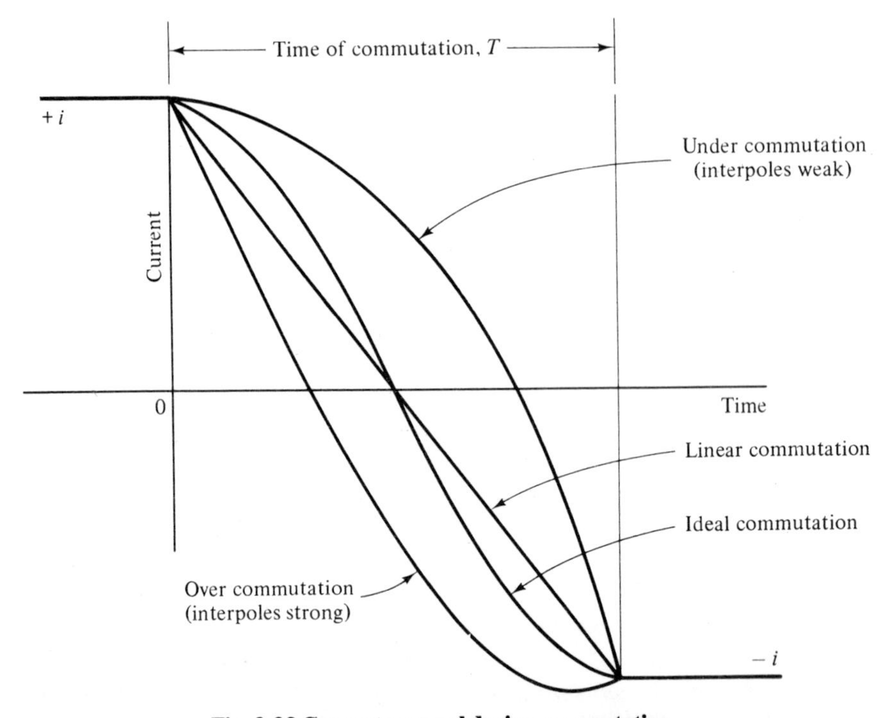

Fig. 3.29 Current reversal during commutation

and how the interpoles must be of the correct strength. There are practical limits to reactance voltage of about 9 volts on full load and 16 volts on overload.

3.4.1.2 Voltage per segment. The other major design factor for successful commutation is the limit on the voltage between segments. This limit again is determined by experience rather than calculation.

$$\text{Average volts/segment} = \frac{V \times p}{C} \tag{3.40}$$

Limit for uncompensated motor approximately 15 volts
Limit for compensated motor approximately 20 volts

These values are derived from the practical limit which is set by the peak volts per segment on load. This is obtained by dividing the average value by the pole arc to pole pitch ratio and then by multiplying by a factor to allow for the flux distortion. See section 3.1.7.1. The limit on peak volts/segment is about 30 volts for large motors to 40 to 50 volts for medium to small motors.

3.4.1.3 Mechanical aspects. Though the basic commutation problem is an electrical one it is considerably affected by the mechanical construction of the motor. The aspects which affect the current collection and switching are balance and vibration, commutator construction and concentricity, stability and accuracy of brushgear. The commutator must run true and total indicator reading (TIR) should not exceed 0·025 to 0·05 mm. Single high or low bars are even more significant and a sudden jump in reading of 0·005 mm could be sufficient to cause trouble. Vibration will cause brush bounce causing poor contact and sparking. The brush must be held by the brushholders in firm and stable contact with the commutator over a wide range of speed. Softening of the segments due to overheating or local stalling must be avoided. A Vickers hardness value of 90 to 95 should be aimed at. Undercutting of the mica must be carefully done to a depth of about 1 to 1·5 mm. Feather edge micas which protrude at the edge of the segments will upset commutation. The edges of the segments should be bevelled with a V form cutter to remove burrs formed during turning of the commutator surface. This can be incorporated in the mica undercutting tool.

3.4.2 *Limit on output and speed*

Due to the limits on reactance voltage and voltage per segment there follow directly limits on output and speed.[18] For simplex lap wound compensated motors the product of output watts and speed in rev/min should not exceed $1{\cdot}8 \times 10^9$ and for a duplex lap 3×10^9. The duplex lap winding does not double the possible output as the effective voltage per segment is more than that given in Eq. (3.40) by a factor of 1·2 to 1·3. The limit for an uncompensated motor is in the region $0{\cdot}4$ to $0{\cdot}8 \times 10^9$ (watts × rev/min) but depends on a number of other factors such as field weakening, peak loads, etc. Another limit which has to be observed is the rate of change of current. For motors with solid magnet and laminated interpoles the limit is about 15 × full load current per second or 15 per unit per second. This can only apply for times much shorter than a second as the peak current is probably limited to 2 to

3 × full load. With a fully laminated motor the current rate of change can increase up to about 80 pu/second. In order to remove some of the above limitations on d.c. motors attempts have been made to replace the switching action of the commutator by thyristors. The most progress towards a practical solution has been made by the thyristor-assisted commutation motor.[19]

3.4.3 *Sparking*

It is possible to design d.c. motors which do not spark but these would tend to be slow speed and expensive. To make them viable economically it is necessary to run them at speeds at which, particularly on thyristor supply, they may spark at the brushes. However, it is now recognized by all the major national specifications that a degree of sparking does not mean unsuccessful commutation (see BS 4999 part 41). What matters is its effect on the wearing life and operation of the commutator and brushes. It is a problem then of deciding whether the sparking is injurious or not.

3.4.3.1 Commutation chart. In order to have some method of indicating the degree of sparking on a motor there are in use various commutation charts of which Fig. 3.30 is an example. There are as yet no official specifications dealing with this subject. Sparkless or black commutation is No. 1. Pinpoint sparking is $1\frac{1}{16}$ and light sparking $1\frac{1}{4}$. Different observers will interpret in different ways. The degree of daylight or artificial light affects the results. However, experienced testers and designers can communicate with one another with such a system to quite a considerable extent. Attempts have been made to assess sparking by measuring one of its physical or electrical effects.[20] This has not yet reached the stage of a simple sparking meter. Most sparking occurs on the trailing edge of the brushes, see Fig. 3.34. Sparking on the entering edge can be caused by flux fringing from the main pole but is more usually caused by mechanical imperfections resulting in the brush bouncing off the commutator.

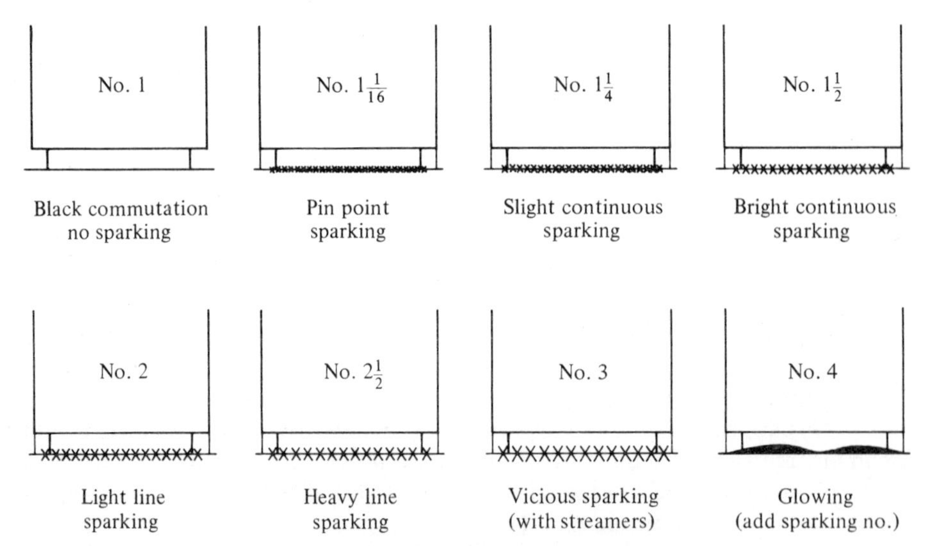

Fig. 3.30 Commutation chart to define sparking

3.4.3.2 Colours of sparking. Apart from a limit on the intensity indicated by the commutation chart how is one to assess if the sparking will be injurious? One method which gives some guide is the colour of the sparking:

Dull yellow sparks are normally harmless.
White sparks unless intense are reasonably non-injurious but may cause some bar marking and increased brush wear.
Red sparks may mean high rate of brush wear as particles of red hot carbon come from the brush.
Blue and green sparks which are usually also brilliant and noisy normally indicate something seriously wrong such as a winding fault.

3.4.3.3 Effects of brushes on commutators. These cover a wide range from a normal healthy commutator skin to bar burning and grooving. In order to illustrate some of the conditions Fig. 3.31 shows the commutator check chart issued by one manufacturer as part of his instruction book. It will be noted that some of the illustrations are also used in IEC 276.

3.4.4 *Motors on thyristor supply*

Many motors are now supplied from thyristor power converters which produce voltage waveforms which are far from smooth. The main factors affected are heating, commutation, noise, and vibration. There are other areas affected such as shaft potential and eddy current effects in cables and terminal boxes. The principal problem is commutation and since the introduction of thyristor controlled supplies the design of the d.c. motor has been improved to meet this challenge.[21]

(a) The fluctuating voltage opposed by a steady back e.m.f. results in a fluctuating armature current. See chapter 11 for typical voltage and current waveforms.
(b) The peak voltage means higher effective volts per segment.
(c) The peak current means a tendency towards saturation in the interpole magnetic circuit. It is also difficult for the interpole flux to follow the current due to damping effects with solid magnetic parts. The black band zone of sparkless commutation is therefore much narrower.
(d) Any current ripple in the field current results in a transformer voltage across the brushes which is another voltage contributing to sparking.

3.4.4.1 Design for thyristor supply. In order to counter the above effects the design of a thyristor-fed d.c. motor should include some or all of these design features.

(a) Full length laminated interpoles. The lamination reduces the damping effect and enables the interpole flux to follow the armature current more closely so providing better compensation for the reactance voltage. The full length together with adequate width ensures a cross section which will not saturate on the peak value of the armature current wave.
(b) Laminated magnet. This reduces the damping again and helps the motor to deal with rapid changes of armature current.
(c) Fully graded main pole gap. This keeps the peak volts per segment down.

COMMUTATOR CHECK CHART

FOR COMPARING COMMUTATOR SURFACE MARKINGS

SATISFACTORY COMMUTATOR SURFACES

LIGHT TAN FILM over entire commutator surface is one of many normal conditions often seen on a well-functioning machine.

MOTTLED SURFACE with random film pattern is probably most frequently observed condition of commutators in industry.

SLOT BAR-MARKING, a slightly darker film, appears on bars in a definite pattern related to number of conductors per slot.

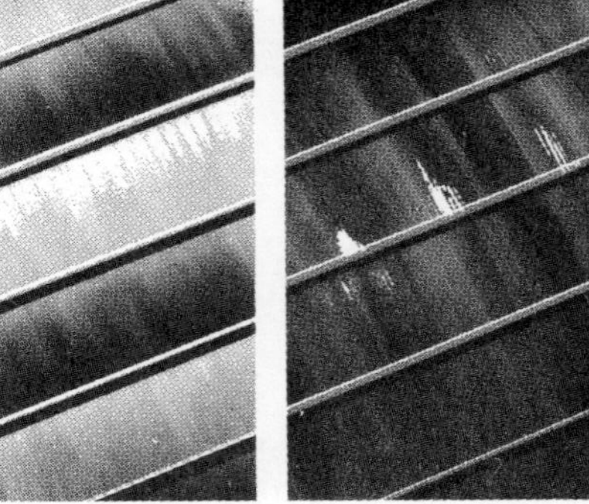

HEAVY FILM can appear over entire area of efficient and normal commutator and, if uniform, is quite acceptable.

CAUSES OF POOR COMMUTATOR CONDITION

Frequent visual inspection of commutator surfaces can warn you when any of the above conditions are developing so that you can take early corrective action. The chart below may indicate some possible causes of these conditions, suggesting the proper productive maintenance.

	Electrical Adjustment	Electrical Overload	Light Electrical Load	Armature Connection	Unbalanced Shunt Field	Brush Pressure (light)	Vibration	Type of Brush In Use		Contamination	
								Abrasive Brush	Porous Brush	Gas	Abrasive Dust
Streaking			X			X		X	X	X	X
Threading			X			X			X	X	
Grooving								X			X
Copper Drag						X	X	X		X	
Pitch bar-marking				X	X	X	X	X			
Slot bar-marking	X	X								X	

How to Get the Most Value from This Chart

The purpose of the Commutator Check Chart is to help you spot undesirable commutator conditions as they develop so you can take corrective action before the condition becomes serious. This chart will also serve as an aid in recognizing satisfactory surfaces.

The box chart at the left indicates the importance of selecting the correct brush and having the right operating conditions for optimum brush life and commutator wear. General Electric offers a complete line of carbon brushes designed to meet all operating conditions and requirements of integral horsepower machines.

For additional information or help with carbon brush application or commutation problems, contact your nearest General Electric Sales Office or Distributor.

WATCH FOR THESE DANGER SIGNS

STREAKING on the commutator surface signals the beginning of serious metal transfer to the carbon brush. Check the chart below for possible causes.

THREADING of commutator with fine lines results when excessive metal transfer occurs. It usually leads to resurfacing of commutator and rapid brush wear.

GROOVING is a mechanical condition caused by abrasive material in the brush or atmosphere. If grooves form, start corrective action.

COPPER DRAG, an abnormal build-up of commutator material, forms most often at trailing edge of bar. Condition is rare but can cause flashover if not checked.

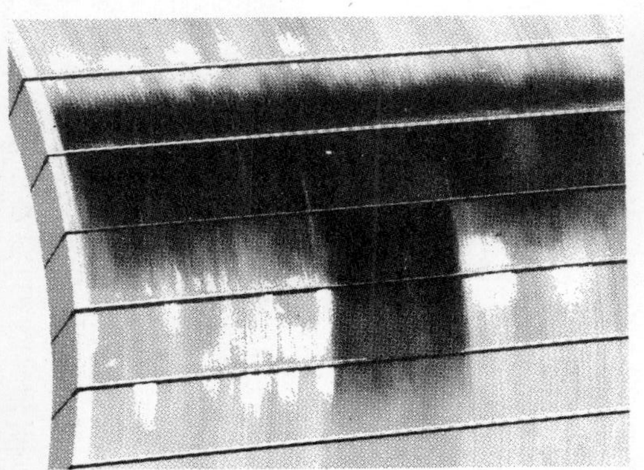

PITCH BAR-MARKING produces low or burned spots on the commutator surface. The number of these markings equals half or all the number of poles on the motor.

HEAVY SLOT BAR-MARKING can involve etching of trailing edge of commutator bar. Pattern is related to number of conductors per slot.

Fig. 3.31 Commutator check chart for surface conditions (Courtesy General Electric, New York)

(d) Reduced pole arc. The ratio of commutation zone to neutral zone is kept down to reduce flux fringing from the main poles interfering in the commutation zone. Neutral zone is the space between the main pole tips.

$$\text{Neutral Zone (N.Z.)} = \frac{\pi D}{p} \times \left(1 - \frac{\text{pole arc}}{\text{pole pitch}}\right) \tag{3.41}$$

(e) Careful choice of armature winding. Aim at maximum number of segments and minimum turns per coil. $e_r \propto t_c^2$. Avoid dummy coils.

(f) Accurate construction. This applies particularly to pole and brush spacing and commutator manufacture.

(g) Good commutating grade of brush.

3.4.4.2 Armature circuit chokes. In difficult cases it would be better to use a three-phase fully controlled thyristor supply as opposed to half controlled supply. The current waveform can be imp.oved by the use of extra inductance in the armature circuit. The current ripple or form factor is obtained from computer calculated graphs which have as the abscissae the per-unit reactance.

$$\text{p.u. } (X) = \frac{(L_a + L_c) I \times 2\pi f}{V} \tag{3.42}$$

L_c = choke inductance, H.

L_a = armature circuit inductance, H. See Eq. (3.25)

f = a.c. supply frequency, hertz

3.4.4.3 Use of existing motors on thyristor supply. In modernizing plant it may be considered desirable to supply an existing d.c. motor from thyristor equipment instead of from Ward–Leonard or other steady d.c. supply. In such cases it is recommended that the motor manufacturer be consulted as the motor may not be suitable or may require fully controlled as opposed to half controlled supply. The derating due to heating with three-phase fully controlled is 0 to 3 per cent and with half controlled 5 to 10 per cent. However, half controlled may result in unacceptable commutation even with additional inductance.

3.5 Specification and Performance

The purpose of standards and specifications, which are discussed in chapter 5, is to ensure that manufacturers and users of motors know what performance is required and can be expected. Guidelines are given to users to assist with ordering the right equipment. When a motor has been delivered it must be properly installed and commissioned and in operation it should be regularly maintained.

3.5.1 *Specifications*

The majority of industrial d.c. motors in the UK were built to comply with BS 2613:1970. This document has been replaced by BS 4999 and BS 5000. Mandatory dimensions apply to a limited extent only, and were covered by BS 3979, which has now been incorporated in BS 4999: Part 10.

The principal specifications dealing with d.c. motors are as follows:

BS 170	FHP Motors (Now BS 5000: Part 11: 1976)
BS 1727	Motors for Battery Operated Vehicles
BS 173	Traction Motors
BS 2949	Marine Motors. See also Lloyd's Rules, Bureau Veritas, Norkse Veritas, American Bureau of Shipping.
BS 4362 sect. B.3	D.c. mill-type Motors
AISE Standard No. 1	Steelworks Motors
NEMA and ASA	American standards
DIN and VDE	German standards
IEC 34 part 1	Rating and performance
IEC 72	Dimensions

3.5.2 *Performance and testing*

It is normal practice for manufacturers of d.c. motors to test them to establish that they are sound and comply with the appropriate specification. To this end, it is essential to test on load. This is different from a.c. induction motors where routine tests can be completed without loading the motor. These tests are of course backed up by full tests on sample motors. A d.c. motor cannot be properly tested unless it is coupled up and put on load to check commutation, speed, and excitation. See BS 4999 part 60.

3.5.2.1 Preliminary inspection. Winding resistances are checked except on some small motors. Field resistances are measured by 'bridge' method or volt drop. Armature resistance is difficult to measure accurately. Dummy brushes with metal point contacts are used to give more accurate volt drop. For accuracy when using the resistance to assess temperature rise it should be taken with the armature in the same position at all times and with the same instruments. Ambient temperature is taken with resistance readings. The brushes are bedded to the commutator surface. If the brushes are not preradiused then some bedding with carborundum paper (not emery) is required. The bedding is completed on test by the use of a brush seating stone which is a very light abrasive stone. The motor is run at reduced voltage at a commutator peripheral speed of at least 7·5 m/s and the stone applied in turn behind each row of brushes. This releases abrasive particles which grind the brushes. After bedding, the motor is blown out with clean dry compressed air at a pressure of not more than 7 kg/cm^2 (100 lb/in^2). The brushes are set in the neutral position, see section 3.1.7.2. There are several methods but the simplest can also be employed quickly on site. The field is separately excited from a low voltage source, e.g., car or torch battery. The brushes should be bedded. The armature is connected to a millivolt meter or universal meter. With the brush rocker slackened the field is flicked on and off and the brushes moved until minimum movement or kick of the needle is observed. If the applied field voltage is higher, then it should be connected by a potentiometer so that there is a discharge path for the field to avoid a peak voltage which may break down the insulation. There is a more accurate kick method which is independent of the brush bedding.[22] Probably the most accurate method is to run the motor on load in both rotations in turn until the same excitation and speed values are obtained. Note that after moving the brushes the brush rocker must be tightened again to prevent movement.

3.5.2.2 Light running loss tests. The loss separation test, often called the Kapp-iron test, consists of running the motor light at constant speed by varying the armature voltage and the excitation. The input current is measured and the results plotted as in Fig. 3.32. The armature voltage is taken down to about one-third depending on stability. The power curve is extrapolated to zero to give the windage and friction. The iron loss is then obtained as shown. The data also gives the basic magnetization curve, see section 3.1.6. The test may be done at several different speeds if accurate assessment of losses is required. The light loss input curve can also be taken at fixed excitation and variable armature voltage. While running light the vibration and noise can also be checked. See BS 4999 parts 50 and 51.

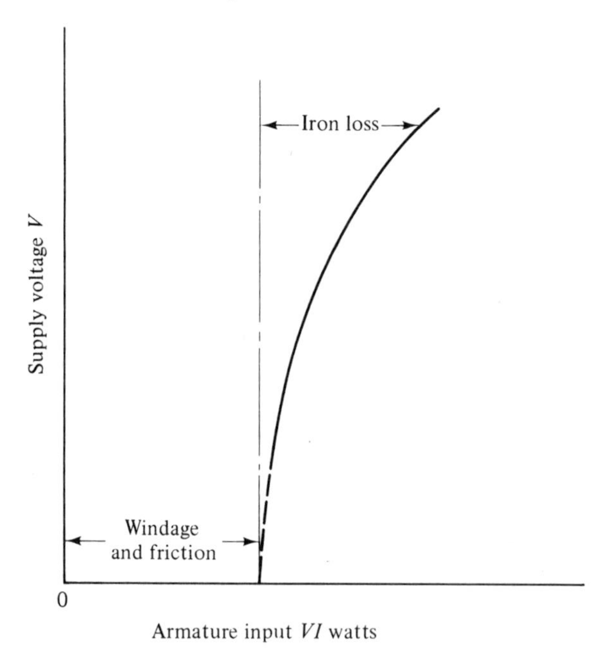

Fig. 3.32 Kapp-iron loss separation test

3.5.2.3 Load testing. The motor is coupled to a test machine and run on load to check commutation, speed, and excitation. On duplicate designs the motor is run long enough to warm up. On new designs a full temperature test is taken. For continuously rated motors this is when temperatures are steady. On short time rated this is for the specified time rating from cold. For copper windings the hot temperature by resistance is

$$t_2 = \frac{R_2}{R_1}(t_1 + 235) - 235°C \qquad (3.43)$$

where

R_1 = cold winding resistance (ohm) at temperature t_1

R_2 = hot winding resistance (ohm) at temperature t_2

For allowable temperature rises see BS 4999 part 32.

If there is significant delay in measuring the hot resistance then a cooling curve against time is taken and the value extrapolated back to zero time. Temperatures are measured by thermometer and resistance. Commutator temperatures are measured by thermometer or thermocouple where applicable, e.g., BS 1727. Inlet and outlet air, frame, and bearing temperatures are also measured. Apart from dead loading in which the power is wasted there are back to back tests, such as the Hopkinson test, in which power is recirculated and only the losses come from the supply. Commutation should be assessed according to a commutation chart, see Fig. 3.30. Acceptable commutation is up to $1\frac{1}{4}$ or slightly higher for overload conditions. If commutation is not satisfactory, then a black band test is carried out at different loads by alternatively bucking and boosting the interpole strength by connecting an adjustable voltage supply across the interpole winding.[23] Readings may be taken when the motor sparks (optimistic method) or more usually when it stops sparking (pessimistic method). Typical black bands are shown in Fig. 3.33. If it is simply incorrect interpole strength then the commutation is corrected by an adjustment of the interpole gap and/or the non-magnetic liners. If the band shows saturation as in Fig. 3.33 curve B, then the designer faces a more difficult problem and more detailed investigation of materials or design may have to be made.

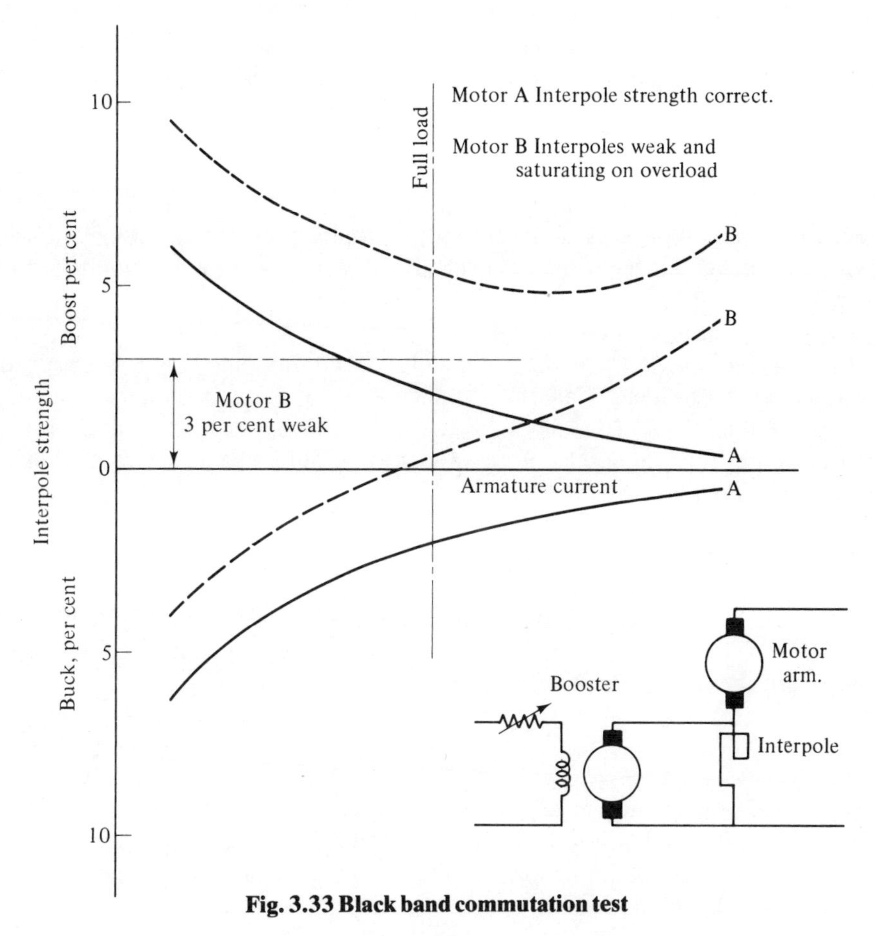

Fig. 3.33 Black band commutation test

When commutation is satisfactory the overload test is carried out. The speed and excitation are checked and also the speed regulation. If the latter is not quite right then a small adjustment to the brush position may have to be made. See BS 4999 part 41 (characteristics) and part 69 (tolerances).

Table 3.6 Effects of brush shift on d.c. motor

Brush shift	Condition	Effect
Back from neutral	Same excitation	Increases speed
	Same speed	More excitation
Forward from neutral	Same excitation	Reduces speed
	Same speed	Less excitation

Brush shift can upset commutation so a compromise may have to be made.

3.5.2.4 Final testing. Includes the overspeed test (20 per cent for 2 minutes) and the high voltage flash test. This is followed by an insulation resistance test at 500 or 1000 volts d.c. After test the motor is painted and prepared for dispatch. A final quality control check is included. Special protection for storage and transport is added. For transport overseas it may be necessary to fit anti-brinelling protection for the bearings.

3.5.3 *Application*

The prospective purchaser of a d.c. motor or a variable speed drive equipment has to provide the manufacturer with sufficient information to enable a quotation to be prepared. BS 4999 part 31 Appendix 31A gives a useful guide.

3.5.3.1 Site and operating conditions. A1(1). Much equipment is for export and conditions can vary widely. State the actual application as this gives a manufacturer, who has probably supplied motors for the particular application, a firm basis on which to quote. Include details of environment conditions and extent of running. Drives on 24 hours a day, 7 days a week running are particularly onerous. For outdoor use indicate if there is any protection from the elements. Detailed information is required if there is any risk of fire or explosion. Low noise requirements should be indicated stating specification and noise level required. This may involve extra cost. It should be noted that the standard specifications, BEAMA 225, BS 4999 part 51 and IEC 34-9 only give limits for no-load uncoupled motor noise. Special arrangements have to be made with the manufacturer if noise guarantees on load are required.

3.5.3.2 Specification to be followed. A1(2). BS4999 or one outlined in section 3.5.1. Some users have their own specifications and where these apply then a copy should be forwarded with the inquiry. It should be stated if the specification is to be strictly applied or if the manufacturer may quote his own standards or if quotations are required for both conditions.

Table 3.7 Extract from BS 4999 part 31 Appendix 31A

INFORMATION TO BE GIVEN BY PURCHASER WITH INQUIRY AND ORDER

A.1 General information for all types of machines

(1) Site and operating conditions.
(2) The number of this British Standard.
(3) The type of protection.
(4) The duty of the machine.
(5) The maximum temperature of the cooling air in the place in which the machine is intended to work in ordinary service.
(6) The altitude of the place in which the machine is intended to work in ordinary service if it exceeds 1000 m.
(7) Values of the voltage, current, frequency, and speed.
(8) Method of earthing the system to which the machine will be connected.
(9) Tests required and where they are to be carried out.

A.3 Particular information for d.c. motor

(1) Rated output (kW).
(2) Rated voltage and voltage range.
(3) Speed at rated output (rev/min). If the speed at rated load is required to be different from the speed at no-load full particulars should be given.
(4) Direction of rotation.
(5) Method of excitation, whether shunt, series, compound or separately excited. State excitation voltage.
(6) Starting torque required.
(7) Any information regarding the driven machine which has a bearing upon the torque required during the accelerating period, e.g., the kinetic energy of the moving parts and the number of starts during a specific period.
(8) Harmonic content of supply.

3.5.3.3 Type of protection. A1(3). The relevant data are to be found in BS 4999 part 20 (Types of Enclosure) and part 21 (Methods of Cooling). These are technically identical with the recommendations in IEC 34-5 and 34-6.

Table 3.8 Enclosures to BS 4999: Part 20 (Extract)

Type of enclosure	Designation
Screen-protected	IP 20
Drip-proof	IP 21 or 22
Splash-proof	IP 24
Hose-proof	IP 25
Totally enclosed	IP 40 or IP 50
Totally enclosed weatherproof	IPW 44 or IPW 54

There is also a designation system for methods of cooling. Examples are ICO1 for self ventilated, ICO6 for force ventilated, and ICO141 for TEFC. The type of enclosure and cooling may also be affected by the mechanical arrangement. See BS 4999 part 22 and IEC 34-7 which give details of the IM designation system on construction and mounting. This includes type of bearings and methods of frame fixing. Any special details of mounting should be advised at the inquiry stage including dimension limitations. Basically a decision has to be taken between a ventilated motor and the very much more expensive totally enclosed. At the 50 kW,

1500 rev/min level, TEFC is about twice the cost of SP. Previous experience of the application is the best guide. The d.c. motor has problems with enclosure which do not occur on a.c. motors where windings can be completely insulated and sliprings put in a TE cover. On a d.c. motor the commutator and brushgear remain exposed and there is always a creepage path to earth. Any atmosphere providing corrosive gases or conducting particles must be kept outside the motor. However, even plastic chippings can cause trouble particularly if any polystyrene solvents are present. The resulting mixture on the hot brushgear tends to burn and ends up as a sticky or hard deposit. This causes the brushes to stick in the boxes. In such cases pipe ventilation can be the answer but the source of air must be clean.

With pipe and forced ventilation, the manufacturer will either have to make an allowance in the blower water gauge for customer's ducting or if the customer is providing the blower he will be advised of the air quantity and the water gauge drop across the motor. It is preferable for the blower to be mounted at the inlet to the ducting so that the whole system is under pressure. This is essential in carbon-black atmospheres. Typical air quantities are 0·038 to 0·057 m^3/s per kW loss.

Plain TE usually applies to short time or intermittently rated motors and for continuous rating of not more than about 30 kW. The temperature rise depends on the watts per square centimetre of the external surface and as the power goes up the percentage of ventilated output comes down to about 20 per cent at 30 kW. TEFC is used up to about 100 kW. Indicate the minimum continuous running speed as there is a derating factor from 1·05 at 500 rev/min to 1·33 at 150 rev/min for constant torque with shaft-mounted fans. With force ventilation of the external air a slightly better output can be obtained. The carcase is often ribbed to improve output. CACA enclosure can give up to 75 per cent of ventilated output and CACW up to 100 per cent. With shaft mounted fans again there are derating factors for low continuous running speeds and the manufacturer will decide when it is cheaper to offer forced ventilation in addition. With force ventilated coolers it is possible to have filters in the internal air circuit though at the expense of size of cooler and added cost. Protection against air failure should be considered where forced ventilation is involved.

3.5.3.4 Motor duty. A1(4). In cases of doubt supply full details of the duty cycle. Motor manufacturers have application or design engineers whose job it is to help customers with these problems. Also involved are items A3(1) rated output; A3(3) rated speed; A3(6) starting torque; and A3(7) acceleration and inertia. Graphs of torque and speed against time are useful for duty cycles. For a simple cycle the engineer will calculate the root mean square (r.m.s.) current. This is because the main losses are proportional to current squared. For a more complicated cycle it is necessary to calculate the losses, and the cooling rate in detail throughout the cycle to arrive at an equivalent output to fix the frame size. Applications such as rubber calenders may require a guaranteed stopping time. Give details of the time or revolutions in which the motor must stop together with the coupled inertia referred to the motor shaft. See section 3.2.5 and BS 4999 part 30.

3.5.3.5 Coolant conditions. A1(5) and A1(6). Normal ambient for BS 4999 is 40°C. Above this, the temperature rises must be reduced. For ambients down to 30°C some increase in temperature rise is allowed. In assessing ambient, allowance should be made for localized heating due to a nearby furnace or hot extruder barrel. In siting motors arrange for air circulation to remove the heat losses to prevent the ambient rising above the specified temperature.

Very low temperatures can cause problems. Equipment may be suitable for storage down to −40°C but not for operation at this temperature due to the risk of brittle fracture. For water-cooled motors, supply water temperature, pressure available, and brief details of the water analysis. This affects the material for the cooler tubes. With altitude over 1000 metres it is necessary to derate due to the lower air density and reduction in cooling. High altitudes over 3000 metres may require special brush grade due to the dryness of the air.

3.5.3.6 Supply conditions. Three items are involved. A1(7) limits of voltage, current, frequency or speed; A3(2) voltage; and A3(8) harmonic content of supply. With a fixed d.c. supply state the voltage and its variation together with any limit on speed variation. With battery supply state maximum voltage range.

With a thyristor drive, if the inquiry is for a motor only, then state the type of thyristor supply, e.g., three-phase half controlled or fully controlled or single phase. Alternatively the wave form would have to be defined but this is difficult as it varies widely. If the inquiry is for a complete drive state the a.c. supply with variations on voltage and frequency and any limitations on harmonics fed back to the supply.

3.5.3.7 Testing. Item A1(9). See section 3.5.2 which covers manufacturer's tests. If special tests such as witness tests are required or commissioning tests then give details to enable an estimate to be included in the quotation.

3.5.3.8 Standards. For industrial d.c. motors apart from special steelworks motors there are as yet no well-defined output, speed, and dimension standards as there are for a.c. motors.[24] The user may require some guidance on motor speed when gearboxes or belt drives are involved. For thyristor drives the majority of motors are ordered at or near 1500 rev/min with smaller numbers at 1800 rev/min. Higher speeds are possible but not so favourable for commutation and brushwear. For field weakening drives 500 or 750 rev/min are common base speeds (n_0) with field weakening to 1000 or 1500 rev/min. Standard thyristor voltages are 1·1 to 1·15× a.c. voltage for three phase, 0·7× for single phase line to line, and 0·40× for line to neutral.

3.5.3.9 Order stage. When an order has been placed and to prevent delays the manufacturer will require some additional information including position of terminal box and cable entry. Cable details may be required particularly on large motors. Direction of rotation is important. Problems relating to the mechanics of the drive including suitability of belt drives and endfloat or endthrust on direct coupled drives must be examined.

3.5.4 *Maintenance*

The subject of maintenance is also dealt with in chapter 17. It is recommended that d.c. motors shoud be regularly maintained.[25] This is largely because of the exposed nature of the commutator and brushgear. For example it only requires brushes to start sticking in the holders for a gradual deterioration in performance to take place. For most motors an instruction book and connection diagram will be provided.

3.5.4.1 Schedule of maintenance. The period between inspections may depend very much on the application, the running time, and the environment. After commissioning, the motor should be inspected after one or two weeks and thereafter every 4 to 12 weeks.

(a) Examine the brushes to ensure they move freely in the holders and there are no signs of large quantities of carbon or copper dust. Check on brush length.
(b) Clean the brushgear. If the motor is very dirty then blow out with clean dry compressed air at a pressure of not more than 7 kg/cm^2 (100 lb/in^2). If the motor is very dirty use a suction cleaner first before blowing.
(c) Check the insulation resistance. This should be several megohms.
(d) Check the sparking level and the condition of the commutator.
(e) Check the bearings and lubricate only if necessary. If any excessive vibration, investigate. This may affect commutation apart from doing mechanical damage.
(f) Check the motor ventilation to ensure it is not restricted in any way. Clean or replace filters.
(g) Check terminals and connections for tightness and cables for fraying.
(h) Check the tacho-generator and its coupling.

3.5.4.2 Operation of d.c. motors. There are certain factors in using d.c. motors which can adversely affect their performance.

(a) Stalling. This should be avoided. On fixed voltage supply it would mean a very excessive armature current and the protection would trip the motor out. On variable speed drives unless an anti-stalling device is fitted there is nothing to prevent operating stalled at low voltage and in current limit. This causes localized heating of the commutator which affects its concentricity. The segments under the brush may burn and cause flats to develop. The situation is worse on self ventilated motors as there is no cooling either. As a rule avoid stalling on full current for more than a few seconds. If the application is likely to involve stalling then consult the manufacturer.
(b) Short circuit. The short circuit current of an uncompensated d.c. motor is up to 10 times full load and on a compensated motor 20 times. Avoid accidental short circuits.
(c) Rapid field change. On motors with field weakening the field strength must not be changed too rapidly particularly with heavy or high inertia load since this will involve high surges of armature current which could cause a flashover.
(d) Rapid voltage change. On drives with voltage control a quick change of voltage will mean a high surge of armature current.
(e) Loss of ventilation. If a blower stops working on a force ventilated motor then unless there is also an internal fan it is necessary to shut down fairly quickly within minutes to avoid overheating or burn out when running on full load. On self-ventilated motors do not run continuously below the specified minimum continuous speed.

3.5.4.3 Fault finding. One of the main problems is finding out which item of equipment is the primary cause of the trouble. A motor is associated with a driven load, control gear, and power supply. If the motor suddenly stops the trouble could apply in any one of these four areas, or a combination of two or more. Sparking on a thyristor d.c. motor may mean there is nothing wrong with the motor but that the thyristor supply waveform is incorrect perhaps due to a fuse blowing. On the other hand it could mean that something in the motor has upset the current distribution or commutation. Table 3.9 shows the main trouble symptoms in a d.c. motor and the action which can be taken.

Table 3.9 Fault finding guide

Symptom	Possible cause	Test or check	Cure
Overheating	Overloading	Armature current	Reduce load if FL exceeded
		Ammeter calibration	Replace if incorrect
	Insufficient ventilation	Air openings	Clean
		Blower rotation	Reverse if incorrect
		Filter	Clean or replace
		Water gauge drop in pipes	Larger pipes or blower
	Incorrect supply waveform	Armature voltage and current waveform	Correct thyristor supply
	High ambient	Air temperatures	Pipe in cool air
		Ambient, inlet, outlet temperatures	Fit heat shield Resite motor Stop recirculation of hot air
	Incorrect voltage	Armature volts Field volts	Set to rating plate values
Low Insulation Resistance	Damp Condensation	Megger windings Check individually also brushgear	Protect from moisture Fit heaters Resite motor
	Dirt or carbon dust or conducting particles	As above	Clean regularly Blow out Wipe brush spindles and commutator overhang
Vibration	Misalignment	Use gauge to check	Realign
	Bent shaft	Clock gauge	Repair
	Armature not balanced	Run uncoupled	Dynamically balance
	Poor foundations	Examine	Resite motor
	Faulty bearing	Examine	Replace
	Unequal airgaps	Measure gaps	Equalize gaps
Noise	Vibration	See above	See above
	Armature fouling	Check windings and fan	Clear obstruction
	Faulty coupling	Examine	Replace worn parts
	Ventilating air	Run light	Resite blower or pipe air
	Incorrect waveform	See above	See above
	Loose mounting	Check all bolts	Tighten
	Bearings	Lubrication and clearances	Lubricate or replace

Table 3.9 Fault finding guide—*continued*

Symptom	Possible cause	Test or check	Cure
Brush chipping or chatter or flexibles pulling out of brushes	Commutator glazing	If running on low load	Change brush grade
	Rough commutator	Check with gauge	Stone or skim
	High mica	Examine	Undercut and bevel
	Holders too far from commutator	Measure distance	Set within 1·5 to 3 mm
Unequal or rapid brush wear Flexibles burning	Brushes sticking in holders	Check clearances	Clean brushgear Renew brushes
Excessive sparking (See section 3.43)	Incorrect supply waveform	Armature voltage and current waveform	Correct thyristor supply Fit choke
	Wrong brush position	Check neutral	Set to mark or neutral
	Wrong brush grade	Brush grade	Consult manufacturer
	Vibration	See above	See above
	Overload	See above	See above
	Control too rapid	Peak current and rate of change	Slow down control
	Interpole strength wrong	Brush drop test Air gaps	Consult manufacturer Correct front and rear gaps
	Wrong spring pressure	Check pressure	Adjust or renew springs
	High/low bars	Check commutator	Stone or skim
Commutator	(See section 3.4.3.3)		
Burning	Sparking	See above	See above
Grooving	Incorrect axial stagger	Check stagger	See section 3.3.2.3
Streaking	Chemicals or fumes	Check air supply	Resite motor or air inlet
Scouring	Light load running	Load	Change brush grade or reduce brushes/arm
Irregular marking	Unstable commutator	Concentricity and profile	Retighten and skim
	Oil on commutator	Examine	Clean. Stop leaks
Hot bearings	Wrong lubrication	Examine	Add or remove grease
	Overload	Belt pull or endthrust	Use larger pulleys Fit suitable coupling
Noisy bearings	Vibration	See above	See above

3.5.4.4 Fault testing. These are some of the tests which may have to be made during fault finding.

(a) Armature drop test or bar to bar test. This checks for some winding faults or bad joints. Pass a d.c. current through the armature at points on the commutator spaced by a pole pitch. Check the voltage drop between adjacent segments all the way round. For most windings the readings will be uniform.

For some mixed turn windings the readings will be non-uniform but in a recurring pattern. There are various forms of this test whereby the current is fed in 180° (electrical) apart or adjacent segments or a commutator pitch apart. Note: Electrical° = Mechanical° × $p/2$.

(b) Impedance test for field coils. If a field coil has some shorted turns a simple resistance test may not show this up. Pass an a.c. current through the field coils and check the voltage across each coil. Due to the mutual inductive effect of shorted turns these will show up as a much reduced voltage reading compared to a healthy coil.

(c) Brush drop curve test. This can be used to get a rough guide to the interpole strength and shows how the current is being picked up by the brushes. From section 3.4.1 and Fig. 3.28 the current density in the brushes depends on the way in which the current is reversed. For straight-line commutation the density is constant and the interpole strength correct. The brush potential curve should preferably be flat or slightly drooping to the leaving or trailing edge.

See Fig. 3.34 for circuit for the test and typical brush drop curves. Care must be observed as the motor may be at a high voltage level.

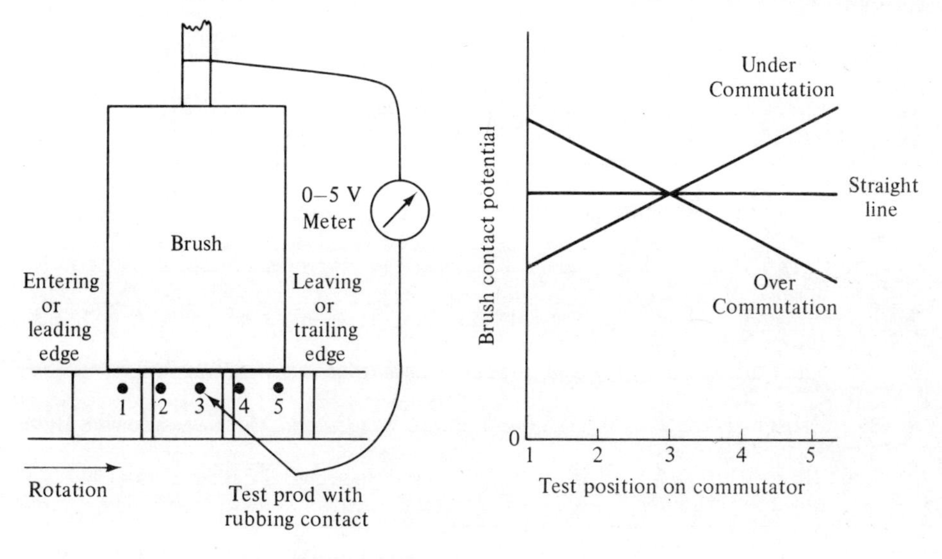

Fig. 3.34 Brush drop test and curves

(d) Brush neutral test. See section 3.5.2.1.

(e) Thyristor waveform test. On a thyristor drive it is preferable to ensure first that the thyristor supply is correct. Check the voltage and current waveforms with an oscilloscope against the standard expected.

(f) Overheating test. Check cold and hot resistance and work out temperature rise by the formula in section 3.5.2.3. Measure when running, temperatures by thermometer of air inlet, outlet, and frame and when shutdown, the windings.

(g) Connections and polarity checks. See section 3.2.6.

3.5.4.5 Repairs. See also chapter 19. For d.c. motors the following points are important.

(a) Windings must be rewound in a similar manner to the original so that the high heat dissipation rates of modern windings are maintained.
(b) If the armature is rewound or a new one fitted or the brushgear changed, check the brush neutral position. See section 3.5.2.1. The brush running position may be slightly displaced from neutral but it is not likely to be more than one bar out.
(c) New brushes should be bedded to the commutator surface before running.
(d) Brush grades must not be mixed except in a few special cases and a grade at least equivalent to the original should be fitted.
(e) Brushholders should be set close to the commutator after skimming. (Within 1·5 to 3 mm).
(f) Brushes must be a good sliding fit in the holders. Check springs and set pressure to the correct value.
(g) Pole spacing and brush spacing accuracy must be maintained.
(h) Liners must be replaced as found.
(i) All queries to the manufacturer should state complete rating plate data.

References

1. A. E. Clayton and N. N. Hancock, Chapter 1, *The Performance and Design of Direct Current Machines*, Pitman, London (1959).
2. A. Draper, Chapter 11, *Electrical Machines*, Longmans, London (1967).
3. A. Tustin, Chapter 2, *Direct Current Machines for Control Systems*, Spon, London (1952).
4. D. R. Shoults and C. J. Rife, Chapter 7, *Electric Motors in Industry*, Wiley, New York (1953).
5. R. L. C. Tessier, 'Present state of Ward–Leonard Control Systems', *Direct Current* (July 1963).
6. Institution of Electrical Engineers, *Conference on Electrical Variable Speed Drives*, Publication 93, IEE, London (1972).
7. B. M. Bird and R. M. Harlen, 'Variable characteristic d.c. machines', *Proc. IEE*, **113**, No. 11 (1966).
8. Morganite Carbon Ltd, Chapter 9, *Carbon Brushes and Electrical Machines*, Morganite Carbon, London (1961).
9. H. B. Ranson and E. T. A. Webb, Chapter 5, *Direct Current Machines*, Cleaver-Hume, London (1960).
10. C. S. Siskind, *Direct Current Armature Windings*, McGraw-Hill, New York (1949).
11. L. Greenwood, Chapter 2, *Design of Direct Current Machines*, Macdonald, London (1949).
12. B. S. Hender, 'Recent developments in battery electric vehicles', *Proc. IEE*, **112**, No. 12 (1965).
13. R. S. Philips, Chapter 6, *Electric Lifts*, Pitman, London (1973).
14. G. O. Watson, *Marine Electrical Practice*, Newnes, London (1962).
15. D. A. Lightband and H. D. A. Bicknell, *The D.C. Traction Motor*, Business Books, London (1970).
16. 'Homopolar Motor with Superconducting Windings', *Electrical Review* (9 February 1968).
17. Institution of Electrical Engineers, *Conference on Commutation in Rotating Machines*, Publication 11, IEE, London (1964).
18. J. Hindmarsh, 'Fundamental ideas on Large D.C. Machines', *Electrical Times*, Series from 7 August 1958 to 2 April 1959.
19. J. J. Bates, 'Thyristor-assisted commutation in electrical machines', *Proc. IEE*, **115**, No. 6 (1968).
20. M. Z. Iqbal and J. Hindmarsh, 'Detectors for Commutation Measurement and Control', *Electrical Machines in the Seventies*, University of Dundee (1970).

21. 'Problems in the Construction of Modern D.C. Machines for Industrial Applications', *The Brown Boveri Review* (Oct/Nov 1968).
22. R. Bourne, Chapter 2, *Electrical Rotating Machine Testing*, Iliffe, London (1969).
23. S. Johnson, 'Commutation Adjustments on D.C. Machines', *Electrical Times* (21 September 1961).
24. D. D. Stephen, 'The evolution of the standard machine', *Electronics and Power*, March, 1970.
25. W. P. Tholen, 'Maintaining Electrical Machines', *Electronics and Power* (3 May 1973).

Acknowledgements

The author wishes to thank the Technical Director of Brush Electrical Machines Ltd, for permission to publish this chapter on direct current machines, and his colleagues for their assistance.

Figure 3.31 is reproduced by permission of the General Electric Company of Schenectady, New York.

Figures 3.16, 3.17, 3.18, 3.20, 3.22, and 3.23 are reproduced by permission of Brush Electrical Machines Ltd of Loughborough, England.

4 Alternating Current Motors

4.1 Types of a.c. Motors

In general, torque and the resulting rotation or linear motion in electric motors results from the interaction of two magnetic fields. One of these fields may be provided by a permanent magnet but more often both are produced by electric currents. The nature of the current supplied to electric motors and the manner in which their windings and magnetic structures are arranged gives rise to a wide variety of designs. Of all electric motors manufactured for industry the three-phase induction motor with a cage rotor has for three-quarters of a century been the type which, for one or more of the following reasons, has been produced in the greatest numbers.

(a) Since the majority of motors are required to produce steady rotation and the rotating magnetic field eliminates many of the design problems associated with providing continuous torque, the polyphase-induction motor is eminently suitable for approximately constant speed application. Only small modifications can make it synchronous, which on public electricity supplies means absolutely constant speed over its full operating range.

(b) Three-phase alternating current supplies have been adopted universally for the distribution of electrical energy, and there is a considerable measure of standardization. In such supplies the three phases peak sequentially in time at intervals equal to one-third of the system periodicity. By arranging the three-phase windings around the airgap so that the poles generated are also in space-sequence a rotating field is set up.

(c) The rotor draws power by transformer action from the stationary windings distributed in the stator slots and converts it into mechanical power.

(d) The absence of brushes makes it possible to operate induction motors with cage or solid rotors under very adverse conditions and ensure considerable reliability.

4.2 Rotating Fields

Figure 4.1 shows how a symmetrical three-phase winding when supplied from a balanced three-phase power supply produces a rotating field. Each phase is represented by two turns spanning one pole pitch of a two-pole motor. The arrow shows the direction of the current in phase A at the instant when it has maximum positive value. In the phasor diagram it is represented by I_A, and I_B and I_C are the other two phase currents at 120 degree intervals. If the phasor diagram is now imagined to rotate in an anticlockwise direction at supply frequency the variation of current or

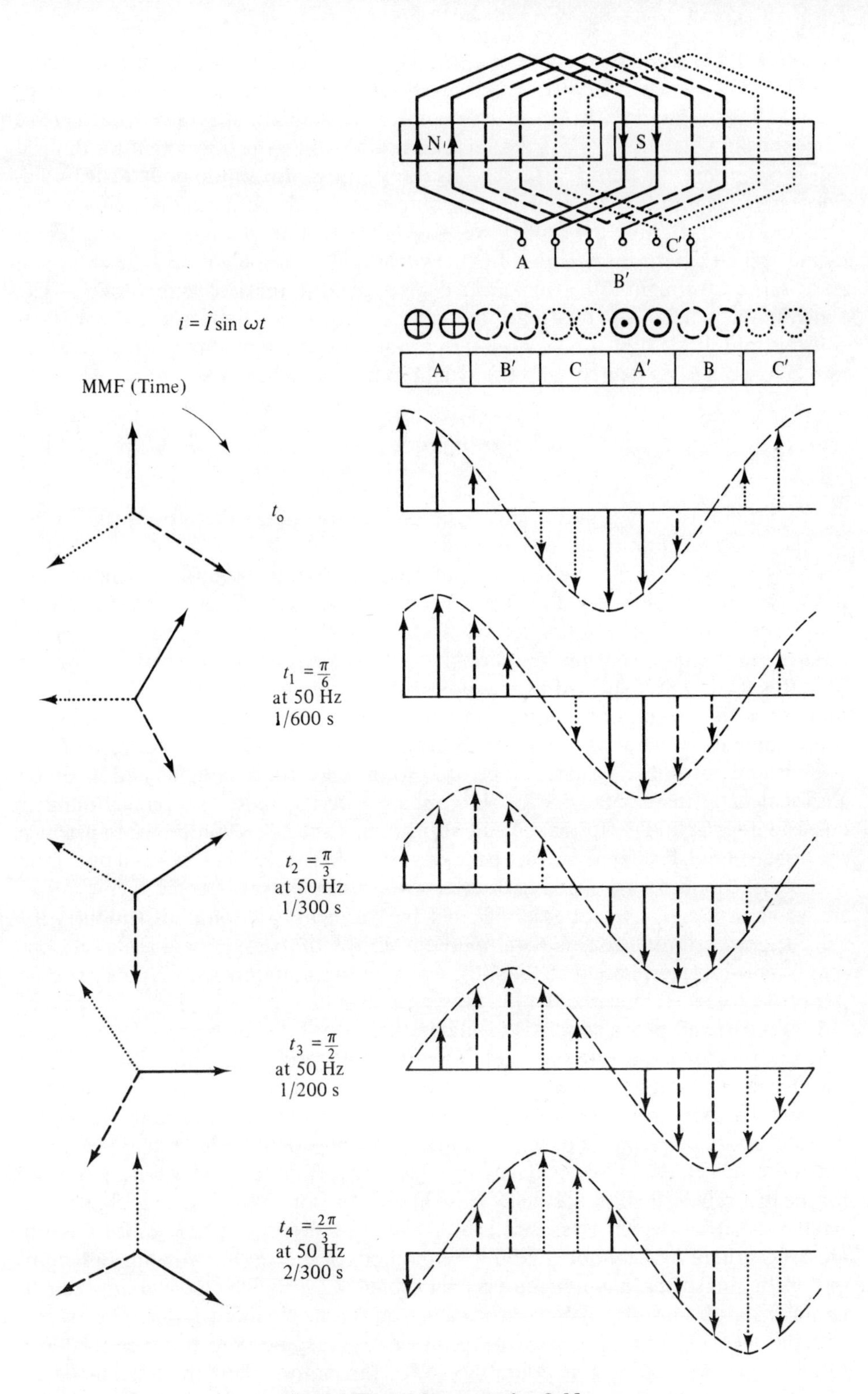

Fig. 4.1 Three phase rotating field

field magnitude with time may be considered. One complete rotation or one cycle is completed in $1/f$ seconds, so that for a 50 Hz alternating current the 30 degree steps occur at one six-hundredth second intervals. If the instantaneous value of each phase current is now shown in its correct spatial position under the developed windings it can be seen that the m.m.f. produced by the stator is also sinusoidally distributed. Successive diagrams show the field distribution at the above-mentioned time intervals and illustrate how the field wave travels across the winding face, which in a cylindrical airgap means rotation. In the synchronous two-pole motor the rotor will have turned through 120 mechanical degrees or one-third of a revolution. The synchronous speed of two-pole induction motors is, therefore, f rev/s or $60f$ rev/min. If a motor is now wound to produce p pole pairs the m.m.f. wave will now move only $1/p$ degrees and the synchronous speed becomes

$$N_S = \frac{60f}{p} \text{ rev/min} \tag{4.1}$$

For 50 Hz motors 4, 6, 8, and 10 pole synchronous speeds are therefore 1500, 1000, 750, and 600 rev/min respectively.

Figure 4.2 shows that two phases are sufficient to produce a rotating field and small power two-phase servo motors have been studied and are used extensively in instrument and instrumentation servo systems. In such applications one phase (called the reference phase) is connected directly to the supply, whereas the other (control phase) is made to either lead or lag by 90° and varied in magnitude. Unbalancing m.m.f.'s of the two phases is detrimental to efficiency, but since the most important characteristics demanded are controllability and stability, this has to be accepted.[1,2]

Thus any polyphase system can be used to produce rotating fields and it can be shown that up to very large sizes three phases are the most economical number consistent with good performance and utilization of the active materials. In practice the airgap m.m.f. distribution is not perfectly sinusoidal since it is produced by a finite number of conductors grouped in discrete slots. Since, however, other factors affect the waveform and it is not essential to have perfectly sinusoidal distribution for satisfactory performance small and medium sized motors may use a relatively low number of slots per phase, i.e., 24 slots, 4 pole, 3 phase stators with 2 slots per pole per phase for small machines would be typical.

If only a single-phase supply is available then Fig. 4.3 shows that it is no longer possible to produce a rotating field. The effect of time variation of current in a single-phase winding produces a pulsating field, and if the magnetic circuit is symmetrical there will be no reason for a motor to develop torque in either direction. The stationary pulsating field can be shown to be equivalent to the sum of two equal and uniformly contra-rotating magnetic fields of half amplitude. This concept is used for the theoretical analysis of single-phase induction motors (reference 3, chapter 5). Another method which was used for the analysis of single-phase motors is the cross-field theory.[3] A number of different methods of starting and running induction motors on single-phase supplies are given in chapter 7. Provided the windings can be reconnected to suit the voltage of the single-phase supply three-phase motors can, with the aid of one or more capacitors, be made to give some 70 to 80 per cent of their rated output. This method is frequently used on fan motors where the starting torque required is usually only some 25 to 30 per cent of rated full-load torque. The fact that

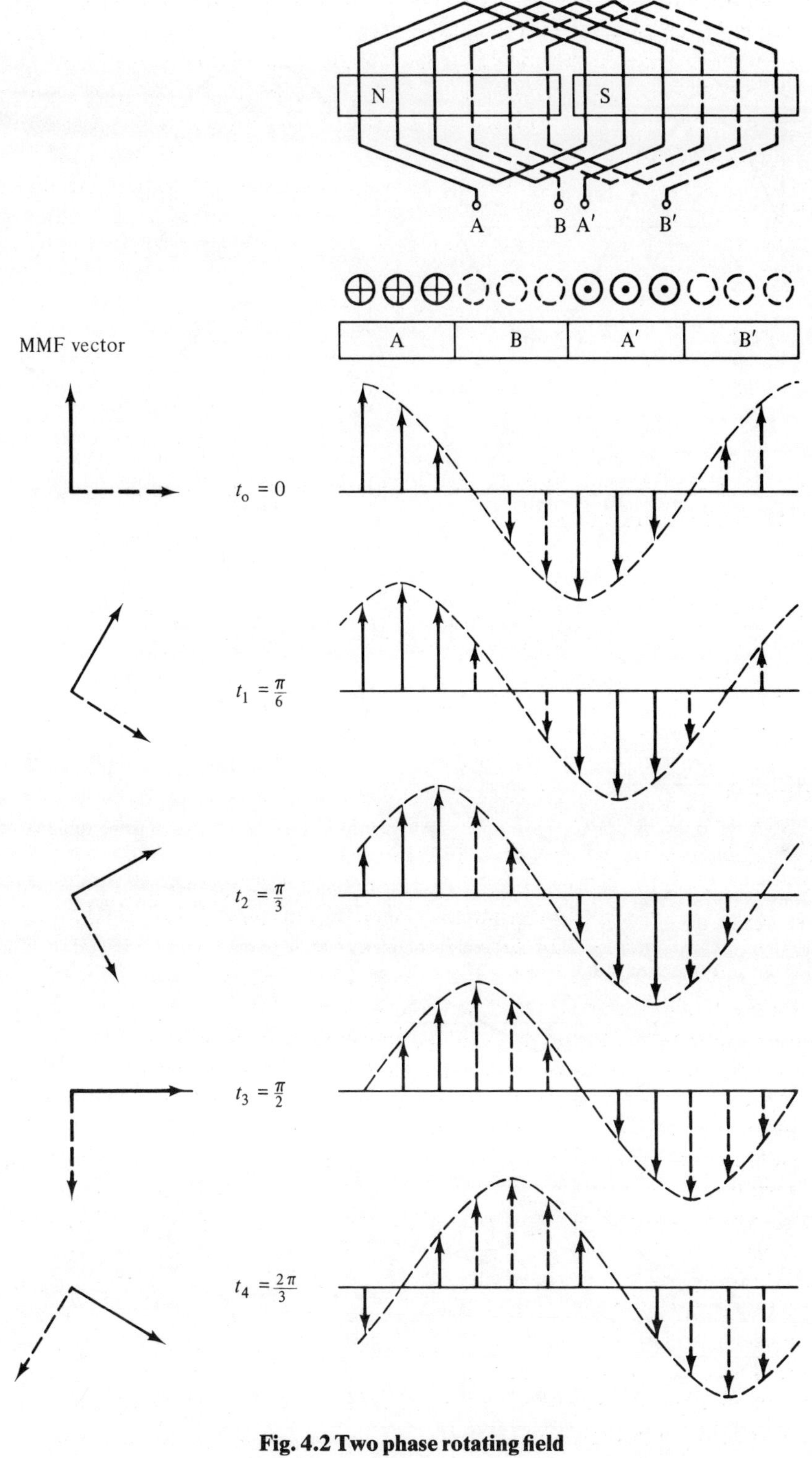

Fig. 4.2 Two phase rotating field

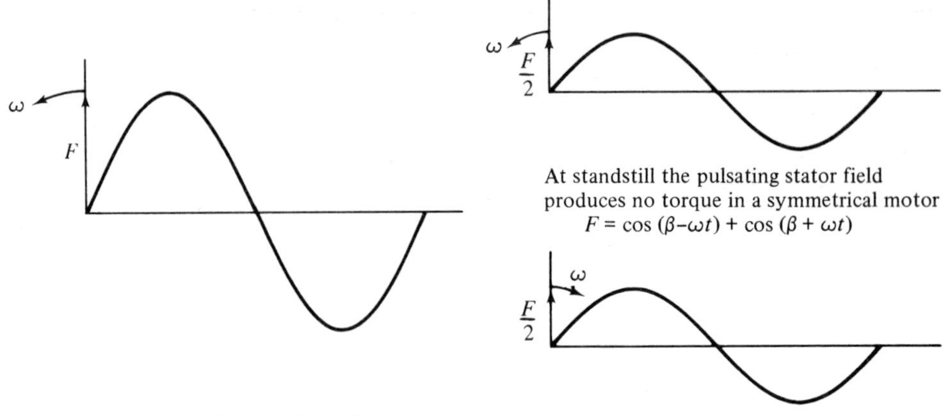

(a) *Two revolving field theory*
When running and considered as two motors with their stators in series opposition acting on the same rotor it can be shown that the forward and backward torque curves shown in (c) below result

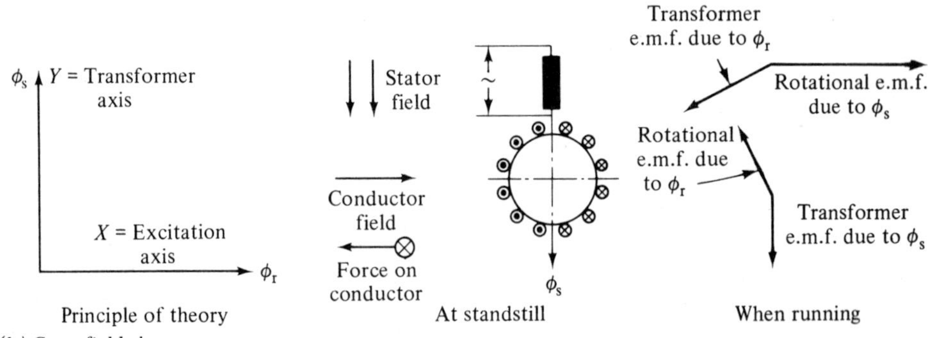

(b) Cross-field theory
Torques are produced by (i) the interaction of ϕ_r and y axis components of current: *the driving torque* and (ii) by the interaction of ϕ_s and the x axis components of current. These torques are the same as those derived by the revolving field analysis.
In both cases calculation of motor parameters is complex and due to deviations from ideal sinusoidal fluxes and currents approximate.

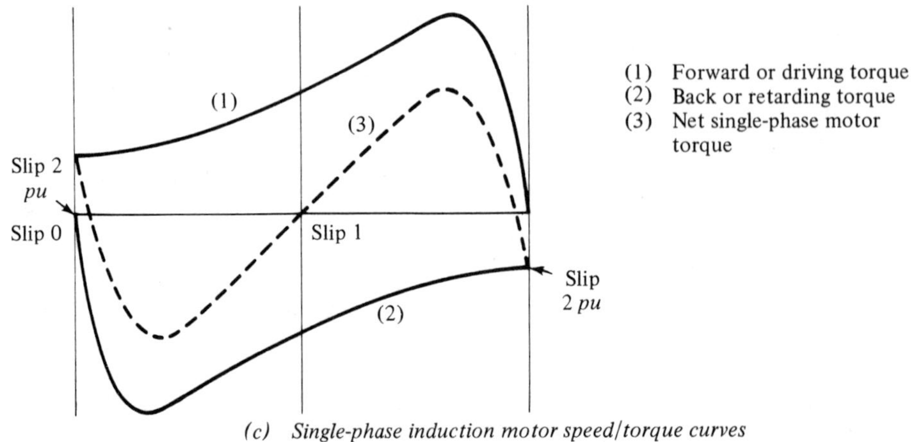

(c) Single-phase induction motor speed/torque curves

Fig. 4.3 Single phase motor theories

a three-phase motor can continue to run when one phase fails may cause overheating or stalling when the load increases and suitable precautions must be taken.

4.3 Speeds of a.c. Motors

The speed of induction motors can be varied by increasing slip, changing the number of poles or varying the supply frequency and voltage.

Special designs of cage rotors can match motor and load characteristics and so reduce the size of motor required. More torque per ampere of starting current can also save on control equipment and generally increase the flexibility of the drive.

If the cage is replaced by a winding brought out to collector- (slip-)rings the rotor resistance can be varied externally and provides not only improved starting characteristics but a measure of speed control. Such control is achieved by dissipating power as heat in the external resistances so that this method can prove not only inconvenient but costly for large powers. Special auxiliary machines which can convert power at slip frequency and return it to the supply are now superseded by commutator motors and electronic controllers which are able to provide a wide range of continuously variable speeds.

Pole changing provides at least one other fixed speed and can, in certain circumstances, prove economical for a third speed. The increase in motor size depends on the torque required at the lower speeds and ratio between the speeds. The pole amplitude modulated windings invented by Professor Rawcliffe are now used by a number of manufacturers to provide cage rotor induction motors with two speeds which need not be in the ratio 1:2 from a single winding. (See section 4.10).

In some applications such as grinding spindles, portable tools, and woodworking machinery high speeds are required or the weight must be as small as possible. To achieve a high power-to-weight ratio the speed must be increased substantially which, for induction motors, means higher frequencies such as 250 and 400 Hz which, in the case of two-pole windings, gives synchronous speeds of 15 000 and 24 000 rev/min respectively.

For aircraft and military applications mainly 400 Hz motors are used and these may be useful in industry where their cost and the provision of a suitable frequency changer for groups of motors can be economically justified.

The speed of induction motors can be controlled electronically either by controlling the power input (varying the firing angle of the thyristors), or by connecting them to inverters, or cycloconverters (frequency changers).

In many applications where frequency conversions cannot be justified economically, a.c. series motors are used. For domestic appliances and portable tools where powers are unlikely to exceed 1 kW, the single-phase series motor (universal motor) consisting of a single laminate two-pole field and 12 to 18 slot armature and, nowadays, operating at load speeds of between 8000 and 18 000 rev/min, provides an ideal power unit. Larger motors with distributed and compensating field windings are used in industry and for electric traction.

The need for variable speeds and 'special' starting characteristics led to much work being done on three-phase commutator motors. These fall into three main categories:

(a) The three-phase series motors.

(b) The three-phase stator fed shunt motors covering mainly the large power range from 150 to 1500 kW.
(c) The three-phase rotor fed shunt (or Schrage) motors limited to about 500 V maximum and power outputs of no more than 200 kW.

Of these the Schrage motor in particular is still manufactured by several European companies to meet variable speed requirements in such industries as paper, textiles, process (stirrers and compressors), and printing, as well as power generation (auxiliaries).

4.4 Rotors, Rotor Currents, Torque, and Slip

The following treatment of the induction motor uses simplified formulae and equivalent circuits sufficient for approximate calculations to illustrate how the

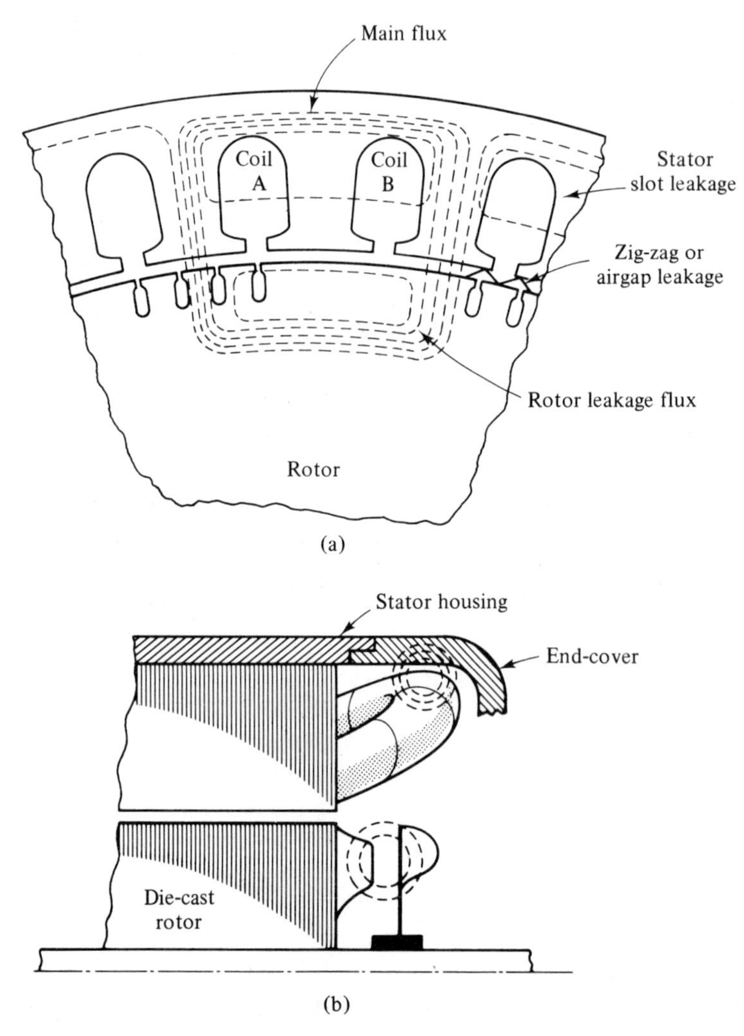

Fig. 4.4 Flux paths in stator and rotor of induction motor. (a) Radial distribution. (b) End-winding flux leakage.

various parameters interact. For more accurate calculations and searching analysis of the various types of motors the formulae used in their theoretical treatment and design must be used. (Bibliography references 3, 4, 5, and 6).

Figure 4.4(a) shows a simplified representation of how the flux produced by two series coils of a primary winding links the windings of the secondary.

As long as there is no relative movement between the rotor (secondary) and the stator (primary), the induction motor is equivalent to a single-phase transformer with an airgap in its magnetic circuit. Secondary induced voltages on starting will have the same frequency as the primary supply. If the primary stator has three-phase winding it will, as shown in Fig. 4.1, produce a field rotating at the synchronous speed determined by the supply frequency and the number of pole pairs (Eq. (4.1)). If the rotor is of the wound type it must have the same number of poles but these may be formed by a different number of phases. In practice rotors are usually wound with three balanced, star-connected, phases brought out to three collector rings. This not only allows the introduction of resistances for starting or speed adjustment but the connection to other smaller machines to form synchronous drives (Ref. 3, chapter 15).

The most common rotor winding, the cage rotor, generally consists of one solid conductor per slot connected to an end ring at each end. Conductors and end rings are sometimes insulated from the core to reduce stray losses caused by currents through the laminations or iron core between adjacent bars. Because of differential expansion between the iron core and the copper or aluminium rotorbars this may be difficult to maintain after prolonged arduous service. Heating during starting or plugging can be considerable and the forces caused by the rapid acceleration of which induction motors are capable are such that even the most carefully manufactured rotor may, after long use, suffer from open circuited or lifted bars.

Cage windings adjust themselves to any number of poles and the exact number and shape of rotor bars used is the result of careful design and development. Compromises in the design of commercial motors leave inherent imperfections, which, under certain conditions, can affect the use of a particular motor. Noise, cogging, and synchronous crawling are usually minimized by skewing the rotor bars[8] and rotors with specially shaped bars are used to obtain the desired starting performance.

Reverting to the rotating field and stationary rotor the rotor current per phase can be expressed as

$$I_2=\frac{E_2}{Z_2}=\frac{E_2}{R_2+jX_2} \tag{4.2}$$

The rotor current vector has a magnitude

$$I_2=\sqrt{\frac{E_2}{(R_2^2+X_2^2)}} \tag{4.3}$$

and lags behind the induced voltage by

$$\cos\phi_2=\sqrt{\frac{R_2}{(R_2^2+X_2^2)}} \tag{4.4}$$

Since E_2 is proportional to the flux per pole, Φ, and is in phase with it, the starting torque is given by

$$T_{st}=k\Phi I_2\cos\phi_2 \tag{4.5}$$

which would be maximum if the rotor had no reactance, i.e., ($\cos\phi = 1$). The effect of rotor power factor can also be shown by drawing the flux and current curves in their correct phase relationship (Fig. 4.5). As the angle of lag increases the number of conductors under a particular pole which carry current in the reverse direction and therefore produce backward torque increases. Since the rotor bars are imbedded or backed by iron, inductance is inevitable and can only be minimized by increasing

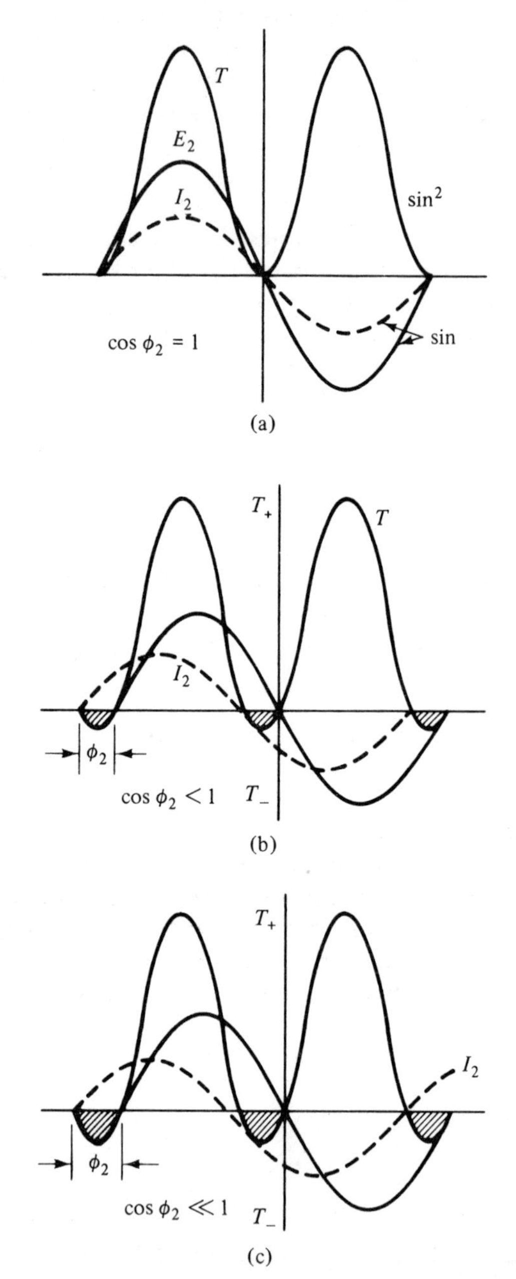

Fig. 4.5 The effect of rotor power-factor on torque in induction motors. (a) Ideal case no reverse torque. (b) Small lagging power factor. (c) Poor power factor

resistance. Maximum torque will be associated with maximum rotor current and by differentiating Eq. (4.3) with respect to R it is found that the current will be maximum when $R_2 = X_2$. Any further increase in resistance though improving the power factor will decrease the current, and therefore torque, to a greater extent.

Once the rotor begins to turn, the e.m.f. in the rotor bar decreases in magnitude and frequency until at synchronous speed the rotor current is zero and no torque is developed. A cage induction motor will, therefore, always run at such speed so that the rotor current can provide sufficient torque to balance the load on the motor.

The slip of a motor is defined as the difference in speed between the rotating field and the rotor (N) expressed as a decimal fraction of the synchronous speed (N_s)

$$s = \frac{N_s - N}{N_s} \tag{4.6}$$

As the magnitude of the e.m.f. produced in the rotor conductors depends on the rate of change of flux linked with the rotor winding, E_2 will decrease proportionally to s. The reactance of the rotor, being a function of frequency will decrease with slip so that the general equation for the secondary current now becomes

$$I_2 = \frac{sE_2}{R_2 + jX_2} \tag{4.7}$$

4.5 The Equivalent Circuit of the Induction Motor

The basic equivalent circuit is shown in Fig. 4.6(a) where the stator and rotor leakage reactances due to leakage fluxes are also introduced. The origin of these reactances is illustrated in Fig. 4.4(b) by the paths of the leakage fluxes which cause them.

The equation for the primary applied voltage is

$$V_1 = R_1 I_1 + jX_1 I_1 + E_1 \tag{4.8}$$

The secondary circuit can be defined by the equation

$$E_2 = R_2 I_2 + j(sX_2) I_2 \tag{4.9}$$

In Eq. (4.9) the slip s is 1 when the rotor is stationary and the circuit could equally represent a short-circuited transformer secondary in which the induced e.m.f. is balanced by the product of short-circuited current and secondary impedance. The standstill e.m.f. E_{ss} induced in the induction motor secondary by the rotating field will be at supply frequency. As soon as the motor begins to turn the induced e.m.f. and the rotor impedance become sE_{ss} and sX_2, respectively. The rotor frequency decreases rapidly until the normally very steep slope of the induction motor's speed/torque curve is reached. The frequency of the rotor current will then usually be of the order of 1 to 5 Hz. In order to be able to continue with the equivalent circuit and phasor representation E_2 when referred to the stator must be considered in terms of its effective value. This concept is justified by the magnetic field which links both stator and rotor.

If the apparently vanished portion $(1 - S)E_{ss}$ is now assumed to be a back e.m.f. due to rotation (similar to that in a d.c. motor armature) then the gross power produced by the rotor per phase will be:

$$P_g = (1 - s)E_{ss} I_2 \cos\phi \tag{4.10}$$

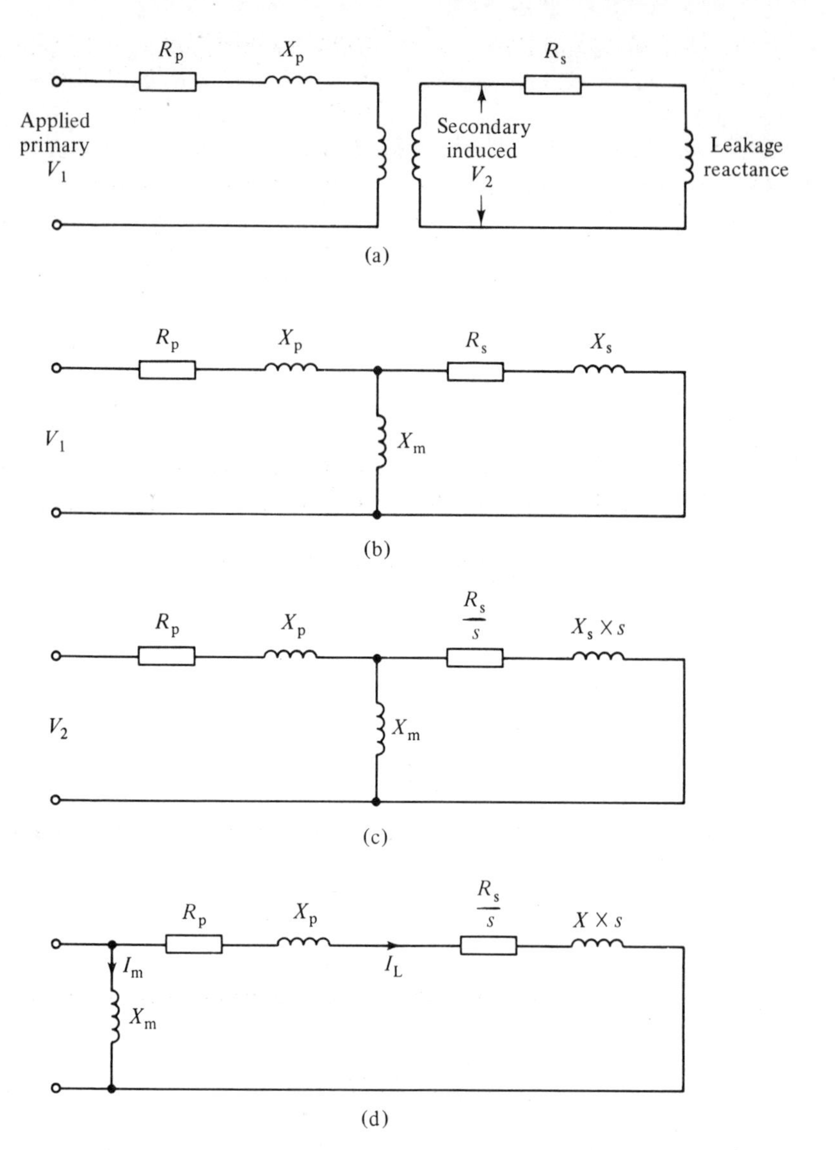

Fig. 4.6 Basic equivalent circuits for induction motors. (a) Introduction of leakage reactances. (b) Rotor stationary. (c) Rotor speed

Substituting for (I_2) from Eq. (4.7) and replacing $\cos \phi_2$ by

$$\cos \phi_2 = \frac{R_2}{\sqrt{(R_2^2 + s^2 X_2^2)}} \tag{4.11}$$

the power equation becomes

$$P_g = \frac{1-s}{s} R_2 I_2^2 \tag{4.12}$$

This equation shows that the output is a fraction of the secondary heat losses.

The equivalent circuit is developed by referring secondary values of resistance and reactance to the primary by multiplying them by the square of the turns ratio and, if applicable, by the ratio of the number of phases.

If C is defined by

$$C = \frac{\text{number of stator phases}}{\text{number of rotor phases}} \times \left(\frac{\text{number of stator turns}}{\text{number of rotor turns}}\right)^2 \tag{4.13}$$

then

$$R_2^1 = CR_2 \quad \text{and} \quad X_2^1 = sCX_2 \tag{4.14}$$

For comparison and general application calculations the simplified equations are more than adequate, and the variation of X_2 with s as well as magnetizing losses are neglected.

In Figs. 4.6(b) and (c) the resistance $(1-s)/s \,.\, R_2$ has been introduced to represent the mechanical load on the motor and the equation for the equivalent circuit is now

$$V_1 = IR + IjX + \frac{1-s}{s} R_2 I \tag{4.15}$$

For the power, Eq. (4.15) is multiplied by current, it being understood that V_1 and I_1 are phasors.

The mechanical power is again represented by

$$\frac{1-s}{s} R_2 I_2^2$$

and this together with the copper loss $I_2^2 R_2'$ represents the total power transferred from stator to rotor

$$\frac{1-s}{s} I_2^2 R_2^1 + I_2^2 R_2^1 = \frac{1}{s} I_2^2 R_2^1 \tag{4.16}$$

Making the appropriate substitution the power output per phase can be expressed as

$$P_g = \frac{R_2^1 V_1^2}{\sqrt{[(R_e + R_2)^2 + X_e^2]}} \tag{4.17}$$

where $R_2^1 = (1-s)/s \,.\, R_2$ which can be shown to be a maximum when the equivalent load resistance equals the standstill leakage impedance. For maximum power output, efficiency, and good speed regulation, resistance and leakage reactance must be as small as possible.

4.6 Synchronous Watts and Torque

Mechanical power output is given by the product of rotational speed and torque. If ω_s is the synchronous speed of the motor in radians/s then the actual speed is given by $\omega = \omega_s(1-s)$ and the output equation is

$$m_2 I_2^2 R_2 \left(\frac{1-s}{s}\right) = T\omega_s(1-s)$$

which reduces to

$$T\omega_s = m_2 I_2^2 \frac{R_2}{s} = \frac{N_s}{N} P_g \tag{4.18}$$

The product of torque and synchronous speed is known as the torque or more correctly the output in synchronous watts. At standstill, however, when $s = 1$ the torque in synchronous watts equals the total rotor copper loss.

The following example illustrates the calculation of synchronous watts and torque.

A 22 kW 4 pole motor runs at 1440 rev/min at rated load. The efficiency claimed is 90·5 per cent and it will be assumed that the mechanical losses are about 25 per cent of the total load losses.

Input to motor	$\dfrac{22}{0\cdot 905} = 24\cdot 31$ kW
Friction and windage	$2\cdot 31/4 = 0\cdot 58$
Input to rotor	22·58 kW
Power in synchronous watts	$= 22\cdot 58 \dfrac{1500}{1440} = 23\cdot 52$ kW
Synchronous speed in radians/s	$= \dfrac{4\pi \times 50}{4} = 157$
Synchronous torque	$= \dfrac{23\cdot 520}{157} = 149\cdot 8$ Nm

The actual maximum value of torque of an induction motor can be shown to be independent of the rotor resistance which, as Fig. 4.7 shows, only governs the slip at which it occurs. The simplified equivalent circuit is used to derive an equation for the synchronous torque by substituting for I_2

$$T_{sw} = \frac{R_2}{s} I_2 = \frac{R_2}{s} \frac{V_1^2}{[(R_1 + R_2/s)^2 + X^2]} \tag{4.19}$$

From this equation it can be shown that the maximum torque will occur at a value of slip

$$s = \frac{R_2}{\sqrt{(R_1^2 + X^2)}} \tag{4.20}$$

and substituting this in the torque equation (4.19) will give the actual numerical value

$$T_{sy \cdot max} = \frac{V_1^2}{2\{R_1 + \sqrt{(R_1^2 + X^2)}\}} \tag{4.21}$$

The maximum torque of a particular induction motor is determined by its stator resistance and leakage reactances, dependent then only on the square of the applied voltage. The nature of the load and the duty rating will now determine the rotor resistance giving the most suitable speed/torque curve. The curves in Fig. 4.7 illustrate that high running efficiency (smallest possible slip on full load) and high starting torque are only possible if the rotor resistance is made large for starting and

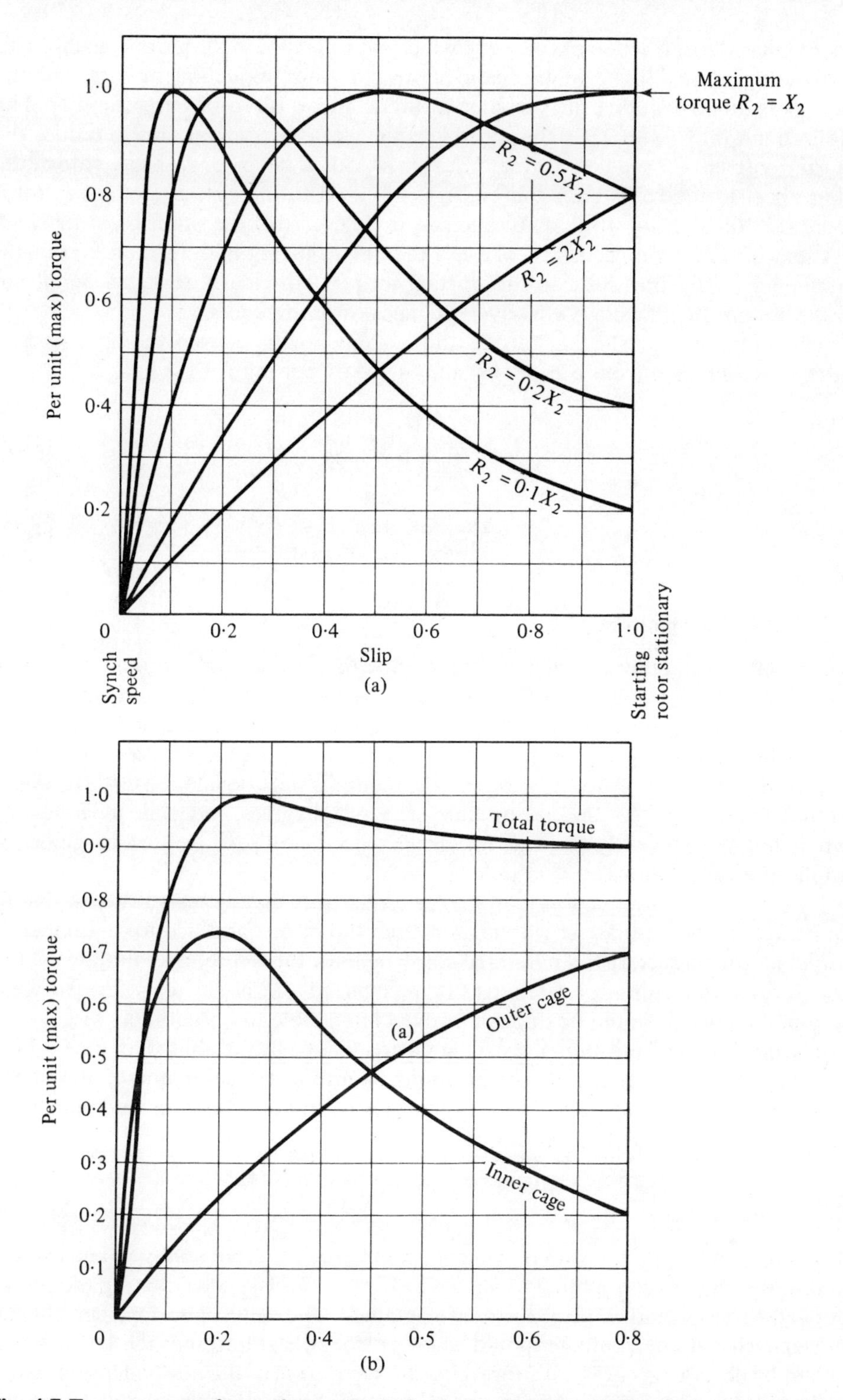

Fig. 4.7 Torque curves for various values of rotor resistance. (a) Standard cage induction motor. (b) Torque curves for typical double cage rotors

then reduced to a fraction of the value which gave maximum torque at starting. This variation is easily made in the case of wound rotor induction motors, but the designers found also that the variation of inductance with rotor frequency could be utilized not only to improve the speed/torque characteristic but also to reduce the starting current for a given torque. Thus, in so-called deep- or sash-bar rotors, the skin effect in the rotor bars and end-rings is used to increase the effective rotor resistance on starting. In double cage rotors (Fig. 4.7(b)) the outer rotor has high resistance to provide curve (a), whereas the inner, having high reactance at mains frequency contributes little to the starting torque, but clearly makes a significant contribution when at slip it effectively reduces rotor impedance.

The equations for starting torque and current obtained from the approximate circuit are obtained from Eqs. (4.19) and (4.15) by putting $s = 1$

$$T_{ss} = \frac{R_2 V_1^2}{(R_1 + R_2)^2 + X_2} \tag{4.22}$$

$$I_{ss} = \frac{V_1}{\surd\{(R_1 + R_2)^2 + X^2\}} \tag{4.23}$$

4.7 The Circle Diagram

The equations and the equivalent circuit require the values of the various parameters to be known or reasonably easy to determine from tests. Until designers could make use of computer programs, and thus obtain a comprehensive and detailed print-out of the motor's performance, the circle diagram was also their easiest means for analysing motor performance. A number of methods were developed using the locus of the current vector for the construction of a circle diagram, and while the following would not be sufficiently accurate for design work it will provide a useful guide for application calculations.

By subjecting an induction motor to two relatively simple tests it is possible to construct an approximate circle diagram from the results and deduce a number of useful motor characteristics by simple constructions. Provided the limitations of the test results and graphical methods are recognized the circle diagram can yield information which would be difficult or even impossible to obtain otherwise.

The diameter and position of the diagram are defined by the phasors of the no-load and the short-circuit current of the motor. Careful measurement of the stator (primary) resistance will enable the copper losses to be evaluated, so defining the very important torque line as shown.

4.7.1 *No load test*

The motor should be run with no load at a number of different voltages, and current and power factor then plotted as shown in Fig. 4.8(a). By extrapolating the power curve the friction and windage losses are obtained. The value of no-load current and power-factor at rated voltage should be taken from the graph since these values will generally be more accurate. A more accurate separation of the no-load losses can be obtained by supplying a small amount of power through another motor to turn the shaft at exactly synchronous speed. The power supplied mechanically will be the

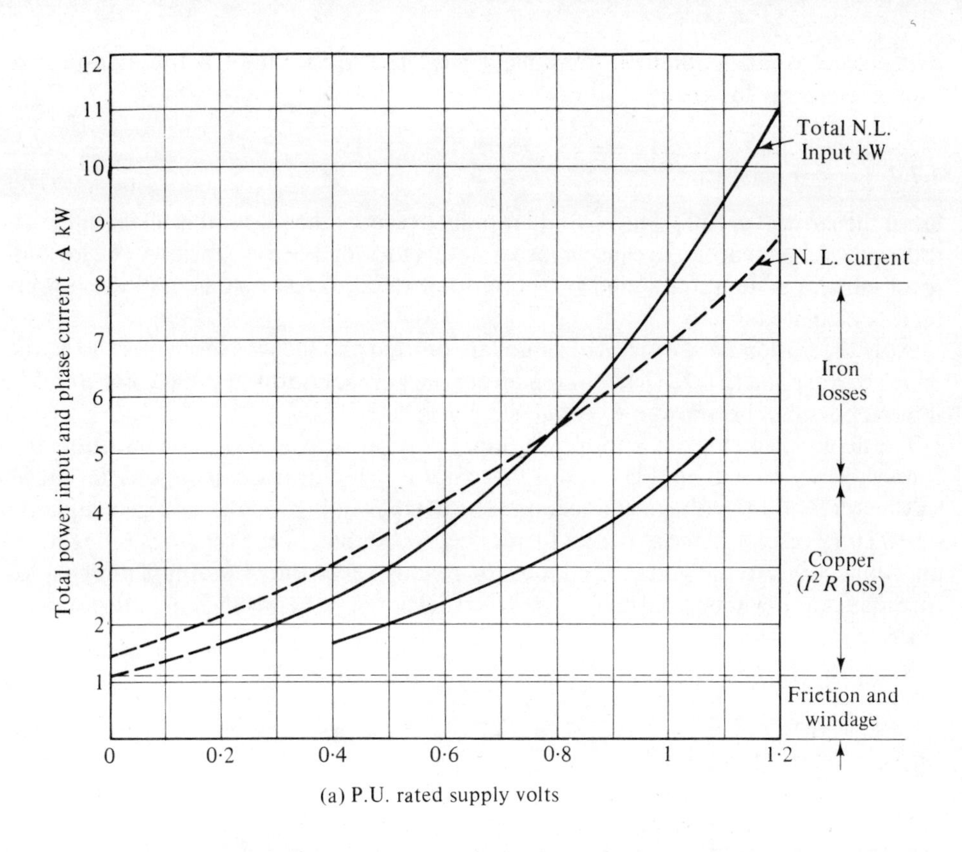

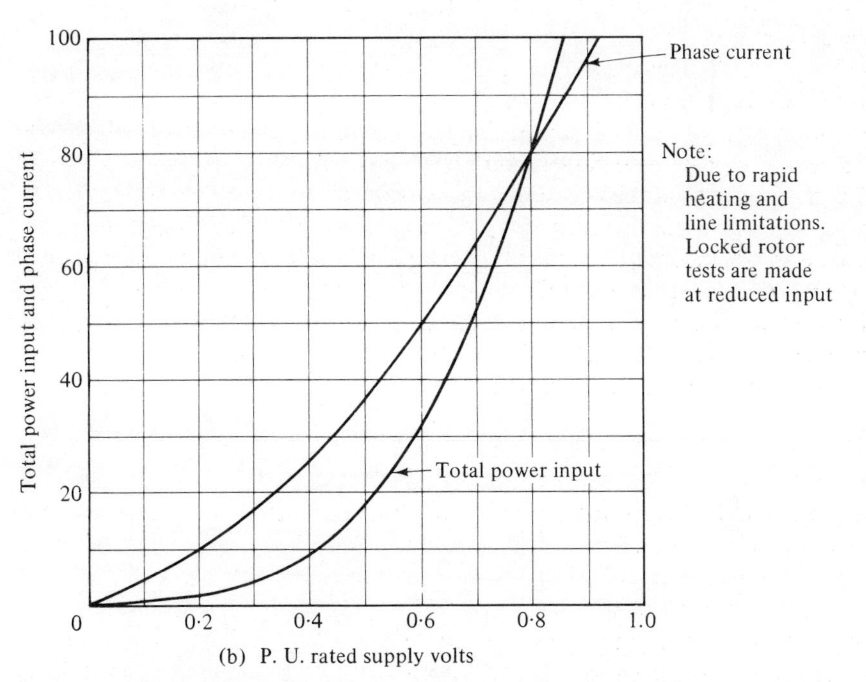

Fig. 4.8 Determination induction motor characteristics. (a) No-load test. (b) Locked rotor test

friction and windage, and the electrical power fed to the stator is the measure of copper and stray losses.

4.7.2 *Locked rotor test*

Since the current at full voltage would rapidly overheat the motor this test is made at reduced voltages chosen to cause approximately full-load current to flow. The locked rotor current is then multiplied by the ratio of rated to test voltage and the power factor calculated.

As in the no-load test a series of values are plotted and the values of volts and watts which correspond to rated current read from the graph. For this test the rotor should, if at all possible, be allowed to rotate slowly.

The no-load and locked rotor power input may be measured with two wattmeters connected as shown in Fig. 4.9(a). As motor windings are balanced a single wattmeter with the voltage connections switched as in Fig. 4.9(b), will give equally satisfactory results, since at power factors below 0·5 one of the readings is negative the connections to the voltage coil must be readily reversible. Careful planning and correctly rated switches and meters are essential to the safe and efficient performance of these tests.

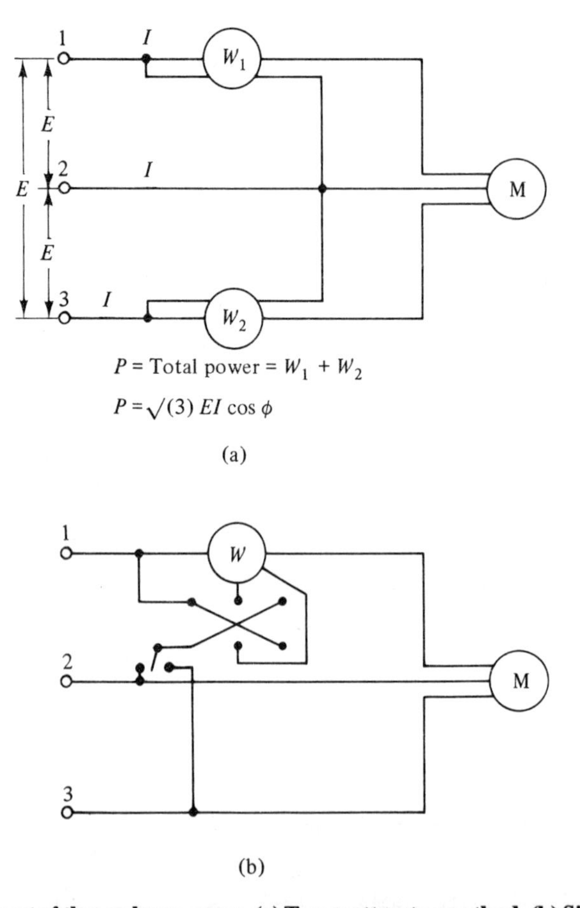

Fig. 4.9 Measurement of three phase power. (a) Two wattmeter method. (b) Single wattmeter

The measured values for current and voltage are line values and the total power input is the arithmetic sum of the two wattmeter readings. The power-factor may either be calculated from

$$\cos\phi = \frac{\text{total power input}}{3\times\text{line volts}\times\text{line current}}$$

or the ratio of the two wattmeter readings, which is likely to give the more accurate result, and justify the extra work involved:

$$\cos\phi = \sqrt{\left(\frac{(1+(W_2/W_1)^3}{1+(W_2/W_1)^3}\right)}$$

$$= \frac{1}{\sqrt{1+3(W_1-W_2/W_1+W_2)}}$$

$$= \frac{(W_2/W_1)+1}{2\sqrt{\{(W_2/W_1)^2+W_2/W_1+1\}}} \qquad (4.2a, b, c)$$

where

$$W_1 > W_2$$

or when

$\cos\phi$	W_1	W_2	W total
0 to 0·5	+	−	W_1-W_2
0·5	+	0	W_1
0·5 to 1·0	+	+	W_1+W_2

Where many calculations of this nature are made, curves, nomograms or tables may be convenient.

The evaluation of the graphs is self explanatory and it is now only necessary to calculate the locked rotor current at full voltage from

$$I_{ss} = \frac{\text{locked rotor current } I'_{ss}\times\text{rated voltage}}{\text{test voltage at which } I'_{ss}\text{ was measured}} \qquad (4.25)$$

The circle diagram, Fig. 4.10, should be at least 250 mm in diameter and the scale for current should now be chosen accordingly. The no-load and standstill current phasors can now be drawn using a practical power factor scale as shown.

Thus

$$\overline{OA} = I_{NL}\cos\phi_{NL}$$

and

$$\overline{OB} = I_{ss}\cos\phi_{ss}$$

A and *B* are two points on the circle diagram and the centre of the circle is found where the perpendicular bisector of the line *AB* cuts a horizontal line parallel to the *X* axis. The perpendicular to the axis *BDE* is taken to represent the total power input

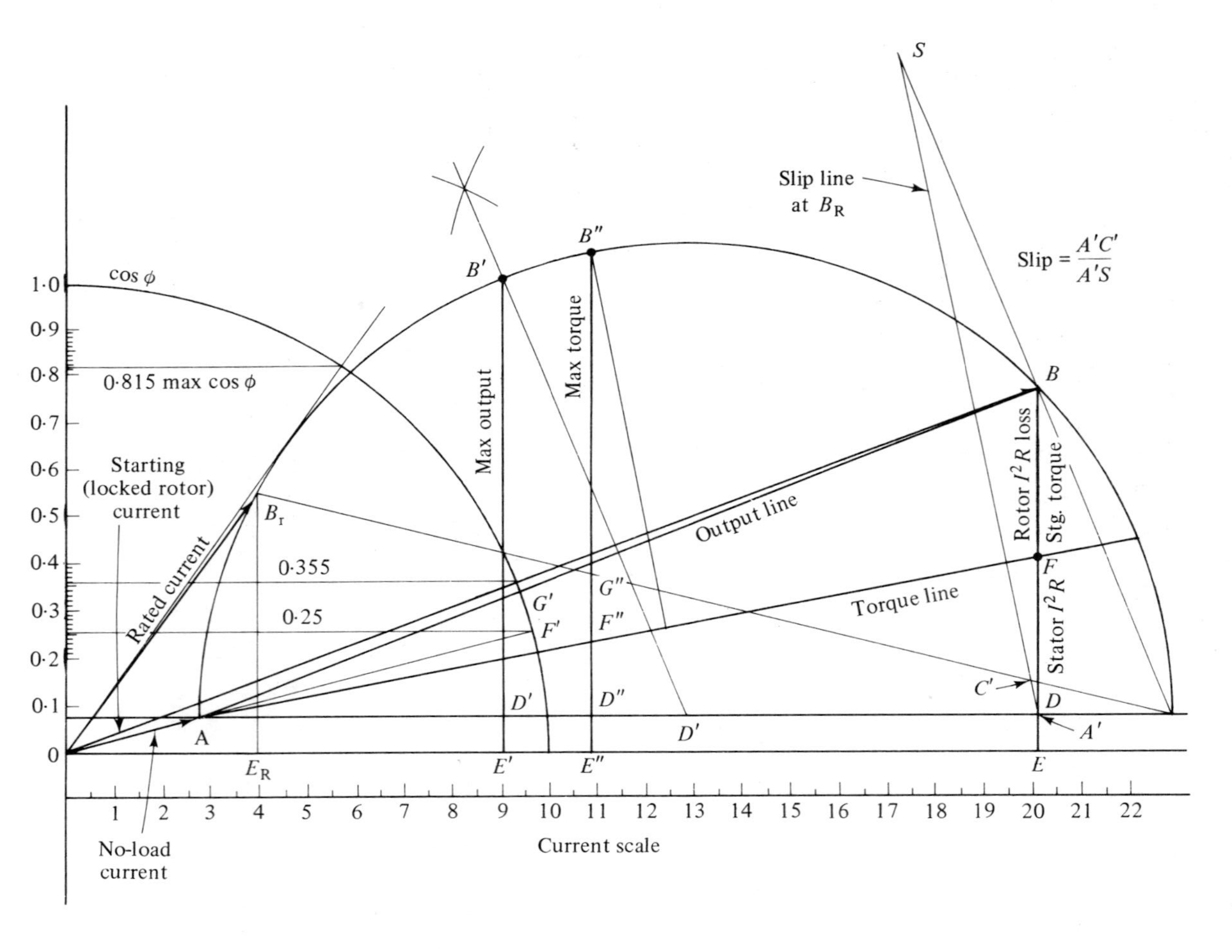

Fig. 4.10 Circle diagram of induction motor

and DE representing the core losses which are assumed to remain constant over the whole of the motor's operating range.

BD on the power scale represents the remainder of the total power supplied which, at standstill, is dissipated as copper losses. For cage induction motors it is not possible to determine the rotor resistance by measurement and the total losses can only be apportioned by subtracting the stator losses, by measuring the resistance and multiplying it by the square of the current. If the resistance has been measured at ambient temperature its value at the motor's normal working temperature must be calculated. A line joining A to F is known as the torque line and all distances BF represent the torque at the respective motor current OF. A tangent to the circle parallel to the torque line thus not only locates the point at which maximum torque occurs, but enables its value to be estimated. At A and B the motor does not convert any of the electrical power supplied into mechanical output, but between those points the distance BG represents the rotor's gross mechanical output and the line AB is referred to as the output line.

The tangent to the circle parallel to the output line thus defines the maximum output obtainable and enables the following values to be determined:

BE = total input per phase (fixes power scale)

DE = represents core losses—the reduction of rotor core losses at full speed is usually compensated by increased stray stator losses

DF = stator copper loss

BF = rotor copper loss = torque in synchronous watts

GB = power output (at no load and locked rotor points G and B therefore coincide)

OB = input current

The point at which OB is a tangent determines the motor's maximum power factor. The perpendicular at that point makes it possible to calculate values of output and efficiency which is given by

$$\eta = \frac{BG}{BE} \times 100 \text{ per cent}$$

$$\text{Torque in Nm} = \frac{\text{Synchronous watts}}{\text{Synchronous speed in rad/s}}$$

or

$$M = \frac{9550 \times \text{k}W}{n} \quad \text{where } n = \text{rev/min}$$

The slip can be obtained from

$$s = \frac{\text{rotor copper loss}}{\text{rotor input}} = \frac{FG}{BF}$$

The length FG, representing the rotor loss, is rather difficult to measure and either the geometric or arithmetic use of similar triangles can be used to obtain a more satisfactory result.

Thus slip for point B is also given by

$$\frac{BF}{AF} \times \frac{AF'}{B'F'}$$

From the circle diagram it is possible to draw an approximate set of performance curves. The curve of particular interest is usually the speed/torque curve since this will show if sufficient torque is available to accelerate or maintain the load at all speeds. The performance of any motor is affected by the magnitude and frequency of the supply, which, whether from a national grid or private supply, will normally be maintained between specified limits. The motor itself will have an effect on the system and a knowledge of the effect of voltage and frequency variations on the characteristics is essential.

4.8 The Effect of Voltage and Frequency on the Characteristics of the Induction Motor

The following relationships are based on the fundamental motor equations and take no account of transients and the variations which are a function of size or special electromechanical design features. Where motors are electronically controlled the behaviour of the complete system must be considered, since supply variations do not necessarily appear at the motor terminals.

From Eq. (4.19) it follows that torque is proportional to the square of the applied voltage and the changes in starting and pull-out torque are usually more significant than the changes around full load. Normally frequency changes are small, but the operation of 50 Hz motors on 60 Hz and vice versa is frequently of interest.

4.8.1 *Change of voltage—constant frequency*

Figure 4.11 shows the speed and torque curves for the motor's rated voltage and those for 80 per cent and 120 per cent of that value calculated on the basis of direct proportionality for current and the square of the voltage ratio for the starting, pull-up, and maximum torque. Once the motor is up to speed when the supply is 20 per cent below nominal value the load torque can be increased until the current reaches its nominal value and the output obtainable, without change of duty rating, will be 80 per cent of its rated value. If the load torque remains constant the rotor current will increase to the required value since the motor can still be loaded right up to pull-out value. The increased rotor copper losses are more liable to cause the motor to overheat. There will clearly be some small changes in speed but these can generally be neglected.

The dangers from voltages substantially higher than nominal values can stem from increased losses and the very much increased accelerating capability.

4.8.2 *Changes of frequency at constant voltage*

If the nominal supply frequency, f, of an asynchronous induction motor is changed to f' without any change in the voltage, the flux and the no-load current will vary as the inverse ratio of the frequency. Assuming that the changes in frequency do not cause

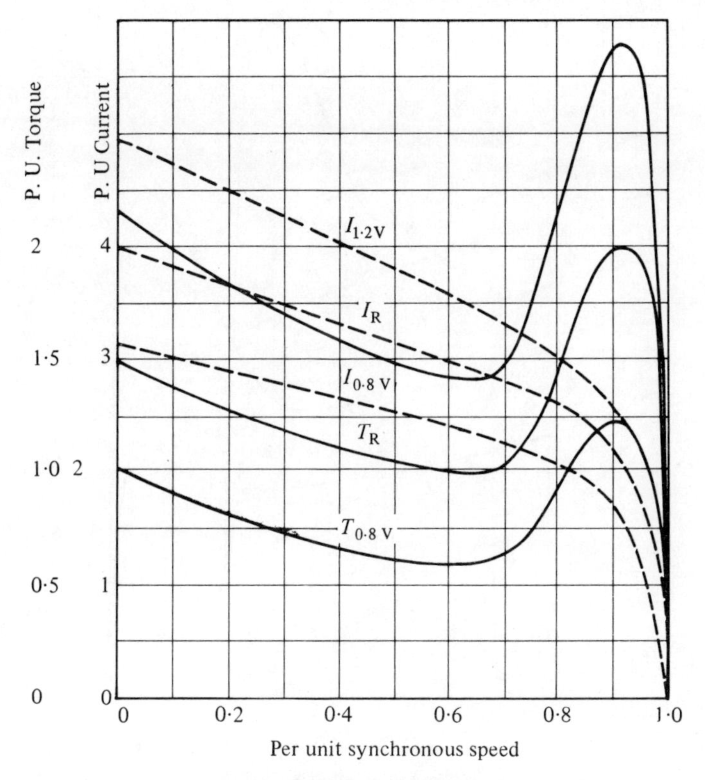

Fig. 4.11 Induction motor current/ and speed/ torque curves for ±20 per cent normal voltage at constant frequency

saturation in the iron then, as shown in Fig. 4.12, the torque will be approximately

$$T' \approx T\left(\frac{f}{f'}\right)$$

Speed is directly proportional to frequency so that the output also increases or decreases with corresponding changes in frequencies. Neglecting the changes in windage the motor output should not be allowed to increase the current beyond its rated values without examining the operating conditions of the motor in detail.

4.8.3 *Simultaneous changes of voltage and frequency*

If voltage and frequency change by the same amount then, using the simplified equivalent circuits, there will be no change in no-load current or motor flux. For moderate changes in frequency the torque developed by the motor can be kept, as Fig. 4.13 shows, substantially unchanged by voltage adjustment. The output power from the motor will be directly proportional to the frequency since the change in slip will be negligible. From the rotor power factor Eq. (4.4), and the torque equation (4.5) it appears that the starting torque must decrease with substantial increases in frequency even when the voltage has been raised to maintain the stator flux and the secondary current. When the motor is running near synchronous speed the stator

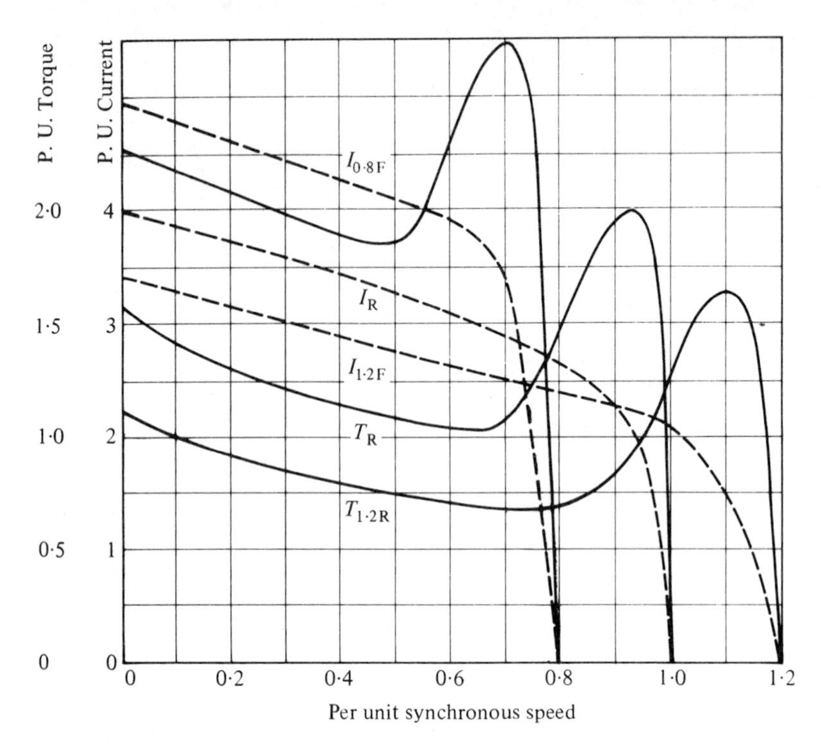

Fig. 4.12 Induction motor current/ and speed/torque curves for ±20 per cent normal frequency at constant voltage

iron losses, friction and windage will increase exponentially with frequency, but again, depending on the particular motor and the amount by which the frequency has been increased, the increased airflow due to higher speed may or may not be sufficient to keep the motor at a safe temperature. At low frequencies the resistive voltage drop across the windings becomes a significant proportion of the supply voltage so that the volt per cycle factor must be increased to maintain the torque. If the variable frequency supply is derived from the static inverter rather than a generator the characteristics of the cage motor will be further modified by the waveform. Under such circumstances it becomes necessary to study the complete drive system and consider the nature of the load.

4.9 Multi-speed Induction Motors

In the cage or solid rotor induction motor the speed is determined by the number of pole pairs of the primary winding. If the stator has two separate windings or a winding which can be reconnected at least two speeds can be obtained. In a single winding motor pole changing is accomplished by regrouping the coils of each phase so that the current in part of the winding is reversed. In order to make use of the whole of the winding and permit reasonably simple connections single windings used in practice had to be restricted to two speeds in the ratio of 2:1 with mostly widely differing outputs. Since the flux paths change with the number of poles, and many of the

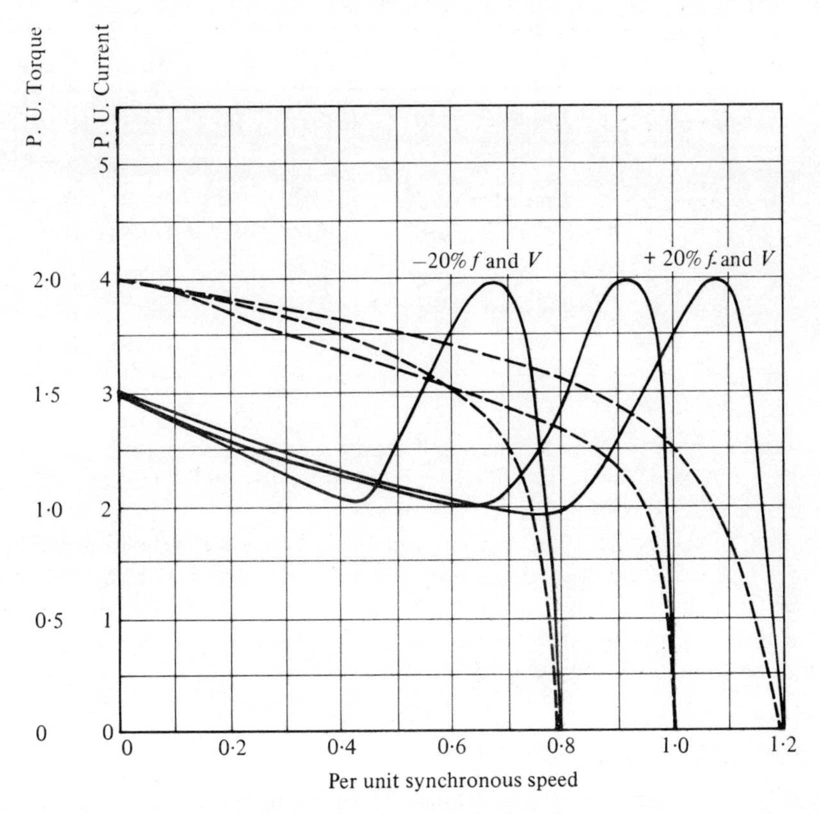

Fig. 4.13 Induction motor current/ and speed/torque curves for simultaneous voltage and frequency changes of ±20 per cent

leakage reactances are very different, the performance at one of the speeds will be very inferior. There are, however, many advantages of being able for example to either run-up, or brake an inertia load with the low-speed winding, and in such applications a motor optimized for the higher speed would be used. A method of pole changing which became well known and is based on the consequent pole principle is the Lindstöm–Dahlander winding.

In practice, multi-speed induction motors catalogued by manufacturers aim to provide one of the following characteristics:

(a) constant torque
(b) square law torque (fan applications)
(c) constant output (torque inversely proportional to speed).

Some of the sacrifices which multi-speed operation necessitates are best illustrated by the following example of a totally enclosed fan-cooled motor in frame size IC 132 when wound for either 2, 4, 6 or 8 poles, with respective outputs of 7·5, 7·5, 4, and 3 kW. Taking these outputs or the respective output of the winding with the smallest number of poles as unit value then multi-speed versions could have the following typical parameters:

		Output		T_{st}	T_{pu}	T_{max}
IC 132—2/4	2 pole	1·07	(8)	1.06	0·91	0·91
	4 pole	0·87	(6·51)	0·67	0·87	0·83
IC 132—4/8	4 pole	0·56	(4·3)	0·34	0·37	0·48
	8 pole	0·87	(2·6)	0·88	0·85	0·81
IC 132—4/8–6	4 pole	0·27	(2·0)	0·22	0·16	0·26
	8 pole	0·41	(1·2)	0·46	0·44	0·56
	6 pole	0·40	(1·6)	0·42	0·38	0·59

In the last example the 6-pole winding is separate and the actual kW output figures given show the reduction in output which is necessary to obtain three speeds from the same motor at approximately constant torque.

Figure 4.14 shows the speed/torque and current curves for the motor.

From these it will be clear that the motor must be started with the lowest speed winding. Considerably more output can be obtained if only two speeds are required, and the load follows a square law. Thus, a 4 and 8 pole motor for fan duty is supplied in the same framesize with respective outputs of 6 and 1·2 kW.

4.10 PAM Winding Induction Motor

The PAM winding represented a major step forward in multi-speed induction motor design and application. (PAM is an abbreviation for 'Pole-Amplitude Modulation'.) It offers a wide choice of speed-ratios, in particular, it gives the close-ratios (4/6; 6/8; 8/10; 10/12 poles, etc.), which are much the most in industrial demand; and, fortunately, they are also the best in technical performance. Since PAM windings were first introduced in the late 1950s, considerable research has been carried out to perfect their design and performance, and to generalize the application of the principle for all sizes, all types of load and many speed-ratios. The PAM motor is now a fully established commercial product, and many tens of thousands of motors, ranging from 0·5 kW to 7000 kW, are in service all over the world. They are made by about twelve licensees, in five leading industrial countries.

The novelty of PAM motors lies in the sequence of coil-connection, which takes place when the motor is being manufactured. The theory of these connections, and the method used to make the actual connections correspond with the theory, are technically very difficult. On the other hand, in practice, for the operating engineer, there is nothing new or difficult to learn. A six-terminal pole-changing winding, for the 2 to 1 speed-ratio only, was first devised in 1897 by Dahlander. The PAM winding uses the same well-known connections and switching sequence, which are shown in Fig. 4.15: parallel-star/series-star (or delta). Standard switches of various ratings are available from most control-gear makers, for this sequence. In 1957 Professor G. H. Rawcliffe of the University of Bristol (where most of the subsequent research has been carried out) invented and patented a method of making the internal connections of three-phase windings by which it becomes possible to obtain almost any speed-ratio, still only using a single winding with six terminals and the

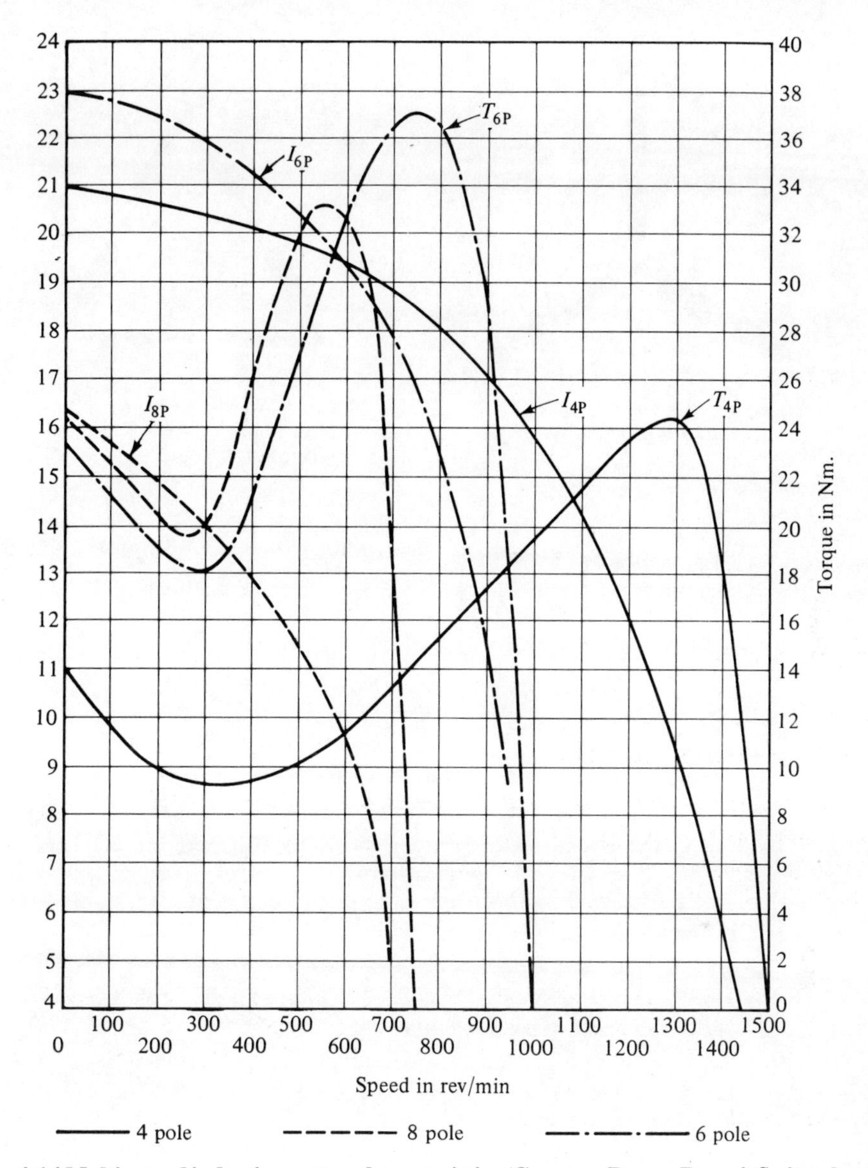

Fig. 4.14 Multi-speed induction motor characteristics (Courtesy Brown Boveri, Switzerland)

same switching sequence. The close-ratio windings give particularly good performance.

The method involves changing the pole-distribution by reversing one half of each phase on the lines simply illustrated in Fig. 4.16. Instantaneous pole-configurations for *one phase only* are shown. The 6-pole case appears unsymmetrical, but the interaction of all three phases results in a rotating field, with equally distributed poles.

Before the development of PAM methods, it was necessary to use separate windings for each speed required, except for speed ratios of 2 : 1, when the Dahlander pole-changing winding could be employed. Currently, the PAM motor, although

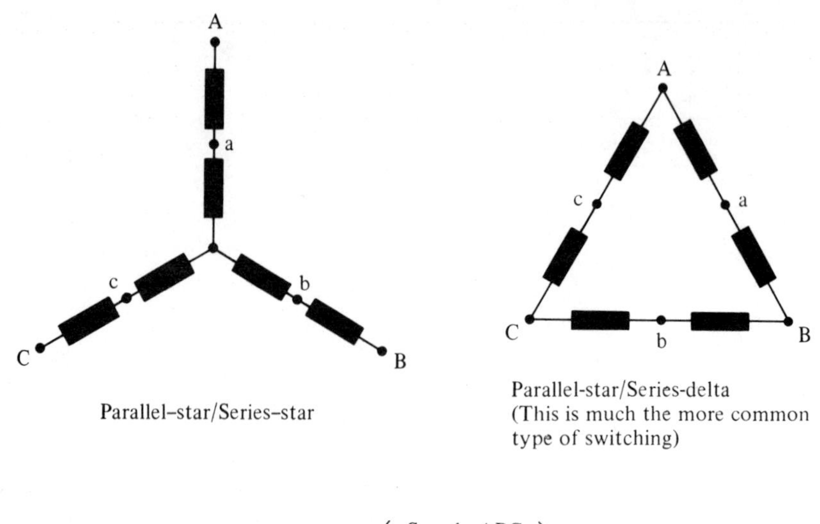

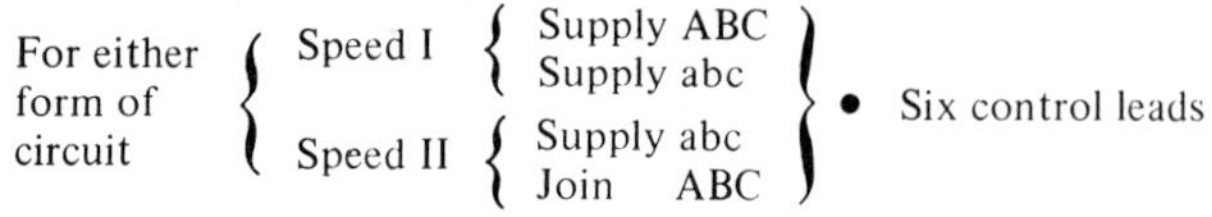

For either form of circuit	Speed I	Supply ABC Supply abc	• Six control leads
	Speed II	Supply abc Join ABC	

Notes: 1. All coils identical and all in circuit at both speeds
2. Switching for all speed ratios is the same as above
3. The invention lies in devising the correct sequence of coil connections. There are millions of possible ways of connecting a few dozen coils

Fig. 4.15 Basic connections and stator winding switching diagrams for pole amplitude modulation

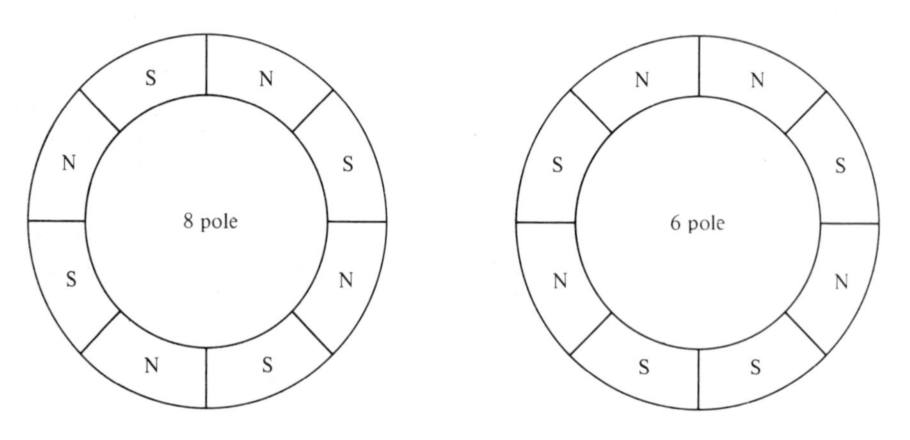

Fig. 4.16 Instantaneous pole configurations for PAM

in essence a single-winding induction motor, is now firmly established as a two-speed motor; and the technical and commercial arguments for its use as a two-speed motor are so overwhelming that it becomes difficult any more to justify the use of a dual-wound motor. Indeed, PAM motors are now being substituted in some applications where a single-speed motor, for the higher speed, is at present being employed. Considerable power savings can often be obtained in this way, by speed-reductions during part of the operating duty.

The stator winding of a PAM motor is essentially similar in construction to a single-speed stator winding. All the coils are identical, and are of the same coil-pitch. Because of this similarity, the high reliability associated with single-speed induction motors is also obtained with PAM motors.

In comparison with single-speed motors the power factor and efficiency of a PAM motor are almost as high for each speed, and the pull-out and starting torques developed are only marginally inferior. In comparison with an equivalent dual-wound motor, every item of the performance of a PAM motor is substantially better.

A leading British company has supplied the following data, as a statistical average, for PAM motors up to 2500 kW. Values given by other manufacturers are sometimes a good deal better, and sometimes rather worse, depending on the basis of calculation.

(a) The output power at the higher speed in the falling-torque designs is at least 80 per cent of the output which could be obtained from a single-speed motor of the same frame size, wound for the higher speed.
(b) The cost of a PAM motor is only 70 per cent to 75 per cent of the cost of a corresponding dual-wound machine for the same duty.
(c) The efficiencies of PAM motors are almost indistinguishable from those of equivalent single-speed machines, for both speeds.
(d) The power factor for the higher speed and lower pole-number is within 0·04 of the power factor of an equivalent three-speed machine. For a falling torque, the power factor for the lower speed diminishes by rather more, (say) 0·15; simply because one is using a large machine for a relatively low load. The same would be true of a corresponding dual-wound machine.

There is a different type of winding for constant-torque motors, the frame size of which is, of course, always determined by the duty and performance at the lower speed. The performance of the PAM motor for constant-torque applications will compare at least equally favourably with that of a dual-wound motor for the same duty. Some manufacturers have said that the comparison between a PAM motor and a dual-wound motor is most favourable for constant-torque applications.

The term Pole-Amplitude Modulation is used, because of the logical similarity between this technique and the concept of amplitude modulation, as used in radio communication, the difference being that PAM involves modulation of the space-distribution of the windings of a stator, whereas the latter involves time-modulation of the current flowing.

The theory of a $2p/2q$-pole PAM winding may be simply expressed as follows:

The m.m.f. of a conventional $2p$-pole winding may be expressed as

$$M_1 = A \sin (p\theta - \omega t)$$

If this m.m.f. is modulated by a wave of $(2p \pm 2q)$ poles, that is by a wave having a number of poles equal to the sum or the difference of the original $(2p)$ and final pole-numbers $(2q)$, the resulting m.m.f. may be written as

$$\begin{aligned} M_2 &= A \sin (p\theta - \omega t) \sin (p \pm q)\theta \\ &= A/2[\cos (\pm q\theta + \omega t) - \cos (\{2p \pm q)\theta - \omega t\}] \end{aligned}$$

Considering the positive condition, i.e., modulation by the sum of the pole-numbers $(2p+2q)$,

$$M_2 = A/2[\cos(q\theta+\omega t) - \cos\{(2p+q)\theta-\omega t\}] \tag{4.26}$$

Considering the negative condition, i.e., modulation by the difference of the pole-numbers $(2p-2q)$

$$M_2 = A/2[\cos(q\theta-\omega t) - \cos\{(2p-q)\theta-\omega t\}] \tag{4.27}$$

From Eqs (4.26) and (4.27) it may be seen that the direction of rotation of the $2q$-pole field, given by the term 'ωt', will be different in the two cases. It will be in the opposite direction to that of the original field of $2p$ poles, for the first type of modulation; and in the same direction for the second type of modulation. The second terms of expressions (4.26) and (4.27) show that there will also be an unwanted harmonic field present. This may be reduced or eliminated by chording of the winding, and by the suitable choice of 'sum' or 'difference' modulation, and other means. Each case is dealt with on its own merits, by those skilled in the art. Almost any two synchronous speeds can be chosen, except a 6/2-pole machine, though the speed-combinations most popular in industry are 1500/1000 rev/min; 1000/750 rev/min; 750/600 rev/min, and other close-ratios (Table 4.1).

Table 4.1 Speed Combinations on 50 hertz supply

	Close speed-ratios		Ordinary speed-ratios		Wide speed-ratios	
	rev/min	Poles	rev/min	Poles	rev/min	Poles
Two-speed for Constant-torque or Fan torque	950/710	6/8	1420/950	4/6	1420/560	4/10
	710/560	8/10	950/560	6/10	2850/710	2/8
	560/470	10/12	710/470	8/12		
Three-speed for Fan torque only	950/710/560	6/8/10	1420/950/710	4/6/8	2850/1420/560	2/4/10
	710/560/470	8/10/12	1420/950/470	4/6/12	2850/710/560	2/8/10

A four-speed motor is obtainable, for any four synchronous speeds, by using two PAM windings; a number of these have been made, for use instead of a motor with continuously-variable speed. Similarly, a number of three-speed motors have been made, using one PAM winding and one standard winding. There is a rapidly growing field of such applications, especially for medium and small-size motors, for machine-tool drives. Such stepped-speed motors are cheaper, and require less maintenance, than motors for continuously-variable speed.

An alternative use for the PAM motor is for power drives where the speed has only to be changed at very long intervals, or even just once in the lifetime of the motor. An example of the first kind of drive is a fan drive, the speed of which is regularly changed twice a year: at the beginning of summer and at the beginning of winter. An example of the second type of drive is a mine ventilating fan, which has to be speeded up when the mine workings become longer. In these cases the speed can be changed in a few minutes by simple alteration of the terminal connections, and a standard single-speed starter, of low cost, can be used.

All PAM motors are perfectly balanced electrically, and take a balanced three-phase current from the supply. All currents are of sinusoidal waveform; and there is no waveform distortion, as there is with some kinds of thyristor control. Above all, PAM motors give great economy in the use of copper—and therefore in the size of punchings, and the motor is lighter, smaller, and cheaper than any other form of change-speed motor.

Although PAM windings have been used mainly for three-phase squirrel-cage motors, the same principle applies to slip-ring and synchronous motors. The PAM principle is also applicable to single-phase machines, but the only such machines for which any substantial demand can be expected are for 4/6 poles and 6/8 poles, of small size. Various simplifications of the switching sequence can be used, with only slight deterioration in performance. The manufacturers of small single-phase induction motors prefer the concentric type of winding on economic grounds. Generally, it has fewer coils and fewer end-winding connections, and may be wound more readily by automatic winding machines; as compared with the double-layer windings commonly used in three-phase motors.

4.10.1 *Conclusion*

The principle of the PAM motor is a permanent addition to the theory of a.c. machinery; and the PAM motor is likely to last as long as (and no longer than) the induction motor itself. The first single-speed induction motors were made in 1888, and the first Dahlander motors in 1898; and the PAM motor is probably at the beginning of another long story.

4.11 Synchronous Motors

The principle of the synchronous motor can be visualized in a number of ways and a classical text-book approach is to link it with the generator. Since it has however, been shown that a three-phase winding creates a powerful rotating field which interacts with induced currents in the rotor to produce torque the transition from asynchronous to synchronous running will be considered first. As the rotor approaches zero slip the induced e.m.f. and, therefore, the current flowing has become very small and of low frequency. If the rotor is of the wound type and is now supplied with current at slip frequency the torque can, according to the direction of the current, be either increased or decreased. If it is increased the motor will accelerate and come even nearer to the speed of the rotating field with the injected current's frequency reducing until at synchronism it becomes zero, that is, a direct current. The rotor has now become polarized and is in step with its magnetic axis more or less coincident with that of the field. It is now expedient to revert to the salient pole representation of the rotor and consider how, once sufficiently near synchronism, it can become polarized other than by d.c. magnetization. These alternative methods are generally only likely to be encountered in relatively small machines, which in this case means machines up to about 10 kW output. Considerable advances have been made in permanent magnets, and for certain applications rotors incorporating a permanent magnet make the most economical synchronous motors with outputs approaching now 100 kW. The combination of salient poles and permanent magnets is particularly effective once the rotor is at synchronous speed, but has little if any inherent starting torque.

The combination of a cage winding with the concept of a salient pole rotor has led to a variety of types known as synchronous reluctance motors. In small motors the same rotor laminations as are used for the asynchronous standard cage version are modified by removing some of the rotor teeth so as to produce a salient pole effect. In those parts of the rotor where the iron has been removed the rotor bars are still fitted, and if necessary are supported in the centre by a few of the unmodified motor laminations. The motor will run up to within a few percent of synchronous speed in the same way as a normal cage induction motor. The stator flux will then cause the rotor to be magnetized and pull it into synchronism with the field.

Such reluctance motors are the most economical to manufacture but have their limitations, particularly with loads having substantial inertias, and when having to operate on single phase with capacitors.

Another type of synchronous motor is the hysteresis synchronous motor whose principles were known before 1900 and dealt with in detail by C. P. Steinmetz in 1917. This makes use of hard magnetic material in the rotor so that as it is magnetized by the rotating field, poles are formed and retained to produce torque. This hysteresis torque is said to be constant down to zero speed and aided by the torque produced by

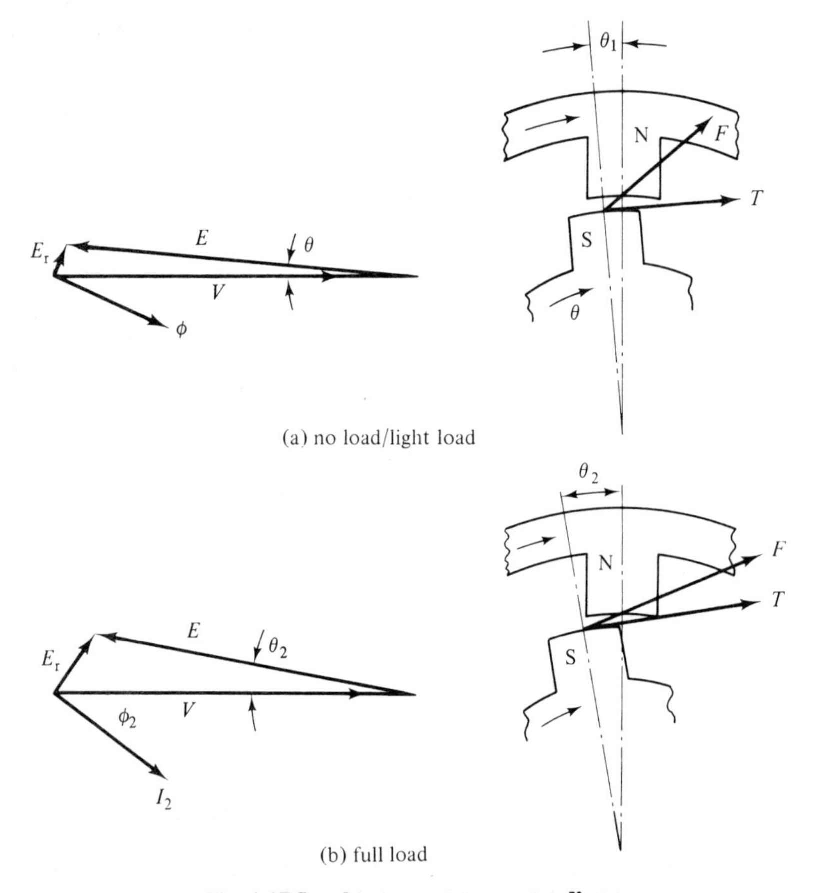

Fig. 4.17 Synchronous motor vector diagrams

the eddy currents is able to run the motor up to speed. The manufacture of rotors from hard steels containing nickel and cobalt while not presenting too many problems in milliwatt output clock motors becomes more difficult in the larger sizes. Mostly a cylinder is built up on nonmagnetic material from a number of rings. Further details are given in chapter 7, section 7.16.

In the non-excited synchronous motors the field strength is near enough constant, and they, therefore, are a particular case of the separately excited synchronous motor which will be used to discuss the general principle of the synchronous motor. Figure 4.17 shows the synchronous motor with no external load applied to it. The magnetic axis of the rotor pole lags slightly behind the rotating field axis (represented by the axis of an imaginary rotating stator pole) by the 'load angle' θ_1 in order to provide the friction and windage losses. The applied voltage V (per phase) is balanced by the voltage induced by the rotor field, the back e.m.f., and the impedance voltage drop E_r. The angle ϕ is the impedance angle and, as the load increases changes to ϕ_2; θ also increases to allow the motor to develop more torque.

If the excitation can now be varied E can be increased or decreased at will. The supply voltage V (Fig. 4.18) which is assumed to be unaffected by the motor is taken as reference and the power input to the motor per phase given by $VI \cos \phi$ can be shown to be proportional to the perpendicular BC. A horizontal line through B will, therefore, represent a constant power input locus. An increase of excitation which makes the phasor E large enough to reach B' results in the motor current now being

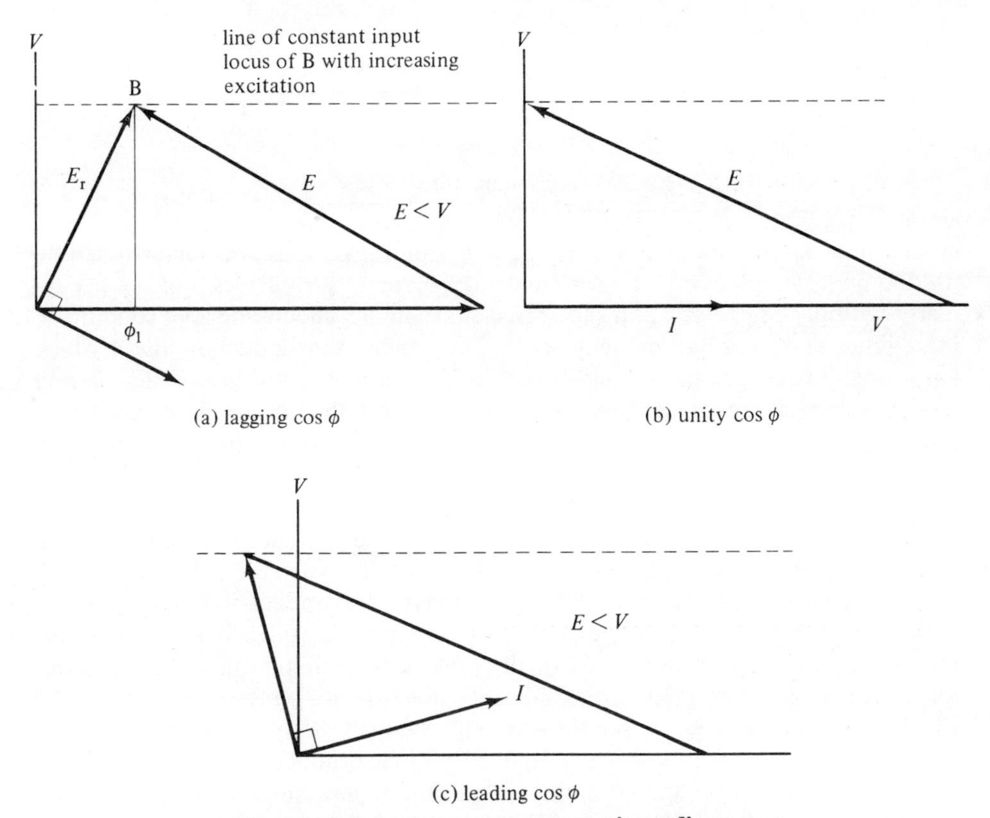

Fig. 4.18 Synchronous motor constant input diagrams

in phase with the supply voltage. Further increase in excitation causes the motor current to lead. Since the excitation can be made to take the current from its full lagging value to its minimum at unity power factor, and then actually cause it to lead the supply voltage, a family of curves can be drawn as shown in Fig. 4.19. Since the

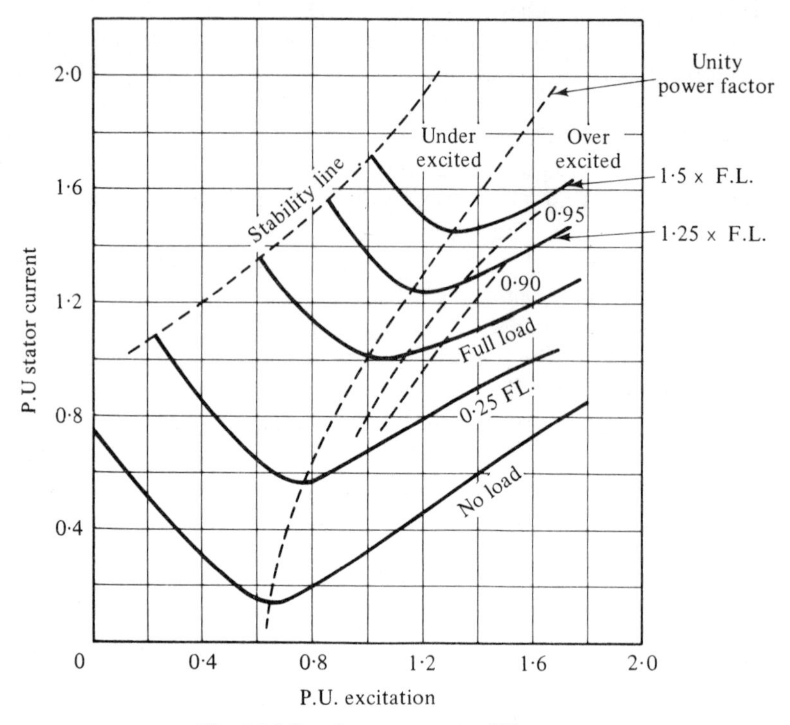

Fig. 4.19 Synchronous motor 'V' curves

synchronous motor can be made to operate with a leading power factor it can be used to improve the overall power factor of a system. Such drives must be particularly carefully engineered and supervised since, under fluctuating load conditions, instabilities known as hunting may occur. The synchronous induction motor which has a wound rotor can be connected so that it is started as a normal collector-ring motor. When all resistance has been cut out and the motor is at maximum synchronous speed, d.c. is applied to the rotor which, if wound for three-phase can have two of the rings connected together. The current divides at the star point, and whereas one phase carries the full excitation current, the other two not only form a parallel return circuit but being a closed loop will also provide a path for damping circuits. If the load of a synchronous motor is increased beyond the pull-out value the motor may have to be restarted; depending on type and application this could be a more or less undesirable incident and must be carefully considered when determining the required size of motor. Normally motors are rated so that their pull-out torque is 1·5 times the rated torque, but since this depends on the power output for which the motor has been chosen the economic value of the synchronous reactance, and therefore the excitation current, must also be determined.

For a synchronous induction motor it is useful to superimpose the locus of the stator current phasor on the diagram, Fig. 4.20. The stator current I_{sy} is the resultant

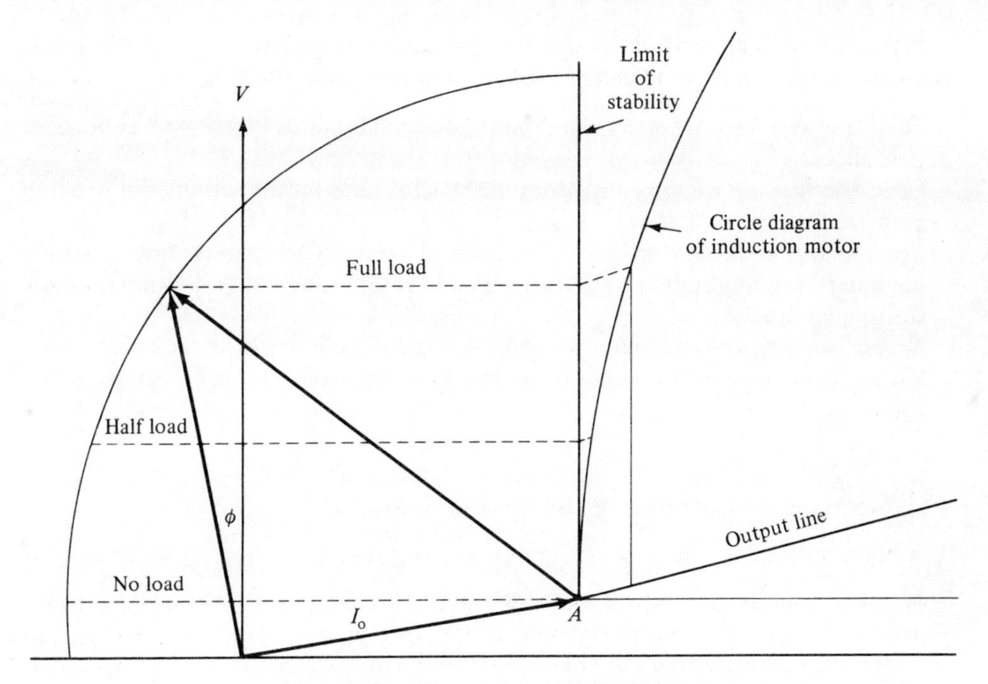

Fig. 4.20 Synchronous motor circle diagram

of two phasors $-E/Z_{sy}$ and V/Z_{sy}, where the latter is the current taken by the motor on rated voltage without excitation. This current is practically the no-load current as induction motor. The current E/Z_{sy} is the short-circuit current in the stator I_x when driven at synchronous speed, and thus depends on the value of excitation.

The locus of the stator current phasor shows that lightly loaded the motor operates at leading power factor and that at constant excitation the power factor becomes unity around full load and lagging on overload. If such an overload reaches pull-out value the motor will continue on its induction characteristic, and automatically resynchronize when the load is reduced. The normal motor protection should, however, operate if the overload is sustained beyond the rated limit.

4.11.1 *Types of synchronous motors*

All synchronous motors are now designed to be capable of breaking away from standstill, accelerating, and pulling into synchronism with the rotating field. The motors start as induction motors and the following types of starting windings are used:

(a) solid pole or rotor machines in which the eddy currents circulating in the steel produce starting torque;
(b) cage machines in which copper or copper alloy conductors are embedded in slots in the rotor, and interconnected at both ends to provide a cage-type winding;
(c) wound rotor machines in which insulated coils in slots are connected as a polyphase winding and brought out to collector rings on the shaft.

Any of these starting windings can be combined with the basic types of synchronous motors and the types in general use may be defined as follows:

Solid pole synchronous motor—a salient pole synchronous motor with solid steel pole shoes.
Cage synchronous motor—a salient pole synchronous motor with a cage winding embedded in the pole shoes
Synchronous induction motor—a cylindrical rotor synchronous motor with a secondary winding similar to that of a wound rotor induction motor and used for starting and excitation.
Salient pole synchronous induction motor—a salient pole synchronous motor which has a coil winding embedded in the pole shoes connected to collector rings on the rotor.

4.11.2 *Starting characteristics of synchronous motors*

To obtain satisfactory starting of a motor it must provide:

(a) a break-away or standstill torque which can overcome any stiction load torque;
(b) an accelerating torque defined as the torque by which the total torque produced by the motor at any speed exceeds the torque required by the load at the same speed. This accelerating torque acts on the total inertia of the load, and the motor and determines the time required to accelerate up to full speed;
(c) acceleration to a small value of slip so that when excitation is applied sufficient synchronizing torque is produced to accelerate the total inertia to synchronous speed against the load torque.

To ensure that the motor characteristic can be matched to the load requirements accurate values of load inertia and torque requirements must be known. It is, furthermore, necessary to know the characteristics of the supply system since the motor torque depends on the terminal voltage which, due to the heavy current drawn on starting, can be affected by a significant voltage drop in the system. In general, a 15 per cent drop in terminal voltage is regarded as an acceptable value, but certain systems which have voltage-sensitive plant connected to the same bus-bars demand the drop to be restricted to 5 per cent or even 2·5 per cent in special cases.

4.11.3 *Excitation systems*

The many forms of excitation systems which have been developed to meet particular requirements mostly fall into one of the following basic groups:

(a) d.c. busbar or distribution systems;
(b) individual d.c. generator or exciter for each particular motor;
(c) a.c. busbar or distribution system with rectifiers;
(d) individual a.c. generator or exciter with rectifier for each motor;
(e) as (d), but windings and rectifiers are mounted on the rotor of the synchronous motor, 'Brushless synchronous motor'.

These groups can be further subdivided according to the means for controlling the excitation:

(i) Fixed excitation;
(ii) manually controlled excitation with a reasonable range of adjustment;
(iii) compensated automatic excitation providing automatic adjustment for load to first order of accuracy;
(iv) compounded automatic excitation which derives its magnitude from the motor input voltage and current in such a manner as to keep the power factor close to the design value;
(v) closed loop automatic excitation measures the quantity to be controlled, e.g., kVAR or power factor, and automatically adjusts the excitation to maintain it within the required limit;
(vi) limited ranges of control using any of the above automatic control schemes or combinations of different schemes.

Motors with fixed rotor windings may be started:

(a) direct on line;
(b) through a series reactor;
(c) by auto-transformer;
(d) with continuously variable or stepped rotor resistance starters.

(a) Direct-on-line starting is the simplest and cheapest method. Normally starting currents are of the order of 4 to 5 times the full load currents but special designs may make them as low as 3 times or as high as 8 times. In all applications the voltage drop must be considered since a high-torque high-current motor may not necessarily produce the best starting torque.

(b) Reactor starting increases the motor impedance as viewed from the supply and thus reduces the starting current. If the value chosen for the reactor is such as to produce K ($K<1$) times normal voltage at the motor terminals, then the current drawn and the torque produced will be K^2 their direct-on-line values. As a result of this starting time will be appreciably increased and the total heat energy produced in the motor may make this a more severe condition for the motor than direct-on-line starting. In applications where the starting load on the motor is modest such as motor generator sets, pumps, and compressors started on no load, reactor starting is satisfactory and limits starting currents to 1·5 to 3 times full load current.

(c) Auto transformer starting systems are more expensive than the reactor method but are more efficient since for the same torque the line current is also K^2 and can, therefore, be kept to about 0·5 to 2 times the rated full load value.

(d) The cost of a wound rotor motor is slightly higher than a cage machine, but the combined cost of motor and resistance starter is comparable to a reactor-started motor system. The effect of the starting resistance is to raise the power factor at starting and, therefore, the ratio of starting torque to starting current. Starting torques between full-load and three times full-load torque with approximately proportional currents can be obtained from such systems.

During the 'induction' motor starting period the amount of heat dissipated in the starting winding is equivalent to the total kinetic energy in the rotating masses at full load speed, plus an amount determined by the load torque. This appreciably heats the secondary conductors, and to avoid insulation failures their temperature must be prevented from exceeding the specified limit. Wound rotor motors with resistors in series with their starting windings have greater thermal capacity and are, therefore, more suitable for starting high inertia drives such as large ventilating fans.

The load inertia which a particular motor can accelerate and the number of starts which it can perform sequentially are, therefore, limited by this temperature rise. Synchronous motors are usually started direct-on-line, as this is the simplest and cheapest method. Large motors are normally installed in large industrial plants with good connection to the power system and a simple study can show if the starting current is likely to have any adverse effects on the system. Motors of twenty or thirty thousand h.p. are started in this way. Smaller motors which are connected to restricted capacity power systems use reactor starting with autotransformer starting used in a few special instances. Where starting current is restricted and torque requirements are high rotor-resistor starting is usually adopted.

Cage-type induction motors have a relatively poor power factor which deteriorates as the number of poles increases. The synchronous motor is not only able to operate at unity power factor but can actually correct the power factor of a plant. Where the excitation of the rotor is provided by a direct supply rather than by induction the airgap can be made larger which makes the synchronous induction motor less affected by unbalanced magnetic pull and supply voltage transients. The synchronous motor, whatever its type, is more costly than its induction counterpart so that a number of economic factors have to be taken into account, particularly the penalty of a poor system power factor. Generally synchronous motors, particularly if they are also used for power factor correction, only prove themselves economical in larger sizes.

4.12 Three-phase a.c. Commutator Motors

Many methods which can be used to control wound rotor induction motors tend to be either wasteful by dissipating power in resistances or require costly auxiliary machines to obtain a very limited range of speeds. The scope of matching the characteristics of the motor to that of a particular load and obtaining more efficient control over a wider range of speeds is considerably increased if the rotor is wound like a d.c. armature, has collecting rings as well as a commutator, and in the case of shunt machines, an additional control winding. The brushgear must incorporate at least three sets of brushes per pole pair and means for rotating them manually or by a small geared motor through an angle of about 30 degrees, preferably either side of the neutral position. The higher cost of winding the commutator and brushgear as well as the need for regular inspection and routine maintenance must be weighed against the saving in power, additional machines, and control equipment. In many applications the improved starting torque, efficiency, and power factor are as important as speed control.

Many designers who found acceptable solutions to the fundamental and difficult problems of commutation gave their names to such motors. The development of better materials and components has given a.c. commutator motors sufficient reliability and robustness to be widely used in process industries.

Like d.c. machines the various polyphase commutator motors can be divided into series- and shunt-wound types, but in this case there are also fundamental differences in their action. Whereas the series motor's action is akin to that of its d.c. counterpart, shunt motors are in essence rotating field induction motors incorporating a commutator to obtain a slip-frequency control voltage.

4.12.1 *Series motors*

The stators of these machines are very similar to those of induction motors but instead of the windings being connected in either star or delta, one end of the phase windings is joined to a brush. The three-phase rotor winding connected to a commutator completes the electric circuit. The inherent series characteristic is particularly suitable for loads whose torque increases with speed such as fans, centrifugal pumps, and compressors. The characteristic high starting torque is useful in hoists or mill applications. Though it is possible to vary the speed by changing the supply voltage the more economical method of moving the brushes is usually adopted. In order to obtain acceptable commutation it is necessary to keep the voltage across the commutator low and the use of a transformer between stator and rotor becomes essential. It can, furthermore, be shown that at speed exceeding 1·5 times the synchronous speed the e.m.f. in the short-circuited coils increases rapidly and results in deterioration of commutation. Practical compensating windings and other measures required to improve commutation have been successfully developed for large single-phase traction motors. The three-phase versions on the other hand appear to have remained mainly in the development stage and two patents are described in reference 3. A detailed treatment of the theory, operating characteristics and equivalent circuits of a.c. series motors can be found in reference 7 of chapter 11.

4.12.2 *Shunt motors*

There have been many and varied designs of three-phase commutator motors with shunt-connected windings. Many started off as separate machines and were referred to by the name of their inventor or most successful designer, but in more general terms the motors in use today can be classified either as rotor- or stator-fed machines.

The rotor-fed three-phase shunt motor which is still manufactured extensively is available for supply voltages up to 500 V and with outputs up to about 150 kW. It is used extensively in the textiles and paper process industries where its stepless and loss-free speed control is found to be particularly useful. The brush shifting mechanism can usually be connected with a servo-motor so that such motors can easily be integrated into fully automatic control systems. Motors with outputs above 200 kW are invariably of the stator fed type to allow them to be wound for a more appropriate supply voltage such as 3·3 kV.

Schrage motor

This motor is typical of the rotor fed (or inverted) variety of the three-phase shunt motor. It uses a tertiary winding on the rotor to feed the secondary winding in the stator. (See Fig. 4.21.) The tertiary or control winding is connected to a commutator by pairs of brushes whose relative position to each other can be varied to obtain the variable second voltage. The theory and principles of operation of the Schrage motor is dealt with in chapter 9 of reference 7 and the operation and a specimen characteristic are covered on pp. 234–7 of reference 2.

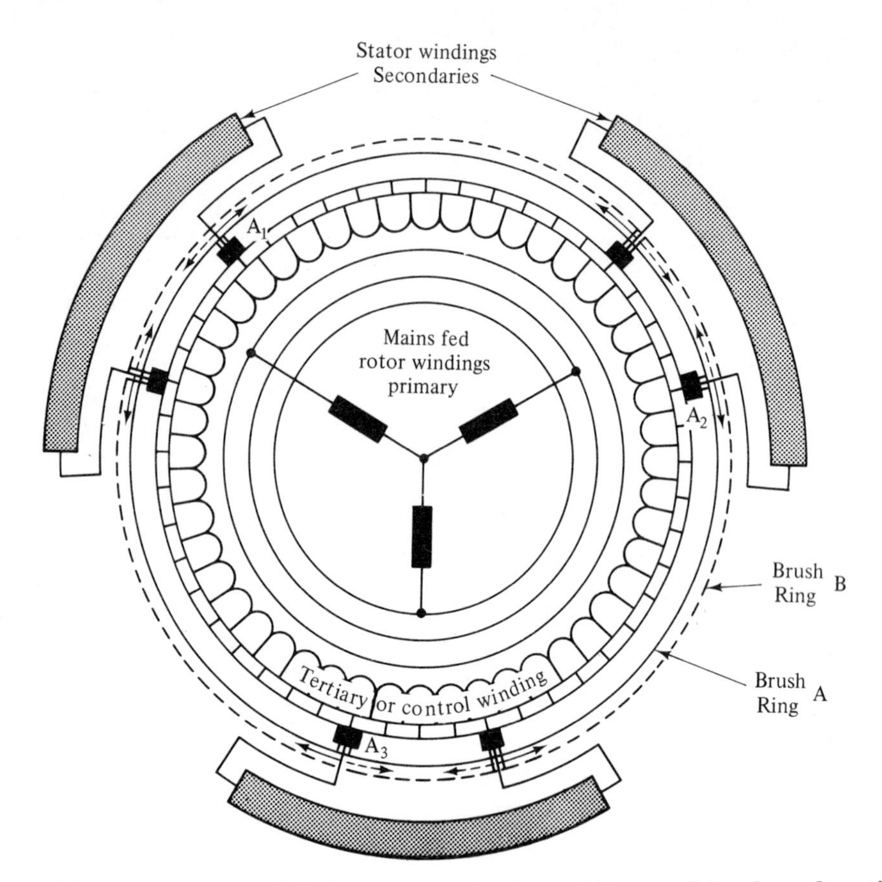

Fig. 4.21 Basic diagram of Schrage motor showing winding- and brush-configuration (Courtesy Elektra Faurndau, Germany)

Figure 4.22 shows the operation of the brush pairs in characteristic positions: (a) when both brushes are on the same commutator segment and the secondary winding is short circuited the motor will behave like a cage induction motor. In (b) the secondary winding has a voltage sE_2 applied to it. This voltage is derived from the voltage induced in the tertiary winding by the rotating flux of the mains frequency fed primary winding. At standstill the rotating field will be moving at f/p rev/s relative to the stator and the brushes, so that E_2 also has mains frequency. When the motor has run up to a speed which causes the field to rotate at a frequency f_r the commutator acts as a frequency converter and E_2 and, therefore the currents and the field of the secondary will have a frequency of $f \pm f_r$, or n and n_r in terms of rev/s. Now, just as in an induction motor with wound rotor, the injection of a current into the secondary at an angle of 180° will cause the speed to drop further below synchronous. In (c) the injected e.m.f. is reversed and the angle of injection is 0° so that sE_2 must also be reversed due to s becoming negative in making the speed rise above synchronism.

If the brush ring as a whole is now moved relative to the secondary winding against the direction of rotation without altering the brush separation, Fig. 4.22(d) shows how at speeds below synchronism this improves the power factor.

The main rotor windings are generally 4, 6, or 8 pole but as in all electric motors the output power decreases with speed and a smaller and more economical solution

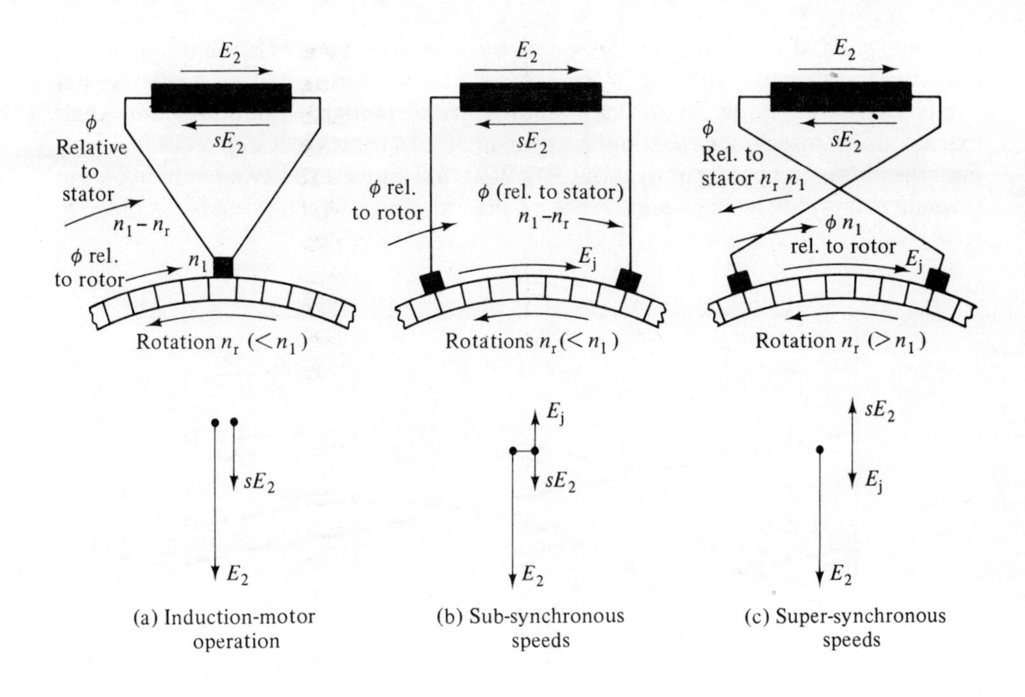

(a) Induction-motor operation

(b) Sub-synchronous speeds

(c) Super-synchronous speeds

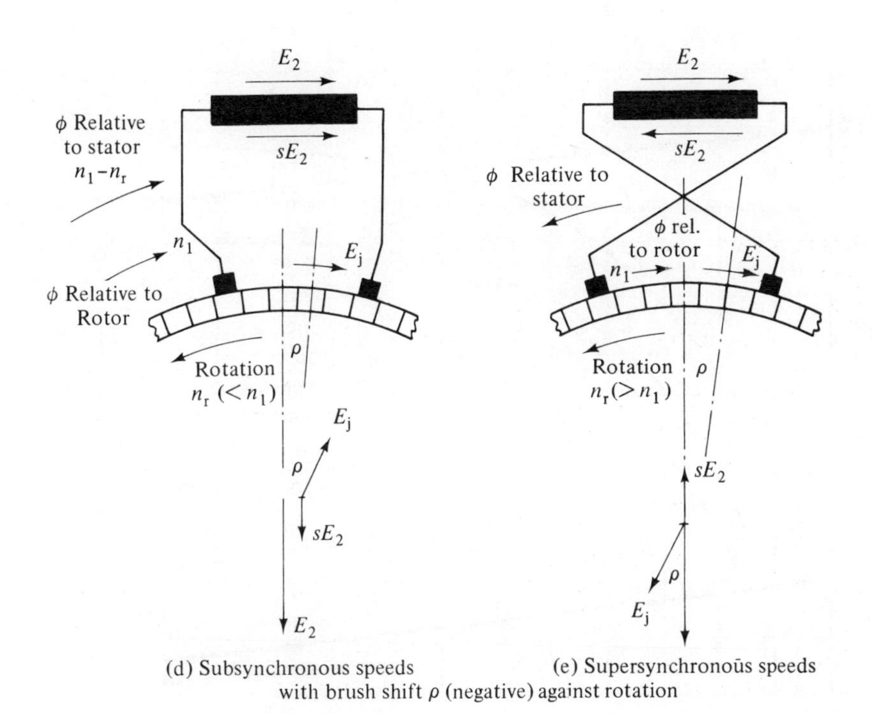

(d) Subsynchronous speeds (e) Supersynchronous speeds
with brush shift ρ (negative) against rotation

Fig. 4.22 Schrage motor fluxes and e.m.f.'s with various brush positions, (from *The Performance and Design of A.C. Commutator Motors*, by E. Openshaw Taylor)

for low-speed drives may be provided by a geared unit. This has become a particularly attractive solution since several manufacturers can now offer either complete geared units, or at least motors with standard mountings and shaft extensions for attachment to standard gear-heads. In practice design considerations limit the output per pole pair to about 30 kW so that above 120 kW maximum output it would in any case be necessary to use a 6 pole machine, which extends the range to

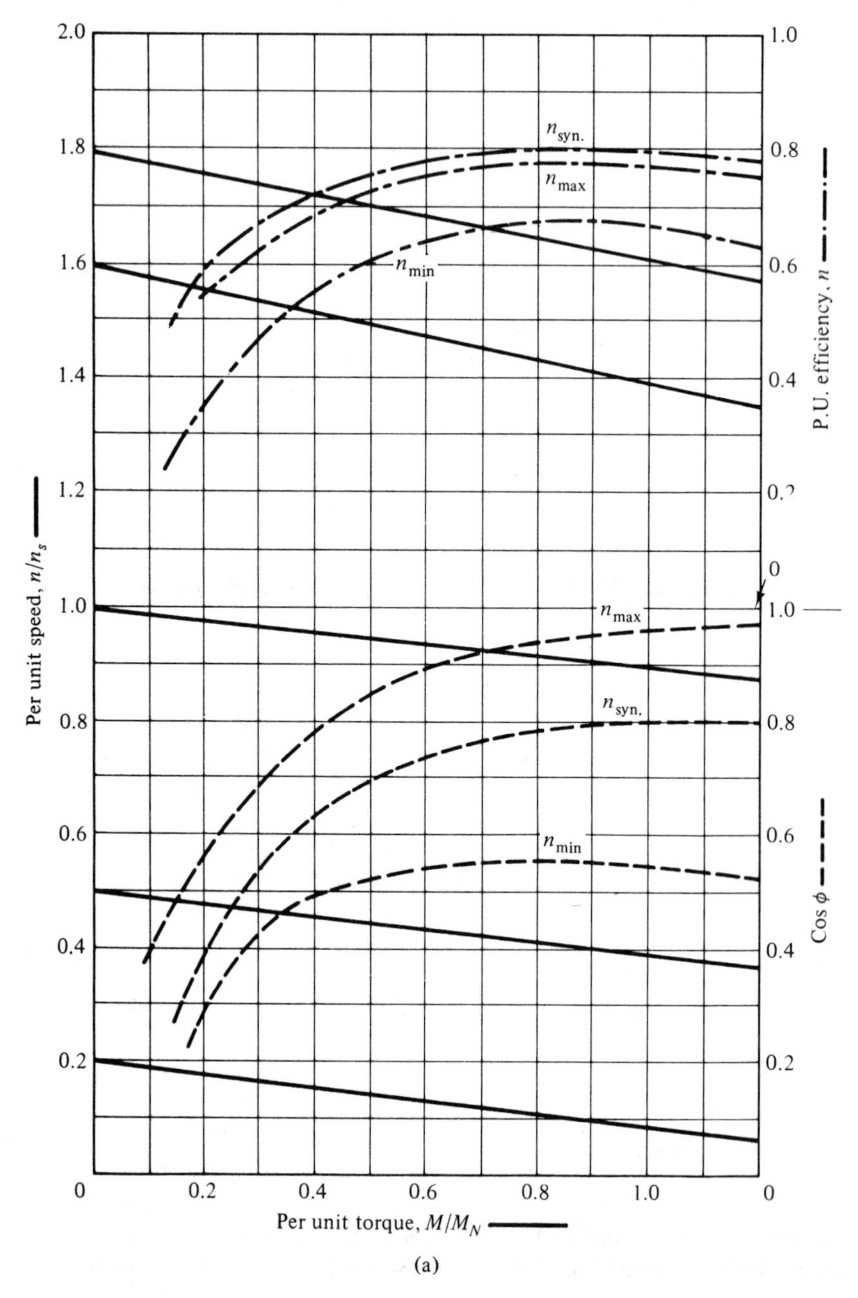

(a)

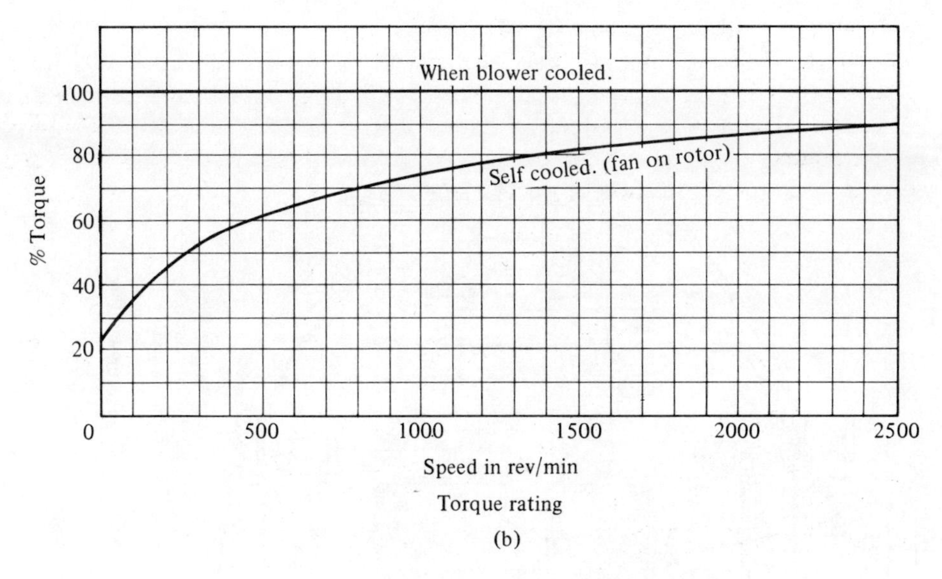

Fig. 4.23 Characteristics of typical small Schrage motor

another limit, namely that of supply. Since it is important to wind rotors for supply voltages above 500 V 50 Hz motors from about 150 kW are made as stator fed machines.

A standard range of speed is 1:3·7 arranged symmetrically about synchronous speed, thus for 4-pole machines typically 600 to 2200 rev/min and for 6-pole versions 375 to 1500 rev/min (1:4). Since self-cooling below 600 rev/min is inadequate most manufacturers supply motors with separate cooling motors. The motors give approximately constant torque so that output is nearly proportional to speed. A 4-pole motor which has an output of 18 kW at 2200 rev/min will give only 4·5 kW at 600 rev/min. Extending the speed range also reduces the maximum output so that the same motor made to work from 375 to 2300 rev/min will have outputs of 2·5 and 15 kW respectively. A further extension of the speed range to 1:20 reduces the maximum output at the same speed by nearly 50 per cent. These figures make it abundantly clear that the larger the range of speeds specified the larger in size and inevitably more expensive the motor will be for a given torque or power loading. The characteristics and construction of a typical small Schrage motor are shown in Figs 4.23 and 4.24 respectively.

Brush shifting is accomplished either manually or by a small geared motor which provides a minimum run-up time of 9 seconds for small machines and 26 seconds for the large ones. Such motors are either controlled from a push-button unit or by a fully automatic preset, programmed or even computer-based system. Such systems must be designed to run the brushes back to the minimum speed position every time the motor is stopped. The starting current of a particular motor depends on its speed range and, as shown in table 4.2, will affect the starting torque in the same proportion. Small and medium sized motors can be started directly on line provided the brushes have been returned to the minimum speed position. Large motors which

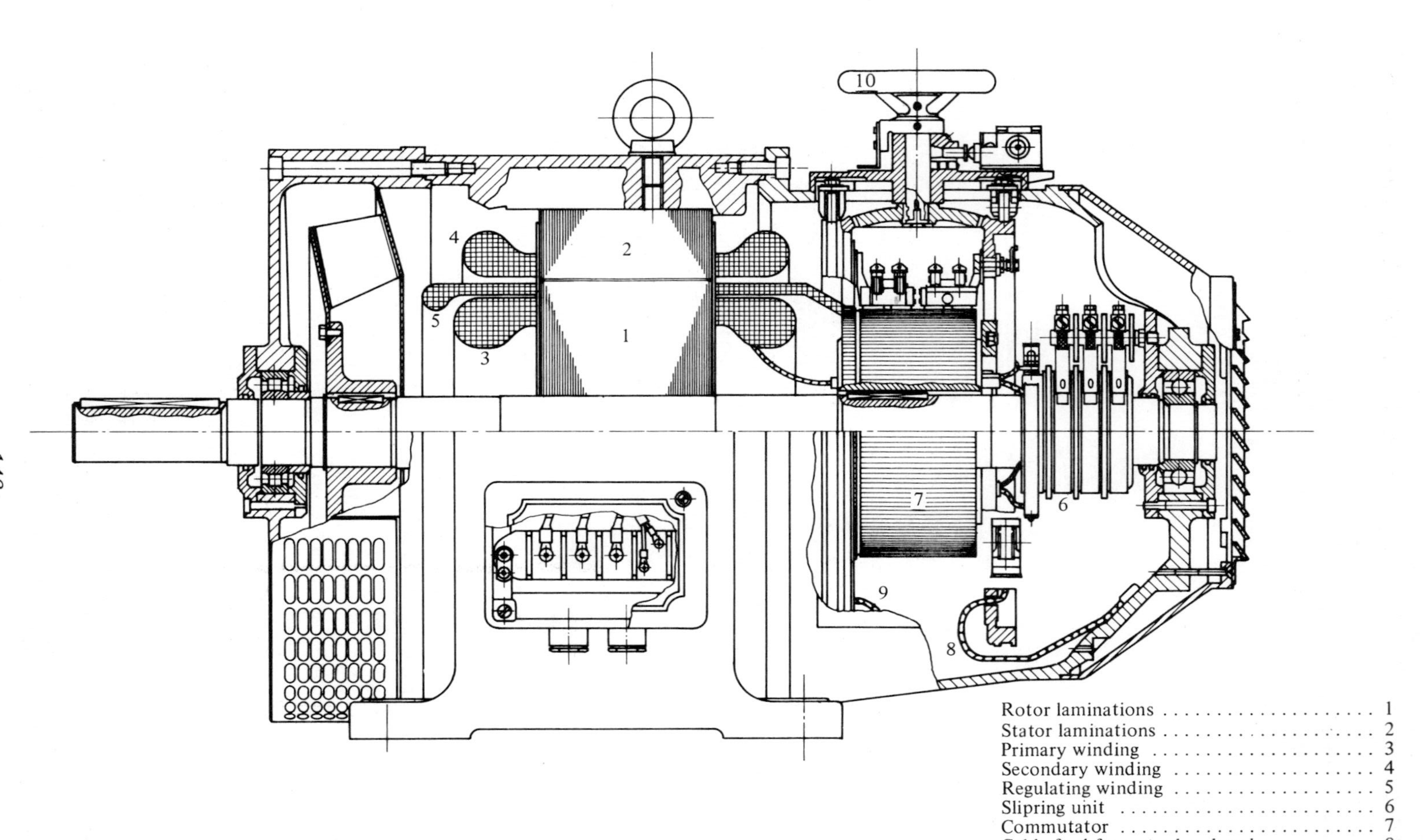

Fig. 4.24 Sectional drawing of a rotor-fed three phase variable speed commutator motor (Courtesy Elektra Faurndau, Germany)

would in any case require a suitable starter can be started with the brushes in any position.

Table 4.2 Effect of speed range on starting torque

Speed Range			Starting torque / rated torque
FL	0·5 FL	NL	
1:3·7	1:3·3	1:3·1	2·0–2·5
1:6·1	1:5·1	1:4·5	1·7–2·0
1:10	1:1·7	1:6·2	1·4–1·6
1:20	1:12·8	1:9·1	1·25–1·4

References

1. D. E. Hesmondalgh and E. R. Laithwaite, 'Method of analysing the properties of two phase servo-motors as tachometers', *Proc. IEE*, **110**, 2039 (1963).
2. D. E. Hesmondalgh and K. L. Driver, 'Design of the a.c. servo-motor', *Proc. IEE*, **113**, 1651 (1966).
3. H. Vickers, *The Induction Motor*, Pitman, London (1953)).
4. P. L. Alger, *The Nature of Polyphase Induction Machines*, GEC, New York (1951).
5. A. Pen-Tung Sah, *Fundamentals of Alternating-Current Machines*, McGraw-Hill, New York (1946).
6. M. G. Say, *Alternating Current Machines*, 4th Edition, Pitman, 1976.
7. E. Openshaw Taylor, *The Performance and Design of A.C. Commutator Motors*, Pitman, London (1958).
8. K. J. Binns, 'Cogging Torques in Induction Machines', *Proc. IEE*, **115**, No. 12 (December 1968).
9. H. Rentzsch, *Handbuch für Elektromotoren. 2.* Auflage, Brown, Boveri, Giradet, 1972.

5 Dimensions and Performance Standards

5.1 Introduction

Standards for electric motors have evolved during the course of this century with the purpose of attempting to specify the 'performance' of the work-horse of industry. This chapter attempts to advise the reader of the situation in the UK and on the Continent of Europe regarding Standards and Codes of Practice existing as at the end of 1975. A major problem in compiling this chapter has been that this situation is a very fluid one with an almost unprecedented standardization activity in BSI and IEC during the early 1970s. As an indication of this, it is only necessary to note that well over half of the Standards listed at the end of this chapter bear dates from 1970 onwards.

The chapter reviews the relevant Standards and selects the more important aspects in respect of the application of electric motors; it presents data from the Standards in descriptive and tabular form and states the relationship between IEC and BSI Standards.

The objective is to introduce the reader to the appropriate Standards by indicating their scope and content in broad terms. It is obviously impossible here to present the full details of any of the current Standards, and hence it must be strongly emphasized that the reader should consult the actual Standard both to cover these points of detail and also to ensure that the latest version of the Standard is used. Consultation of the BSI year book is recommended for this latter purpose.

A number of interesting papers[1,2,3,4] give a useful insight into the very high level of activity in standardization of electrical machines in Europe. The surge of activity in recent years has been largely brought about by the need to harmonize the multitude of National Standards in Europe, and also by the BSI decision to reorganize the various electrical machine Standards under two main Standards, BS 4999 and BS 5000. This re-grouping will be virtually complete by 1976–7 when BS 2613 will be withdrawn. It has attracted considerable interest in Europe, and it is likely that a similar scheme will be set up for IEC Standards.

The chapter covers all general-purpose and special-application motors for which BSI or IEC Standards exist with the exception of motors for use in hazardous atmospheres. Standards for such motors are discussed in chapter 9.

5.2 General Requirements of Electric Motors

5.2.1 *Duty and rating*

The various standard duties and ratings of electric motors are specified in BS 4999: Part 30. There are four basic classes of rating:

(a) Maximum continuous rating (MCR), namely the load at which the motor may be operated for an unlimited period while complying with the normal performance requirements.
(b) Short time rating (STR), namely the load at which the motor may be operated for a limited period (preferably 10, 30, 60 or 90 min) starting at the ambient temperature, while complying with the normal performance requirements.
(c) Equivalent continuous rating (ECR), namely the load and conditions at which the motor may be operated until thermal equilibrium is reached while complying with the normal performance requirements, and which is considered to be equivalent to one of the duty-types listed below.
(d) Duty-type rating (DTR), namely the load and conditions at which the motor may be operated while complying with the normal performance requirements, in accordance with one of the standard duty-types S3–S8 listed below.

5.2.1.1 Standard duty-type. There are eight standard duty types, as follows:

(a) S1—continuous running duty-type; equivalent to maximum continuous rating (MCR).
(b) S2—short-time duty-type; equivalent to short-time rating (STR).
(c) S3—intermittent periodic duty-type, consisting of repeated cycles of a period (N) under rated load, followed by a period (R) at rest and de-energized. The cyclic duration factor (see below) is given by $N/(N+R)$.
(d) S4—intermittent periodic duty-type with starting. This is similar to S3, but with a period (D) during which the motor accelerates to full-speed. The cyclic duration factor is then $(D+N)/(D+N+R)$.
(e) S5—intermittent periodic duty-type with electric braking. This is a further extension of S3 where a significant period (F) of electric braking occurs. The cyclic duration factor is then $(D+N+F)/(D+N+F+R)$.
(f) S6—continuous operation duty-type, consisting of a period (N) under rated load, followed by a period (V) on no-load. The cyclic duration factor is given by $N/(N+V)$.
(g) S7—continuous operation duty-type with electric braking. This consists of repeated cycles of a period (D) of starting, a period (N) of operation at rated load, followed by a period (F) of electric braking. The cyclic duration factor is considered to be unity.
(h) S8—continuous operation with related load and/or speed changes. This consists of a complex sequence of periods (N_1, N_2, N_3, etc) of operation at rated loads, interspersed both with periods (F_1, F_2, etc) of electric braking and also with periods (D_1, D_2, etc) of acceleration. Various cyclic duration factors can be expressed in terms of a period of acceleration or braking followed by the subsequent period of load in relation to the total cyclic repetition time (duty-cycle time).

The preferred duty-cycle time is 10 min and the cyclic duration factor can be either 10 per cent, 25 per cent, 40 per cent or 60 per cent. For duty-types S4, S5, S7, and S8 it is necessary to specify the motor and load stored energy constants, H_m and H_l respectively, referred to rated speed. Alternatively the factor of inertia, FI, equivalent to $(H_m + H_l)/H_m$, may be used in place of H_l.

5.2.1.2. Examples. Duty-type S2 is indicated by the abbreviation and the duration, e.g.:

S2 60 min.

Duty-types S3 and S6 are indicated by the appropriate abbreviation and the cyclic duration factor, e.g.:

S3 25 per cent or S6 40 per cent

Duty-types S4 and S5 are specified:

S4 25 per cent $H_m 2\ H_l 4$

or

S4 25 per cent $H_m 2$ FI3

For duty-type S7, the cyclic duration factor is taken as unity and is not specified.

Duty-types S8 require a statement of the combinations of load, speed, and cyclic duration factor as follows:

S8	$H_m 1\ H_l 9$	24 kW	740 rev/min	25 per cent
	$H_m 1\ H_l 9$	60 kW	1460 rev/min	25 per cent
	$H_m 1\ H_l 9$	45 kW	980 rev/min	40 per cent

It is obviously essential that the purchaser should specify the duty of the motor as accurately as possible. If the actual loading/time sequence is indeterminate, it is necessary to select a fictitious sequence no less onerous than the true one. If a test is required, it is usually sufficient to test to an equivalent rating. Where a test to an actual or estimated duty is required, this should be arranged specially.

5.2.1.3 Preferred outputs and voltages. The output available at the motor shaft is expressed in watts (W). The preferred outputs for motors with standard metric dimensions (BS 3979, but being transferred to BS 5000: Part 10) are given in table 5.1 in relation to the standard frame sizes under specified conditions. For motors outside the scope of BS 3979 (i.e., non-metric dimensions, or with outputs exceeding 132 kW), the preferred outputs are given in table 5.2.

There are three basic categories of voltage ratings:

(a) Single voltage, e.g., 415 V motors.
(b) A limited voltage range, e.g., 380–420 V motors, with a normal maximum voltage range of 10 per cent of the mean, such motors designed to be suitable for operation at all voltages in the range and to meet the performance requirements at the mean voltage.

Table 5.1 Preferred outputs

Single speed, TEFC, cage rotor, MCR motors, Class E or Class B insulation. Suitable for a three-phase, 50 Hz, 415 volts supply

Output kW				
Synchronous speed (rev/min)				Frame number (D—)
3000	1500	1000	750	
1·1	0·75	0·55	—	80
1·5	1·1	0·75	0·37	90 S
2·2	1·5	1·1	0·55	90 L
3·0	2·2 & 3·0	1·5	0·75 & 1·1	100 L
4·0	4·0	2·2	1·5	112 M
5·5 & 7·5	5·5	3·0	2·2	132 S
—	7·5	4·0 & 5·5	3·0	132 M
11 & 15	11	7·5	4·0 & 5·5	160 M
18·5	15	11	7·5	160 L
22	18·5	—	—	180 M
—	22	15	11	180 L
30 & 37	30	18·5 & 22	15	200 L
—	37	—	18·5	225 S
45	45	30	22	225 M
55	55	37	30	250 M
75	75	45	37	280 S
90	90	55	45	280 M
110	110	75	55	315 S
132	132	90	75	315 M

NOTE: Somewhat higher ratings are allocated to each frame for enclosed ventilated motors, and lower ratings for wound-rotor motors.

Table 5.2 Preferred outputs for larger motors

above 75 kW up to and including 750 kW				above 750 kW	
kW				kW	
80	140	250	450	800	2 800
85	150	265	475	900	3 150
90	160	280	500	1000	3 550
95	170	300	530	1120	4 000
100	180	315	560	1250	4 500
106	190	335	600	1400	5 000
112	200	355	630	1600	5 600
118	212	375	670	1800	6 300
125	224	400	710	2000	7 100
132	236	425	750	2240	8 000
				2500	9 000
					10 000

(c) A dual voltage, e.g., 220/380 V or 220–240/380–420 V motors, obtained by winding reconnection, such motors designed to meet the performance requirements at the two voltages or at the means of the two ranges.

If a single voltage motor is required to be rated for a limited voltage range, then the ambient temperature should be restricted to 30°C. The preferred voltages for three-phase, 50 Hz motors are 415 V, 3·3 kV, 6·6 kV, and 11·0 kV. The minimum recommended outputs at these voltages are as follows:

3·3 kV—150 kW

6·6 kV—300 kW

11 kV—1000 kW

5.2.2 *General characteristics*

These are specified in BS 4999: Part 41.

5.2.2.1 Starting characteristics. The main variations in performance of cage induction motors relate to the starting characteristics. In this respect, there are six different standard designs for direct-on-line starting, specified as A to F as follows:

Design	*Starting current*	*Locked rotor torque*
A	High	Normal
B	Normal	Normal
C	Normal	High
D	Low	Normal
E	Low	Low
F	High	High

In addition, Design G can be used to specify other starting characteristics.

5.2.2.2 Pull-up torque. For single-phase cage induction motors, the pull-up torque is normally not less than 0·3 times rated load torque. For single speed three-phase motors rated at less than 100 kW, the pull-up torque is normally not less than 0·5 times either rated load torque or locked rotor torque; at and above 100 kW, the corresponding factors are 0·3 and 0·5 respectively.

For large motors (e.g. $>$1000 kW), it may be more economical to match the motor and load characteristics, so that the developed motor torque (allowing for a 10 per cent drop in supply voltage) exceeds the load torque by not less than 10 per cent of rated load torque.

5.2.2.3 Frequency of starting. The normal design criterion is that the motor shall be capable of two starts in succession at rated load torque, inertia and temperature, followed by a cooling period of 30 min before attempting a further starting sequence. It may be necessary to change the value of 30 min for special applications and designs.

5.2.2.4 Momentary excess torque. This torque is the extra loading which a motor is capable of withstanding without injury, stalling or abrupt change in speed, having previously attained its normal running temperature. The specified values of momentary excess torque in per unit of rated load torque are:

d.c. motors	0·6
a.c. induction motors	0·6

synchronous (salient pole) motors	0·5
synchronous induction motors (cylindrical rotor or salient pole)	0·35

For cage induction motors designed for a starting (locked rotor) current less than 4·5 times rated load current, i.e., basically designs D and E, the momentary excess torque can be less than 0·6 but not less than 0·5 times rated load torque.

5.2.2.5 Overspeed. Motors will normally be designed to withstand an overspeed of 1·2 times either the maximum rated speed or the maximum speed in normal service. If the motor can be over-driven by the load, then the motor should be designed to withstand this runaway speed.

5.2.3 *Service and operating conditions*

These conditions are specified in BS 4999: Part 31. The basic conditions for which the motor should be suitable are as follows:

(a) *Altitude*—not exceeding 1000 m above sea level.
(b) *Coolant temperature*—not exceeding 40°C for motors cooled by air or other gas at atmospheric pressure.
—not exceeding 25°C at the inlet to liquid-cooled heat exchangers where these are fitted.
(c) *Voltage variations*—not exceeding ±5 per cent of rated voltage (±6 per cent if for use where the power supply is subject to UK Electricity Supply Regulations), at rated frequency for single voltage motors. For a motor rated for a limited voltage range, the voltage variation should be restricted to within 95 per cent of the minimum and 105 per cent of the maximum voltages of the range. If the motor is intended for prolonged operation at these limits, then the temperature rise limits in table 5.3 may be exceeded by 10°C for motors up to and including 1000 kW, and by 5°C for larger motors. In such a case, a motor rated for a more suitable voltage or voltage range should be specified.
(d) *Unbalanced or non-sinusoidal supplies*—negative phase sequence and zero-phase sequence components of voltage individually not exceeding 0·02 per unit of the positive phase sequence components.
—instantaneous values of phase voltages not differing from the fundamental by more than 0·05 per unit of the amplitude of the latter.

These conditions are the same as in IEC 34-1. It should be noted here that a possible change in specifying the supply unbalance is under consideration, namely to consider current rather than voltage components. The suggested limits are that the negative and zero phase sequence components of current should not exceed 0·05 per unit of the positive phase sequence component.

5.2.4 *Limits of temperature rise*

Table 5.3, from BS 4999: Part 32, specifies the limits of temperature rise for electric motors with rated voltage not exceeding 11 kV. These limits are based on the service conditions specified in section 5.2.3 above for continuously rated motors (S1). For

Table 5.3 Limits of temperature rise for machines

	Temperature rises in °C														
	Class of insulation (see Note 1)														
	A			E			B			F			H		
	Method of measurement (see Note 2)														
Part of machine	T	R	E	T	R	E	T	R	E	T	R	E	T	R	E
A.C. winding of a.c. machines rated for 11 000 V *or less*															
Windings of air cooled machines having an output of 5000 kW (or kV A) or more or having a core length of 1 m or more	—	—	60	—	—	70	—	—	80	—	—	100	—	—	125
Windings of air cooled machines smaller in output or size than those designated above	—	60	—	—	75	—	—	80	—	—	100	—	—	125	—
Permanently short-circuited insulated windings	60	—	—	75	—	—	80	—	—	100	—	—	125	—	—

NOTE 1: This classification is in accordance with BS 2757, 'Classification of insulating materials for electrical machinery and apparatus on the basis of thermal stability in service'.
NOTE 2: T = Thermometer method
R = Resistance method
E = Embedded temperature detector method
} see section 5.2.11.8

motors with a short-time rating (S2) these limits may be increased by 10°C by agreement. For duty-types S3–S8, the temperature rises specified apply to the middle of the period of greatest load. If this cannot be measured easily, then the peak temperature attained must not exceed those values in table 5.3 by more than 10°C.

At rated voltages above 11 kV, the limits are lowered if the temperature is measured by embedded detectors. The amount of lowering is 1°C for each 1 kV between 11 and 17 kV with a further lowering of 0·5°C for each 1 kV above 17 kV.

The adjustment of the temperature rise limits for service conditions differing from those in section 5.2.3 above are given in Table 5.4.

Table 5.4 Adjustments for various service conditions

Coolant Temperature °C	Altitude 0 m–1000 m
<30	+10°C
30–39	+(40 − coolant)°C
40	0
Above 40–60	−(coolant − 40)°C
>60	by agreement

5.2.5 *Vibration*

Vibration amplitudes are specified in BS 4999: Part 50, with the allowable amplitude increasing as the rotational speed decreases. These limits are set out in table 5.5, and are for general-purpose motors. If high precision balancing is required for a particular application, then lower limits can be agreed. As the measured amplitude will depend on the stiffness of the foundation and the degree of unbalance of a coupled load, then amplitudes measured on site may exceed those in table 5.5, but not by more than 25 per cent.

Table 5.5 Limits of vibration amplitude (peak to peak in millimetres)

Speed rev/min	limit	Speed rev/min	limit	Speed rev/min	limit	Speed rev/min	limit
3600	0·020	1800	0·036	900	0·059	450	0·085
3400	0·021	1700	0·038	850	0·061		
3200	0·022	1600	0·040	800	0·063	400	0·090
3000	0·024	1500	0·041	750	0·065		
2800	0·025	1400	0·043	700	0·068	350	0·094
2600	0·027	1300	0·045	650	0·071		
2400	0·029	1200	0·048	600	0·074	300	0·100
2200	0·031	1100	0·051	550	0·077		
2000	0·033	1000	0·055	500	0·080	250	0·100

NOTE: Where the actual test speed is not tabulated interpolation is permitted.

These vibration limits are for the radial direction. It is accepted that, for machines with frame sizes above 180, the amplitude of vibration in the axial direction may be significantly higher. Also, machines other than horizontal foot-mounted machines may generally have higher levels of vibration.

It is worth noting at this point that the UK practice is to balance rotors with a half-key fitted. The IEC practice is to balance without a key.

A new international Standard, ISO 2373, issued in November 1974, sets limits for vibration severity in terms of maximum r.m.s. values of vibration velocity. It uses three grades, namely 'N' (normal), 'R' (reduced) and 'S' (special). Normal stock motors would have a grade N.

The allowable limits of vibration severity are invariant with speed for grade N between 600 and 3600 rev/min, but increase in steps with increasing shaft height between 80 and 400 mm, from 1·8 mm/s to 4·5 mm/s. Comparison with the values in table 5.5 is therefore complicated in that the latter are independent of frame size and are quoted as a vibration amplitude and will hence be related to the vibration velocity by the vibrating frequency.

5.2.6 *Noise levels*

Three degrees of noise level are specified in BS 4999: Part 51, namely 'normal,' 'reduced,' and 'specially low.' The noise levels apply to all types of motor, except airborne equipment, traction-vehicle motors, and small-power machines, up to ratings of 16 MW. The upper limits of noise level are lower than the maximum levels specified in IEC 34-9, 1972 by between 1 and 3 dB(A), the greatest difference occurring at the higher speeds and powers.

The noise level is classified in terms of dB(A) sound power level on no-load. The allowable limits of sound power level depend on the rotational speed, rating, and type of enclosure. Table 5.6, from BS 4999: Part 51, gives the upper limits of sound power level. These limits apply to motors of normal sound power. The limits for motors of reduced sound power level are lower than those in the table by 5 dB(A). This can be achieved by modifications to a standard design, but these may not be practicable below 132 kW. Motors of specially low sound power will have special design features to give noise levels lower than the reduced sound power level. The actual value will be subject to agreement between manufacturer and purchaser.

Table 5.6 Limiting mean sound power level for enclosed machines

Rating kW (or kVA)		Rated speed (rev/min)					
		960 and below	961–1320	1321–1900	1901–2360	2361–3150	3151–3750
Above	Up to	Sound power level dB(A)					
	1·1	76	79	80	83	84	89
1·1	2·2	79	80	83	87	89	91
2·2	5·5	82	84	87	92	93	95
5·5	11	85	88	91	96	97	99
11	22	88	91	94	97	100	102
22	37	91	94	96	99	103	104
37	55	93	96	99	101	104	106
55	110	95	99	102	103	106	108
110	220	98	102	105	106	108	109
220	630	101	104	108	109	109	111
630	1100	104	107	111	111	111	112
1100	2500	107	110	113	113	113	113
2500	6300	108	112	115	115	115	115
6300	16 000	110	113	116	116	116	116

NOTE: Sound power levels for open ventilated machines are generally 3–4 dB(A) lower

5.2.7 *Dimensions*

The standardization of dimensions in the UK has been a prolonged affair mainly because of the change from inch to metric units and also because of the need to avoid preferences for existing schemes of dimensions as used by a particular manufacturer or country. The fundamental UK document was BS 3979, giving metric values for a wide variety of mountings, enclosures, and types of motor. This has been replaced by BS 4999: Part 10. In addition, two documents are still in being, concerned with inch units. These will be considered first.

5.2.7.1 Flameproof motors. Dimensions of three-phase, 50 Hz cage induction motors rated between $\frac{1}{2}$ and 100 h.p. (0·37 to 75 kW) with flameproof enclosures are given in inch units in BS 2960: Part 3: 1964 (Note: Parts 1 and 2 have been withdrawn). This Standard is not in wide use and hence only minimal details will be given here; it will probably be withdrawn in 1976–7. The Standard assigns ratings to each frame size depending on the synchronous speed of the motor. The frame size is designated by the letter 'E', followed by three numerals, the first two giving the height of the shaft centre from the underside of the feet in quarter-inch units, and the third numeral indicating the axial length. A letter S indicates a short shaft extension. The absence of a further letter indicates foot mounting, while the use of a final letter D or V indicates flange or skirt mounting respectively.

The standardized dimensions cover the fixing dimensions, shaft size, and keyway, and shaft height (for foot-mounted motors). In addition, the maximum overall dimensions (total length, diameter, etc) are specified to ensure that a given frame size will fit an existing space for any make of motor. Finally, for foot mountings, the terminal box is positioned on the right-hand side looking at the non-driving end.

5.2.7.2 Fractional horse-power motors. The dimensions of fractional horse-power motors in terms of inch units are specified in BS 2048: Part 1: 1961 (Note: subsequent parts were not issued). These dimensions are in common usage and, for common frames, the shaft and mounting dimensions are interchangeable with those in the appropriate part of NEMA Standard MG1 (except for frame 42 and all flange-mounted motors). The specified dimensions cover d.c. motors and single and polyphase a.c. motors. Frames are designated by the letter 'B' followed by two numerals giving the shaft centre height in $\frac{1}{16}$ inch units. A final letter from A to F indicates the type of mounting, if other than plain foot-mounting. The relationship between frame and rating is not specified because of the wide variety of types of motor and enclosure covered by the standard dimensions.

The standardized dimensions are those relating to fixing, shaft, keyway, and the shaft centre height. Reference can be made to the Standard for the actual dimensions relating to any frame size (Tables 3A and 4A in the Standard). The entry for the supply lead should be on the right-hand side looking from the non-driving end.

5.2.7.3 Metric dimensions. The dimensions of a wide variety of electric motors in metric units are specified in BS 3979: 1966. This Standard is in general agreement with practice in Western Europe; in particular the allocation of shaft numbers and ratings to particular frame sizes is identical. It is most important in using this Standard to ensure that the most recent copy is available, due to the considerable changes and extensions to its content since its introduction in 1966. As an indication of this, eight Amendments were issued up to April 1973. The Standard is mainly concerned with a.c. induction motors. The dimensions of d.c. machines (up to frame 500) are not fully covered, nor are outputs and shaft numbers assigned to particular frame sizes.

The system of frame nomenclature is internationally agreed up to and including frame size 400, i.e., primarily for low-voltage induction motors. The nomenclature starts with a letter indicating the enclosure, namely 'C' for enclosed ventilated, 'D' for totally enclosed (not flameproof) or 'E' for flameproof. This is followed by a two- or three-digit number giving the shaft centre height in millimetres. Next, a letter indicates the length, S, M, or L for short, medium, and long, respectively. A final letter follows this indicating the method of mounting, namely 'D' for flange, 'V' for skirt, 'C' for face-flange, 'P' for pad, and 'R' for rod. The absence of a final letter indicates foot mounting. This system of symbols will be transferred eventually to BS 4999: Part 2.

Fixing dimensions are standardized for foot-mounted, flange-mounted, and skirt-mounted frames up to frame number 400 for all three types of enclosure. For face-flange frames, dimensions are given up to frame 160 for all enclosures except flameproof. For pad or rod-mounted frames, dimensions are given for totally enclosed motors only, up to frame 250. Each frame has one specific set of mandatory fixing dimensions, except for pad or rod-mounted frames, where the radius of the circle to which the mounting pads (or faces or rods) are tangent is not a specific value. For large foot-mounted motors, with high or low shaft centres from frame 355 to 1000, the distance between centre lines of fixing holes (side view) is given a median value for each frame, and a choice from twelve other values is allowed to accommodate the large variety of types of motor associated with large frame sizes. In addition, two values of the distance from the centre line of the fixing holes at the driving end to the shoulder of the shaft are allowed to accommodate either ball and roller or sleeve bearings. The particular dimensions for these motors are most extensive and are not included here. The reader should refer to BS 3979 for detailed data.

Each standard-dimensioned frame is allocated an output rating (table 5.1) and a shaft number, depending on the synchronous speed, type of enclosure, and type of rotor. The shaft number relates to a specific set of dimensions giving length, diameter, and key-way detail. The allocation is generally made for all frames from 80 to 315 inclusive, but some motors have allocations over a smaller range.

For totally enclosed frame-surface cooled motors, e.g., motors cooled by the air stream from the driven fan, each frame from 80 to 250 has a minimum specified value of average air velocity for a particular synchronous speed.

For foot-mounted frames, the terminal box is positioned at the left-hand side of the motor, looking at the non-driving end.

Standardized sets of dimensions for slide-rails for foot-mounted motors are given for frames 80 to 315.

Finally a nomenclature is given for mounting arrangements. This consists of a letter, either B for horizontal shaft or V for vertical shaft, followed by a number indicating a particular combination of stator support arrangement, shaft configuration, and flange or mounting arrangement.

IEC 34–7, 1972, gives a much more extensive nomenclature in terms of two codes. Code I is similar to that in BS 3979 but prefixed by the letters IM and with a greater variety of mounting arrangements. Code II is a general code, again prefixed IM, but followed by four numerals. The first numeral indicates the type of construction (foot mounted, flange mounted, etc). The second and third numerals indicate the mounting arrangement, while the fourth numeral indicates the shaft extension (e.g., one or two shaft extensions, cylindrical or conical, etc.). Over 150 different configurations are tabulated, indicating the extent of the system.

When BS 4999: Part 10 is published, dimensions for frame sizes 56, 63, and 71 will be included for all enclosures for foot-mounted frames, for totally enclosed and flameproof enclosures for flange-mounted frames, and for enclosed ventilated and totally enclosed enclosures for face-flange frames.

5.2.8 *Rating plate markings*

The detailed information to be given on the motor rating plate is specified in BS 4999: Part 4. This consists of a list of general information for all types of machine, plus additional information for particular types. The general information includes the class of rating, insulation class and (for duty-type ratings), the stored energy constant (H) or factor of inertia (FI). The additional information includes speed, current, rated voltage, rated output, excitation (for d.c. motors), supply frequency and number of phases (for a.c. motors), and primary winding connection and design letter (for induction motors).

5.2.9 *Enclosures*

A system of classification of types of enclosure for electric motors has been agreed internationally in IEC. It is given in IEC 34-5 and BS 4999: Part 20. The system accounts for protection by the enclosure in three ways:

(a) To persons against contact with live and moving parts;
(b) to the motor against ingress of solid foreign bodies;
(c) to the motor against harmful ingress of liquid.

The first two of these are inter-related and the system groups them as one. The classification does not include special systems of protection as suitable for example for hazardous atmospheres or corrosive vapours. The classification consists of the letters IP followed by two numbers. The first number indicates the degree of protection regarding contact with live or moving parts and regarding ingress of solid foreign bodies. The second indicates the degree of protection against harmful ingress of water. The two numbers may be followed by a letter (S or M) indicating whether the motor was tested against ingress of water with the motor stationary or running respectively. The absence of a letter indicates that the motor was tested under both conditions. The first number has the following significance:

0 non-protected
1 protected against solid bodies greater than 50 mm diameter
2 protected against solid bodies greater than 12 mm diameter
4 protected against solid bodies greater than 1 mm diameter
5 protected against dust

The second number indicates the following degree of protection against ingress of water:

0 non-protected
1 dripping water
2 drops of water 15° from vertical
3 spraying water
4 splashing water
5 water jets
6 conditions on ship's decks
7 immersion of limited duration
8 submersible.

In addition, the letter W between IP and the number indicates weather protection, i.e., when the motor can withstand operation in rain or snow and in the presence of airborne particles. The possible combinations of first and second number are extensive in theory, but in practice only a few combinations are necessary. As examples, the following equivalents relate to some of the more common enclosures:

IP 11S ventilated, drip-proof
IP 44 totally enclosed
IP 54 dust proof.

Discussion took place during 1974 in IEC and BSI concerning the need for an intermediate degree of protection between IP4 and IP5 to cover totally enclosed machines. The argument is that IP4 allows ingress of solid bodies of 1 mm diameter, while IP5 is a dust-tight enclosure. A totally enclosed machine can be said to lie between the two and a classification IP4–5 giving protection against solid bodies greater than 0·1 mm diameter has been suggested.

5.2.10 *Method of cooling*

In order to facilitate understanding between manufacturer and purchaser, an internationally agreed system of classifying methods of cooling electric motors is described in IEC 34-6 and BS 4999: Part 21. Two codes of classification are given, a simplified one to cover the more usual cooling methods, and a complete code to cover all methods. Only the simplified one will be described here. Reference to BS 4999: Part 21 will give the complete system.

In the simplified code for air-cooled motors, the method of cooling is described by the letter IC and two numbers. The first number indicates the cooling circuit arrangement and the second gives the method of supplying power to circulate the coolant. The first number follows the following code:

0 free circulation
1 inlet duct
2 outlet duct
3 inlet and outlet duct
4 frame surface cooled
5 integral heat exchanger using surrounding medium
6 machine mounted heat exchanger using surrounding medium
7 integral heat exchanger not using surrounding medium
8 machine mounted heat exchanger not using surrounding medium
9 separately mounted heat exchanger.

The second number has the following code:

0 free convection
1 self circulation
2 integral component mounted on separate shaft
3 dependent component mounted on the motor
5 integral independent component
6 independent component mounted on the motor
7 independent and separate device *or* coolant system pressure
8 relative displacement.

The component referred to is the fan or blower producing the air movement, and the dependence referred to is in regard to the electrical supply, i.e., an independent component is a fan with a separate supply. For motors with self circulation, i.e., second number 1, the code may be abbreviated by using only the first number.

By means of example, some of the more common methods of cooling are classified below:

IC 01 or ICO	normal ventilated motor
IC 11 or IC1	inlet duct ventilated
IC 41 or IC4	enclosed fan cooled
IC 05	normal ventilated with separately supplied fan
IC 17	inlet duct ventilated with separately supplied fan in the duct
IC 48	surface cooled by air-stream.

5.2.11 *Tests*

In any system of standardization, it is not only necessary to specify performance parameters, etc, but also to be specific in describing test methods whereby the various parameters are determined. Most of the parameters are very highly dependent on the method of test, e.g., temperature rise, noise, and vibration. This section outlines the various test procedures as specified in British Standards and comments on the degree of international agreement on the methods.

The classification of tests is described in BS 4999: Part 60, as follows:

(a) Basic tests—to establish compliance with selected, widely applicable requirements.
(b) Duplicate tests—tests on a motor of the same design as one previously tested and sufficient to indicate accordance with the original design.
(c) Routine check tests—tests to check correct assembly and workmanship.

Separate tables for induction machines, synchronous machines, and d.c. machines set out the relevant tests under the appropriate headings of basic tests, duplicate tests, and routine check tests. As examples, the high voltage test applies to all three classifications of test, while the temperature rise test applies to basic tests only. Basic tests are normally made on the first machine on any order, except that the manufacturer may submit routine check tests if these are supported by earlier basic tests on a machine of the same major external dimensions, enclosure, method of cooling, and insulation class and with similar electrical and magnetic loadings. Routine check tests are normally performed on all other machines.

5.2.11.1 Overspeed tests (BS 4999: Part 60). The test duration is 2 min and the motor is considered to pass the test if no permanent abnormal deformation is apparent, if no damage is sustained sufficient to prevent normal operation and if the windings withstand the specified high-voltage test subsequently.

5.2.11.2 Commutation tests (BS 4999: Part 60). These tests are made on no-load after the temperature rise test. For series motors, the no-load condition is replaced by the lowest test load as agreed.

5.2.11.3 Vibration tests (BS 4999: Part 60). These tests are made on the completely assembled motor on no-load and at rated voltage and speed (or the highest speed of its speed range). The vibration is measured on the bearings on a horizontal plane through the shaft centre line of horizontal machines. If measurements on site exceed the allowable amplitudes by more than 25 per cent, then the balance of the coupled equipment should be improved. If this is not successful, then it may be necessary to stiffen the foundations or supporting structure.

5.2.11.4 High voltage tests (BS 4999: Part 60). These tests are made between the windings under test and the motor frame, with the core and any other windings all connected to the frame. The test is made only once at full test voltage immediately following the temperature rise test. If the test is made on site on an assembled group of several pieces of new apparatus, each of which has previously passed the test, then the test voltage should be limited to 80 per cent of the full value. Similarly, for a repeat test on site, the test voltage is limited to 80 per cent of the full value, and the windings should be previously cleaned and dried. If a winding is completely replaced, e.g., following failure in an existing motor, then the full test voltage is applied. After a partial repair the test voltage should be arranged with the repairer, but should not exceed 75 per cent of the specified value. (IEC 34-1 quotes the same values, but states that the test voltage after partial repair should be 75 per cent of the normal value. In addition, it states that the test voltage after an overhaul should be $1{\cdot}5V_r$, with a minimum of 1000 V if $V_r \geqslant 100$ V and with a minimum of 500 V if $V_r <$ 100 V. The test voltage can be of any convenient frequency between 25 and 100 Hz, and approximately sinusoidal so that the peak value is not more than 1·45 times the r.m.s. value. Alternatively, by agreement a d.c. supply can be used such that its magnitude is not greater than 1·6 times the r.m.s. value of the a.c. test voltage. It is recommended that the applied voltage is increased from about one-third the test voltage as rapidly as is consistent with correct measurement, is then held at the test voltage for 1 min, when the applied voltage is rapidly reduced to at least one-third before switching off. The applied voltage is either increased steadily or in increments of not more than 5 per cent of the final value. It is recommended that the duration of increase should not be less than 10 s. For routine testing of motors rated up to 75 kW, the 1 min period can be reduced either to 5 s at the specified test voltage or 1 s at 120 per cent of this value. The full details regarding the appropriate test voltage are given in BS 4999 part 60, Table 60.8.5. It is convenient here simply to summarize the test voltages for stator windings only, as follows:

Rating	Rated Voltage, V_r	Test voltage (r.m.s.)
<1 kW	<100 V	$500\ \text{V}+2V_r$
<1 kW	⩾100 V	$1000\ \text{V}+2V_r$
1 kW up but <10 MW	—	$1000\ \text{V}+2V_r$ but $\nless 1500$ V
10 MW and above	up to 2 kV	$1000\ V+2V_r$
10 MW and above	>2 kV up to 6 kV	$2{\cdot}5V_r$
10 MW and above	>6 kV up to 17 kV	$3000\ \text{V}+2V_r$
10 MW and above	>17 kV	subject to special agreement

5.2.11.5 Momentary overload test. This test relates only to the electromagnetic design of the motor and not to the thermal properties. Hence the test has to be made as quickly as possible consistent with obtaining an accurate reading, but the test duration should not exceed 15 s. For d.c. motors, the brush setting should remain as for normal operation.

5.2.11.6 Noise tests. Measurements of noise emanating from an electric motor have to be made under carefully controlled conditions. Hence a most elaborate and detailed set of instructions is given in BS 4999: Part 51, regarding both the test method and the evaluation of the test data. The reader should refer to this Standard for the full test methods which are selected from ISO Recommendation R/1680. The test methods specified apply to all electrical machines except small power machines and machines for traction vehicles. The basic method relates to the determination of the (A) weighted sound power level. If this value exceeds 93 dB(A), or if one or more tones are prominent, then a second test method has to be used based on a frequency band analysis of sound radiation.

The motor is tested on no-load, and (for synchronous motors) at unity power factor, while supplied at rated voltage (and frequency for a.c. motors). The motor has to be fully assembled and uncoupled, and running at or near rated speed, or the highest speed of a speed range. The test methods describe corrections to be made to take account of any background noise, the corrections being applied if the difference between measured sound power with and without the motor running is less than 10 dB. If this difference is less than 3 dB then no significance can be attached to the results. The tests are intended to be carried out under free-field conditions, and a further check is described regarding the reverberation qualities of the room. This is done by checking that the measured sound pressure level at the prescribed distance from a source of broad-band noise and at half that distance differ by at least 5 dB. If this test shows that free-field conditions do not exist, then the measured sound power level will be slightly higher, but no correction is made. Alternatively, a special test-method is given for semi-reverberant conditions, indicated by a half-distance difference of less than 5 dB but greater than 2 dB. If it is desired to estimate the noise level of a motor on load, then it is necessary to measure the sound pressure level around the motor, sufficiently close to it to minimize the effects of background noise or reflecting surfaces. (It is suggested that directional microphones would assist in this.) The difference between the no-load and on-load values is then added to the no-load sound power level as measured in the usual way, to give the approximate load value.

5.2.11.7 Starting performance. On large motors, it will probably be necessary to measure the starting current and locked rotor torque at reduced voltage. In such circumstances, the full-voltage torque will be slightly greater than that estimated from the reduced-voltage value assuming a square-law relationship with voltage. The correction factor for the saturation effect will depend on the particular motor design.

5.2.11.8 Temperature rise. Three methods of temperature rise are specified in BS 4999: Part 32, with recommendations regarding the particular method to be used for any given motor or part of a motor. The specification also gives instructions in measurement of coolant temperature and regarding estimation, e.g., by extrapolation from a cooling curve, of the actual working temperature.

The three methods described are as follows:

(a) Embedded temperature detector method. These detectors (ETD) are either thermocouples or resistance thermometers inserted into the motor during construction. Six detectors are required, positioned as closely as possible to the likely hot spots, with not less than three detectors positioned between coil sides in the slot portion of the winding.

(b) Resistance method. This method relies on the increase in resistance of a winding when hot. The winding temperature is given by the expression:

$$t_2 = (R_2/R_1)(t_1 + 235) - 235$$

where R_2 and R_1 are the hot and cold resistances at the temperatures t_2 and t_1 °C respectively. The constant 235 applies only to copper windings; for other materials it is replaced by the reciprocal of the temperature coefficient of resistance at 0 °C for that material. Before starting the test, the motor should be at rest for sufficient time for the winding temperature, as checked by a thermometer, to have stabilized at the coolant temperature.

(c) Thermometer method. This method requires thermometers to be placed on the hottest accessible surfaces of the stationary parts of the motor when running under test, and for others to be placed similarly on the rotating parts as soon as the motor stops after the test run. The thermometers can be either mercury or alcohol bulb thermometers, but it is recommended that an alcohol one is used in the presence of a strong or moving magnetic field. It is also recommended that the bulb of the thermometer should be in direct contact with the surface to be measured and that it should be covered by a pad of non-conducting material 3 mm thick and at least 20 mm in radius to prevent heat loss.

The allowable temperature rises take account of the fact that the thermometer method only measures temperatures of accessible surfaces and hence the thermometer should not be inserted into narrow gaps between windings, etc, where the temperature will be greater. For a motor with a continuous maximum rating, the test is continued until thermal equilibrium is attained. This is defined as the condition when the rate of rise of temperature does not exceed 2°C/h. For a short-time rated motor, the temperature of the machine at the start of the test should be within 5°C (4°C for motors for battery operated vehicles, BS 1727: 1971 and 2°C for small-power motors, BS 5000: Part 11) of that coolant. The duration of the temperature rise test for short-time rated motors is the time specified in the rating. For motors with duty-type ratings S3–S8 the test has to be continued until thermal equilibrium is reached.

The tests are carried out at the rated voltage of the motor, or at the mean of the voltage range.

The temperatures are measured immediately before and after switching off the motor following the test run. The highest value, either as measured directly or as extrapolated, is taken as the final temperature. For motors on S3–S8 duty-cycles, the temperature is measured in the middle of the period of greatest load in the last cycle of operation. The maximum temperature during this last cycle should also be recorded. If measurements cannot be taken until the motor stops, then whether or not extrapolation is necessary is shown by the following table:

Motor	Time to first reading	
	Take actual value	Extrapolate
0–50 kW	<0·5 min	0·5–1·0 min, extrapolate to switch-off
>50–200 kW	<1·5 min	1·5–3·0 min, extrapolate to switch-off
>200–5000 kW	<3·0 min	3·0–6·0 min, extrapolate to 3·0 min
>5000–15 000 kW	<4·5 min	4·5–9·0 min, extrapolate to 4·5 min
small-power motors	<20 s	⩾20 s, extrapolate to switch-off
battery-traction motors	⩽45 s	>45 s–2 min, extrapolate to switch-off

A method of extrapolation is given in the Appendix to BS 4999: Part 32. This suggests plotting temperature on a logarithmic scale against a linear time scale. Three alternative plots will result, with the following interpretation:

(a) If the plot is linear, then extrapolation is made linearly to the specified time.
(b) If the plot is initially curved, but with a slope of constant sign, then the straight part is extrapolated linearly to T_1 say. The differences between the linear extrapolation and the measured temperatures are then plotted on a log/linear graph, and extrapolated to give T_2 say. The final temperature is then $T_1 + T_2$.
(c) If the plot has a slope which changes sign initially, then a similar double extrapolation is made, and the final temperature is $T_1 - T_2$.†

A method of superposition is specified in IEC Publication 279 to allow the measurement of winding resistance and hence winding temperature while a motor is energized and on load. Either a bridge method or a volt-ampere method can be adopted, using a principle of d.c. injection. Combinations of capacitors, reactors, and transformers are used to separate the paths for the a.c. power and the d.c. test signals. It should be noted, however, that as no extrapolation is required, because measurement of resistance is made before switch-off, then the temperature rise recorded will be higher than that associated with any of the other basic methods of measurement. The IEC Standard states that the difference might be between 5°C and 25°C. Hence while the method is of value and interest, the recorded value cannot be directly compared with those from any other method.

5.2.11.9 Enclosures. Two basic sets of tests are necessary, one to test for contact with live or moving parts and for ingress of solid foreign bodies, and one to test for harmful ingress of water. Both sets of tests are described in BS 4999: Part 20.

The first set of tests regarding contact with live or moving parts and for ingress of foreign solid bodies, to be carried out with the motor stationary, is as follows (the numbers referring to the enclosure classification scheme):

1st number	*Test*
0	No test.
1	Sphere 52·5 mm diameter with a force of 50 N cannot enter enclosure.

† For case (c), IEC 34-1 recommends that the maximum temperature rise attained after switch-off should in general be taken as the final value.

1st number	*Test*
2	Live—test finger at 40 V or more, no circuit. moving (a) test finger not able to touch moving parts. (b) sphere of 12·5 mm diameter cannot enter enclosure.
4	Wire of 1 mm diam cannot enter enclosure.
5	Talcum powder must not enter the enclosure in harmful quantities under specified conditions of pressure, time and dust concentration.

The second set of tests regarding harmful ingress of water is as follows:

2nd number	*Test*
0	No test.
1	Dripping water at 3 mm/min for 10 min (normal operating position).
2	Dripping water at 3 mm/min for 10 min (motor at 15° from normal).
3	Spraying water at up to 60° from vertical for 10 min, at 8×10^4 N/m^2, <9 litres/min, 1 m distance.
4	As 3, but splashed in every direction.
5	Splashed in all directions from a nozzle of 12·5 mm inside diameter, 3×10^4 N/m^2 at 3 m, 50 litres/min, $\not<10$ min.
6	As 5, but at 10^5 N/m^2 and 100 litres/min.
7	Total immersion for 30 min. *or* internal pressure of 10^4 N/m^2 and check for leaks.
8	By agreement.

Tests for 1–4 may be replaced by suitable examination of drawings. The tests are normally carried out with the motor stationary, except for tests for 4–6 when the motor may be running. After the appropriate test, a check has to be made of the amount of water in the motor. All live parts must be dry (bearing in mind that some internal condensation may have occurred). Then, if the motor was tested when stationary, it is necessary to run it on no-load at rated voltage for 15 min, and then subject it to the high voltage test at 50 per cent of the normal test voltage (but not less than 125 per cent rated voltage). If the motor was tested running, then only the high voltage test is required.

The specification gives useful sketches of apparatus suitable for subjecting motors to the various tests, and these sketches should be referred to for guidance.

5.2.11.10 Losses and efficiency. The determination of motor efficiency by an output/input method is recognized as unreliable, except for low efficiencies, because small percentage errors in power measurements can have important effects on the recorded efficiency. Hence in BS 269, it is recommended that a method of summation of losses be used, in which any error in measurement of losses has a less significant effect on the recorded efficiency. This specification simply itemizes the various losses in a.c. and d.c. motors, defines the derivation of the losses and makes certain recommendations regarding estimation of stray load loss.

IEC 34–2: 1972 describes in detail the necessary tests to be performed in order to determine the losses and efficiency. This Standard will be taken as the basis for BS 4999: Part 33 at some future date, to replace BS 269.

5.2.12 *Tolerances*

The permissible tolerances on motor performance are given in BS 4999: Part 69. Tolerances on dimensions are given in the appropriate Standards (see section 5.2.7).

Performance tolerances are quoted for:

(a) Efficiency—by summation of losses and by input–output test.
(b) Declared total losses.
(c) Induction motor power factor and speed.
(d) d.c. motor speed and variation of speed (shunt, separate, series, and compound excitation).
(e) Speed of a.c. motors with shunt characteristics.
(f) Locked rotor torque.
(g) Pull-out torque.
(h) Starting current of induction and synchronous motors.
(i) Moment of inertia or stored energy constant.

5.2.13 *Brushes and brush gear*

Two British Standards are in existence regarding brushes and brush gear namely BS 96: 1954 (inch units) and BS 4362: A1: 1968 (metric units). The latter is similar to IEC 136.

Inch units for brush and brush gear dimensions are decreasing in usage and will presumably eventually be discarded. Hence it is only necessary to consider BS 96 in outline. The Standard defines the brush dimensions as length (at 90° to commutator axis), width (along commutator axis), and thickness (tangential to commutator surface). It specifies dimensions and tolerances of brushes, brush holders, flexibles, and terminal shoes. Brushes are designated by the letter C (for commutator) or S (for slip-ring), followed by four digits; the first two give the width in sixteenths of an inch and the second two similarly give the thickness. To limit the number of standard sizes, brushes for slip-rings are tabulated in five groups with increasing width to suit a range of five slip-ring widths.

The metric dimensions in BS 4362: A1 in general correspond to those in IEC 136. The British Standard in addition specifies a range of longer brushes suitable for constant-force brush-holders (not yet agreed internationally). The main dimensions of a brush are defined as tangential, axial and radial dimensions, corresponding to thickness, width, and length respectively in BS 96. Dimensions and tolerances are specified for brushes, brush holders, flexibles, and brush terminations (terminal shoes). The standard values for tangential and axial dimensions range from 3·2 to 50 mm with radial dimensions from 10 to 64 mm. Square brush sections, i.e., equal tangential and axial dimensions, are not recommended. For constant force brush-holders, a range of brushes with radial dimensions up to 125 mm is specified. Values are also specified for chamfer dimensions, contact bevel angles, top bevel angles, and spring pressure area. Finally a detailed nomenclature is given, relating to dimensions, angles, surfaces, brush shapes and types, types of connections and terminals, slip ring and commutator types and irregularities, commutator wear and marking patterns, brush disposition, and voltage drops through the brush system.

A recent IEC Standard 467 (1974) describes test procedures for determining certain physical properties of carbon brushes. A voltmeter-ammeter method is described to measure the electrical resistance of the brush/flexible connection. A test method is given to measure the pull strength of tamped or moulded connections. The Standard describes the test equipment, procedure for measurement and interpretation of the results. There is no equivalent British Standard.

5.2.14 *Information required by supplier*

To assist the purchaser in obtaining the correct motor for a given application, BS 4999: Part 31 lists particulars to be supplied to the manufacturer when inquiring for or ordering an electric motor. In addition, other items are mentioned in various other specifications. The general data required are as follows:

(a) Site and operating conditions, i.e., maximum cooling-air or cooling-water temperature, altitude, limits of voltage, frequency, current, and speed (section 5.2.3).
(b) British Standard number.
(c) Type of enclosure (section 5.2.4).
(d) Motor duty (section 5.2.1).
(e) Method of earthing the system to which the motor will be connected.
(f) Particulars of tests (section 5.2.11).
(g) Voltage drop at starting, where starting kVA or torque is important (section 5.2.2.1).
(h) Sound power level (section 5.2.6).
(i) Mounting arrangement (section 5.2.7.3).
(j) Direction of entry to terminal box for frame size above 200L.

For an induction motor, the following additional information is required:

(a) Supply frequency, number of phases, and voltage or voltage range.
(b) Rated output (kW) and speed at rated output.
(c) Direction of rotation and phase sequence.
(d) Type of rotor and collector ring gear.
(e) Method of starting and any limitations on starting current.
(f) Nature of the load, inertia of moving parts, number of starts during a specified period.

In addition, for d.c. motors information is required on rated output, rated voltage and range, speed at rated output, direction of rotation, excitation (shunt, series, compound or separate), starting torque, kinetic energy of load, starting frequency, and harmonic content of the supply.

5.2.15 *Terminal identifications*

The internationally agreed system of terminal markings in IEC 34-8, 1972, differs radically from that presently in use in the UK, namely BS 822: Part 6: 1964. The former is less extensive as it excludes a.c. commutator motors. A quick comparison of the two systems can be obtained from the following table.

		IEC 34-8	BS 822: Part 6
	Three-phase	stator U, V, W; rotor K, L, M	Primary A, B, C; secondary D, E, F
	Neutral	Q	N
a.c.	Two-phase	stator U, V; rotor K, L	Primary A, B; secondary D, E
	Single-phase main winding auxiliary excitation	 U Z F	 A Z X
d.c.	Armature Series Shunt Separate Commutating	A_1A_2 D_1D_2 E_1E_2 F_1F_2 B_1B_2	A AA Y YY Z ZZ X XX H HH

Both documents describe the methods of determining the direction of rotation depending on the a.c. phase sequence or the d.c. polarity.

These radical differences between current British and European terminal identifications will be resolved when BS 4999: Part 3 is published. This will supersede BS 822: Part 6 and will be based on IEC 34-8. In addition it will include provision for the colour marking of leads of small power a.c. and d.c. motors as an alternative or in addition to the symbols.

5.3 Motors for Particular Applications

In order to describe the performance characteristics and requirements of a particular type or design of motor, or of a motor for a particular application, it is necessary to combine various features of performance, etc, from section 5.2 above and to add specific special requirements for the type of motor or application in question. The unique requirements of a particular motor are determined by the nature of the supply, the design of the motor, the characteristics of the driven load, and the total environment. These together determine the variations necessary in terms of temperature rise, enclosure, starting conditions, etc.

A considerable number of Standards exist for motors for particular applications. The following sections describe the salient features of these.

5.3.1 *Motors for general purposes and miscellaneous applications*

The basic Standard for industrial-type motors at present is BS 5000: Part 99. This covers the range of motors previously specified in BS 2613: 1970 and is based on IEC 34-1 with the addition of some features such as classified starting performance. BS 2613 is to be withdrawn in 1976–7. The basic performance features, etc, are selected from BS 4999, as summarized in section 5.2 above. Where alternatives exist, e.g., starting characteristics, enclosure, noise level, etc, it is obviously essential that these are stated and agreed at the time of inquiry and order.

BS 5000: Part 99, will be replaced by Part 10 eventually, to specify motors for general purposes. This will state a specific combination or alternative combinations of parameters, to remove the need to describe these individually in each instance.

A draft of BS 5000: Part 10, General Purpose Induction Motors with Standard Dimensions and Outputs, was under review at BSI during 1974. This draft covered motors for three-phase 50 Hz supplies at rated voltages up to and including 660 V, for frame sizes as set out in BS 4999: Part 10. Outputs were allocated as in BS 3979 (see table 5.1) with the addition of frame numbers D56, D63, and D71 for ratings from 60 W to 550 W for 3000 and 1500 rev/min t.e.f.v. cage motors. General performance to BS 4999: Part 99 was specified (see above) with starting performance according to Design B and temperature rises in accordance with BS 4999: Part 32 (see table 5.3).

The position of the terminal box for foot-mounted induction motors was centered on the motor centre line on the right-hand side looking at the drive end for frame sizes 80 and above; for frame sizes 71 and below, the terminal box was shown on top of the motor. The draft stresses the need for the supplier to be advised of the shaft axis (whether horizontal or vertical) and, for frame sizes above 200L, the direction of entry to the terminal box.

5.3.2 *Small-power motors*

Small-power motors are defined as those with a continuous rating (or equivalent continuous rating) up to and including 0·75 kW per 1000 rev/min, with a rated voltage up to 250 V d.c. or single-phase a.c., and up to 650 V three-phase a.c. Such motors are specified in BS 5000: Part 11 (previously in BS 170). The type of enclosure is limited to either open, protected, drip proof, totally enclosed, hose proof, weather proof, submersible or flameproof. Also the method of cooling is restricted to either free convection; self ventilated; totally enclosed free convection; totally enclosed fan ventilated; drip proof, air over motor; or totally enclosed, air over motor.

If the motor is designed for general-purpose applications, then it will have a continuous rating. Alternatively, the motor might be matched to a particular application, and be assigned a short-time rating, preferably either 5, 10, 15, 30 or 60 minutes. The other preferred parameters are:

d.c. voltage	100–250 V
a.c. voltage	240 V single phase 415 polyphase
a.c. frequency	50 Hz
number of phases	1 or 3
output	2·5, 4, 6, 10, 16, 25, 40, 60, 80, 90, 100, 120, 180, 250, 370, 550, 750, 1100, 1500 or 2200 W
speed	Approximately 950, 1425 or 2850 rev/min (except for series motors), somewhat slower for shaded-pole motors.

5.3.2.1 Temperature rise. The allowable temperature rises of windings, iron parts, slip-rings, and commutators for ventilated motors are similar to those given in table

5.3. The main differences are an extra 5°C allowed for Class B and F insulated windings (measured by resistance), 5°C or 10°C lower for iron parts in contact with insulated windings, and up to 15°C extra for commutators and slip rings. In addition, an extra 5°C is allowed for windings with Class A, E or B insulation if the motor is totally enclosed. The corrections for maximum coolant temperatures other than 40°C are as given in table 5.4, except that the temperature range above 40°C is taken as 40–100°C rather than 40–60°C. Corrections for altitudes between 1000 and 4000 m are not necessary if the test altitude differs from the site altitude by 1000 m or less. If the difference is greater, then the allowable temperature rise is reduced by 1 per cent for each 100 m in excess of 1000 m difference between site and test altitude.

5.3.2.2 Momentary overload. The momentary overload capability of small-power motors depends considerably on the type of motor. The specified values are given in table 5.7.

Table 5.7 Momentary overload for small power motors

Type of machine	Minimum excess torque for 15 seconds
	percentage of full load
split-phase	60
capacitor start induction run	60
capacitor start capacitor run	60
capacitor start and run	25
shaded pole	10
a.c./d.c. and series commutator	50
polyphase cage induction motors wound for 8 poles or more or rated 180 W or below	60
polyphase cage induction motors wound for fewer than 8 poles and rated above 180 W	100
d.c.	60

NOTE: Greater excess torques may be obtained for many of the types of motors listed.
These depend upon the number of poles, speed, output, etc. and should be subject to agreement between the manufacturer and the purchaser.

5.3.2.3 Tests. The specified tests are basically as set out in section 5.2.11, with the exception that the HV test voltage has to be between 25 and 60 Hz, and that for rated voltages of 100 V or more, the test voltage has a minimum limit of 1500 V. Also, for measurement of torque, the tests are started with the motor at ambient temperature ($\nless$ 15°C).

5.3.3 *d.c. mill-type motors*

In British Standard Specifications for mill motors, that published by the Association of Iron and Steel Engineers of the USA (AISE Standard No. 1) has been the basis for dimensions for many years. The present British Standard, BS 4362: Section B3: 1969, aligns closely with the 600 series of the American Standard. It relates to foot-mounted motors rated between 5 and 500 h.p. (3·75 to 375 kW). The rating is assigned on a short-time basis for one hour for a totally enclosed motor. This rating is available continuously if the motor is forced ventilated.

The maximum allowable temperature rise of the windings above a 40°C ambient temperature is 75°C, measured by thermometer for Class B, F or H installations. The rated voltage is either 230 V or 460 V. The motor will either be totally enclosed or duct-ventilated.

The frame sizes are designated by the letter 'M' and a number, not directly related to the frame dimensions or the assigned rating. The latter are tabulated in table 5.8.

Table 5.8 Horsepower outputs assigned to frames for mill motors

Frame size	Output
	h.p.
M 2	5
M 602	$7\frac{1}{2}$
M 603	10
M 604	15
M 606	25
M 608	35
M 610	50
M 612	75
M 614	100
M 616	150
M 618	200
M 620	275
M 622	375
M 624	500

For each rated output, a rated base speed is assigned for series, shunt or compound-wound motors along with a speed range for shunt-wound motors. Separate values are quoted for 230 V and 460 V motors, the latter being somewhat higher.

No specific performance parameters are quoted, e.g., noise and vibration, starting torque or current. The main purpose of the Standard is to ensure interchangeability of dimensions. These are tabulated for fixing, shaft, and key and key-way. In addition, overall dimensions are quoted. Motors are required to be double-ended, i.e., an output shaft at each end, with a tapered shaft extension, and the terminal arrangement on the left-hand side looking at the commutator end.

A similar specification has been under consideration in IEC for some years and a draft has been debated during 1974–75. An IEC Standard on mill type electric motors is therefore possible during the late 1970s.

5.3.4 *Motors for battery-operated vehicles*

Direct-current motors for use in battery electric vehicles, either for the main drive or for auxiliary applications (hydraulic compressors, ventilation fans, etc) are specified in BS 1727: 1971 (there is no IEC equivalent). This covers industrial trucks and tractors, and road vehicles. Motors to this Standard are designed for operation in an ambient temperature between −10°C and +40°C with a nominal voltage of either 12, 24, 36, 48, 72, 80 or 96 V. Motors for road vehicles have an enclosure type IP24, while those for industrial trucks and tractors have type IP20, IP44 or IP54 (see section 5.2.9). The motor rating is either continuous or on a short-time basis (60, 15 or 5 minutes), or a one-hour rating followed by a 200 per cent torque-overload for 5 minutes (for driving motors only).

The allowable temperature rises of windings in battery-vehicle motors on test are considerably higher (by between 30 and 40°C for armature windings, and by between 40 and 55°C for field windings) than the corresponding windings in normal industrial-type motors (see table 5.3). Experience has shown that these higher temperatures result in an adequate insulation life, bearing in mind the particular application (i.e., operating life, ambient temperature generally much lower than the allowable 40°C, test carried out without cooling corresponding to that produced by the motion of the vehicle).

5.3.4.1 Tests. Characteristic curves of motor performance are plotted showing the variation of speed, output, torque, and efficiency against armature current. Tolerances on speed are given depending on rating and loading.

Checks on commutation quality are made at normal working temperature, both over the full working range and also at minimum field strength/maximum speed, one-hour rated current, and finally at the maximum current, each condition being maintained for 30 s. The commutation quality will be sufficient to prevent sparking of a level sufficient to damage the commutator or brushes. An overspeed test requires that the motor withstand 1·2 times maximum service speed for two minutes when hot.

The high voltage test is made at 1000 V r.m.s. for 1 minute between 25 and 100 Hz, or at 1500 V d.c. The insulation resistance to earth should exceed 2 MΩ at 500 V.

5.3.5 *Motors for marine applications*

The current British Standard for motors for use in ships (BS 2949: 1960, there is no IEC equivalent) does not differ greatly in content from the general requirements discussed in section 5.2 above. Indeed, this Standard is not in extensive use, the general philosophy being to utilize good-quality general-purpose industrial motors.

The main differences or additions are outlined below, relating in general to particular constructional features.

Types of ships are divided into Group 1 (ocean-going and tropical climates) and Group 2 (all other than Group 1). Two service-condition classes are considered; service-condition 1 relates to motors in machinery spaces and on weather decks in Group 1 ships; service-condition 2 relates to Group 1 machine-tool motors, all others in Group 1 except those in Condition 1, and all motors in Group 2. The significance of these conditions is simply in terms of the allowable temperature rise. A motor with service-condition 1 has to withstand higher ambient temperatures, e.g., on deck in the tropics.

Motor ratings are either continuous or short-time (1 hour or $\frac{1}{2}$ hour). To avoid ridging of commutators after prolonged use, it is recommended that brushes should be staggered longitudinally in multi-polar d.c. motors, with a positive brush in line with a negative brush on an adjacent brush-arm.

The Standard pays particular attention to bearing lubrication under rolling and trim conditions at sea. It also recommends consideration of design features to prevent overlubrication, leaking of lubricant on to insulation, and bearing currents, and to achieve simple bearing replacement. Insulation materials are recommended to have non-hygroscopic and anti-tracking properties, with windings being treated to withstand the effects of moisture, sea air, and oil-vapour.

Temperature rise limits are about 10°C less than those in table 5.3, but are based on a 50°C ambient for service-condition 1. An extra 10°C rise is allowed for service-condition 2 if the ambient temperature is never greater than 40°C. For water-cooled motors, with a water temperature not greater than 30°C, then the limits are increased by 20°C.

Momentary overload limits are greater than those given in section 5.2.2.4 above for motors of 50 h.p. or less. Depending on the rating, the percentage excess torque ranges from 50 to 100 for d.c. motors and from 60 to 100 for a.c. induction motors.

5.3.6 *Motors for road and rail vehicles*

Electric motors employed in electrically propelled rail vehicles and large road vehicles (trolley-buses and trams) are specified in IEC 349, 1971 and BS 173: 1960. These do not cover motors in small battery-supplied road vehicles (see section 5.3.4) or small ancillary motors such as windscreen wiper motors. IEC 349 replaces IEC 48, 101 and 102. The British Standard is very similar to IEC 349; the latter is outlined below, and the differences from BS 173 are indicated.

Four main types of supply are recognized, namely:

(a) direct current (including rectified three-phase a.c.);
(b) pulsating current (i.e., rectified single-phase a.c.);
(c) single-phase a.c.;
(d) multi-phase a.c.

Motors are classified either as traction motors (used for propulsion) or auxiliary motors (e.g., for compressors or blowers).

The four types of rating considered are continuous, one-hour, intermittent (a sequence of identical loading cycles) and equivalent rating (continuous or short-time rating equivalent to intermittent as regards temperature rise). (In BS 173, ratings are specified as continuous for ventilated traction motors, 1 hour for totally enclosed traction motors, and either continuous or short-time for auxiliary motors.)

A precise series of type tests and routine tests are specified. It is only possible here to mention certain aspects of these and the reader should refer to IEC 349 for complete details. Some of these require agreement between manufacturer and purchaser regarding test conditions. The allowable temperature rises of windings (as measured by resistance) are tabulated below for different classes of insulation:

Insulation Class	E	B	F	H
Armature (rotating) winding	105	120	140	160
Field (generally stationary) winding	115	130	155	180

The maximum temperature rise of commutators or slip-rings is 105°C (90°C in BS 173). These rises correspond to test values obtained with normal ventilation, but not including that due to the motion of the vehicle, except for totally enclosed traction motors.

If industrial-type motors of standard construction are used for auxiliary applications, their temperature rises should be in line with those outlined in section 5.2.4.

(BS 173 allows temperature rises up to 10°C greater than those tabulated above for totally enclosed motors. It also specifies maximum limits of temperature in service. The normal maximum winding temperatures as obtained under the most adverse loading conditions related to normal service are the tabulated values increased by 25°C. The peak temperatures as obtained under emergency conditions are 15 to 20°C higher still.)

For routine tests, the measured temperature rises can be within 8 per cent (with a maximum of 10°C) of the average values obtained from initial type tests. The average is taken from the first four tests for up to ten motors, and from the first ten tests for more than ten motors.

There is a slight difference in the determination of the final temperature. IEC 349 recommends that a cooling curve should be plotted, on a log/linear scale, taking the first reading within 45 seconds of switch-off and this curve is extrapolated to switch-off. BS 173 states that a similar cooling curve should be obtained and a straight line drawn up to two minutes from switch-off and then extrapolated to switch-off.

Tests are described to determine the ability of traction commutator-motors and auxiliary motors to withstand the appropriate frequency of starting. Direct-current traction motors and pulsating-current motors are tested four times at up to 1·7 times the continuous rated current for 15 s at standstill with a 5 min interval between tests. Single-phase traction motors are tested at reduced speed for 1 min in each direction of rotation. Auxiliary motors have to withstand five successive starts at minimum voltage and five at maximum voltage with 2 min intervals. (BS 173 has two sets of ten starts.)

Checks on commutation performance are made under steady-state conditions and also during supply interruption and restoration (for motors supplied from a contact system). The steady-state tests are made over a range of currents and directions of rotation, holding each test condition for 30 s The transient performance is checked by interrupting the normal rated current for 1 s, once for traction motors and five times for auxiliary motors. (BS 173 requires the test to be made six times. In addition it specifies a short-circuit test for compound-wound auxiliary motors running on no-load; the supply is disconnected and the motor short-circuited, the sequence being repeated twice at 5 min intervals).

High-voltage tests are made at somewhat higher test voltages than those specified in section 5.2.11.4, namely 2·25 times nominal voltage plus 2000 V for motors fed from the line, and twice nominal voltage plus 2000 V for other motors.

The general performance of motors for road and rail vehicles is described by a series of characteristic curves giving the variation of torque, speed, and efficiency (and power factor for a.c. motors) with current. These curves are related to a reference temperature of 115°C. Precise details are given regarding the performance limits within which test values are required for type tests and routine tests. Useful details are given regarding methods of determining efficiency. (BS 173 has a reference temperature of 75°C for Class A and E insulation, and 110°C for Classes B, F, and H. Also, it states that characteristic curves are not generally required for auxiliary motors.)

5.3.7 *Motors for power station auxiliaries*

The performance requirements of motors for driving essential power station auxiliaries, e.g., boiler feed pumps, draught fans, pulverizers, etc (excluding ventilation fans, drainage pumps, etc) are somewhat more exacting than for normal industrial motors. They are required to continue operating satisfactorily on supply variations outside the normal industrial limits, such as might occur during a severe system disturbance or fault condition. The particular requirements are specified in BS 5000: Part 40: 1973 (there is no IEC equivalent); this itemizes the special requirements additional to those in BS 5000: Part 99 (see section 5.3.1 above).

The type of enclosure for indoor motors has to be not less than IP 12 (see section 5.2.9). For outdoor use, weather protected motors are required.

Motors have to be able to supply the starting and normal load requirements at 0·8 times rated voltage for a duration of at least twice the normal starting period.

Motors have to be able to operate continuously at any frequency between 0·96 and 1·02 per unit (p.u.). They also have to be able to withstand short emergency conditions with frequency between 0·94 and 0·96 p.u., allowing a pro rata reduction in voltage and load.

If simultaneous variations of voltage and frequency occur, then steady-state operation has to be satisfactory at the following combinations:

p.u. frequency	*p.u.voltage range*
1·02	0·95–1·07
1·01	0·94–1·06
0·98	0·93–1·05
0·96	0·91–1·03

These voltage and frequency variations also apply to d.c. motors operating from rectified a.c. supplies.

Regarding starting characteristics, Design B is specified, i.e., normal current and torque at start. The momentary excess torques are greater than those specified in section 5.2.2.4, being as follows:

a.c. induction motors	0·85
d.c. motors	0·85
synchronous motors and synchronous-induction motors	0·65

5.3.8 *Motors for other applications*

The performance and principal design features and characteristics of turbine-type machines are specified in IEC 34-3 and BS 5000 Part 2. While these cover synchronous motors as well as synchronous generators, synchronous compensators, and exciters, the details of these are generally outside the scope of this chapter.

5.4 Application Requirements

Electric motors constitute one of the major loads on industrial power systems. The interaction of a motor and a particular supply will affect the performance of the motor and the quality of the supply to other users. Because of these two factors the UK Electricity Council has drawn up important Engineering Recommendations relating to electric motor installations. These are referred to in the following sections. The reader is required to discuss particular circumstances with his local Electricity Board. The following sections merely indicate the scope of the Recommendations.

5.4.1 *Starting conditions*

Electric motors draw their largest current at starting and this inevitably results in a reduction in supply voltage. The magnitude of the voltage dip depends on the type and size of the motor, the method of starting, and the supply impedance. It is important that this magnitude is kept within limits to avoid annoyance to other consumers. The degree of annoyance of a given voltage dip will of course depend on the frequency of occurrence. The reference point is taken as the point of common coupling with other consumers. Recommendations regarding starting conditions are set out in document P13.

Two degrees of frequency are considered, namely frequent starts with less than two hours duration between consecutive starts and infrequent starts with two hours minimum between consecutive starts. The voltage dip allowable with frequent starts is naturally the smaller.

To avoid the need for calculating the voltage dip caused by every motor, the Recommendation gives maximum ratings of motors which can be connected to any particular supply, depending on the nature of the supply and the load application. The main factor in this assessment is the rated kVA of the transformer if the installation is supplied from an on-site substation. The maximum motor rating for infrequent starting allowable without calculating the dip is 100 h.p. for infrequent starting with a 1 MVA transformer, and 30 h.p. for frequent starting (assuming direct-on-line connection). Allowable motor ratings for smaller transformers are reduced approximately in proportion to the transformer rating.

For motors with greater ratings, it is necessary to calculate the actual voltage dip during starting at the point of common coupling, from a knowledge of the starting current and the supply impedance. To determine the latter, the Recommendation tabulates impedances for cables, overhead lines, and transformers of various sizes, ratings, and voltages. If such a calculation indicates that the motor rating is too great for the particular system, then it will be necessary to consider the use of star/delta or auto-transformer starting methods.

5.4.2 *Limitation of harmonic currents*

Modern forms of electronic control of electric motors result inevitably in harmonic currents and voltages being generated both in the motor windings and also back into the main supply. The harmonic currents in the supply can result in:

overloading of power factor correction capacitors;
series resonance between such capacitors and transformer reactance;

noise in telephone circuits;
harmonic volt drops in the supply which can lead to maloperation of ripple control systems.

Because of these supply disturbances, two Engineering Recommendations have been drawn up. These are G5/2 relating to control systems operating with 6 or more pulses, and G11 for 1, 2 or 3 pulse systems. These Recommendations set limits of harmonic current based on the largest value of r.m.s. current at a number of harmonic frequencies. If the harmonic currents occur only in short bursts then the limits can be increased. A second set of limits is given for the maximum load (based on a half-hour maximum demand) for a naturally commutating convertor.

If more than one convertor operates at the same supply point, then the harmonic currents are initially summed arithmetically to check against the current limits. If these limits are found to be exceeded, then the phase shift between the harmonic currents can be allowed for in the estimation. This phase shift is particularly relevant when there is a reactance between the convertors or with large resistive loads.

For large convertors rated greater than 1 MW, then it is likely that measurements will be required both of the harmonics already existing in the supply at the point of common coupling and also of the harmonics generated by the convertor. For very large convertors, e.g., greater than 30 MW 24-pulse, then a full network study of the harmonic content would be required.

5.5 Other Related Standards and Codes of Practice

It is virtually essential that any motor installation includes motor starters to switch the motor on and off and to protect it against overloads. The special requirements of such equipment are that they are able to withstand the high current surges that occur during switching-on or re-switching a motor (typically 4 to 8 times rated current) and that a motor-current or motor-temperature monitoring device be incorporated with appropriate trip facilities. It is most important that the motor and its starter are matched both in terms of the currents to be switched and the thermal analogue of the motor. A number of BSI and IEC publications exist for motor starters and these go some way towards helping the purchaser achieve a satisfactory matching of equipment.

The basic British Standard which has served for many years is BS 587: 1957. This is now in process of being replaced by a series of new Standards under the title of BS 4941, 'Motor starters for voltages up to and including 1000 V a.c. and 1200 V d.c.' The 'equivalent' IEC Publication is IEC 292. Parts 1 of both documents deal with direct-on-line (full voltage) a.c. starters, and they are quite closely aligned. IEC 292, Part 2 is concerned with reduced voltage and star-delta starters for a.c. motors. An equivalent Part of BS 4941 has not yet been published.

Of these documents BS 587 covers the widest range of equipment. It applies to hand-operated and automatic equipment for starting and/or controlling a.c. or d.c. motors rated up to 11 kV a.c. and up to 650 V d.c. It also applies to resistors and auto-transformers included in starters or controllers. It specifically excludes field rheostats and equipment for traction and aircraft applications. Liquid starters and controllers as applied to rotor circuits are dealt with separately in BS 140: 1957. The latter document covers rheostatic equipment to be used in rotor circuits of induction motors and synchronous-induction motors, including equipment for d.c. dynamic braking, reverse current braking, and slip regulation. It excludes similar equipment for use in stator circuits of a.c. motors and for use with d.c. motors. Part 3 of IEC 292 covers similar equipment.

BS 4941: Part 1 deals exclusively with direct-on-line a.c. starters, including reversing starters, rated up to 1000 V. Such starters are not designed to provide interruption of short-circuit currents, and an Appendix comments on the requirements of co-ordination with short-circuit protective devices. The Standard does not apply to static starters or to motor-protection equipment operating with built-in thermal devices in the motor.

Although thermostats have been used for many years to protect motors, they were not previously covered by a British Standard. The increasing use of thermistors for this purpose emphasized the need, resulting in the preparation of BS 4999: Part 72. This Standard sets out the requirements to be met by motors with built-in protection from thermistors and thermostats. (There is no equivalent IEC Publication.) The Standard applies only to cage induction motors other than those complying with BS 170, and does not apply to devices which are designed to carry the motor current. It includes provision for using thermal devices either to operate a warning or to trip the motor. The Standard describes various tests to be carried out on thermistors and thermostats as components, on thermistor control units, and on the motors when fitted with the devices.

Two codes of practice issued by the BSI are relevant in the industrial application of electric motors. These are CP 1003 dealing with electrical apparatus and associated equipment for use in explosive atmospheres of gas or vapour (other than mining applications) and CP 1011: 1961 dealing with the maintenance of electric motor control gear. A third code of practice CP 321.102 describing the installation and maintenance of electrical machines and other equipment has been withdrawn for a number of years, mainly on the grounds of obsolescence; most of the aspects are covered by the UK IEE Regulations for the Electrical Equipment of Buildings (Fourteenth Edition, 1966, reprinted May 1974). These Regulations relate to positioning of the motor starter, automatic restart prevention, excess current protection, and cable ratings.

5.6 Definitions

A considerable number of British Standards contain a section giving the definitions of certain special terms used within that Standard, e.g., BS 1727, BS 4362 A1, BS 4999: Part 51. To avoid this need in general and to prevent possible conflict of definitions, a single glossary of terms has been drawn up under BS 4727. Part 2 of the Standard lists terms particular to electrical power engineering, of which Group 03 relates to rotating electrical machines. The terms and definitions in this Glossary are in close agreement with those agreed internationally in IEC 50 (411).

A number of well-known and commonly-used terms are deprecated, and the more important of these are listed below:

Term previously used	*Term in BS 4727*
Alternator	Alternating current generator
Squirrel cage induction motor	Cage induction motor

Term previously used	*Term in BS 4727*
Slip-ring induction motor	Wound-rotor induction motor
Variable-speed motor	Adjustable-speed motor
Kramer system	Kraemer system
Inter (pole)	Commutating (winding)
Squirrel cage	Cage winding
Mush winding	Random winding
Distribution factor	Spread factor
Slip ring	Collector ring
Commutator bar	Commutator segment
Baffle	Air guide
Totally enclosed fan cooled	Totally enclosed fan ventilated
Standstill torque or current	Locked-rotor torque or current

All the important performance parameters are listed and defined and hence the possibility of confusion should be reduced by the use of these terms by both manufacturer and purchaser.

5.7 Summary of Probable Future Changes in Motor Standards

1. BS 2960: Parts 3 and 4 will be replaced by the metric dimensions in BS 3979.
2. BS 2613 will be replaced by BS 5000: Part 99.
3. BS 3979 will be replaced by BS 5000: Part 10 for ratings, by BS 4999: Part 10 for dimensions and by BS 4999: Part 2 for symbols.
4. BS 269 will be replaced by BS 4999: Part 33, the latter being based on IEC 34-2.
5. BS 96 will probably be replaced by BS 4362 A1 with metric dimensions, or a new BS 4999: Part 70.
6. BS 822: Part 6 will be replaced by BS 4999: Part 3, the latter being based on IEC 34-8.
7. BS 5000: Part 99 will be replaced by BS 5000: Part 10.
8. The IEC Standards relating to electric motors will probably be regrouped eventually along similar lines to BS 4999 and BS 5000.

5.8 Complete List of Relevant British Standards as at 31 March 1977

BS 96: 1954	Carbon brushes (parallel-sided for use on commutator and slip-ring machines (1 amendment)
BS 140: 1957	Liquid starters and controllers
BS 229: 1957	Flameproof enclosure of electrical apparatus (last amendment July 1974)
BS 269: 1927	Rules for methods of declaring efficiency of electrical machinery (excluding traction motors) (Confirmed 1957) (latest amendment August 1961)
BS 587: 1957	Motor starters and controllers (latest amendment June 1974)
BS 741: 1959	Flameproof electrical motors for conveyors, coal-cutters, loaders, and similar purposes for use in mines
BS 822	Terminal markings for electrical machinery and apparatus
Part 6: 1964	Terminal markings for rotating electrical machinery (1 amendment)

BS 1259: 1958	Intrinsically safe electrical apparatus and circuits for use in explosive atmospheres (last amendment October 1964)
BS 1727: 1971	Motors for battery operated vehicles (latest amendment July 1974)
BS 2048:	Dimensions of fractional horsepower motors
Part 1: 1961	Motors for general use (latest amendment December 1972)
BS 2613: 1970	Electrical performance of rotating electrical machines (1 amendment) (Proposed for withdrawal)
BS 2949: 1960	Rotating electrical machines for use in ships (1 amendment)
BS 2960	Dimensions of 3-phase electric motors
Part 3: 1964	Flameproof motors
Part 4: 1964	Slide rail dimensions.
BS 3979: 1966	Dimensions of electric motors (metric series) (latest amendment April 1973)
BS 4362:	Rotating electrical machinery
Part A—	General data. Section A1: 1968. Dimensions of brushes and brush holders for electrical machines
Part B—	Applications. Section B3: 1969. D.C. mill-type electric motors (1 amendment)
BS 4727: Part 2	Group 03: 1971. Rotating machinery terminology
BS 4941:	Motor starters for voltages up to and including 1000 V a.c. and 1200 V d.c.
Part 1: 1973	Direct-on-line (full voltage) a.c. starters
BS 4999: 1972	General requirements for rotating electrical machines
Index: 1976	Issue 3—General introduction and index
Part 1: 1972	Definitions
Part 2: 1976	Symbols
Part 4: 1972	Rating plate markings
Part 10: 1976	Standard dimensions
Part 20: 1972	Classification of types of enclosure (1 amendment)
Part 21: 1972	Classification of methods of cooling
Part 22: 1977	Symbols for types of construction and mounting arrangements
Part 30: 1972	Duty and rating (1 amendment)
Part 31: 1972	Service and operating conditions (1 amendment)
Part 32: 1972	Limits of temperature rise and methods of temperature measurement (1 amendment)
Part 41: 1977	General characteristics
Part 50: 1972	Mechanical performance—vibration (1 amendment)
Part 51: 1973	Noise levels
Part 60: 1976	Tests (1 amendment)
Part 61: 1977	Tests of insulation of bars and coils of high-voltage machines
Part 69: 1972	Tolerances
Part 71: 1974	Winding terminations
Part 72: 1972	Built-in thermal protection for electric motors rated at 660 V a.c. and below
BS 5000	Rotating electrical machines of particular types or for particular applications
Index: 1973	Issue 2—General introduction and index

Part 2: 1973	Turbine-type machines
Part 11: 1973	The electrical performance of small power electric motors and generators
Part 16: 1972	Type N electric motors (1 amendment)
Part 40: 1973	Motors for driving power station auxiliaries
Part 99: 1973	Machines for miscellaneous applications
CP 1003	Electrical apparatus and associated equipment for use in explosive atmospheres of gas or vapour other than mining applications
Part 1: 1964	Choice, installation and maintenance of flameproof and intrinsically safe equipment (1 amendment)
Part 2: 1966	Methods of meeting the explosion hazard other than by the use of flameproof or intrinsically safe electrical equipment
Part 3: 1967	Division 2 areas
CP 1011: 1961	Maintenance of electric motor control gear

5.9 Complete List of Relevant International Standards as at 31 December 1975

IEC Standards

IEC 34	Rotating electrical machines
34-1 (1969) Part 1	Rating and performance (Seventh edition)
34-2 (1972) Part 2	Methods of determining losses and efficiency of rotating electrical machinery from tests (excluding machines for traction vehicles) (Third edition)
34-2A (1974)	First supplement. Measurement of losses by the calorimetric method
34-3 (1968) Part 3	Ratings and characteristics of three-phase 50 Hz turbine-type machines (Third edition)
34-4 (1967) Part 4	Methods for determining synchronous machine quantities from tests. (First edition, 1 amendment)
34-4A (1972)	First supplement
34-5 (1968) Part 5	Degrees of protection by enclosures for rotating machinery (First edition)
34-6 (1969) Part 6	Methods of cooling rotating machinery (First edition)
34-7 (1972) Part 7	Symbols for types of construction and mounting arrangements of rotating electrical machinery (First edition)
34-7 (1974)	Erratum
34-8 (1972) Part 8	Terminal markings and direction of rotation of rotating machines (First edition)
34-9 (1972) Part 9	Noise limits (First edition)
IEC 50	International Electrotechnical Vocabulary
50(10) (1956)	Machines and transformers (Second edition)
50(411) (1973)	Rotating machines (First edition)
IEC 72 (1971)	Dimensions and output ratings for rotating electrical machines—Frame numbers 56 to 400 and flange numbers F55–F1080 (Fifth edition)
72A (1970)	Dimensions and output ratings for foot-mounted electrical machines with frame numbers 355 to 1000

IEC 136:	Dimensions of brushes and brush-holders for electrical machinery
136-1 (1962) Part 1	Principal dimensions and tolerances (First edition)
136-1A (1972)	First supplement
136-2 (1967) Part 2	Complementary dimensions of brushes—Terminations of brushes (First edition)
136-2A (1972)	First supplement
136-2B (1973)	Second supplement
136-3 (1972) Part 3	IEC technical questionnaire for users of carbon brushes (First edition)
IEC 279 (1969)	Measurement of the winding resistance of an a.c. machine during operation at alternating voltage (First edition)
IEC 292	Low-voltage motor starters
292-1 (1969) Part 1	Direct-on-line (full voltage) a.c. starters (First edition)
292-1A (1971)	First supplement
292-1B (1973)	Second supplement
292-1C (1975)	Third supplement
292-2 (1970) Part 2	Reduced voltage a.c. starters: Star-delta starters (First edition)
292-3 (1973) Part 3	Rheostatic rotor starters (First edition)
IEC 349 (1971)	Rules for rotating electrical machines for rail and road vehicles (First edition)
IEC 356 (1971)	Dimensions for commutators and slip-rings (First edition)
IEC 413 (1972)	Test procedures for determining physical properties of brush materials for electrical machines (First edition)

ISO Standards

ISO 2373 (1974)	Mechanical vibration of certain rotating electrical machinery with shaft heights between 80 and 400 mm. Measurement and evaluation of vibration severity.

References

1. —. 'Motors for Europe', *Electrical Equipment*, 28–30 (March 1973).

2. H. Bradshaw, 'Trends in new standards for rotating machines', *Electrical Review*, 237 (15 March 1974).

3. L. G. Roth, 'Difficulties beset standards compilers', *Electrical Times*, 7 (23 May 1974).

4. H. A. Moncur, 'Reflections on changing standards in motor control gear', *Electronics and Power*, 817–820 (17 October 1974).

6 Linear Motors

6.1 Essential Differences Between Linear and Rotary Machines

6.1.1 *Definition*

For the purpose of restricting the coverage in this chapter a linear machine is considered to be a structure which results from the imaginary process of splitting and unrolling a rotary electric machine, as shown in Fig. 6.1. It is further to be understood, however, that although the rotary machine so treated is normally

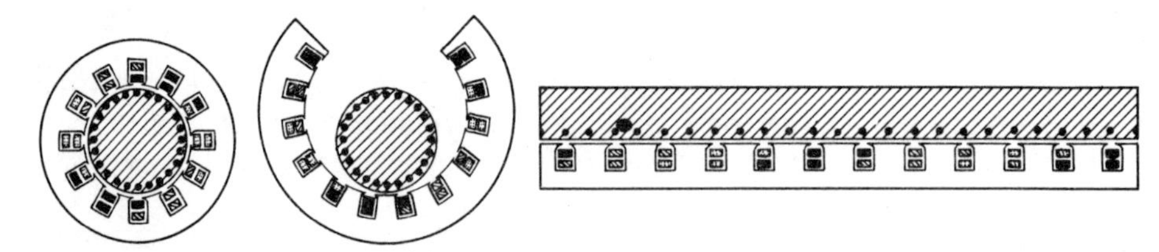

Fig. 6.1 A linear motor can be described by imagining a rotary machine to be split and unrolled

designed for continuous motion, its linear counterpart may also be designed in a form which produces reciprocating motion. Certain recent developments in linear machines have involved topological changes which result in their having no rotary counterpart which has ever been built or is likely to be built for to do so would be to incur severe economic penalties due to the topological differences between linear and rotary forms. Nevertheless, the very fact that a rotary form can be envisaged, whether or not it would be profitable, qualifies its linear form for inclusion here.

6.1.2 *Historical*

The first linear motor to be patented and built is undoubtedly that of Charles Wheatstone, better known for his electrical 'bridge' circuit and for his invention of the five-needle telegraph. Wheatstone's machine was built in 1845, only 14 years after Faraday's discovery of the principles of electromagnetic induction. It is also of interest to note that less than 10 years later, another inventor, likewise better known for his other work, reproduced almost identically the machine of Wheatstone without knowledge of the earlier work. The experimenter was William Henry Fox

Talbot, the first man to take a photograph using a negative from which prints could be made and he is often referred to as the 'Father of photography'.

Until the invention of the induction motor by Nicola Tesla (1888) all the linear machines were of the reluctance type, but inventors were not slow to exploit Tesla's machine in linear form and it is said (albeit so far without confirmation) that a patent was granted to the Mayor of Pittsburgh in 1890 for a linear motor to propel shuttles across weaving looms. Whether confirmed or not, a patent was certainly granted to the Weaver, Jacquard and Electric Shuttle Company in 1895, for a linear motor for this same application.

In 1903–05 the German Zehden[1] made important developments in linear induction motors with a view to their application to railway traction and although his ideas were not taken up commercially until the 1950s he was certainly the inventor of what is now known as the 'double-sided sandwich motor', described fully in the next section.

Between 1905 and 1946 there were many patents on linear motors and their applications although over 90 per cent of these were intended for the textile industry and again, mostly for shuttle propulsion. Many of these patents contained much ingenuity in linear motor topology, many of the modern forms effectively being invented and discarded because their inventors only saw them in the context of weaving looms. One particular arrangement is worthy of note in that it represents the first attempt to combine linear motion by induction with the phenomenon of induction levitation. In 1914 a Frenchman Emile Bachelet suggested a single-sided primary arrangement of coils which would both levitate and propel, and although his system was not laterally stable and needed mechanical side guides, Monsieur Bachelet was able to interest many famous people in his machine, including Hiram Maxim and Winston Churchill, and had the confidence to launch two companies to exploit the idea, one for shuttle propulsion and the other for railway traction. Neither succeeded, and certainly no-one else tried for the next 50 years, for the application to high speed transport is an expensive development requiring perhaps £400m just to try! There is not enough space in one chapter to do justice to all the large-scale developments which are currently in progress in at least seven major countries on this aspect of linear propulsion. In writing a summary of the theory and shapes of various families of machine, reference will, however, be made to the fact that certain new topologies are being developed almost exclusively for the high-speed transport market, where this is the case.

The unfruitful attempts of textile engineers to apply the linear motor which occupied nearly half a century were terminated in 1946 with the building and testing of an aircraft launching linear motor by the Westinghouse Corporation. Known as the Electropult,[2] this machine had a top speed of 225 miles/h ($\approx$362 km/h) and produced a starting thrust of nearly 5 tons ($\approx$5 tonnes). Half speed with aircraft attached was attained in 4·2 seconds from rest.

6.1.3 *Physical phenomena peculiar to linear motors*

At first sight there appear to be very few differences between linear and rotary machines as the result of the splitting and unrolling process illustrated in Fig. 6.1. The linear equivalent of the cast-cage rotor of an induction motor is a conducting 'ladder' of rungs and side bars with the rungs sunk into slots in a laminated steel core. Primary coils can be arranged in slots also, producing a linearly travelling field across the pole surface instead of the more familiar rotating field, and the exact linear counterpart is then as shown in Fig. 6.2.

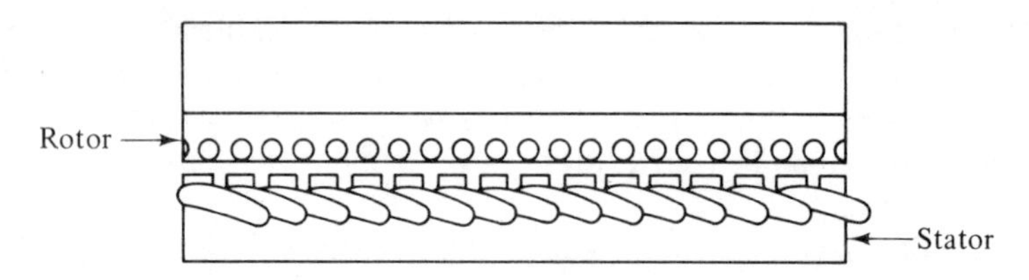

Fig. 6.2 Exact linear counterpart of a cage-type induction machine

A glance at this figure is enough to show perhaps the simplest and certainly the most fundamental difference introduced by the fact that the linear motor has ends—a start and a finish to both primary and secondary members defined in terms of the direction of travel. As soon as movement is allowed to take place a part of the secondary moves away from the area of influence or 'active zone' whilst an equivalent area of primary is left 'bare' to send its magnetic field into free space. The situation is as shown in Fig. 6.3. If the motion continues further, primary and

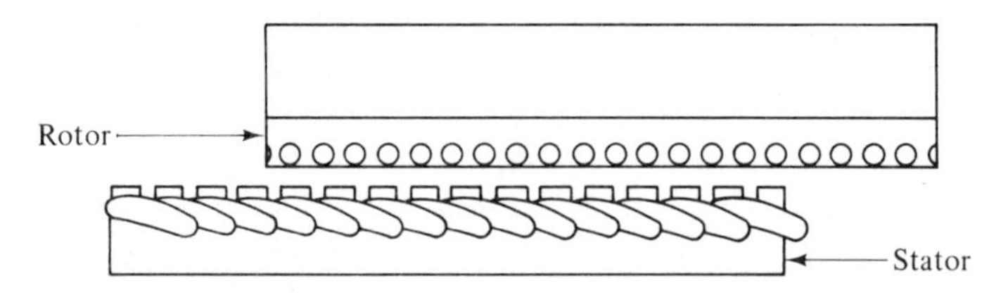

Fig. 6.3 Relative movement of the motor shown in Fig. 6.2 at once reduces the 'active zone'

secondary members separate completely after which no further electromagnetic action can take place. For continued motion it is therefore necessary to elongate either primary or secondary member (but not both) and this at once gives rise to the two classes of linear motor shown in Fig. 6.4, depending on which of the two members is the shorter.

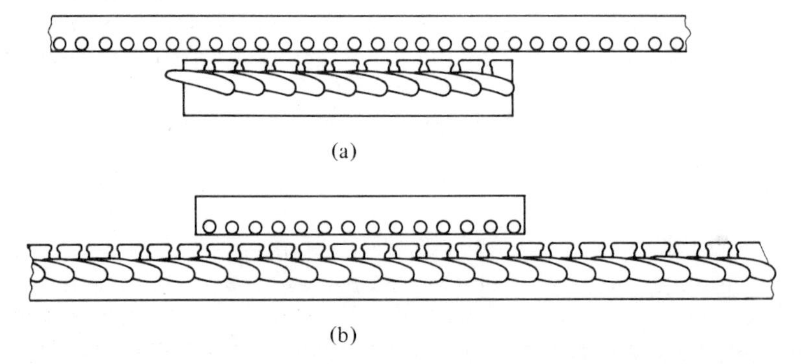

Fig. 6.4 Basic division of linear motors into two classes. (a) Short primary. (b) Short secondary

Everything which has so far been described and illustrated in relation to the induction type of machine applies equally well to any other type of linear motor and it is obvious that there are as many types of linear motor as there are rotary since each of the latter can be split and unrolled. The cheapness and inherently robust nature of the solid-metal secondary of the induction type has made it dominant in the world of electric rotary drives. There is no reason to suggest that this situation will not obtain also in the linear field, in fact the necessity to elongate one member almost demands that such member should be an induction type secondary.

It is also to be noted at this point that either the primary or the secondary can be the fixed member and while a fixed-secondary rotary machine is rarely seen in commercial practice, the linear version is much more regularly used since the secondary is not only the cheaper to produce but is not required to be energized all along its length at all times. Perhaps the only disadvantage of the moving primary system is the need to feed the moving part from a polyphase supply. For short-stroke actuators this is possible using flexible leads but for longer distances a system of brushes and slip tracks (corresponding to slip rings in a rotary motor) is needed. Thus the Electropult was a short-primary, moving primary system with the secondary member (in one of the two machines built) some $\frac{3}{8}$ mile in length, and in the case of the later motor, over a mile long.

The second fundamental change due to linearization arises directly from the need to elongate. Figure 6.4 shows that when there is relative motion between the members, the portions of the longer member which are not at one instant under the active zone will suffer transient phenomena as the edges of the shorter member pass over any such particular portion. As will be shown in the theoretical section, such transients are usually detrimental to performance.

The third main difference between rotary and linear electric machines is that the cylindrical symmetry of the former tends to balance all forces *normal* to the direction of motion, i.e., radial forces, against each other and in a machine with perfect symmetry the net result of all these forces is seen to be zero, no matter how large the force/unit area at any given point may be. Such radial forces arise as the result of at least a part of the primary current being used to drive a magnetic flux across the airgap into the secondary steel. The situation differs little from that in which a horseshoe magnet attracts its keeper, and is, in a sense, not related to any *electromagnetic* action which may take place, e.g., the induction of secondary current, as in an induction machine, the reaction with secondary current as in d.c. or a.c. commutator motors. So large is this magnetic 'pressure' that even minor asymmetry causes the rotary machine designer to worry lest the resultant 'unbalanced magnetic pull' (UMP) should bend the shaft. Shaft sizes are often dictated by such apparent trivia as a shaft keyway, the direction in which the sheet steel of the primary punchings has been rolled, etc.

In a linear machine of the form shown in Fig. 6.2, the whole of the magnetic pull is seen to operate in the same direction and its value may be such as to multiply the apparent weight of the supported member (in a horizontal airgap arrangement) by a factor of 10 or more. This feature was undoubtedly a strong deterrent to those engineers who seriously considered the manufacture of linear motors between the years 1905 and 1945. As will be shown in the next section, however, designers of linear machines can change the shape of component parts to eliminate or at least to minimize this force.

6.1.4 *Linear motor topology*

The first important changes in design between a linear motor designed for a specific application and the machine which results from a simple splitting and unrolling (Fig. 6.1) exercise were undoubtedly those by Zehden mentioned earlier. Although specifically aimed at transport applications, later commercialization was seen to exploit the same designs for quite different and generally lower-speed applications. Ironically, the double-sided sandwich motor of Zehden was finally discarded as unsafe for high-speed transport applications between 1967 and 1973. The fact that its philosophy was progressive entitles it to a full treatment here as the forerunner of many new topologies which were subsequently to establish linear motors as successful commercial machines in their own right.

The arguments leading to the sandwich motor (sometimes called the 'sheet-rotor motor') are illustrated in Fig. 6.5, in which (a) represents the unrolled cylindrical machine with extended ladder-type secondary. The first and major step is shown in (b) in which the secondary conductor has been removed from its slots and accommodated in the airgap. Such relocation, however, demands that the airgap be made much larger to allow for the space occupied by the conductor as well as the normal

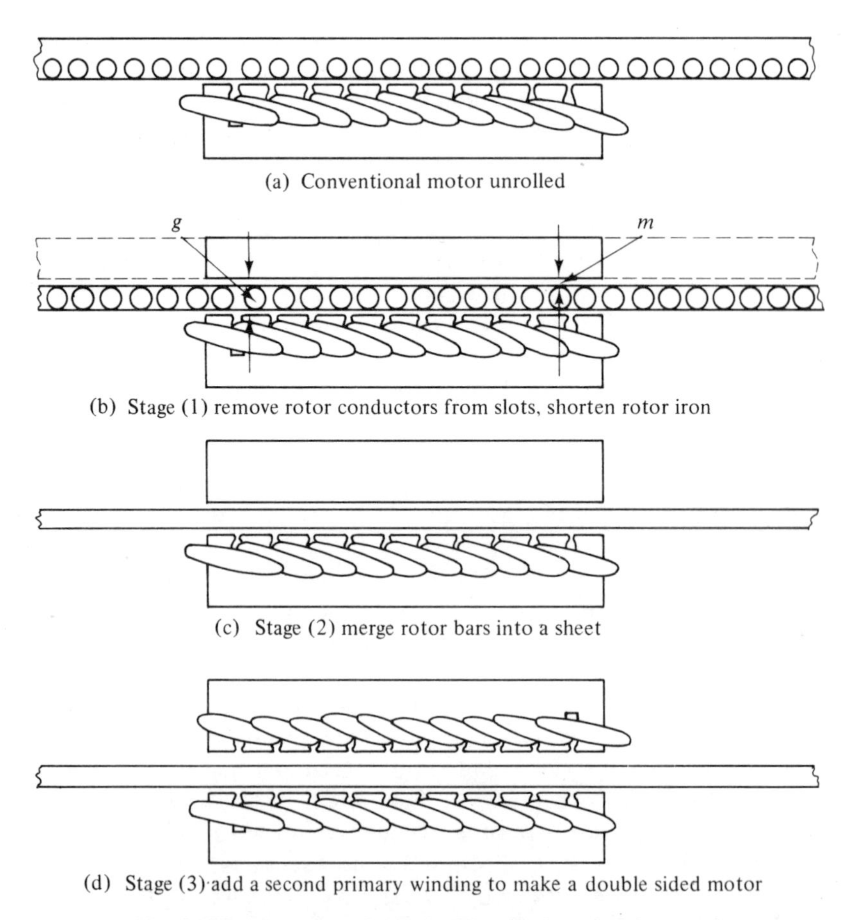

Fig. 6.5 Development of the 'double-sided sandwich motor'

mechanical clearance demanded for safe relative motion between primary and secondary.

At this point it is important to distinguish between this clearance and the total effective gap in the magnetic circuit which contains only material of low permeability. For example, the distance m is the mechanical clearance, which might rightly be termed the 'airgap', yet the distance g is the one to be used in magnetic circuit design as that part of the circuit having relative permeability approximating to unity. The French word *entrefer* (literally the 'between-iron') is a better and totally unambiguous word describing g and will be used in this chapter wherever the word 'airgap' would otherwise be ambiguous.

Returning to Fig. 6.5, there is a second step also included in (b), for having dispensed with secondary slots it can easily be demonstrated that the desired horizontal force of induction now acts directly on the secondary conductor, as opposed to an arrangement as in (a) in which virtually the whole of this force can be shown to have its points of application on the sides of the slots in the secondary steel. There is no need therefore to *elongate* the secondary steel in (b), since its only purpose is now to close the magnetic circuit as effectively as possible. In this role, the secondary steel in effect 'changes sides' so far as the entrefer is concerned for it can now, indeed *must*, be fixed to the primary and regarded as a permanent 'keeper' for the row of magnets which constitute the primary.

The secondary conductor is the next component to receive attention in the development. Being no longer in slots it is not necessary to retain its ladder-like construction. It can, in effect, be melted down and re-cast into a solid sheet as in (c). In the past, experts in 'stray-load loss' have worried about excess secondary ohmic loss now being incurred by the ability of the secondary current to flow, in part, in longitudinal paths under the active pole, instead of in the equivalent of conventional cage-rotor 'end rings'. It has been suggested that it is essential to design sheet-rotor motors with a multiplicity of parallel slits in the sheet to ensure the bar-like distribution of current in a 'cage rotor' cylindrical machine. This is not so, for although some current will flow in roughly elliptical paths entirely in the active zone of the pole area, as shown in Fig. 6.6(a), the proportion of such current is a measure of the Goodness factor, G, of the machine, G being a quantity derived in 1965[3] for the rapid assessment of most properties of electromagnetic machines. In any machine having a high value of G, the pattern of primary current (imposed entirely at the will of the designer because conductors can be insulated) should be almost perfectly mirrored by the secondary current distribution, even if the latter is an uninsulated, indefinite mass. The current distribution of the sheet rotor shown in Fig. 6.6(b) is therefore indicative of a higher Goodness factor. An alternative 'physical' explanation is that longitudinal secondary currents under the active zone can find no mirrored current in the insulated, and therefore pre-routed primary circuits, and thus face the same task as a magnetizing winding which tries to set up transverse flux. In a good machine the magnetizing reactance component of the equivalent circuit may be many times the secondary resistive component (in fact G times it) and even at speed corresponding to slip s, an induction motor may have $sG > 1$. The 'local' currents inside the active zone face a 'stifling' impedance of sG compared with a full linkage flow impedance of unity.

Finally, to carry a purely rectangular block of laminated steel as permanent 'keeper', as shown in Fig. 6.5(c) is wasteful of steel, for at a cost of very little extra

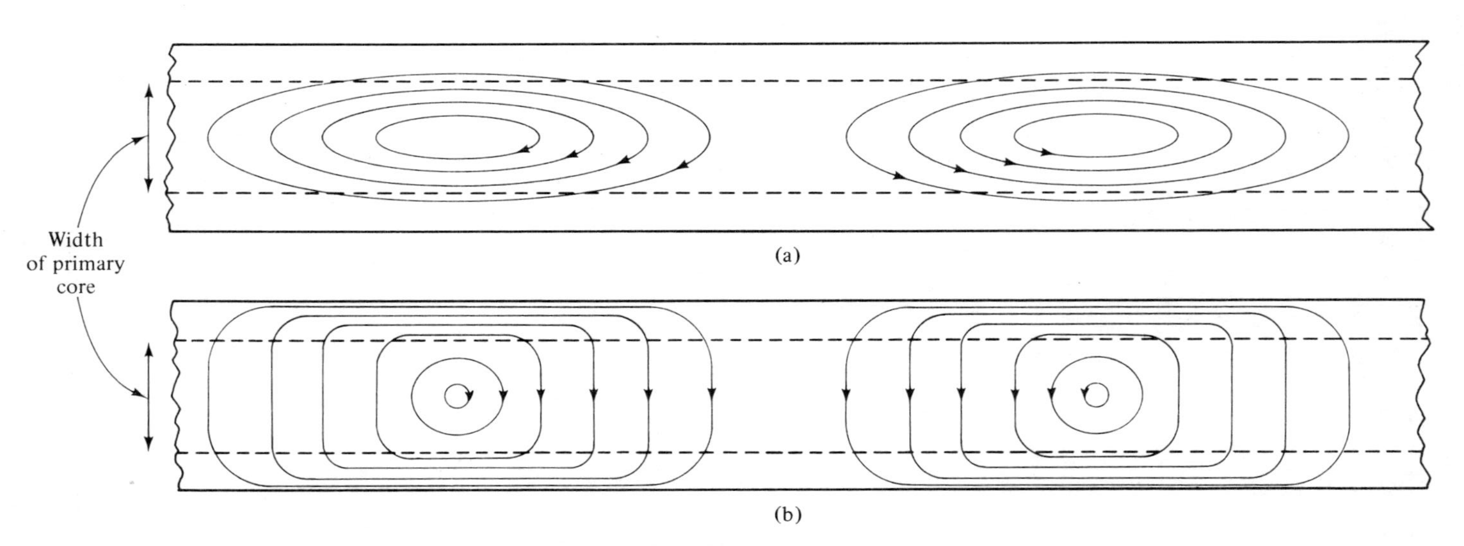

Fig. 6.6 Current flow lines in solid sheet secondaries having (a) A low Goodness Factor. (b) A high Goodness Factor

steel, the block which began as 'secondary steel' in Fig. 6.5(a) can indeed become a full member of the primary components by the fitting of a second winding, in slots, as shown in Fig. 6.5(d). This last development can be considered either as a means of halving the primary ohmic loss for a given output or, what is a better coin in which to take the advantage, as a means of doubling the output for much less than double the primary weight.

The reason that this second alternative is possible highlights the greatest advantage of all short-primary systems which places them a whole order of magnitude better than rotary machines, where suddenly to double the primary current loading of a well-designed machine would almost certainly burn out the rotor (of limited size). In linear machines in which the secondary is many times the length of the primary, the secondary power loss is confronted with an almost infinitely large heat sink, and is never likely to melt the secondary, unless stalled for a long period.

While the double-sided sandwich motor is seen to minimize the secondary cost in a short-primary machine and at the same time to eliminate magnetic pull, it should be noted that when a sheet of conductor lies between oppositely magnetized surfaces which are excited by a.c., there exist lateral electromagnetic forces which, like their magnetic counterparts in a ferrous-cored secondary, tend to attract the secondary to whichever of the two faces is the nearer, i.e., the system is naturally de-centralizing. In a large range of linear motors of small or medium size, however, this force is but a small percentage of the useful thrust, rather than, as with magnetic pull, many times the drive force.

6.1.5 *Theoretical considerations*

Most of the theory which must be added to that of rotary machines could be said to arise as the result of 'edges', i.e., sharp discontinuities which either do not arise in cylindrical machines or are more easily treated by rigorous mathematical techniques. This statement is illustrated in Fig. 6.7, which shows a portion of single-sided, sheet-rotor secondary in which the dotted lines indicate the active zone opposite the short primary block.

The edges BC and DA might be thought to constitute a similar problem to that of calculating end-ring resistance and leakage reactance in a conventional machine.

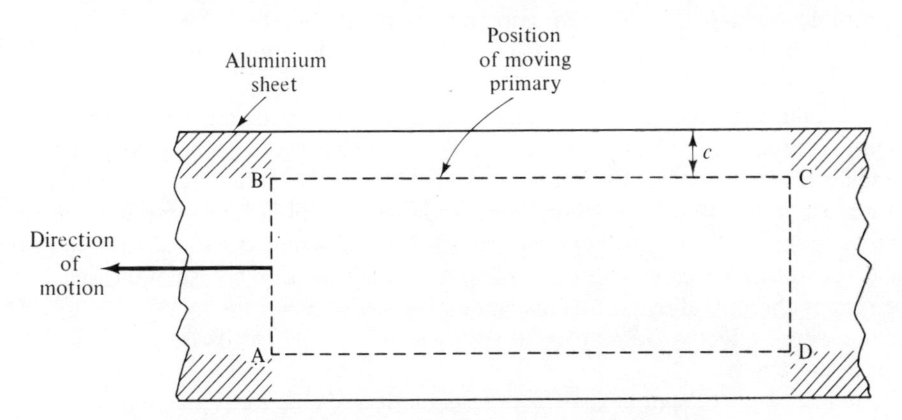

Fig. 6.7 Edge effects in a double-sided sandwich motor

That this is not so is due to the compact nature of end rings where a single bar can be assumed not to couple with the much extended end windings of the primary. The fact that the calculation of *primary* end-winding leakage in a conventional machine has long been considered a rather hazardous affair by designers, even though the currents are clearly specified, indicates the difficulty of performing the calculations for secondary end-winding leakage in a sheet rotor which, for purposes of lowering the effective secondary resistance, has been extended well beyond the limits of the primary steel, i.e., the distance c in Fig. 6.7 is comparable with half a pole pitch.

Computation of the secondary resistance component to be used in the equivalent circuit is also difficult, but as with most problems, there are different levels of approximation which may be used to simplify the problem on the understanding that accuracy is being sacrificed for simplicity. Perhaps the most sweeping assumption that can be made in this problem is to imagine that a non-distributed current pattern of the form shown in Fig. 6.8 exists at all times and that the resistance of its end paths

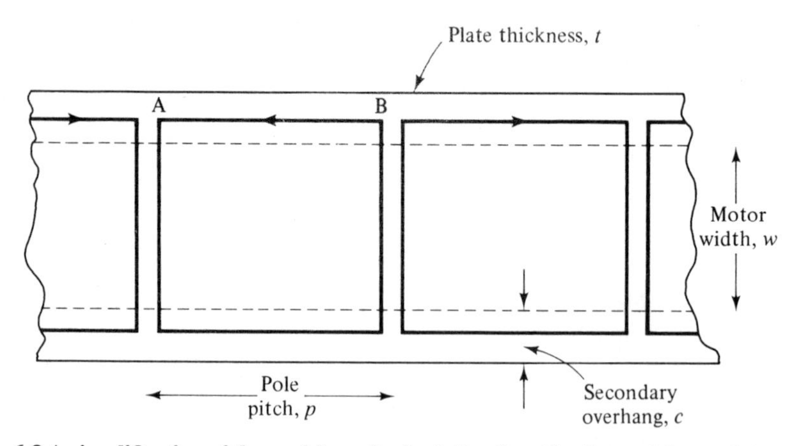

Fig. 6.8 A simplification of the problem of calculating the effective end-bar resistance

(outside BC and AD in Fig. 6.7) can be related to those in the active zone by assuming the latter to flow in a conductor of length w and area $p/2 \times t$ (thereby assuming *uniform* distribution of the concentrated current along AB in Fig. 6.8) while the end current flows in a path of length p and area $c \times t$. The ratio of end resistance to active zone resistance is then $p/c \div 2w/p = p^2/2cw$ and the more accurately calculable active zone resistance can then be multiplied by $(2w + p^2/c)/2w$. Even this formula can be simplified further if the overhang distance c is roughly equal to $p/2$ when the multiplying factor is simply $(p+w)/w$. This formula is not to be despised for the inelegance of its origins, for it gives remarkably good results over a wide range of values of c and is useful in enabling quick calculations (of G for example) to be made in first assessments of a problem. The reason for this is that as c is reduced, the value of G decreases (because effective secondary resistance decreases) and more end current is thereby allowed to flow under the active zone (as in Fig. 6.6(a)). This increases the effective value of c to compensate in part for its reduction in actual measured length.

The next stage of complexity for the calculation of effective end winding is due to Russell and Norsworthy[4] who assumed a uniformly-travelling wave of flux which was of constant amplitude across the width of the machine and then calculated accurately the flow in the end regions on the assumption of no magnetic coupling with primary end windings. Their results may be interpreted as indicating that the effective resistivity of the secondary may be obtained by calculating the resistivity of the active zone and multiplying by a factor $1/(1-k)$ where

$$k = \tanh \varepsilon/[\varepsilon(1+\tanh \varepsilon \tanh \phi)]$$

in which

$$\varepsilon = (\pi/2)\left(\frac{\text{Primary width}}{\text{pole pitch}}\right)$$

$$\phi = \pi\left(\frac{\text{secondary overhang/side}}{\text{pole pitch}}\right)$$

Still greater accuracy is obtainable by formulae due to, firstly H. Bolton,[5] and secondly Preston and Reece.[6] The former assumes only that there is no secondary leakage and that there are no fields in the end winding region, while the latter assume that the end-winding fields which they do take into account are unaffected by a notional increase in stator width up to the full secondary plate width. What no analysis so far has taken into account are the 'no-man's land' areas (shaded in Fig. 6.7) where edge effects due to AB and CD interact with those from BC and DA.

The effect due to the edges AB and CD is quite different, being entirely due to transient currents. A first approach to the problem is more easily understood by returning temporarily to the ladder type of secondary conductor embedded in slots, as shown in side cross-sectional view in Fig. 6.9. A tooth A approaching the active zone carries no flux and on entry is faced with the primary excitation at its full value (assuming all coils of the primary in each phase to be connected in series). The tooth A, surrounded by a thick conducting loop, (bars X, Y and their associated end conductors) is subjected to similar treatment to that imparted to a transformer with short-circuited secondary, whose primary is suddenly connected to a d.c. source. Since there cannot be an infinitely large initial rate of rise of flux (which would demand infinite applied voltage) the very first reaction of the secondary conductor is to pass ampere-turns identically equal and opposite to those of the primary it faces, and to allow zero flux to thread the secondary loop, as in tooth B (Fig. 6.9).

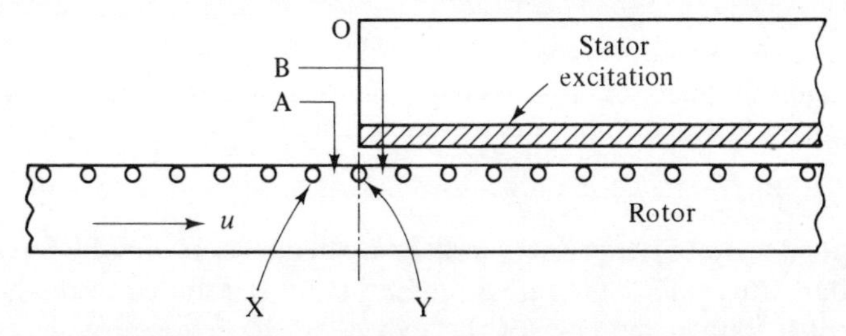

Fig. 6.9 Use of a 'conducting ladder' concept to simplify study of the entry and exit edge transient

As with the problem of edges BC and DA (Fig. 6.7) the problem of the subsequent action is easiest to appreciate if taken in successive stages of decreasing simplicity. The first step then is to assume that the machine is an induction motor with the secondary travelling at the field speed v_s (the condition known as 'running light' in conventional motors). It is further assumed at this stage that the secondary conductor has *zero* resistivity. In this case no flux could ever penetrate the loop surrounded by bars XY and end rings and the passage of any such loop across the active zone would sweep it clean of flux. This is apparently a trivial solution but is important to the understanding of the second stage in which the speed of the secondary is still v_s, but the secondary bars have an effective surface resistivity ρ. The flow of secondary current is now required to have a secondary voltage ρJ_2 to maintain it (where J_2 is the secondary current loading) and the only source of such voltage is a rate of change of flux, so that if such a flux change could be produced without demanding magnetizing ampere-turns to drive it, the flux density along the block will rise linearly with distance, as shown in Fig. 6.10. In the next stage, the magnetic circuit can be regarded as imperfect and when magnetizing current is included, the linear rise of flux density gives way to an exponential rise as shown dotted at OA in Fig. 6.10.

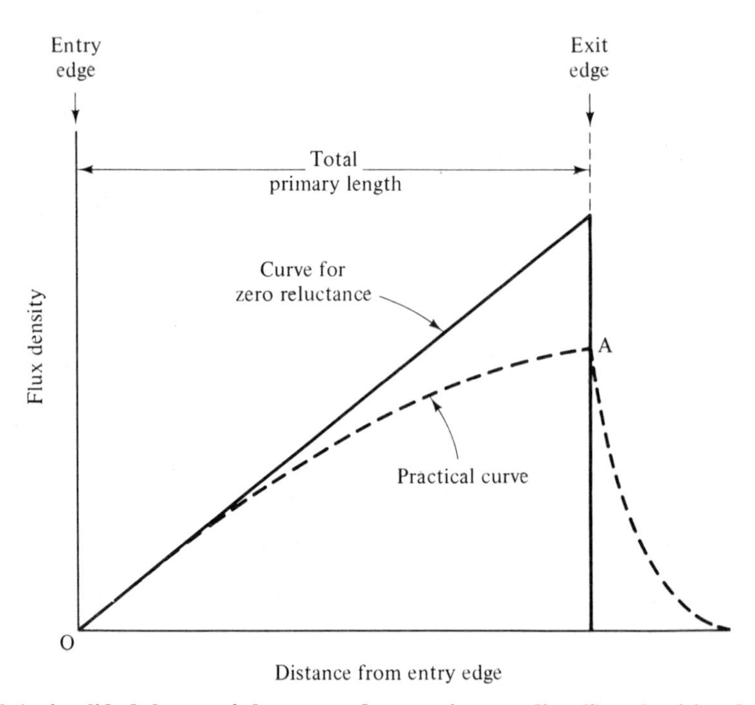

Fig. 6.10 A simplified theory of the entry edge transient predicts linearly-rising flux. When magnetizing current is included, the flux pattern is modified to that shown dotted (including exit edge effects)

At the opposite end of the primary a reverse effect occurs and the flux-laden teeth of the secondary are reluctant to release their flux for the same reason that they were slow to accept it, but in the case of the exit edge, the current which flows in the secondary to maintain the flux occurs outside the active zone of the primary where the Goodness factor is much lower and the resulting exponential decay of flux beyond the exit edge (also shown dotted in Fig. 6.10) is much more rapid than was the build-up under the active zone. The exit edge flux tends to enter the primary steel (at least in part) along the vertical surface of the back edge and, as such, exerts a reversed thrust on the secondary even though its relative speed to the primary has not exceeded synchronism. A short-primary motor may therefore be seen not to run light at indicated field speed, but at a somewhat lower speed.

This phenomenon should not be confused with a comparable one in which a linear motor secondary runs light at slightly *greater* than the indicated field speed. This latter is observed in motors with a large airgap and low G-value (often caused *by* the large gap). In such cases the flux builds up to its full value almost instantaneously at the leading edge and decays in the same way at the exit edge. But flux *fringing* at both edges gives the impression to the secondary that the effective zone length is greater than it actually is, i.e., that the pole pitch and hence the synchronous speed is increased.

This detailed examination of the running-light condition of a linear motor has revealed a number of new features not encountered in a symmetrical, cylindrical induction machine which, for easy reference, will now be listed.

(a) There is secondary current and hence secondary ohmic loss under the active zone even when field and secondary speeds are equal. Such loss does not give rise to a proportionate forward thrust as is usual in the case of conventional motor secondary ohmic loss. While the effect is most marked at zero slip, it occurs at other speeds, but has been shown to be zero at slip values $1/(n+1), 2/(n+2), 3/(n+3) \ldots$, where n is the total number of poles contained in the active zone. Between the values $1/(n+1)$ and $2/(n+2)$, $3/(n+3)$ and $4/(n+4) \ldots$ the ohmic loss under the active zone is actually less than would be calculated on conventional induction motor theory, a typical graph of 'extra ohmic loss' plotted as a function of speed being shown in Fig. 6.11.

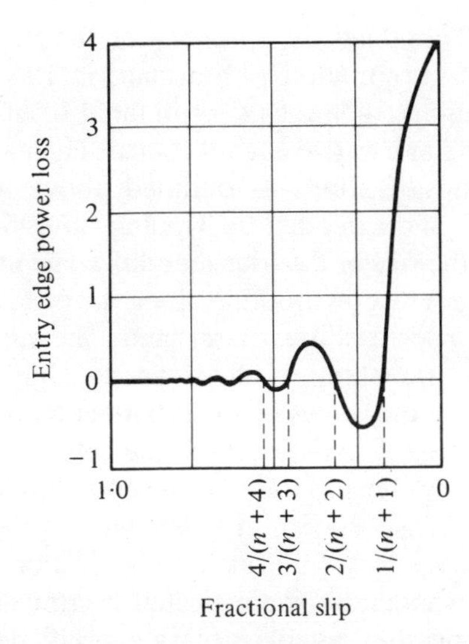

Fig. 6.11 Variation in secondary loss, due to entry edge effect, with slip

(b) The continuous creation of flux under the active zone in the running-light condition gives rise to reactive volt-amps not accounted for either by conventional magnetizing current or by leakage flux. This effect also occurs in lesser degree at speeds other than synchronism.

(c) Secondary current and associated power loss occur outside the active zone but such loss is seen to result in a loss of synchronous power (Fv_s, where F is the forward thrust) rather than in excess power drawn from the supply. The variation of this loss with speed is indicated for a typical motor in Fig. 6.12.

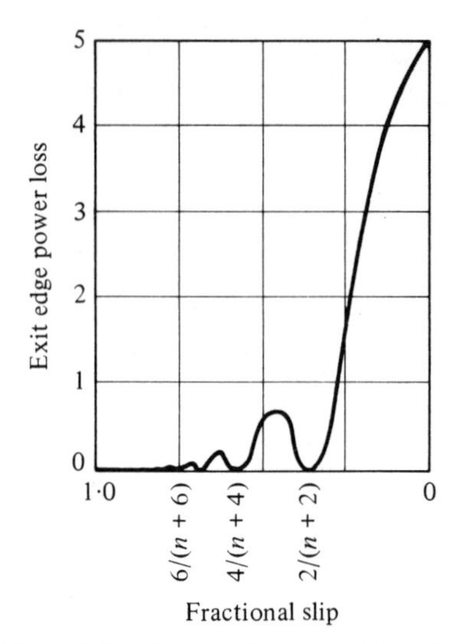

Fig. 6.12 Variation in exit edge loss with variation in slip

These results can be obtained by a variety of theoretical methods. The first calculations of this kind were made by Shturman[7] in 1946 by solving third-order partial differential equations, themselves formulated from a consideration of field theory which included secondary leakage reactance. The method of approach to the problem by physical considerations as outlined above was first formulated by Williams et al.[8] in 1957, and extended by Tipping[9] in 1964. Basically the method consisted of calculating the size of the transient flux wave initiated at the entry edge and observing the resultant flux distribution along the primary as the transient wave travelling at secondary speed, slipped alternately in and out of phase with the conventional flux wave, travelling at field speed. Tipping's extensions include allowance for reflections of the waves set up at both entry and exit edges and for the fact that a small input step in flux did occur at the actual entry edge itself.

Later, an elaborate computer program produced by the Garrett Corporation for the US Department of Transportation in Washington[10] took into account leakage reactance, but not exit edge loss. Still further elaboration appears in the book by Yamamura.[11] The disadvantage of sophisticated treatments of the entry and exit edge effects is that while they enable accurate predictions to be made of the performance of any given design, they give little or no guide as to how the design can be improved, except by trying other parameters and running the computer program over and over again. The main advantage of Williams' method was that although it is not especially accurate it provides quick 'rules of thumb' which give guides to the designer regarding the direct influence of this or that parameter on ultimate performance.

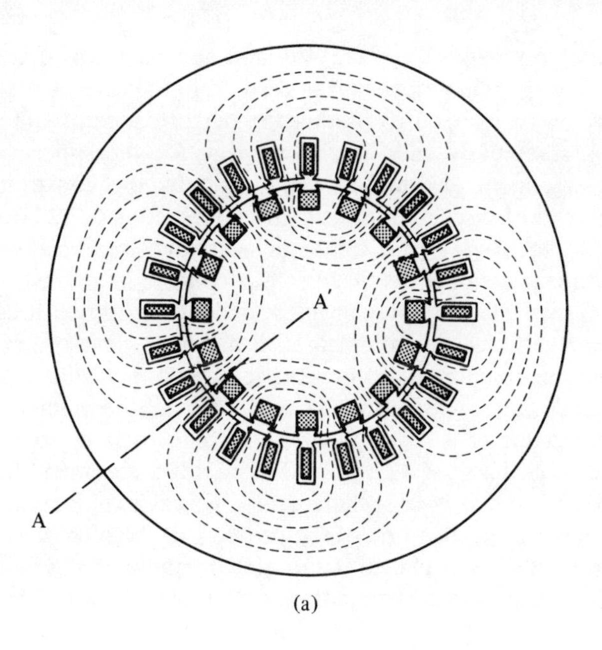

(a)

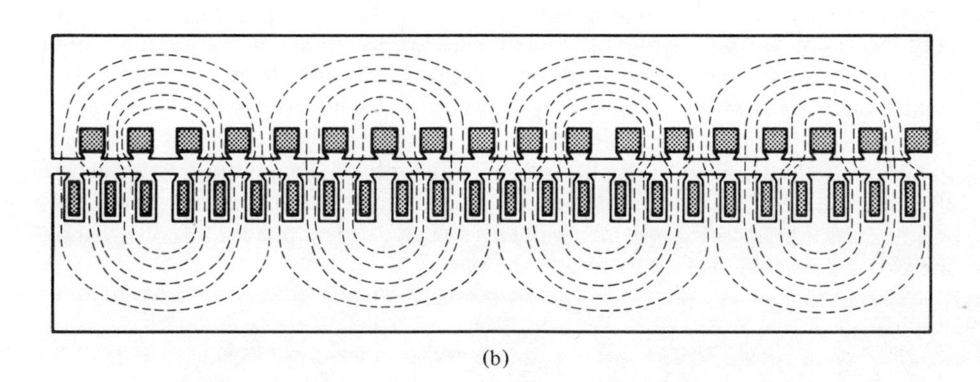

(b)

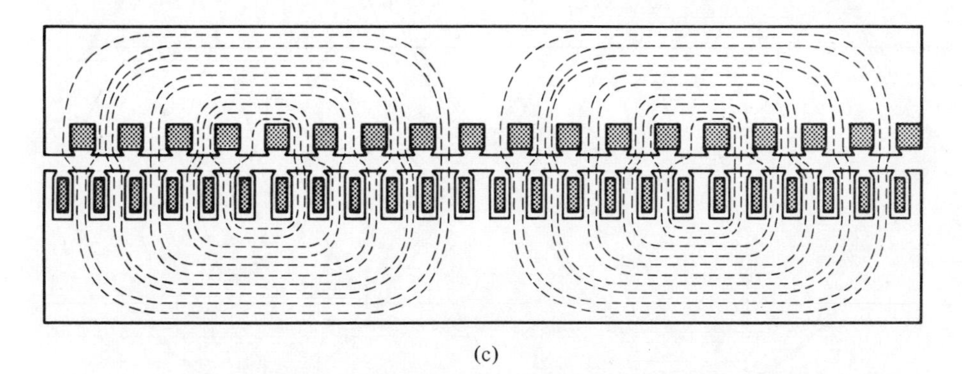

(c)

Fig. 6.13 Instantaneous flux patterns in linear and rotary machines. A section AA′ produces the diagram (b), but $\frac{1}{4}$ cycle later results in flux doubling as at (c)

One of the main omissions from the Williams approach was that of ignoring the fact that the total airgap flux must sum to zero, as in the case of rotating machines. Since his theory was advanced to predict the performance of spherical motors,[12] however, it was completely valid in this respect, for the spherical motor had a transverse flux arrangement which is not bound by the constraint $\int$(flux/input length) $dx \equiv 0$ as will be discussed in the high speed section 6.2.2 later. It will suffice here to say that the spherical motor, although itself not commercially viable, can now be seen as the forerunner of the transverse flux concept in at least two aspects.

For those who like their complex problem divided up so that it can be seen as a number of distinct physical phenomena each having a clear origin, the following approach to core flux in a linear motor, coupled with the Williams approach to the edge transients, will have appeal. The Garrett program, the much earlier Shturman papers, and the much later work of Yamamura are all inclined to 'eat the cherry in one bite'. That is no criticism of their work. It is merely a statement that the author of this chapter prefers inductive teaching and physical explanations to deductive teaching and complete specification of the problem at the outset.

Figure 6.13 illustrates at a glance the origins of standing waves in the core of a linear motor. The instantaneous flux pattern in the rotary motor shown at (a) retains its form at all times, the whole merely rotating at synchronous speed. If the splitting and unrolling process is performed along the radius AA′ (a), the linearization is completed as at (b) and there appears to be no *essential* difference in the pattern of core flux. In particular the flux from each pole divides equally to the left and right in the core, so that the core depth required is equal to the sum of the tooth widths in *half* a pole pitch (plus some allowance for leakage flux and bolt holes). It should be obvious, however, that the choice of AA′ as the splitting line was particularly fortuitous for had the flux been allowed to rotate for but a further quarter of a cycle of the mains supply, the flux pattern necessitated would have been that shown at (c) where the required core depth is now *double* that at (b). This picture, of course, shows the worst condition, as Fig. (b) shows the best.

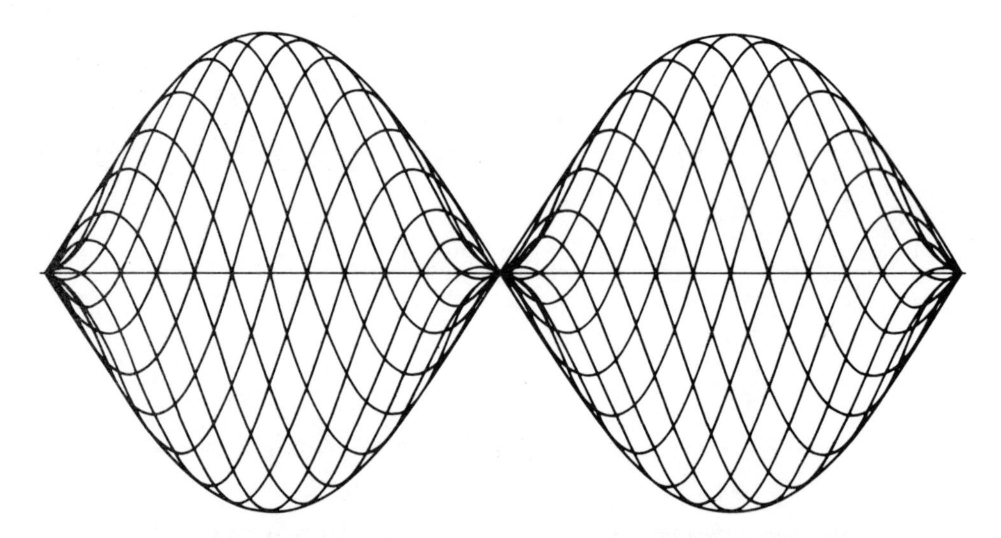

Fig. 6.14 Standing wave pattern in the core of a four-pole machine

When this problem is analysed mathematically, albeit in isolation from other phenomena[13] (what such analysis assumes is that there is no secondary conductor) the core flux is seen to consist of travelling waves which reflect to produce a standing wave envelope, as shown in Fig. 6.14, in which the travelling waves are contained. Were it not for the fact that moving secondary conductor throws the whole core flux distribution back into the 'melting pot', the designer might have shaped his core iron to match the wave envelope and used no more iron than in the rotary machine at (a), Fig. 6.13. In general, however, the core depths of what are now called 'longitudinal flux machines' to distinguish them from the transverse type (6.15) should be larger than those of equivalent cylindrical structures, and generally by a factor approaching the value 2·0.

The most recent developments in linear motor topology, especially that concerned with high-speed motors, have necessitated a theoretical re-appraisal in order to predict the performance of the new machines. It would seem unhelpful to include such new theory at this point merely for the appearance of tidiness, and it is therefore proposed to mention the theoretical modifications as and where they occur in a topological discussion.

6.2 Forms of Construction

The single-sided and double-sided sandwich types of motor, either of which may have moving primary or moving secondary, have already been described. All species in this 'genus' (to use biological notation which is a useful way of classifying topological differences) have flat pole faces and entrefers. The motors of the second genus which were also second historically have essentially cylindrical symmetry although no part of their physical action is rotational. The topological transformation is obvious from Fig. 6.15.

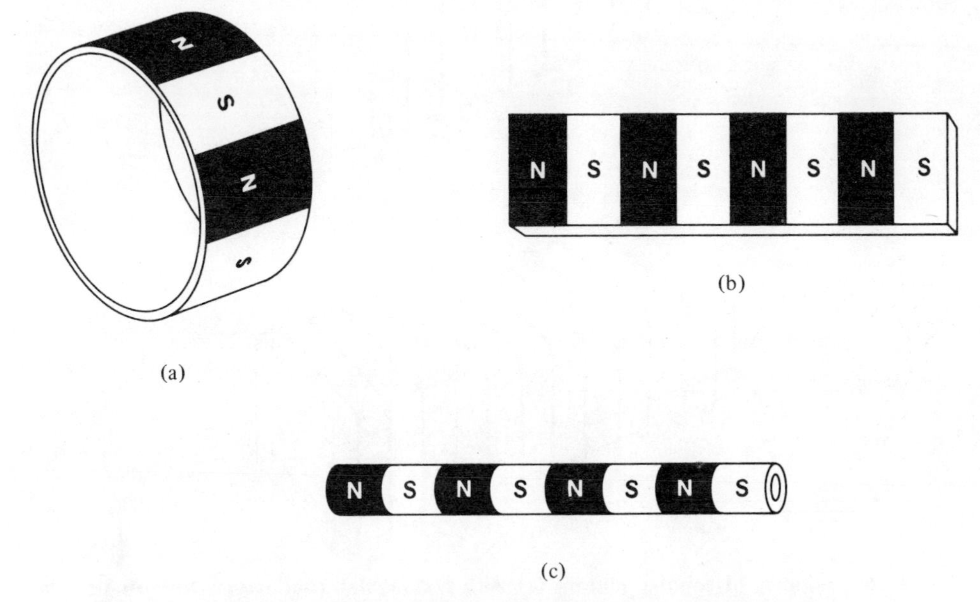

Fig. 6.15 Tubular motor seen as a re-development of a flat linear motor

6.2.1 *Tubular machines*

The 'parent' rotary motor at (a) is seen to have been split and unrolled to give the flat machines typified by (b). Re-rolling as in (c) produces a tubular arrangement in which the direction of travel of the moving field system and of the moving part (be it primary or secondary) is along the axis of the cylinder. While many species of this motor are possible, few are used commercially, the others either being difficult to manufacture or offering little commercial advantage. Nevertheless all possibilities are here listed for the sake of completeness and in the hope that one of the hitherto unused forms will suggest to a reader a new application, more suited to its shape.

Double-sided tubular motors are seldom used, but clearly possible. It is not so much manufacturing difficulty which dissuades industry from applying this type, as the very small benefit that would accrue from the inner winding, which on a decreasing radius would be so confined that its contribution to the total current loading would be but a few per cent.

Then, as with flat machines, either the primary or the secondary can be made the moving member. This alternative is then to be crossed with the idea that either the inner or the outer can be the primary member, giving rise to four sub-families of motor. (These are sub-families rather than species, for each is now seen to be available in a variety of shapes in other respects. The word 'genus' used by analogy with biological nomenclature should therefore perhaps, as in biology, be replaced by 'super-family'.)

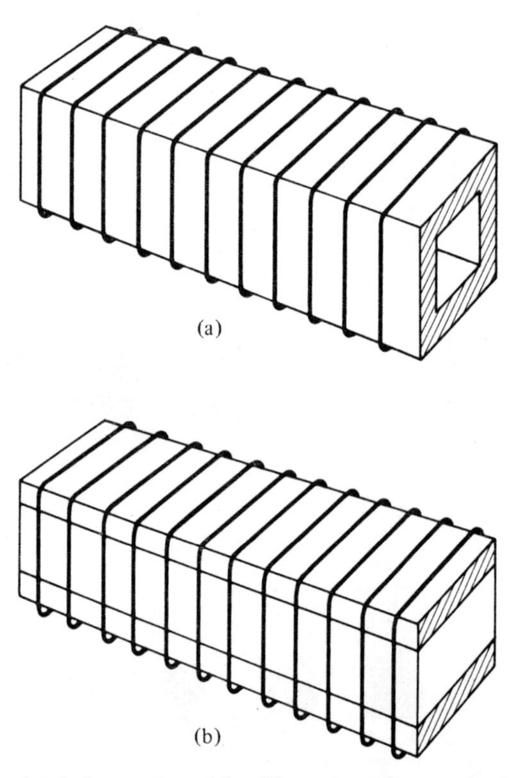

Fig. 6.16 Examples of tubular motors (a) with rectangular conductor; (b) with two flat conductors

The word 'tubular' is not to be taken too literally and any machine in which the flux from pole-to-pole travels parallel to the direction of motion and then fans out radially throughout the full 360° in a plane normal to the direction of motion (its distribution being not necessarily uniform) is seen to be a tubular motor, for its primary coils lie essentially in planes perpendicular to the direction of travel. Thus the square or rectangular cross sections of Fig. 6.16 are tubular types. Before proceeding further with pure topology it is necessary to state the three features of tubular machines that make them less attractive commercially than the flat machines. The first is the obvious mechanical difficulty of mounting either primary or secondary if the required stroke is very long. A short, inner, moving-secondary machine has its moving part inaccessible to the job except by a long push rod along the axis of the tube. A long, short-primary (or short-secondary), fixed-inner-member machine, mounted for horizontal motion can only be end mounted and is therefore liable to bend under its own weight, and so on.

The second, less obvious disadvantage is that of incorporating ferromagnetic material into the construction to ensure a good magnetic circuit. At first sight this seems a relatively easy difficulty to overcome. Figure 6.17(a), for example, shows a four-block construction which is no more than two double-sided (but singly-excited) flat motors arranged in a hollow square. But note the wasted space. A better arrangement would appear to be that shown at (b), yet the inner core is incorrectly laminated for the outer blocks B and D. The flux from these blocks must enter the faces of the steel punchings and thereby induce large circulating currents. A feature of the process of lamination is that it is always possible in conventional rotary machines to laminate in such a way that all the useful flux is contained in paths essentially in the plane of the laminae. The condition that this should be so is a completely topological one, namely that the magnetic circuit of the machine should lie entirely in parallel planes. This has been a most restricting requirement in electrical machine development inhibiting the inventor by implying that such constraint was laid down by the nature of the phenomenon of electromagnetism, and this is not so. In particular, insufficient use has been made of the fact that a stack of steel punchings can be butted against a second set, the planes of the two sets being perpendicular to each other, without incurring iron loss due to eddy currents in the planes of the laminae. Thus the magnetic circuit shown in Fig. 6.18 incurs no more iron loss than one of the same effective length and cross section, carrying the same flux at the same frequency, in which the whole of the laminae are parallel to each other. More discussion of the non-planar magnetic circuit will occur later in this section.

Returning to the tubular motor, it is no comfort to know that the core of the inner member may consist of a bundle of iron wires rather than a stack of laminae as in Fig. 6.17. Although a bundle of wires constitutes, in one sense, two-dimensional lamination, wherever the flux is not being transmitted directly along the wires, it must cross from wire to wire through varying degrees of airgap, as shown in Fig. 6.19, which leads to local bunching near the lines of contact between wire and wire, which in turn leads to local saturation and increased hysteresis loss, and to the introduction of a large effective airgap, being the sum of all distances such as AB, CD, etc (see Fig. 6.19). Note also that Fig. 6.19 represents an ideal case where all wires are parallel to each other and packed as closely as is possible. Practical cores may be much less effectively packed.

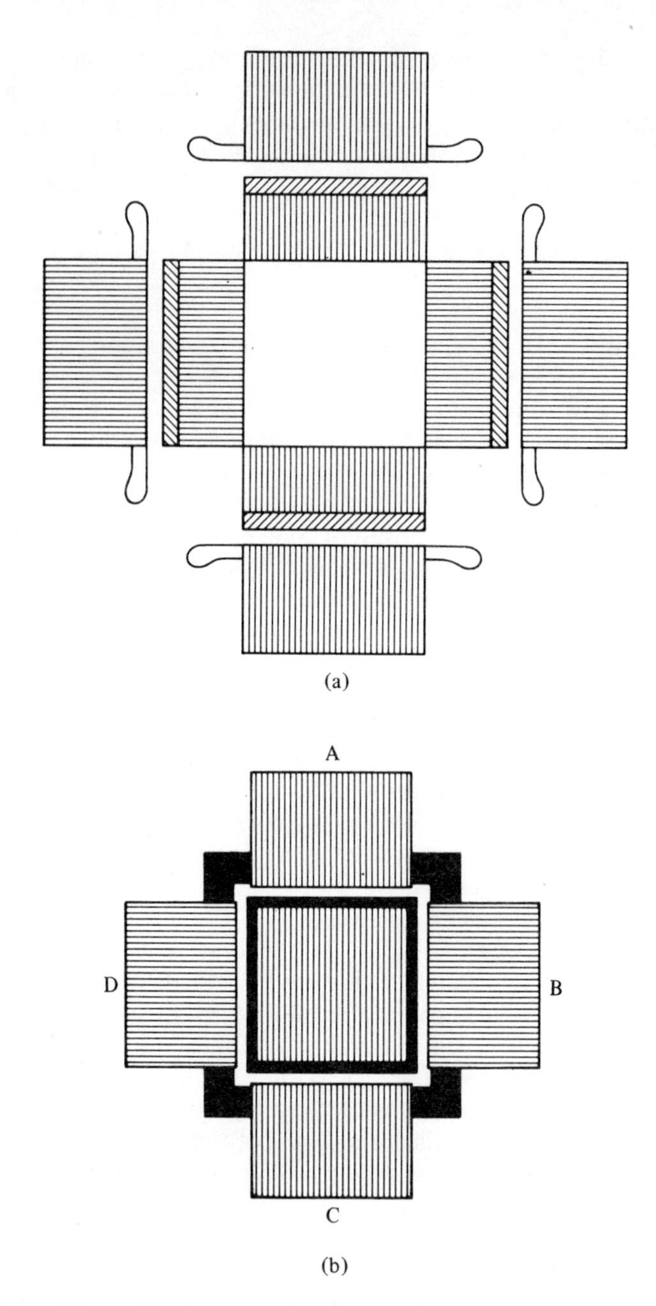

Fig. 6.17 Four-block constructions for tubular motors, shown 'end-on'

The third disadvantage of the tubular motor is that all the flux from one particular pole, in the worst case, which enters the core of the inner member must be passed axially through the core to the neighbouring pole so that the relationship between core flux density B_c and maximum airgap flux density (useful flux) B_g is given by

$$(2\pi rp)B_g = (\pi r^2)B_c$$

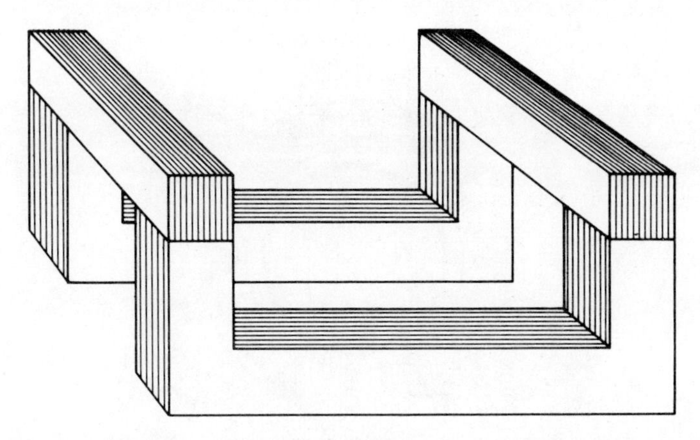

Fig. 6.18 An example of a non-coplanar magnetic circuit consisting of laminated steel

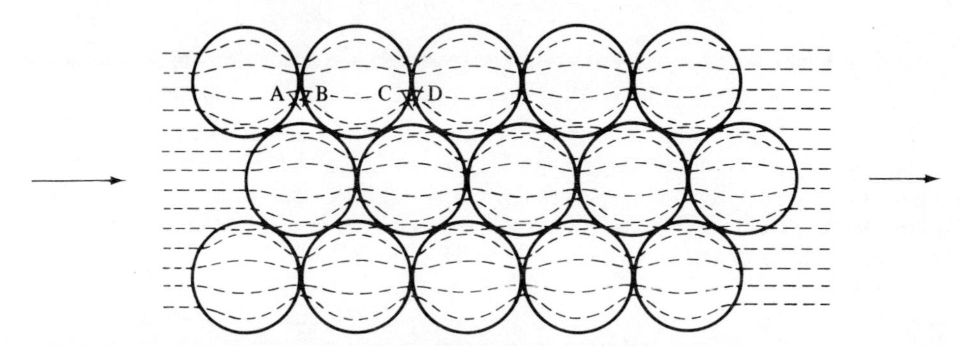

Fig. 6.19 Flux paths involved in penetrating a two-dimensionally laminated core, i.e., a bundle of iron wires

that is

$$B_g = \frac{r}{2p} B_c$$

where r is the core radius. Assuming the core flux density to set the limit due to its saturation of the core, the working gap density is seen to be much lower than that used in flat machines, even taking into account the wide slots and narrow teeth of the latter, unless the radius of the tubular motor is large, making the moving part (whether inner or outer) very heavy.

Nevertheless commercial tubular motors have been successfully exploited as described in the next section, by suitable re-shaping of parts and relaxation of constraints, e.g., the need to laminate the steel.

The constraint imposed on the flux by the contracting geometry of the inner member is in part compensated by the magnetic circuit of the outer member. For example it is not necessary to surround the outer electric circuits entirely with radially-laminated steel (wedge-shaped punchings which would be very difficult to produce). Machines with the outer magnetic material divided into packets in an otherwise truly tubular structure, as shown in Fig. 6.20, have been manufactured, but little reduction in performance results from the complete removal of the outer steel packets. The physical reason for this can be appreciated in two steps.

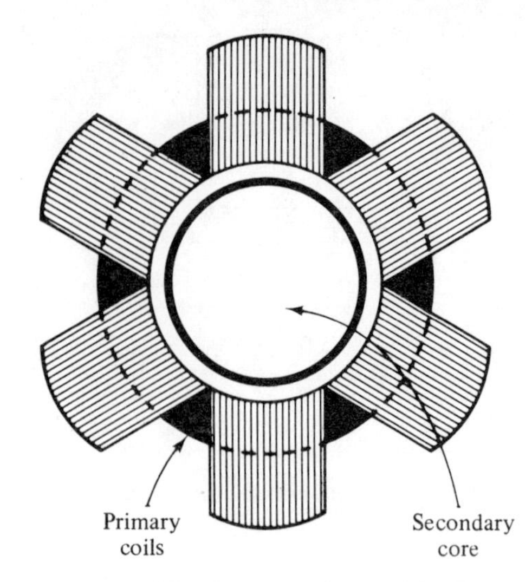

Fig. 6.20 A tubular construction with primary steel punchings divided into separate packets

First it is instructive to calculate the effective airgap of a single-sided flat motor in which there is no secondary ferromagnetic material. Figure 6.21 shows a side view of such a machine on the assumption that the secondary conductor is to consist only of non-ferrous structures. The no-load (no secondary) situation is shown in which the flux per pole is allowed to spread into the space above the pole surface (in

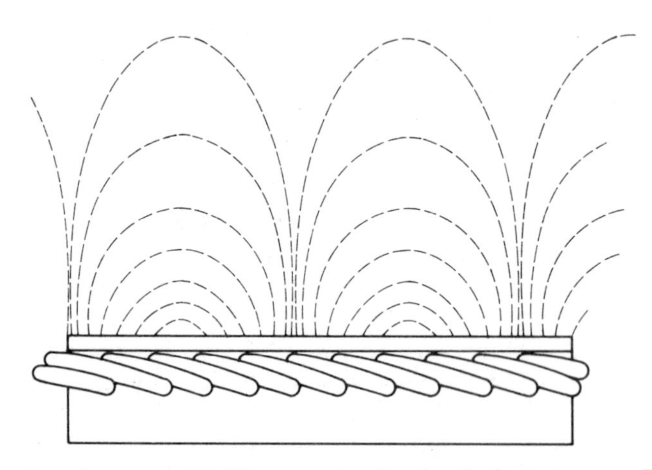

Fig. 6.21 Side view of an open-sided linear motor showing the instantaneous field form in the space above the pole surface

exponentially reducing quantities to infinity). The integrated effect of this spreading results in the magnetizing reactance of the primary winding being of such a value as to suggest[13] that the machine was a double-sided magnetic structure with an airgap of p/π. When the system is re-rolled to make a tubular structure, a much greater spreading is possible and the additional factor by which the effective gap is multiplied is illustrated in Fig. 6.22 in which equivalent gap is plotted (as a multiple of p/π) against r/p.

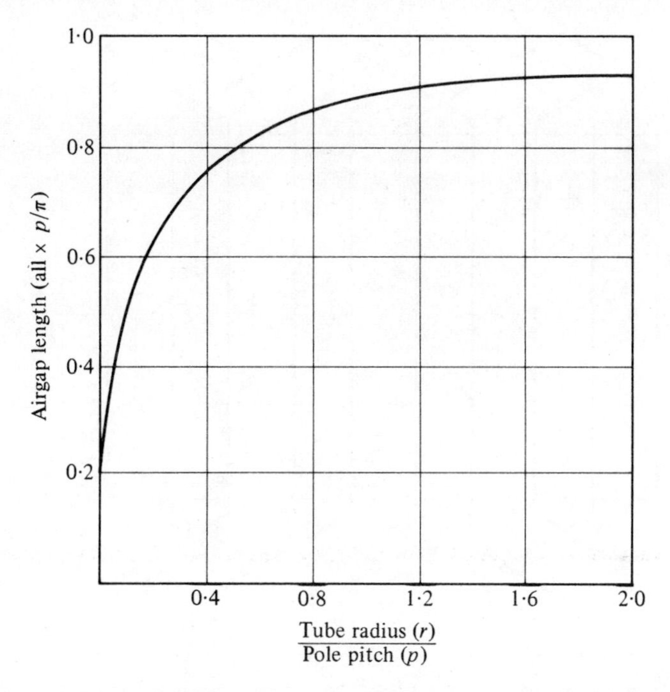

Fig. 6.22 Effective airgap outside a tubular motor surrounded entirely by air

If there is real benefit in a tubular topology it undoubtedly lies in the simplicity and effectiveness of its electric circuits. Figure 6.23(a) shows a plan view of a flat linear motor (or a 'developed' diagram of a rotary motor) and it is usual to regard conductors such as AB and CD as 'useful' and the portions BC and DA as end windings (implying *wasted* copper and ohmic loss). If this diagram is re-rolled to form a tubular motor, as shown in Fig. 6.23(b), the active or useful parts of the coils are seen to complete circular loops and the end windings become totally redundant. The primary windings of simple tubular motors may therefore consist of a row of simple drum-shaped coils or a multi-layer, continuously wound primary, one layer of one phase of which is shown in Fig. 6.23(c).

As in biology, classification is always made difficult by the 'borderline' cases, and after examination of the motor shown in Fig. 6.17(b), one may ask whether the structure shown in Fig. 6.24 is of tubular construction or flat, for it is seen to remove the inner core lamination problem, but leave the inaccessibility to the inner member problem unsolved. If such a structure be indeed accepted as a tubular motor it is fair to ask whether modification to the form shown in Fig. 6.25(a) removes it to the category of flat machines, for the rotary form of Fig. 6.25(a) is shown in (b) to be a very old friend—the 'Gramme-ring' windings of the Victorian era which were soon supplanted on the grounds of difficulty of manufacture and excessively large airgap, neither of which is seen to apply necessarily to linear forms. In a recent book[14] the author attempted to classify tubular motors absolutely by the statement that: 'A tubular machine is one in which the outer electric circuit is contained in a plane perpendicular to the direction of motion, and in which this electric circuit surrounds the other member entirely.'

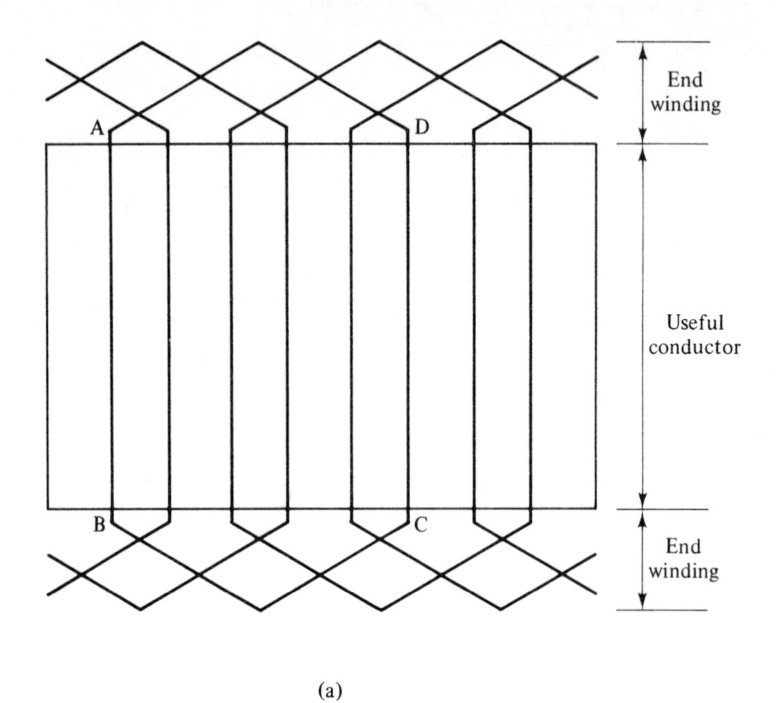

(a)

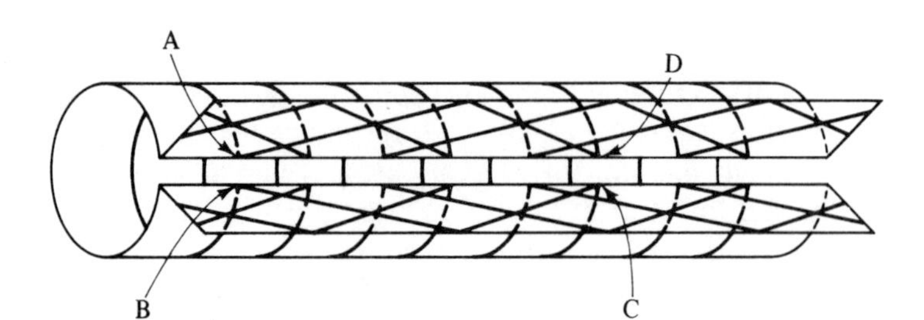

(b)

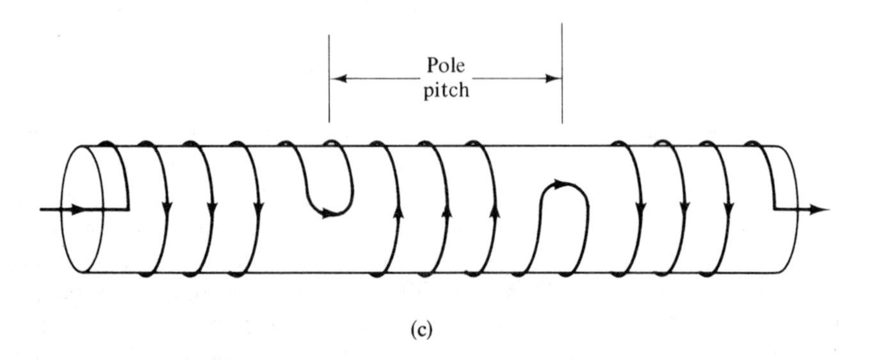

(c)

Fig. 6.23 The advantage of a tubular motor in respect to its absence of 'end conductors'

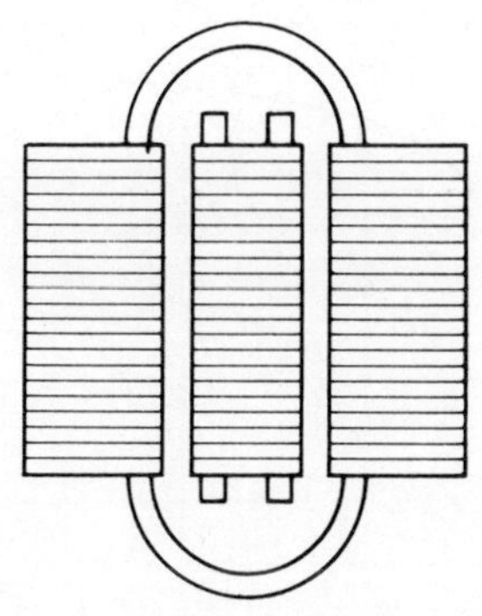

Fig. 6.24 A question of classification as between 'flat' and 'tubular' types

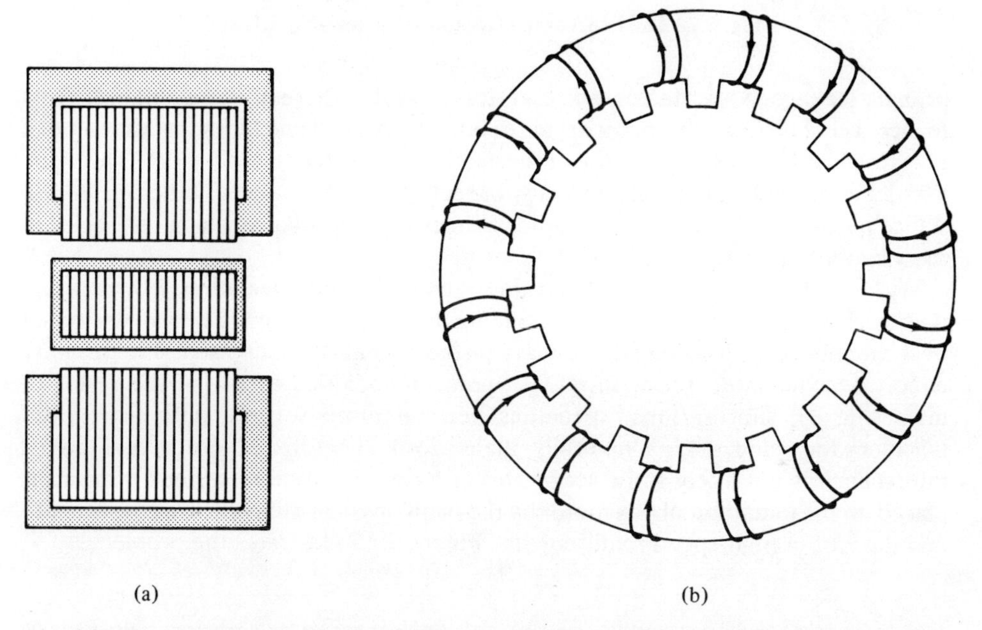

Fig. 6.25 The question posed by Fig. 6.24 is answered: the machine there illustrated is tubular. That in (a) above is flat; cf. its rotary counterpart at (b)

6.2.2 *Flat high-speed motors*

The speed of the travelling magnetic field, v_s, in a linear machine is given simply by $v_s = 2pf$ where f is the frequency of the supply. The Goodness factor is proportional to p^2f, indicating that for a given value of v_s, better machines are to be made by using a large value of pole pitch, p, and low value of f. The question often facing the design engineer is just how far he must go in this process and just how high the value of G needs to be for a motor to be acceptable. The second question is the easier to answer and probably best answered by the use of an example. Figure 6.26 shows the efficiency of a machine of given rating and size plotted against G, it being assumed that the value of the latter is varied by changing the resistivity of the secondary conducting material, so as to interfere as little as possible with other factors, such as

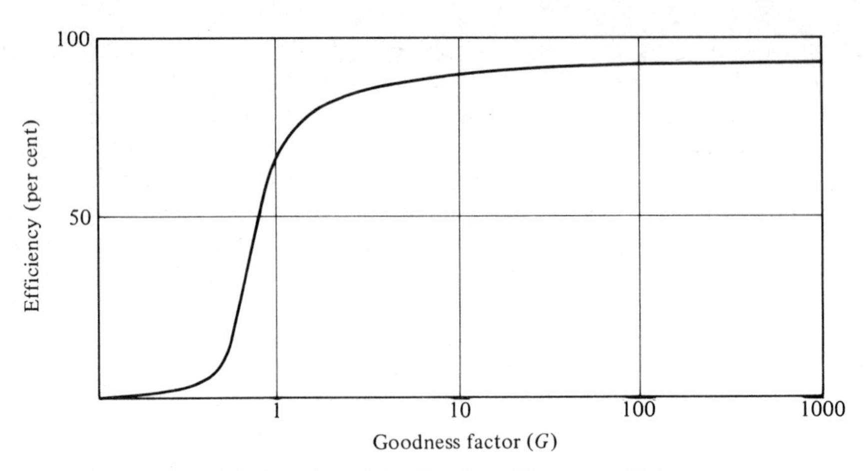

Fig. 6.26 The effect of the Goodness Factor on efficiency

primary leakage, power factor, and cost. It is clear that there is little point in striving to increase G beyond 100, nor in trying to build such small machines that the value of G is restricted, by the very nature of the elements, to values less than 0·1. The value $G = 1$ has special significance in many aspects of rotating machine design. It will suffice here to note only the very rapid changes in performance which take place in the vicinity of $G = 1$.

With small- or medium-speed machines, the limit on the technique of increasing pole pitch and reducing frequency (assuming that the latter facility is available) is set by the number of poles along the primary surface, for as G is increased, short primary effects become more pronounced. As pole number decreases (as the result of increasing p), short primary penalties become more severe. With high-speed machines the value of G is 'naturally' higher for a given size of motor, and the p/f interchange is not likely to be set by this criterion, for a more serious restraint is placed on the longitudinal flux motor by the plain physical difficulties in construction and the cost of high-speed requirements. Figure 6.27 illustrates the problem.

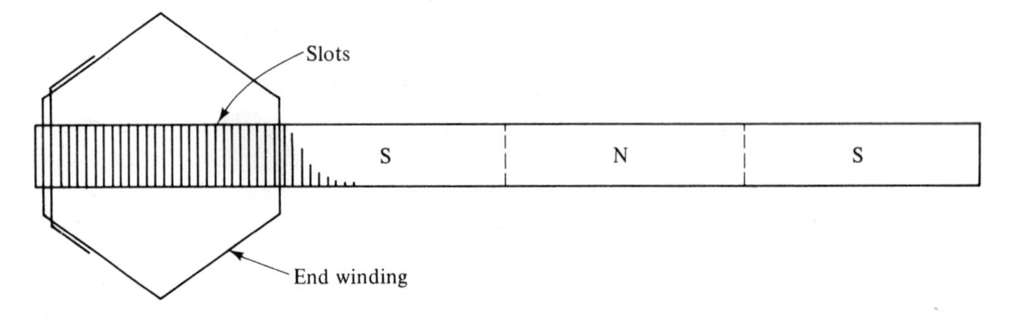

Fig. 6.27 The problem of long-pole pitches in longitudinal-flux machines

Only one primary coil is shown in the plan view of a four-pole, 50 Hz motor designed for a synchronous speed of 300 miles/h which, after allowance for slip, might reasonably be expected to have a full-load speed of 250 miles/h ($\approx$400 km/h). The problem of packing in the primary end windings from a 33 slot/pole machine with a pole pitch of 5 feet is acute, requiring a region perhaps 3 feet wide on either side of the one-foot wide pole area. Since the value of G ($\propto p^2$) is naturally high for this speed, the restrictions that p be large and f small can be relaxed and the reverse technique of increasing f introduced. There are, however, two possible deterrents to the use of this latter technique. The first occurs because a high-speed motor is often a high-*power* motor and the cost and weight of a solid-state frequency converter for a power handling capacity of perhaps 5 MVA may each be 4 or 5 times that of the motor itself. If generation is to take place on the moving part of the system, there may then be a sound case for attempting to design the linear motor so as to benefit from the higher frequency. The second deterrent, however, is ever present in the fact that the slot depth of the primary is quite severely restricted by the leakage reactance and the latter rises in proportion to frequency for any given geometry. A high-frequency motor is therefore a shallow-slot motor and the value of current loading is accordingly small. The only way in which the thrust can be made up to the desired value is by increase of pole area over that required for a mains-powered machine.

If the demands of a high-speed vehicle should be such as to prohibit on-board generation and make the cost and weight of conversion equipment too high it may be necessary to modify radically the whole topology of the linear motor primary so that it can accept raw mains frequency of 60 Hz (50 Hz in Europe).

The problem of the electric circuit can almost certainly be solved by resorting to the use of very short-chorded coils and accepting a greater proportion of primary ohmic loss (for $\frac{1}{3}$ chording the primary loss is doubled compared with a full-pitch winding). But the magnetic circuit cannot be similarly treated. The whole of the flux per pole must, at two instants per cycle, have space to pass from one pole to the next through both primary and secondary core. Even allowing for the fact that linear motors are designed with wide slots and narrow teeth, as discussed also in section 6.4, a tooth width only $\frac{1}{4}$ of a tooth pitch demands core iron 15 in deep which is quite unacceptable in terms of weight and cost.

In 1969, a way to redesign the shape of the magnetic circuits of linear motors was invented at the Imperial College, London.[15] The technique is available for a wide range of linear machines, including tubular types, but the description was not included until this point because the invention had its origins in high-speed machine design and to emphasize this fact is to help indicate to the designer the applications for which it offers greatest benefits.

Instead of the flux being compelled to pass along the core in the direction of motion, as is the case with all designs which represent split-and-unrolled rotary machines, the pole area is effectively divided down the middle, as shown in Fig. 6.28. Now all the flux from a N-pole can pass *transversely* through a core which is laminated transversely, for a S-pole exists alongside it, as shown. The core depth is now seen to be independent of pole-pitch and dictated rather by the *width* of the motor and the ratio of tooth width to tooth pitch. In the example chosen earlier the reduction in core depth is from 15 in to 1·5 in assuming no saturation. The number of magnetizing ampere-turns required to overcome core reluctance is basically reduced by $(p/w)^2$, where w is the width of the motor. The designer may elect to forgo this benefit in return for the ability to reduce the core depth still further and run the steel into such a degree of saturation as to reduce its effective permeability to $(w/p)^2$ times its unsaturated value. The core depth in the example above now falls to a fraction of an inch and the designer may now consider dispensing with the luxury of a laminated

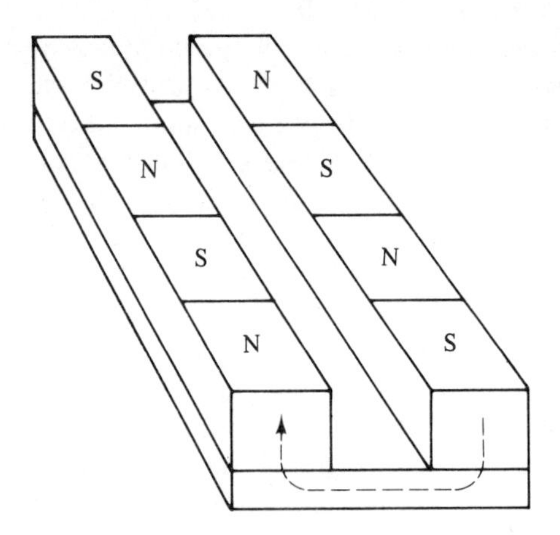

Fig. 6.28 Principle of the transverse flux concept

secondary core, if the latter is much longer than the primary, and using a far cheaper steel sheet.

The tubular versions of the transverse flux machines ('TFMs') also show considerable advantages.[16] The windings are as shown in Fig. 6.29. In (a), the helical three-phase windings form a six-start, screw-thread shape, each passing across a diameter at one end of the tube and having its feed point at the other. Alternatively,

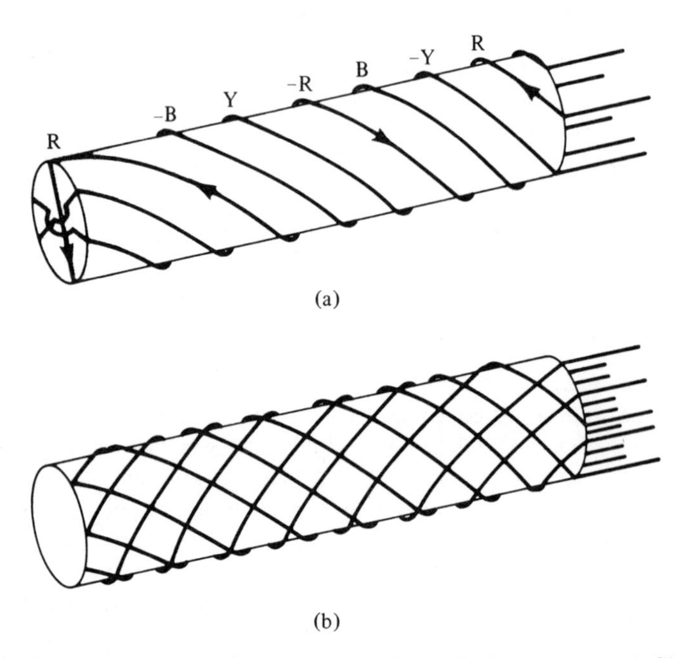

Fig. 6.29 Winding arrangements for transverse-flux, tubular motors. (a) Simple helices produce both rotation and axial thrust. (b) A double-helix system produces thrust only

windings could be fed from both ends. This system will both propel and spin the secondary, which must of necessity be a sheet-rotor construction. But the pole-pattern of the primary is seen to be such that the flux always passes through the central core in diametral planes and all the disadvantages of two-dimensional core lamination are eliminated, the inner core consisting of a stack of circular laminae, exactly like the structure of conventional rotating machines. Outside the primary winding the flux has a two-dimensional spreadability, reducing the effective reluctance values below those shown in Fig. 6.22. If the component of field producing spin in the transverse flux tubular motor is not required, a double-layer, counter-wound helix can be used as shown in Fig. 6.29(b). TFMs represent, in biological terms, a whole new sub-family of machines, some species of which have been described elsewhere,[17] but there are clearly many topologies still unexplored.

6.3 Application Considerations

The history of technology teaches us that there have been fashions in engineering as has been the case with clothing. In the early part of the nineteenth century, 'Can it be done at all?' was the popular question. As the century progressed, efficiency began to exert a greater and greater influence on the designers of all types of machine (electrical or not). This quantity, and for a.c. machines power factor, held the lead until increased transport facilities and regular air transport in particular, forced the emphasis to change from efficiency to power/weight ratio, for power/weight went hand-in-hand with power/cost and the 'age of the accountant' was at hand, where *overall* economics were replacing the short-sighted 'efficiency only' criterion. This process is continuing at the present time, except that low maintenance charges, reliability, absence of pollution, and absence of noise are supplanting the weight and cost criteria. The 'Can it be done at all?' criterion is, to some extent, back in fashion. 'Ask not what it costs to put a man on Mars. Ask only what are his chances of return.'

Since only the Electropult (abandoned by the 1950s) and a few liquid metal pumps (used by those who researched into nuclear power station possibilities) had ever been seen to be useful, the doubts about the more general use of linear motors in industry were great, despite the changing fashions. In the 1950s the linear motor was still in the 'Can it be done at all?' stage, but with this advantage. Modern tools and materials enable modern research to advance at a much greater rate and only ten years after the first linear motor paper since Shturman (Laithwaite, 1957)[18] linear motor applications were seen to be settling down to form three natural groups, which can be described thus:

(a) *Force machines* in which the full extent of the movement is usually small. Here efficiency, power/weight and power/cost are replaced by force/input, force/weight and force/cost as criteria of excellence. The benefits are to be found in absence of moving parts, especially wheels, giving rise in turn to easy maintenance and a high reliability. The flexibility both in design and in use is often the reason for the replacement of hydraulic or pneumatic actuators by linear motors, since the demand of a longer steel rod (in the case of a tubular motor) is less expensive than the requirement of a whole new hydraulic ram. A linear motor can be fed from a flexible cable which can be reconnected in a new location simply by removal and replacement of a plug in a socket. Electric machines cannot compete on a force/weight or force/input basis with either pneumatic or hydraulic thrust units, especially the latter, for compared with the thousands of pounds per square inch available as oil pressure, ordinary electric motors do not attain more than *one* pound per square inch of pole area. Their performance in this respect can often be improved by a factor of 10 or more, if they are effectively cooled and only required to be used for perhaps 2 seconds every 20 minutes. Such requirements frequently arise in industrial processes, for example in the traverse mechanism for cranes and for conveyor operation or sliding doors. The best designed actuators for use on direct mains supply can usually provide thrust at the cost of 30–40 watts per lbf ($\approx$5 Newtons). The advantage which should not be forgotten is that while other types of actuator are required to make physical contact with the workpiece to be moved, a linear induction motor need have no physical or electrical (via electric arc) contact with the load. This fact is exploited in using linear motors for uniformly back-tensioning aluminium sheet during processing, without scratching the surface.

Force machines work well in closed-loop automatic control systems. Both a.c. and d.c. types are readily adaptable and if great accuracy is required the continuous conductor of a sheet-rotor motor is sufficiently homogeneous to allow the position accuracy to be no worse than that of the sensing device. In one optical experiment designed at Imperial College, London, a two-phase linear motor measuring only 4 in $\times$ 1 in pole area, monitored from a laser sensing device, achieved an accuracy of $\pm$20 Å ($\approx$0·000 000 08 in).

(b) *Energy machines* are used, as their name implies, as producers of kinetic energy. They are accelerators and the appropriate criteria associated with input, weight, and cost are all measured as energy efficiency, energy/weight, etc. Experience with this type of machine has shown that the most likely advantages over mechanical or other means of propulsion are in first cost, repeatability of operation to great accuracy, and reliability. Chemical means, such as cartridges, are obviously better in energy/weight ratio but can only guarantee a relatively low accuracy in terminal velocity and are both noisy and polluting. Following the Electropult, which was a machine of this class, the next large-scale linear motor built has been working at the Motor Industry Research Association Laboratories at Nuneaton, England since 1968. It is used to accelerate automobiles containing human dummies rapidly up to one of a number of fixed speeds, and *at* the steady speed the vehicle is crashed into a concrete block. On impact, the linear motor primary is automatically decoupled and plunges below the crash block into a tunnel where it is stopped by aircraft-type arrestor gear. The cost of the energy per launch is approximately $\frac{1}{2}$p. The cost of the crashed car may be £2000. A similar machine has been in use for almost as long a period for rope testing at the National Engineering Laboratories at East Kilbride, Scotland.

(c) *Power machines* are almost restricted to high-speed applications, not so much because the lower speed machines may waste a sizable proportion of the input power, but because of the difficulty of removing that wasted power which, in the main, appears as heat in the secondary conductor. The only possible exception occurs where flat linear motors are used to operate on the face of a disc when the area of disc available to be blown by cooling air may be many times the pole area. The use of multiple disc arrangements on a common shaft is attractive in that several standard, relatively small, primary units can operate on each face of each disc. This has a twofold advantage:

(a) If a greater load is demanded extra primary units and discs can be fitted. If the load is reduced, the appropriate number of primary units can be switched off.

(b) If one primary unit is damaged or fails, its loss until it can be replaced is such a small percentage of the total power that the main drive need not be stopped even whilst the new unit is being fitted.

The application of linear motors to high-speed transport and its inter-relationship with the topic of electromagnetic levitation are subjects needing special treatment and only a few references are included here.[19,20,21]

6.3.1 *Oscillating linear machines*

Industrially, linear motion is much more usually a 'go-and-return' requirement than it is one of continuous motion. Yet self-oscillation is seldom used even though both synchronous and asynchronous mechanisms are known and have been investigated. It should be noted, however, that this general statement refers only to what might be termed 'real' machines, for very small self-oscillating linear devices of the synchronous type (both moving conductor and moving iron forms) belong in the electronics world and are generally known as 'loudspeakers', which of course are manufactured in very large numbers.

There are fundamental reasons why linear self-oscillators are not acceptable commercially. To understand these reasons analogous quantities to those in successful rotating machines must be appreciated. Centrifugal force is the 'enemy' of the rotating machine designer. It prevents him from increasing the power/weight ratio of a given machine by simply increasing its speed, for it is only the *force* per unit pole area which is limited by saturation of the magnetic circuit and temperature rise of the electric. Composite rotating parts including insulated windings in slotted steel are more liable to fly to pieces at lower speeds than truly solid rotors and it is solely for this reason that reluctance-type aircraft alternators were developed in which the rotor carried no winding. (In some designs the single exciting coil was mounted on the stator concentric with the shaft, thus making a 'transverse flux' cylindrical machine.)

Naturally the linear motor designer is not tormented by '$(mr\omega^2)$' but discovers a comparable 'enemy' in the kinetic energy at the centre of the oscillation that must be created and destroyed twice per cycle, even though the machine is running on no load. If this 'circulating energy' is supplied electrically, the losses associated with it can quite easily exceed those associated with the load forces, and the resulting efficiency is likely to be low. Nor does the linear motor designer escape the possibility of the moving part breaking up, for the maximum linear acceleration it must withstand is given by $\omega^2 a$, where a is the semi-amplitude of oscillation. Comparing longitudinal forces in a linear machine with radial forces in a rotary, a is seen to correspond to the rotor radius r. This comparison extends to other aspects also, for example the kinetic energy of the rotary machine is $\frac{1}{2}I\omega^2$, where I is the moment of inertia $(=\Sigma\, mv^2)$, while that of the oscillating member as it crosses the centre of its travel is $\frac{1}{2}mv^2 = \frac{1}{2}ma^2\omega^2$.

The second fundamental consideration is that it is difficult to construct multipolar linear synchronous devices so that 3600 strokes per minute are almost a necessity in the oscillating, mains-fed machine. The circulating energy of a machine with more than a small stroke ($>0{\cdot}1$ m) will be prohibitively large unless the design is such as to allow secondary core steel to be stationary. One design of such a machine is shown schematically in Fig. 6.30. A and B are coils carrying direct current that set up flux in

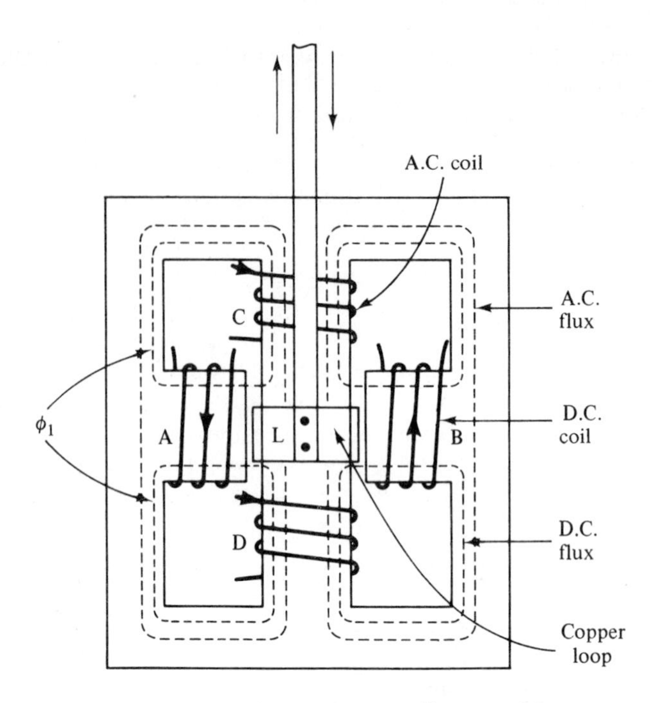

Fig. 6.30 As oscillating synchronous linear machine

paths denoted ϕ_1. Coils C and D carry alternating current and constitute the primary of a transformer whose moving secondary is the solid loop L. Even with this arrangement the efficiency was found to be low for a stroke of 3 in, for the maximum linear velocity at 3000 strokes per minute (this was a British experiment with $f = 50$ Hz)[22] was less than 40 ft/s which, even if it were maintained throughout the motion would be too slow to give a high value of Goodness factor (a large value of airgap having been introduced by the necessity for the conducting loop to be the sole moving part).

Relaxation of the 'only two-pole systems are practical' rule seemed to be possible using two linear induction motors placed back-to-back so that their travelling fields were directed towards the centre of the track. It has been shown[13] that the nature of an induction motor is such that, provided $G > 1$, the system is capable of building up an oscillation in a short secondary whose amplitude and frequency are functions of many parameters. The designer can thus easily arrange for an amplitude of say, 6 feet, with a frequency of two cycles per second. Unfortunately the nature of an induction machine is also such that in accelerating the secondary, of mass m, from rest up to synchronous speed v_s, the heat loss that appears in the secondary is equal to the full-speed kinetic energy $\frac{1}{2}mv_s^2$. Deceleration to rest in a 'plugging' field incurs a further secondary loss of $\frac{3}{2}mv_s^2$ so that in one complete cycle of operations, the secondary heat loss is 8 times the peak kinetic energy, and is seen to be prohibitively large for most practical purposes.

The only purpose in including considerations of linear oscillators is that relief from the burden of the circulating energy can be sought by 'tuning' the moving mass with the use of mechanical springs. Once the primary coils are required only to make up the spring and frictional losses, the possibility of commercial machines is revived and work continues to this end in several parts of the world at the present time.

6.4 Typical Applications

While it is relatively easy to make comparable cost calculations as between a linear and a rotary drive doing the same job on the basis of efficiency, power factor, and first cost, the advantages of most of the applications for linear motors to date have been seen at once to lie in other aspects such as convenience, reliability, 'self-containedness', and so on. It is much more difficult to put a cash value on such things as absence of noise, portability of apparatus, and similar features than it is on first cost or efficiency. The main proof that linear motors are a success must lie, at the present time, in the large number of satisfied customers who have bought and used linear machines. Over 50 000 units have been installed in the UK alone by the British company, Linear Motors (Lintrol) Ltd. The reliability is such that a single service engineer, only 70 per cent occupied in this capacity, is well able to maintain all the units in working order. Because of the element of 'adventure' which has often accompanied initial orders, the machines purchased have tended to be restricted to small 'force' machines on the grounds that, if they failed to satisfy, not much capital would be lost.

Some of the earliest commercial machines available were of tubular construction with a relatively short traverse (a few inches maximum) and it is interesting to note that their 'selling point' was that they could achieve a constant thrust over a much greater distance than could a simple electromagnet and keeper. Accordingly, they were advertised as 'Long-stroke Actuators', and found almost immediate application in post office parcel-sorting equipment. One of the attractive features of such an

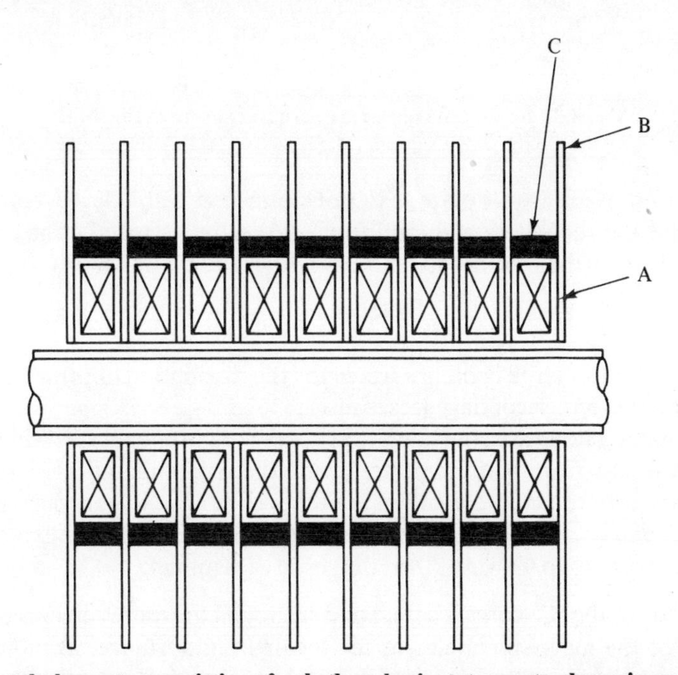

Fig. 6.31 A tubular motor consisting of only three basic stator parts, shown in cross-section

application was that unlike hydraulic actuators there was zero possibility of a burst oil pipe, an event that was almost intolerable in the vicinity of goods wrapped in paper and cardboard. Other applications in the early 1960s included the opening and closing of curtains and the operation of pen-recorders and X-Y plotters. It should be noted how, at that time, the size of motors was tending to become smaller even though as a consequence the force/input, force/weight, and force/cost were all reduced.

Ingenious methods for reducing manufacturing costs were developed at this time; two examples will illustrate this point. Figure 6.31 shows a cross-section of a tubular motor whose primary consists of only three basic elements: coils, core plates, and 'C'-cores. Figure 6.32 shows an exploded view of this type of thruster. The core

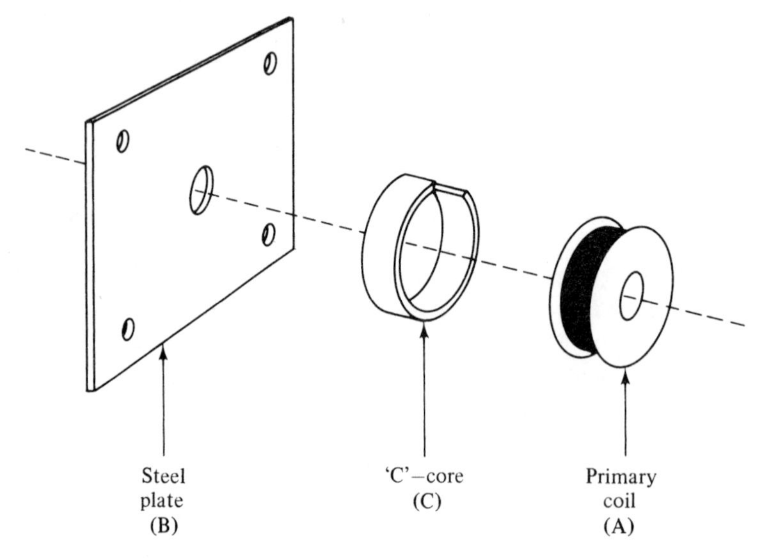

Fig. 6.32 Exploded view of the actuator shown in Fig. 6.31

plates consisted of mild steel plate $\frac{1}{16}$ in thick punched with a large, central hole to accommodate the secondary rod, and four smaller holes to take the locating and clamping rods when the whole is assembled. These core plates serve a triple purpose:

(a) They constitute what would be termed the primary 'teeth' which project right to the surface of the primary bore.
(b) They provide a large cooling surface for the dissipation of primary losses (and to some degree, secondary losses also).
(c) They allow a large area outside the primary core proper across which magnetic flux can pass from plate to plate, giving the combined effect of a very small primary core reluctance even without the 'C'-cores. So beneficial is this effect that the input power per pound thrust was only raised (in one typical machine tested) from 45 to 48 watts/lb as the result of removing the 'C'-cores entirely.

The purpose of the 'C'-cores is a dual one, although the more obvious one, namely the closure of the magnetic circuit, is the lesser in importance, as indicated at (c) above. The other use is to provide mechanical protection for the primary coils.

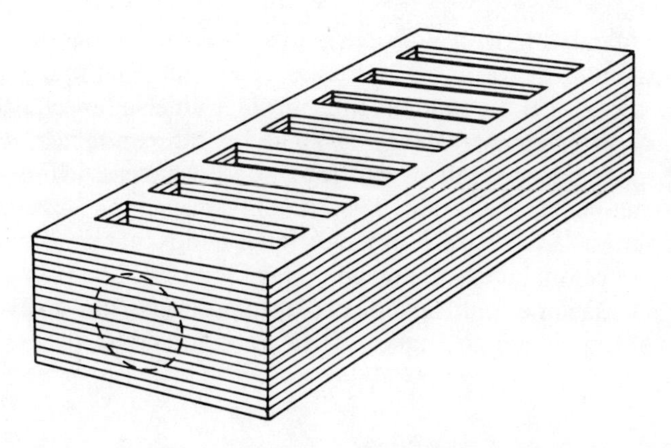

Fig. 6.33 A tubular motor construction in which the stator requires only two basic parts. (The steel portion only is shown)

The second form of construction is illustrated in Fig. 6.33. A stack of punchings each containing a row of rectangular slots is constructed, being impregnated and baked to produce a mechanically solid block. The block is then drilled from end to end, in the position indicated by the dotted circle in the figure, to take the secondary rod. Individual coils are then inserted into the slots and a locating notch (not shown) locates each coil with its central hole in line with the hole in the steel core. This construction thus limits the number of primary assembly parts to two—coils and steel punchings. It should be noted that the steel punchings unaffected by the drilling operation, i.e., those on either side of the hole, do not constitute in any way a magnetic short-circuit, for all can share in the 'transportation' of flux from a cylindrical N-pole in the bore to the neighbouring S-pole. The primary magnetic circuit of this type is generally of lower reluctance than is that of the type shown in Figs. 6.31 and 6.32. The power required for that earlier machine was of the order of 40 watts/lb, that of the later construction 30 watts/lb. The earlier model was the lighter of the two and it provides an excellent example of the rule that where the ratio of pole pitch to airgap (entrefer) is small the machine should be designed to have wide slots and narrow teeth. The axial length of the coils in the earlier type was of the order of $\frac{3}{8}$ in compared with a tooth width (core plate thickness) of $\frac{1}{16}$ in. Yet the five-second rating of these machines gave ten times the thrust obtainable on continuous rating, indicating that in the latter condition the flux in the core plates was below $\frac{1}{3}$ that which it was possible for them to contain without saturation.

6.4.1 *Liquid metal pumps*[23]

There is one notable exception to the statement that the revival of interest in linear motors began in 1946 with the Electropult. The earlier indication that nuclear power stations might become feasible led to the development of liquid metal pumps with no moving parts, which might be used to transport the heat from the centre of a reactor core to the outside. Several types of electromagnetic pump were developed including two linear types—the *induction* pump (a double-sided sandwich type) and the *conduction* pump (a double-sided linear d.c. motor). The metal to be pumped was a sodium/potassium mixture with almost as high a conductivity in its hot liquid state as that of solid aluminium at normal temperatures. Even though the speed of the liquid was relatively slow (20–30 ft/s) and the tube through which it flowed had to consist of a thick wall of stainless steel (a reasonably good conductor, virtually only paramagnetic), giving rise to high losses due to induced currents in the wall, efficiency values over 30 per cent were obtained with the induction type. A tubular induction pump had been tried earlier but discarded on the grounds that if a winding failed, the pipeline must be broken before a new unit could be fitted. The primary blocks of a double-sided, flat machine could be replaced by remote control—an essential feature for a motor operating in a highly radio-active contaminated zone.

6.4.2 *Applications for 'force' machines*

The largest numbers of individual motors employed at the present time are those used for the traverse motions of travelling cranes and for the opening and closing of sliding doors. Removal of objects for conveyor belts has effectively already been mentioned in connection with post office applications. To this can be added the driving of conveyor belts themselves. For several years some of the baggage handling facilities at Heathrow Airport, London have been linear-motor operated. Among the medium-sized machines (several hundred pounds thrust) are included those for back-tensioning of aluminium sheet (already mentioned), positioning of X-ray equipment in hospitals, production line handling in two dimensions, mobile racking (as in Patent libraries), and programmed drilling jigs. In many of these applications automatic electronic units operate on the linear motors to provide position, speed or acceleration programmes. Other similar units are in operation commercially for powering the trolleys in extrusion processes and for rotary drives such as are used on roll-out tables in steel mills. Some of these applications are illustrated graphically in Fig. 6.34. In the world as a whole, Linear Motors (Lintrol) Ltd have supplied machines for over 5000 different applications.

In smaller sizes, machine tool positioning, adjustment of optical systems, and road barrier operation are typical examples of the exploitation of the short-rated motor where a one-second operation every 20 minutes may enable thrusts of the order of 20 times the motor weight and perhaps 100 times the weight of the moving secondary to be achieved. It is interesting to record that the maximum linear acceleration produced by induction is of the order of 1200 times that due to gravity, a figure which approaches that produced by explosive cartridges.

6.4.3 *Other applications*

High-speed transport systems apart, larger machines have been developed for the cold forming of metal (impact extrusion) and for ship-testing tanks and similar applications where the acceleration required excludes the use of wheels on account of wheel slip.

Looking to the future it is interesting to speculate on other possible uses for direct linear drive. For example one could envisage a ski-run where the chair lift is replaced by a pair of linear motor tracks into which the skis could be fitted (having first had aluminium strips clipped on to them if they are not made of metal). The enthusiast

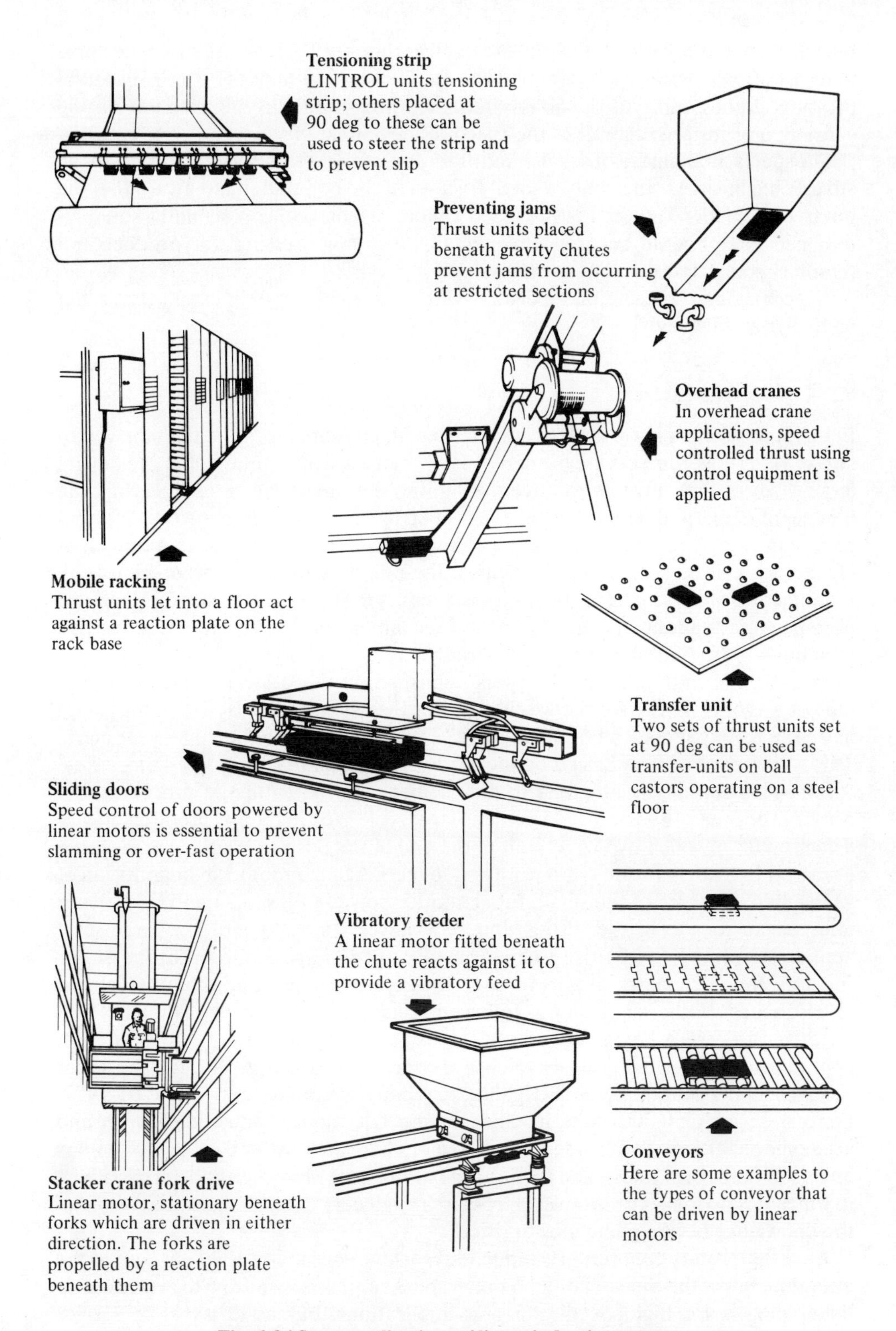

Fig. 6.34 Some applications of linear induction motors

may then effectively 'ski' up the slopes as well as running freely downhill. In the home a push-button-operated smoothing iron could be used to take the drudgery out of pressing clothes and with a suitably designed pressing table the pressure on the garment is adjustable as well as the speed of movement of the iron.

The spectacular nature of self-levitating motors makes them an ideal subject in the advertising business since only a small fraction of the population are aware that the forces of induction require neither electrical nor mechanical contact and a suitably-designed model could be used effectively as 'window dressing' for products not remotely connected with linear motors!

A recent comprehensive paper dealing with the application of linear motors is that by Sadler and Davey.[24]

6.5 Commercial Interest

In the pure sciences knowledge accumulates with no more incentive than the will to know. Many governments are prepared, as of course are many individuals, to devote large slices of their income to adventuring into the unknown. In engineering, the pioneer of modern times is severely constrained by the need to demonstrate financial gain in the very short term, or he goes hungry. Evidence of success in a branch of technology can be seen in two activities, the one academic, the other industrial. Manufacture throws up its problems for academics to solve. Academics for their part become curious about a particular facet of technology and write papers for learned societies—try to 'sell' their ideas industrially.

In 1960, the total number of papers published on the subject of linear electric motors was only sixteen for the whole year, for the whole world. By 1968, this figure had risen to 10 papers a month, and in 1977 it may well reach the 200 a year mark. This is the growth as evidenced by the academic community.

Following spasmodic attempts to adapt linear motors to textile processes that occurred between 1895 and 1945, and the large scale Westinghouse aircraft launcher 'Electropult' of which two were built at the end of the Second World War, Skinner Precision Industries Inc. of Ohio were first to market linear motors in quantity, all of which, initially, were of the small, tubular kind shown in Figs. 6.31 and 6.32. Their sales began to rise in the 'fifties, but they never ventured into the large-scale applications. In 1966, a British company made a medium-sized motor for the Motor Industry Research Association who used it to crash test cars and other vehicles at precise speeds. Other applications followed rapidly. By 1974, it was estimated that linear motors were being used on a global scale for more than 3000 different applications, only a handful of which were for transport, even though the latter received 95 per cent of the publicity that frequently surrounds innovation. By 1977, there were over 50 000 units installed in the UK alone. Industrial growth and academic paper growth exploded side-by-side. There are now well over 3000 papers on linear motor technology and the subject has taken its place alongside many others in which it is no longer profitable for research engineers to try to digest the whole of the knowledge before going into practice.

As in their rotary counterparts induction machines occupy over 90 per cent of the attention, but at the time of going to press, some emphasis is shifting to synchronous linear motors for high-speed transport applications, but none is yet in service commercially.

References

1. A. Zehden, US Patent No. 782312 (1905).
2. 'A wound rotor motor 1400 ft long', *Westinghouse Engineer*, **6** (1946).
3. E. R. Laithwaite, 'The goodness of a machine', *Proc. IEE*, **112**, 3 (1965).
4. R. L. Russell and K. H. Norsworthy. 'Eddy currents and wall losses in screened-rotor induction motors', *Proc. IEE*, **105A**, 20 (1958).
5. H. Bolton. 'Transverse edge effect in sheet-rotor induction motors', *Proc. IEE*, **116**, 5 (1969).
6. T. W. Preston and A. B. J. Reece. 'Transverse edge effect in linear induction motors', *Proc. IEE*, **116**, 6 (1969).
7. G. I. Shturman. 'Induction machines with open magnetic circuits', *Elektrichestvo*, No. 10 (1946).
8. F. C. Williams, E. R. Laithwaite, and L. S. Piggott. 'Brushless variable-speed induction motors', *Proc. IEE*, **104A**, 14 (1957).
9. D. Tipping. 'The analysis of some special-purpose electrical machines', Ph.D. thesis, Manchester University, 1964.
10. 'Study of linear induction motor and its feasibility for high-speed ground transportation', Final Report No. 67-1948 (Study Contract No. C-145-66/NEG) by Garrett AiResearch Division for the Office of High-speed Ground Transportation, US Department of Transportation (1967).
11. S. Yamamura. *Theory of Linear Induction Motors*, Halsted Press, division of John Wiley, New York (1972).
12. F. C. Williams, E. R. Laithwaite, and J. F. Eastham. 'Development and design of spherical induction motors', *Proc. IEE*, **106A**, 30 (1957).
13. E. R. Laithwaite. *Induction Machines for Special Purposes*, Newnes, London (1966).
14. E. R. Laithwaite. *Linear Electric Motors*, Mills and Boon, London (1971).
15. E. R. Laithwaite, J. F. Eastham, H. R. Bolton, and T. G. Fellows. 'Linear motors with transverse flux', *Proc. IEE*, **118**, 12 (1971).
16. J. F. Eastham and J. H. Alwash. 'Transverse-flux tubular motors', *Proc. IEE*, **119**, 12 (1972).
17. J. F. Eastham and E. R. Laithwaite. 'Linear-motor topology', *Proc. IEE*, **120**, 3 (1973).
18. E. R. Laithwaite. 'Linear induction motors', *Proc. IEE*, **104A**, 18 (1957).
19. E. R. Laithwaite and F. T. Barwell. 'Application of linear induction motors to high-speed transport systems', *Proc. IEE*, **116**, 5 (1969).
20. E. R. Laithwaite. 'Electromagnetic levitation', *Proc. IEE*, **112**, 12 (1965).
21. J. F. Eastham and E. R. Laithwaite. 'Linear induction motors as magnetic rivers', submitted to *Proc. IEE*.
22. E. R. Laithwaite and R. S. Mamak. 'An oscillating synchronous linear machine', *Proc. IEE*, **109A**, 47 (1962).
23. L. R. Blake. 'Conduction and induction pumps for liquid metals', *Proc. IEE*, **104A**, 13 (1957).
24. G. V. Sadler and A. W. Davey. 'Applications of linear induction motors in industry', *Proc. IEE*, **118**, 6 (1971).

7 Small Power (Fractional Horse-power) Motors

7.1 Introduction

Since the kilowatt has been designated as the standard unit of mechanical power output the term 'fractional horsepower' must now be replaced by 'small power motors' as used in the British Standard 5000. The definition in both cases specifies motors with outputs below 0·75 kW per 1000 rev/min so that the watt, which avoids decimal points and vulgar fractions, is the most practical unit for output. In very small motors of special design, such as those used in electric clocks, the input has dropped to a few watts and the output well below one watt. Such motors are mostly specified in terms of their rated speed and permissible torque loading, but for output power the milliwatt is the most appropriate unit.

Small motors are produced in a large variety of electrical and mechanical types. The IEC frame sizes 56, 63, 71, and 80 which are also included in most national standards are mainly used for induction motors suitable for industrial applications, and are similar in construction to the larger integral or kW output motors. Though the dimensions which are standardized are not sufficient to ensure interchangeability in all cases, the chances of obtaining a quick replacement or reducing the number of spares which must be stocked are substantially increased.

In many applications the required characteristics and quantities to be produced are sufficient to warrant the design of a special motor and possibly even new laminations. More moderate quantity requirements restrict the degree to which an existing design can be economically modified and only close liaison between motor and equipment manufacturer can ensure that the best compromise has been found. Complete motors and skeleton type motors which are protected by the equipment they drive are generally easier to build in than stator rotor units where much more care is required to ensure that the alignment and the airgap are maintained. Attempts have been made to standardize appliance, skeleton, and stator/rotor motor units but requirements of applications and manufacturers were too diverse to enable any worthwhile standardization to be agreed.

Clock motors are one typical example of a special design where low cost and ease of production were of such value that an efficiency of the order of 1 per cent is considered to be acceptable. However, the input power is only about 3 W for the full rated output of 30 mW, and production quantities are sufficient to warrant investment in appropriate tooling. These low-cost synchronous motors also find application in timers, programme controllers, recorders, and a variety of instruments. In contrast, small d.c. motors for portable tape recorders, cine cameras, and the like, though still requiring to be economical to manufacture must have as high an efficiency as possible. The small permanent-magnet motors which are manufactured for such applications are typically between 40 per cent and 70 per cent efficient, and larger motors are now powering electric drills, lawn mowers, and the like, making them not only considerably safer but also mobile and easier to use.

Where a motor has to be selected for a new product which is only likely to be produced in small quantities the choice is restricted to motors in current production and requiring the least possible modification. A number of small motors are produced by several manufacturers simply because they are commonly used in certain types of equipment (fans, record-players, tape recorders, etc) and these should, if possible, be given preference since procurement and subsequent replacement are made very much easier.

Such dimensional and output standards which apply to small motors have fewer types of mountings, enclosures, and shaft variants than their larger counterparts, and their number of combinations and permutations still exceeds economic stockholding levels by manufacturers and stockists. Since availability is invariably one of the selection criteria, manufacturers' stock lists should be consulted at an early stage and preference be given to those motors which are most likely to be held in stock. The scope for compromise in small motor applications is usually considerable and should be thoroughly explored to ensure that the cost of the motor is minimized, while ensuring that it has the required performance and reliability. Though the preferred dimensions and frame sizes of BS 2048: Part 1 (see chapter 5, section 5.2.7.7) which are in inches are now obsolescent, the specification will remain of interest as long as the American NEMA standards, with which it substantially agrees, remains in force.

A very large number of small-power motors reach the end user in equipments such as office machines (typewriters, copiers, and dictating machines), cash registers, accounting machines and small computers, machine tools, instruments, items of plant (heating and ventilating, processing, and process control), domestic, and other industrial appliances. Here the only safe source of replacement is the manufacturer of the equipment or his authorized agent or stockist.

7.2 Induction Motors

A summary of the essential theory and characteristics of induction motors is given in chapter 4 and, as has been pointed out in the introduction to this chapter, the smallest frame sizes of the IEC's new standard range (BS 4999: Part 10) house 'small power' induction motors. Their construction will be basically the same as the large motors in the range and consist of the following main sub-assemblies:

(a) a stator unit with windings distributed in a number of slots (typically 36, 24, and 12);
(b) a rotor mostly cage type die-cast unit with one or both ends carrying straight fins to act as fan blades for cooling;
(c) a shaft;
(d) bearings, predominantly single-row deep grooved ball races;
(e) end frames, one of which may be used to locate and fix the motor.

In the smaller sizes a number of special stator configurations with only one or two coils are used. Though electrically rather inefficient the single-coil shaded-pole motor provides low-cost driving power not only for record players and fans but innumerable other motor-driven devices. There are also a number of variations on rotors which are either of the conventional cage-, solid- (slotless-), drag-cup-, or, for synchronous motors, variable reluctance type. Wound rotors are, except for special types such as synchronous drive motors (Selsyns) not normally used as there are other more economical methods for speed control and starting. Some cage rotors may still be made with individual copper or brass bars brazed or hard soldered at each end to end-rings, but pressure or centrifugal casting of aluminium and its alloys lends itself more to less labour-intensive mass production. The polyphase induction motors will have their characteristics modified by rotor resistance as outlined in chapter 4, but in single-phase versions it is not only the slip at which maximum torque occurs, but the actual value of that maximum which is affected by rotor resistance.

In certain applications and motor types such as slow-speed fans, synchronous and high frequency motors the construction is 'inverted' and the rotor is made as a rotating cylinder around the central stationary stator.

7.3 Polyphase Induction Motors

The characteristics of these motors have been dealt with in detail in chapter 4, and it is only necessary to summarize the main reasons why, wherever possible, they should be used in preference to single-phase or commutator types.

Three-phase motors are inherently self-starting with high values of starting and pull-out torques. The windings are balanced and produce a better distributed rotating field than can be obtained from any windings supplied from a single-phase source and incorporating means for phase splitting. The absence of special starting switches or relays makes for better overall reliability and in machine-tool applications the ability to be brought to rest quickly by reversing 'plugging' or even to run in the reverse direction. The high starting torque will make the motor run up in fractions of a second and permit many more starts per hour than the single-phase version.

Unfortunately the three-phase voltages which are generally used in most countries using 50 Hz lie between 380 and 440 volts, which causes winding and safety problems in very small sizes. The lower output limit for motors in this range is around 150 watts. Because of the fine wires which have to be used, the insulation problems, and smaller production runs, the small polyphase induction motor may, despite its greater simplicity, be more expensive than the single-phase version with the same output.

The electricity supply in offices, shops, and homes was for many years almost exclusively single-phase. Even in factories there were many areas where it was not economical to provide three-phase supplies. The more extensive use of electricity for heating, lifts, and ventilating plant is likely to make three-phase supplies more generally available.

7.4 Single-phase Induction Motors

Millions of small motors are now used in equipment ranging from typewriters, cash registers, copiers, and dictating machines to fans, pumps, vacuum cleaners, and hair-driers, to name but a few. Most of these motors are either made by the equipment manufacturer or supplied to him in bulk by a manufacturer specializing in small motors. Such motors are often of the skeleton type, that is without proper enclosure, as they are mostly well protected by the machine they drive and only become accessible when covers, bearing appropriate safety warnings are removed. Though these motors are generally specially designed for particular applications some find many other varied uses.

The standard $\frac{1}{8}$, $\frac{1}{4}$, $\frac{1}{2}$ and $\frac{3}{4}$ h.p. motors of the now obsolete BS 170 are now replaced by metric sizes and preferred output ratings of 100, 120, 180, 250, 370, and 550 watts given in BS 5000: Part 11 (see chapter 5, section 5.3.2). The standard supply frequency and voltage recommended are 50 Hz 240 V single phase and 415 V polyphase. Outputs in watts additional to those given above are 2·5, 4, 6, 10, 16, 25, 40, 60, 80, 90 at the low end, and 750, 1100, 1500, and 2200 at the top. The classification 'small power' implies a motor with a continuous (or equivalent continuous) rating of 750 W per 1000 rev/min, so that a four-pole 400 Hz induction motor with an output of about 8·5 kW would still be included, though it would run from a 'special' supply.

For outputs above 100 watts as mentioned above it will usually be possible to obtain motors in a more or less standard frame size and, where necessary, have the

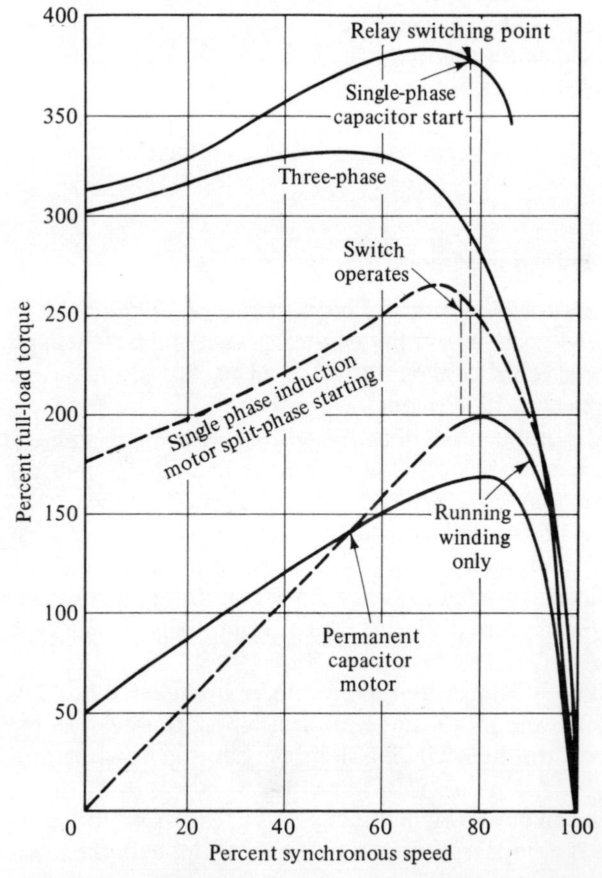

Fig. 7.1 Speed/torque curves for cage induction motors

choice of fitting either a three-phase or single-phase motor. Care must be taken in such cases to ensure that the single-phase motor will not be overloaded. In fan applications requiring only 0·25 to 0·4 p.u. starting torque the three-phase winding is also used for single-phase supplies by employing a start and run capacitor in what is sometimes known as Steinmetz connection. A 415 V star connected three-phase motor, when reconnected in delta to a single phase 240 V supply of the same frequency, with a capacitor across one of the other phases, may be expected to have an output rating of 70 to 80 per cent of its three-phase value. The exact value of capacitor will depend on the size of the motor and its electrical design, but for 230 V motors with three-phase ratings of 180 to 750 W will be of the order of 10 to 60 μF. The single-phase starting torque may, however, be as low as 10 per cent and the pull-out torque down to 60 per cent of their three-phase values.

Single-phase induction motors can be further classified according to the method of starting, which, according to the size and actual value of the torque required, will usually fall within one of the following:

(a) split-phase (with or without disconnection after running up) motors;
(b) repulsion-start induction motors;
(c) repulsion induction motors;
(d) capacitor-start motors;
(e) capacitor-start- and run-motors;
(f) two value capacitor motors;
(g) shaded-pole motors.

Figure 7.1 compares the various types of single-phase motors with a three-phase type.

7.5 Split-phase Induction Motors

The split-phase motor is one of the oldest types of single-phase motor built and currently still finds service as a utility motor in a host of different applications such as fans, machine tools, centrifugal pumps, washing machines, and business machines, in various sizes up to about 375 watts.

Essentially the motor consists of two separate and distinct windings on the stator; a main or running winding having a low resistance and high reactance and an auxiliary or starting winding having a high resistance and low reactance.

The difference in the inductance of the two windings causes a split in the single-phase power supply giving rise to a phase displacement between the currents in the two windings of somewhat less than 90° and thus producing a rotating field which reacts with the bars of the rotor, conventionally a cage type, to produce a starting torque.

At standstill both windings are required to be in the circuit to develop torque, as described, but after the motor has run up to approximately 75 to 80 per cent of synchronous speed, the main winding alone can develop nearly as much torque as the two combined.

Furthermore it can be shown that for speeds in excess of 90 per cent synchronous speed the motor develops less torque, for a given slip, with the auxiliary winding in circuit.

Consequently a centrifugal switch is connected in series with the auxiliary winding, to disconnect it automatically when 75 to 80 per cent synchronous speed has been reached. The motor then continues to run on a single oscillatory field in conjunction with the rotation of the rotor.

With the auxiliary winding so disconnected there ceases to be any rotating field in the motor and consequently it is not possible to reverse the direction of rotation until either the motor is brought to a standstill or the speed reduced sufficiently for the centrifugal switch to close and bring the auxiliary winding back into operation.

In practice the starting torques of split-phase motors range from around 130 per cent to 200 per cent of full running torque depending upon the class of insulation used. They should be used where starting is infrequent and the inertia is low and with Class B insulation they can run with a 35 per cent overload indefinitely.

A typical torque/speed curve and connection diagram are shown in Fig. 7.1 and Fig. 7.2(a) respectively.

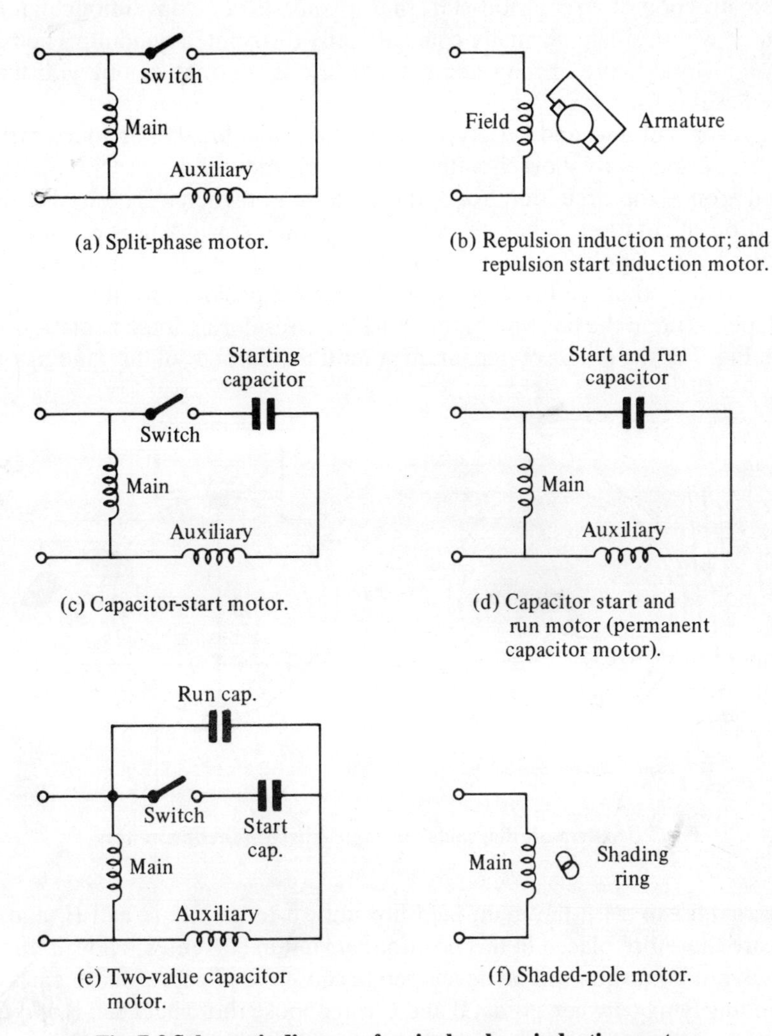

(a) Split-phase motor.

(b) Repulsion induction motor; and repulsion start induction motor.

(c) Capacitor-start motor.

(d) Capacitor start and run motor (permanent capacitor motor).

(e) Two-value capacitor motor.

(f) Shaded-pole motor.

Fig. 7.2 Schematic diagrams for single-phase induction motors

7.6 Repulsion-start Induction Motors

Although also ranking as one of the oldest forms of single-phase induction motor with a very wide popularity in the mid 1930s, the repulsion-start induction motor has now virtually been superseded by the capacitor-start and two-value capacitor types. Many of them, however, are still operating satisfactorily for such applications as pumps, compressors, and miscellaneous general purpose applications where the desired parameters are a high locked rotor torque and a low locked rotor current.

In dealing with the repulsion-start motor it should not be confused with the repulsion motor which is a different type and covered in section 7.19 of this chapter.

Basically the repulsion-start motor may be defined as a single-phase motor having the same windings as a repulsion motor but at a pre-determined speed the rotor winding is short circuited or otherwise connected to give the equivalent of a cage winding. This type of motor starts as a repulsion motor but operates as an induction motor with constant-speed characteristics.

The construction of a repulsion-start motor consists of a conventional laminated stator stack with a single normally concentrically distributed winding. The rotor is wound on a slotted core in the same manner as a d.c. armature but with the core laminated as it is for use on a.c.

As the rotor is of a wound variety, a commutator and brushes are necessary. The brushes are permanently short circuited and suitably mounted in order that they may be moved around the circumference of the commutator when it is desired to reverse the direction of rotation. When the stator (or primary) winding is connected to a single-phase a.c. supply, current is induced in the rotor by transformer action providing, that is, that the brushes are in the correct position, for the action of the motor depends upon the position of the brushes. Considering the schematic diagram shown in Fig. 7.3, the points of maximum potential difference of the windings will be

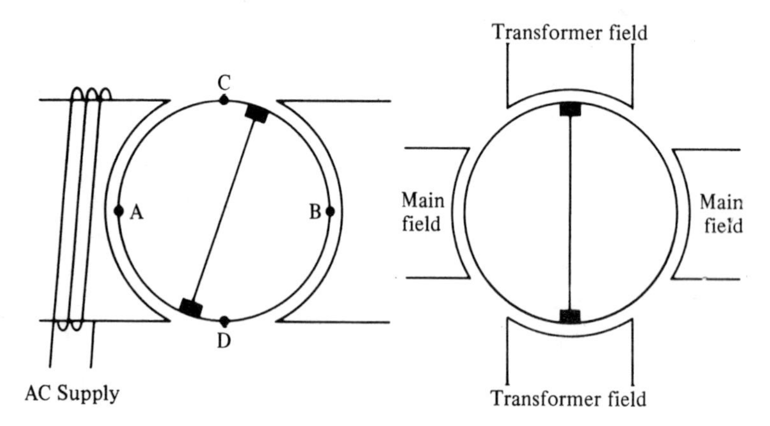

Fig. 7.3 Schematic diagrams for single-phase induction motors

in a horizontal axis with the main field flux shown by points A and B, and if the brushes are therefore placed in this position maximum current will flow in the short circuit link but no torque will be developed because the effects of those parts of the rotor winding lying between points B and C will oppose those between B and D, and similarly for A to C and C to D. If the brushes are now moved to a position at right angles to the horizontal axis (i.e., points C and D) there is no transformer action and hence no current will flow. An intermediate position is therefore sought where some unbalance of rotor torques will occur causing the rotor to rotate. The optimum brush position is found to be in the order of 17° from the horizontal field axis.

The speed/torque curve for an induction-start repulsion motor is similar to that of a d.c. series motor in that once starting torque has been produced the speed will rise indefinitely until limited by the load torque and losses. However, as previously stated, to prevent this at a predetermined speed of the order of two-thirds synchronous speed, a centrifugal switch short circuits the entire commutator and allows the motor to continue running as a normal induction motor. The centrifugal mechanism is also frequently designed to lift the brushes from the commutator and thus reduce friction and wear.

When running at speed the induced harmonic currents are very small due to the distributed rotor winding, and hence the torque curve is quite smooth. As the armature winding currents are produced by induction the motor can be designed for a low voltage to assist commutation. In addition because of the favourable ratio of high starting torque to low locked rotor current these motors are often able to handle their loads satisfactorily even when operated on low voltage.

Repulsion-start induction motors are made in many small sizes up to about 4 kW.

A typical torque/speed curve is shown in Fig. 7.4 (and a connection diagram in Fig. 7.2(b)).

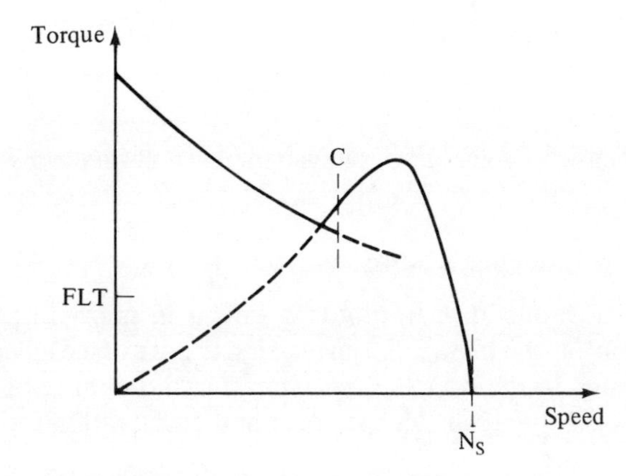

Fig. 7.4 Torque/ speed curve, repulsion-start motor

7.7 Repulsion Induction Motors

These may be considered as a variation of the repulsion-start induction motor already discussed in the previous section.

In order to avoid using the centrifugal mechanism to facilitate the shorting of the rotor windings at a pre-determined speed, the repulsion induction motor utilizes a rotor which has a permanent, deeply buried, low resistance cage with leakage slots between the bars and the repulsion winding. The high reactance of these slots ensures that the induced currents in the cage bars are small in the locked rotor condition and hence the motor has the normal repulsion motor characteristic up to about two-thirds synchronous speed. At higher speeds there is a marked increase in cage conductor current and the induction torque becomes greater than the repulsion torque, thus limiting the no-load speed to a little above synchronism.

Unfortunately as there is no centrifugal mechanism in this type, the brushes cannot be lifted from the commutator as for the repulsion-start induction motor and hence the brushes have to carry current continuously, which creates problems regarding radio interference.

By varying the wound and cage portions of the rotor, these motors can be custom built to suit many definite purpose applications, such as printing presses, textile machines, and laundry extractors.

A typical torque/speed curve is shown in Fig. 7.5 (and a connection diagram in Fig. 7.2(b)).

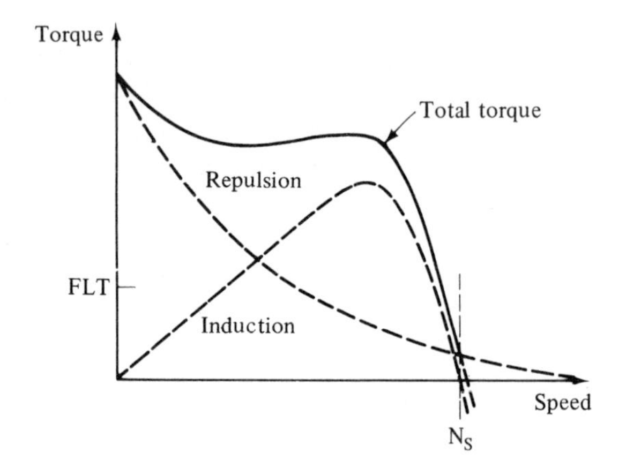

Fig. 7.5 Torque/speed curve, repulsion induction motor

7.8 Capacitor-start Induction Motors

The capacitor-start, induction-run motor is similar in many respects to the split-phase induction motor, the main difference being the use of a capacitor in series with the auxiliary winding. In addition, the capacitor-start motor has a better power factor resulting in considerably higher locked rotor and accelerating torques per ampere than a split-phase motor.

After starting, the auxiliary winding is removed from the circuit and the motor gives a performance almost identical to that of a split-phase motor running with constant speed characteristics. Consequently the capacitor-start motor should be used for requirements where the locked rotor and accelerating torques are beyond the capabilities of the split-phase type.

By varying the value of capacitor and rating of the auxiliary winding a wide range of starting torques can be designed into capacitor-start motors. The most usual value offered commercially is around 300 per cent full load torque which is in the same order as that for three-phase versions.

In common with split-phase motors, they can only be reversed at standstill or at a speed low enough for the centrifugal switch to be in circuit, unless special switches have been incorporated. If suitable provision is made for the motor to be reversed while running at speed the capacitor motor is found to be more suitable for this purpose than the split-phase variety, as it provides greater reversing ability at less watts input.

Capacitor-start motors come in many convenient sizes ranging around 125 to 746 watts and they may be used in virtually all applications. They are particularly suitable for duties that may involve overcoming high inertia loads as refrigeration compressors, hand tools, machine tools, and conveyer belt drives.

Types having a high starting torque are ideal for compressors, loaded conveyor belt systems, and reciprocating pumps, etc. Normal and low starting torque types are best suited to centrifugal pumps, machine tools, and fans, etc.

Although, normally, they are of a rugged construction, accurate machining, quality bearings, and the choice of a centrifugally or relay operated starting switch ensures quiet dependable operation and freedom from radio interference.

The capacitor used may be of the electrolytic type and mounted on the side of the motor.

Typical characteristics for the capacitor start motor are shown in Fig. 7.6 (and a connection diagram in Fig. 7.2(c)).

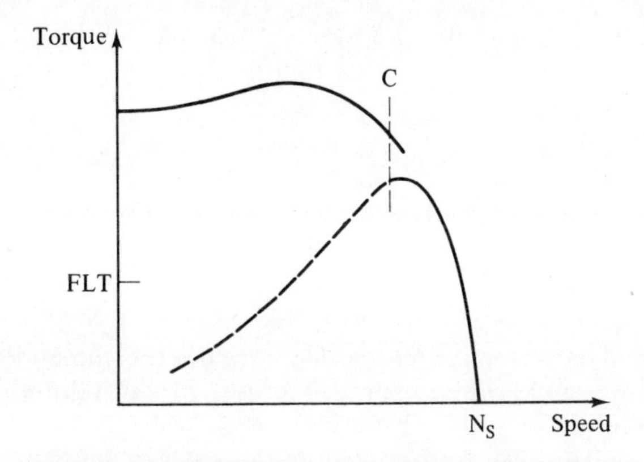

Fig. 7.6 Torque/ speed curve, capacitor-start motor

7.9 Capacitor Start and Run Induction Motors

In a similar manner to the capacitor-start motor, the capacitor start and run version has an auxiliary winding with a capacitor connected in series. However, in this case the capacitor and auxiliary windings are designed to be continuously energized and assist in producing a higher power factor than other designs. The two windings, main and auxiliary, can be made identical or unlike, the phase shift being achieved entirely by the capacitor. The operation of a capacitor start and run motor simulates very closely that of a true two-phase motor. The efficiency is comparable to, and sometimes better than, that of other single-phase motors.

Reversal may be achieved by changing the polarity of the windings by means of a simple single-pole double-throw switch and as both windings are always in circuit this may be carried out while the motor is rotating. In addition, owing to the absence of a centrifugal switch this type of motor may be shorter than other single-phase versions.

Locked rotor torque is low compared to other capacitor-start or split-phase motors and therefore capacitor start and run motors are generally applicable only to

Fig. 7.7 B56 frame size capacitor start motor; solid base and spigot mounting. (Courtesy GEC Ltd)

direct drive fan and blower applications. They are not recommended for belt-driven loads. Power ratings and normal operating speeds are also different from those of capacitor-start motors.

In installations where space around the motor is at a premium the phase-shift capacitor may be located in some remote position and in types where the two windings are identical only three leads or terminals need be used, one common and one for each of the two windings.

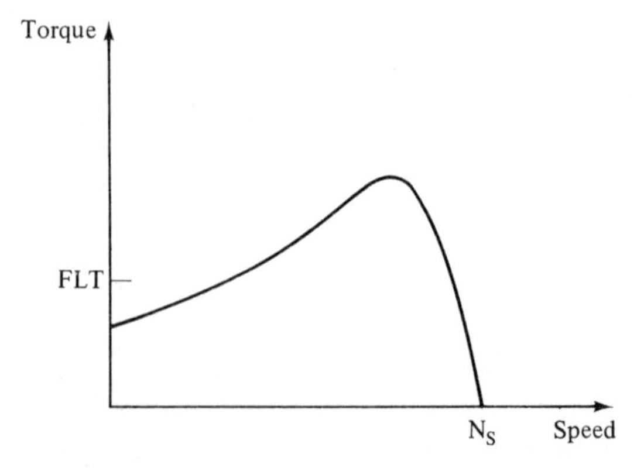

Fig. 7.8 Torque/ speed curve, capacitor start and run motor

More often than not capacitor start and run motors are designed and custom built for a specific application.

Typical torque/speed characteristics are shown in Fig. 7.8 (and a connection diagram in Fig. 7.2(d)).

7.10 Two-value Capacitor Motor

In construction the two-capacitor induction motor is very similar to the capacitor-start motor, only in this case the motor has two capacitors of different values in parallel with each other but connected in series with the auxiliary winding. One of these is designated as the starting capacitor and the other as the running capacitor. The starting capacitor is only in the circuit during the starting period, being cut out by means of a centrifugal mechanism, as in the case of the capacitor-start motor, when a predetermined speed is reached.

In effect the two-capacitor motor may be considered as a capacitor-start motor with a running capacitor connected permanently in circuit. The operation of the running capacitor may be understood by referring to ch. 4 p. 110 where the derivation of the single-phase rotating field was discussed. Here the resultant field was considered as being comprised of two component pulsating fields, a main field and the cross field separated by 90° both in time and space.

When the auxiliary winding is now maintained in the circuit by means of a running capacitor of the correct value it draws a current which leads the main field current by 90°. The net result is that the auxiliary winding assists in setting up a part or all of the cross field, thereby reducing or eliminating the magnetizing currents in the rotor and the inherent copper losses.

The resulting effect of this running capacitor on the motor is to give a marked improvement in performance and a reduction in losses. Full load efficiency and power factor are improved and the double frequency torque pulsations which tend to be inherent in single-phase motors are also reduced.

Increases in the order of 5 to 30 per cent for breakdown torque, and 5 to 20 per cent for locked rotor torque may also be obtained.

The impression may be gathered that the presence of the running capacitor will result in this type of motor giving a performance comparable to that of a true two-phase motor. It must be remembered, however, that the capacitor value chosen is normally 'tuned' for one particular value of load, it not being possible to duplicate two-phase motor performance at all values of load with a single capacitor. Different values of capacitor would be required for each different load.

Standard motors are normally designed for continuous operation and the reversing characteristics are similar to those of the capacitor-start motor. They are most used in applications requiring a high locked rotor torque.

A typical torque/speed characteristic for the two-value capacitor motor is shown in Fig. 7.6 (and the connection diagram in Fig. 7.2(e)).

7.11 Shaded-pole Induction Motor

The advantage of this type of motor is that it is simple to manufacture, low in cost, and extremely rugged and reliable. Like the polyphase induction motor it employs no commutator, brushes or switches.

The main disadvantage is the low efficiency but as power inputs are low this is not usually regarded as important. However, the low efficiency does basically prevent it from being built in the larger horsepower ratings.

The shaded-pole motor is adaptable to three types of construction, i.e., salient pole, skeleton, and distribution winding. Figure 7.9 depicts a salient pole version having two windings on the stator displaced at an angle of less than 90 electrical degrees. One of these is the main or excitation winding and the other is an auxiliary winding in the form of short circuited turns, otherwise known as the 'shading' winding, from whence the motor derives its name.

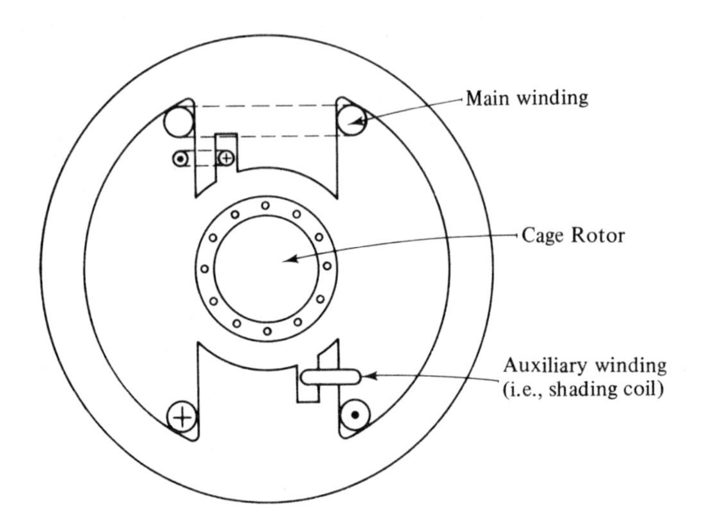

Fig. 7.9 Shaded-pole motor. Salient-pole type

The main winding acting by itself will produce a pulsating flux wave in the air gap and the motor like any other single-phase induction motor will not have any inherent starting torque. However, the introduction of the auxiliary coil splits the main flux into two components which will be displaced both in time and space; this it does by setting up a component field along an axis in space, different from that of the main winding. The field of the auxiliary winding is produced by transformer action of the main winding inducing a voltage in it, hence the need for the displacement angle of the windings to be less than 90 electrical degrees.

The projection of a rotating diagram on a vertical axis will give instantaneous values of the fluxes of two halves of the pole at different instants. These fluxes together produce a rotating elliptical field, this in turn induces currents in the cage rotor bars. The interaction of these currents with the rotating field produces the torque.

The fact that the flux in the shaded pole always lags behind the main flux fixes the rotation of the shaded-pole motor in one direction only, which in some applications is a disadvantage. These motors however, may be reversed by using separate windings for each direction of rotation or by connecting alternate sets of shading coils. These complex methods of reversal normally add too much to the cost of the motor and hence are not considered practical.

Other disadvantages of the shaded pole motor may be listed as follows:

(a) With a concentrated winding an approximate square wave is produced in the air gap which is equivalent to an infinite number of harmonic fluxes. The action of some of these fluxes on the rotor is to impair the motor performance.
(b) The leakage fluxes of the motor are high. In particular the rotor bars must be skewed to reduce the effect of harmonic fluxes and the result on the motor is reduced magnetizing reactance and increased skew leakage reactance. This plays an important part in determining the performance of the motor.
(c) In order to reduce the effect of harmonic fluxes a relatively small number of bars is used. This results in an increased harmonic leakage reactance and the rotor is of inherently high reactance.

Fans, record players, hair dryers, displays, and many other devices do not require high starting torques or very powerful motors and so still provide a very large market for the shaded-pole motor. For fan applications, in particular, the shaded-pole motor has the further advantage that it can be speed controlled over a fairly wide range by a simple series resistance.

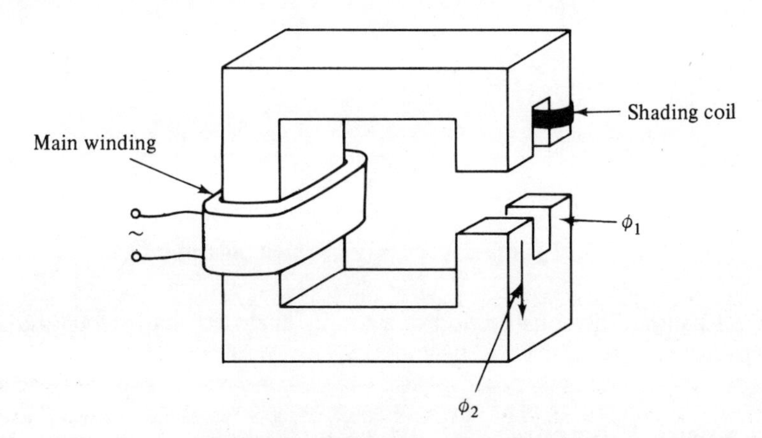

Fig. 7.10 Shaded-pole motor, simplified

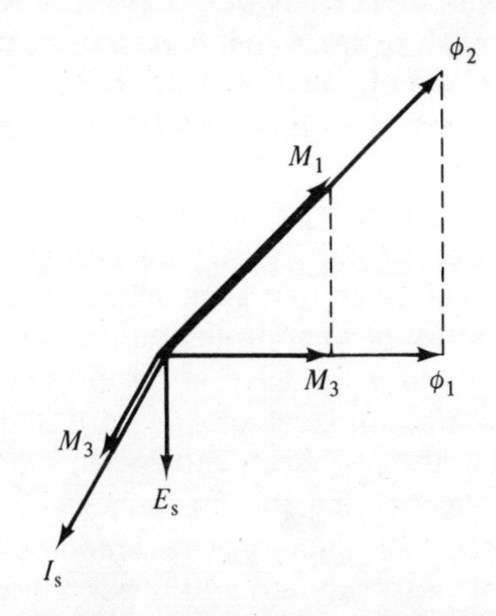

Fig. 7.11 Shaded-pole motor—phasor diagram

By referring to the simplified drawing of a shaded-pole motor, Fig. 7.10, the phasor diagram Fig. 7.11 may be developed as follows.

Flux ϕ_1 links with the shading coil and E_s is the resultant e.m.f. induced. I_s is the shading coil current lagging E_s by a small angle due to the leakage reactance of the coil. The total m.m.f. M_3 driving the flux ϕ_1 is assumed in phase with ϕ_1 and has two components, the m.m.f. of the main coil, M_1, and that of the shading coil, M_2. The unshaded flux component ϕ_2 is driven by M_1 alone, so that ϕ_2 and M_1 can be assumed in phase. From Fig. 7.11 it can be seen that the phase difference between ϕ_1 and ϕ_2 depends on the effect of shading.

A typical torque/speed characteristic is also shown in Fig. 7.12 (and a connection diagram in Fig. 7.2(f)).

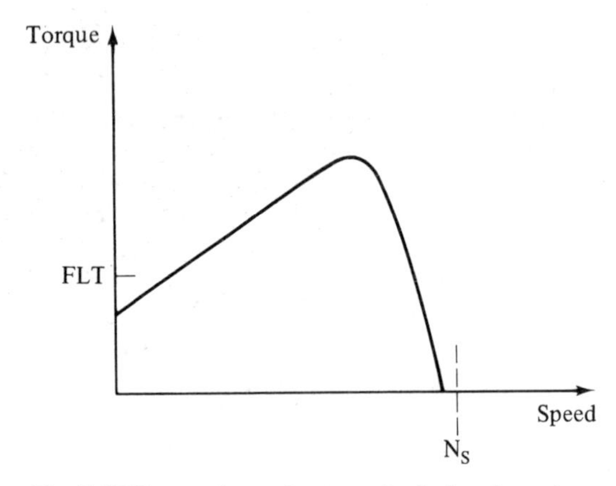

Fig. 7.12 Torque/ speed curve, shaded-pole motor

Figure 7.13 shows the construction of a totally enclosed shaded-pole fan motor, with air stream ratings of 3 W to 10 W at 1275 rev/min.

7.12 Synchronous Motors

Synchronous motors are built in a variety of types and constructions having a stator structure and winding similar to their induction motor counterparts in which a revolving magnetic field is set up when energized with alternating current. The exception to this is the single-phase or impulse type of motor which will be discussed later.

Their primary use is for constant speed requirements although the efficiency is considerably less than that of a comparable induction motor and their use should therefore be limited to critical speed tolerance application only.

Small synchronous motors for power outputs in the range from a few fractions of a watt up to several hundreds of watts have at the present time wide and various applications. Typical uses are, for example, in wire and radio communications installations; self-recording devices; electrical clocks; synchronous servo-systems, and other special fields of engineering.

Three different types of synchronous motor predominate; they are hysteresis motors, reluctance motors, and a sub-variety the permanent magnet (or polarized) synchronous motor.

Fig. 7.13 Totally enclosed shaded-pole fan motor (Courtesy GEC Ltd)

Reluctance and hysteresis motors are the simplest as regards their construction, among all the existing types of electro motors for they have a winding only on the stator, and have no sliding contacts. In addition, the hysteresis motor requires no special provision for starting and synchronizing.

Most synchronous motors are self-starting except some varieties (e.g., as used for clock motors) which have a salient pole stator and one field coil and which require to be started manually.

7.13 Reluctance Motors

Reluctance synchronous motors resemble conventional squirrel cage induction motors except that the rotor has salient poles (without permanent magnets or exciting windings) which lock into step with the rotating field, the speed of which is dependent upon the supply line frequency.

The general construction of stators for reluctance motors normally takes one of two forms:

(a) Ordinary non salient pole stator with slots and conventional windings distributed in them.
(b) Salient pole stators with one or several field coils.

Rotor construction can be of one of the following types:

(a) Salient pole rotor of common design.
(b) Sectional salient pole rotors.
(c) Salient pole rotor with cage.
(d) Toothed rotor.

The toothed rotor version is applicable to the stepping motor and these are dealt with separately in chapter 18.

Of the above variations in rotor design, the commonest is that with the cage winding (or bars) centred around a magnetic core which is so designed as to allow the passage of magnetic flux in a preferred direction only. The rotor will thus align itself with the rotating field, produced by the energized stator winding, when its position is such as to offer the minimum reluctance to the main flux. Therefore, because of the salient poles the motors will pull into step at some definite angular position depending on the number of poles, i.e., a four-pole motor will exhibit four discrete rotor-to-stator positions.

The purpose of the cage is, of course, to provide starting, and hence upon energizing the stator winding the reluctance motor behaves very much like an ordinary induction motor, accelerating up to about 98 per cent of its synchronous speed; although the saliencies cut efficiency due to the large average air gaps and also introduce torque pulsations.

As the rotor approaches synchronous speed, the salient poles of the rotor pass (or 'slip') the poles of the stator's rotating field at a constantly slower rate. As the pole passes and begins to lag behind the pole of the rotating field, a torque is exerted which tends to make the rotor pull into step with the field. Thus, this reluctance torque accelerates the rotor above its induction motor speed into synchronism because it is pulling the rotor in the direction of rotation.

7.13.1 *Reluctance motor performance*

Reluctance motors are less efficient than induction motors and are sensitive to inertia loads, therefore the key to synchronizing is to approach it as closely as possible (as in a low-slip induction motor) and to maintain a torque margin above load requirements. In addition, as the speed change from slip speed to synchronous speed is abrupt, the load inertia must be minimized and as a consequence, reluctance motors are not suitable for synchronizing high inertia loads.

With an inertia load present, the pull-in torque, which is the maximum load torque the motor can pull into synchronism, may be far lower than the pull-out torque which is unaffected by inertia. Pull-in torque even under worse conditions must therefore be capable of synchronizing the load. Pull-out torque is the maximum load torque that the rotor can hold in synchronism and under normal conditions this is slightly in excess of the pull-in torque. As a comparison the hysteresis motor is unaffected by inertia and has a characteristic as shown in section 7.16.

Although the synchronizing or pull-in torque is not directly dependent upon rotor resistance, none the less, for good pull-in torque characteristics, the rotor resistance should be low. The reason for this is that as the rotor approaches the synchronous speed by induction motor action (as previously described) the lower the rotor resistance, the higher will be the torque at this pull-in speed and hence higher will be the pull-in torque.

Power factor and efficiency are marginally lower than that for the same motor as an induction motor at one-third full load, and considerably less at full load for an induction motor of the same power and speed rating.

7.14 Single-phase Reluctance Motors

Single-phase reluctance motors with a salient pole stator having one field coil and also a salient multi-polar rotor are often used as drives in various small mechanisms. Power output is in the order of a few watts and starting is obtained by manual operation.

Interaction takes place between the poles of rotor and stator and the synchronous rotation of the rotor depends upon the periodic pull between rotor and stator poles and the inertia of the rotor.

Basically, the principle of operation is that every pair of poles of the rotor are periodically pulled by stator poles with the mechanical inertia of the rotor overcoming any reverse pull between them when the axis of a rotor pole pair passes the axis of the stator poles. At this instant, the flux of the stator coils passes through zero and then grows in the opposite direction. The rotor has to possess considerable inertia to overcome the effect of the reverse pull on the rotor poles by those of the stator until the next pair of rotor poles pass, and are acted upon by, those of the stator. The manual starting is effected by spinning or pushing the rotor by hand.

Single-phase reluctance motors of the type described above have a tendency to hunt.

7.15 Single-phase Capacitor Reluctance Motors

In order that starting torque might be obtained in single-phase reluctance motors, it is necessary to produce the rotating magnetic field as for the polyphase motor. To achieve this, two single-phase windings with a mutual displacement of half a pole pitch are placed on the stator. One of these is the main winding and the other is an auxiliary winding. A capacitor is also included in the circuit of the auxiliary winding as shown in Fig. 7.14. The rotor is again a salient-pole version having a cage winding to provide synchronizing torque and starting torque respectively.

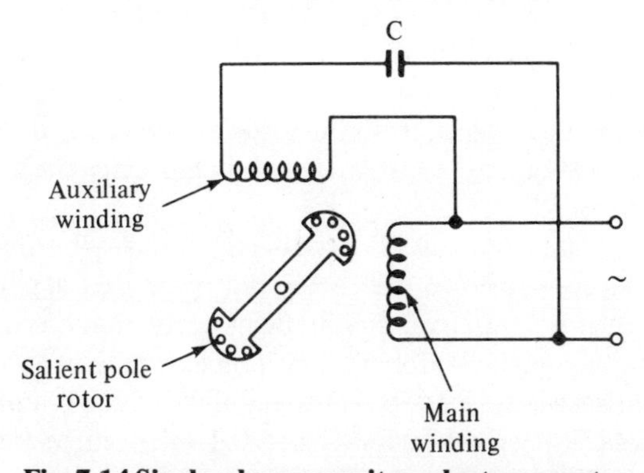

Fig. 7.14 Single-phase capacitor reluctance motor

The object of the capacitor is to produce a 90° phase shift between the currents flowing in the two windings and hence develop a rotating magnetic field as discussed in chapter 4 on the single-phase induction motor.

7.16 Hysteresis Motors

As far back as 1908 C. P. Steinmetz showed that a motor principle could be obtained by using the production of torque by magnetic hysteresis. However, at this, and other subsequent stages, the motor remained little more than a scientific curiosity until around the mid 1940s when the hysteresis motor became firmly established in the field of economical fractional horsepower motors. Since that time a steady development and expansion of the hysteresis motor has taken place, not just in keeping with the tremendous advances in new magnetic alloy materials but also to answer the industrial demands for high efficiency synchronous motors.

In its simplest form the hysteresis motor has a conventional slotted and laminated stator wound with a single or three-phase winding. The active part of the rotor consists of a homogeneous cast sleeve of hardened permanent magnet material secured to the shaft over a non-magnetic insert (Fig. 7.15). Unlike other synchronous motors, the hysteresis motor has a perfectly round and symmetrical rotor with no salient poles and hence it has no preferred direction for synchronizing.

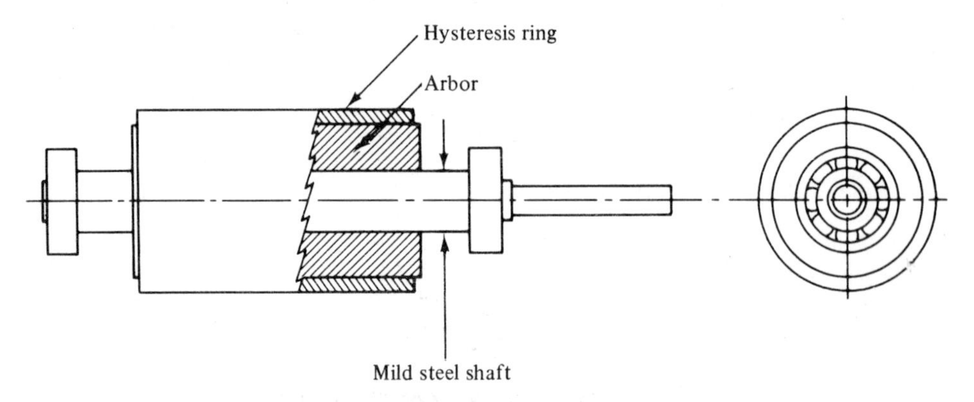

Fig. 7.15 Hysteresis rotor assembly

While the construction of the hysteresis motor is simple, the same cannot be said for its theory of operation, but briefly, several effects interact to produce torque and hence self starting.

When the stator is energized from an alternating current supply a rotating magnetic field is established in the air gap which magnetizes the steel cylinder of the rotor and induces poles.

Below synchronous speed, torque is produced primarily by the hysteresis effect. As the stator field sweeps around the rotor, the rotor flux cannot follow along in phase because the rotor is made from magnetic material that has high hysteresis loss. (The rotor flux, in fact, lags the stator field by a phase angle proportional to the area of the rotor material's hysteresis loop.) This angular displacement of rotor flux with respect to stator field is precisely what is needed to produce torque, so the rotor receives an acceleratiı g force (Fig. 7.16).

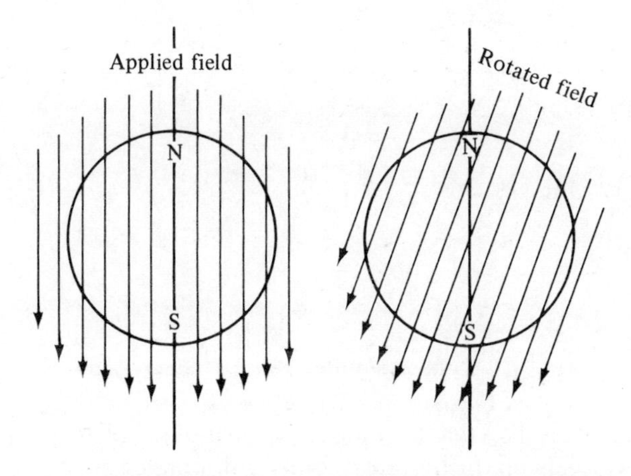

Fig. 7.16 Hysteresis motor, principle of torque production

In addition, the rotor flux induces eddy currents as it sweeps through the rotor at speeds below synchronism. These eddy currents produce additional torque for acceleration, which is why the motor's torque is greatest at lower speeds.

Once the rotor reaches synchronous speed, the rotor flux ceases to sweep around it (Fig. 7.17). Both eddy-current and hysteresis torque disappear, leaving the rotor magnetized and the motor running essentially as a permanent-magnet motor.

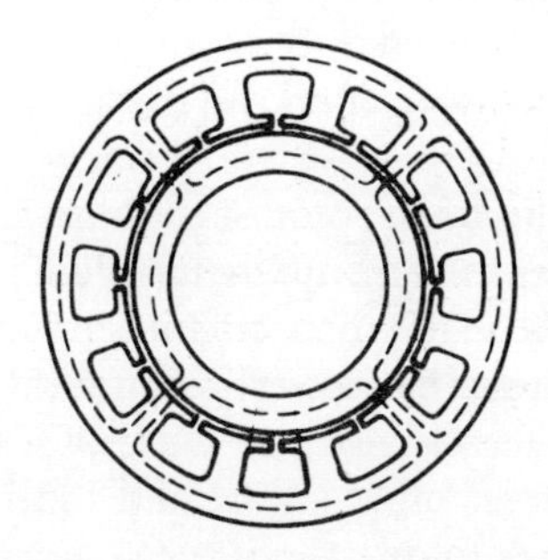

Fig. 7.17 Hysteresis motor, magnetic path

Motor designers have optimized both subsynchronous and synchronous operation by experimenting with different rotor materials and configurations (Fig. 7.18).

Starting torque is proportional to the product of the induced rotor flux, the stator m.m.f. and the sine of the hysteretic lag (or torque angle) and may be represented by the equation

$$T = \pi/2(P/2)\phi F \sin\theta$$

where

P = number of poles

ϕ = Air gap flux due to applied voltage

F = Peak value of induced m.m.f.

θ = Torque angle.

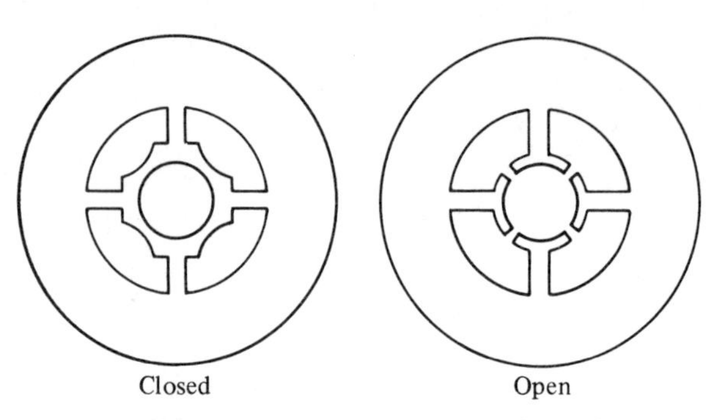

Fig. 7.18 Hysteresis motor, stator configurations

If the load torque is less than that developed by the motor, the rotor accelerates. The lag angle depends solely on the hysteresis of the rotor material and not on the rate at which magnetization is changed, so motor torque is constant up to the synchronized speed.

With a salient-pole motor there is a definite fixed time period when the poles are suitably aligned to pull into synchronism. With such a fixed time a high inertia load requires more synchronizing torque than a low inertia load. This is why all synchronous machines with salient poles are greatly affected by load inertia.

With a uniformly round rotor, this limitation does not exist. In both theory and practice, there is no discontinuity whatever in hysteresis-motor operation below synchronism and at synchronism. Since no discontinuity exists, the motor has no concern for the load inertia connected to it. Since there is no phase orientation the rotor may be synchronized in any position and the abrupt acceleration into step does not occur.

Because of the smooth cylindrical rotor of the simple hysteresis motor, there are no fixed positions for the poles as there are in the reluctance motor. Hysteresis motor poles can assume any position on the rotor circumference, and if the load exceeds the torque developed by the motor, the rotor slips smoothly as the pole location drifts. The lack of rotor pole saliences is also the cause of a deficiency in damping. The loaded rotor assumes a torque angle lag behind the rotating field. If the load is removed, the rotor accelerates to overshoot the stator pole and is then retarded by generation action, developing a new load torque. Depending on motor dimensions and other factors, this hunting can be damped out or persist.

In the ideal hysteresis motor, torque, current, and power remain uniform from stall to synchronism. In real hysteresis motors, eddy currents in the special hysteresis rotor material superimpose induction motor torques. Stall torque and stall current and power, therefore, differ from the synchronous values. Operation as a capacitor motor results in current and power variation with speed and leads to unbalanced operation and reduced torque at certain speeds. Thus, in motors designed for balanced output at synchronism, stall operation is unbalanced, reducing stall torque. For this reason capacitor-driven hysteresis motors often have less stall torque than synchronizing torque. In motors optimized at stall, the reverse holds true.

Relatively sensitive to non-ideal performance conditions, hysteresis motors lose more efficiency than induction motors from unbalanced voltages or from the presence of supply harmonics (as with square-wave excitation). Low supply voltage sharply reduces torque output. Inside the hysteresis motor, slot openings and

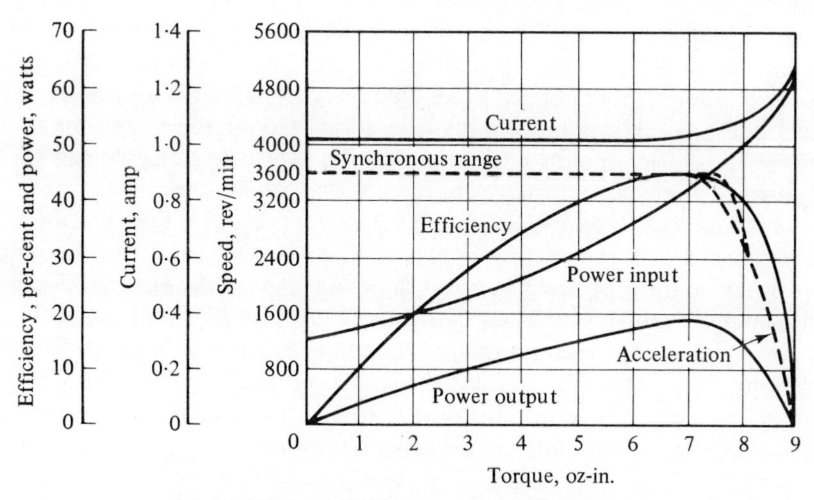

Fig. 7.19 Hysteresis motor, performance curves

imperfect windings also reduce efficiency. Performance, however, is generally good for most practical conditions (Fig. 7.19).

Optimum values of synchronous torque may be obtained at the expense of stall torque, or vice versa by suitable heat treatment of the rotor. If efficiency is the dominant factor it may be obtained by using special closed slot stator laminations with uniform air gaps.

Several performance features combine to make the hysteresis motor an important factor in the small synchronous motor field. Because it develops torque up to synchronism and has the ability to bring very high inertia loads up to synchronous speed, it is ideal for loads such as tape reels or gyrorotors. The motor's advantage on high inertia loads permits the use of a smaller frame for a given application and in smaller sizes its efficiency can be as good as, or better than, conventional induction motors.

In miniature high-speed fans, the hysteresis motor can provide more air at higher pressure than an induction drive. Hysteresis motors can be designed for very high shaft power in small size, important in intermittent duty applications which can tolerate high peak powers.

Another characteristic of the hysteresis motor, associated with its rotor design, is the very low surge currents. This is due to the accelerating torque being produced by hysteresis and eddy current action rather than by induction motor action hence the high starting currents associated with cage rotor windings are not present. In fact, the rotor currents in the hysteresis motor are limited to a low value by the high reactance and resistance inherent in the rotor configuration and the special magnetic materials used.

Motors using cage rotors can have starting currents of up to 600 per cent full load current and with conventional synchronous motors, other design factors can increase starting current as much as 10 or 15 times normal full load current. Thus, a hysteresis motor with 150 per cent surge current may take only one-tenth as much starting current as other synchronous motors. This is especially significant in adjustable-frequency systems, where hysteresis motors are supplied from relatively small alternators.

Since the hysteresis rotor is continuous, the motors are quieter and smoother in operation than reluctance motors and are therefore more suitable for applications having low flutter requirements such as tape drives and record turntables.

Single-phase hysteresis motors may be connected to the supply in the same manner as the capacitor start and run motor, and reversing may be accomplished either at standstill or, under favourable conditions, while rotating.

Hysteresis motors were primarily used for clocks and other timing devices but in modern times the cheaper type impulse motor is now used for this purpose. However, due to the progress made over the years in magnetic materials, such as the higher content of cobalt steels and the Alnico range, it has been possible to design more power into the hysteresis motor. As a consequence it has now found wide applications, in addition to those already mentioned, in recording equipment, aircraft gyros, aircraft precision instrumentation, and other precise speed applications where a very reliable and robust drive is required.

7.17 Polarized Synchronous Motors

This sub-variety of both reluctance and hysteresis synchronous motor types has a permanent magnet in the rotor. The permanent magnet alone would provide no starting torque for the motor, thus the rotor may take the form of a cage with the permanent magnet embedded in it, or as is more common practice, a permanent magnet and a cage assembly are placed in line on a common shaft. Then, as with the reluctance motor, the cage winding brings the rotor up to near synchronism when it abruptly accelerates from slip speed into synchronism under the influence of the permanent magnet.

Thus this type of motor has the qualities of either a reluctance or hysteresis motor, plus something more. The permanent magnet retains its direction so that the polarized synchronous motor always maintains a fixed orientation with respect to the rotating magnetic field subject of course to the displacement of the power angle.

Because of the strong magnetic field produced by the permanent magnet, this motor has higher lock-in torques than the other types and although synchronizing high inertia loads is a problem it can drive ordinary loads synchronously at high efficiency.

Another characteristic of this type of unit is its ability to lock in at certain definite phase angles which relate to instantaneous line voltages and shaft positions. Because a north pole of the rotating flux field can only lock in with the permanent magnet south pole, only one lock-in position exists for a two-pole motor, two lock-in positions for a four-pole motor, and so on. At no load, rotor phasing is very accurate, but as load is applied the rotor tends to lag. This lag angle remains constant for a constant load.

Unlike the hysteresis motor the polarized motor has a much higher pull-out than pull-in torque; consequently pull-in torque must clearly be adequate to handle load.

Because of the characteristics discussed above, the polarized synchronous motor is usually specified for applications where a constant angular position must be maintained as well as synchronism; for example, when controlling remotely located devices that must interact with each other as though they were mechanically coupled, or as when two or more shafts need to be controlled in a definite phase relation.

By using an integrally wound generator winding, sensitive to changes in the power angle, any number of polarized synchronous motors can be identically matched by means of feedback for extremely precise process control, optical sensing, and similar applications.

7.18 a.c. Commutator (Universal) Motors

In modern times the instances when domestic appliances, machine tools, and similar apparatus actually need to operate on both a.c. and d.c. supplies are few and far between and as a consequence, applications formerly requiring the use of universal motors now use motors specifically designed for a.c. operation only, otherwise known as a.c. commutator motors.

In construction both types are similar, any differences being in the design, (e.g., winding modifications) in order to obtain a better performance. Therefore to appreciate the design, operation, and performance of the a.c. commutator motor it is considered worthwhile to start by comparing the universal motor as such with its purely d.c. counterpart, the d.c. series motor.

7.18.1 *The universal motor*

If the direction of current flow to a d.c. series motor is reversed, the direction of the magnetic field is reversed and therefore the motor continues to exert a torque in the same direction as before. If a single-phase a.c. is now applied to the same motor the current in the armature will reverse 100 times per second (for 50 Hz supply) and, as they are in phase with the armature current, so will the field excitation and flux. Once again the torque will continue to be exerted in the same direction and thus the motor will start and run on a.c. in the same manner as on d.c.

However, although the universal motor is therefore basically similar in construction to the d.c. series motor there are some effects present on a.c. operation which are not present on d.c. operation. These necessitate certain modifications to the design before a true universal motor is obtained, which will give essentially the same performance and operating characteristics on alternating as on direct current.

7.18.1.1 Laminated field core. Owing to the alternating nature of the stator flux it is essential that a laminated core is used for the stator as well as for the armature in order to reduce eddy current and hysteresis losses.

7.18.1.2 Reactance voltage. On d.c. the current is limited only by resistance, whereas on a.c. it is limited by the impedance which comprises the two components, resistance and reactance. It therefore follows that on a.c. operation a reactance voltage will be present which will be absorbed from the main line voltage thus reducing the voltage supplied to the armature. This reduction in effective voltage results in the speed for any given value of current being lower on a.c. than d.c. for the universal motor.

7.18.1.3 Saturation effect. A counter effect to that outlined in 7.18.1.2 above is also exhibited on a.c. supply. It is that an r.m.s. value of current will produce less r.m.s. flux than a direct current of the same value, owing to the saturation effects of the iron.

At low currents and high speeds the reactance voltage is relatively unimportant, and the saturation effect usually causes the motor to operate at a higher no-load speed on a.c. than d.c. It can be shown that this characteristic is most prominent around 25 Hz.

7.18.1.4 Winding inductance. On a.c. operation the power factor tends to be very bad because of the large inductance of the windings, but this can be improved by reducing the number of field turns and by designing the motor for a low magnetic flux density.

7.18.1.5 Commutation and brush life. The commutation on alternating current is substantially poorer than on direct current and the brush life is correspondingly less. The reasons are that the commutator and brushes not only have to contend with the reactance voltages (see 7.18.1 above) but in addition, with voltages induced in the short-circuited coils undergoing commutation by the transformer action of the alternating main field. No such transformer action exists on d.c.

Clearly then the torque and speed will be somewhat greater on d.c. than a.c. for any given winding, with the ratios depending upon the particular winding design and the applied load; however, at the higher speeds better universal characteristics (i.e., nearer the same performance on both alternating and direct current) can be obtained as well as more output per given weight of armature.

Having established some of the fundamental problems and differences of the universal motor, and bearing in mind that, as previously stated, modern design tends to err in favour of a.c. performance rather than a universal characteristic, we can now look at the a.c. series commutator motor as a whole.

7.18.2 *Operation and performance of a.c. series motor*

One of the most important characteristics of an a.c. series motor is that it has the highest power-to-weight ratio of any a.c. motor because of its ability to operate at speeds many times higher than that of any other mains-frequency motor. For example, a typical 375 W, 1750 rev/min induction motor weighs 13·6 kg and has a 16 mm diameter shaft, whereas a 375 W, 19 000 rev/min a.c. series motor weighs only 1·2 kg and has an 8 mm shaft.

The vast majority of domestic appliances use the a.c. series motor. The type of motor used for this application requires carbon brush replacement after 200 to 1200 hours depending on the type of appliance. However, when required, a series a.c. motor can be designed for a brush life as long as 3000 to 5000 hours, but at considerable sacrifice in size and cost.

This type of motor is ideally suited for operation at a given rated output, where an occasional overload or intermittent heavy load occurs; stall torque may be from 4 to 10 times the continuous rated torque. Small size and economy give the a.c. series motor a distinct advantage over induction motors.

Series motors for a.c. can be built in a range of speeds, generally 4000 to 24 000 rev/min with rated power of 75 to 750 W. Peak efficiency of these motors ranges from about 30 per cent for small ones to about 75 per cent for the larger sizes. For example, a vacuum-cleaner motor rated at 420 W and 17 500 rev/min may have a peak efficiency of 70 per cent. Peak efficiency of a typical 75 W, 12 000 rev/min motor is about 50 per cent.

When operated without load, the rotor tends to 'run away', speed being limited only by windage and friction. Large motors are therefore nearly always connected directly to a load to limit speed. On portable tools such as electric saws, the load imposed by the gears, bearings, and cooling fan is sufficient to hold the no-load speed down to a safe value. In addition, speed control of a series motor is simple, since motor speed is sensitive to both voltage and flux changes. With a rheostat or adjustable autotransformer, motor speed can be readily varied from top speed down to zero. This ease of speed control permits a speed governor to be used when required. Such a governor consists of a set of speed-sensitive contacts which insert a fixed resistor into the circuit to limit the motor speed to a predetermined value.

Frequency variations have less effect on a.c. series motors with distributed field windings than on motors having the more commonly used salient pole field winding. The distributed winding is occasionally used on longer life, higher powered (and more expensive) a.c. series motors and has inherently better commutation qualities.

The commutation of a series motor can cause radio interference, but when critical this can be minimized by the use of radio noise filters in the supply line.

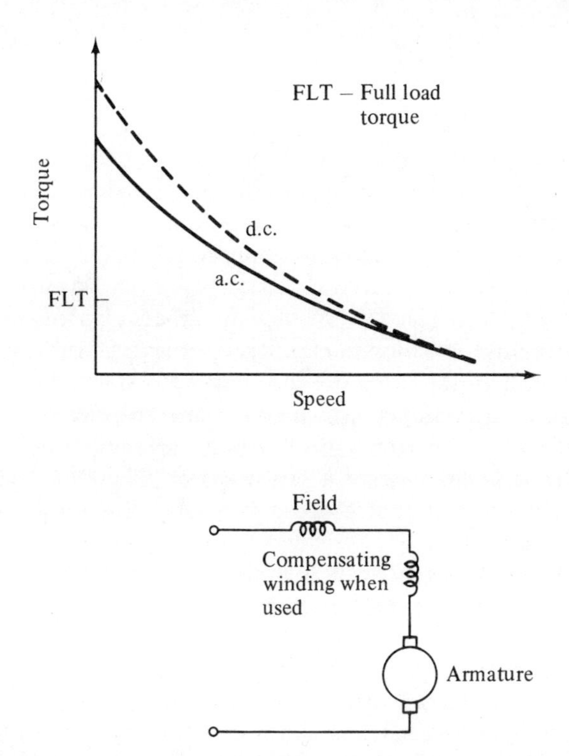

Fig. 7.20 Alternating current commutator motor, torque/speed curve and connection diagrams

A typical speed/torque characteristic and a connection diagram are shown in Fig. 7.20. Figure 7.21 shows a typical universal motor with two-stage impeller, for vacuum cleaner applications. Typical full output ratings are 300 W at 18 000 rev/min.

Fig. 7.21 Alternating current commutator (universal) motor

7.19 Repulsion Motors

In many respects the repulsion motor is very similar to the d.c. series motor and exhibits the same type of torque characteristics. Like the series motor, it has a conventional armature and brushes except that the brushes are permanently short circuited upon themselves. The stator carries the normal single-phase winding but in addition there is a second (or inducing) winding connected in series but displaced 90 electrical degrees from it.

When energization takes place the main winding sets up a magnetic field but this, by virtue of its relative position to the armature and brushes, is incapable of inducing any currents in the armature and hence no starting torque is produced. However, the inducing winding acts in the same manner as the primary of a transformer and induces currents to flow in the armature in the same predetermined paths as in the case of a series motor. The short-circuited armature winding acts as the secondary of the transformer. Thus torque is produced and rotation takes place.

It therefore follows that as the main field winding and the inducing winding are in series the torque characteristics of the repulsion motor must be similar to those of the d.c. series motor, as previously stated.

A further consideration of the similarity of d.c. series and repulsion motors is that in the series motor the armature current is equal at all times to the field current due to the series connection; in the repulsion motor the armature current is proportional to the current in the inducing winding, the latter in turn being in series with the main field winding. By displacing the two stator windings of the repulsion motor 90 electrical degrees apart, neither winding can induce a voltage in the other by transformer action and only the inducing winding can induce a voltage in the armature. The repulsion motor may be reversed by interchanging the connections of the main winding or the inducing winding.

The simplified phasor diagram, shown in Fig. 7.22 for a repulsion motor may be developed as follows:

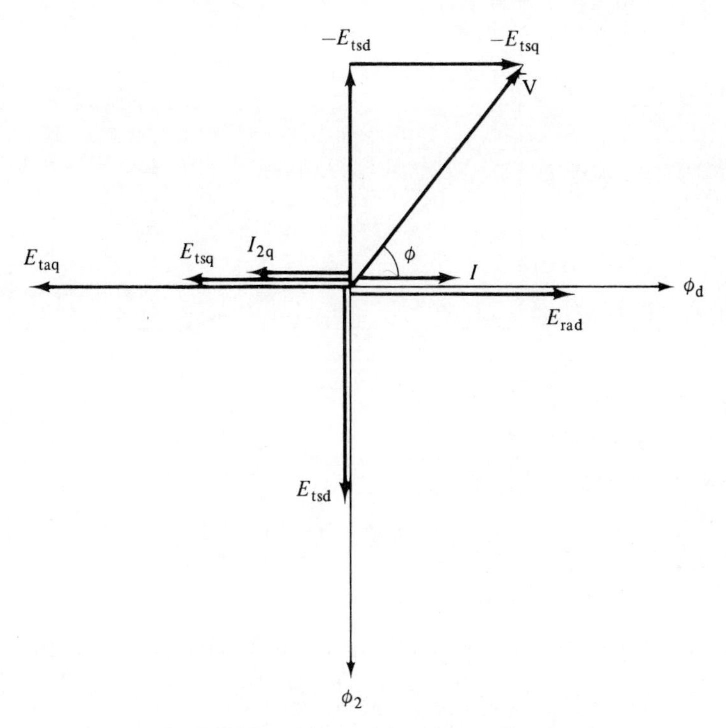

Fig. 7.22 Repulsion motor, phasor diagram

For simplicity, resistance and leakage reactance drops and the magnetizing current required to set up the quadrature flux ϕ_q have been neglected. Let ϕ_d represent the exciting flux, then the rotor conductors rotating in this develop a rotational e.m.f., E_{rad}, in phase with ϕ_d. The flux ϕ_d sets up a transformer e.m.f. in the rotor, E_{taq}. If impedance drops are ignored there are no other e.m.f.s in the rotor and therefore $E_{taq} = E_{rad}$ and is opposite in phase. The transformer e.m.f. must lag the flux ϕ_d by 90°.

Assuming for further simplicity a double-wound motor the flux ϕ_q induces a transformer e.m.f. in the quadrature stator winding E_{tsq} which lags ϕ_q by 90°. Also the exciting flux induces a transformer e.m.f. E_{tsd} in the direct axis stator winding lagging ϕ_d by 90°. Neglecting impedance drops, the applied voltage must be equal and opposite to the sum of these two vector e.m.f.s as shown.

Since, in the direct axis, there is no m.m.f. other than that produced by the main current I, the m.m.f. due to this sets up the exciting flux and therefore I is in phase with ϕ_d.

In the quadrature axis, the flux is produced by the resultant of the m.m.f.s due to I and the rotor current I_{2q}. Since the magnetizing current to set up this flux is neglected; I and I_{2q} (referred to the stator) must be equal and opposite as in short-circuited transformers.

Drawing I_{2q}, it is clear that it is opposed to E_{rad}, which is correct for motor operation. $E_{rad}I_{2q}$ gives the rotor power, which neglecting losses is equal to the mechanical power. Figure 7.23 shows a typical speed/torque characteristic and a connection diagram.

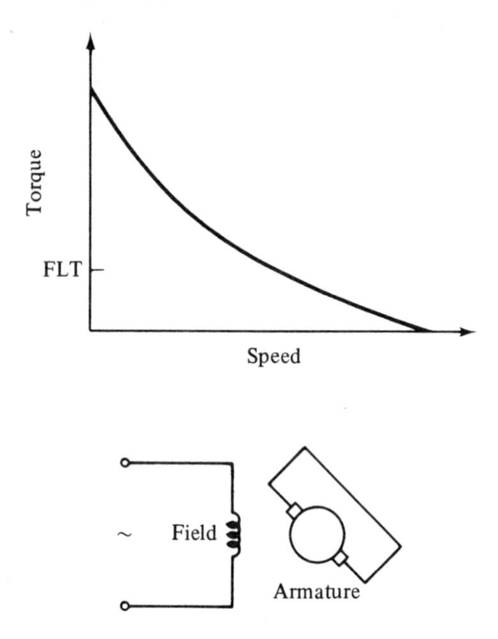

Fig. 7.23 Repulsion motor, torque/ speed curve and connection diagram

7.20 The Deri Motor

It has already been pointed out that the torque developed in a repulsion motor is directly influenced by the position of the brushes, with a small movement of the brushes producing a large variation in torque. Furthermore the rotor coils are subjected to an excessive short circuit current when the stator and rotor axis are coincident. However, in the Deri type of repulsion motor these problems are overcome by employing a double set of brushes.

One set is fixed permanently in line with the stator axis whilst the other can be moved around the commutator. This has the effect of placing short-circuited rotor coils in both axes, and for a given brush movement the movement of the resultant m.m.f. is much smaller, thereby allowing a finer speed control. (See Fig. 7.24.)

7.21 Motor Selection Factors

7.21.1 *Basic requirements*

Apart from the all-important problem of cost there are other technical aspects which must be carefully considered in specifying a motor for a particular application. Some of these points are discussed below together with a summary of some of the salient features of the various types of motor studied in the previous sections of this chapter.

Since the power supply to an installation is normally already fixed the initial choice between a.c. or d.c. motors is pre-determined, but in the case of a.c., further limitations are presented by the frequency and number of phases available. Where special frequencies and voltages exist or where variations in either of these two parameters is likely to occur special care must be exercised.

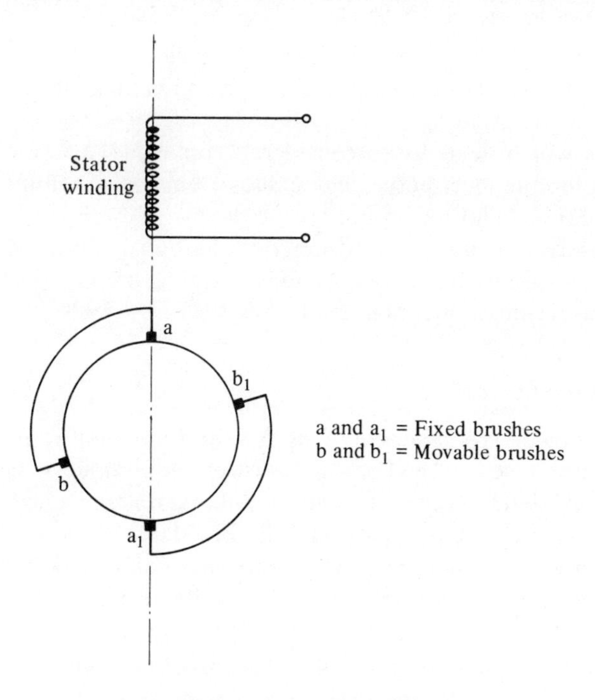

Fig. 7.24 Deri motor, brush configuration

The function the motor is expected to perform should next be considered, such as, would the shaft be turning at constant speed under constant load? or, would both these factors be varying together perhaps with a frequent period of stopping and starting? If the speed is likely to vary, provision must be made for a suitable speed control and the permissible amount of variation must be determined. If the motor is to be reversed it is desirable to know whether this needs to be done with the motor running at speed or at standstill.

Power considerations pose several questions that need to be answered in the process of selection; such questions as: what power is required to operate the machine? In what manner is the power to be applied? Will the torque required be maximum at starting, running, or accelerating? Will the torque applied to the shaft stay constant or will it fluctuate at different points during the operation of the machine?

One of the most important considerations in regard to the above is an analysis of the starting conditions. In some machines, the initial load is primarily one of friction: as speed increases, the power requirement increases rapidly.

In certain applications the motor is brought up to speed before the load is applied. Here, friction and inertia comprise the initial load but if the motor is of sufficient capacity, trouble is not likely to be encountered in meeting the requirements of the load.

Some applications require a motor that is equipped to handle a heavy starting torque. The heaviest demands on the motor are therefore during starting and accelerating; but, once full speed has been achieved, less power is needed to keep the unit operating.

The type of operation the motor is destined for also needs to be known, that is whether for continuous or intermittent operation and in the case of the latter what is the length of the duty cycle.

Other factors which need to be considered come under a general heading of 'mechanical requirements'; these include such items as mounting requirements, space available for installation, limitations on the overall weight of motor, environment (i.e., exposure, climatic and atmospheric conditions), vibration and any limitation on noise, accessibility for servicing and lubricating, and the type of connections required, such as terminal boards or leads.

7.21.2 *Comparison of types*

If a.c. and d.c. power are available, consider the functional requirements of the application in motor selection. For small power requirements, choose universal motors where considerable speed regulation is necessary, and a high starting torque is essential. Generally, d.c. and universal motors start, accelerate, and run small power devices easier and smoother, and reverse easier and quicker, than a.c. motors of equivalent power. However, they cannot normally be used in hermetically sealed systems because of the commutator and air flow requirements. They also require more maintenance because of brush wear and armatures which are exposed to the elements.

Small power a.c. motors generally run at constant speed; they have moderate to high starting torques and are reversible only from rest or during rotation of high frictional, low inertial loads.

Series wound universal motors may have an unadjusted speed drop of about 50 per cent at full load. The slip of a.c. motors, with load, largely depends on the mode of operation. Synchronous motors have no speed drop whereas split-phase machines may slip up to 10 per cent at a 35 per cent overload, the slip being approximately linear with load.

7.21.2.1 Universal v. induction. Universal motors are series-wound commutator type motors that are operable on a.c. and d.c. but not necessarily with the same performance. Usually these motors are not available as stock items, rather they are custom made, although custom making in lots of 50 or 100 is not a problem if similar tooling is available or if the motor can use parts common with an existing mass-produced model.

These motors are usually built with outputs from 7 W to 3 kW and various operating speeds from 5000 to 25 000 rev/min. They have a marked speed drop with load and they attain high starting torques of the order of 350 per cent of rated torque. They can be designed to be used for either continuous or intermittent operations over much of their entire speed and load range.

Universal motors are preferred over induction motors where one or more of the following characteristics are desired:

Ability to operate on a.c. and d.c.
Ability to operate at high speeds.
High starting torque.
Shape of speed/torque curve.

Greater speed variation with load.
Ability to operate with temporary torque overloads.

These characteristics are usable only if the application can tolerate the generally comparatively short brush life (50 to 2000 hours) of most universal motors.

The maximum speed of induction motors is determined by the line frequency (3600 rev/min for 60 Hz). Universal motors have no such limit and so high-speed versions benefit from very high ratios of power/weight of motor, and low ratios of cost and size/power. The load size or cost also frequently can benefit from high-speed operation. High-speed universal motors with gearing take advantage of the high power/weight ratio even in slow-speed applications.

Approximately 30 per cent of the small power motors used today are of the universal type.

7.21.2.2 Split-phase v. capacitor-start. Among utility motors capable of long rugged service, the split-phase types cost the least. They can be specified for light duty, such as driving low-inertia fans, small machine tools and the like, up to about 400 W ($\frac{1}{2}$ h.p.). Their starting torques range from 130 per cent to 200 per cent of full running torque depending upon class of insulation. Capacitor-start motors develop starting torques ranging from 250 per cent to 350 per cent of full running torque, but cost from 35 per cent to 50 per cent more than power-equivalent split-phase types. Split-phase motors should be used where starting is infrequent and the inertia is low, whereas capacitor-start motors operate in intermittent as well as continuous service. As a group, split-phase and capacitor-start motors represent about 50 per cent of all electric motors being sold. With Class B insulation, these motors can run with a 35 per cent overload indefinitely. Split-phase motors can be reversed from rest or occasionally during rotation only under high frictional, low-inertia load conditions, providing that the starting winding is not controlled by a centrifugal switch or current-type relay sensitive to speed.

Capacitor-start motors come in convenient sizes ranging from about 100 W to 1 kW ($\frac{1}{6}$ h.p. to 1 h.p.). They may be used in virtually all applications, being particularly suitable for duty that may involve overcoming such high inertia loads as refrigeration compressors, hand tools, machine tools, and conveyer belt drives.

7.21.2.3 Single-phase v. polyphase (three-phase). In the small power range, the choice between single-phase or polyphase motors almost invariably depends on the kind of a.c. power available, but there are exceptional circumstances where the application determines the choice. Most small power a.c. motors are single-phase, because single-phase power is usually available. Since they are simpler and generally the less expensive of the two types, single-phase motors should be specified unless polyphase motors are essential. Two special conditions make polyphase motors mandatory: when the power supply is limited to 230 or 460 V three-phase a.c. current, or where the need is for virtually instant reversal, as in some machine tool applications. Three-phase motors can be reversed during rotation and under load.

Polyphase (almost invariably three-phase) motors formerly in $\frac{1}{4}$, $\frac{1}{3}$, $\frac{1}{2}$, $\frac{3}{4}$, and 1 h.p. sizes, and 50/60 Hz speed steps of 950, 1140, 1425, 1725, 2850, and 3450 rev/min, are included in BS 5000, part 11 (see ch. 5, page 152). The basic parts of this type of motor should not cost more than a 1 h.p. capacitor-start induction motor. The price of any small-power polyphase motor approximately equals that of a totally-enclosed, fan-cooled 1 h.p. split-phase capacitor-start motor equipped with precision ball bearings.

7.21.2.4 Shaded-pole, permanent-split-capacitor, hysteresis or reluctance synchronous. These types are popular in applications requiring power ranging from 0.5 to 370 W. They can be inexpensive as motors for toys and small fans, or relatively expensive for driving instrument packages. Reluctance synchronous motors have precise speed regulation and lock into phase in a relatively limited number of positions—an advantage in driving many recording devices. These types of timing motors should be used in tape recorders, computers, small pumps and blowers, typewriters, recording decks, vending machines, cash registers, and movie projectors—in applications requiring low torques and relatively constant-speed operations. Where load inertia is high, a hysteresis motor should be specified. Remember, these motors are normally not reversible; in most instances they have to be redesigned to reverse on demand.

References

1. P. L. Alger, *Induction Machines*, Second Edition, Gordon and Breach Science Pubs, New York (1951).
2. C. G. Veinott, *Fractional Horsepower Motors*, McGraw-Hill.
3. M. G. Say, *The Performance and Design of Alternating Current Machines*, Pitman, London (1976).
4. M. Jevons, *Electrical Machine Theory*, Blackie, London (1966).
5. D. E. Knights, 'Variable Speed Drives for Industrial Applications', *ERA Report No. 74-1189* (December 1974).
6. D. E. Knights, 'Drives for Domestic Appliances', *ERA Report No. 74-1195* (December 1974).
7. *Machine Design*, Electric Motors Reference Issue (April 1974).
8. *Electro-Mechanical Design*, Volume 3, Benwill Publications (1973).
9. A. Bekey, 'New High Performance Designs for Small Synchronous Motors', *Electro-Technology* (November 1961).
10. S. K. Pal, 'A Review of Hysteresis Motor Technology', *ERA Report No. 72-64* (July 1972).
11. D. R. Driver, 'Magnetic Alloys for Hysteresis Motors', *Electrical Times* (August 1967).
12. J. L. Tasker and M. Bradford, 'The Effect of Stator Slot Openings on the Performance of Hysteresis Motors' *ERA Report No. ERA 75-39* (November 1975).
13. *Phillips Technical Review*, **33**, No. 8/9 (1973).

8 Definite-Purpose Motors

8.1 Introduction

Standard general-purpose motors have several advantages of importance to the user of electric motor drives. These arise due to the high degree of standardization in terms of rating, duty, dimensions, electrical and thermal characteristics, and construction and also because they are manufactured in large numbers. The main advantages are their relatively low cost, they are easy to replace directly with another motor without any modifications, they are easy to repair, and are readily available. However, where there are special requirements for particular applications 'special' motors are inevitable. In some cases standard frame sizes can be used with special shafts, finishes, or mounting arrangements. In others, either special performance characteristics or space constraints demand a completely new design.

BS 4727 defines a definite-purpose motor as a motor designed, listed, and offered in standard ratings with operating characteristics and mechanical construction suitable for a particular type of application.

These motors, which are not fully covered in other sections, are generally rated about 1 kW and above. While they are generally built to special specifications, they are nevertheless important and are applied in considerable quantity. Application of any type of definite-purpose motor for a duty other than that for which it is intended must be preceded by careful consideration. Although some types are similar to general-purpose motors others are not and their electrical and thermal characteristics may be considerably different from those of general-purpose machines.

8.2 Low-voltage Motors for Battery-powered Vehicles

8.2.1 *Introduction*

The d.c. series-wound battery-powered traction motor is, at present, almost universally employed as the main propulsion motor for road and industrial electric vehicles, although the d.c. compound-wound motor is used in some industrial trucks to provide electric braking.

The conventional d.c. series motor has been found to have suitable torque-speed characteristics for the majority of traction applications. One of the main objectives of the development of traction motors is reduction of the specific weight, which is somewhat high at about 5·7 to 9·6 kg/kW for a continuous rating, improving to about 3·3 to 4·8 kg/kW for peak rating. It should be noted that other motor systems can give a higher specific output than the d.c. series motor. The possible weight-to-power ratios including the gearbox, invertor, and cooling arrangements of a high-speed cage induction motor (10 000 to 20 000 rev/min) could be about 2·1 kg/kW at continuous rating. Possible weight-to-power ratios for d.c. homopolar motors, including auxiliaries, could be similar.

For traction purposes, the short-term rating, which governs acceleration and hill-climbing ability, is possibly as important as the continuous rating. By virtue of its overload capacity the electric motor is superior to the petrol engine in performance because the latter cannot utilize its full rated power all the time. For this reason an 11 kW electric motor is able to replace a 30 kW petrol engine, for almost the same performance. Several different types of d.c. motor are used for battery vehicle applications.

8.2.2 *Permanent magnet d.c. motors*

The d.c. permanent magnet motor has a conventional wound armature, and a commutator, on the rotor, with the field on the stator supplied by either Alnico or ceramic permanent magnets. With this type of motor the pole flux remains basically constant at all motor speeds and hence the speed/torque curves are linear (Fig. 8.1). Due to the limitations of permanent magnet materials these motors are normally limited to relatively small powers, usually below 1 kW.

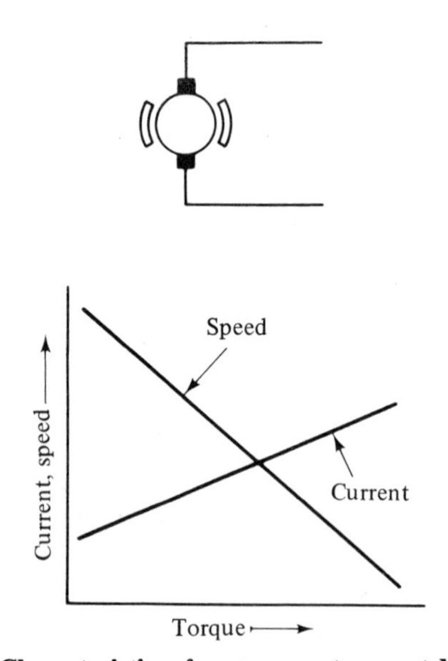

Fig. 8.1 Characteristics of a permanent magnet d.c. motor

Permanent magnet motors are useful for giving savings in cost, space, weight, and power consumption, all of which are important in battery vehicle applications. Permanent magnet motors in the range of 375 W to 1 kW have found wide acceptance in battery electric lawn mowers, sweepers, and electric mopeds. Torque requirements, similar to those of shunt motors, allow permanent magnet motors to be used for low-power applications in fork lift trucks, such as for power steering pump drives.

8.2.3 *D.C. series motors*

Series wound d.c. motors create the field flux by coils which are electrically in series with the armature. Torque/speed characteristics are shown in Fig. 8.2. Most often, the traction motors are series motors. Such motors have a high starting torque, and, during start-up, the torque is directly proportional to the current. As the torque requirement decreases, the speed rapidly increases, and the current gradually decreases. At higher speeds, the torque is proportional to the square of the current.

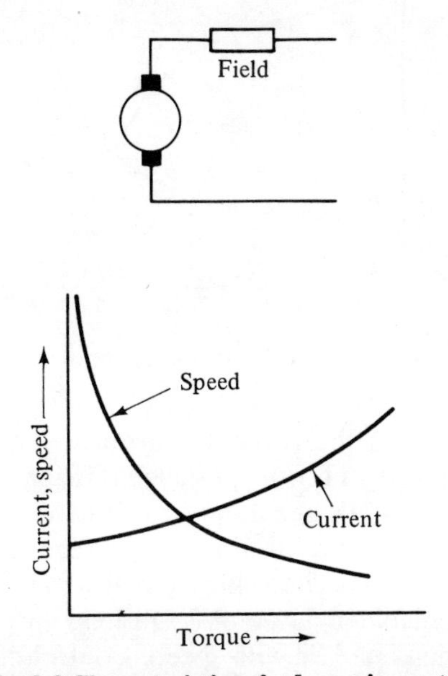

Fig. 8.2 Characteristics of a d.c. series motor

The speed of the series motor is dependent on the applied armature voltage. If the series motor is switched on to full battery voltage, when at rest, a high current flows through the motor, and the motor develops a high starting torque causing it to start suddenly. In order to limit the starting current and torque, reducing the jerkiness and strain on the transmission, it is necessary to limit the applied armature voltage, the voltage being increased gradually as the back-e.m.f. builds up due to the acceleration of the motor.

Series motors are widely used for delivery vehicles and industrial trucks. These vehicles generally operate at low speed and, except during starting, are usually driven with full battery voltage applied to the motor.

8.2.4 *D.C. shunt motor*

Shunt motors have characteristics similar to those of permanent magnet motors, except for the effect of armature demagnetizing flux. The motor characteristics are shown in Fig. 8.3. The shunt winding can either be connected to the same power supply as the armature or it may be separately excited.

The shunt motor has the advantage of controllability over the permanent magnet motor. Speeds can be controlled by varying the field current, which varies the field

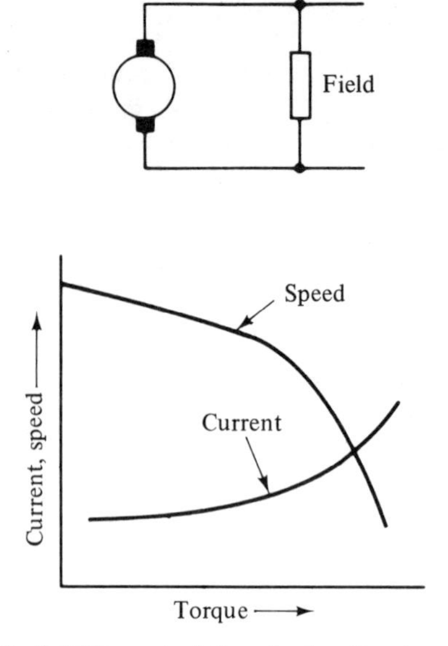

Fig. 8.3 Characteristics of a d.c. shunt motor

strength without operating on the armature current. Hence, control currents are smaller for the shunt motor than for the permanent magnet motor.

The speed of a d.c. shunt or separately excited motor may be changed by controlling either its field current or its armature voltage. The characteristics of the motor are shown in Fig. 8.4. By controlling the armature voltage with fixed field excitation, the motor is capable of developing full-load torque over the entire speed range. The power will then increase with speed. Control of the motor field, which usually occurs when the armature voltage is at the maximum rated value, will give a constant-power characteristic. As the field current is reduced the motor speed will increase and the torque will reduce. Modern closed-loop systems enable the motor speed to be controlled virtually to standstill by armature-voltage control. Motor-field control is usually limited to a speed range of about 3 : 1 or 4 : 1, mainly because of commutation problems due to the increased effect of armature reaction in the weak field condition.

The field of a separately excited d.c. machine can be controlled in such a way that the field current and the armature current are held in a fixed relationship. This type of control is most suited to solid-state controllers. A wide range of torque/speed characteristics is possible and one arrangement, specially developed for battery traction, allows the separately excited machine to behave as a series machine when operating as a motor, and as a shunt machine when braking, to allow efficient regeneration. Shunt machine characteristics could be used when operating at relatively high speeds.

For operation as a series machine the field current $I_f = I_a/K$ where I_a is the armature current and K is a constant, which for maximum efficiency is shown to be $K = (R_f/R_a)$ where R_f and R_a are the field and armature resistances respectively. For shunt motor characteristics $I_f = C/I_a$ where C is a constant.

At present the use of a separately excited motor for a battery traction drive has not found many applications. It has mainly been applied to a few battery buses operating

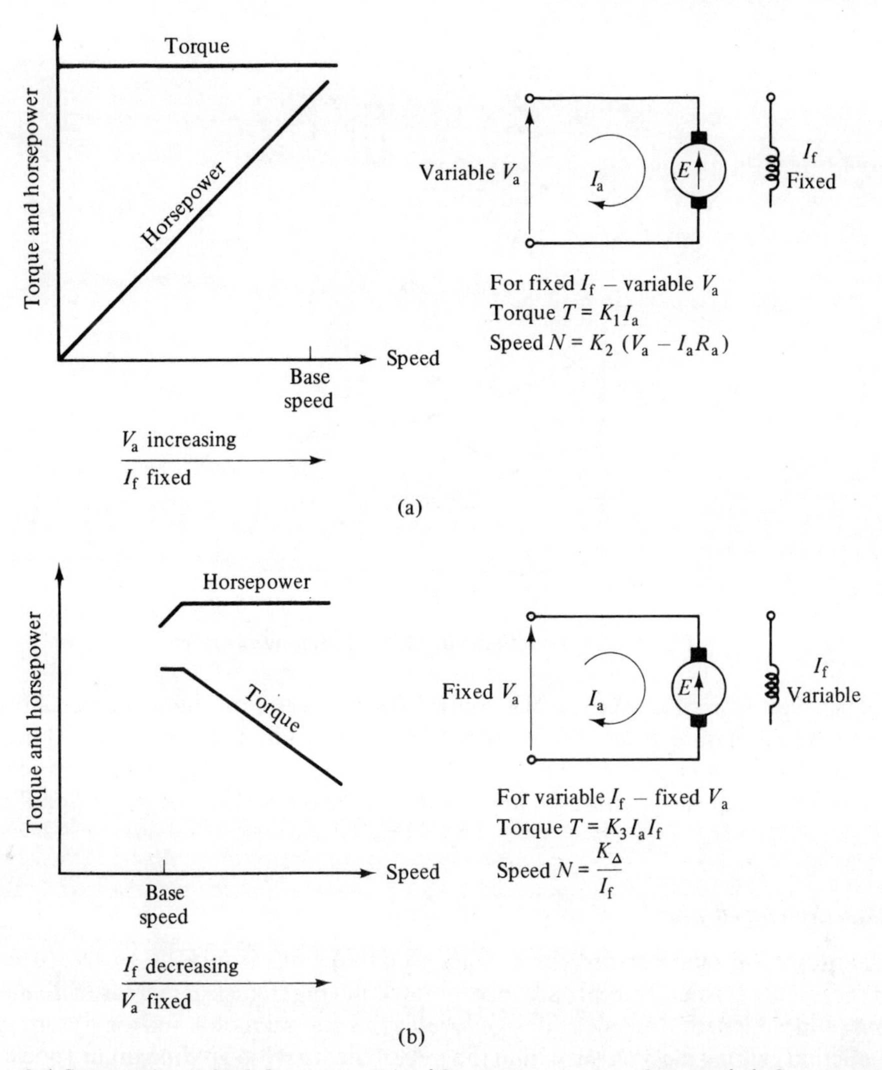

Fig. 8.4 Separately excited d.c. motor (a) with constant-torque characteristic by armature voltage control and (b) with constant horse-power characteristics by field-current control

on the continent, but it is expected to become an important drive motor when relatively high road speeds are necessary, for applications such as battery buses, cars, and vans.

8.2.5 *D.C. compound motors*

Compound motors have both shunt and series fields. The disadvantage of the series motor overspeeding at light loads can be avoided by use of a compound-wound motor. At no load, there is little current in the series field and the speed is determined by the shunt field. At higher loads, the speed depends on the sum of the two fields. The series field gives a reasonably high starting torque. The speed/torque characteristics are shown in Fig. 8.5.

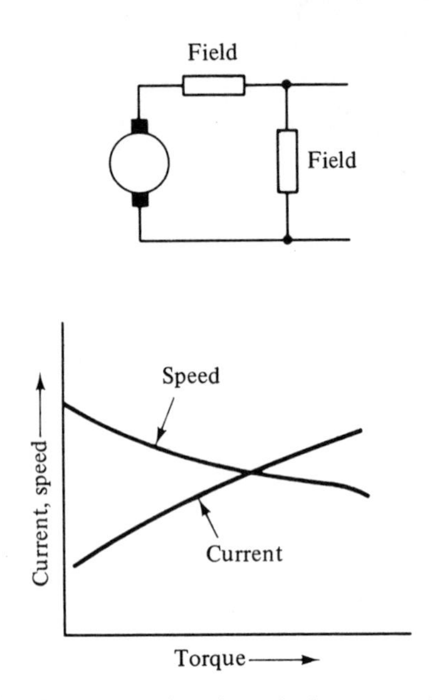

Fig. 8.5. Characteristics of a typical compound motor

This type of motor gives softer control than a series or shunt motor. Abrupt changes in armature volts or field current will not result in abrupt torque changes. Compound motors are especially useful in drives for hydraulic lifts, where a maximum pump speed must be held over a reasonably close range of 'load' and 'no-load' speeds.

8.2.6 *Motorized wheels*

Electric motors built into drive wheels are used frequently in fork-lift trucks. In many of these, slower-speed motors are incorporated having frames of increased diameter and reduced length. Ingenious design locates this special motor and a gyratory type reduction gearing mechanism within the hub of the wheel to give maximum power in the minimum of space.

The speed reducing mechanism is generally so designed that power is delivered to the wheel by a rolling action and with an oil film separation of the driving surfaces. The gyrator gearing provides a single-stage gear reduction that eliminates the frictional losses imposed by the conventional gearbox, and hence improves the efficiency of the drive system.

The armature shaft of this type of motor is usually provided with double eccentrics phased 180° from each other. Each eccentric revolves within a roller race forming the hub of a large planet gear wheel which engages with internal gear teeth formed in the end roller. (See Fig. 8.6.) The gear wheels are prevented from rotating by pins fixed in the non-rotating motor housing. Surrounding each pin are two rollers located within the circular holes in each gear wheel. The pins are eccentric to the roller centre by the same amount as the throw in the eccentrics on the motor shaft. Because of this, as the motor shaft rotates, the planet gear wheels are oscillated by the eccentrics in a gyratory motion. This gyratory movement of the planet gear rotates the annular gear

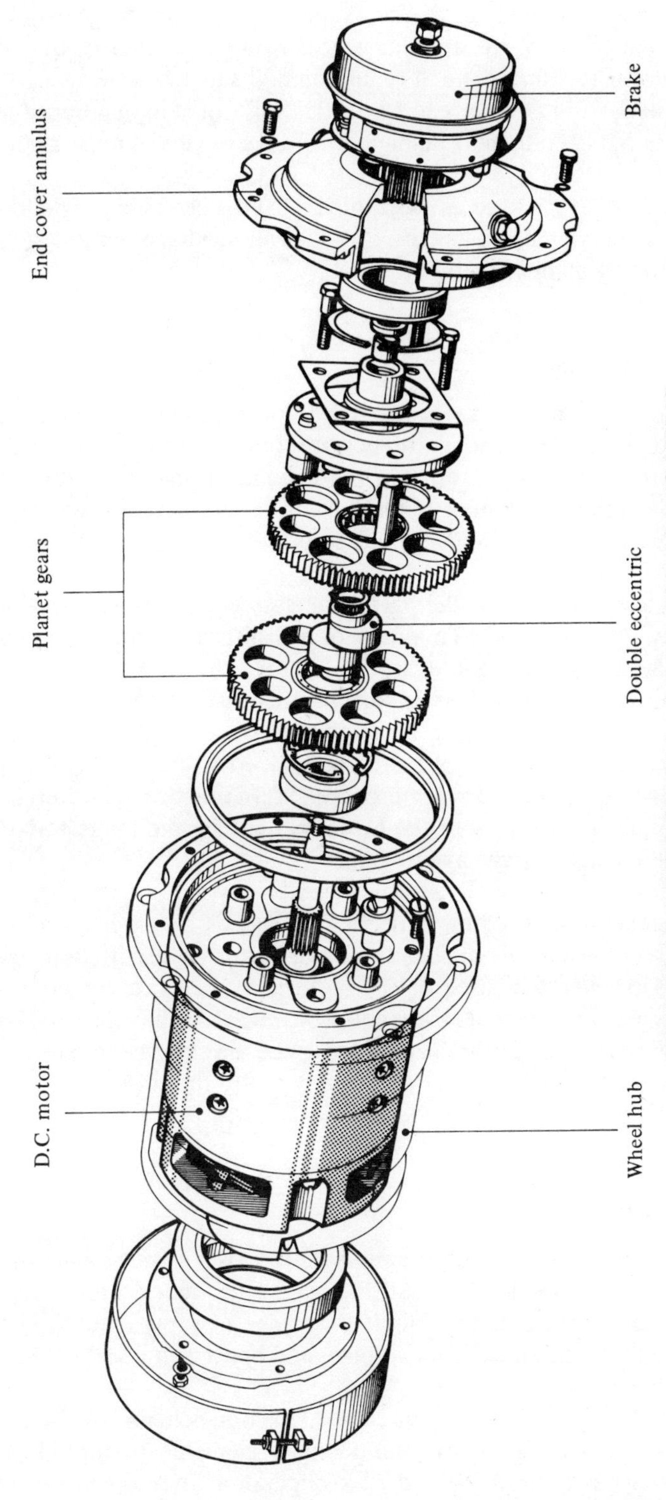

Fig. 8.6 Exploded drawing of wheel motor and gearbox (CAV Ltd)

in the end cover by an amount equivalent to the difference in the number of teeth of the planet gear and the annular gear. For example, if the annular gear has 92 teeth and the planet gear 85, then the annular gear is rotated a distance equivalent to 7 teeth for each motor revolution, i.e. a reduction of 13 to 1.

Because the planet and annular gears are so nearly equal in number of teeth, the sliding movement between them is negligible, and energy loss due to friction is very low.

Motorized wheels can be used as steerable or fixed wheels for single, double, or multiple wheel drives. In most instances, these powered wheels are used to steer the truck, thus improving manoeuvrability.

8.2.7 *Traction motor ratings*

The nominal kilowatt rating for traction drive motors is usually related to a one hour operating period plus an overload of 200 per cent full load torque for a period in the order of 5 minutes, with temperatures of the windings remaining within the limits prescribed by the relevant standards for the insulation system used. Usually the motor is rated in practice for the normal continuous operation of the electrical vehicle involved.

Pump motors are designed for the particular duty cycle required, having ratings ranging normally between 5 and 15 minutes for fork-lift trucks and continuous rating for hydrostatic drive systems.

Vehicle application dictates their rating, but because of battery characteristics vehicle designers aim to obtain the best possible tractive effort per ampere. To achieve this, the ratio of motor speed to road speed is kept as high as possible. Designing for high ratios keeps down motor weight. In practice, it is often found that the limit to the thermal capacity of the motor is fixed by the longest and steepest gradient which the vehicle may have to climb.

The weight/power ratio of a d.c. motor depends very much on the basic speed at full load so the tendency would be to use higher speeds. At high speed the motor efficiency may tend to fall off due to the higher iron losses, friction losses, and windage losses. However, the latter may be reduced by the use of forced ventilation. The power to drive a small blower motor could be much less than the windage loss on a high speed self-ventilated motor. The motor speed may also be determined by the gear ratio it is possible to include in the transmission to the road wheels. High gear ratios could mean higher losses and more noise.

8.2.8 *System voltage*

For a given motor peak power requirement there exists a wide range of possible voltages which would provide for a satisfactory battery pack. However, sensible values of maximum currents tend to limit this range to some extent. Also, safety considerations tend to limit maximum voltage, and hence some sort of compromise has to be arrived at.

The nominal voltages most used in the battery traction field are 12, 24, 36, 48, 72, and 80 V. The system voltages for commercially available battery-electric road vehicles tend to vary between 24 V and 72 V depending upon the vehicle capacity.

Low capacity vans and milk floats tend to operate on 24, 30 or 32 V while the large capacity type operate 72 or 80 V systems.

Small electric mopeds use 12 or 24 V systems, while prototype electric cars have batteries at 24 to 432 V. The trend seems to be about 100 V or even up to 200 V, although the only commercially available electric car in the UK used a 48 V system.

Several experimental electric buses have been built in Europe, the drive systems varying from about 72 V to 380 V. However, the trend seems to be for systems of about 350 V or more. High voltages are possible in vehicles such as buses and coaches as they are serviced by skilled personnel who will be aware of the dangers of high voltages.

Within the sphere of industrial electric trucks there is a trend to high-voltage drive systems. In a general sense, the term 'high-voltage' applies to the high side of the available voltage spectrum for a particular style of truck. The recent evolution of high-performance rider-operated fork-lift trucks has tended to 48 and 72 V drive systems. Both are considered to be high-voltage trucks and have a better performance than the 24 or 36 V trucks. For example, a change-over from a 36 V to a 48 V battery in a 'dual-voltage' fork-lift truck can boost travel speed by 34 per cent and lifting speed by 37 per cent.

However, there is a different range for a variety of pedestrian-operated trucks. In general, this range is from 12 V to 24 V and applies to high-lift counterbalanced trucks, low-lift pallet and platform trucks, and several types of narrow-aisle trucks. Thus the 24 V system is the high-voltage drive and solves critical handling problems, such as long steep ramps, frequent high lifts, and numerous heavy loads.

Also, there are groups of rider-type trucks in which the term 'high-voltage' would include both 24 and 36 V. For example, the voltage spectrum for rider low-lift pallet and platform trucks, and for pedestrian/rider models, ranges from 12 to 24 V, 30 or 36 V.

High-voltage systems for the same output as low-voltage systems obviously require less current. This is advantagous as far as the controller is concerned as the cost goes up with current capacity as well as the problems of thyristor commutation.

The output power per amp of d.c. traction motors has been increased through a combination of improved insulation, higher rotational speed, and increased motor voltage. With d.c. series motors there is a reduction in speed with an increase in load, but this reduction is less with a high-voltage system.

High-voltage, high-performance power systems tend to reduce electrical power losses throughout the entire circuit from the battery to the motor armature. For example, it is expected that there would be a reduction in power losses of about 10 to 15 per cent when a truck, equipped with a solid-state controller, is changed from 24 to 36 V. The main reasons for this are that high-voltage motors require less current for the same output than lower-voltage motors and that power losses in various parts of the electrical circuits are generally directly proportional to the square of the current.

8.2.9 *Efficiency*

Clearly one of the most important factors is efficiency, or more accurately losses, to obtain maximum energy and range while under battery propulsion. Typically, 1 kWh of lost energy requires the carriage of an additional lead-acid battery weight of some 25 to 30 kg.

The peak efficiencies of vehicle traction motors generally lie in the range of 75–90 per cent. Higher voltages and higher speeds lead to more efficient motors; higher speeds lead to lighter motors. With high speeds, however, there are the disadvantages of higher gear ratios, more noise, more costly motor construction, and a reduction in the time that the motor can thermally tolerate overloads.

8.2.10 *Motor enclosures and ventilation*

The type of enclosure is normally chosen to suit the specific requirements of the vehicle, taking into account the conditions and environment under which it must operate. Standard enclosures are available to give alternative protections against ingress of:

(a) medium solid bodies;
(b) medium solid bodies and splashing water;
(c) dust and splashing water.

Watertight and flameproof enclosures are also normally available.

Three alternative cooling arrangements are generally used for battery traction motors.

(a) Natural cooling, relying on natural heat dissipation without the aid of cooling fans. This type usually has a totally enclosed frame.
(b) Self cooling, normally using a fan on the motor shaft for passing external cooling air through the machine. This type can have a ventilated or totally enclosed frame.
(c) Forced cooling, where the supply of cooling air is trunked to the motor from an external source.

If the motor is to be where explosive gases and liquids are present a totally enclosed explosion-proof motor is used. The enclosure is designed and constructed to withstand an explosion of a specified gas or vapour which may occur within it and to prevent the ignition of the gas or vapour surrounding the machine by flashes, sparks, or explosions of the gas or vapour which may occur within the motor casing.

The continuous output rating for an open-frame motor may be 70 per cent or more, higher than the totally enclosed frame motor of the same construction. The choice between open and totally enclosed is therefore an environmental consideration and the problems of dust, dirt, water, and other debris affecting the motor must be evaluated. Table 8.1 lists the parameters of several typical d.c. battery traction motors. The table shows the variations in the one-hour rating as a result of ventilating and not ventilating the motor.

Improving the ventilation increases the power-to-weight ratio. For closed frame motors the specific power rating is about 13 kg/kW and is reduced to about 9 kg/kW or a little less for a ventilated motor operating at maximum voltage conditions. With the ventilated motor this fraction is reduced to 3·5 kg/kW when considering the peak motor rating.

Table 8.1 D.C. Motor comparisons

	1 hr power rating kW	0·5	1·3	1·5	2·5	7·5	12·5
Type	Excitation	PM	PM	CW	CW	SW	SW
	Frame	E	V	E	V	E	V
Performance	Voltage, V	36	36	36	36	72	72
	rev/min	3700	3200	3600	3000	1900	1900
Size	Frame Dia. mm	141	141	168	168	283	283
	Length mm	152	203	292	292	410	410
	Weight kg	6·4	10	20	20	96·2	115·6
	Kg/kW	12·8	7·7	13·3	8·0	12·8	9·2

NOTE: PM = permanent magnet; CW = Compound wound; SW = series wound; E = totally enclosed frame and V = ventilated frame.

With thyristor control the motor temperature does not decrease as much with decrease in load as it would with resistor control, because the form factor increases as the voltage decreases. As the speed or load decreases the magnitude of the harmonics in the armature current increases and produces correspondingly higher copper and iron losses. This has brought about the need of forced ventilation by a separate fan motor for some applications with thyristor controllers. This allows the motor to be run at any speed, virtually down to zero, with the separate ventilation fan keeping the winding below the maximum permissible values. For a self-ventilated motor the maximum permissible torque would have to be reduced with a reduction in speed to prevent the motor from overheating.

8.2.11 *Motor components*

Generally four-pole motors are used for d.c. battery traction applications, as they are considered to be the most economic choice both in terms of energy and cost. A two-pole motor would have inferior weight and performance; a six-pole machine would have adequate performance, but the 50 per cent increase in the number of poles, field coils, and brushgear compared with a four-pole motor results in increased cost.

Many electric-vehicle d.c. motors are at present manufactured without interpoles to keep cost and weight down to a minimum. This tends to restrict the armature speed of conventional motors to a maximum of about 3500 rev/min. for commutation reasons. With interpoles, commutator speeds of up to about 50 m/s (of the order of 5000 rev/min) become possible with conventional motors.

The action of the interpoles is to develop an m.m.f. which is in direct opposition to the armature reaction in the commutating zone and which produces the required field to neutralize the reactance voltage. This results in satisfactory commutation being achieved at a higher specific electrical loading than would be possible without the interpoles.

For the main poles and the interpoles there is a choice between laminated or solid poles. The choice is mainly dictated by the type of d.c. supply used. For example, with smooth d.c. supply, such as for a d.c. motor supplied from a battery with a resistance controller, solid poles would be adequate, but for a d.c. motor controlled by the thyristor chopper it is advisable to use a laminated pole. Interpoles will probably be necessary for good commutation, especially if regenerative braking is used. However, not all traction motor manufacturers use interpoles.

Some motors use an ingenious field system designed for a d.c. machine which operates on thyristor power supplies. The poles used consist of stacks of 'C' shaped laminations which divide the yoke into four separate parts (for a four-pole motor) which are integral with the opposite halves of adjacent poles. This is called the Dynalam* field system with split main poles as shown in Fig. 8.7. The advantage of this split-pole arrangement is illustrated in Fig. 8.7 which gives a diagrammatic comparison of flux distribution in the field system of Dynalam motors and conventional d.c. machines. For purposes of explanation, the interpole flux has been omitted in each case.

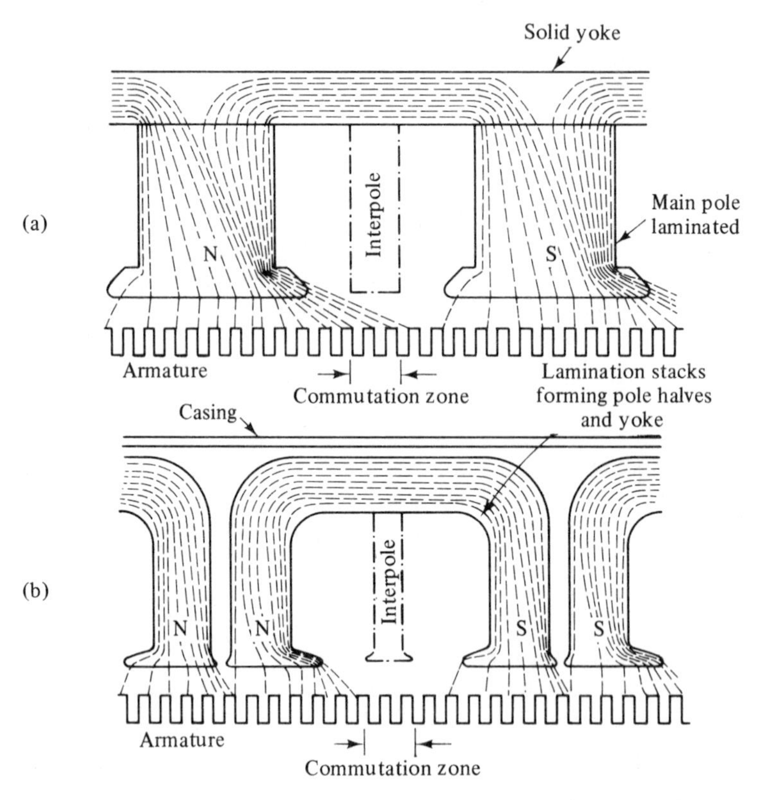

Fig. 8.7 Diagrammatic comparison of typical main pole flux distortion due to armature reaction on load in (a) conventional and (b) Dynalam motors. (Interpole flux has been omitted)

In the Dynalam field system, Fig. 8.7(b), the flux is divided into magnetically-isolated fully-laminated paths in the field system and, by virtue of the air space between the two halves of each pole, the diagonal drift of flux from one half of the pole to the other half is prevented, apart from minor leakage. In this way magnetic

* Developed by the Electro Dynamic Construction Co Ltd

flux distortion is reduced considerably when compared with a d.c. machine with conventional one-piece poles.

To increase the ability of the motor to withstand overload tungsten inert gas (TIG) welding is usually used to join the armature conductor to the commutator risers. This eliminates the use of solder with its low melting point. The welding is usually carried out on the front face of the risers and penetrates approximately 1 mm. Should a rewind be necessary this is quickly achieved by turning 1·5 mm off the front face of the risers enabling the coil ends to be lifted out.

The commutator should be cleaned periodically with a soft fluff-free cloth to remove all traces of oil, moisture, and dust. Any accumulation of dust in the mica recesses between the commutator segments should be removed by brushing along the grooves with a stiff brush. Difficult oil or dirt on the commutator may be removed with a clean rag moistened with a suitable electrical cleanser. Solvents such as acetone or alcohol should not be used as they may soften the insulating varnish.

An eccentric, ridged or flatted commutator can only be trued up by skimming or grinding. The process is best carried out with armature in situ in the motor which is driven at full speed with the brushes lifted, but alternatively the armature may be removed and the commutator skimmed in a lathe. After skimming, the mica between segments should be undercut by 0·5 mm ($\frac{1}{32}$ in) and the edges of the segments slightly bevelled with a smooth file followed by a piece of smooth hard steel to remove the sharp edges.

The motor windings are manufactured from enamelled copper wire or copper strip. The dimensions used will depend upon the motor power rating, current rating, number of poles, and type of winding. Armature and field coils are often wound with copper strip for improved space factor.

Because of the low-voltage operation the armature winding is of heavy section and usually of the wave wound type. It is usually a two-circuit winding irrespective of the number of poles. The field coils are mounted on the main poles, and because of the small confined space in many low-voltage motors the field coils are wound on a flat former and then bent on a press to obtain a curvature which allows them to fit against the motor frame. This contact with the frame also allows good heat transfer from the coil to the frame and to the surrounding atmosphere. The design of the field coil depends upon the characteristics expected from the motor, i.e., shunt or series. The series field carries full motor current, as it is connected in series with the armature, and hence is of a heavy copper section. A shunt field is connected across the armature, taking only a small proportion of the motor current. The copper section is considerably less than in the series case and with a much higher resistance. The interpole winding also carries full motor current and is again of a heavy copper section.

Some motors have two series windings giving a choice of two motor speeds. They can be connected so that only one is in circuit and then switched connecting both in series with the armature, or alternatively they could be switched from both fields connected in parallel to both connected in series. This dual series field arrangement reduces the number of control contactors and resistors in the accelerating circuit.

For battery vehicle applications, where vibration may be a problem, the brush holder should be close fitting and its support should be substantial and rigid in order to avoid any relative radial movement of the brushbox to the commutator. Radial type brush holders have been very successful in service, being suitable for both directions of rotation. However, the brush tends to tilt in the brushbox due to the couple created by the friction force and the reaction force on the brushbox face. Hence the pressure across the brush face is not even, being concentrated towards the leaving edge. It is therefore important to minimize both the brush clearance (consistent with freedom from brushes sticking) and also the clearance between brushbox and commutator face.

The brush gear is mounted on the brush rocker, which is a steel ring clamped onto the motor endframe. The manufacturer sets the brush rocker to locate the brushes in the magnetic neutral position, and this setting is usually marked by a pointer mounted inside the endframe aligned with a notch on the rocker.

On a 1 hour rated traction motor for material handling trucks, it is usual to provide sufficient brush area to obtain a brush density at rated load current of 70 to 80 per cent of the brush manufacturer's recommended current density. This is necessary to have the capacity to carry the heavy peak currents imposed during acceleration, ramp climbing, and braking. Overloaded brushes cause excessive heating, poor commutation, and rapid brush wear. Short duty (3 to 5 minutes) rated motors may use brushes which operate at 150 per cent recommended brush density at rated motor current.

The brushes should be examined regularly to see that they are bedding properly and not sticking in their holders for any reason; the most likely cause is an accumulation of carbon dust. The flexible connections should be kept clear of the commutator risers and any earthed part of the motor. Carbon dust should be regularly blown out of the motor with clean, dry compressed air.

Materials, impregnations, and techniques are generally used for the effective insulation of armatures and field systems to Class E, B, F or H requirements as defined in BS 2757, in accordance with BS 1727 for the particular duty specified.

The temperature rises at rated output should not exceed the values shown in table 8.2 from BS 1727: 1971. Experience has shown that machines to these insulation specifications have an adequate insulation life greater than those associated with the temperature classes in BS 2757.

Table 8.2 Temperature rise

Part	Method of measurement	Temperature rise (°C)			
		Class E insulation	Class B insulation	Class F insulation	Class H insulation
Armature winding	Resistance	105	120	140	160
Field winding	Resistance	115	130	155	180
Commutator	Electric thermometer	105	115	115	115

8.2.12 *Typical motor ratings for various applications*

Table 8.3 shows the rating of low-voltage motors used in various types of industrial trucks. Drive motors vary from about 2 kW to 14 kW while pump motors cover a range from 2·5 kW to 18·75 kW. Table 8.4 lists the rating of traction motors for battery road vehicles, and in this case the range is from 1·1 kW to about 8·25 kW.

Table 8.3 Typical motor ratings for industrial trucks

Type of truck	Capacity (kg)	Motor	Duty (h)	Output range (kW)
Rider operated	less than	Drive	1	2 to 4·2
fork-lift truck	900	pump	$\frac{1}{4}$	2·5 to 6
	900 to	Drive	1	2 to 8·25
	2699	pump	$\frac{1}{4}$	3 to 18·75
		steering	$\frac{1}{4}$	0·4 to 1·5
	2700	Drive	1	4·5 to 14
	and above	pump	$\frac{1}{4}$	9 to 18·75
		steering	$\frac{1}{4}$	0·4 to 1·5
Pedestrian operated	900 to	Drive	1	0·5 to 1
pallet truck	2699	pump	$\frac{1}{4}$	1·2 to 1·5
Rider operated	900 to	Drive	1	1·5
pallet truck	2699	pump	$\frac{1}{4}$	1·2
Rider operated	900 to	Drive	1	1·5
stacker truck	2699	pump	$\frac{1}{4}$	3·6

Table 8.4 Typical drive motor ratings for electric road vehicles

Gross vehicle weight (tonnes)	Duty (h)	Output range (kW)
Less than 2	1	1·1 to 2·2
2 to less than 3	1	2·1 to 5·1
3 to less than 4	1	2·2 to 8·0
4 and over	1	3·4 to 8·25

8.3 Crane, Hoist, Lift, and Conveyor Motors

8.3.1 *Cranes and hoists*

The term 'crane motors' can usually be applied to motors which carry out the three primary crane functions:

(a) load lifting and lowering (hoist motors);
(b) motion along the workshop;
(c) motion across the workshop.

Generally one motor is employed for each of these functions, except in the case of bucket or grapple cranes where two motors work together to provide the opening and closing motion of the bucket or grapple and the hoisting motion.

Because of the precise control requirements (variable torque loads, acceleration, stopping, and reversal) crane motors are either d.c. or a.c. wound-rotor induction motors, although cage motors are used for small hoists where precise control is not necessary. Heavy-duty steelworks cranes which have wide load variations are equipped with d.c. series motors supplied from a constant voltage d.c. power supply.

The basic speed control is inherent in the motor speed regulation. Series-connected tapped resistance banks are switched to provide current limiting on starting and low-speed operation.

In cases where fine speed control is essential, Ward–Leonard or thyristor controlled d.c. motor drives are usually the first choice. Wound-rotor a.c. induction motors are also applied successfully in some instances, usually below 200 kW, but they tend to have control difficulties at low creep speeds, which is limited by their acceptance and dissipation of the slip energy.

Other important factors are ambient conditions, such as outside or inside service, clean or dusty conditions, and ambient temperatures. A general-purpose motor is almost never satisfactory when applied to crane service. Another factor, more difficult to analyse, and which is obscure in application is the duty-cycle operation. For ordinary engineering shop service, crane motors rated to develop their full power for a period of half an hour with a temperature rise of 55°C are suitable. For steelworks and other situations where the duty is severe, the rating should be for one hour, and on heavy continuous grabbing duty it has been found necessary to have the motors continuously rated. Cage-rotor induction motors need to be of the high torque type as starting torque must be high to avoid stalling.

Total enclosure of the motor is desirable to exclude dust, and in the case of outdoor cranes it is essential to afford protection from the weather. At the same time, covers over collector rings, or commutator and brushgear, should be easily removable for inspection or adjustment of the brushes, etc. The use of nuts or other loose parts should be avoided in fixing the covers, as they are easily lost, or may be dropped from the crane, and even a small nut may do considerable damage or cause severe injury if dropped from a height. Some manufacturers anchor the covers themselves with a short length of chain, as a precaution against dropping.

Crane motor requirements are often very severe, and require a high standard of reliability to assure uninterrupted service as well as safety to personnel and equipment. Motor construction must therefore be of the optimum quality, including ample metal sections with adequate insulation and bracing of the coil windings and connections, particularly in rotor windings. Plug stops and overspeed conditions are frequently involved, requiring the rotors to be insulated for at least twice rated voltage. Thermal motor protection should function to warn the operator of motor overheating and impending failure to enable him to take the necessary action. Thermal cut-outs are not used as this may cause dropping of loads or the stopping of a load in an undesirable position.

8.3.2 *Lift motors*

The choice of motor is determined, to some extent, by the available supply. Direct-current motors give the advantages of high starting torque, smooth acceleration, and readily variable speed. However, the majority of lifts depend on a.c. supplies, although thyristor or rectifier bridges enable the d.c. motor to be used. The a.c. motor is probably the most common, and they are often of the two-speed cage rotor type or the wound-rotor type.

Lift motors comply with BS 2655: Part 2, in which two grades are required:

Grade 1 covers motors when silent running is essential, and are fitted with sleeve bearings

Grade 2 covers motors which are not required to comply with Grade 1, fitted with ball bearings.

There are also two types of rating:

(a) 90 starts per hour. This is the standard maximum rating of cage motors.
(b) 180 starts per hour. In the case of wound-rotor induction motors, this may be specified as an alternative to (a).

High smooth accelerating torque is required. The figures specified are 225 per cent minimum and 275 per cent maximum starting torque for cage motors and 225 per cent for wound-rotor motors, in conjunction with a suitable rotor resistance starter. Typical examples of lift motor ratings for passenger lifts are listed in Table 8.5.

Table 8.5

Maximum No. of passengers	Speed m/s	kW	Maximum No. of passengers	Speed m/s	kW
2	0·5 to 0·8	$1\frac{1}{2}$	8	0·5 to 1·0	$5\frac{1}{4}$ to $7\frac{1}{2}$
4	0·5 to 0·8	$2\frac{1}{4}$ to 3	15	0·5 to 0·8	$7\frac{1}{2}$ to $11\frac{1}{4}$

Similar safety requirements, where applicable, are necessary for lift motors as for crane motors. Square ended shaft extensions are usually fitted as standard for passenger lift motors to enable them to be operated manually in case of a mains failure or emergency.

8.3.3 *Conveyor motors*

Belt conveyors constitute the best method for handling materials in bulk, such as coal, gravel, sand, and ores. Initial starting loads are usually heavy and so high-torque cage motors, or alternatively wound-rotor induction motors should be considered. They are generally exposed to dusty and grit-laden conditions and hence totally enclosed, fan-cooled motors are preferred for most applications.

Two or three discrete speed steps may be satisfied by pole-change motors, but when wide speed variation is required, with frequent positioned steps, it may be necessary to employ a closely controlled d.c. motor drive to produce the specified requirements.

8.4 Motors for Fans, Blowers, and Pumps

8.4.1 *Fans and blowers*

The vast majority of small shaft-mounted fan and blower motors utilize the air from the fan to cool the motor. They are generally of a totally enclosed construction, and are usually split-phase or capacitor start and run motors up to about 500 W. Small shaded-pole motors are also common. Three-phase cage motors are used above about 100 W.

Two-speed capacitor start and run, dual voltage wound motors are used to drive shaft-mounted fans. A second speed of the order of two-thirds of the synchronous speed is obtainable by operation of a switch when driving a fan load. With the aid of a suitable voltage control several speed steps are available down to above half synchronous speed. Two- or three-speed shaded-pole motors are used, the speed steps usually being obtained by tapped stator windings.

The rear bearing for shaft-mounted fan motors is designed to take the axial thrust of the fan mounted near the front bearing.

Belt-driven fan and blower motors have relatively low starting torque and relatively high performance at full load. Special care is generally taken in the design and manufacture of these motors to ensure quiet operation. The drive motors for small fans are conventional split-phase and capacitor-start, one or two speed motors, usually rated between about 100 and 600 W with speeds of 1500 rev/min or 1500/1000 rev/min. They are generally drip proof and foot mounted with a resilient base.

The oil burner motor could be included either in a fan and blower section, or a pump section, because it performs both functions. The oil burner pump and a cage blower are both close coupled to the motor shaft in a typical home oil burner system. The motor is generally a totally enclosed, single-phase, split-phase motor of about 60 W to 130 W at 1500 rev/min. They are flanged motors with clearance holes provided in the end frame outer diameter to bolt the motor to the burner. The motor name plate carries the words 'oil burner motor'. It has a manual reset thermal-overload protector, with directions for resetting usually displayed on the motor.

Pole-amplitude-modulated (PAM) type speed changing motors have found wide-spread use for driving fans and pumps. The quantity of air pumped is not normally so critical that it cannot be met by two choices of speed. Figure 8.8 shows the torque/speed characteristics of a typical fan with two typical motor torque/speed

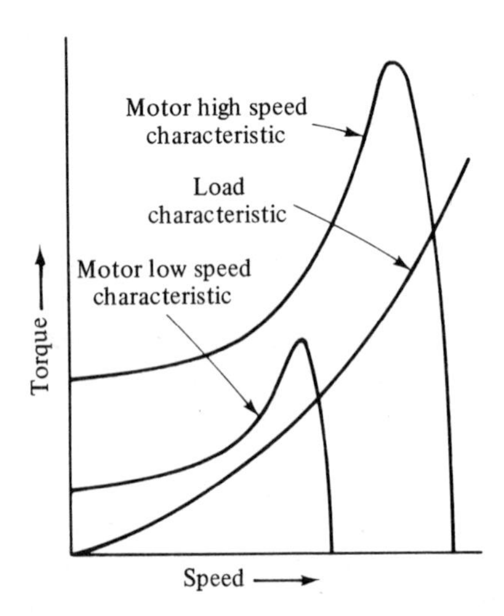

Fig. 8.8 Typical motor and load characteristic for a fan having a speed ratio of 1·5 : 1

characteristics superimposed upon it. In this case the motor characteristics are for a 4/6 pole winding giving a ratio of synchronous speeds of 1·5 : 1. It is evident from these characteristics that the motor torque required at low speed is significantly less than that at high speed and consequently the motor windings need to have higher reactances for the low-speed condition. This dictates the basic connection system for the winding as a parallel-star/series-star connection, as shown in Fig. 8.9.

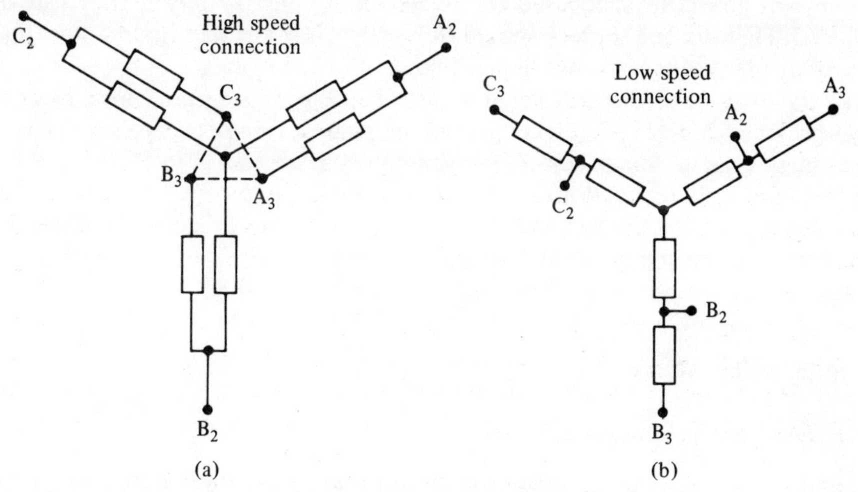

Fig. 8.9 Basic PAM connection for a fan or pump drive; shows (a) the parallel star connection and (b) series star connection

Many PAM motors have been built and are running successfully as fan and pump drives. They are in widespread use over the whole range of motor sizes and with various speed combinations. At the lower end of the size scale 4/6 poles are the most popular, but many 6/8 pole and 8/10 pole motors have also been built. Three speed- and four-speed fan drives are in service, in for example a tunnel ventilation scheme. These are 6/8 / 12/16 pole motors, wound with a large 6/8 pole PAM winding and a much smaller 12/16 PAM winding.

8.4.2 *Pumps*

For large pumping scheme drives consuming several hundreds of kilowatts with a requirement for variable speed, a.c. commutator motors have been successful, although the increasing popularity of slip energy recovery schemes makes them a strong competitor. The main drawback of such schemes is that their operation is limited to sub-synchronous speeds whereas the maximum pump efficiencies may occur at a super-synchronous speed which is within the range of the commutator motor. Direct-current motors and thyristor convertors may also be used on water pumping applications and although the overall efficiency is comparable with an a.c. commutator motor, the power factor at reduced speeds is significantly worse. Where several pump units are necessary to meet the output requirements, and the base load is high, a hybrid installation of constant-speed induction motors and a wide speed range variable-speed motor can be justified.

For small pump applications drip-proof and totally enclosed fan-cooled a.c. motors are the most frequently used types. They are often mounted on a common bedplate with, and directly coupled to the pump. Complete motor pump units often employ flange mounted or vertically mounted motors.

Drip-proof motors may be used for centrifugal pumps and compressors, but if the surroundings are excessively damp or dusty, totally enclosed fan-cooled types are more suitable.

Standard motors may be used for reciprocating pumps only if they can start unloaded. High-torque motors should be used if they are required to start under load, particularly where ratings higher than 4 kW are involved.

Motors for gasoline-dispensing pumps are of an explosion-proof construction and may be directly coupled or belt coupled to the pump. Four-pole capacitor-start and three-phase cage motors of 350 W are generally employed.

Motors for coolant pumps are enclosed ball-bearing motors built for horizontal or vertical operation for direct connection, usually on a support pipe, to the direct drive centrifugal coolant pump. These pumps are generally driven by two- or four-pole capacitor-start or three-phase cage motors from 37 to 750 W.

8.5 Submersible Motors

8.5.1 *Submersible motor construction*

Submersible motors must obviously differ from conventional motors in both mechanical and electrical design because of their unique operating environment. There are generally three basic types of construction, and the advantages of each must be properly understood for proper selection.

All motor components exposed to sea water are made from corrosion resistant materials, for instance, the stator windings may be hermetically sealed with a suitable resin in an all-welded stainless steel case, and the rotor shaft and motor end plates are of stainless steel.

For undersea operation special designs of labyrinth shaft seals prevent sand from getting into the motor.

Submersible motors are usually relatively long and slender and usually of the induction motor type. Their mechanical structures fall into three distinct categories:

(*a*) *The 'all-wet' submersible motor.* This motor, Fig. 8.10(a), employs a heavy waterproof insulation on the field winding wire, since water fills the inside of the motor and surrounds both the stator windings and the rotor. Successful performance depends upon the reliability of the insulation. The principal advantage is simplicity, but it is difficult to ensure against weak or damaged spots in the insulation. The type of insulation available for such service is bulky, and is usually only useful for low temperatures, and is therefore easily damaged by high or abnormal load conditions. Such motors are sometimes used for water pumping applications in small bore pipes for central heating schemes.

(*b*) *The oil-filled submersible motor.* This motor, shown in Fig. 8.10(b), has conventional motor windings treated with an oil-resistant varnish. The motor is filled with an oil having good dielectric properties. A shaft seal is used to keep the oil in and the water out, the effectiveness of this seal being of critical importance. However, no rotational seal is perfect and eventually some water will pass into and mix with the oil, ultimately resulting in failure of the insulation. Simplicity of design and manufacture are primary advantages.

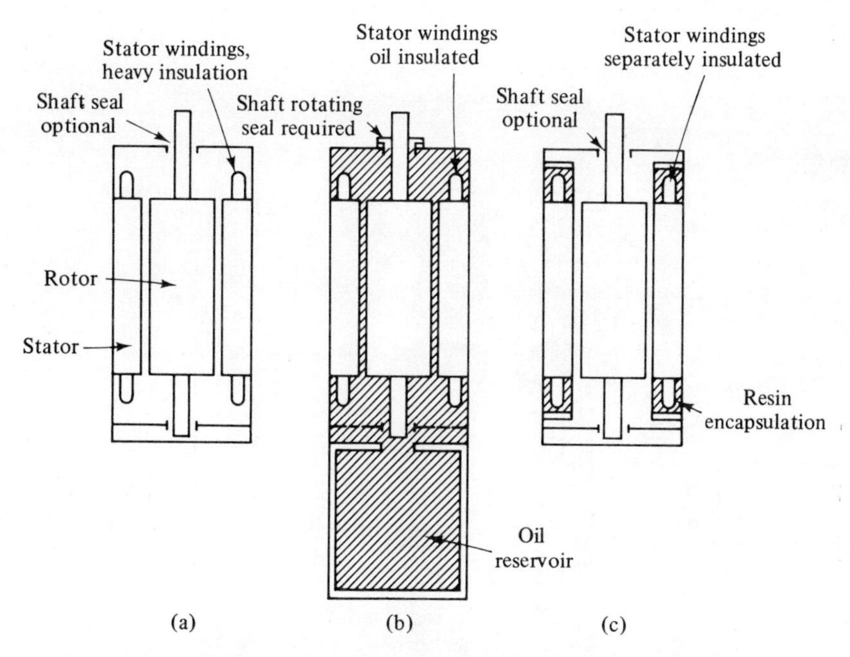

Fig. 8.10 Simplified cross-sections of three types of submersible motor

Some d.c. motors have been developed to run as oil-filled motors. The current switching by the brushes and commutator takes place submerged in oil.

(*c*) *The hermetically sealed motor*. This motor, shown in Fig. 8.10(c), employs conventional insulation and field winding wire. The stator is completely enclosed in a steel casing, with a thin tube or liner in the motor air gap. With proper design and material, this liner can be made very thin and electrical losses due to it kept low. If the enclosure is welded then the leakage of gaskets is eliminated. The enclosure is filled and the windings impregnated with a resin mixture which provides insulation, anchors the windings, and conducts heat.

8.5.2 *Submersible pump motors*

The extreme difficulty of constructing satisfactory seals for pump driving-shafts entering systems containing liquid at high pressure and temperature, or liquids of a highly volatile or toxic nature has led to the development of an electric motor and pump unit enclosed within a common pressure casting.

Figure 8.11 shows a diagram of a typical motor pump unit with a high-pressure and low-pressure cooling circuit.

Submerged pump motors cover such duties as forced power circulation, heater extraction pumps, and deep well pumps. The motors are in some cases designed to operate under water pressures up to 200 kg/cm^2.

The highly specialized field of submerged motors has increased considerably in importance, particularly in nuclear and fossil-fuelled power stations. The elimination of mechanical seals for rotating shafts operating in the working fluid is attractive

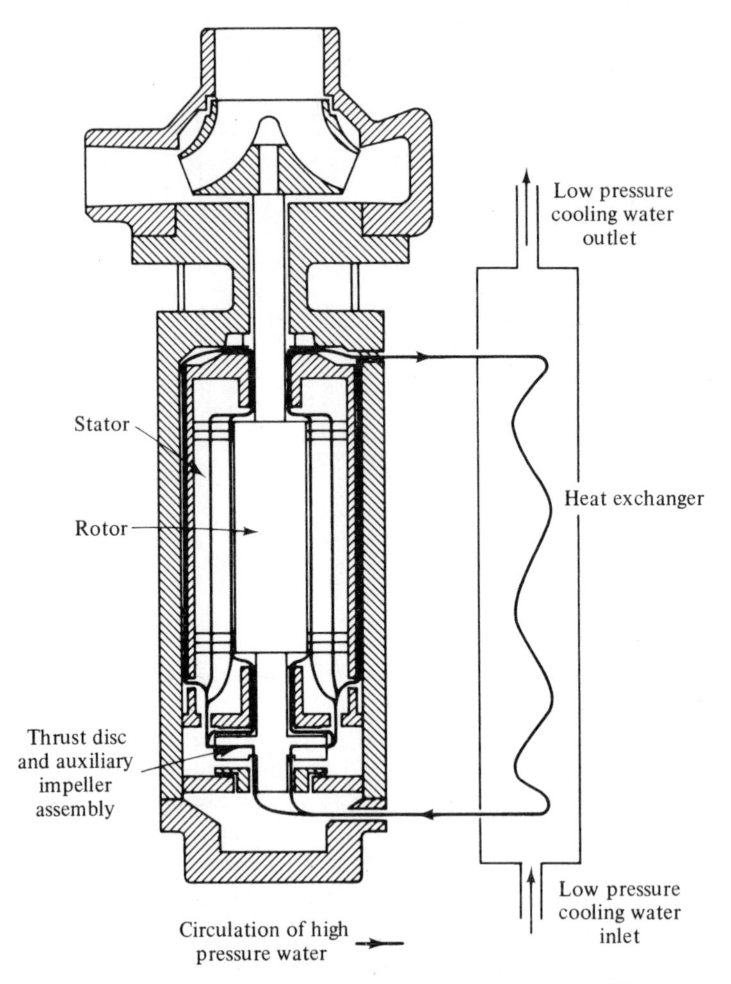

Fig. 8.11 Diagram showing the high-pressure and low-pressure cooling circulation in a zero-leakage pump unit (Hayward Tyler & Co. Ltd.)

because of their reliability and simplicity. They are becoming especially important as gas circulator motors in nuclear fuelled power stations.

There are basically two types of submersible 'zero-leakage' motors; those with stator enclosures, known as dry or hermetically sealed units, and those in which the liquid pumped is circulated through the stator windings, known as wet stator units. The pump–motor units are mounted vertically, and since the motor is invariably of the cage induction type there is no electrical contact with the rotating assembly. In all these units the liquid pumped is usually circulated through the rotor-stator gap as well as acting as a lubricant for the thrust and journal bearings. The rotor is carried by a film of liquid, often water, generated within the journal and thrust bearings. This film provides an almost frictionless support for the motor and for the end thrust transmitted from the pump.

For some applications, particularly in connection with nuclear reactors, all metallic parts including the motor core plates are manufactured from stainless steel. In some cases, for very large units installed in primary circuits of water-moderated reactors, carbon steel casing clad internally with stainless steel may be used. For boiler-circulation duties, and for pumping non-corrosive chemicals, ordinary carbon steels are used.

When handling hot liquids, such as steam and water mixtures the motor temperature has to be reduced by one of three methods:

(a) By special design to restrict conduction and convection between pump and motor.
(b) External heat exchangers.
(c) A cooling jacket surrounding the motor containing cooling water.

The ratings of submersible deep-well pump motors range from about 10 kW to several hundred kW, and typically cover deep well diameters from 8 in to 16 in. Submersible pump motors are usually two- or four-pole machines and are typically 230 or 430 V up to about 100 kW, 460 V up to 300 kW, and 2·3 kV, 3·3 kV or 6·6 kV for ratings above this.

Ratings for gas circulator motors for nuclear power stators probably range from about 50 kW to about 1500 kW. These are discussed further in section 8.7 dealing with power-station auxiliary motors.

8.6 Marine Motors

Marine motors are manufactured to comply with the following specifications and rules: BS 2949, AIEE No. 45, Lloyd's Register of Shipping, American Bureau of Shipping, Bureau Veritas, Germanischer Lloyd, and Norske Veritas. These can be either drip-proof or hose-proof for use between decks, or deck watertight.

Industrial motors are often mounted on rigid, vibration-free foundations in a relatively low humidity atmosphere. When a motor is installed onboard a ship the foundation is a steel deck subject to vibration, shock, pitch, and roll, and the air is usually very humid and salt-laden. The environment in which the motor is operating as well as the function it is performing are the main reasons for the important differences between industrial and marine motors.

8.6.1 *Non-propulsion motors*

Deck auxiliaries include cargo winches, cranes, capstans, warping winches, windlasses, and hatchcover winches. For some classes of ship variable speed for deck auxiliaries is still preferred, but in general two- or three-speed pole-change or PAM induction motors are suitable. For cranes a high speed facility is needed for rapid return of the empty hook. Light-hook speeds on cargo winches are 3 to $4\frac{1}{2}$ times normal full-load speed. Where adjustable speed is essential in a.c. systems, wound-rotor induction motors or a Ward–Leonard drive may be employed. As a general rule the number of brush-type motors is kept to a minimum.

Ships are usually classed for either Unrestricted Service or Restricted Service, the latter relating to service only outside the tropics. There are different rules and regulations regarding ambient temperatures and permissible temperature rises for the two different classes.

Drip-proof marine motors are suitable for applications between decks where hose-proof conditions are not required. They have ventilated enclosures designed to prevent entry of falling liquid or solid particles.

Totally enclosed, fan-cooled, hose-proof marine motors usually have bearings protected by metal water throwers and all joints are sealed against water entry from hose washing.

Deck watertight motors are totally enclosed and non-ventilated; the fan cooling is omitted because of the possibility of ice formation. They have sealed bearings and a watertight terminal box, and are capable of complete submersion in shallow water for short periods. Sealing washers are fitted under all screws and a coat of special corrosion resistive paint is generally applied to all internal and external surfaces.

There are special requirements set out in the Lloyd's Register of Shipping for petroleum tankers and for ships carrying oil in bulk having a flash point less than 65·5°C. On open deck, within 3 m (10 ft) from an oil tank outlet or vapour outlet a flame-proof motor is necessary. Elsewhere on the open deck a watertight enclosure is generally used.

The metal frame, end shields, terminal box assembly, and all hardware are treated for corrosion resistance or manufactured from a corrosion resistant material. Cage rotors are protected by a coating of insulating varnish. Shafts of corrosive material are also varnished, but in extremely corrosive environments non-corrosive materials, such as stainless steel, are generally used. Motor windings are treated to resist moisture, sea air, and oil vapour.

Grease lubrication is usually employed as it is not sensitive to the roll and pitch of the ship. Oil lubrication is permitted where the speed of ball-bearings requires this type of lubrication or where the application requires sleeve bearings.

The motor end play is usually reduced to a minimum, and adequate provision, where necessary, is made for taking the end thrust due to the motion of the ship. Wherever possible motors are placed with their axes of rotation in the fore-and-aft direction.

Marine motors are designed so that all parts which may require repair, servicing or replacement during the life of the motor are readily accessible. Motors employing brushes should have access openings of sufficient size to allow inspection of the brushes while the motor is running and to enable the brush to be easily replaced.

Motors weighing about 70 kg or more are generally provided with removable lifting eyes.

Motors essential to the safety of the ship in the event of damage, such as emergency bilge pumps, must be capable of running for a reasonable period after the compartment in which they are situated has been flooded.

Consideration should be given to the union of two dissimilar metals, particularly in a salt-laden atmosphere, as the corrosion rate may be very high. To minimize this electrolytic corrosion reference to the galvanic series of dissimilar metals should be made to avoid contact of two metals that are far apart in the series. If the only medium separating two parts is a layer of paint then a metal-to-metal contact should be assumed as far as corrosion is concerned.

8.6.2 *Propulsion motors*

Variable-frequency a.c. drives are common for ship propulsion schemes. The variable frequency, over a 3:1 range, is supplied from an alternator driven by a variable-speed gas turbine. This type of system employs either direct or gear-coupled induction or synchronous motor drives for the propellers.

The type of drive used often depends upon experience with regard to the reliability of gearboxes. The choice is also dictated by the type of a.c. machine used. With a large direct-coupled low-speed induction motor the power factor is of the order of 0·5, while a similar synchronous motor can operate at unity, the efficiencies being more nearly equal at about 85 per cent and 95 per cent respectively. For higher speed gear-coupled motors the efficiencies are comparable and there is only a 10 per cent difference in power factor. However, the marginally better performance of the high-speed synchronous motor does not compare with the greater simplicity and reliability of the induction motor.

Variable-frequency propulsion schemes are normally considered viable for large power requirements from 8 MW upwards, with 3·3 or 6·6 kV supplies.

Constant speed or pole-change induction motors have been applied with variable-pitch propellers, but because of the concern regarding the efficiency and maintenance requirements they are not common.

Direct-current propulsion motors are used, driven by thyristor convertors, supplied from a gas turbine alternator set. However, in these systems, because of the concentrated power systems on ships, there can be problems with harmonic generation, interaction between parallel convertors, and the radiation of harmonic voltages at radio frequencies.

Diesel motors driving d.c. generators to supply the d.c. propulsion motors are used, but are normally restricted to the smaller ships such as trawlers and survey vessels requiring a propulsion power of about 3 MW.

8.7 Power-station auxiliaries

As boiler/turbine units in power stations become larger, auxiliary electrical machines of increased size are being used in great numbers in modern power stations. These drives can absorb of the order of 120 MW of the total power generated and the most common types of drive motor used are the cage-rotor induction motor, the wound-rotor induction motor, and the a.c. shunt commutator motor. The table below lists some of the larger sizes of motor associated with a 500 MW turbine-alternator unit.

Table 8.6 Larger sizes of motor associated with a 500 MW unit

Auxiliary	Output kW	Motor type
Starting and standby boiler feed pumps	7125	WRM
Circulating-water pump	3900	CM
Forced-draught fan	746	CM
Induced draught fan	1650/1106	CM
Primary air fan	476	CM
Coal mill	705	CM

WRM = Wound rotor motor; CM = Cage motor

The largest motor in a thermal power station is generally that driving the boiler feed pump. Because of its power demand it has a great effect on the electricity supply system for the auxiliaries. The feed pumps are usually driven electrically for starting up requirements, and they generally also serve as standby pumps when the steam-turbines are running at full speed. Typical sizes for a 500 MW set would be about 6000 to 7000 kW at 11 kV, and usually two such motors are necessary. Often a further booster feed pump is provided with about 50 per cent of this rating.

Motors of the 750 to 2000 kW range are mostly used for driving induced—and forced— draught fans, circulating water pumps, and condensate pumps, while 375 to 750 kW motors are common for powering milling plant. A typical large modern boiler-turbine-alternator unit in a power station may have about twenty motors of over 750 kW associated with it.

8.7.1 *Motor types*

Wherever possible, motors in power stations are cage-rotor motors arranged for direct-on-line starting and switched by an air or oil circuit breaker.

The induced-draught fans driven by cage motors are of the two-speed design and in general these have been of the two-separate-winding type, with a separate terminal box for the high-speed and low-speed winding. Recently the introduction of PAM motors has resulted in two close-ratio speeds being obtained from a single winding, and a number of these machines are now installed for this particular duty. The fine control of draught is by fan-vane adjustment.

Alternating-current commutator motors have been used in considerable numbers where variable speeds are required, for example on fans associated with coal-milling plant. This type of motor usually has fixed brush gear, and speed control is by means of an induction-regulator which feeds the armature with a variable voltage of 50 Hz. Induction-regulator movement can be controlled by a pilot motor operated either from the control room, or from an automatic-boiler-control loop. Outputs of several thousand kilowatts are in use at 3·3 kV.

For the very large motors for starting and standby feed pumps, which have to cover a fairly narrow speed range, wound-rotor induction motors are generally used. The variable speed is obtained from a liquid resistance tank. The electrode drive motor, which is generally included in an automatic control loop, raises or lowers the electrodes in order to control the main motor speed. The minimum speed has to be carefully controlled because of the high rotor losses and the danger of overheating the rotor. These machines have the advantage of a low starting current, usually about $1\frac{1}{2}$ to 2 times full load current, and a fast starting time. This time is critical when such pumps are on autostandby duties covering the loss or failure of the main steam-driven feed pump.

Power station auxiliaries are required to continue to operate satisfactorily on supply variations outside the normal limits, such as might occur during a severe system disturbance or fault condition.

8.7.2 *Motor enclosures*

Motor enclosures are usually of the closed-air-circuit type with air/air or air/water heat exchangers. Extensive fabrication of frames incorporating a box-shaped structure is used.

The high and increasing costs of civil works has tended to make the motor room very expensive, and the enclosures of large machines have been developed to permit the installation of motors out of doors with little, if any, additional protection from the weather in the form of buildings. Motors used out of doors are classed as 'weather-protected' machines. The weatherproof motor is an open machine with its ventilating passages so constructed as to minimize the entrance of rain, snow, and airborne particles to the electrical parts and having its ventilated openings so constructed as to prevent the passage of a cylindrical rod 19 mm ($\frac{3}{4}$ in) in diameter. These motors are usually constructed to the standards of the (USA) National Electrical Manufacturers Association, type I or II. These standards are met by means of an inlet/outlet air housing fitted to the top of an otherwise standard motor.

8.7.3 *Submerged gas circulator*

In nuclear power stations two-pole induction motors have been used to drive gas circulators in Advanced Gas-cooled Reactors to circulate the carbon dioxide cooling medium. Some have been used to circulate helium. The combined motor and impeller unit is generally horizontally mounted (although some have been mounted vertically) in the side of the reactor, and submerged in carbon monoxide at a static pressure of 40 atmospheres. The circulator motors are usually rated between 5 and 7 MW operating at 11 kV.

The motor insulation techniques and requirements are substantially the same as those for air insulated equipment, as carbon monoxide has substantially the same dielectric properties as air. Carbon monoxide is non-oxidizing with a consequent advantage to insulation life.

8.8. Motors for Vibratory Applications

Electric motors are normally built to be as free as possible from vibration, but there is one type of motor which is built to allow a shaft unbalance for the specific purpose of producing vibration. These are known as vibrator, shaker, or exciter motors.

Shaker motors are used to produce the desired exciting force at the required frequency for various kinds of vibrating equipment, such as feeders, conveyors, hoppers, grinders, finishers, and mixers. Vibrating equipment provides a more positive flow of the material being handled than would be provided by a gravity feed, especially when the material is sticky. It can often perform jobs that would be difficult for other kinds of conveying and feeding equipment. For example, the equipment can readily be designed to handle very hot materials, such as fresh iron castings, and abrasive materials such as sand, steel scrap, rock, and coal.

8.8.1 *Motor construction*

A shaker motor is basically a totally enclosed, non-ventilated cage induction motor, but with its shaft extended at each end, and an eccentric weight attached to each shaft extension. Also special bearings and bearing fits are used. These special mechanical features along with the electrical design are dictated by the loads on the shaft.

The motor rotating parts, including the eccentric weights, do not vibrate with respect to the bearing supports. They actually move about an imaginary axis at the shaft rotational speed, hence the principal load on the motor is the centrifugal force providing a steady unidirectional, but rotating load on the shaft. Under these conditions it has been found best to fit the outer bearing ring tightly in its housing, while the inner ring is fitted loosely on the shaft to permit assembly and to allow thermal expansion of the shaft. If the outer ring was fitted loosely in the housing the load due to the centrifugal force would be such that it would tend to make the outer ring first creep in the housing and then rotate, generating heat and drastically reducing the bearing life.

Bearings are generally grease lubricated because of the design advantages and because of the operating conditions of most shaker motors. The principal design advantage of grease lubrication is the simpler bearing-housing construction that it permits. Shaker motors are often mounted vertically or at an angle to the horizontal. This does not matter with grease, but it would be important with oil lubrication. Bearing lubrication is the most difficult and frequently occurring problem in the design and application of these motors.

Table 8.7 lists the typical characteristics of a range of vibrator motors.

Table 8.7 Characteristics of typical vibratory motors

Current consumption at 400/440 V three-phase 50 Hz		Centrifugal force kg
Starting A	Running A	
0·90	0·20	177
1·00	0·25	281
1·50	0·30	454
2·30	0·45	544
2·50	0·50	794

8.8.2 *Application factors*

There are basically two approaches to the design of vibratory equipment and these can be labelled as non-resonant and resonant designs.

In the non-resonant designs all of the force necessary to vibrate the equipment is derived from the shaker motor. The motor is attached directly to the vibrating equipment, which is supported on springs that isolate the equipment from the floor or support structure. When only one shaker motor is used the vibrating equipment has a circular or elliptical motion, and this is used in gyratory hoppers and grinders.

In a partially restrained system, motion is made rectilinear by rotating two eccentrically loaded shafts in opposite directions as shown in Fig. 8.12. The shafts may be geared to a single drive motor, or they may be the shafts of two synchronized shaker motors.

In the resonant design approach coupling springs are employed, and isolation springs may or may not be used. Resonance or near resonance is achieved by designing the spring system in such a way that the natural frequency of the vibrating mass with its spring system is equal to the frequency of the driving force. The principal advantage is that the driving force is amplified and thus produces the

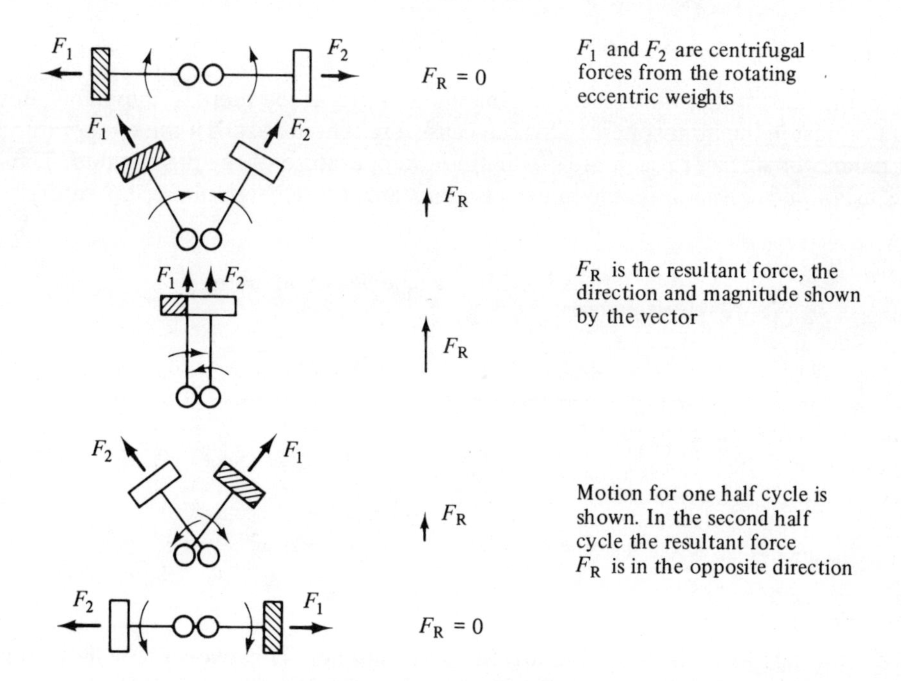

Fig. 8.12 Diagram to show how a linear force is obtained from two contra-rotating eccentrically weighted shafts

desired amplification of motion at a lower applied force. Therefore, lighter-duty motors can be used than would be needed for an equivalent non-resonant system.

Single mass resonant systems have the vibrator motor mounted directly on to the vibrating equipment, as in the non-resonant system, but the unit is fixed to a stiffer reaction system instead of on the soft isolation springs. Two mass-resonant systems employ two vibrating masses. The vibrating equipment and the motor are coupled with reactor springs and the entire system is supported on isolation springs, normally attached to the vibrating equipment.

The most important motor consideration is starting and accelerating torque. The vibrator motor must overcome considerable resisting torque due to the unbalanced weight and then come up to speed in a matter of seconds so as to pass quickly through system resonances due to the isolation springs.

Analytical determination of the power required to handle materials in vibration applications is extremely difficult, if not impossible, mainly because it is difficult to evaluate the effective mass of the working material, i.e., the quantity of mass in contact with the vibrator system. Power consumption is usually determined by measuring it in actual service.

8.9 Textile Motors

In the textile industry the two outstanding characteristics which must be considered are those of the environment and the necessary accelerating torque. The majority of applications use three-phase cage induction motors.

8.9.1 *The environment*

In textile mills where fibres are being handled or processed there is, inevitably, lint, fluff or dust in the atmosphere. This can cause a potential hazard if the temperatures of the motor surfaces reach the self-ignition temperature of the dust or fluff. Table 8.8 lists some common textile materials and states the self-ignition temperature.

Table 8.8 Self-ignition temperatures of pure, clean, material

Material	Temperature (°C)
Cotton, absorbent, rolls	266
Woollen blanket, roll	205
Viscose rayon, roll	280
Nylon roll	475
Silk roll	570
Pure scoured wool	525

The materials listed in the table are for pure samples. However, the self-ignition temperature is generally dependent upon the length of exposure and the amount of impurities present. Both these effects generally tend to lower the self-ignition temperature.

The motor total temperature is governed by the total temperature which the motor insulation system can withstand continuously without any deterioration in motor life. The insulation total temperature is set by the relevant standards at 105°C for Class A, 130°C for Class B, 150°C for Class F and 180°C for Class H. These temperatures actually represent the maximum continuous hot-spot temperatures in the machine and are well below the self-ignition temperatures listed in the table. However, these temperatures are internal and due to the cooling effect of the motor surfaces will be at a much lower temperature. The highest surface temperatures will be with totally enclosed, non-ventilated motors, but even these will be safe.

The safety margin does not mean that textile motors do not need further attention. They should be checked at regular intervals to ensure that there are no problems causing the motor to overheat due to a motor failure or due to a blanket of fluff or lint on the motor surface. Lint or dust build-up on the motor should be prevented by brushing and blowing, as it will form a good insulating jacket causing the motor surface temperature to increase. This may not necessarily cause self-ignition but it will reduce the motor life. Lubrication oil or grease on the motor will greatly help an insulating blanket to form and should be removed from the motor surfaces.

8.9.2 *Enclosures*

The presence of dust, fluff, and fibres requires careful attention to the cooling arrangements. Motor enclosures for textile applications are often described as 'lint-free'. The totally enclosed, fan-cooled and totally enclosed, non-ventilated motor is generally used where lint and combustible fibres may be present in the atmosphere. For motors below about 1·5 kW textile motor manufacturers try to provide smooth surfaces and contours where possible to minimize the accumulation of lint and to omit grids and screens upon which fluff or dust may accumulate.

Motors with outputs above 1·5 kW tend to be fitted with external fans and cooling fins since their surface is insufficient to allow the frame sizes to be rated economically. The protecting grill at the fan end of the motor is usually removed and replaced by a sheet-iron end-shield with a protecting plate, as shown in Fig. 8.13. This arrangement avoids restriction of ventilation due to blockage of an air inlet grid and the accumulation of fibres between the fan and frame and between the frame ribs.

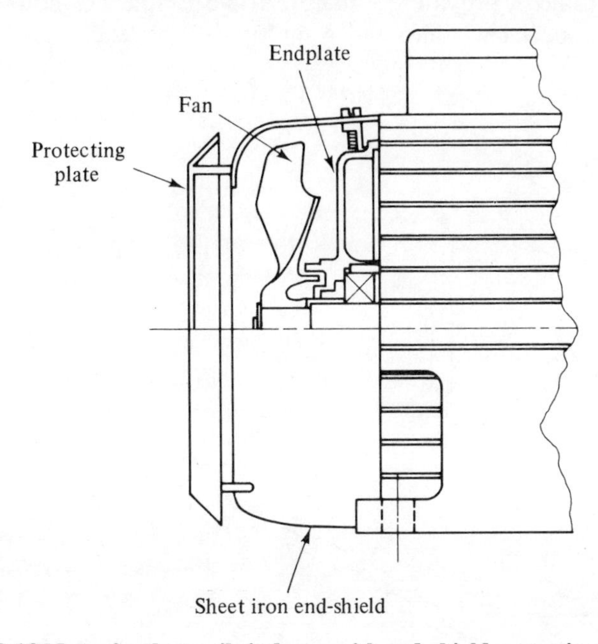

Fig. 8.13 Motor for the textile industry with end-shield protecting plate

Motors which are driving auxiliary equipment, such as fans, pumps, and compressors, located outside the production area, may be of conventional design.

8.9.3 *Torque requirements*

The electrical design must take into account the mode of operation of the loom, the action of the shuttle, and movement of the frame. Since in large factories many thousands of motors may be used, reliability and efficiency are important, and motors are specially designed to meet these requirements. Textile machinery usually uses individual drives integral with the machine.

Torque requirements, in particular accelerating torque, are a matter of prime importance in textile motors. Although many applications use a standard performance motor there are applications where special requirements are necessary.

On a card drive the inertia of the card is high compared with that of the motor. In this case the motor is allowed to accelerate nearly to full speed before the card load is imposed on the motor. This can be done either by belt drives which permit belt slippage on starting or by a clutch which can be mechanically or electromagnetically controlled or centrifugally governed. This method of allowing the motor to accelerate reduces the need for a highly specialized motor, although the motor should have high torques and thermal capacity because of the driven inertia. Some individual drives employ gear motors, eliminating the need for belt drives. However, in these cases a motor with a very high starting torque, about 400 per cent full-load torque, and with a high thermal capacity is required.

Roving-frame motors are often referred to as soft-start motors. The problem with these motors arises because of the lack of strength of the yarn in the process. This problem is overcome by providing a dual- or triple-torque a.c. motor. The motor is provided with a special winding and a multiplicity of leads, see Fig. 8.14. It is

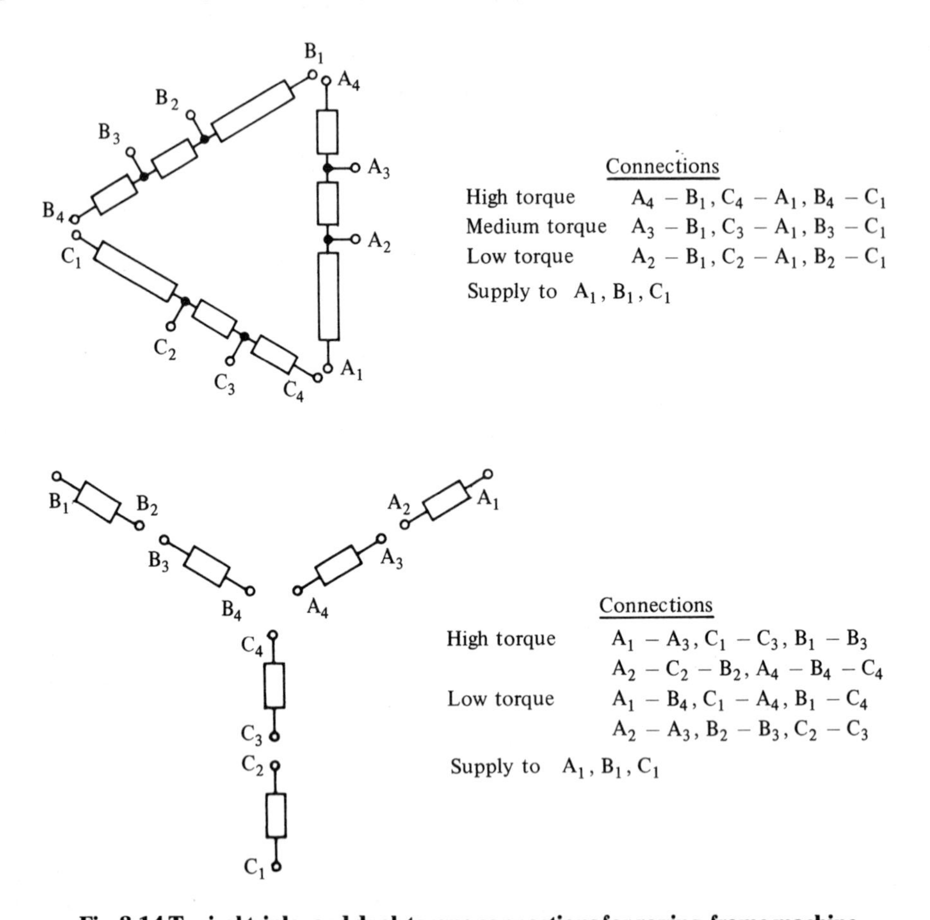

Fig. 8.14 Typical triple- and dual-torque connections for roving-frame machine

reconnectable in the motor terminal box to permit the selection of that torque characteristic which will no more than start and accelerate the frame without damage to the material in process. The dual-torque motor is connected in a two-circuit parallel star for high torque and a series delta for low torque. The high torque is approximately 35 per cent greater than the low torque. The triple-torque motor has a delta winding that is provided with taps and usually results in torques of 100, 84, and 69 per cent. Figure 8.15 shows the soft-start principle.

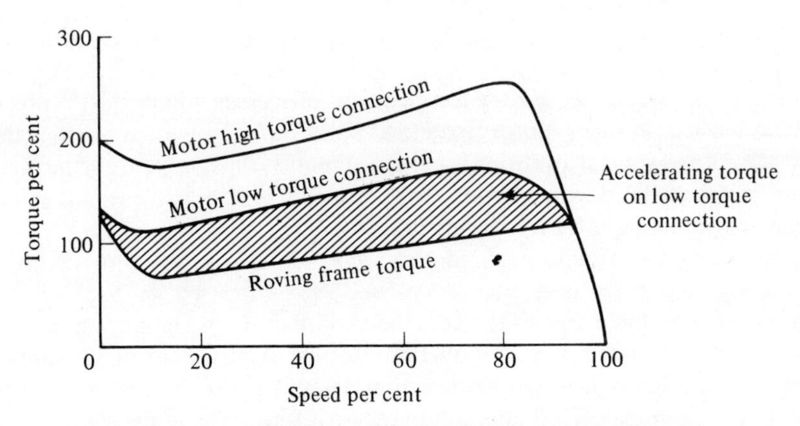

Fig. 8.15 Low and high torque relationships for roving-frame drive

The fluctuations of the load can be such that intermittently the load torque becomes negative and the motor must work as a generator. The torque fluctuations are partially smoothed out by adding a flywheel on the motor shaft, so that motors with normal values of slip (4 to 5 per cent) are usually found to be satisfactory. This type of drive is mechanically arduous and leads to wear of mechanical parts of the system.

Where variable-speed drives are required the trend is to use variable-frequency a.c. systems. The normal frequency range is about 25 to 150 Hz, although with invertor-driven induction motors the range can be extended. Advantages of a variable-frequency system are that a stepless speed range, inching, threading, and controlled acceleration are easily obtained and the simple reliable cage or reluctance motor is used. Also for multimotor drives there is the assurance of synchronized operation when synchronous reluctance motors are used.

The electrical connections should be in a flexible conduit and control gear should not, because of vibration, be mounted on the machinery.

Close coordination with the machinery manufacturer is necessary for a successful application, particularly when existing equipment is to be modified for speed-up, increased load, or other factors not originally included in the specifications.

8.10 Motors for the Chemical Industry

Simple cage induction motors are relatively resistant to the corrosive atmosphere which may occur in chemical works. Acid, alkalis, dust, grit, oil, and moisture may be present, and moisture combined with corrosive chemicals in the air are a particular threat to the life of the motor. Motors with carbon brushes and sliding contacts, whether collector rings or commutators, rely on the maintenance of a suitable oxide film on the sliding surfaces for satisfactory current collection. These motors give rise to special problems when employed in the chemical industry.

8.10.1 *Application factors*

Totally enclosed, fan-cooled motors are the most frequently employed type of motor. The totally enclosed, non-ventilated motor may be necessary in applications where it is undesirable to have fan cooling which may, under certain conditions, upset chemical processes.

Dust-tight motors are necessary for chemical processes where dust represents a dangerous hazard. In these motors extended paths on bearing caps are machined to fine tolerances and special attention is paid to all joints, thus preventing the ingress of airborne dust particles down to about one micron in size. Slingers on shaft extensions repel dust at points of entry to bearing housings.

Explosion-proof and flame-proof motors are available for conditions involving the use of inflammable materials.

Many companies have tended towards the manufacture of aluminium, rather than cast iron, motor frames for cage induction motors. Apart from the advantages of lightness, good appearance, strength, and economy, aluminium is not affected by most common chemicals. Cast aluminium cage rotors also have the same corrosion-resistant properties.

Where the atmospheric conditions are severe, standard motors are often employed with epoxy resin encapsulation of the end-windings and slots. The majority of acids and alkalis are prevented from entering the windings hence preventing a premature insulation breakdown.

Motors for the chemical industry are often supplied without final painting so that a special paint suitable for the surroundings can be applied direct on the undercoat. Painting is often not carried out until after final installation of the motor in order to avoid any danger of damage to the protective coating during transportation or when the motor is being fixed to its bed plate.

8.10.2 *Special problems with brush machines*

It is important that a thin oxide layer should adhere firmly to the collector surface to achieve low-level sparking and low brush and collector wear. As the skin is formed by a chemical process, the presence of other chemicals can upset its formation or retention. Typical chemicals found in the chemical industry which upset the correct film formation are hydrogen sulphide, sulphur dioxide, chlorine, and carbon tetrachloride. These chemicals attack the metal under the normal skin on the current collector or commutator surface forming copper salts which have little resistance to friction and the electrical properties of which are very different from the normal skin. The consequence of this is a relatively rapid wear of the commutator, generally together with excessive brush wear. The rapid wear is usually accompanied by brush sparking, often coloured green by the copper salts. If this situation is not corrected the conditions will cumulatively worsen, ending with the final destruction of the brushes.

There are generally three possible remedies for motors operating in these conditions:

(a) Total enclosure of the machine or the compartment containing the brushes is the normal method of protecting the collector surfaces from relatively low concentrations of gases. Where contamination is heavy the machine may need to be pressurized from a clean source of air.
(b) Employ a specially treated brush which will deposit a protective film on the commutator or collector ring.

(c) Employ brushes having a high cleaning action.

As a precaution, brush flexibles are usually protected from contaminated atmospheres by tinning them or by the use of a protective sleeving made from plastics (PVC) or silicone varnished glass-fibre braiding. In commutator machines the risers are often tungsten-inert-gas (TIG) welded, and all soldered joints are encapsulated. Collector rings are often manufactured from a cupro-nickel alloy because of its resistance to wear and chemical reaction.

8.11 Motors for the Steel Industry

The steel manufacturing and process industry is one of the largest single users of electrical drives. Apart from drives for unloading, pouring, shearing, and ladle manipulation the main applications for motors are mill motors and roller table motors.

8.11.1 *Mill motors*

The essential operational characteristics of primary rolling mills demand a high overload, slow-speed operation at the start of the rolling programme, but after the steel has been passed backwards and forwards through the mill rollers several times the load requirements are reduced, accompanied by an increase in speed. The only machine to readily satisfy these wide operational requirements is the d.c. motor, which, if suitably designed, can successfully for short periods develop heavy overload torques of about $2\frac{1}{2}$ times the continuous rating at the lower operating speeds, and also be capable of fulfilling the higher speed requirements. A large mill motor of this type would typically be rated continuously at several thousand kilowatts at 11 kV with a speed range of 75 to 200 rev/min. Single-pass mill motors are generally smaller, sometimes down to a few kilowatts, and their overload requirements are generally less stringent than a reversible multi-pass mill motor. There are, however, several types of mill for both hot rolling and cold rolling applications and many of these use a d.c. drive motor.

Mill motors are usually manufactured to the Association of Iron and Steel Engineers (AISE) standards. These standards cover in detail the outputs, dimensions, types of enclosure, ratings, armature inertias, class of insulation, temperature rises, etc.

Mill motors are of rugged construction to withstand the severe duties encountered in steel works, and to minimize maintenance requirements. A heavy duty shaft of high strength steel is used and to permit driving from either end of the motor, two identical shaft extensions are provided, each with a standard taper and keyway to allow couplings or pinions to be fitted easily. Owing to the arduous nature of the drive, sometimes resulting in extensive damage to the shaft and its keyway, the shaft has to be easily replaceable, and generally the armature shaft can be pressed out of the armature core without dismantling the core or commutator.

The commutator is usually of convential vee ring construction, but some manufacturers use the glass-banded type. Tungsten-inert-gas (TIG) welding is used to connect the armature conductors to the risers. This increases the ability of the motor to withstand overloads, and also means that, if necessary, an armature rewind can be carried out quickly. Machining $1\frac{1}{2}$ mm off the front face of the risers enables the conductors to be removed, as the TIG welding only penetrates about 1 mm.

Both the main and commutating poles are of laminated construction to improve both transient performance and commutation of the machine particularly with thyristor supplies. The main field coils have to be capable of continuous operation at standstill with the machine totally enclosed.

The modern method of providing a controlled supply for d.c. mill motors is the utilization of thyristor convertors connected directly to the incoming a.c. supplies. A regenerative convertor is necessary for a reversing mill, but for a single-pass mill this is not necessary. The thyristor control methods have generally superseded the Ward–Leonard system, which has inherent regenerative characteristics.

8.11.2 *Roller table drives*

For high-speed roller tables and for the entry tables at both sides of a primary mill, where high response is an essential requirement, d.c. motor drives of a few kilowatts each are used. These motors now generally incorporate permanent magnet field systems and are totally enclosed. These allow savings in both initial cabling costs and overall capitalized costs, an important consideration when two or three hundred motors are arranged in roller table sections.

All other roller table drives are powered by a.c. induction motors, because of the reduced maintenance requirements. Speed variation is of the order of 5 : 1 and is obtainable by frequency control, over a range of 2 to 10 Hz, from either a rotating frequency convertor or a static frequency invertor or cyclo-convertor.

8.12 Motors for Agricultural Machinery

Many farms receive only single-phase supplies, and the capacitor-start induction motor is the most common for driving agricultural machinery. The motors used must be robust, reliable, and require little maintenance, as once installed they are often neglected until a fault condition arises.

For sheltered and dust-free situations the standard drip-proof motor is suitable. Where dust or grit is present, or if the motor is used out of doors, totally enclosed, fan-cooled motors are generally applied. In certain extreme conditions weather-proof or dust-tight motors may be necessary. In the dust-tight motor extended paths on the bearing caps are machined to fine tolerances and special attention is paid to all joints, preventing the ingress of airborne dust particles. On some motors protection can be fitted to prevent the motor from being damaged by vermin.

8.13 Motors for Woodworking Machinery

The woodworking industry generally uses a.c. cage motors as individual drives to every machine. The motors are usually 'built-in' to the equipment but where normal types are used these are sometimes drip-proof, fixed on the machine, and driving through a short vee belt. Totally enclosed fan-cooled motors are a better alternative because of the sawdust in the atmosphere, and are essential where there is a likelihood of the motors being buried in sawdust. The motors should be regularly cleaned to prevent the cooling air passages being clogged with sawdust. Where motors with higher speeds than 3000 rev/min are required it is usual to use frequency changers or invertors to supply cage motors running up to 24000 rev/min.

8.14 Rotor-stator Units

Rotor-stator units are the essential parts of an a.c. motor designed for building into machines. The rotor is fixed to the shaft of the machine and the stator into a prepared cavity, the motor becoming an integral part of the machine. These are largely used by the metal and woodworking industries. Special units for refrigerator compressors, known as hermetic rotor-stator units, are produced.

8.14.1 *Hermetic rotor-stator units*

A hermetic motor consists of a stator and rotor without a shaft, end shields or bearings, for installation in refrigeration compressors of the hermetically sealed type. The hermetically sealed motors in refrigeration compressors have over the past 10 to 20 years come to encompass the majority of all refrigeration and air-conditioning compressor units below about 75 kW. The compressor manufacturer presses or bolts the stator into his housing and shrinks the rotor onto the compressor crankshaft. The refrigerant in the air conditioning system passes through or around the motor, and because this is an excellent cooling medium the motors are not rated in terms of the full-load power, but rather in terms of breakdown torque, locked rotor torque, and current.

The motor design is matched to that of the compressor, and because of the varying requirements of different compressors there is very little standardization apart from the motor diameter. A wide variety of electrical designs is available. Single-phase motors are usually split-phase, capacitor-start, two-value capacitor, or permanent capacitor motors. Three-phase induction motor designs can be found up to about 100 kW. There is generally a trend towards two-pole motors although four-pole motors are common.

References

1. B. M. Bird and R. M. Harlen, 'Variable-characteristic d.c. machines', *Proc. IEE*, **113**, 1813–1819 (1966).
2. B. S. Hender, 'Recent developments in battery electric vehicles', *Proc. IEE*, **112**, 2297–2308 (1965).
3. A. J. Graumlich and C. V. Kern, 'ABC's of small electric vehicles', *Automotive Eng.* (USA) 58–69 (Sept. 1974).
4. 'Why the trend to high-voltage Trucks', *Mod. Mater. Handl.* (USA), 42–45 (Sept. 1973).
5. M. A. Thompson and L. A. Watlers, 'The design of d.c. commutator motors for high performance electric vehicles' SAE Paper No. 740169 (March 1974).
6. D. E. Knights, *Drives for electric vehicles* No. 74–1163, ERA Ltd., Leatherhead, Surrey (1974).
7. *Machine Design*, Electric Motors Reference Issue, 13 April 1972.
8. K. K. Schwarz, 'Submerged gas circulator motors for Hinkley "B" and Hunterston "B" AGR power stations', *LS Engineering Bulletin*, **13**, 6–12 (1974).
9. D. E. Knights, *Variable speed drives for industrial applications*, No. 74-1189, ERA Ltd, Leatherhead, Surrey (1974).
10. D. G. Searle, 'Pole amplitude modulated motors for fan and pump drives', *Electrical Review*, 407–409 (April 1974).

11. J. R. Hazel, 'Power-station Auxiliaries', *Electronics and power*, 430–436 (Dec. 1965).
12. Z. A. Tendorf, 'Motors for vibratory applications', *Westinghouse Engineer*, 74–80 (May 1971).
13. J. Midwood, 'Developing mill motors with increased output', *Electrical Times*, 9–10 (Feb. 1973).
14. F. Kappins and M. Liska, 'Electronic motors for industrial applications', *Siemens Review*, **XXXVIII**, 453–456 (1971).

9 Electric Motors for Flammable Atmospheres

9.1 Introduction

Flameproof motors at one time formed the only type approved for use in flammable atmospheres; there are now, however, other types of motor available. The purpose of this chapter is to describe the various types of construction, and the factors affecting their choice to suit particular conditions.

Selection and specification of electric motors is largely determined by the environment in which they are to be located. For the purpose of defining the hazard, two factors are important, namely the classification of the area, and certain characteristics of the flammable gas or gases which may be present.

9.2 Classification of Hazardous Areas

The classification of hazardous areas according to the likelihood of flammable gases being present is given in IEC 79.10 and described more explicitly in the British Code of Practice CP1003 Part 1 as follows:

(CP1003 will eventually be superseded by BS 5345, of which part 1 is published already.)

Zone 0 or Div. 0

An area or enclosed space within which any flammable or explosive substance, whether gas, vapour or volatile liquid is continuously present in concentration within the lower and upper limits of flammability.

Zone 1 or Div. 1

An area within which any flammable or explosive substance, whether gas, vapour or volatile liquid is processed, handled or stored, and where during normal operations an explosive or ignitable concentration is likely to occur in sufficient quantity to produce a hazard.

Zone 2 or Div. 2

An area within which any flammable or explosive substance, whether gas, vapour or volatile liquid, although processed or stored, is so well under conditions of control that the production (or release) of an explosive of ignitable concentration in sufficient quantity to constitute a hazard is only likely to occur under abnormal conditions.

The responsibility for classification lies with the owners of the plant.

An area falling within the category of Zone 2 is sometimes known as a 'Remotely Dangerous Area'. Although any of the approved types of motor can be used in Zone 2 areas, advantage can be taken of the less stringent conditions by using motors developed specially for use in Zone 2, known as Type 'N' motors, and which are of lower cost than the other types.

In Zone 1 areas, Type 'N' motors are not applicable, and one of the other approved types must be used, the factors governing the choice being discussed later.

No motors are approved for Zone 0, and the use of electric motors within a continuously ignitable environment is not recommended.

9.3 Grouping of Gases

The types of flammable gas present in the environment also affect the specification of motors, and two characteristics are relevant.

Table 9.1 Enclosure suitable for a particular flammable gas or vapour

Group of enclosure	Gas or vapour
I	Methane (firedamp)
IIA	Ammonia
	Industrial methane
	Blast furnace gas
	Carbon monoxide
	Propane
	Butane
	Pentane
	Hexane
	Heptane
	iso-Octane
	Decane
	Benzene
	Xylene
	Cyclohexane
	Acetone
	Ethyl methyl ketone
	Methyl acetate
	Ethyl acetate
	n-Propyl acetate
	n-Butyl acetate
	Amyl acetate
	Chloroethylene
	Methanol
	Ethanol
	iso-Butanol
	n-Butanol
	Amyl alcohol
	Ethyl nitrite
IIB	Buta-1,3-diene
	Ethylene
	Diethyl ether
	Ethylene oxide
	Town gas
IIC	Hydrogen

Firstly, gases are grouped according to the dimensions of maximum safe gaps which have been found by experiment to be capable of preventing transmission of a flame through enclosures. This is of importance only in the case of flameproof motors, and the grouping is shown in table 9.1.

Secondly, the ignition temperature of the flammable mixture should be higher than the highest surface temperature likely to be attained in service. Table 9.2 shows examples of gas ignition temperatures, and their grouping into the six internationally recognized temperature classes. (BS 5345 Part 1 covers these groupings in detail.)

Table 9.2 Temperature classes: gas ignition

T1 450°C		T2 300°C		T3 200°C	
Acetic Acid	485	Acetylene	305	Crude oil	220
Acetone	535	Amyl acetate	325	Cyclohexane	259
Ammonia	630	*iso*-Amylacetate	390	Heptane	215
Benzene	560	Amyl alcohol	343	*n*-Hexane	233
Carbon monoxide	605	*n*-Butane	365	Hexane	
Ethane	515	*n*-Butyl acetate	370	(Commercial)	247
Ethyl acetate	460	*n*-Butyl alcohol		Petrol	220
Ethyl chloride	510	(*n*-Butanol)	340	Phenyl glycol	225
Hydrogen	560	*iso*-Butyl alcohol			
Methane	595	(*iso*-Butanol)	426		
Methyl acetate	475	Ethyl alcohol			
Methyl alcohol		(Ethanol)	425		
(Methanol)	455	*n*-Propyl alcohol			
Methyl ethyl ketone	460	(N-Propanol)	405		
Naphthalene	528	Ethylene	425		
Propane	470				
n-Propyl acetate	455				
Toluene	535				
Town gas	647				
Water gas	570				
Xylene	464				
T4 135°C		**T5 100°C**		**T6 85°C**	
Acetaldehyde	140	Carbon		Ethyl nitrite	90
Ethyl ether	170	disulphide	100		

9.4 Types of Motor Available

Four basic types of motor are available for hazardous atmospheres, namely Flameproof, Pressurized, Type 'e', and Type 'N'. It is expected that these will eventually be internationally designated (Ex)d, (Ex)p, (Ex)e, and (Ex)N respectively.

The principles involved in these different types of construction are based upon three quite distinct philosophies.

In the case of flameproof motors, the object is to contain any explosion which may occur inside the motor by designing the motor casing strong enough to withstand the resultant pressures, and by ensuring that flames cannot pass through from inside the casing so as to cause ignition of the surrounding atmospheres. Thus, the electromagnetic parts of the motor may be of orthodox design to normal standards, without any special precautions to prevent internal arcing or sparking, nor are any measures taken to prevent the ingress of flammable gases to the interior of the motor. The essential feature is the flameproof enclosure.

In other types of motor, no flameproof enclosure is used, the safety of operation being achieved by preventing or minimizing the chance of an explosion occurring inside the motor.

Thus in the case of pressurized motors, flammable gases are prevented from being present inside the motor during operation, so that any internal arcs or sparks can occur only in the presence of non-flammable gases.

In the case of Types 'e' and 'N' motors, the object is again to prevent an internal explosion occurring, but by designing the motor to minimize the chances of arcs, sparks or dangerously high temperatures occurring during service. The special features are therefore included in the electro-mechanical parts of the motor, and in the electrical and mechanical clearances. The enclosure is usually quite orthodox, indeed the motor may be of the ventilated type.

Later in this chapter the above four basic types are described in greater detail.

9.5 Certification

In most countries, users of motors in areas where there are recognized explosion hazards generally require that the motors carry some form of certification as assurance of their suitability. The administration of approval in the UK is a function of the Department of Trade and Industry, the responsibility for mining and other industries being divided, however, within the Department as follows:

The Department of Energy's Safety and Health Division, London, issue certificates for flameproof motors for use in mining.

The Department of Prices and Consumer Protection's British Approvals Service for Electrical Equipment in Flammable Atmospheres (BASEEFA) at Buxton, is responsible for certification of all types of explosion-protected motors for use in industries other than mining.

Both certifying authorities employ the test facilities of the Department of Energy's Safety in Mines Research Establishment, (SMRE) at Buxton. The certificates are recognized not only in the UK, but in many overseas countries, especially those which have no authorized approvals service of their own. They are not, however, universally accepted, and exporters of equipment incorporating flameproof and other types of motor for use in hazardous atmospheres should check on the recognition extended to the BASEEFA and S & HD certificates in the country of destination.

Two of the foreign approval services which deserve particular mention, because their influence extends beyond the frontiers of their own countries are the German Physikalisch—Technische Bundesanstalt, (PTB) and the US Underwriters Laboratories (UL).

9.6 Flameproof Motors

Flameproof motors have been used for many years and because of the amount of development and successful experience which has been accumulated, are likely to retain a degree of popularity in spite of their relatively high cost. In principle, the

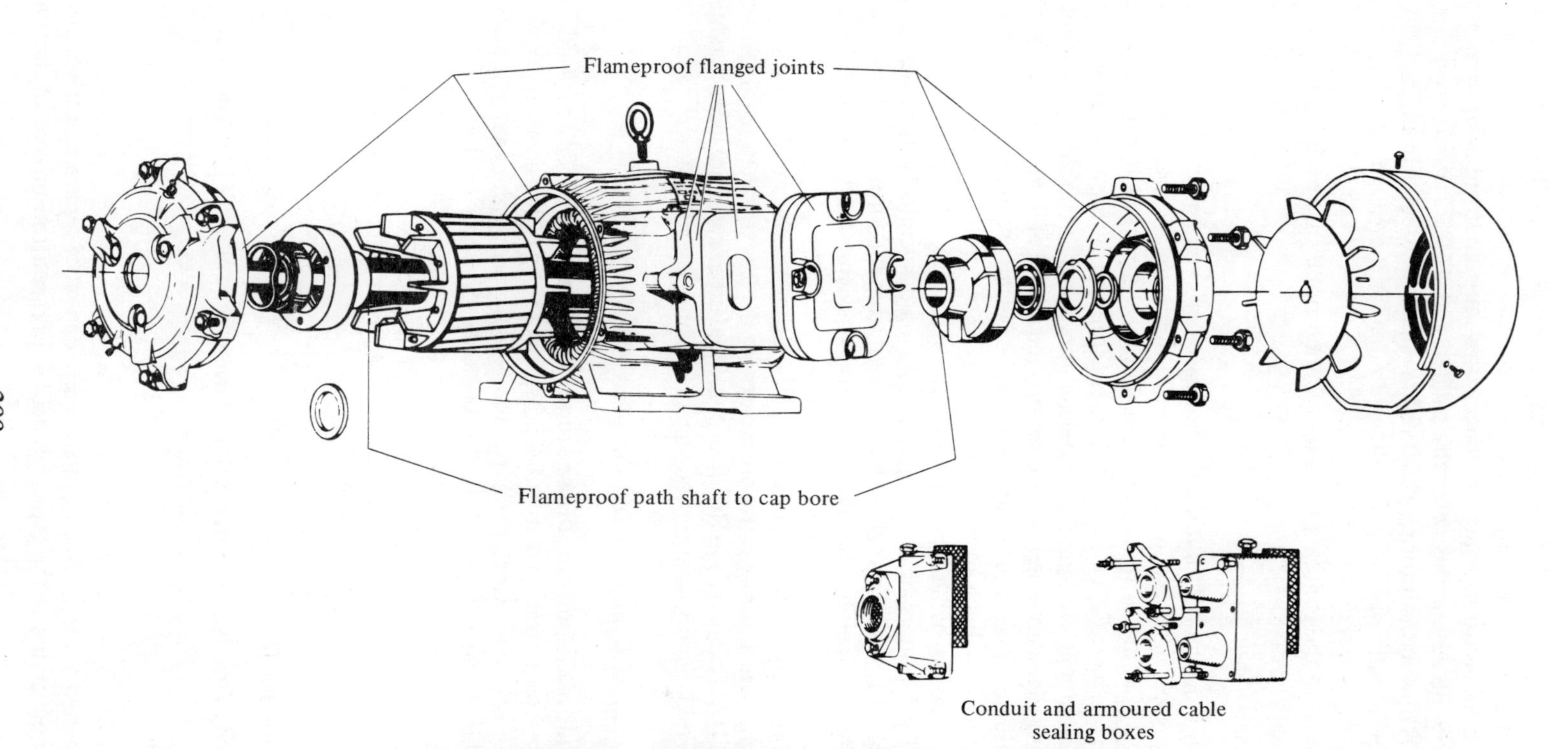

Fig. 9.1 Main features of a typical flameproof motor. (Brook Motors Ltd)

motors are of normal electrical design and performance; it is the enclosures which distinguish them from general-purpose motors.

The British Standard to which the enclosures must conform is BS 4683 part 2, which replaces BS 229, and is generally aligned with IEC 79-1. However, a European Norm is now in draft form, which differs in certain details from BS 4683, and which will eventually precisely align the national standards of all EEC member countries.

Some aspects of the general problem of flameproofing are outlined below.

9.6.1 *Flameproof gap*

If an explosion occurs within a motor, hot gases or even flames may be projected through any gaps or holes in the enclosure, and could cause explosion of any flammable gases in the surrounding atmosphere. By experiment, it has been found that gaps can be dimensioned so that the gases projected during an internal explosion are not sufficiently hot to ignite the external atmosphere. The safe gap dimensions vary according to the pressure, concentration, and nature of flammable gas, and have been established by repeated tests on a number of gases. For practical purposes, flammable gases have been arranged in groups. The old grouping in Britain is given in BS. 229, the groups being designated I, II, III, and IV. This, however, is being superseded by the internationally agreed grouping designated I, IIA, IIB, and IIC shown in table 9.1. Although somewhat different from the former, the new grouping follows a similar principle, the higher the group designation, the smaller being the safe gap.

It is customary for motor manufacturers to design motors suitable for gas group IIB, so that they are automatically satisfactory for the less stringent requirements for groups I and IIA. For gas group IIC (hydrogen) and other gases not listed in table 9.1, there are formidable obstacles in the way of obtaining certification.

9.6.2 *Explosion pressures*

The magnitude of the pressure attained during any explosion inside a motor depends on the nature, concentration, and initial pressure of the gas, the geometry of the free space, and the point at which the explosion is initiated. A phenomenon known as 'pressure piling' can be caused by the presence of small openings or passages communicating between separate free spaces in a motor. Thus an explosion inside one space causes a high pressure wave through the communicating passage, causing pre-compression of the gas inside the other space prior to its ignition. Designs should whenever possible minimize this possibility.

9.6.3 *Pressure rise time*

This is related to the dynamics of pressure during the explosion, and is defined by Fig. 9.2.

9.6.4 *Certification tests*

These are carried out by SMRE at Buxton, and consist of separate explosion tests to verify firstly the ability of the gap clearances to prevent transmission of the internal

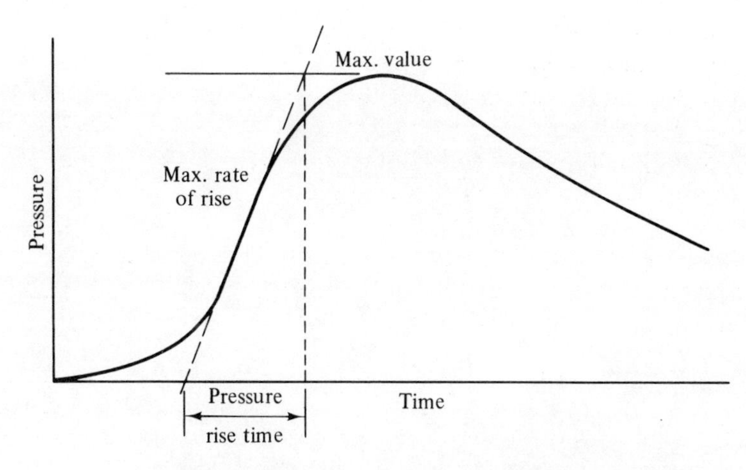

Fig. 9.2 Definition of pressure rise time

explosion, and secondly the strength of the structure. For each gas group, representative gases or mixtures are used, which by experience have been found to produce the most onerous conditions. The gases used for the two types of test are different, as the mixture which produces the most onerous conditions for transmission of the flame is not the same as that which produces the highest pressure.

The internal mixture is ignited by a sparking plug, and in the strength-of-structure test the internal pressure and pressure rise times are measured. In order to establish the factor of safety, similar motor casings with sealed joints are subjected to static pressure tests, and providing the pressure can be raised to four times the value measured during the explosion test without causing damage, routine testing of enclosures of other than welded construction may be waived. If, however, the structure is incapable of withstanding this pressure, routine tests on every motor casing must be conducted, but at a pressure of 3 or $1\frac{1}{2}$ times the explosion pressure, depending on whether the pressure rise time was greater or less than 5 ms. All enclosures of welded construction are subjected to routine static pressure tests.

Both the strength-of-structure tests and the transmission tests are repeated a number of times, and may be conducted with the motor at rest or running, at the discretion of SMRE.

9.6.5 *Performance*

The performance of flameproof motors will in due course be specified in BS 5000: Part 17, at present in draft form. The general electrical performance does not differ from that of standard motors for general application, but there is a particular requirement relating to external surface temperature of motors, which should not exceed 135°C when operating in the maximum ambient temperature (normally 40°C) when operating at normal load, or with recognized overload. This requirement prevents auto-ignition of any flammable gas in temperature classes T1 to T4 inclusive, but should any motor be required for operation in the presence of gases or ignitable dusts of lower auto-ignition temperatures, then temperature class T5 or T6 should be specified as appropriate. This may, however, require motors of increased physical size for equal output ratings.

Fig. 9.3 2·2 kW flameproof motor with sealing box and armoured cable gland: terminal box cover removed. (Brook Motors Ltd)

9.6.6 *Dimensions and output ratings*

Flameproof induction motors are available, conforming to the same internationally agreed mounting dimensions as are used for electrical machines generally. In Western Europe, the outputs allocated to particular frame sizes are standardized, and are the same for flameproof and non-flameproof motors of standard performance. These are at present specified in BS 3979, but will eventually be transferred to BS 5000: Part 17. The range covered extends from 0·37 to 150 kW output, at speeds from 750 to 3000 rev/min. Larger flameproof motors are not readily available, owing to the high explosion pressures, and associated problems and costs.

9.6.7 *Constructional details*

Joints between endshields and yokes, terminal facings, boxes and covers, etc, all present possible flame paths. They must not be less than 6 mm wide in small motors, but are more typically 12·5 mm or 25 mm wide in larger motors. The wider the joint, the greater is the permissible gap for each gas group, and similar rules apply to other flame paths such as the radial clearances between shafts and glands. Maximum gaps are tabulated in BS 4683: Part 2.

In a motor with sleeve bearings, the radial clearance between the shaft and the gland should preferably be greater than the motor air-gap, so that excessive bearing wear would result in rubbing of the rotor and stator within the safety of the flameproof enclosure, rather than the generation of sparks by friction between the shaft and the external gland. If the gland clearance cannot be made greater than the motor air-gap, then it must be designed with easy access to facilitate systematic measurements of the clearance.

Holes for screws, etc, should not penetrate right through the wall of the enclosure. The heads of screws for securing removable covers must be shrouded or otherwise prevented from being unfastened except by special keys. In some countries this requirement extends to screws for holding together component parts of the enclosure.

Cable entries to the flameproof terminal box (if provided) or direct to the main enclosure must be made through either suitable approved glands, or through sealing boxes, or by an approved and certified plug and socket. (See Fig. 9.4.)

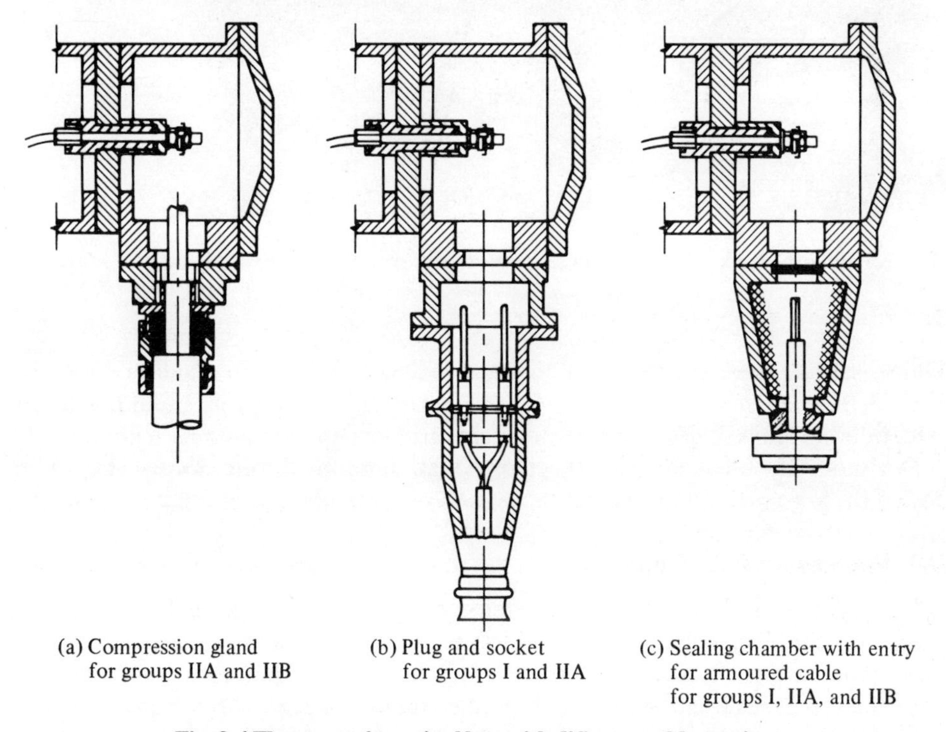

(a) Compression gland for groups IIA and IIB

(b) Plug and socket for groups I and IIA

(c) Sealing chamber with entry for armoured cable for groups I, IIA, and IIB

Fig. 9.4 Flameproof terminal box with different cable entries

9.6.8 *Condensation problems*

For obvious reasons, drain holes as normally provided in non-flameproof totally enclosed motors are not permitted, and although flameproof draining devices are available, opinions differ regarding their effectiveness. Most manufacturers prefer not to fit such devices. In consequence, flameproof motors are particularly susceptible to the problems associated with the progressive accumulation of internal

Fig. 9.5 200 h.p. 3·3 kV 100 rev/min flameproof slip ring motor. (Brush Electrical Machines Ltd)

condensation. These problems do not occur when motors are in regular operation, but it is recommended that motors for infrequent or standby duty should be fitted with internal anti-condensation heaters, energized when the motors are not operating. Motors which have been subjected to high humidity during storage should be tested for low insulation resistance and dried out if necessary before installation.

9.7 Motors for use in Mines

The foregoing applies to flameproof motors generally, but because of the particularly rough handling to which flameproof motors from Group I (mining) may be subjected, the National Coal Board have their own additional requirements. Mechanical robustness is paramount, and for many applications normal grey-iron castings are not acceptable, SG iron, cast steel or welded steel constructions being preferred. The terminal boxes must be designed to withstand excessive force from the pull of trailing cables. Plug and socket connections are usually required in all motors.

Watercooled motors are preferred for many underground applications, when either the location or available space precludes the use of orthodox fan cooling.

9.8 Pressurized Motors

As the name implies, the internal spaces inside these motors are maintained at a higher pressure, so that the surrounding atmosphere cannot enter.

Fig. 9.6 120 h.p. 1500 rev/min flameproof water-cooled motor for use in mines. (Brush Electrical Machines Ltd)

Pressurized motors are usually adaptions of standard forms of totally enclosed motors, e.g., totally enclosed fan-cooled motors, or in the case of larger sizes, closed-air-circuit machines. In order to minimize leakage, particularly through the bearing housings, it is necessary to seal all apertures except those provided for the inlet pipe, and for the controlled outflow of the pressurizing medium. It goes without saying that the latter must be non-flammable, dry clean air being quite satisfactory.

To ensure that all parts of the interior are at higher pressure than the atmosphere, the value of inlet pressure would typically be 50 mm water gauge, but may have to be greater to exceed the pressure head generated by the motor's own internal circulating fan. Unnecessarily high pressures merely increase the leakage, and can cause displacement of bearing lubricants.

It is essential to ensure that a motor can be adequately purged while at rest, and this requires care in the design, and verification by type testing.

The positions of the inlet and outlet should be at opposite ends of the machine, the choice of upward or downward displacement being dependent on the density of the flammable gas. In large motors, the inlet usually connects to a tube having graded nozzles spaced along its length, in order to increase the spread of the inlet air. The possibility of stationary gas pockets is reduced by providing cut-outs in offending baffles, etc, as necessary.

Purging the space inside the terminal box could present a particular inconvenience, to avoid which it is customary to employ a different type of protection for this

Fig. 9.7 120 h.p. 1500 rev/ min flameproof air-cooled motor driving a conveyor in a coal mine. (Brush Electrical Machines Ltd)

part of the motor, say flameproof or type 'e', or to dispense with a separate terminal box altogether, and have direct entry.

Careful consideration must be given to the auxiliary apparatus needed for the supply of the pressurizing air, and safety devices must be included for tripping or raising an alarm whenever the inlet pressure falls below the required value. It is important to purge the motor before switching on, to sweep out any flammable gas which may have previously entered.

A suitable control sequence for the pressurizing equipment is as follows, and would preferably be automatic.

(1) Turn on pressurizing air.
(2) Pressure switch on motor operates.
(3) Open purge outlet on motor.
(4) Check rate of flow.
(5) Allow purge time to elapse.
(6) Close purge valve.
(7) Clear to start motor.

The purge time should allow five volume changes or more, and should be based on type tests on the motor to check the effectiveness of the design. The following is the

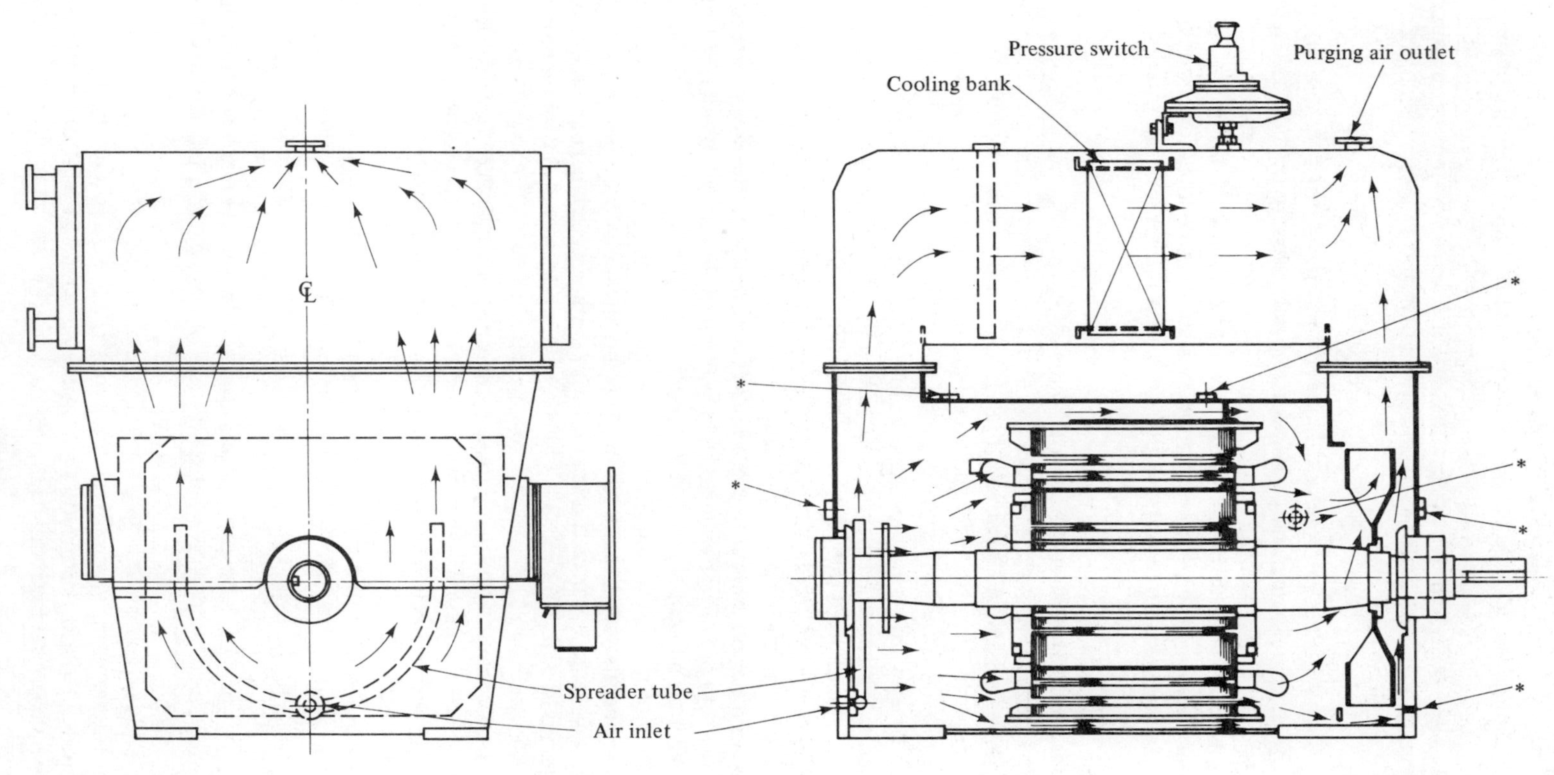

Fig. 9.8 Purging air flow diagram for 400 kW 10 pole pressurized closed air circuit water-cooled induction motor. (Brush Electrical Machines Ltd)

testing procedure carried out by one British manufacturer. In preparing the motor for test, tapping points are provided at various positions around the motor casing.

Leakage check. This consists simply of a preliminary injection of 50 mm water gauge from a smoke generator, and with the purging outlet closed, to give a visual check that there are no significant leakages.

Pressurization Test. With the purging outlet closed and the motor first stationary, then running, the inlet pressure is increased by stages and pressure readings taken at the inlet and at the various tapping points. The air flow to the motor is also measured. The object of this test is to confirm that when the inlet pressure is at the design value (normally 50 mm water gauge) the minimum pressure inside the motor at any point does not fall below 25 mm water gauge, also that the air leakage rate is not excessive.

Purging Test. Ideally this test should be carried out with the motor initially filled with either the actual flammable gas, or (for safety) with a non-flammable gas of equal density. Results have shown, however, that filling with nitrogen represents a generally more onerous test than with the actual flammable gas. This is because the system design initially includes a choice, according to the relative density of the flammable gas, of upward or downward purging displacement to exploit gravity assistance, whereas with nitrogen having approximately equal density to air, gravitation is negligible. The motor is therefore completely filled with nitrogen, and with the purging outlet opening into an outlet pipe of the specified length, and the air inlet pressure adjusted to the design value, the oxygen percentage is monitored at various tapping points for a period and plotted graphically against time. Air flow measurements enable the rate of volume changes to be measured. Figure 9.9 shows the test results on a particular machine designed for operation in the presence of hydrogen. Note that at seven minutes, the oxygen content inside the machine had reached the specific level indicating 4 per cent residue of the invading nitrogen gas. With hydrogen mixed with air, 4 per cent is the lower flammability level. The number of volume changes up to that point was 3·1 which is well inside the normal minimum of five.

The safety factor would be even greater than this in practice, because the concentration of the flammable gas inside the motor at the commencement of purging would always be much lower than the 100 per cent concentration of invading gas used in the purging test.

As in the case of flameproof motors, the maximum attainable temperature of the outer casing is of importance, particularly in the presence of gases in temperature Classes T5 and T6, see table 9.2.

Pressurized motors, though arguably the 'safest' of all available types, are the least used, probably because of the complications involved by the auxiliary pressurizing apparatus. Nevertheless, they have a useful area of application, particularly for motors of large size.

There is currently no British Standard for this type of motor, but a European Norm is in course of preparation, generally aligned with IEC 79-2. In the absence of a national standard in Britain, certification is a problem, but BASEEFA are prepared to issue letters of approval for individual motors.

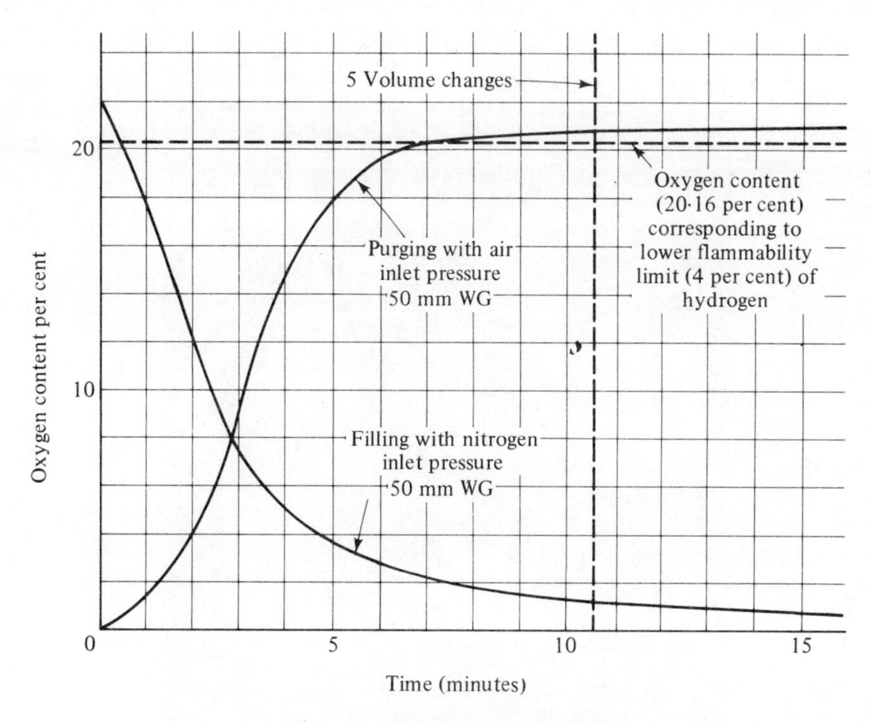

Fig. 9.9 Results of purging test on 400 kW 10 pole pressurized induction motor. (Brush Electrical Machines Ltd)

9.9 Type 'e' Motors

Often described together with type 'N' as non-sparking, type 'e' motors are based on an entirely different concept from flameproof or pressurized motors. The enclosure is of normal construction, usually totally enclosed fan-cooled, but the interior or electrical parts are so designed and constructed that arcs or sparks or dangerous temperatures cannot be produced in normal service. For practical purposes, this rules out commutators, and in fact type 'e' motors are almost invariably cage type induction motors.

The letter 'e' is an abbreviation of the German 'erhöhte Sicherheit' meaning 'increased safety', and the concept of the type 'e' originated in Germany, where its use has spread at the expense of flameproof equipment. The advantage lies in the cost, which in the case of motors is substantially less than flameproof of equal output rating, particularly in small sizes. The principle can be applied, moreover, to very large motors which are outside the upper limit of practicability for flameproof designs. They have not yet an established place in UK practice, although they are permitted by HM Factory Inspectorate for use in Zone 1 areas.

A European Norm is in draft form, generally aligned with IEC 79-7, and will eventually align all existing national standards in EEC member countries, including BS 4683: part 4.

In order to meet the requirements of the standards, type 'e' motors must include the following safety features:

Fig. 9.10 Terminal box of 2·5 kW type 'e' motor with cover removed. (Brook Motors Ltd)

9.9.1 *Windings*

All winding wire must be either double enamelled or single enamelled with an additional insulation, and approved materials employed for all components of the insulating system. The complete winding is double impregnated, employing an approved insulating varnish. Cage rotors if other than die-cast must employ tight-fitting bars, and the joints between bars and end-rings and the general design and construction must ensure reliability, and must reduce as far as practicable the risk of sparking during starting and running.

9.9.2 *Terminal arrangements*

The actual terminals must be able to accept stranded line cables, without the need to form an 'eye', and the clamping parts must not turn or slide when the terminal is being tightened. Clearance and creepage distances between terminals, and to earthed parts must be not less than specified values, and the terminals constructed so that loose strands cannot reduce the distances. If cable lugs are used on the winding leads, they must be connected by an approved method, and means must be provided to prevent them turning on the terminals so as to reduce the clearance distances. The insulating blocks or boards on which terminals are mounted must have a known value of comparative tracking index, compatible with the creepage distance between terminals.

The terminal box itself must be of dust- and splash-proof construction, and must not be open to the interior of the motor unless the latter is totally enclosed. If the supply cables are of flexible type, they must be suitably packed and clamped to withstand tensile stresses, and the entry shaped so as to prevent the cable having too sharp a bend at the entry point.

9.9.3 *Mechanical running clearances*

The air-gap dimensions, mechanical clearances between shaft and glands, and between fans and adjacent parts are all important safety factors, and the manufacturer must organize his quality control so that none of these clearances can be less than the minimum values specified.

9.9.4 *Mechanical strength*

All parts of the motor must be capable of withstanding a degree of mechanical shock, as defined by a specific impact test, without impairing serviceability, or sustaining damage to a degree which reduces clearances below the specified values, or impairs the effectiveness of the terminal enclosure.

9.9.5 *Limiting temperatures*

In order to reduce the risk of burn-out within the normal expected life of the motor, the temperature rise of the windings attained at rated output must be at least 10°C lower than the values permitted for similar insulating materials in normal motors.

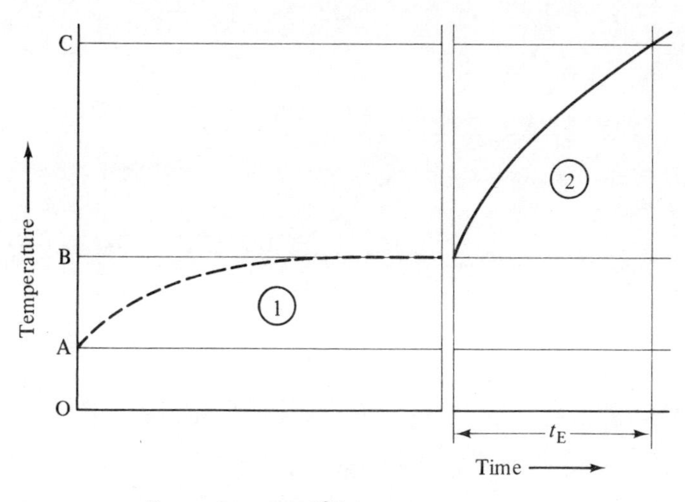

O = temperature 0°C
A = maximum ambient temperature
B = temperature in rated service
C = limiting temperature

① = temperature rise in rated service
② = temperature rise during stalled motor test

Fig. 9.11 Schematic diagram explaining the significance of time t_E

For example, Class E and Class B insulating materials are limited to 65°C and 70°C respectively.

In the previous sections of this chapter relating to flame proof and pressurized motors, mention has been made of the necessity to limit the outer casing temperatures, in order to prevent auto-ignition of flammable gases in the surrounding atmosphere. This is important to all types of motor for use in flammable atmospheres, but in the case of type 'e' motors, we are concerned not only with outer frame temperatures, but with the surface temperatures of all internal parts of the motor with which the atmosphere is in contact. The particular condition which causes concern is a failure to start, or stalled condition, when a motor has already attained maximum operating temperature, after prolonged operation at full load in maximum rated ambient temperature (normally 40°C). Under these conditions, and with a stalled time of t_E, the following temperature limitations must not be exceeded.

Firstly, the hottest internal or external surface temperature must not exceed the maximum safe temperature appropriate to the temperature class (see table 9.2).

Secondly, the rotor cage temperature must not in any case exceed 300°C.

Thirdly, the winding temperature must not exceed the limiting values shown in table 9.3.

The criterion which determines the value of the stall time t_E may be any one of the above limitations, all three of which must be met. Figure 9.11 shows t_E graphically defined in terms of the temperature of the critical part of the motor.

The minimum acceptable value of t_E is a function of the per-unit value of locked rotor current (see Fig. 9.12) but should never in any case be less than 5 seconds.

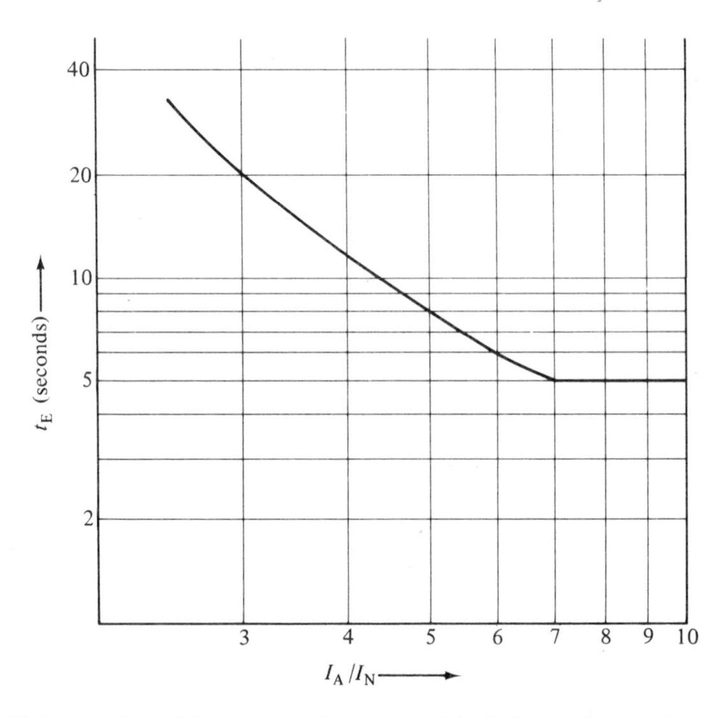

Fig. 9.12 Minimum value of the time t_E of the motor in relation to the starting current ratio I_A/I_N

Table 9.3 Winding temperatures: Short-time limiting values at the end of time t_E

Insulation class	Temperature classes					
	T1	T2	T3	T4	T5	T6
A	160	160	160	135	100	85
E	175	175	175	135	100	85
B	185	185	185	135	100	85
F	210	210	200	135	100	85
H	235	235	200	135	100	85

9.9.6 *Dimensions and output ratings*

In order to meet these temperature specifications, it is necessary to reduce the output ratings of all frame sizes below the output ratings of general purpose motors, by 10 per cent or more, depending on the temperature class of the flammable gas.

For example, a 1500 rev/min motor of 18·5 kW (25 h.p.) rated output is normally in frame D180M. However, a type 'e' motor in the same frame size is rated at only 17 kW for temperature classes T1 and T2, at 15 kW for class T3, and is de-rated even further for lower temperature classes. If a particular application requires a motor having the full 18·5 kW output rating, a larger motor is required. In contrast, a flameproof motor of 18·5 kW is available in frame size E180M, which is dimensionally similar to a general-purpose motor.

Thus we see that one disadvantage of type 'e' motors is that they tend to be larger than flameproof motors of equal output, and that the difference in frame size can be considerable if the gases happen to place the motors in one of the lower temperature classes.

Output/frame size relationships for motors with standard dimensions are given in the W. German standard DIN 42673 Blatt 2, and are expected to be included in BS 5000 part 15.

9.9.7 *Certification procedure and tests*

In certifying type 'e' motors, BASEEFA first scrutinize all drawings and design details, to ascertain that the design complies with the requirements of the specification with regard to clearances, suitability of materials, etc. They then either conduct, commission, or witness impact tests on such components as are deemed vulnerable, also electrical performance tests including conventional full-load heat run, and measurement of the t_E time.

The heat run is carried out on conventional lines, by loading the motor against a suitable dynamometer. Load measurement is by summation of losses in accordance with BS 269, or by direct mechanical load measurement, and full load is applied for several hours until the rate of rise of frame temperature falls below 1°C per hour, when it is assumed that the ultimate rise of temperature of the internal parts of the motor has been attained.

The heat run serves two purposes. Firstly, to ensure that the full-load temperature rise of the insulated windings does not exceed the permissible limit appropriate to the insulation class, and secondly to measure the surface temperature of the hottest parts of the motor, as a preliminary to the test for determining the t_E time.

In a cage motor, the cage winding is often the hottest part, therefore at the end of the heat run both the stator and rotor temperature rises are measured. The specified method of stator winding measurement is by rise-of-resistance, which should be effected as quickly as possible after the rotor is stopped.† Rotor temperatures may be measured by heat-sensitive paints or indicators, or by thermocouple or thermistor type probes through access holes in the casing, or by other suitable methods. It is usually sufficient to measure only the end-ring temperature, as all parts of the cage winding attain practically equal temperatures during a full-load run.

Fig. 9.13 Die-cast rotor of 16 kW type 'e' motor, prepared for locked-rotor test. (Brook Motors Ltd)

The locked-rotor test for measurement of t_E time calls for special techniques. The cage rotor is first prepared by embedding thermocouples in one end-ring, and in one or more bars. Unlike the full-load heat run, the locked-rotor test is likely to produce widely different temperatures between bars and rings, and even along each bar. In large rotors, particularly those with ducted rotor cores, it is difficult to predict the point of maximum temperature, and several thermocouples should be embedded at different points. In a small rotor with die-cast aluminium cage, it is considered sufficient to check only the mid-point of the bar. The current density in the rotor bar is always highest at the outermost part of its section, so the thermocouples should not be deeply embedded. They should be of light section wire to minimize the heat conducted away from the junction. The latter should be kept small, and embedded by caulking into small drilled holes.

Although the definition of t_E time refers to a stalling condition in a motor previously heated by a full-load run in the maximum ambient, for practical reasons the test is carried out with the motor initially at room temperature, and the limiting

† BS 4999: Part 32 requires the temperature measurement to be completed within a specified time limit, otherwise cooling curves must be plotted and extrapolated to the instant of shut down.

temperatures arithmetically reduced by the sum of the difference in ambient temperature and the recorded temperature rises on full load of both stator and rotor.

There is another important difference in temperature measurement between heat run and the t_E test. In the former, the temperature falls relatively slowly after the motor is switched off, as the whole motor is hot. In the t_E test, which is of short duration, the current-carrying conductors heat up rapidly, resulting in a large temperature differential in relation to adjacent iron cores, and these act as a heat sink after switch-off, quickly reducing the conductor temperatures. It is therefore not permissible to take the first measurement after switch-off as being the maximum temperature, and techniques must be employed which enable accurate assessment of the peak value to be made. The following procedure has been successfully used, and has the approval of BASEEFA.

To measure the stator temperature by rise of resistance, a stabilized direct-current source is first applied to the winding terminals when cold, and the voltage drop measured by a digital precision millivoltmeter. After the motor has been energized at rated voltage for a minimum of five seconds, the measurement is repeated several times at short intervals as the windings cool down and the digital display photographically recorded. This enables an accurate cooling curve to be plotted, and extrapolated to the instant of switch-off.

To measure rotor temperatures, the embedded thermocouples are used, the temperature of each being simultaneously recorded on a UV recorder.

Figure 9.14 shows typical results. Only the cooling curves are recorded or plotted from measured values. It is convenient to assume that the heating curves are linear,

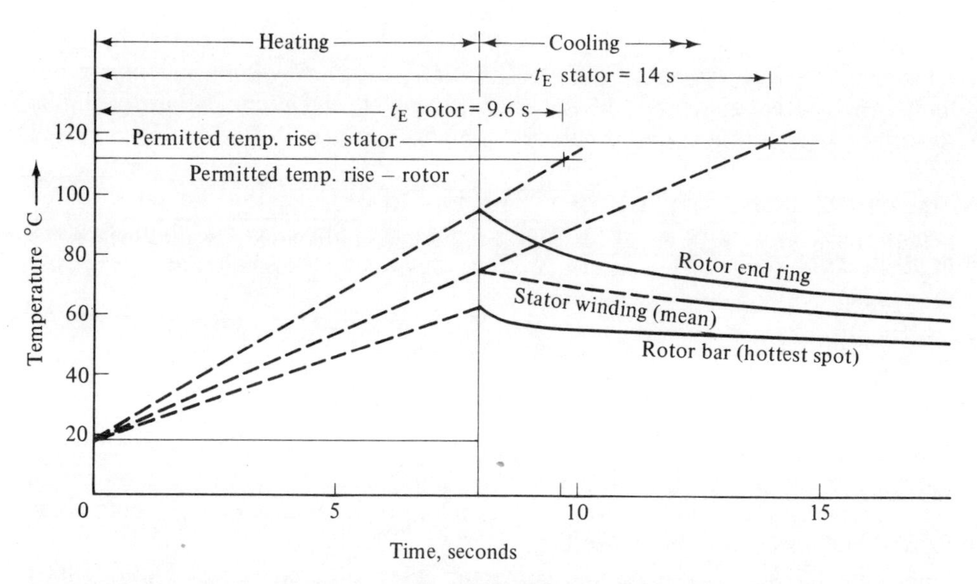

Fig. 9.14 Locked-rotor test curves for 12·5 kW 3000 rev/min type 'e' motor, temperature class T3, $t_E = 9{\cdot}6$ s

the time for each part to attain the limiting rise being obtained by extrapolation if necessary. Generally the times for each measured part will be different, and the lowest is the t_E time for the motor. The assumption that the heating curves are linear is not strictly correct, and to minimize the error the duration of the locked-rotor test should be as nearly as possible equal to and not greater than the t_E time.

For large motors, locked-rotor tests may be impracticable, in which case it is permissible by agreement with the certifying authority to resort to methods of calculating the t_E time from the design parameters.

9.9.8 *Control equipment*

A type 'e' motor is not safe to use unless controlled by suitable equipment. In particular, protective devices must be used which will trip within the t_E time when the motor is stalled, or which will always prevent the limiting temperatures for the various parts of the motor being exceeded.

9.10 Type 'N' Motors

Developed in Britain, type 'N' motors are similar in basic concept to type 'e' motors, but with some relaxations in requirements because of the lower risk in Zone 2 areas, to which their application is limited. They were originally made to an OCMA specification, but are now covered by a British Standard, Part 16 of BS 5000. There is as yet no international or foreign equivalent.

The specification for type 'N' motors is similar to that for type 'e' in the matter of non-sparking of cage rotors, minimum permissible running clearances, and mechanical shock tests.

The requirements are less stringent than for type 'e' in the design of terminals and their clearances, in the specification of insulation materials, and in temperature limits. The concept of t_E time does not apply to type 'N', and although there are limits to surface temperatures, the permissible temperature rise of the windings at rated load is the same as for general-purpose motors, i.e., 10°C higher than for type 'e'. This enables frame sizes for type 'N' motors to be generally the same as for general-purpose motors, at least when the flammable atmosphere is limited to gases in temperature classes T1, T2, and T3. Thus they tend to be smaller and less costly than type 'e' motors of equal output rating.

9.11 Combinations of Different Types of Protection

It is possible and sometimes expedient to combine different types of protection in one motor, particularly in large-size units. One example has already been given in the section on pressurized motors, where it was mentioned that either flameproof or type 'e' terminal boxes were often used.

It is not possible to guarantee complete absence of sparking at the contact faces of brushes and sliprings, therefore in type 'e' or type N slipring induction motors and synchronous motors, it is essential to enclose the sliprings in a flameproof casing.

To similarly segregate a commutator inside a separate enclosure, with its multiple connections brought out to the armature, poses a formidable design problem, and the author knows of no example of type 'e' or type 'N' commutator motors.

9.12 Choice of Motors

In mining, flameproof motors are still exclusively used, but in other industries, there is now a variety of types of electric motor construction, and it is worth considering the factors influencing their choice.

Undoubtedly, from purely cost considerations, type 'N' has the advantage over all others, and has encouraged engineers in endeavouring to ensure whenever possible that areas are scheduled as Zone 2. The result is that Zone 2 plant areas in Britain now outnumber Zone 1 areas in the ratio of 5 : 1. Type 'N', or Div. 2 motors as they were previously known, are now well established, having been used with success since 1963.

It is in Zone 1 areas that the user is now facing a proliferation of choices. The traditional flameproof motor is challenged by the type 'e' newcomer, and the latter, although often physically larger, has the advantage of lower initial cost. One factor which occasionally can be decisive is the category of the hazardous gas. Thus, if the gas has an ignition temperature low enough to place it in temperature classes T4 to T6 the frame size of a type 'e' motor is likely to be prohibitively large, and the choice would be for flameproof. Conversely, if the gas is in group IIC or is excluded from the groups of table 9.1, then there are considerable difficulties in obtaining certified flameproof motors, and the choice would be either type 'e' or pressurized.

Pressurized motors even of small to medium size have been used in the past for those gases for which flameproof motors were not certified, but their use for groups IIA and IIB gases is limited mainly to large sizes, where the cost of auxiliary pressurizing apparatus is relatively small.

In the majority of cases, however, for outputs up to about 150 kW the above factors do not influence the choice, which must then be a matter of judgement. The many years of successful experience in the use of flameproof motors must be set against the less costly but fundamentally different safety concept of type 'e', which although relatively untried in Britain, has more than a decade of Continental experience behind it.

References

1. J. Haig, H. C. Lister, and R. L. Gordon, *The Testing of Flameproof and Intrinsically Safe Apparatus*, IEE Conference Report (1962).
2. D. I. A. Grainger, 'Standards and Types of Protection applicable to Electric Motors used in Hazardous Atmospheres', *Mining Technology* (June 1973).
3. P. B. Greenwood, 'Selection of Electric Motors for Flammable Atmospheres', *Chemical Processing* (February 1974).
4. H. Dreier et al., P.T.B. Testing Memorandum: 'Explosion Protected Machines of the type of Protection "Increased Safety" (EX)e' (1969). English Translation published by BASEEFA.

10 Small Special-purpose Motors

10.1 Introduction

BS 4727 defines a special-purpose motor as a motor with special operating characteristics, or special construction, or both, designed for a particular application, and not falling into the definitions of general-purpose or definite-purpose motors.

These motors are generally small, possibly rated up to a few kilowatts, but usually much less. The following sections cover quite a wide variety of motors some of which are considerably different from a general-purpose motor. Some of these motors, such as the Steromotor are produced in relatively small numbers, while others, such as the electromagnetic vibrator motor and timing motor are manufactured in large quantities.

10.2 Brushless d.c. Motors

10.2.1 *Introduction*

Requirements for long life, high reliability, low radio noise, and low power consumption have been the main reasons for recent developments in brushless d.c. motors. In small, fast-rotating machines the wear caused by friction and sparking limits the life of the brushes to a few thousand hours under normal conditions, and possibly only to a few minutes at very low atmospheric pressures. A longer life without maintenance becomes possible if electronic switching elements are used instead of mechanical commutators.

Some of the important applications of these motors are to be found in the aerospace industry where great reliability and exceptionally long life without maintenance is very important. The very high overall efficiency may also be of advantage in this field, since it permits a considerable saving in weight in the energy source. Other typical applications for brushless d.c. motors are drives for implanted artificial hearts, high-speed prism drives for infra-red seekers, cryogenic coolers, gyro spin motors, video tape recorder head drives, high reliability tape recorders, and stereo record player turntables.

These motors are not simply a.c. motors powered by an electronic invertor, as they require a position feedback of some kind so that the input voltage waveforms are kept in the proper timing with the rotor position. Their performance in terms of motor output closely allies itself to that of mechanically commutated d.c. motors, and in almost all cases, when supplied with the proper electronic control system, can be substituted directly. Although generally brushless d.c. motors are more expensive than conventional d.c. motors, they are available in a wide range of designs from the most sophisticated type used in space applications, to high-volume low-cost devices used in industrial equipment. The weight of the motor with its control system is not appreciably greater than a conventional d.c. motor with brushes.

10.2.2 *The basic brushless motor*

A commutatorless d.c. motor comprises four basic elements: electronic commutator, rotor position sensor, the stator, and the rotor. In a commutatorless d.c. machine, the roles of the stator and rotor are reversed when compared to a conventional d.c. machine. The stator now becomes the armature, and resembles a typical induction motor stator. The rotor now provides the magnetic field, which is generally of the permanent magnet type.

The commutator/brush system in a d.c. motor performs two important functions:

(a) It senses the position of the rotor coils relative to the pole pieces, by virtue of its physical position.
(b) It reverses the rotor coil current at the appropriate instant of time.

These two actions ensure that the motor torque is always unidirectional. It is feasible to substitute each commutator segment by a semiconductor switch, so that the motor coil current can be individually switched. However, this may lead to an expensive machine, as some d.c. motors have a large number of commutator segments. Since an electronic commutator can normally stand a much higher voltage than the mica between conventional commutator segments and is not troubled by switching inductive loads, it is possible to interconnect some of the armature coils in series so as to reduce the number of switching devices.

The thyristors or transistors which make up the electronic commutator need signals or firing pulses, which are dependent upon the rotor position, to enable them to conduct at the correct time. This of course is done automatically with a conventional mechanical commutator, but with an electronic commutator an auxiliary mechanism called a rotor position sensor is required. Several different rotor position sensors are possible.

There is, at present, a considerable range of brushless d.c. motors available. Table 10.1 shows the performance data for a typical range of brushless motors.

Table 10.1 Typical range of transistor brushless d.c. motors

Rated voltage,	V	12	12	24	60
Rated speed,	rev/min	6000	6000	6000	6000
No-load speed,	rev/min	11 600	8400	7000	6700
Rated torque,	$Nm \times 10^{-2}$	0·2	1	5	20
Rated current,	A	0·3	1	2	2·8
Mean starting torque,	$Nm \times 10^{-2}$	0·5	3·7	36	200
Starting current,	A	0·6	3·3	13	27
Moment of inertia of rotor,	$Nms^2 \times 10^{-6}$	0·45	3·3	41·6	230
Run-up time at rated current,	s	0·14	0·21	0·53	0·72
Efficiency,	%	30	50	65	75
Weight of motor,	kg	0·07	0·21	0·75	1·5
For ambient temps. under rated conditions,	°C	−20 to +45°C			

10.2.3 *Rotor position sensors*

One method for determining the rotor position which has gained favour, for small motors, in the last few years, is the use of photo-transistors and LEDs or lamps. A small stationary light source is mounted within a shield attached to the shaft. The shield has a small aperture which allows illumination to fall on only one of the equally spaced photo-sensors which are attached to the stator. The number of photo-sensors used depends upon the number of stator windings and hence the number of power transistors. A simplified schematic diagram of a transistor motor with photoelectric commutation is shown in Fig. 10.1

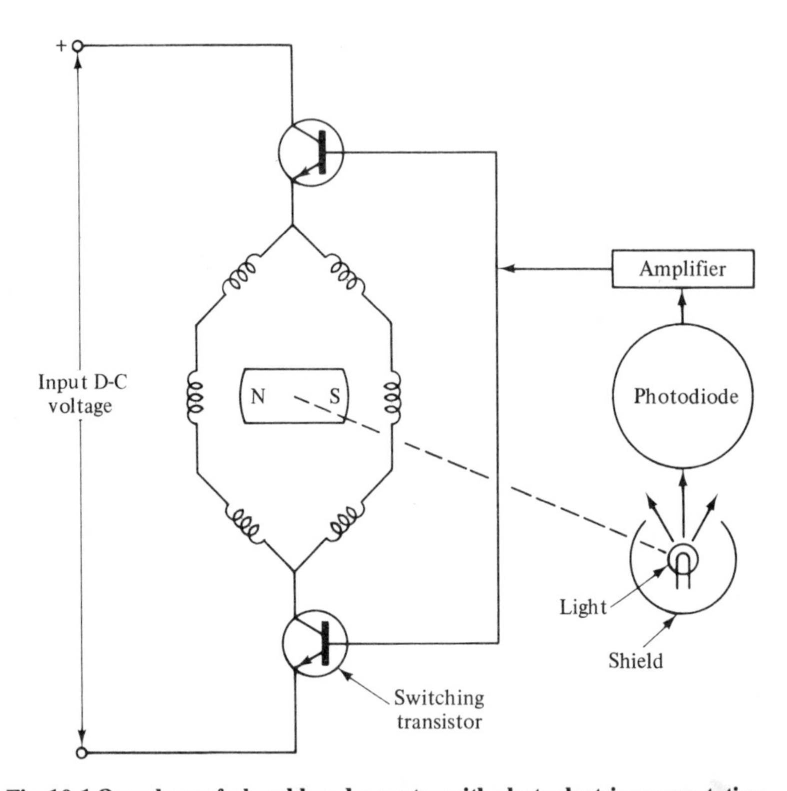

Fig. 10.1 One phase of a brushless d.c. motor with photoelectric commutation

Another method uses a wheel with a proper sequence of windows inserted between the light source and the photo-transistor.

The illuminated photodiode produces a current which is amplified and fed to the bases or firing gates of the switching transistors or thyristors. These signals allow the supply current to be applied to the appropriate stator windings.

This system is lighter than equivalent magnetic methods and can be made to produce a sharply changing d.c. signal. It is, however, less robust than other methods, and generally used only on very small motors.

A few companies have developed brushless d.c. motors using magnetic-sensing commutation, usually using the Hall effect. These motors use the main magnetic poles, or an auxiliary rotating permanent magnet that provides a rotating field for the stationary Hall generator. To define the commutating axis, the control signal can be derived either from the point of intersection of two Hall voltages induced by the rotor flux from Hall generators positioned 90 electrical degrees apart or from the point of passage through zero of the Hall voltages obtained from two generators positioned 45 electrical degrees apart. The latter method is preferable since, because of the temperature dependence and manufacturing tolerance of Hall generators, the Hall voltages may differ greatly. The consequent displacement of the point of intersection with respect to the optimum commutating axis causes an additional fluctuation of the output torque and a reduction in efficiency. Although the sensor output is d.c. this system also has a residual signal due to flux leakage, and requires an auxiliary d.c. source and amplification of the Hall voltages.

Magnetic coil position sensors are widely used in industrial motors and can take several forms. Figure 10.2 shows a typical system. Its output is capable of directly

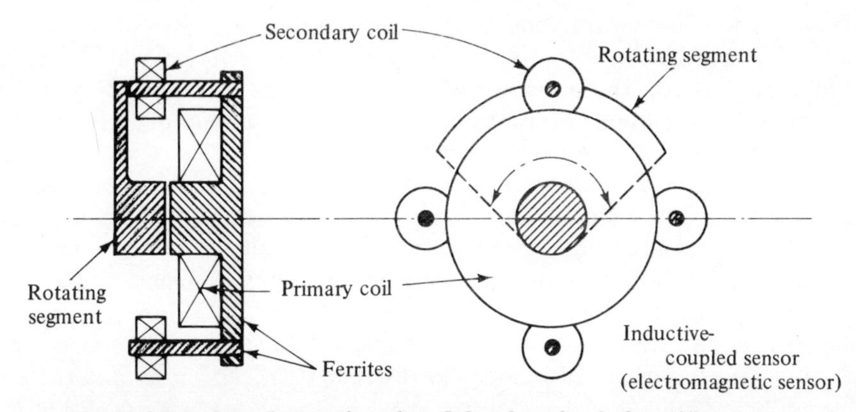

Fig. 10.2 Principle of operation of modulated carrier shaft position sensors

driving the switching circuitry thus dispensing with the need for amplifying circuits. It consists of a high-frequency oscillator which energizes a primary coil mounted statically adjacent to the end of the rotor. A rotating segment on the rotor couples this primary coil to a series of secondary sensor coils. During motor operation, oscillator current is induced in each pick-up coil in succession as the rotating segment moves into position to couple energy from the oscillator coil. As the voltage induced in the pick-up secondary coil rises, the power transistor or thyristor is switched on, supplying current to the armature coil. This current is switched off as the control segment moves away, approaching the next secondary coil, and turning on its supply semiconductor. Thus the armature is commutated as the motor turns. This direct-driving capability enables the switching circuitry to be considerably simplified and hence more cheaply constructed.

One brushless d.c. motor uses four small toroidally wound ferrite cores mounted at angular distances of 90° around the stator. Depending on the position of the rotor, the cores are either exposed or not exposed to the magnetic field of a small permanent magnet rotor attached to the end of the main rotor. Hence the appropriate power transistor switching times can be determined.

Electromagnetic sensing devices are robust and produce signals which can be made to operate the commutator without auxiliary amplification. They suffer from the disadvantage of having a relatively high residual signal caused by leakage in the motor.

Other methods of sensing the rotor position have been used. These include the use of capacitor probes, metal detectors, and reed switches with a rotating magnet. These methods are not very common, as the photo-electric, magnetic sensing, and magnetic coil methods appear to be more reliable and hence are used for most applications.

Once the signal from the rotor position sensor has been removed from the base of the transistor it is switched off. The transistor remains in this state until the firing signal is applied again. For machines greater than about 750 W thyristor switches may be used. These require forced switching off or forced commutation, usually by forcing the current to zero by discharging a capacitor across the thyristor.

10.2.4 *Armature windings for small commutatorless d.c. motors*

The commutator in a conventional d.c. motor is placed on the rotor to avoid the necessity for rotating brushes, but this construction is inverted when electronic commutation is used. The armature windings are now on the stator and if the conventional d.c. armature is transferred as a multi-phase armature, the machine can have a similar construction to a synchronous motor.

In the conventional d.c. motor, a multi-segment commutator produces a rectangular distribution of armature current which reacts with the main field flux to produce electromagnetic torque. Conventional d.c. armature windings may be adapted for solid-state commutation by connecting the armature tappings to pairs of transistors or thyristors rather than commutating segments. Figure 10.3 shows a simple lap winding with a solid-state commutator. With this winding the armature m.m.f. wave does not rotate uniformly but moves in discrete steps, and produces the usual

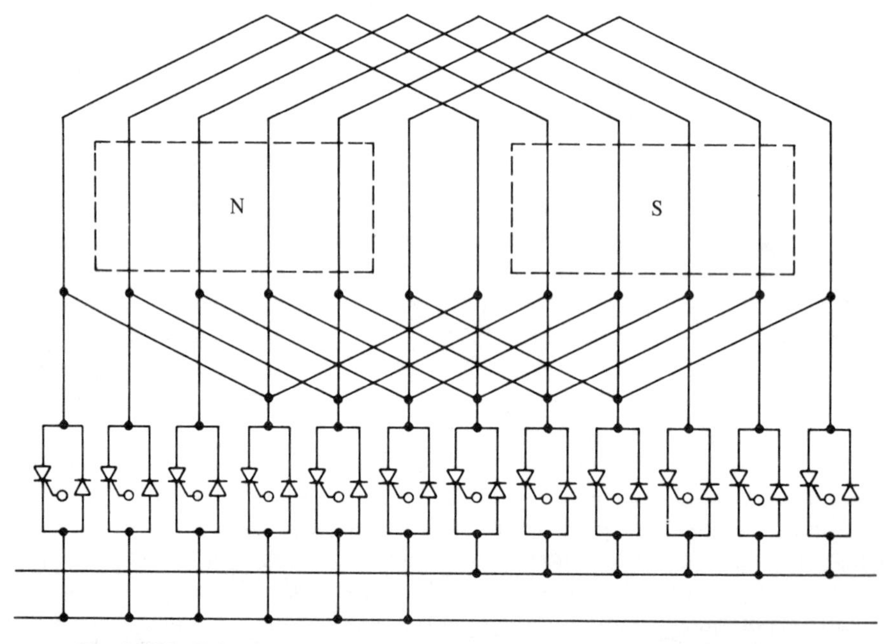

Fig. 10.3 Individual armature coils controlled by an electronic commutator

triangular armature m.m.f. distribution. A larger number of armature coils and thyristors will give a smoother field rotation and so reduce the rotor losses and undesirable torque pulsations. On the other hand, as the number of thyristors is increased, the conduction periods become shorter and the device utilization is poorer. For applications such as cassette recorder drives the number of transistors or thyristors is kept to a minimum.

A typical example of a commercially available, low power, brushless d.c. motor with a transistor commutator, which could be used in tape recorders, measuring instruments and medical equipment is shown in Fig. 10.4. This motor uses reed

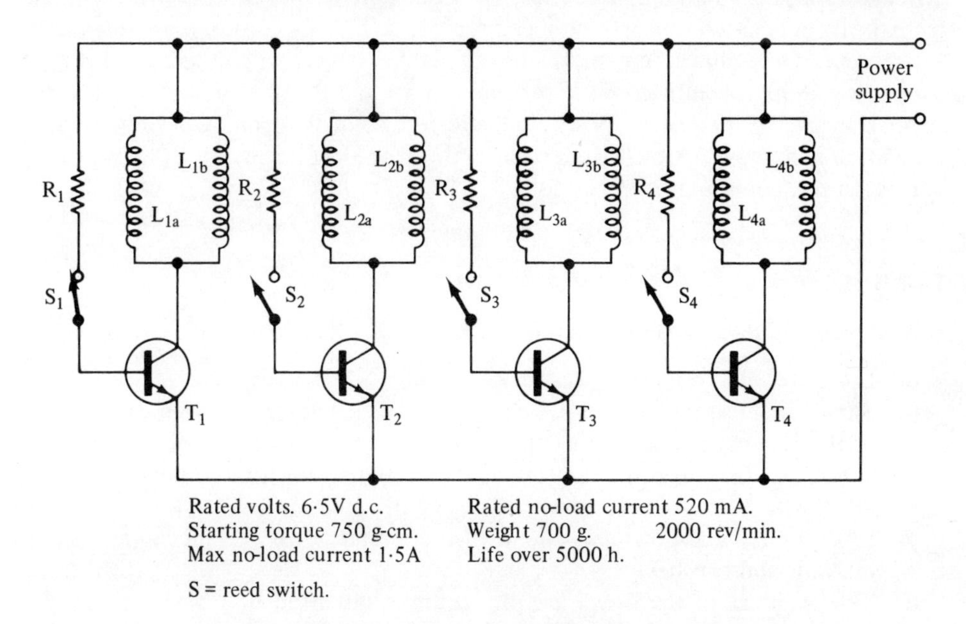

Fig. 10.4 Diagram of a typical small brushless d.c. motor

switches and a rotating magnet for the rotor position sensor. Normally small brushless d.c. motors are restricted to three or four phase windings on the stator, keeping the number of power transistors to a minimum.

10.2.5 *Stator construction*

With a brushless d.c. machine the stator now contains the armature. The armature windings are usually housed in a laminated core similar to an induction motor stator. However, there is one commercially available motor which has a stator with no slots, the winding being placed in the magnetic air gap between the rotor and stator.

A slotless stator arrangement permits a very economical winding technique. At high speeds the iron losses are low; in addition motors with this type of winding do not produce magnetic residual torques and therefore generate no magnetically excited structure-borne noise. Because of the slotless winding arrangement and the large air gap, the commutating reactances are smaller by an order of magnitude than for motors with slotted stators. Conditions are therefore very favourable for the design of the electronic commutator.

10.2.6 *Speed regulation*

The shaft position sensor, although usually considered a disadvantage, can be used for speed measurement because the frequency of the signal from the position sensors is proportional to the angular velocity of the shaft. The conventional method is to convert the angular frequency to a signal voltage by signal integration.

For very small motors, a speed control system often uses a centrifugal governor that disconnects the motor from the supply source as the speed increases, and reconnects it when the rotor speed drops below a predetermined level. Using a governor, speed regulation is in the range of 2 per cent to 5 per cent. Using the signals from the rotor position sensors, the speed regulation for small motors can be less than 1 per cent. One motor, commercially available, which will develop 14·4 g cm torque at 3000 rev/min for 180 mA and 6·5 V at the motor, claims a speed regulation of 0·5 per cent. With sophisticated control techniques better speed regulation is possible. A tachogenerator is therefore not necessary for closed-loop control of d.c. commutatorless motors.

10.2.7 *Advantages*

It is convenient here to summarize the advantages of these motors:

(a) Long life, high reliability, fast response.
(b) Small brushless d.c. motors are capable of efficiencies in excess of the usual 30 to 50 per cent obtained from conventional motors with powers of the order of a few watts. Brushless d.c. motors can be designed for a rated power of up to 20 W with efficiencies reaching 80 per cent.
(c) Little or no maintenance, due to the elimination of the mechanical commutator and brushes.
(d) The presence of the brush and the commutator itself in a conventional d.c. motor causes electrical and mechanical noise. However, brushless d.c. motors are much quieter and emit very little electrical noise.
(e) A tachogenerator is not required for speed control.
(f) Very high speeds and high power densities possible. Speeds up to 100 000 rev/min have been claimed.
(g) Speed of motor not limited by the supply. Wide speed range.
(h) Good speed regulation; good wow and flutter characteristics possible.
(i) Conventional d.c. motor characteristics without the limitations of a mechanical commutator.
(j) Simple speed control.

10.2.8 *Disadvantages*

(a) Expensive because at present production quantities have been rather small. This will reduce with increasing demand.
(b) Needs a rotor position sensor to fire semiconductor switching.
(c) Difficulties with starting and low-speed running can arise with thyristor commutator, but not usually a problem with a transistor commutator.
(d) Overload capacity low; about 150 to 200 per cent.

10.3 The External Rotor Motor

10.3.1 *Construction*

The 'inside-out' motor or the external rotor motor was patented by a German, Hermann Papst, who started producing them after the end of the Second World War. The basic features of these motors are their low noise level, high efficiency, and constant speed. Figure 10.5 shows a cross-section of such a motor.

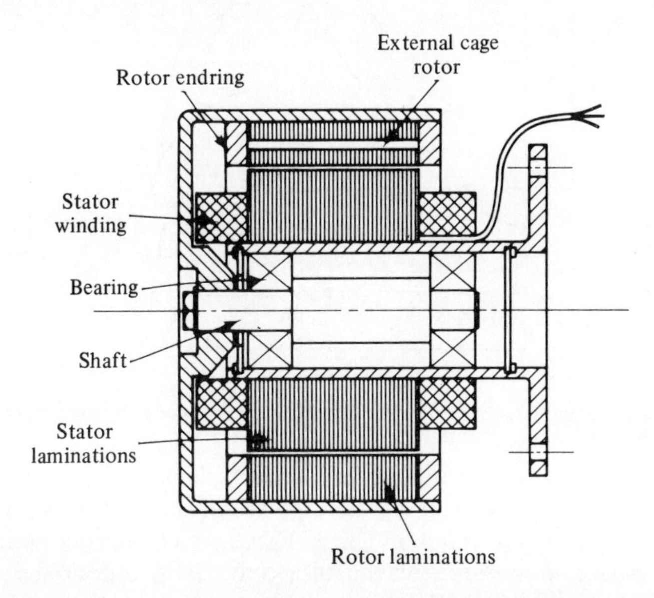

Fig. 10.5 Cross-section of a typical cage external rotor induction motor

External rotor type motors generally have a better performance than conventional motors of comparable size and weight, because of their higher efficiency. The outer housing, the rotor, rotates and carries the secondary winding, while the internal stator contains the primary winding, giving substantially increased air gap area and hence reduced field densities, which in turn reduces the magnetic losses. The internal location of the stator practically halves the magnetic losses, and reduced magnetizing current and reduced iron loss together result in reduced winding losses.

The external rotor motor can be one of several different types, such as a cage induction, hysteresis, eddy current, or reluctance motor.

The use of sintered sleeve bearings, reduced field density in the air gap and virtually closed or completely closed slots allow extreme quietness of operation. Ball bearings are available, and with these bearings a long operational life of 20 000 to 30 000 running hours is possible.

The flywheel effect of the rotor overcomes slight pole-to-pole magnetic changes, resulting in operation at low speeds without cogging. It also gives improved speed holding through transient load changes.

The rotary housing provides effective screening of the internal stator and its rotating field.

10.3.2 *Performance*

External rotor motors are usually equipped for either three-phase or single-phase operation for standard voltages and frequencies. Speeds with 50 Hz supplies generally vary between 500 and 3000 rev/min. Figure 10.6 shows the torque/speed curves

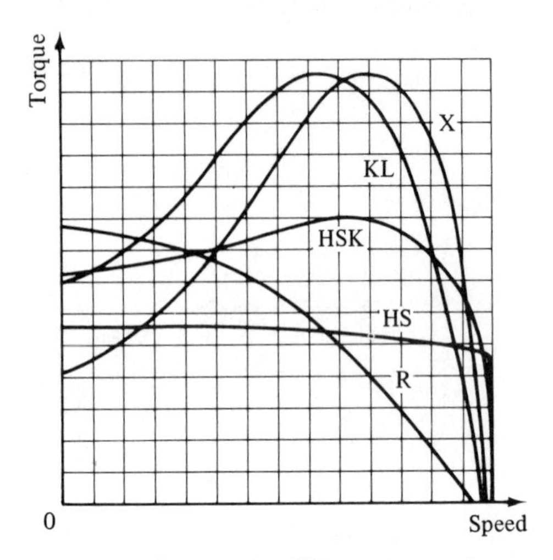

Fig. 10.6 Typical torque speed curves for different types of external rotor motor

for various types of external rotor motors. Cage induction motors, for both three-phase and single-phase, are available for high-speed stability (curve K) and where increased starting torque is required (curve KL), up to an output of about 250 W. Pole-change induction motors are available up to 120 W output and synchronous induction motors up to 70 W. Three types of hysteresis motors are available up to about 30 W output. The first type has a high starting torque, silent running, with a smooth transition to synchronous running (curve HS) while the second type contains a squirrel-cage giving high damping properties against hunting effects, with a high starting torque and silent running (curve HSK). In case of overload this motor pulls gradually out of synchronism without faltering. The third type also contains a squirrel-cage, giving flutter-free running, and at overload falls abruptly out of synchronism. Slotless eddy current rotor motors with a load-dependent speed

Table 10.2

Motor type	Voltage (V)	Frequency (Hz)	Speed (rev/min)	Rated input power (W)	Rated output power (W)	Starting torque (g cm)	Weight (kg)
AS	240	50	1300	28	4·5	360	0·8
AS	240	50	1325	33	6·0	432	0·9
HS	42	50	3000	13	0·7	50	0·2
HS	240	50	1500	27	1·6	252	0·7
HS	115	60	1800	51	15	800	1·9
E	220	50	2500	18	2	200	0·7
E	220	50	750	34	5	1200	1·3

NOTE: AS = Asynchronous motor; HS = Hysteresis synchronous motor; E = Eddy current motor.

characteristic are available up to about 80 W output (curve R). This type of motor gives high starting torque, minimum noise, and a power consumption which increases only slightly at standstill.

Typical motor characteristics are listed in table 10.2.

10.3.3 *Applications*

The external rotor motor can take the place of conventional motors in applicatons where the flywheel effect of the rotor may be of value in providing a smoother drive and more gentle starting conditions, and where the larger rotor inertia is not a disadvantage.

External rotor motors are generally fairly high precision motors and find quite a wide range of applications; some typical examples are studio and domestic tape recorders, dictating machines, office calculating machines, electric typewriters, Hi-Fi turntables, Video tape recorders, medical, optical, and chemical instrumentation systems, precision machine-tool control, and cooling fans for electronic equipment.

Induction motors with external rotors have been used for fan-applications with outputs from 3 W to 10 kW. The fan blades are directly attached to the external rotor.

Although the vast majority of external rotor motors are low-power motors British Rail are developing a motor of this type to power railway locomotives. They have mounted an induction motor within a large diameter tubular axle, with the stator mounted in the centre of the axle. The rotor is external to the stator and is attached to the inside diameter of the tubular axle.

10.4 Clock and Timer Motors

A clock or timing motor is an electric motor which operates at constant speed and has a usable output power. They fall into an entirely different classification from other types of motor. The output powers are very low, usually less than 1 W. Normally they are used as prime movers for timing devices, such as timed cam switches, clocks, and chart drives, from which the exact speed of the output shaft can be utilized, and can be a.c. or d.c. motors.

The decision to use an a.c. or d.c. motor usually depends upon the available power supply. Where both types of supply are available the choice depends upon the merits of the motors.

10.4.1 *Direct-current timing motors*

Direct-current motors have several advantages making them useful as timing motors. The main ones are high efficiency, typically between 30 and 70 per cent, linearity between speed and voltage of up to ±5 per cent, and a starting torque of up to 10 times greater than running torque. However, their main disadvantage as timing motors is that variation of the supply voltage results in variations of the motor speed. As a result d.c. motors by themselves are only practical where speed tolerances of ±10 per cent or greater are permissible. Better speed-holding accuracy can be obtained by the use of a governor.

There are basically three principal types of governor in common use, namely centrifugal, mechanical or electrical oscillator, and electronic governors. The centrifugal governor has spring-loaded contacts which open at a set speed of rotation. This type of governor is cheap and accuracies of ±1 to 10 per cent can be maintained. With a mechanical or electrical oscillator governor the frequency of an oscillator is compared to the speed of the motor, and corrections are made as required. The oscillator may be a tuning fork, crystal, or vibrating reed, and accuracies of ±0·05 per cent or better are possible. The electronic governor compares the back-e.m.f. of the motor being controlled to a reference voltage in the governor. This type of governor provides accuracy of the order of ±3 per cent with 20 per cent voltage variations, zero to full-load. More sophisticated electronic methods can provide better accuracy.

Direct-current motors used in timing applications are generally of the permanent magnet type, and are usually low voltage type of 6, 12, or 24 V.

The hollow-rotor motor type is common. The permanent magnet is stationary and is located at the centre of the rotor. The housing is made of soft steel and provides a return path for the magnetic flux. The rotor coils are mounted on a non-magnetic cage to form a shell, or an external rotor. Epoxy rotor mountings increase rotor rigidity and resistance to shock and vibration as well as protecting the winding. The rotor diameter is relatively large compared to the motor outside diameter, resulting in greater torque at low speeds, and producing less mechanical noise and longer bearing life. Cogging is eliminated as there is no iron in the armature. An aluminium rotor cage is often used to provide eddy current damping. This provides a more constant speed, even under load conditions. The aluminium cage also provides temperature compensation.

On small d.c. motor timers the commutator is often manufactured from a silver alloy with gold alloy brushes. Sometimes an etched copper-clad laminated commutator is used, which is rhodium plated to ensure maximum life and low potential drop across the brush-commutator interface. With this commutator silver-graphite brushes are used.

Printed-circuit motors are also used. Their construction and characteristics are dealt with elsewhere.

10.4.2 *Alternating current timing motors*

The most common type of a.c. timing motor is the synchronous permanent magnet motor, often called the synchronous inductor motor. Figure 10.7 shows an exploded view of such a motor.

The stator consists of two high-precision stampings of a high flux density soft steel which surrounds the excitation coil; two copper phase rings and an outer sleeve. The coil is encapsulated to ensure a high degree of environmental protection and a long life. The use of the phasing rings ensures a consistent unidirectional start, accelerating rapidly to synchronous speed in about a half to a whole cycle of the supply frequency. Some motors do not have phasing rings and rely on a non-reversing mechanical system, or a flick of the shaft for starting.

The stator is formed like a cage; that is the stator winding and the rotor are completely surrounded with iron. Thus the external magnetic field is very small and the danger of interference with sensitive instruments or electronic circuits is greatly reduced.

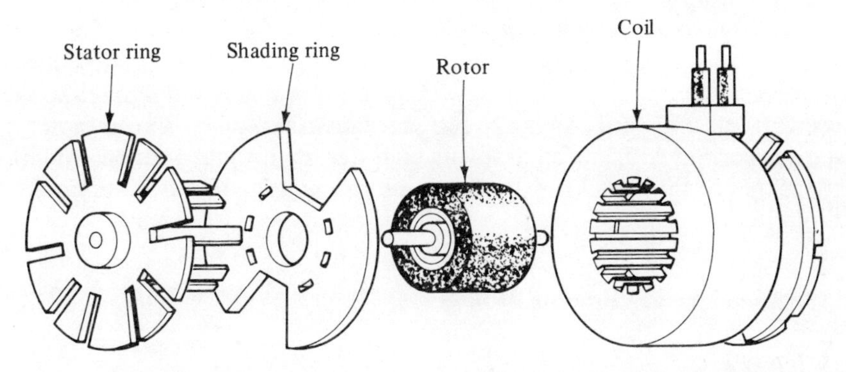

Fig. 10.7 Exploded view of a basic permanent magnet synchronous timing motor

The rotor usually consists of a ceramic ferrite magnet of high coercivity and high field strength, moulded on to a hardened steel shaft. The rotor is magnetized radially with about 12 to 18 poles of alternate polarity. This design ensures a high locking torque when the motor is de-energized giving virtually instantaneous stopping, important for stop clocks measuring precise time intervals.

The permanent magnet rotor is driven by impulses from the alternating field of the stator once the motor has been started by shaded-pole or mechanical methods. The motor will then continue to run in synchronism unless the 'pull-out' torque is exceeded, when it will stall.

The rotor bearings are generally manufactured from a plastics material or nylon to ensure quiet running, with long life and little maintenance.

Most small a.c. clock motors use no fan cooling and depend on free radiation and convection for cooling. The expected life of an a.c. timing motor is usually around 3 to 5 years of continuous running under rated torque and normal ambient conditions. Intermittent duty will extend this expected life. Improvements in the heat dissipation characteristics of the device in which the motor is mounted will also improve the life. These improvements would take the form of increasing the motor ventilation and/or increasing the weight of metal attached directly to the motor, i.e., increasing the size of the heat sink.

These a.c. timer motors are generally manufactured with 'built-in' gearboxes with ratios of up to 1 000 000:1, although a 375:1 ratio is the most common. Typical permanent magnet synchronous characteristics are listed in table 10.3

Table 10.3 Typical permanent magnet synchronous motors for timing applications

Motor	A	B	C	D
Voltage (V)	48	117	48	220
Frequency (Hz)	50	60	50	50
Speed (rev/min)	250	300	250	250
Current (mA)	12·5	8	40	7·5
Power consumption (W)	0·6	0·9	1·6	1·6
Starting torque (g cm)	5	5	25	25
Running torque (g cm)	5	5	30	30

The synchronous hysteresis motor is also used as a timing motor. The stator is virtually identical to that of the permanent magnet synchronous motor described, but the rotor is manufactured from a magnetic material which is soft enough to permit induced currents and hard enough to retain magnetism. The hysteresis motor has a low starting current with a comparatively high starting torque. There are no torque pulsations, hence the speed is extremely uniform, and the motor is very quiet and vibration free.

10.5 Electromagnetic Vibrating Motors

10.5.1 *Introduction*

When a reciprocating motion has to be produced electrically the use of a rotary electric motor with a suitable transmission is really a rather roundabout way of solving the problem. It is generally a better solution to look for an electromechanical system which produces a to-and-fro movement directly. Such systems, called vibrator motors, can be based on various principles. One such approach is to use the forces exerted on a piece of soft iron or a permanent magnet in a magnetic field produced by an a.c. current. The piezoelectric effect can also be used to produce the required motion from an a.c. input voltage. Magnetostriction and the variation in forces between two plates of a capacitor when fed with an a.c. voltage are other alternatives.

All the possibilities mentioned above require an a.c. supply voltage. The choice of operating principle depends on the particular application; the amplitude of the necessary reciprocating motion is very important. So far, electromagnetic drives have given the largest amplitudes; magnetostrictive and capacitor drives give a very small amplitude, severely limiting their application. Piezoelectric drives produce an amplitude greater than the latter types but smaller than the electromagnetic type. In most cases the amplitude required makes it necessary to choose an electromagnetic drive.

10.5.2 *Construction of electromagnetic vibrators*

There are basically two types of electromagnetic vibrator; one type uses a soft iron armature and the the other a permanent magnet armature. The choice depends upon cost, necessary force and movement required, size, and speed. Figures 10.8 (a) and (b) show simplified diagrams of two soft iron armature electromagnetic vibrators. In Fig. 10.8(b) the armature can either be mounted for linear movement or attached to a pendulum transmitting the alternating force through a lever. With a soft iron vibrator supplied with 50 Hz the armature oscillates at a frequency of 6000 cycles per minute, as there are two attractive forces per cycle. However, if a half-wave rectifier is inserted in series with the magnetic coil the resulting speed will be 3000 cycles per minute, as there is now only one attractive force per cycle.

Some types of magnetic vibrators use permanent magnets in their design. Figure 10.9 shows a simplified diagram of this type of motor. The poles of the permanent magnet react with the right-hand side coil to produce an attractive force and the permanent magnet reacts with the left-hand side coil producing a repulsive force during the first half of the cycle. During the second half the forces are reversed hence

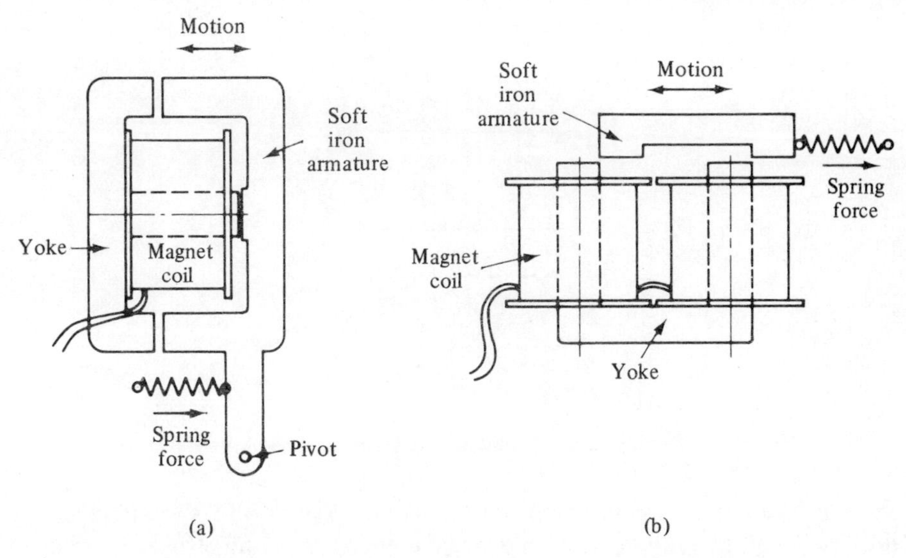

Fig. 10.8 Types of soft iron armature electromagnetic vibrator

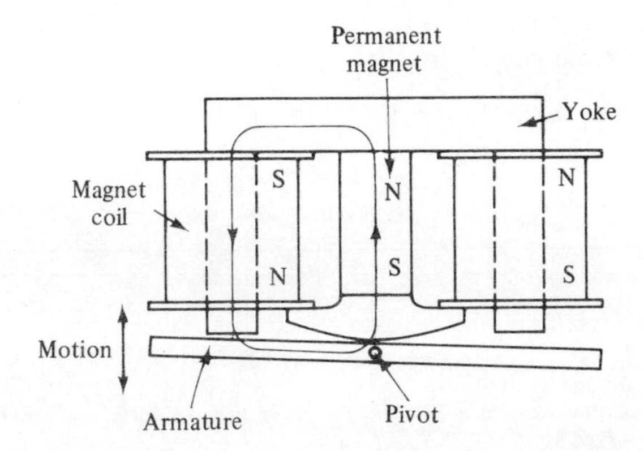

Fig. 10.9 Permanent magnet electromagnetic vibrator

performing one mechanical cycle to one electrical cycle. This results in a vibrating motion of 3000 cycles per minute on a 50 Hz supply. The vibrator in Fig. 10.8(b) would oscillate at half the frequency of the soft iron armature motor if the armature was a permanent magnet.

Another permanent magnet vibrator is shown in Fig. 10.10. This type is often used to drive a refrigerator compressor, and here a coil carrying a current moves in the field of a permanent magnet. The magnetic circuit is provided with an air gap in which the coil can move. The vibrator operates exactly the same way as a loudspeaker cone and oscillates at the same frequency as the supply.

It is common practice to make the mechanical resonance frequency of the armature, determined by its mass and the stiffness of the suspension springs, approximately equal to the speed of operation of the vibrator to ensure the vibration amplitude will be a maximum.

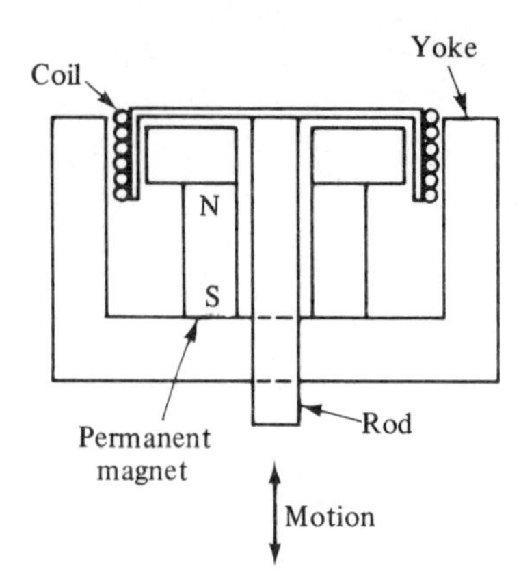

Fig. 10.10 Moving coil, permanent magnet electromagnetic vibrator

The vibrator motor is very simple with few moving parts and is virtually maintenance free. The pendulum type motors usually contain a pivot bearing made of sintered bronze saturated with oil, which requires very little lubrication. Some simple motors do not contain bearings.

10.5.3 *Performance and characteristics*

Typical characteristics of electromagnetic vibrators are shown in table 10.4.

Table 10.4

Type of vibrator	Permanent magnet armature	Soft iron armature
Continuous kW/kg	0·01	0·007
Continuous kW/cm^3	$0{\cdot}1\times10^{-3}$	$0{\cdot}07\times10^{-3}$
Maximum developed kW/kg	0·03	0·02
Starting torque Nm/kg	4·1	3·2
Watts input/kg at continuous duty	24	37

The motor efficiency depends upon the design configuration and how well the mechanical spring system is tuned to the supply frequency. Permanent magnet motors are more efficient than soft iron armature types.

The speed of the electromagnetic vibrator will remain constant, and as the load increases the movement is reduced. These motors are designed for specific applications, and to produce a given force required to overcome the load while oscillating with the required stroke.

Variations in the supply voltage from the nominal voltage, of say ±10 per cent, must not affect the operation of the vibrator too greatly. Too high a voltage should not make the armature exceed the limits of its free path, and too low a voltage should not make the speed too low.

The electromagnetic vibrator motor lends itself to applications where a vibratory motion at constant speed is required. Because of their simple design they lend themselves to mass production techniques with resulting low costs. Other main advantages are that they are maintenance free, and since there are no brushes or commutator they do not generate radio or television interference.

Vibrator motors do, however, suffer from some disadvantages, the main ones being their relatively low efficiency, their low output to weight ratio, and their inherent magnetic hum and vibration. Speed control is limited and can only be carried out satisfactorily with expensive electronic frequency control. Typical examples of their applications are drives for hair clippers, shavers, pumps, and sanders.

10.6 Direct-current Axial Air-gap Motors

Direct-current axial air-gap motors are often referred to as disc motors and there are basically two types. The first and most common type is the printed-circuit motor and this is discussed in chapter 18. The second is the wound-rotor type disc motor; Fig. 10.11 shows a typical cross-section of such a motor. The only real difference between the two types of motor is the rotor construction.

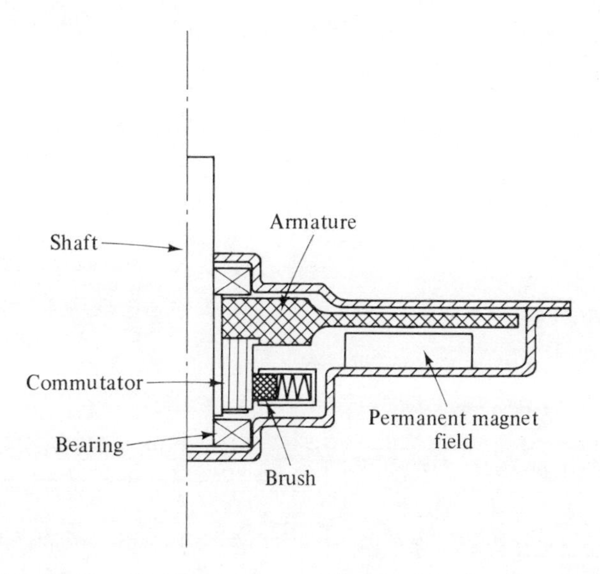

Fig. 10.11 Simplified sectional view of a wound-rotor type d.c. disc armature motor

10.6.1 *Construction and characteristics of the d.c. wound-rotor disc motor*

In the disc wound motor, the armature coils are wound into a flat disc, which is stiffened by reinforcing it with a thermosetting epoxy resin. The rotor coils are wound on to a frame to the correct number of turns, the frame is then removed, and the coils hot pressed to form a disc. The resin used must not cause the insulation to deteriorate and must also display minimal size variation under severe conditions of high temperature, high speed, centrifugal force, and stresses caused by repeated stops and starts. Because of the narrow clearance of the armature and the permanent magnet field any armature distortion would damage the motor. These rotors can be constructed with barrel-type or face-type commutators.

The disc motor has a lightweight armature with a low inertia and no iron core. Because of the iron-less construction the inductive influence of the armature coils is decreased, assuring good commutation and hence increasing the brush life over that of conventional motors. The iron-less construction also results in no core loss, and the permanent magnet field does not require electrical power, hence this motor is of higher efficiency than conventional wound-field types.

The d.c. disc motor is extremely thin, and only 25 to 33 per cent the size of conventional motors. It can be operated on pure d.c. or rectified a.c.

These motors have the capacity to deliver high instantaneous torques, allowing very low reversing and stopping times. They also have a high power to weight and volume ratios. Motors up to 4 kW are standard items, but outputs up to 20 kW are available. The characteristics of a typical range of motors is shown in table 10.5

Table 10.5 Typical characteristics of d.c. wound-rotor disc motors

Rated speed (rev/min)	6000	3000	3000	3000
Rated voltage (V)	40	88	40	245
Rated output (W)	152	630	1120	3150
Rotor inertia (m^2 kg)	$0{\cdot}04\times10^{-3}$	$0{\cdot}68\times10^{-3}$	$2{\cdot}4\times10^{-3}$	$11{\cdot}5\times10^{-3}$
Efficiency (%)	76	84	86	92
Weight (kg)	1·7	7	12	25·5

The motors in this table are totally enclosed; however, forced ventilation will increase the output by about 40 per cent for the same motor frame.

10.6.2 *Applications*

These motors have a wide variety of possible applications; some typical ones are:

(a) Domestic appliances, such as polishers and washing machines.
(b) Automobile equipment such as car heaters, windscreen wipers, power window winders, and radiator fans. In the last two applications the flat shape is a particular advantage.
(c) Industrial equipment such as wire feeders, machine tools, and blowers and pump units.
(d) Automatic control equipment such as computer hardware and office equipment.

10.7 High-frequency Motors

High-frequency motors were originally designed for military applications where communication interference cannot be tolerated. These motors are generally of the three-phase cage induction motor type and high output powers can be obtained in small sizes. They are used, for example, in ultra-high speed centrifuges for uranium enrichment running at about 60 000 rev/min, nylon thread spinning machines at 100 000 rev/min, and textile machinery drives and high-speed grinders running at 25 000 to 180 000 rev/min. Typical ratings for high-speed grinder drives are 0·9 kW at 84 000 rev/min and 0·45 kW at 42 000 rev/min.

Operating conditions for these motors can be extremely severe. The limit in terms of peripheral speed is at present 200 m/s, and under these conditions the need for reduced weight and accurate balancing of the rotor is especially critical. Other conditions include frequent start-stop operation, wide temperature variations, operation submerged in liquids or in a radioactive environment, and operation at low pressures.

The air gap is generally very small, possibly down to 0·1 mm, and the rotor surface is smooth to reduce windage losses to a minimum. The cage bars are embedded in closed rotor slots. Special rotors can be made to suit particular drives; solid steel rotors for inexpensive, simple, and robust designs, laminated or sintered permanent-magnet rotors for hysteresis or reluctance type synchronous motors. Solid steel rotors offer reliable operation at ultra-high speeds and may be of a flat disc shape for an axial air-gap motor. The rotor is vacuum impregnated to prevent humidity from impairing the balance.

Because of the high-frequency supply the laminations are very thin. Effective insulation on the laminations is necessary and they are usually glued together, as the use of rivets would considerably increase the iron losses. The stator is often encapsulated in epoxy resin.

The choice of bearings is determined by several factors including shaft stiffness and loading, maximum rotational speed, expected service life, and operating temperature. Grease lubrication is not recommended at speeds above 60 000 rev/min. Special ball-bearings and closed-circuit 'oil-mist' lubrication methods are generally used for speeds much lower than this up to a maximum of about 200 000 rev/min, although at this speed the bearing life may be low. Air bearings may be used for special applications.

High-frequency motors require a high-frequency supply and the extra cost for this extra equipment limits its use to special applications.

10.8 The Steromotor

The conventional method of producing a high torque at low running speeds has usually been the use of a standard high-speed motor together with a reduction gear train. There is, however, one motor, called the Steromotor, which will provide a high-torque low-speed characteristic without the use of a gearbox. The name, Steromotor, comes from the first syllables of the inventors' names. M. C. Rosarn and M. G. Stcherbetcheff, two French engineers, produced the first of these motors at the Paris Mesucora Exhibition in 1961. Full-scale production began in mid-1962.

The special characteristics or advantages of the Steromotor are listed below:

(a) High torque at low speed without gear reduction.
(b) No commutator, slip rings, or brushes.
(c) No high-speed bearings and no lubrication.
(d) Continuous operation possible under stalled condition at rated voltage.
(e) Negligible rotor inertia.
(f) Reversibility of rotation in approximately $\frac{1}{50}$ second.
(g) Rated torque and speed available in $\frac{1}{50}$ second.
(h) Inherent braking torque without energization; there is no overrun.
(i) High positional accuracy in synchronous operation.

(j) Reliability, robust construction, immunity to humidity and to most chemical atmospheres.

However, even though the Steromotor has many useful advantages it has not found widespread use.

10.8.1 *Construction and principle of operation*

Electrically the Steromotor is a conventional a.c. motor, but mechanically it has an unusual design feature in that a permanent magnet rotor is free to roll round inside of the stator bore without the usual restraint of bearings on a fixed axis. There is no metal-to-metal contact of the rotor and the stator as the rotor is fitted with two resilient tyres at each end, which run in two annular tracks in the stator. The epicyclic motion of the rotor is transferred to the output shaft through a flexible linkage and suspension system. This type of motor is also referred to as a hypocycloidal motor, so named because a point on the rotor surface moves along hypocycloidal trajectories.

One type of motor proposed by R. Schön, in 1964, contains a stationary d.c. excitation field wound in the stator, which induces the d.c. field in the rotor. Hence a permanent magnet rotor is not required with this type of motor.

The stator windings are conventional and are completely embedded in a potting compound, making the motor impervious to humid or chemical atmospheres.

Both asynchronous and synchronous versions of the Steromotor are available. The basic asynchronous design uses flat tyres which allow negligible slip between tyre and track. Where synchronous rotation is essential toothed wheels and tracks are used.

The stationary stator coils are arranged to produce a rotating magnetic field, its angular velocity depending upon the supply frequency and the number of stator poles. The specially constructed permanent magnet rotor, having no fixed axis, is allowed to roll freely in the stator bore and to align itself with the instantaneous magnetic flux. Figure 10.12 shows the permanent magnet field and the a.c. rotating field in the Steromotor. From this it can be seen that the rotor is drawn to the bottom of the rotor. As the field moves round, so the rotor follows. The rotor movement results in a slow output shaft rotation, which is predetermined by the constructional parameters of the rotor and stator.

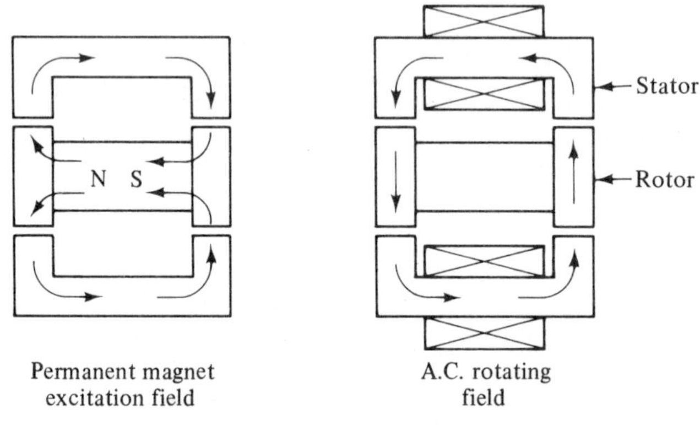

Note. Circuit diagram shows single pair of coils in one plane only.

Fig. 10.12 Magnetic circuit diagram for a Steromotor

10.8.2 *Performance*

Steromotors cover a torque range from about 9 kg cm to 58 kg cm, and speeds are available from 2 rev/min to 200 rev/min for asynchronous types and 20 rev/min, 30 rev/min or 60 rev/min for synchronous types.

Standard Steromotors may be operated from single, two- or three-phase supplies as well as from suitable pulsed d.c. supplies to provide a stepping action. They are usually available with standard voltages of 115 V or 230 V, 50 Hz.

Figure 10.13 shows a set of torque/speed curves for a typical asynchronous Steromotor at various speeds. The crosses on the curves show the maximum power points which occur at about 75 per cent of rated speed. Asynchronous Steromotors are rated for continuous operation at stall or rated speed. They may be reversed at full load at a rate of up to 30 reversals per second, if the load inertia and the control circuit permit.

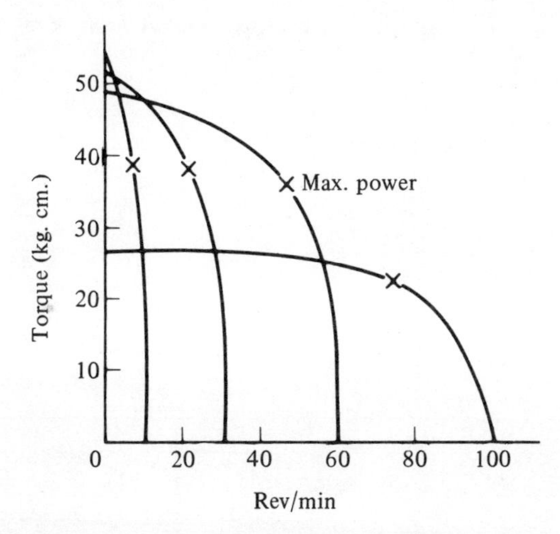

Fig. 10.13 Typical torque/speed curves for Steromotors

Synchronous versions are particularly suitable for step-by-step application, in which pulses are fed in sequence to the stator coils. The full rated torque may be maintained only for low frequency applications; say up to 5 pulses per second.

Steromotors may be used for braking and holding applications or as controlled torque devices. With the stator coils supplied continuously with d.c. the rotor locks in one position, the current value determining the holding torque. They also have inherent braking properties even with the stator de-energized, preventing any overrun.

10.8.3 *Applications*

Applications are diverse, and range from simple high-torque drives through various fast-reversing actuators to step-by-step indexing or positioning devices. The most suitable applications for Steromotors are when some of the motors' special characteristics, listed previously, are required or when size or cost considerations preclude the use of reduction gearing.

Some typical examples of Steromotor applications are: Gas and liquid valve regulation, position control of vent flaps and stopcocks, control of wire and cable reelers, automatic wrapping and cutting machines, cigarette making machines, position control of paper feed drums, nuclear reactor equipment, overload torque devices, and control of tool heads on machine tools.

References

1. *Machine Design*, Electric Motors Reference Issue, 13 April 1972.

2. D. E. Knights, *Variable speed drives for industrial applications*, No. 74-1189, ERA Ltd, Leatherhead, Surrey (1974).

3. F. Kappins and M. Liska, 'Electronic motors for industrial applications', *Siemens Review*, **XXXVIII**, 453–456 (1971).

4. W. Radziwill, 'A highly efficient small brushless d.c. motor', *Philips Tech. Rev.*, **30**, 7–12 (1969).

5. J. Timmerman, 'Two electromagnetic vibrators', *Philips Tech. Rev.*, **33**, 249–259 (1973).

11 Large a.c. Motors

11.1 Introduction

11.1.1 *General*

In an environment in which the purchaser is increasingly finding himself being offered standardized packages, around which he is expected to design his plant, the large electric motor is still (possibly wrongly) a product where manufacturers offer a wide and sometimes bewildering range of alternatives from which to choose. To exercise this choice to the best advantage requires some knowledge of the way motors behave, and more than a superficial knowledge of the problems which confront the motor designer.

The economic penalties which accrue over the lifetime of a plant as a consequence of an indifferent choice of motor, possibly due to an inadequate understanding of the problems involved and the possible solutions available, can be frighteningly high. Once made, fundamentally wrong decisions can seldom be rectified, and whatever palliatives are subsequently adopted by those concerned with the operation of plant, they can seldom do more than mitigate the worst of the effects of such decisions.

In what follows no attempt is, or could be, made to provide facile solutions or simple rules to enable a layman to arrive at 'correct' solutions. Rather an attempt is made to describe the available alternative solutions in as much detail as possible, in a form which a project or applications' engineer can use as a basis for weighing the pros and cons of the various alternatives in the light of the special and often unique circumstances which exist for each and every project. The better informed the decision makers, the better will they be able to select an effective solution.

11.1.2 *Scope*

Large a.c. machines are commonly available in a variety of general types, viz., squirrel-cage or slipring induction motors, salient pole or cylindrical rotor synchronous motors (including both 'conventional' and 'brushless' designs), and variable speed a.c. motors of the stator-fed commutator type.

As 'large' is a purely relative term, it is necessary to define, even if only in broad rather than precise terms, the limitations which can conveniently be ascribed to this term. The two relevant parameters are 'power' and 'physical size'. One thousand kilowatts has been somewhat arbitrarily chosen to represent a convenient bottom end to what would generally be accepted as a 'large motor'. Power output, however, cannot be the only criterion, as it is the output torque rather than the output power which determines the physical size of an electric motor. Hence, the well-known fact that machines required to develop high torques at low speeds, though having relatively low power outputs, can nevertheless be physically very large. There is, therefore, a need also to specify a lower limit for torque, which can conveniently be taken to be about 0·5 kW per rev/min (or 5000 Newton metres).

At the other end of the scale, a cursory examination of large installations reveals that motors are not normally required with outputs which exceed either 20 megawatts, or 30 kW per rev/min.

Using these limits to define 'large' will, of course, effectively exclude many well-defined types and constructions, for instance, single-phase machines, squirrel-cage machines with die-cast aluminium rotors, synchronous-reluctance motors, and rotor-fed a.c. commutator motors, as well as the 'blowover' type of totally-enclosed, fan-cooled motor, all of which are typical of small motor construction and are, for various reasons, seldom built in the sizes being considered herein.

It is, of course, conceivable that motors having ratings approaching those of the present generation of turbo-generators (say, 660 megawatts at 3000 rev/min) could be manufactured, though in fact no demand exists at present for such machines. On the other hand, very large slow-speed motors do have a practical application in connection with pumped storage hydro-electric installations where reversible motor/generator sets are in operation with ratings in excess of 100 megawatts or 200 kW per rev/min. Turbo-electric propulsion of large ships has been another specialized field for very large slow-speed motors.

11.2 General Considerations

11.2.1 *Performance standards*

Very often one of the first duties of a project engineer is to prepare a specification. Done well, many future problems can be avoided at this stage. However, the writing of specifications is an art not easily acquired, as is evidenced by the numbers of manifestly poor documents currently in circulation. Two principal pitfalls exist, the temptation to assume that the requirements of the particular drive in question are so unique that the specification has to be written from scratch, and the almost diametrically opposed view that the writing of the specification should consist of no more than the addition of a few personal idiosyncrasies to a venerable existing document. The first ignores the amount of basic expertise which has been built into national and international standards, while the latter perpetuates the type of document, still all too common, which is replete with the jargon inherited from, and much beloved by, an earlier generation of engineers who, faced with the difficulty of quantifying their requirements, were forced to rely on such generalities as 'low loss', 'non-hygroscopic', 'best quality materials', or 'trouble-free operation'. As the specification will eventually form part of the contract documents, it is essential that its requirements are stated clearly and unambiguously. Its aim should be to specify what is to be achieved rather than how it is to be achieved; it is not the correct vehicle for the propagation of views more appropriate to a code of practice. The specifying of one particular solution to a problem is often a realistic reflection of an individual's or organization's unhappy experience, but progress presupposes a freedom of action which badly written specifications may well inhibit. The protection which the purchaser rightly requires should be obtained by clearly indicating the objectives to be achieved (which are not always as easy to set down as might appear at first sight).

The technique of monitoring a project by keeping a close watch on 'exceptions' ('management by exception') is widely practised. Similarly, the preparation of a specification should start from the assumption that the basic national and international specifications, prepared by experts with widely differing backgrounds, provide the basic framework within which most machines properly fall. Starting from such specifications the project engineer should consider where and why his particular project requires special or non-standard solutions, and only in these areas should his specification differ from that of the general run of machines.

The above should not be assumed to reflect any complacency about the state of national or international standardization. Much is being done and much yet remains to be done in this field. As these standards gain wider acceptance so will work in this area gather momentum, work which will be of growing importance as multi-national and international trading expands.

The specification of performance standards is clearly very closely allied to the requirements of the driven machine, the requirements of safety and reliability, and last, but not least, with the integration of the motor within the overall supply system design. The economics of motor drives are really the basis of any proper cost evaluation, and the correct assessment of costs, both capital and running, is a worth-while exercise in this field. Here, national as well as company and personal preferences play a large part. For instance, European practice pays great attention to running economics, both in terms of kW and kVAr, while American practice is inclined to attach much less weight to the running costs in comparison with the initial capital outlay. This difference in philosophy can be ascribed, at least superficially, to the relative costs of energy. Typical kWh costs in the United States, for instance, are likely to be half those in Western Europe.

Again, the synchronous motor is likely to be the 'norm' for a large motor drive in the United States, while in Europe, other things being equal, it is much more likely to be an induction motor. Technical reasons for these preferences are hard to find. Large slow-speed synchronous motors are certainly cheaper and more efficient than corresponding induction motors, but for high-speed machines the simplicity of the cage rotor construction outweighs all other considerations. In the large area which falls between these extremes, the final choice will be dictated partly by prejudice, and partly by the fact that when differences are marginal, constructors tend to be able to offer their most advantageous terms in the field in which they are manufacturing in the greatest quantities.

11.2.2 *Design considerations*

In view of the internationally recognized R20 centre height series (see table 11.1), it is convenient to use centre heights as a yardstick with which to compare outputs. Figure 11.1 shows a logarithmic plot of the kW per rev/min (i.e., motor torque) against the centre height for a 'typical' range of squirrel-cage motors. The figures are typical only for ventilated or CACW motors (wound for 3·3 or 6·6 kV supply at the lower end, and 6·6 or 11 kV at the upper end). Other voltages and enclosures will modify the actual figures without substantially changing the general picture.

Table 11.1 Recommended dimensions for foot-mounted electrical machines with frame numbers from 355 to 1000

	R20	R20	R20	R40	R20	R40	R20	R40	R20	R40	R20	R40	R20	R40	R20	R40	R20	R40	R20	R40	R20
H	250	280	315	335	355	375	400	425	450	475	500	530	560	600	630	670	710	750	800	850	900
355	Ⓒ**	B C	B C	C	B C	C	B C	C	A B Ⓒ	A C	A Ⓑ C	A C	A Ⓑ C	Ⓐ** C	A B C	A C	A B C	A C	A B C	A	A B C
400	C	Ⓒ	B C	C	B C	C	B C	C	B C	Ⓒ	A B C	A C	A B C	A C	A Ⓑ C	Ⓐ*** C	A B C	A C	A B C	A	A B C
450	C	C	Ⓒ	C	B C	C	B C	C	B C	C	B Ⓒ	C	A B C	A C	A B C	A C	A Ⓑ C	Ⓐ C	A B C	A	A B C
500	C	C	C	Ⓒ	C	C	B C	C	B C	C	B C	Ⓒ	B C	C	A B C	A C	A B C	A C	A Ⓑ C	Ⓐ	A B C
560	C	C	C	C	Ⓒ	C	C	C	B C	C	B C	C	B Ⓒ	C	B C	C	A B C	A C	A B C	A	A Ⓑ C
630	C	C	C	C	C	Ⓒ	C	C	C	C	B C	C	B C	Ⓒ	B C	C	B C	C	A B C	A	A B C
710	C	C	C	C	C	C	Ⓒ	C	C	C	B C	C	B C	C	B Ⓒ	C	B C	C	B C	—	A B C
800	C	C	C	C	C	C	C	Ⓒ	C	C	B C	C	B C	C	B C	Ⓒ	B C	C	B C	—	B C
900	C	C	C	C	C	C	C	C	Ⓒ	C	B C	C	B C	C	B C	C	B Ⓒ	C	B C	—	B C
1000	C	C	C	C	C	C	C	C	C	Ⓒ	B C	C	B C	C	B C	C	B C	Ⓒ	B C	—	B C

R40	R20	R40	R20	R40	R20	R40	R20	R40	R20	R40	R20	R40	R20	R40	R20	R40	R20	R20	R20
950	1000	1060	1120	1180	1250	1320	1400	1500	1600	1700	1800	1900	2000	2120	2240	2360	2500	2800	3150
—	B C	—	B	—	—	—	—	—	—	—	—	—	—	—	—	—	—	—	—
A	A B C	—	B	—	B	—	—	—	—	—	—	—	—	—	—	—	—	—	—
A	A B C	A	A B	—	B	—	B	—	—	—	—	—	—	—	—	—	—	—	—
A	A B C	A	A B	A	A B	—	B	—	B	—	—	—	—	—	—	—	—	—	—
Ⓐ	A B C	A	A B	A	A B	A	A B	—	B	—	B	—	—	—	—	—	—	—	—
A	A Ⓑ C	Ⓐ	A B	A	A B	A	A B	A	A B	—	B	—	B	—	—	—	—	—	—
A	A B C	A	A Ⓑ	Ⓐ	A B	A	A B	A	A B	A	A B	—	B	—	B	—	—	—	—
—	A B C	A	A B	A	A Ⓑ	Ⓐ	A B	A	A B	A	A B	A	A B	—	B	—	B	—	—
—	B C	—	A B	A	A B	A	A Ⓑ	Ⓐ	A B	A	A B	A	A B	A	A B	—	B	B	—
—	B C	—	B	—	A B	A	A B	A	A Ⓑ	Ⓐ	A B	A	A B	A	A B	A	A B	B	B

NOTES: (a) All values in millimetres, based on IEC 72A and BS 3979, and the preferred numbers of the Renard series, ISO R3 and R17; (b) Ⓐ values: BS 3979 recommendation; (c) Ⓑ values: BS 3979 median value; (d) IEC 72A lists additional C values: 0, 100, 200 and 224, Ⓒ values: BS 3979 recommendations, lower range for rolling bearing, higher range for sliding bearings; (e) To line up with IEC 72, BS 3979 and IEC 72A, list A** as 610, A*** as 686 and C** as 254.

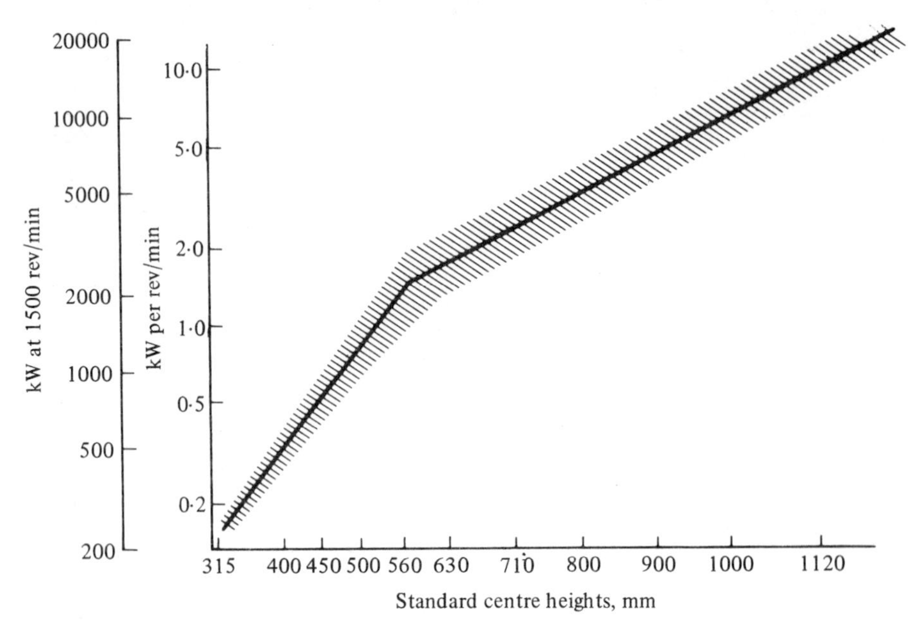

Fig. 11.1 Typical ratings (kW per rev/min) for standard squirrel-cage induction motors 315–1120 mm centre height

Specific outputs, though often quoted in textbooks, are only meaningful for a particular construction at specified voltages, speeds, and enclosures. However, other things being equal, the output of a motor is proportional to the product of its core length and the cube of its core diameter (i.e., the product of its specific electrical and magnetic loadings). The closer a range of designs approaches this figure, the better will be the utilization of the active material (i.e., core and winding). Figure 11.1 also illustrates how mechanical constraints affect the outputs obtainable from motors. For instance, doubling the diameter from a 315 mm to a 630 mm centre height machine gives a 10:1 increase in output (i.e., somewhat more than as the cube of the diameter) though a similar doubling from the 630 to the 1250 frame results in an increase of about half this figure. This reflects varying limitations to the length-to-diameter ratio of machines. Mechanical considerations such as shaft stiffness and thermal considerations in connection with the cooling of long core lengths, have the effect of significantly reducing the diameter to core-length ratio of large machines in comparison with corresponding figures for the smaller sizes.

Figure 11.2 is a similar presentation, this time of the variation of efficiency against motor size; corresponding figures for the losses of machines, illustrated in Fig. 11.3, highlight the progressive increase in efficiency with size, as well as emphasizing at the high and low limits of the curve the effect of low and high poleage machines respectively.

In the field of thermo- and aerodynamic[73,74] design much work has been carried out in recent years with a view to improving the cooling of machines. The magnitude of the problem is perhaps most clearly illustrated by the fact that as yet no single solution has emerged, though this might well be less surprising than appears at first sight when one considers the complexity of the overall design problems.

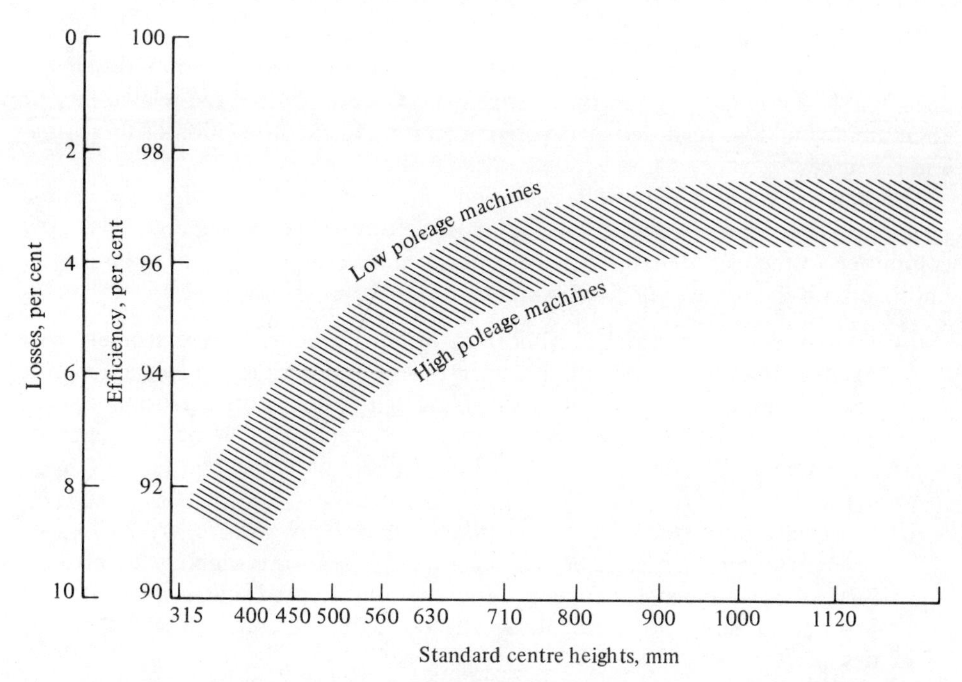

Fig. 11.2 Typical efficiency levels for standard squirrel-cage induction motors 315–1120 mm centre height

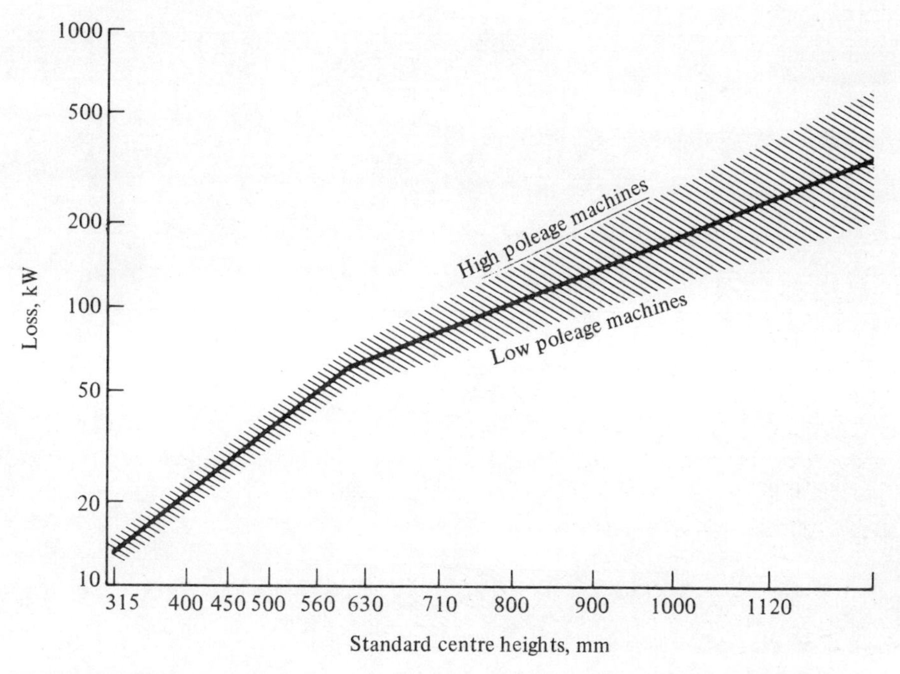

Fig. 11.3 Typical loss levels for standard squirrel-cage induction motors 315–1120 mm centre height

The general availability of computers has helped tremendously in the speeding up of design investigations, and while this in itself will not produce any 'better' designs, it does enable designers to undertake investigations which before the availability of computers would be regarded as occupying an unrealistic proportion of their time and resources.

The emphasis which is nowadays placed on aerodynamic and thermodynamic problems is very largely due to the fact that the designer of a large motor is now committed to the use of materials whose properties are unlikely to be significantly improved. Of the three major components:

(a) The conductor material is almost invariably copper,[45] an extremely bad construction material which, if its electrical properties are to be retained, can be only marginally improved by alloying. High-temperature performance can, therefore, only be obtained by sacrificing conductivity. The only practicable alternative, aluminium, has made little impact as yet in the field of large motors.

(b) Electrical sheet steels have not changed materially with regard to loss or magnetization characteristics in recent years, and are unlikely to provide further easy means of making progress. Grain-oriented silicon steels are, for practical reasons, not particularly attractive for the common run of motor designs.

Fig. 11.4 16·5 MW cage induction motor for driving a compressor. (Courtesy of GEC Machines Ltd)

(c) There have been major improvements in insulation over the last 25 years, to such an extent that it has now overtaken the properties of the conductors which it covers in terms of stability of performance, particularly constructional, at elevated temperatures. Insulating materials are now available whose thermal properties are hard to exploit to the full without incurring major problems elsewhere in the machine.

In spite of the limitations enumerated above, significant steps forward are still being made, not so much by the employment of spectacular new concepts, but rather by the painstaking application of known techniques which in the hands of expert and experienced designers has ensured steady rather than spectacular progress. An indication of the magnitude of the progress can be gained from a consideration of the 16·5 MW induction motor illustrated in Fig. 11.4 which is a typical example of the fruits of this form of progress. Compared with a similar machine manufactured only five years earlier, it is some 30 per cent lighter as well as offering a significantly better performance.

Bearing designs, both of the rolling and sliding type,[85] have increased their scope and reliability in recent years, and a real understanding has been obtained of the vibration problems of rotating systems. In particular, the stiffness and damping coefficients of sliding bearings can now be used to compute, with some confidence, the vibration characteristics of a rotor subject to both mechanical and electrical unbalance forces,[123] and this has significant impact on the behaviour of machines in their working environment where, for instance, the foundations, or other supporting structure may well be far from ideal.[28,86]

11.3 High-voltage Stator Windings

11.3.1 *Background*

Machines of the sizes being considered will require primary windings suitable for operation from high-voltage supplies, normally either 3·3 kV or 11 kV in the UK, and 6 kV abroad.

Insulation technology is a vast, complex, and rapidly changing field,[29] and the following paragraphs can do no more than present a bird's eye view of the present situation. Large motors will inevitably be connected to modern high fault level, high operational security supply systems, and these may in their turn impose very onerous conditions on the primary windings of motors connected thereto. These windings will therefore need to be designed for maximum reliability from many different points of view, dielectric, electro-mechanical, and environmental. All these requirements are interrelated, generally very difficult to measure or quantify and hence to specify precisely, and, generally speaking, not very well understood (as is exemplified by the complex state of specifications, or the lack of same, dealing with many aspects of winding insulation).

11.3.2 *Dielectric requirements*

The dielectric requirements for a winding arise directly from the steady state and transient voltage levels normally encountered in modern supply systems. Table 11.2

Table 11.2 Summary of stator-winding insulation design and test levels

1 System	2	3 Earth insulation	4	5
Nominal voltage 50 Hz	Surge-voltage level	1 min routine test voltage 50 Hz <10 MVA (≥10 MVA)	Breakdown-type test voltage 50 Hz	Impulse-withstand type test 175/2500 μs
kV r.m.s.	kV peak	kV r.m.s.	kV r.m.s. (peak)	kV peak
3·3	18	7·6 (−)	13·2 (18·7)	18
6·6	31	14·2 (16·2)	28·4 (40·0)	31
11·0	49	23·0 (25·0)	46·0 (65·0)	49
Derivation	4V+5	2V+1 (2V+3)	2(2V+1)	4V+5
Reference	29	13–15†	15	29
	32	22		
		25–27		

† Reference 15 specifies 2V+1 above 10 MVA

6	7	8 Interturn insulation	9	
			High-frequency withstand routine coil test 80–100 kHz for 10 s	
1 min type test voltage 50 Hz	Breakdown-type test voltage 50 Hz	Impulse-withstand type test 0·25/0·8 μs	before final processing	after final processing
kV r.m.s.	kV r.m.s.(peak)	kV peak	kV	peak
2·1	4·2 (6·0)	4·5	6	11
3·2	6·4 (9·1)	7·8	12	18
4·7	9·4 (13·3)	12·3	19	27
V/3+1	2(V/3+1)	(4V+5)/4		
29	29	29	32	32
27				
15*				

* Reference 15 specifies 0·3 V

summarizes the requirements to be found in specifications related directly or indirectly to machine windings.[29] In attempting to understand the implications, it is important to realize that the impulse wave fronts specified, which are relevant specifically to test procedures, are nevertheless chosen to provide realistically onerous tests. Moreover, although these are working overvoltage levels with an unspecifiable recurrence frequency, the test levels adopted take into account practical field experience. Similarly, it is essential to obtain a clear grasp of the essential difference between insulation to ground (earth) and interturn insulation. Insulation to ground is traditionally, and in practice, effectively monitored by the conventional high-voltage test procedures.[37] Interturn insulation is difficult to check in practice, although the test levels in table 11.2 are increasingly being accepted as realistic in the light of experience.

For the quality control of windings at the higher voltages, monitoring of the ground insulation by quantitative, non-destructive testing is common. Normally, this is done by measuring the loss angle of the winding insulation by means of a Schering bridge,[15] although more recently various methods of measuring discharges, for instance by the use of a dielectric loss analyser,[24] have been introduced. The results of such tests on completed windings still require interpretation by skilled observers, particularly if an attempt is being made to use such tests as a device for monitoring and detecting long term trends in the deterioration of insulation. To a large extent, the problem revolves around the fact that all such measurements are bulk measurements or, expressed in another way, represent the sum of a large number of discharges. There is no way of differentiating whether a particular reading consists of a very large number of very small discharge currents, or, alternatively, of a relatively few heavy discharges. In the one case the windings may be considered to be in good condition, in the other the winding may well be at significant risk.

In any case, these measurements relate only to the insulation between the winding and earth, and say nothing about the state of the interturn insulation. Operational stresses on interturn insulation lie in the range of 10 to 100 volts, but a voltage surge on the supply system may well expose the interturn insulation to peak voltages of between 5 and 20 kV. Conventional surge testers have neither the capacity nor the rate of rise of voltage to produce stresses of this order in a stator winding. While it would be wrong to exaggerate the importance of surge voltages, nevertheless, the magnitude and frequency of switching and other surges which large motor windings must withstand, is increasing rather than diminishing with the increase in size and capacity of supply systems, and the development of modern switching devices such as vacuum contactors. Windings designed to withstand the test procedures indicated in table 11.2 have nevertheless proved exceptionally reliable in operation. As with all statistics, interpretation is difficult, but evidence from an analysis of breakdowns generally shows that the incidence of winding faults due to unknown causes, i.e., after the elimination of manufacturing errors, accidents, and indirect damage, is now very small. Perhaps equally significant is the fact that windings, which for various reasons are not or cannot be designed on these principles, do seem to show relatively high failure rates due to unknown causes.

11.3.3 *Mechanical requirements*

The fact that large machines will be connected to supply systems capable of providing high fault currents has a direct influence on winding design. During the short period that the fault current flows prior to the operation of the protective system the windings must be mechanically stable, i.e., should not contribute further to the fault damage by failing themselves. It is an important objective that the design of a motor should utilize the full capacity of the supply system for direct-on-line starting, reswitching between alternative supplies, and safety under 'normal' fault as well as some abnormal fault conditions.

11.3.4 *Thermal requirements*

The mechanical usable temperature of copper lies below about 130°C to 160°C. At these temperatures serious mechanical (particularly creep) and thermal (expansion) problems arise, and windings designed to operate at these temperatures require special consideration.

11.3.5 *Execution*

It is now substantially universal practice to utilize epoxy/mica paper insulation systems,[31] which have largely replaced the traditional 'black bond' and shellac-bonded micafolium systems except at the bottom end of the 3·3 kV range. This is not to say that these older systems have not given and are not capable of giving excellent service, but the advantages of the newer systems have led to their adoption for economic reasons. The basis of these insulations is glass-backed mica paper, bonded with an epoxy resin, the mica being in the form of very fine splittings and the resin in the 'B' stage. The requisite number of layers of insulation are used on the conductor to provide the required interturn dielectric properties for the coil which, after shaping, is taped to provide the chosen wall thickness for the ground insulation, and then pressed to size. Pressing to size ensures consolidation to minimize the amount of air inclusions in the coil.

The overhang portions of the coils tend to be dealt with in different ways to retain the necessary flexibility required by the winder. It may, for instance, be taped with the same material as used for the slot portion, but as it will not be heated during the pressing operation it will remain sufficiently flexible for the coils to be inserted into the slots, and will then be consolidated only after having been fully processed in the machine. Alternatively, the overhang may be taped with various types of 'flexible' tapes generally containing mica and often based on silicone elastomers.

With a wide variety of satisfactory insulating systems of comparable technical merit from which to choose, it is perhaps not surprising that individual constructors will have come to conflicting conclusions as to the technical and economic merits of one or the other system. Considerations which strongly influence the choice of system for an 11 kV winding may well be of much less significance at 3·3 kV, and continental manufacturers, whose main output is likely to be confined to 6·0 kV, may arrive at different solutions to UK-based companies whose output will be divided between 3·3 and 11 kV. As an example, the reduced space requirements of an epoxy/mica paper system at 6·0 kV and above, represents a significant economic gain; at 3·3 kV where the minimum insulation thickness is dictated much more by mechanical considerations, micafolium systems are still, and probably will remain, relatively common. Similarly, interturn insulation at this level does not require the use of mica, and synthetic conductor coverings are used with complete success.

A further indication of how the choice of winding system is influenced by inter-related requirements, is revealed by only a cursory look at the differences between epoxy/mica and shellac-bonded micafolium systems at 3·3 kV. The latter are significantly cheaper to manufacture (provided the requisite relatively high quality mica splittings continue to be available), but this advantage may be counter-balanced by the higher thermal classification of the epoxy/mica system (Class F rather than Class B), thus permitting more output to be obtained from the same active materials by running the windings at higher temperatures. Special environ-mental conditions may well also have an influence; for instance, the epoxy/mica system lends itself to the production of a 'sealed winding' with a substantially homogeneous dielectric which will comply with the American (NEMA) specification[22] for sealed windings intended for use in damp or other difficult environmental conditions.

11.3.6 *Cable and terminal boxes*

The effects of high surge voltages and high prospective fault current capabilities (see table 11.3) have focused attention on the means of connection between the motor and its supply cables. The mechanical stability required by windings must be matched

Table 11.3 Typical system parameters

System voltage	Surge-voltage level	Short-circuit fault level		Induction-motor load-stability limit (approximate)
kV r.m.s.	kV peak	MVA	kA r.m.s.	MW
3·3	18	57	10	2·5
		150	26	7
		250	44	11
6·6	31	115	10	5
		350	31	16
		500	44	22
11·0	49	190	10	9
		500	26	22
		750	39	35

by the connecting cables, as must the dielectric security of the connections themselves,[36,38] i.e., reliability coupled with a mechanical robustness sufficient to eliminate explosion risks as a consequence of a short-circuit fault within the box. Traditional designs employ post-type terminals[37] and incorporate pressure relief vents to deal with the explosion risk. Alternative connector-type, phase-segregated terminal boxes are available, where special attention is paid to sealing and desiccation of the enclosure, and where bare live conductors can be eliminated by fully taping all such parts. Such terminal boxes can be relied on to operate as 'pressure contained' boxes with fault currents up to about 10 kA, and as 'pressure relief' boxes up to the maximum short-circuit currents likely to be encountered.

There are also a number of designs of boxes available for use in hazardous (Division I) areas,[39] using phase separation to limit the prospective fault to a phase to earth fault.

The problem of containing short-circuit faults is, of course, eased where fuses are available, by making use of the inherent limitation of the fault energy input which is available with fuses. Such techniques are particularly favourable in hazardous areas where fused supplies enable flameproofness[38] to be achieved even in the worst fault conditions.

11.4 Squirrel-cage Induction Motors

11.4.1 *General*

In spite of the apparent simplicity of the squirrel-cage motor (Fig. 11.5) it is at least as difficult to design well as any other type of machine. Since its inception in 1889, it has frequently been said that no further development is likely except in detail, although

Fig. 11.5 Squirrel-cage motor for a power station soot blower compressor drive (3120 kW, 333 rev/min, 11 kV)

nothing can in fact be further from the truth, either in general or in relation to large motors.[43] To appreciate the role which this motor plays in modern industry it is necessary to concentrate attention on the salient points of its real advantages (and its very real limitations):

(a) It is simple, provided it is used as a single-speed direct-on-line start motor.
(b) It is highly reliable and robust, provided it is correctly designed for known operating conditions.
(c) It requires no sliding contacts or equivalent semiconductor devices for its normal operation as described in (a).

Incorrect application may easily attenuate these advantages to such an extent that they become substantial disadvantages. To use such machines to solve excessively difficult starting conditions, or to emulate variable speed operation is a temptation which should be avoided, or at least looked at with a critical eye.

11.4.2 *Rotor design*

The design of a cage motor is only as good as that of its cage.[44,45] While every possible shape of squirrel-cage bar (see Fig. 11.6) from plain rectangular to complex

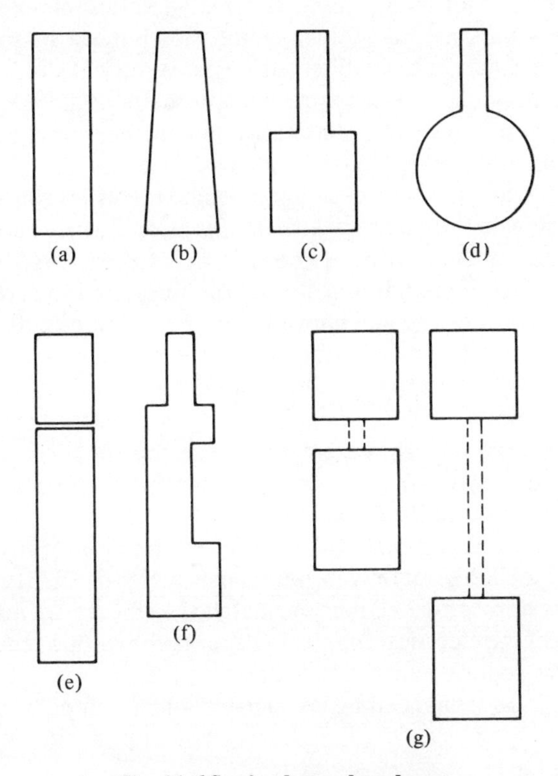

Fig. 11.6 Squirrel-cage bar shapes

machined or extruded sections seem to have been used there are in fact only a small number of logical standard solutions for given ranges of machines. Hence, though the precise shape of the bar is not a matter of principle, it does to a large extent determine the problems associated with the permanent fit of the bar in the slot, and also has a major influence on the magnetic circuit due to saturation effects in the narrow parts of the rotor tooth.

Of paramount importance is the method of applying and jointing the bars to the short-circuiting endrings. The design of the bar/ring structure must take fully into account the properties of the materials, the thermal and centrifugal stresses, and the degradation of the material during the jointing process. For these reasons, plain copper is not generally acceptable for machines of any size, and is usually replaced by copper alloys.[45] Where mechanical forces are such that the ring can no longer be expected to be self-supporting, non-magnetic high-tensile alloy steel shrink rings are used to provide the necessary support while still allowing free axial and radial movement. For the control of starting current larger machines utilize sash bars (Fig. 11.6(c) or (d)) in preference to plain rectangular sections. For more onerous starting conditions, double cage machines provide significant advantages. In particular, the use of high resistance cupro-nickel starting cages enables a high thermal capacity to be built into the starting cage, while still permitting the use of a copper running cage for low slip, and therefore low loss running.

Considerable controversy persists regarding methods of mechanically fixing the bars within the rotor slots, and many alternative solutions exist. For instance, wedging with steel bars at the bottom of the slot is used in some areas, while high-temperature insulation techniques, using conventional slot liners and wedges reinforced with impregnation, is a contrasting alternative. The use of idle bars (Fig. 11.6(e)) at the bottom of the slot is illustrative of a further construction intended to alleviate the connection problem at the endrings.

In all cases the starting problem of a cage machine revolves around the fact that an amount of heat equivalent to the total kinetic energy of the drive is generated within the cage winding during the starting period. The heat so generated is far greater than that which has to be dissipated during normal running, and the corresponding rapid rise of temperature of the cage and endring during the comparatively short starting period has to be absorbed by the winding without exceeding relatively low temperatures, say, 150–200°C for copper, or 300–400°C for copper alloys such as cupro-nickel or chrome copper.

The shape of the rotor bar has a great influence on the electrical performance, though this must clearly take second place to constructional problems. All large cage motors will have rotors designed to exhibit substantial current displacement effects in order to improve the starting performance. The starting performance of the motor (and this of course includes not only its performance at the instant of breakaway but also throughout its run-up period) must be matched to the driven machine, as well as to the capabilities of the electrical supply. In general, the torque requirements will be determined by the requirements of the driven machine, while limitations to the starting current will be influenced by the capacity of the supply system.

11.4.3 *Various performance parameters*

The overall cost of a motor depends critically on the interaction between the heat to be dissipated (losses) and the aerodynamic and thermodynamic design which is intended to dissipate these losses without incurring an excessive temperature rise. Secondary phenomena have a vital effect; noise, vibration, and torque irregularities are associated with flux harmonics generated and influenced by the winding disposition and slotting of the stator and rotor. As is usual, many of the design requirements are mutually incompatible and design compromises are essential. Inherent noise levels must be limited to acceptable figures, and torque requirements at reduced speeds and voltages play a vital part in determining the final design of machine. Reduction of stray losses[48–51] are greatly eased by the use of insulated cages, where they are structurally permissible, by allowing a greater freedom in the choice of slotting, particularly in permitting the use of high numbers of rotor slots compared with those in the stator.

The choice of slot numbers, both absolutely and relatively, is one of cost against performance. Magnetic noise is particularly critical with respect to the correct choice of slot ratios,[70–72] though slot numbers in themselves are no guarantee of freedom from trouble in this respect. Large airgaps are generally helpful in reducing harmonic levels and increasing mechanical stability (for instance in relation to unbalanced magnetic pull).

11.4.4 *Multi-speed motors*

The traditional doubly-wound stator, though still retaining a field of application for a few limited speed ratios, has now been generally replaced by machines with pole amplitude modulated (PAM) tapped windings[41,42] (Fig. 11.7). These windings have significantly extended the scope of the well-known Dahlander winding which is limited to a 2:1 speed ratio. PAM windings are available for a comprehensive,

Fig. 11.7 PAM squirrel-cage motor for an induced draught power station drive fan (2380/1720 kW, 1493/750 rev/min, 6 kV)

though not quite complete, range of speed ratios. The PAM two-speed machine is some 10 per cent to 15 per cent larger than its single-speed counterpart and requires an additional multi-pole reconnection switch (which is usually designed for off-load operation). More than two speeds can be obtained, though only at the cost of considerable additional complication. The additional restraints imposed on the designer by the requirements of two-speed operation clearly involve more substantial compromises than are necessary for single-speed designs but, nevertheless, multi-speed machines have an important place in the economic application of squirrel-cage motors in certain fields.

11.5 Slipring Motors

11.5.1 *General*

Though historically the slipring motor was the precursor of the squirrel-cage machine, it is with some reason regarded as the next step in sophistication of the squirrel-cage motor. The fact that the secondary winding is accessible through the sliprings means that a much greater degree of control can be exercised over the performance of the machine by external means, generally but not always in the form of resistance. The slipring motor is generally applied in those cases where:

(a) An improved starting performance is required, using external resistances to increase the torque/current ratio at start.
(b) To permit repeated and very difficult starts to be carried out, the bulk heat generated at start being external to the motor in resistance banks, and therefore easier to dissipate.
(c) To obtain a limited amount of speed control, at the expense of additional losses dissipated externally to the machine in external resistances.
(d) To permit the injection of a secondary e.m.f. to obtain 'loss free' speed control, the e.m.f. being obtained from rotating or static frequency changers.

11.5.2 *Rotor winding design*

Rotor windings are similar in concept to the primary windings of motors, except that voltages are generally kept down to no more than 2 kV or 3 kV because of the problems of external apparatus and indeed of the collectors themselves. The dielectric requirements of the windings are generally mechanically biased in the sense that preferred methods of winding and insulating the conductors are those which permit the introduction of the conductors singly into semi-closed slots, thus permitting simple wedging procedures to be used. The overhang portion of the windings is supported on winding drums, and modern glass banding techniques make all but the very largest and highest speed machines possible without the necessity of using shrink rings.

The design of the sliprings themselves is not particularly difficult, bronze, cupro-nickel, and occasionally steel being used for the ring material, with helical grooves frequently being employed to improve the reliability of current collection. To avoid the additional problems of brush dust, particularly where metal–graphite brushes are involved, sliprings are often accommodated in separate slipring enclosures which may well be separately cooled. While there are many advantages associated with the use of electro-graphitic brushes, metal–graphite brushes have frequently to be used for thermal reasons in view of their higher current-carrying capabilities and lower losses. Modern brushgear tends to use constant force springs, and constructors are beginning to pay more than lip service to the adoption of national and international standardization of brush dimensions.

11.5.3 *Resistance control of secondary voltage*

The resistance control of slipring machines is appropriate when the combination of supply system limitations with those of a squirrel-cage motor ceases to be viable for economic or technical reasons. Furthermore, whenever the requirements of the driven machine are such that frequent high torque starting is necessary, the slipring machine is ideal and may be used with a permanent slip resistance to improve the overall stability. Moreover, where a drive is intermittent the losses incurred at reduced speed, which are proportional to slip and load, are not of great significance. On the other hand, where continuous low-speed operation is likely, these losses may well prove prohibitively expensive—not only must the energy be paid for, it must also be dissipated.

It should also be noted that the steepness of the slipring motor speed/torque curves, all of which pass through synchronous speed, may be disadvantageous, for instance, speed control at very light loads is hardly possible. In this respect the slipring motor is completely equivalent to a slip coupling, eddy-current or fluid, and there are certainly occasions where it can be argued that an electromechanical combination, i.e., a direct-on-line started squirrel-cage motor used in conjunction with a fluid coupling, provides a better compromise solution. This is another area where American and European practice differ, in that it is common in the United States to rely very heavily on slip coupling drives, whereas European practice, being much more loss conscious, tends to adopt different solutions.

11.5.4 *Cascade connections for speed control*

Putting aside the possibility of multi-speed motors using PAM techniques, control of the secondary voltage of slipring motors has long been possible with various cascade connections. The importance of these solutions lies in the fields where the application of commutator motors (either a.c. or d.c.) is impossible because of the inherent limitations in commutator design. Modern modifications of the classical Kramer Cascade are illustrated in Figs. 11.8 and 11.9, using static (semiconductor) techniques. Substantially loss-free regulation, i.e., only incurring losses due to the constituent parts, is obtained by varying the voltage injected into the rotor, which is achieved via rectifiers and smoothing reactors, using a conventional a.c./d.c. convertor.

As the conversion equipment has a high capital cost, it is economically necessary to limit its output to the required continuous speed range; starting is therefore accomplished by means of conventional resistance starters. Such static slip energy recovery systems, though having advantages over the conventional Kramer or Scherbius systems in which the slip power is wholly handled with rotating machines, nevertheless differ from them in one important aspect, the generation of harmonics.

Static frequency convertors, in common with all systems using semiconductor devices such as thyristors, generate harmonics whose effects are manifest in the motor in the form of additional losses and heat. They are also fed back into the supply system, and may, as a consequence, give rise to undesirable side effects on other apparatus connected to the same supply system, particularly capacitors (resonant effects at the harmonic frequency) and other thyristor-controlled devices (spurious firing of thyristors).

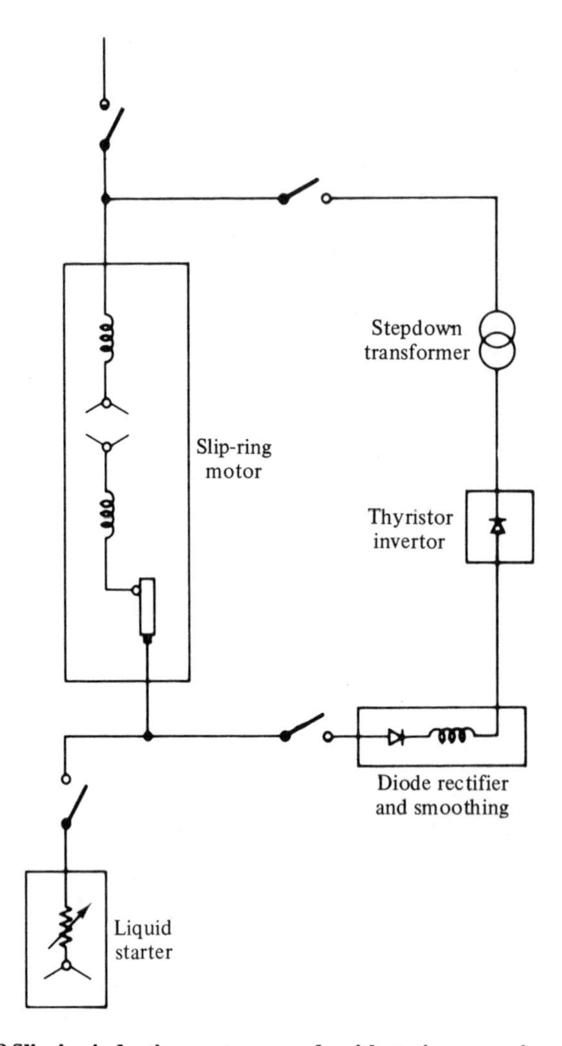

Fig. 11.8 Slipring induction motor cascade with static conversion equipment

The slipring cascade has a mechanical equivalent, the variable speed epicyclic gear, normally used in conjunction with a variable speed hydraulic motor to inject a slip speed into one of the epicyclic gear systems. Some use has been made of this arrangement for applications similar to those where the slipring cascade is viable on its own account, i.e., large and/or high-speed drives, particularly where the driven

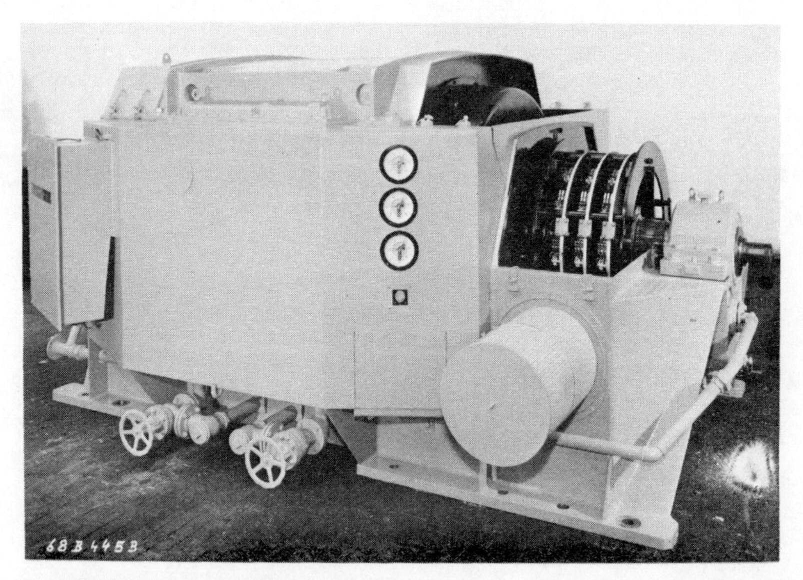

Fig. 11.9 Slipring motor cascade set for a power station boiler feed pump drive (8200/1100 kW, 1493/750 rev/min, 6 kV)

machine, for instance a compressor, can be economically designed for speeds above that of a two-pole motor.

11.6 Synchronous Motors

11.6.1 *General*

Because small synchronous motors are more expensive than corresponding induction motors, their application is limited to the few drives where their special characteristics can justify the use of the more expensive machine. However, in the larger sizes the cost differential reverses; hence the high proportion of synchronous machines amongst the very largest sizes of motors (Fig. 11.10). Designed to operate at unity or slightly leading power factors, such synchronous machines have rather higher efficiencies than those of corresponding induction motors, as well as being somewhat smaller and lighter. These differences are particularly pronounced on low-speed machines such as are commonly required for the drive of large reciprocating compressors.

All large synchronous motors require a source of direct curent for the exciting windings, which may be obtained from a d.c. exciter, a static excitation system, or, more commonly nowadays, from a 'brushless' excitation system mounted on, and revolving with the motor rotor (Fig. 11.11).

A simple synchronous motor develops no significant starting or running-up torque, and, as a consequence, steps must be taken to accelerate the machine up to substantially synchronous speed by other means so that the machine can synchronize. Pony motors are occasionally used for this purpose, but, more usually, the motor relies on accelerating torques developed by slip frequency currents induced in pole face windings or in the pole itself. Differing methods of obtaining the required

Fig. 11.10 Steelwork centrifugal compressor drive utilizing a conventional salient-pole synchronous motor (8600 kW, 1000 rev/ min, 11 kV)

breakaway and run-up torques are reflected in variations in the basic rotor construction of synchronous motors.[57,60,61]

11.6.2 *Salient pole rotors*

11.6.2.1 Solid pole construction. Most large high-speed synchronous motors are of the solid pole type, the solid pole shoes, pole body, and pole hub or shaft providing paths for the eddy currents induced therein during the starting period which develop the starting and run-up torques.

The thermal capacity of the solid poles makes this type of rotor construction particularly suitable for starting and accelerating high inertia loads. Since the electrical and magnetic properties of the pole shoe materials cannot be varied to any substantial extent, the inherent starting characteristics of this type of motor (say, 100 per cent full-load torque at breakaway when taking around 5 times full-load current) cannot be varied to any great extent either, so that matching the machine and drive starting characteristics is not always easy. During run-up the exciting winding is normally closed through a starting resistor (which, in the case of a brushless machine, has to be carried on the rotor). This resistor serves the dual purpose of limiting the slip frequency voltage appearing across the exciting windings, as well as assisting the development of additional run-up torque, particularly close to synchronous speed.

Fig. 11.11 Salient-pole rotor for a brushless synchronous motor (3540 kW, 18 poles)

The centrifugal forces on the exciting winding and poles of these machines are likely to be high; where possible the solid pole tips are bolted to the pole body, but where the forces are such as to prohibit this construction, the pole, complete with pole tip, is dove-tailed into the pole hub or shaft.

Field windings for these machines are usually of the strip-on-edge type, each turn being built up from rectangular copper strips welded or brazed together at the corners. Cooling is assisted by the introduction of wider strips at regular intervals to obtain a finned coil profile. The field winding coils require proper support to counteract the effects of high centrifugal forces, particularly on the long core-length machines currently being manufactured, and some form of non-magnetic interpolar coil support wedging between coil sides is almost always necessary.

11.6.2.2 Laminated pole construction. On slow-speed motors or where the starting characteristic of the solid pole machine does not adequately match that required by the driven machine, a cage-type winding is used for starting.

To accommodate such a winding, the poles are built up from punchings in which suitable slots for the starting cage can be easily arranged. The poles are either bolted (slow-speed machines) or dove-tailed (high-speed machines) to the pole hub. The starting cage (which also acts as a damping winding when the motor is running synchronously) may well be fairly rudimentary, in that only the bars within each pole face are shorted together. Where starting requirements are particularly onerous, the pole face may have to be deepened to accommodate a three-phase distributed starting winding brought out to sliprings so that external starting resistors can be employed (and a starting performance, comparable to that of a slipring induction motor obtained).

11.6.3 *Synchronous induction motors*

Although by far the majority of large synchronous motors have salient pole rotors, a few large synchronous induction motors are manufactured from time to time. These machines have laminated cylindrical rotors and distributed field windings similar to those of slipring induction motors. The ability of these machines to operate satisfactorily as either synchronous or slipring motors has attractions, and has led to their adoption for certain specialist drives, for instance where the majority of running is required at top speed (where the advantage of synchronous running is the high power factor of the synchronous motor), but which must be combined with shorter periods of reduced speed running (using the machine as a slipring motor with external rotor resistances).

11.6.4 *Excitation systems*

The starting and synchronizing of synchronous motors with d.c. exciters required nothing more elaborate in the way of starting equipment than a changeover switch (to connect the exciter to the field for synchronous running and to isolate it from the slip frequency voltages induced in the field winding while the motor was running up to speed), and a resistor connected across the field winding to limit the magnitude of the voltage which appears across the field winding terminals to the volt drop across the resistor. Static excitation systems (which are rapidly superseding the d.c. exciter) provide the required d.c. excitation by static rectification from the a.c. supply system. These static systems, in common with the 'brushless' system (where the sliprings are eliminated by the use of an a.c. exciter, together with diodes carried on the motor shaft), introduce additional problems during synchronization as, in contrast to the

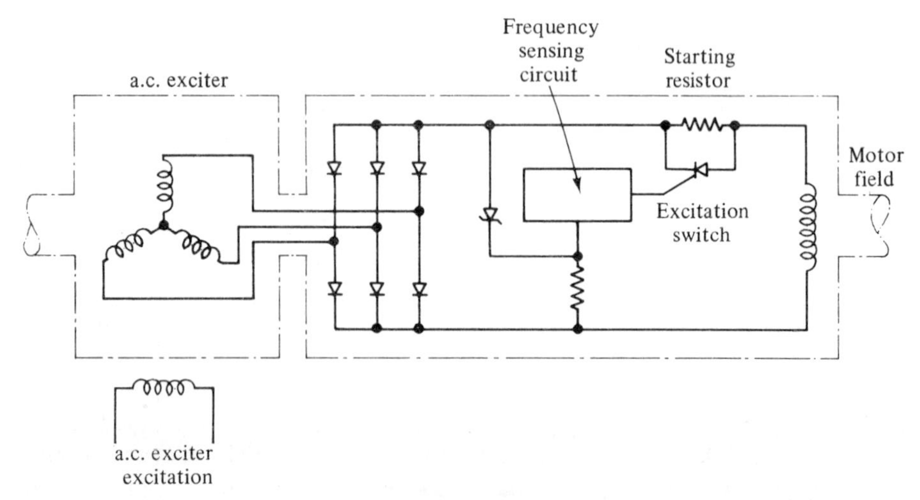

Fig. 11.12 Typical excitation system for brushless synchronous motor

d.c. exciter, the diodes allow slip frequency currents to flow only in the positive direction, thus introducing a superimposed pulsating torque component[58] during run-up.

The desire to eliminate sliding contacts completely has led to the development and widespread adoption of 'brushless' designs in which an alternating-current exciter mounted on the motor shaft feeds the excitation winding via shaft-mounted diodes. A starting resistor connected across the motor excitation winding is still necessary to limit the slip frequency voltages applied to the diodes during the run-up period. Additionally, semiconductor switching devices mounted on the rotor may be used to protect the diodes during the run-up, and to disconnect the starting resistor once the machine has synchronized. Most constructors have their own patented circuits for carrying out these switching procedures. Figure 11.12 is typical of one such basic system.

11.7 A.C. Commutator Motors

11.7.1 *General*

The special suitability of thyristor-fed d.c. motors for wide speed range drives, and 'Modified Kramer' Cascade Slip Recovery systems for very large high-speed

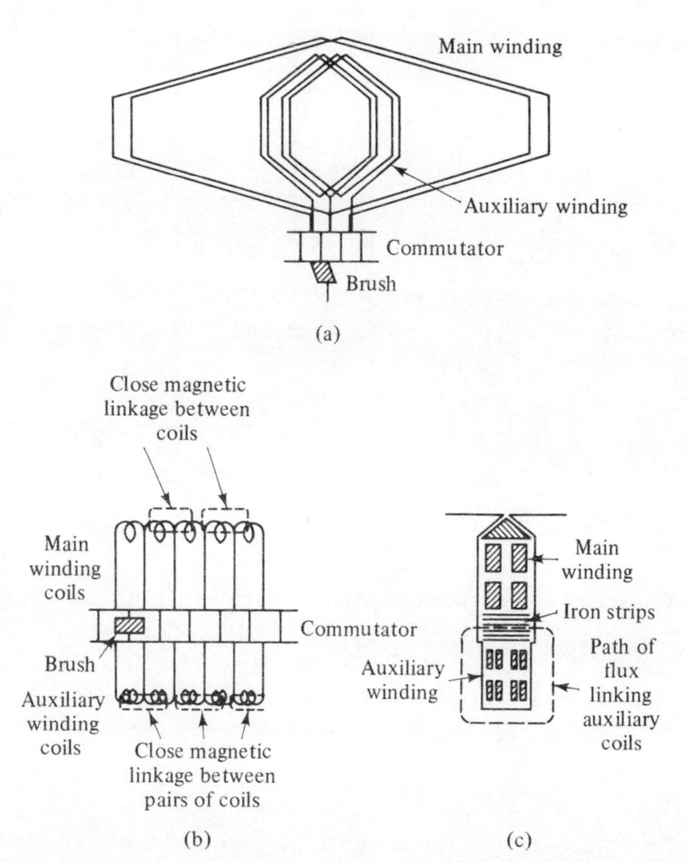

Fig. 11.13 Stator-fed a.c. commutator motor commutating winding arrangement

machines, indicates the limits of the modern field of application of variable speed a.c. commutator motors which, at least in the United Kingdom, are the normal solution for variable-speed drives requiring narrow to medium speed ranges (say, $1\frac{1}{4}:1$ up to $3:1$).

The voltage levels appropriate for large machines, i.e., 3·3 kV, 6 kV, and 11 kV, are such that the stator-fed rather than the rotor-fed version of the a.c. commutator motor is almost exclusively used, generally with induction regulator[52] control. Effective auxiliary commutating windings (Fig. 11.13) have been an integral part of standard practice for many years; successful upward extension of outputs has been dependent on the use of multiplex rotor windings. Duplex windings on d.c. machines have not been without their problems, but on a.c. commutator motors triplex, quadruplex, and quintuplex windings are commonplace on large machines.[53] Success with such complex windings is critically dependent on many factors, perhaps the most important of which is to ensure proper equalization between the various windings. Output limitations for these motors are similar to those of d.c. motors, limitations associated with the mechanical design of the commutator itself generally setting the limits rather than problems of commutation. Examples of large high-speed a.c. commutator motors in service are 1550/900 kW at 1025/875 rev/min, typical of a high-speed waterworks pump drive (Fig. 11.14) and 2350/94 kW at 750/250 rev/min, typical of a large power station draught fan drive.

Fig. 11.14 Stator-fed a.c. variable-speed commutator motor waterworks pump drives 1550/900 kW, 1025/875 rev/min, 3·3 kV and 1075/660 kW, 1025/900 rev/min, 3·3 kV

11.7.2 *Performance characteristics*

Most a.c. commutator motors are designed as 'shunt' characteristic machines (i.e., are machines in which the speed/torque curve is relatively flat for a fixed regulator position). 'Series' characteristic motors are also made, particularly for falling torque drives such as power station draught fan drives requiring speed ranges around 3 : 1, where the variable flux characteristic of the series motor results in higher efficiencies, particularly at low speeds, and enables somewhat simpler induction regulators to be used.

There are few variable-speed drive applications where some form of automatic speed control is not required. Controls based on all the usual systems have been used—electric, hydraulic or pneumatic, controlling the induction regulator position via a pilot motor or slave cylinder so that the motor speed is varied in accordance with the requirements of the input control signal. Although very sophisticated control systems are used where justified, surprisingly often it is found that quite simple systems are all that are required.[54]

Some type of power factor correction is an integral part of the design of all large a.c. commutator motors, generally obtained by injecting a compensating voltage into the secondary circuit of the equipment. On the largest sizes, the correction may be made adjustable by the use of a compensating induction regulator so that the magnitude and/or phase position of the injected compensating voltage can be automatically controlled. By suitable design, in addition to controlling the power factor, this injected voltage may be used to modify the inherent speed/torque characteristics of the machine to suit, for instance, the special drive characteristics of a reciprocating compressor.

11.7.3 *Application*

The a.c. commutator motor still retains many advantages over alternatives utilizing semiconductor devices. Generally cheaper for narrow and medium speed ranges, and with an inherently higher thermal and overload capacity, it also has the additional advantage of not drawing significant harmonic currents from the supply, a feature which could well be of increasing importance in the future as the capacity of existing supply systems to absorb further harmonic currents is progressively reduced, forcing supply authorities to impose increasingly expensive limitations on the design and installation of thyristor drives.

11.8 Enclosures

11.8.1 *General*

The enclosure of a motor serves two distinct functions, one of which is associated with 'protection' and the other with 'cooling'. Personnel need protection from live or moving parts and, in addition, the machine itself may need protection from the environment in which it is to work. Appropriate enclosures for these purposes also provide or may be adapted to provide, guidance for and control of the machine cooling medium (which is generally, but not always, air). The increasing acceptance of internationally agreed codes for degrees of protection and types of cooling (IP and IC codes)[14,16] has helped to emphasize the distinction between these two separate aspects of the machine enclosure, and has highlighted the fact that many of the conventional descriptions of motor enclosures are an illogical mixture of degrees of protection and types of cooling.

Although most types of enclosure have been applied to large machines, ventilated machines have in the past tended to be in the majority (Fig. 11.15) because of the inherent problems of cooling large totally enclosed machines. The bigger the machine, the more unfavourable becomes the ratio of surface area to losses, so that on large motors the simpler methods of providing the necessary cooling surfaces (such as the use of a ribbed frame) are difficult to apply, and more elaborate cooling systems are much more common.

Fig. 11.15 Ventilated squirrel-cage motor power station circulating water pump drive (6150 kW, 750 rev/ min, 11 kV)

Most of the smaller sizes of electric motor are installed indoors, not so much for their own protection but because, in many cases, they are being coupled to driven machines which need the protection of a building, either for the convenience and protection of personnel, or because the manufacturing process itself makes this essential. With the increase in size of driven units there has been a tendency to concentrate them in central areas leading, in many cases, to the provision of special 'motor rooms'. Such rooms were a simple way of controlling the environment in which the machines ran, and provided suitable working conditions for operating and maintenance staffs (as well as encouraging the use of ventilated enclosures for the motors).

The high and increasing costs of civil works is beginning to make motor rooms an expensive luxury, and the enclosures of large machines have therefore been developed to permit the installation of motors out of doors with little if any additional protection from the weather in the form of buildings. There is, consequently, a significant movement away from ventilated machines with simple drip-proof enclosures and towards machines with closed ventilating circuits which are more readily adapted to operation in the open by relatively simple 'weatherproofing'. Typically, these employ air/air heat exchangers, though machines with air/water heat exchangers have been installed in the open, either in locations where the risk of frost is negligible, or where suitable precautions can be taken to protect the cooling water system from sub-zero temperatures.

11.8.2 *Weather-protected machines*

North American practice tends to differ somewhat from European, in that whereas in Europe the tendency has been to use totally enclosed machines for outdoor installations, in North America, and areas influenced by North American practice, it is usual to use ventilated machines fitted with suitable structures at the machine inlets and outlets to discourage the entry of dust, rain or snow into the interior of the machine.[22] The provision of suitably designed baffles, settling chambers, and possibly dust filters (Fig. 11.16), provide in themselves a high degree of protection for the machine; modern insulating systems are in any case much better able to withstand operating under damp conditions, and can be readily given additional protection by the use of 'sealed' windings.

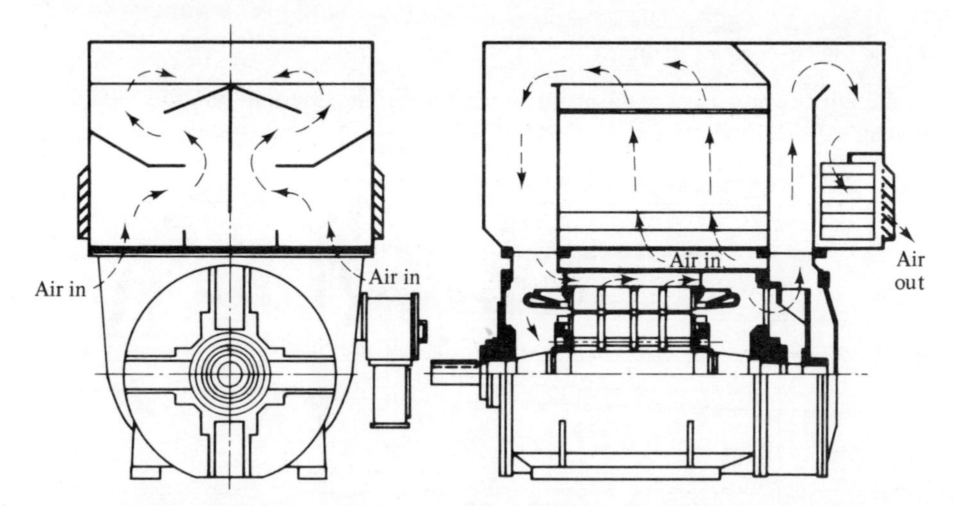

Fig. 11.16 Basic ventilation circuit for a weather-protected (NEMA 2 specification) motor design

11.8.3 *Closed ventilated machines with air/air heat exchangers*

By far the commonest method of cooling large totally enclosed machines is by means of separate air/air heat exchangers, of the tube or plate type, with either cross or contra-flow paths for the two cooling circuits (Fig. 11.17). These coolers have many attractions when compared with the corresponding system using air/water heat exchangers, not least of which is that the cooling system is self-contained. Such machines, however, always require more active material, i.e., are likely to be physically larger than corresponding ventilated machines, due to the additional temperature differential across the heat exchanger.

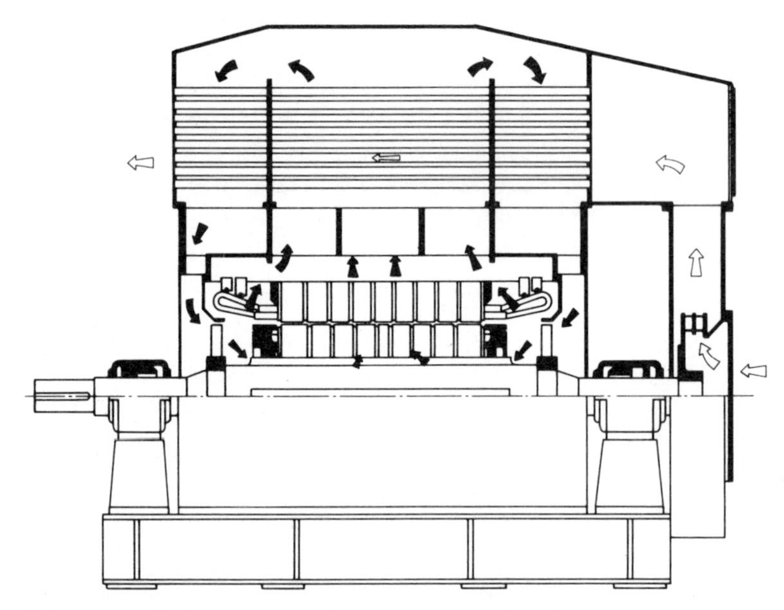

Fig. 11.17 Basic ventilation circuit for closed air circuit air/ air heat exchanger motor design

11.8.4 *Tube-cooled machines*

Although the use of a separate heat exchanger is essential for the very largest machines, the tube-cooled machine (Fig. 11.18), though presenting considerable

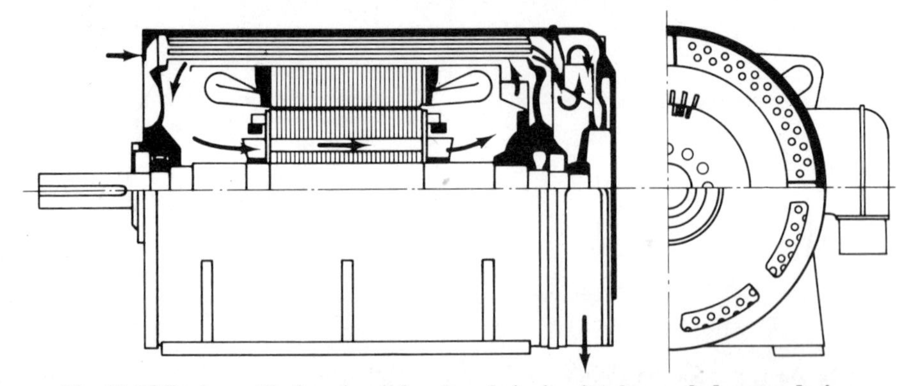

Fig. 11.18 Basic ventilation circuit for closed air circuit tube-cooled motor design

problems for the designer, has certain fundamental attractions; the distribution of the heat exchanging surfaces around the outer periphery of the carcase is attractive from several points of view, not the least being its aesthetic appeal (Fig. 11.19). European practice had tended not only to retain but develop this construction for the smaller end of the 'large' machines in spite of the acknowledged difficulty in accommodating sufficient surface area on the machine periphery. To compensate for these limitations, very high air velocities have to be used within the cooling tubes. Careful design of the air paths is essential to minimize the pressure drops through the air circuit which are otherwise likely to be unacceptably high, and, of course, large well-designed fans must be used.

Fig. 11.19 Tube-cooled squirrel-cage motor (1500 kW, 1500 rev/min, 6 kV). (Courtesy Heemaf)

11.8.5 *Closed-ventilated machines with air/water heat exchangers*

Because of the very much better heat transfer across metal-to-water interfaces compared with corresponding metal-to-air interfaces, the water cooler is a relatively compact object. In addition, when cooling water is available in the necessary quantity (and this may not always be the case) it is likely to be at significantly lower temperatures than those conventionally assumed for ambient air (for instance, 25°C instead of 40°C). With cooling water at these temperatures it is not difficult to design air/water heat exchangers to maintain the machine ambient (i.e., the temperature of the air entering the machine) below 40°C. Hence, compared with ventilated machines, no derating is required.

The provision of sufficient cooling water for large machines often presents something of a problem. Three litres per minute per kilowatt loss is a typical requirement, so that once-through cooling systems can be prohibitively expensive unless the machine is installed close to sources of abundant water. Alternatively, recirculatory systems using cooling ponds or other forms of cooling are used, although generally only if such elaborate installations are in any event required for other installations on the same site. The quality and cleanliness of the cooling water may have a significant influence on the service life of an air/water heat exchanger. Water velocities tend to be high to promote good cooling, and hence erosion as well as corrosion may take place unless the cooler has been designed with a full knowledge of the water analysis.

11.8.6 *Noise*

The noise emitted by a motor increases with its size and power (see table 11.4) and, as a result, the noise levels of large and/or high-speed machines may well be quite unacceptable[67,68,72] unless the noise can be confined within a restricted area where the level is already high, or where personnel are unlikely to have to work for any length of time. As the majority of noise emitted by a properly designed motor is associated with the motor cooling air at exit and entry, changes in enclosure (for instance, from a ventilated to a machine with an air/air heat exchanger) will have only marginal effects on the machine noise level (with the notable exception of the machine with air/water heat exchanger which, by eliminating the external air circuit,

Table 11.4 Noise–power ratings of electrical machines

Rating		Noise–power rating Rated speed		
Above	Up to	3000–1501	1500–1001	Below 1001
kW	kW	rev/min	rev/min	rev/min
	2·5	78	78	73
2·5	6·3	88	83	78
6·3	16	93	88	83
16	40	98	93	88
40	100	103	98	93
100	250	103	103	98
250	630	108	108	103
630	1100	108	108	103
1100	2500	110	110	108
2500	6300	113	113	108
6300	16 000	115	115	110

Corresponding noise-rating numbers, based on pressure and referred to a 3 m hemisphere, will be approximately 18 dB less than the above noise–power-rating figures.

offers the possibility of a significant reduction in noise, sometimes by as much as 15 dB). Mounting criticism of ambient noise levels, even in traditionally noisy industrial environments, has led to increasing attention being focused on the noise emitted by electric motors.[69–71] The increasing tendency to site large machines out of doors aggravates the problem, in that brick structures have in the past provided quite effective sound-insulating barriers between motors and an increasingly noise-conscious and hence hostile world. Modifications to machine design in the interests of quiet running are becoming part of the normal machine design (trailing bladed fans for instance) and increasingly the inlets and outlets of machines are being made suitable for the addition of sound-absorbing cowlings (see Fig. 11.20). The weather-protected machine with its elaborate cowlings designed to restrict the entry of the

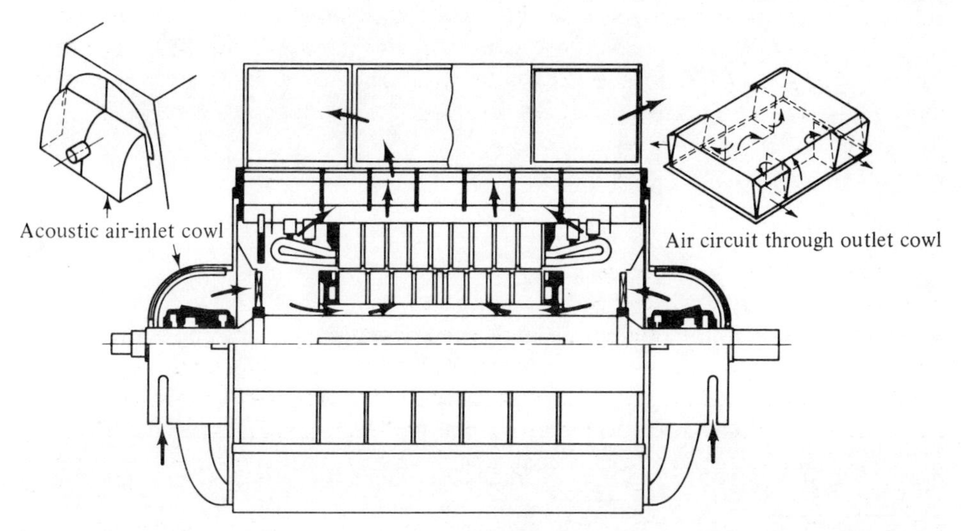

Fig. 11.20 Basic ventilation circuit for a ventilated machine with inlet and outlet noise attenuators

weather to the interior of the machine, also serves to inhibit the escape of noise. Such cowlings, often with the addition of sound-absorbing materials on the internal surfaces of the ducting, are one of the most effective methods of cheaply reducing noise levels of machines to those likely to be acceptable for outdoor installations (Fig. 11.21).

Noise levels rise when the motor load is applied, but it is not usually difficult to design to keep this increase in magnetic noise to a relatively low value. Where very low noise levels are required lagging of the machine carcase may be necessary, although the neatest solutions in such circumstances are often those in which the complete machine is built into its own self-contained, free-standing, sound-insulating enclosure.

11.8.7 *Hazardous atmospheres*

11.8.7 General. The increasing size and complexity of petrochemical complexes has introduced completely new thinking relative to the use of electric machines in

Fig. 11.21 Weather-protected oil refinery compressor drive (with noise attenuation). (Courtesy Siemens)

hazardous atmospheres. The traditional flameproof enclosure was originally developed for underground use in mines in which there is a significant risk of explosive gases being present during normal working, where machine sizes are relatively small, and where the fault capacity of the supply system is low. The situation in large petrochemical installations is almost exactly the opposite, i.e., there will be a very large electrical supply system, with a high fault capacity, supplying numbers of motors, many of which may be very large, mostly situated in locations where the atmosphere will normally be perfectly safe and hazardous gases are unlikely to occur except in the event of an accident of plant failure.[84] This has resulted in the development of further types of machines for use in such areas, for instance type N machines ((Ex)n), pressurized machines ((Ex)p), and increased safety machines ((Ex)e), see chapter 9.

11.9 Standardization and Optimization

11.9.1 *Performance standards* (see also chapter 5 section 2.11)

Not many years ago a section on standardization would have been deemed inappropriate in a chapter dealing with large machines. In the smaller sizes the benefits of standardization were being accepted, albeit reluctantly, but large machines were considered to be custom-built and therefore outside the scope of any but the most basic of standards. Few would now take up such a position, in an environment where even in the largest sizes of machine the benefits of an appropriately drafted standard are recognized.

Another discernible trend, as important in the field of large motors as elsewhere, is the increasing acceptance of the necessity for national specifications to be broadly in line with each other and in accordance with IEC recommendations. The greater breadth of modern motor standards (and British specifications were in the past open to criticism in this respect relative, for instance, to corresponding American and German documents) has significantly reduced the necessity for over-long and detailed individual user specifications. A purchaser can now safely confine himself, on most machines, to calling up a national specification, to which he may add a relatively short schedule specifying those additional features which are peculiar to his specific application. An example of an extension of this form of thinking into the format of British Standards is the recent publication of BS 4999 and BS 5000. BS 4999 specifies requirements which have a relevance to the generality of rotating machines, while the various parts of BS 5000 deal with the special requirements of a particular type or application of machine, selecting from BS 4999 those parts which are deemed appropriate to the type of machine in question. Part 41, for instance, covers the special requirements of large motors for driving power station auxiliaries; and other parts deal with such varied classes as general purpose motors, type 'N' motors, and small power motors.

11.9.2 *Dimensional standards* (see also chapter 5 section 2.7)

The degree of dimensional standardization appropriate to a large machine is very different from that which is widely accepted on smaller ones. Nevertheless, it is essential that the primary dimensions, for instance centre heights, should be consistent with and form a coherent whole with those adopted for the very tightly standardized smaller machines. The basic international standards upon which modern dimensional standards are based are IEC 72 and 72A.[19] Because of the variety of constructions and enclosures which are commonly used on large machines, constructors must always be permitted to exercise some choice. Nevertheless, dimensional standards of the type appropriate to large machines are useful in providing each individual designer with the same guide as is available to all other designers, by offering tables of preferred numbers for many of the leading machine dimensions (see Fig. 11.22 and table 11.1). Although it is unlikely that there will ever be a sound economic case for the dimensions of a large machine to be as closely controlled as those of, say, a 280 size, nevertheless, there are positive advantages in standardization reaching a point at which a project engineer will have a general idea of the overall space occupied by a given machine of known centre height and enclosure. Similarly, a standardized range of shaft ends and flange or skirt dimensions for vertical machines will certainly extend into at least the bottom end of the large motor field.

11.9.3 *Output allocations* (see also table 5.1)

Output allocation is an integral part of the standardization of small motors, so much so that there is a tendency to forget that although dimensional standardization is an essential preliminary to output allocation, the two exercises are essentially quite

Fig. 11.22 Outdoor installation in a Division I/hazardous petrochemical area employing pressurized CACW squirrel-cage motors (4750 and 2600 kW, 1500 rev/min, 11 kV). (Courtesy GEC Ltd)

separate. With large motors, the diversity of duties and the relatively small number of any one type of machine manufactured, means that it will be seldom possible to justify, on economic grounds, the inevitable compromises inherent in the allocation of standard outputs, compromises which in the smaller sizes are easily offset by the ability to introduce series or mass-production methods into the manufacture of the relatively much larger numbers of smaller machines. Nevertheless, standards incorporating output allocations do exist for use in certain limited areas; one such example is the CEGB Standard 44231 which covers a selected range of machines up to about 1500 kW.

11.9.4 *Application of standards*

In spite of the undoubtedly wide variety of specialist machines in the large motor field, manufacturers have directed considerable efforts towards developing a range of basically standard units with sufficient built-in versatility for them to be easily adapted for a wide variety of specialist applications. One particular construction of this type which has been adopted, with minor modifications, by several manufacturers, comprises a simple basic unit consisting of little more than the wound stator core which is inserted in or onto a rudimentary form of bedplate or cradle which serves to support the rotor and bearings. The unit is completed by the addition of a separate sheet steel or fibreglass enclosure (see Figs. 11.23 and 11.24). Most designs have a suite of enclosures to cover the complete range of common cooling systems (for instance, ventilated, pipe-ventilated, NEMA 2, closed-ventilated circuit with air/air or air/water heat exchangers).

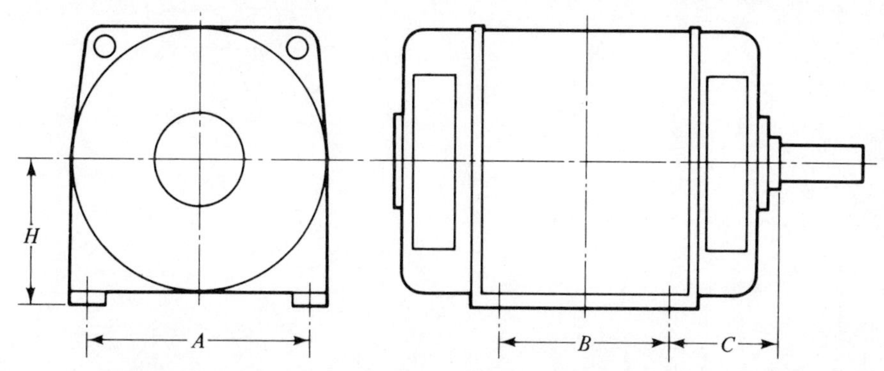

Fig. 11.23 Standard symbols for the principal fixing dimensions of horizontal shaft machines

Up until quite recently, performance standards for machines tended to be drafted for and to be primarily concerned with machines at the smaller end of the spectrum. In recent years there has been a gratifying swing away from this type of thinking; BS 4999 for instance contains tables of standard starting performance figures for cage motors up to 10 000 kW, and lists noise rating numbers for machines up to a similar output. Few contracts for large machines are based on the uncritical acceptance of such standard figures; their usefulness is that they provide a realistic starting point from which users and manufacturers can discuss and evolve a specification to match the motor to the drive. Limitation of the winding temperatures has long been accepted as an effective method of ensuring adequate insulation (and therefore winding) life; although it will never be possible to guarantee the life of a winding on anything more than a statistical basis because of the random nature of the mechanism of breakdown, the thermal performance of insulating systems is well understood (in spite of the problems of carrying out realistic life and development tests).

A similar situation exists with respect to the life of rolling bearings used at the lower end of the large motor spectrum. Users are increasingly asking for nominal bearing lives to be calculated in accordance with ISO Recommendation R281-1962E,[28] and specifying B10 lives up to 40 000 hours. Although it can be argued, rightly, that the calculation in the present document is too crude to represent accurately the probability that 90 per cent of the bearings will survive the period calculated, ignoring as it does many factors which contribute significantly to the bearing life, nevertheless, it does provide an internationally agreed basis for comparison between bearings on the basis of their loading, speed, and design.

11.9.5 *Optimization*

The optimization of machine designs has attractive overtones on large machines where the motor may well be one of the most important components of an installation. In the first flush of computer availability, optimistic claims were made

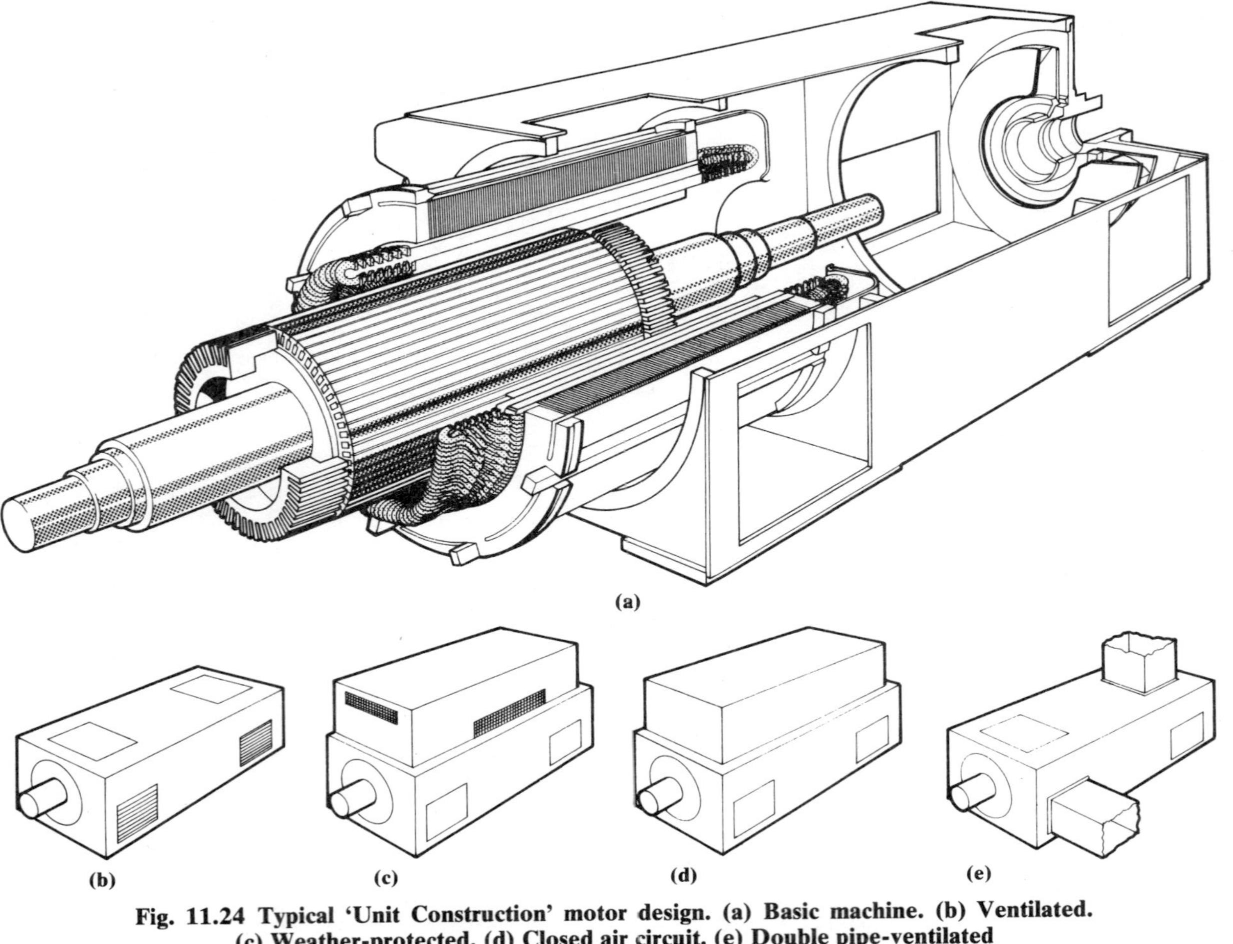

Fig. 11.24 Typical 'Unit Construction' motor design. (a) Basic machine. (b) Ventilated. (c) Weather-protected. (d) Closed air circuit. (e) Double pipe-ventilated

for the complete 'computerization' of the design process. Practically all calculations associated with machine design are now carried out by computers, but the resultant design still reflects very much the designer's individuality, simply because any design is a very delicate balance between a considerable number of inter-related, and often mutually incompatible, requirements. It is still (perhaps disappointingly) difficult to quantify the mental processes which lead the experienced designer towards achieving this balance.

The availability of the computer as a working tool has reduced the necessity for many of the simplifying assumptions which were part of the stock-in-trade of an older generation of designers. Increasingly detailed computer programs provide much more accurate calculations for such traditionally difficult areas as machine reactances under saturated conditions at start, thermal design relative to the heat flow within the machine, the calculation of fans and ventilating circuits, and flux and current distribution in solid pole synchronous motors during starting.

One of the problems in formulating an optimization programme is that any investigations in this area quickly show that the optimization of the machine alone can be a relatively meaningless exercise. The impact on the motor design of the requirements of the driven machine on the one hand and the supply system on the other, is such that a well-designed machine in one environment may be completely inappropriate in another. Similarly, the emphasis placed on any particular feature of the machine design, for instance its losses, will be very dependent on the local cost of electric power, the economic climate, interest rates, and the availability of capital, all factors which vary from country to country and time to time.

11.10 Motors and Systems

11.10.1 *Basic economic considerations*

However interesting, sophisticated, and even revolutionary a motor design, it has always to be borne in mind that the motor is only a means to an end, be it the provision of an adequate water supply for a city, the manufacture of chemicals or the refining of oil. The applications engineer must consider the overall efficiency of the installation as his primary objective. It may well be more realistic to talk of the good or bad application of a particular motor rather than imply any possibility of defining the intrinsic merit of a particular design. It is becoming increasingly difficult to formulate a balanced view, partly due to the proliferation of choices, but also partly due to the delicate and complex balance that is involved between technical and calculable factors and the many imponderable though, nevertheless, equally vital factors.

However simple a drive may appear at first sight, the proper specification of its individual requirements is, and probably always will remain a difficult problem. Typically 'simple' questions are—constant versus variable speed; if the latter, how is it to be varied and over what range?—how significant is it to know accurately the actual operating conditions with regard to load, frequency of starting, ambient temperature?—what is a realistic economic weighting for kW losses, maximum kVA or kVAr demand, mains interference due to harmonics or oscillating loads? It is even realistic, on the largest sizes, to query whether the electric motor is the proper driving means, though this question is rarely posed for industrial systems in Europe.

Fig. 11.25 Partially erected motor illustrating one form of unit construction (7000/ 1500 h.p., 1200/ 720 rev/ min with PAM winding)

11.10.2 *System design*

Whatever the drive it is essential to investigate the interaction between the supply system and the motor load.[79] There are heavy cost penalties involved if the limitations of standard switchgear are exceeded.[81] This will set a fixed limit to the system stability, i.e., the voltage reduction versus time from which a system will eventually recover from disturbances on the supply line.[76,80] Table 11.3 indicates the sort of levels of induction motor load which can be stably supported based on the underlying assumptions that squirrel-cage motors are designed for a rather low nominal starting current of 450 per cent full load current, that they have speed/torque curves which allow acceleration of the driven machine at 80 per cent of the nominal voltage, and have a standard overload capacity of 160 per cent full load torque. Interruptions of supply, causing a complete stoppage of motors, can be very expensive, not to mention dangerous, in many continuous process plants. There is much that the motor designer can do to assist in this matter, both from the point of view of providing the necessary design information, and in introducing design variations to suit the system problems. While recovery currents can seldom be influenced to any great extent, improvements in the torque/current characteristics or increased inertia in the drive will generally increase stability. Squirrel-cage motors, with significant current displacement effect rotors, by reducing the starting currents, are a specific example where the appropriate choice of machine can ease stability problems. Synchronous motors on the other hand, particularly solid pole machines, are inherently more difficult to finesse in this respect, and in any case have the added complications of excitation systems and automatic synchronization control. On the other hand, a.c. commutator motors, mainly due to their low starting currents, show a natural advantage in this respect over both squirrel-cage and synchronous motors.

There is, furthermore, the converse problem in so far as under fault conditions all motors contribute to the short-circuit current of the system[77,82] and this effect needs to be taken into account when designing cables and switchgear.

Motor torques produced during normal and abnormal switching operations[75,78] require some consideration in the context of shaft/coupling systems, and it is obviously advantageous to have compatible fault and dielectric properties throughout the complete installation; thus transformer, switchgear, fuses, cables, terminal boxes, windings, shafts and couplings, may all play a part in providing the proverbial weak link in the chain.

Voltage unbalance problems rarely affect large installations during steady-state running, but are, of course, well known during transient conditions where unbalanced instantaneous currents are inevitable during the starting of motors. With the exception of some slipring cascade connections, none of the motors considered are in themselves generators of harmonics, i.e., they will not be expected to produce any problems, and can generally be assumed to contribute less than 2 per cent individual and 3 per cent total harmonic content to the system.

CHECK LIST TO AID SYSTEM DESIGN

The detailed design of a supply system for a large installation may well require the attention of a specialist in the field. There are, however, many simple checks which the project engineer can employ to ensure that, in the early stages, an installation is planned in such a way that problems are foreseen and circumvented before irrevocable decisions are taken with respect to such fundamental matters as system voltage level, MVA capacity of switchgear, and type and characteristics of motors.

The following check list should enable a project engineer to simply and quickly assess whether the preliminary system design concept is likely to be satisfactory or not, and to suggest alternatives which may be worth exploring before opinions have hardened and decisions become difficult to alter.

Information

(a) Obtain line diagrams of supply system, with sizes of supply transformers and their respective percentage reactances.
(b) Obtain speed/torque and speed/current curves of motors, speed/torque characteristics of driven machines, and inertias of drives.

Investigation

(a) Check the system voltage drop when supplying motor starting currents.
(b) Check that the motor starting torques are sufficient to start and accelerate the drives up to normal running speeds at these reduced voltages.
(c) In the case of synchronous motors check that there is sufficient voltage at the motor terminals, when up to speed, for the machines to synchronize.

(d) Check that the voltage drop, due to the starting of any one motor, does not involve a risk of other motors on the system pulling out (i.e., ensure that machines have sufficient overload capacity to accommodate this reduction in volts).
(e) Consider voltage depressions of various magnitudes and durations, particularly those which are sufficient to allow machines to pull out. Check that upon restoration of the voltage the supply system is capable of supporting the corresponding restoration currents of all the machines simultaneously, without such an excessive volt drop that a return to full speed is not possible. Also check that proper interlocks and sequencing is introduced to deal with any machines which are not designed for direct-on starting (slipring machines, a.c. commutator motors, auto transformer or reactance started induction motors, or synchronous motors for instance). Ensure that for all these machines an appropriate re-starting sequence is introduced.

Alteration

If, as a consequence of the above investigation, it is clear that the system design requires modifications, consider the following:

(a) Increasing the inertia of the drives to reduce the speed drop during supply depressions by, for instance, specifying a large-diameter short machine rather than a conventional small-diameter long machine.
(b) Improve the starting torque characteristics of the motor to provide more accelerating torque or a higher pull-out torque by, for instance, the usc of sash bar or double cage induction motors.
(c) Alternatively, or in addition, investigate the possibilities of reducing the starting current by similar means.
(d) If there are problems associated with the re-synchronizing of synchronous motors, consider whether some or all might be replaced by induction motors.
(e) If the system consists of a combination of fixed- and variable-speed motors, consider whether a case can be made for increasing the proportion of variable-speed a.c. commutator motors in relation to fixed-speed motors.

If, in spite of considerations along the above lines the system stability still seems marginal, the supply system is probably not stiff enough for the size of the installation. Consider the use of low reactance supply transformers or reinforce the system with larger or additional input transformers. If by so doing the short-circuit MVA capacity of the system then exceeds recognized standard fault levels for switchgear (for instance, 150 MVA for 3·3 kV, and 500 MVA for 11 kV), after allowing for the fault contribution of the motors themselves, then the voltage level of the supply is probably too low; consider replacing a 3·3 kV system by 6·6 kV or 11 kV system (or, alternatively, splitting the system into two parts).

Finally, check local supply regulations relating to the permissible harmonic levels at points of common coupling with the supply authority system with respect to any thyristor-controlled motors on the system, and if high consider reducing the harmonic levels by increasing the number of phases (3 pulse to 6 pulse, or 6 pulse to 12 pulse) or, alternatively, replacing some of the thyristor drives with other non-harmonic producing forms of variable-speed drive (variable-speed a.c. commutator motors or slipring motors, for instance).

11.10.3 *Environmental design*

This has to be judged from two aspects, the effect of the motor on its environment,[118] and the effect of the environment on the motor. Generally speaking, this problem largely revolves around that of motor enclosures. The chemical contamination which a machine will accept may well require close investigation and, in addition, the heat generated by the machines and the consequent problems of its removal from confined spaces may prove difficult. Noise is becoming an important aspect of machine design, particularly with large high-speed machines, although, in general, once the necessity for quiet running is appreciated, appropriate action can be taken at the design stage.

In many ways the design of a machine with respect to its environment is closely connected with that of reliability. Reliability[87] is very difficult to quantify; it involves not only good initial design but equally good supervision and clear instructions for the operation and routine maintenance. The objective of manufacturers is to build machines to provide long, reliable life. It is, however, impossible economically to design a machine to be proof against everything, and consequently it is better for operators to realize where the limitations lie, and not be misled by such facile and untrue statements such as 'indestructible squirrel-cage rotors'. Reliability is best achieved by the closest collaboration between constructor and operator, and this covers the whole life of the project from the initial feasibility studies right through its operating life.

11.11 Conclusions

If it is accepted that progress in engineering is largely an economic problem, whether it be in the area of savings in first cost or in running cost, or in the reduction of outages, then, though much has been done, there is nevertheless still much to be learnt, much to be investigated, and much improvement possible within the traditional lines of large motor design, (quite apart from the possibilities, however remote, of fundamental basic innovations which may or may not introduce new concepts into large motor designs).

11.12 References

The following references, generally post 1960, have been selected from the very large available literature[4,5] with a view to indicating where the reader might usefully direct his further reading with the aim of pursuing any particular subject in greater detail. As an aid to clarity the references are arranged in subject groups, in date order, under six main headings.

Where a reference has a direct bearing on the text, this is indicated by adding the reference number at the appropriate point; most, though not all, are so referenced. The references marked with an asterisk quote a significant number of further references.

11.12.1 *Reviews*

1. G. A. Juhlin, 'Electrical plant and machinery', *JIEE*, **86**, 193–202 (1940).

2. W. N. Kilner, 'Rotating electrical machinery', *Proc. IEE*, **99-1**, 280–285 (1952).

3. R. D. Ball, 'Alternating-current machines', *Proc. IEE*, **109-A**, 33–44 (1962).
4. Bibliography of rotating electric machinery for 1948–1961, *Trans. IEEE*, **83**, 589–606 (1964).
5. Bibliography of rotating electric machinery for 1962–1965, *Trans. IEEE*, 87-*PAS*, 679–689 (1968).
6. *The application of large industrial drives*, IEE Conf. Publ. 10 (1965).
7. 'Motoren für industrielle Antriebe', *Siemens-Zeitschrift*, **40**, Beiheft, 6–101 (1966).
8. G. Zaar, 'Elektrische Maschinen', *ETZ-A*, **89**, 479–481 (1968).
9. K. F. Raby, 'Motors for modern industry', *Electronics & Power*, **18**, 374–379 (1972).
10. 'Large electrical machines', *LSE Eng. Bull.*, **12**, 1–49 (1972).
11. J. C. H. Bone, 'Motors in Industry', *Electronics & Power*, **18**, Suppl. S3-S16 (1972).
12. Zueva, et al., 'A new series of large AC machines', *Electrotekhnika* No. 7, 10–17 (1972).

11.12.2 *Specifications*

13. BS 2613/70 'Performance of rotating machinery'.
14. IEC Publ. 34, Parts 1–9—1967/72, 'Rotating electrical machines'.
15. VDE 0530 (1972), 'Bestimmungen für elektrische Maschinen'.
16. BS 4999, 'General requirements for rotating electrical machinery'.
17. BS 5000, 'Rotating electrical machines of particular types or for particular applications'.
18. BS 3979/71, 'Dimensions of electric motors'.
19. IEC Publ. 72 (1971), IEC 72A (1970), 'Dimensions and outputs of electric motors'.
20. BS 4683: Parts 1–3, 1971/2, 'Specification for electrical apparatus for explosive atmospheres'.
21. IEC Publ. 79, Parts 1–10—1962/72, 'Electrical apparatus for explosive gas atmospheres'.
22. NEMA Publ. MG1-1972, 'Motors and generators'.
23. BEAMA Publ. No. 225 (1967), 'Measurement and classification of acoustic noise from rotating electrical machines'.
24. BEAMA Publ. REM500 (1970), 'Recommendations for dielectric discharge energy and loss tests for new coils of rotating electrical machines'.
25. BS 2949—1970, 'Rotating electrical machines for use in ships'.
26. IEC Publ. 92, Parts 1–6 (1964–5), 'Electrical installations in ships'.
27. Lloyd's Register of Shipping, 'Extracts from the rules for the construction and classification of steel ships', No. 7 (1972) and Notice No. 2 (1972).
28. ISO Recommendation R281 (1962), 'Methods of evaluating dynamic load ratings of rolling bearings'.

11.12.3 *Insulation and allied topics*

***29.** K. K. Schwarz, 'Performance requirements and test materials for high-voltage AC motor insulation', *Proc. IEE*, **116**, 1735–1743 (1969).
30. H. Rotter and L. Eckbauer, 'Isolierung für Hochspannungsmaschinen kleinerer und mittlerer Leistung', *Siemens-Zeitschrift*, **46**, 878–885 (1972).
31. 'Windings', *LSE Eng. Bulletin*, **12**, 20–25 (1972).
32. D. Krankel and R. Schuler, 'A method of checking the turn insulation of form-wound coil windings for high voltage rotating machines', *BBC Review*, **57**, 191–196 (1970).
33. IEC Publ. 216 (1966), *Guide for the preparation of test procedures for evaluating the thermal endurance of electrical insulating materials.*
34. B. J. Chalmers and J. Richardson, 'Performance of some magnetic slot wedges in an open slot induction motor', *Proc. IEE*, **114**, 258–261 (1967).
35. H. Keuth, 'Magnetischer Protofer—Nutverschluss für elektrische Maschinen', *Siemens-Zeitschrift*, **44**, 736–740 (1970).
36. K. K. Schwarz, 'The design and performance of high- and low-voltage terminal boxes', *Proc. IEE*, **109**, 151–165 (1962).

37. 'Development of the BEAMA/CEGB motor terminal box', *BEAMA J.*, **69**, 39–44 (1962).
38. K. K. Schwarz, 'Further developments in the design and performance of high-voltage terminal boxes', *Proc. IEE*, **112**, 957–964 (1965).
39. W. E. Beck and G. R. Small, 'A motor terminal bushing', *The Engineer*, 12.8.66, 237–239.
40. BS 159: 1957, 'Busbars'.

11.12.4 *Asynchronous motors*

41. G. H. Rawcliffe and W. Fong, 'Speed changing induction motors, *Proc. IEE*, **107**, Part A, 513–528 (1960).
42. G. H. Rawcliffe and W. Fong, 'Speed changing induction motors. Reduction of pole number by Sinusoidal Pole Amplitude Modulation', *Proc. IEE*, **108**, Part A, 357–368 (1961).
***43.** E. C. Andresen, 'Drehstrom—Käfliglaufermotoren grosser Leistung', *ETZ-A*, **86**, 213–220 (1965).
44. G. Leroy, R. Chaverney and M. Simon, 'L'experimentation enterprise au banc d'essai d'EDF dans le demaine des rotors à cage', *RGE*, **75**, 1047–58 (1966).
***45.** K. K. Schwarz, 'The design of reliable squirrel-cage rotors', *LSE Eng. Bulletin*, **9**, 1–10 (1967).
***46.** 'Glandless pumps for power plant', *Proc. I.Mech.E.*, **184**, Part 3K, 64–69 (1970).
***47.** K. K. Schwarz, 'Submerged gas circulator motors for Advanced Gas Reactors'. *Proc. IEE*, **120**, 777–785 (1973).
48. B. M. Bird, 'Measurement of stray load losses in squirrel-cage induction motors', *Proc. IEE*, **111**, 1697–1705 (1964).
***49.** K. K. Schwarz, 'Survey of basic stray losses in squirrel-cage induction motors', *Proc. IEE*, **111**, 1565–1774 (1964).
50. 'Les gros moteurs asynchrones à haute tension', 1re Partie: 'Les pertes supplimentaires', *RGE*, **77**, 137–172 (1968).
***51.** H. Jordan and W. Raube, 'Zum Problem der Zusatzverluste in Drehstrom-Asynchronmotoren', *ETZ-A*, **93**, 541–545 (1972).
***52.** B. Schwarz, 'The Stator-fed AC commutator machine with induction regulator control', *Proc. IEE*, **96**, Part II, 755–767 (1949).
53. B. Schwarz, 'The design of armature windings for AC commutator motors', *IEE Conference on Commutation in Rotating Machines*, Conference Publication No. 11, 178–183 (1964).
54. J. C. H. Bone and R. Mederer, 'Applications and control of large stator-fed variable speed AC motors', *IEE Conference on The application of large electrical drives*, Conference Publication No. 10, 76–79 (1965).

11.12.5 *Synchronous motors*

55. J. Walker, 'Generator/Motor problems in pumped storage installations', *Proc. IEE*, **107A**, 157–165 (1960).
56. L. W. W. Graham, 'The AEI propulsion system for the liner *Canberra*', *AEI Eng.*, 150–187 (1961).
57. W. Fong, J. French and G. Rawcliffe, 'Two-speed single winding salient pole synchronous machines', *Proc. IEE*, **112**, 351–358 (1965).
58. I. Zborovski and A. Malevinskaya, 'The starting of a synchronous motor with connected rectifier', *Electrichvestro*, No. 10, 17–22 (1966).
59. G. Widger and B. Adkins, 'The starting performance of synchronous motors with solid salient poles', *Proc. IEE*, **115**, 1471–1484 (1968).
60. J. Walker, 'Output coefficient of synchronous machines', *Proc. IEE*, **115**, 1801–1807 (1968).

*61. B. Chalmers and J. Richardson, 'Steady-state asynchronous characteristics of salient pole motors with rectifiers in the field circuit', *Proc. IEE*, **115**, 987–995 (1968).
62. E. Blauerstein, 'The first gearless drive of a tube mill', *Brown, Boveri Review*, **57**, 96–105 (1970).
63. D. Richlein, 'Getriebeloser Antrieb für Zement Muhlanlage', *Siemens–Zeitschrift*, **45**, 189–191 (1971).

11.12.6 *General*

*64. W. Schuisky, *Berechnung elektrischer Maschinen*, Springer-Verlag (1960).
65. H. Chalmers, *Electromagnetic problems of AC machines*, Chapman & Hall, London (1965).
66. P. Hammond, *Applied electromagnetism*, Pergamon Press, Oxford (1971).
67. J. Walker and N. Kerruish, 'The open circuit noise in synchronous machines', *Proc. IEE*, **107**, Part A, 505–512 (1960).
68. R. Muller, 'The problem of noise in rotating electrical machinery', *ACEC Review*, No. 1, 10–25 (1964).
69. A. J. King, *The measurement and suppression of noise*, Chapman & Hall, London (1965).
70. W. Acton, T. Bull, R. Hore and K. Schwarz, 'Noise reduction in the circulating Water Pump House at Blyth "B" Power Station', *Proc. I.Mech.E*, **181**, Part 3C (1966). Symposium on 'Noise from Power Plant Equipment'.
71. 'Les gros moteurs asynchrones à haute tension', 2e Partie: 'Le Bruit', *RGE.*, **77**, 377–391 (1968).
*72. B. Pomfret, 'Noise of Electric Motors', *LSE Engineering Bulletin*, **12**, 21–29 (1973).
73. H. Rentzsch, 'Systems for cooling electrical machines from the point of view of noise generation', *ETZ-A*, **87**, 450–453 (1966).
74. 'Les gros moteurs asynchrones à haute tension', 3e Partie: 'La ventilation aérodynamic interne des gros moteurs', *RGE*, **77**, 753–894 (1968).
75. M. R. Chidambara and S. Ganapathy, 'Transient torques in 3 phase induction motors during switching operations', *Trans. AIEE*, **59**, 47–53 (1962).
76. D. A. Dewison, 'Effect of system disturbances on generating station auxiliaries', IEE Conference Publication No. 8, *Abnormal loads on power systems*, 39–44 (1963).
77. D. Fabrizi, 'The contribution of motors and generators to a short circuit', *Trans. IEEE*, **111**, 337–343 (1964).
78. W. Wood, F. Flynn and A. Shanmugasundaram, 'Transient torques in induction motors, due to switching of the supply', *Proc. IEE*, **112**, 1348–1353 (1965).
79. K. K. Schwarz, *Effect of motor choice on supply system design*, IEE Conference Pub. No. 10, *The application of large industrial drives*, 28–35 (1968).
80. D. D. Stephen, 'Effect of system voltage depressions on large AC motors, *Proc. IEE*, **113**, 501–505 (1966).
81. R. A. Hore, *Advanced studies in electrical power system design*, Chapman and Hall, London (1966).
82. W. Wagner, 'Short circuit contribution of large induction motors', *Proc. IEE*, **116**, 985–990 (1969).
83. M. Rychtera and B. Bartakova, *Tropic proofing electrical equipment*, Leonard Hill, London (1963).
84. W. Fordham Cooper, 'Electrical safety in industry', *Proc. IEE*, **117**, 1509–1545 (1970).
85. M. C. Shaw and F. Macks, *Analysis and lubrication of bearings*, McGraw-Hill (1949).
86. P. A. Boto, 'Detection of bearing damage by shock pulse measurement', *Ball Bearing Journal*, No. 167 (1971).
87. G. W. Ireson, *Reliability Handbook*, McGraw-Hill (1966).

12 Electronic Controls for Electric Motors

12.1 Introduction

Electronic power control for electric motors probably started in 1930 with a simple power triode or thyratron system. However, electronic control has only relatively recently taken off since the advent of semiconductors.

It is probably true that present technology is sufficiently advanced to solve practically any power control problem, the only real limitation being cost. Until recent years capital cost was of greater importance than such things as ease of maintenance, reliability, and efficiency. However, since the realization that the energy content of the world is finite, efficiency is gaining in importance and may eventually overtake capital cost. Since electronic power control circuits are highly efficient, there is likely to be a gradual increase in the use of solid-state devices in all areas of power control.

The thyristor and transistor are not the ideal power control devices, but they go a long way towards providing ideal control systems. This is because circuit technology has overcome most of their disadvantages. The basic solid-state power control circuits have not changed much over the last ten years. They have improved in detail and the devices have been improved, but the foundations were laid in the early days of thyristors and transistors. In some cases solid-state circuits are simply modified thermionic valve circuits. Because of the greater reliability, life, and efficiency of solid-state devices, circuits that were hitherto impracticable with thermionic devices, are now in common use.

As a leader into solid-state power control of electric motors, there follows a description of the solid-state devices that are widely used in power control circuits. There are three types of device suitable for wide scale use in motor and other power control systems, namely:

The Thyristor
The Triac
The Transistor

Of these, the Thyristor is the most commonly used in applications ranging from simple electric drill speed controllers to complex high power invertors. The Triac is in fairly wide use, but it is limited in use to the control of a.c. systems and is, therefore, not so versatile as the thyristor and the transistor. The Triac is essentially two thyristors connected in reverse parallel.

The transistor as described later is the most promising of semiconductor current-control devices. Providing the present limitations of the power handling capabilities of the transistor are overcome, this device will probably supersede all other semiconductor devices for power control applications.

12.2. Power Control Devices

12.2.1 *The thyristor*

The thyristor is a controlled rectifier. It will almost completely block current in one direction, but will allow current to pass in the other direction, providing a gate signal is applied. Basically the current flow situation is as shown in Fig. 12.1. The thyristor

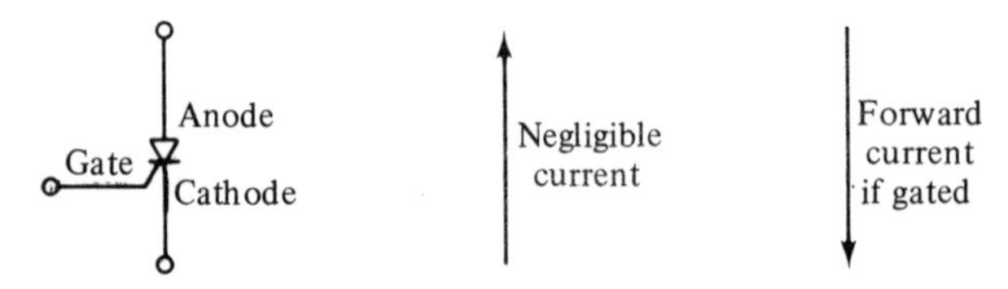

Fig. 12.1 Thyristor symbol and direction of currents

will conduct current from anode to cathode if a signal is applied to the gate. This current will continue to flow even if the gate signal is removed, providing the anode is positive with respect to the cathode. The thyristor can only be reset to the nonconducting state by reducing the anode current to a very low value (typically 60 mA) for a specified time (typically 50–100 μs). A curve of anode–cathode voltage against

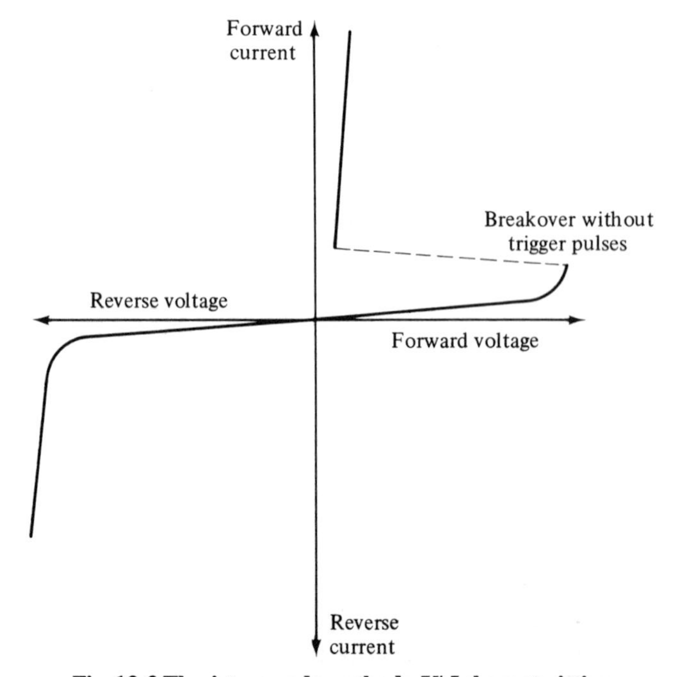

Fig. 12.2 Thyristor anode–cathode *V*/ *I* characteristics

current is shown in Fig. 12.2. The main characteristics and ratings of the thyristor are listed here:

V_f forward breakover voltage without trigger pulse on gate;
I_A maximum average anode current;
V_r reverse blocking voltage;
dV/dt maximum positive rate of change of anode volts not to cause conduction;
dI/dt maximum rate of change of forward current not to damage the device.

The gate characteristics are best explained with the aid of a diagram. The gate voltage/current characteristic for a typical thyristor is given in Fig. 12.3.

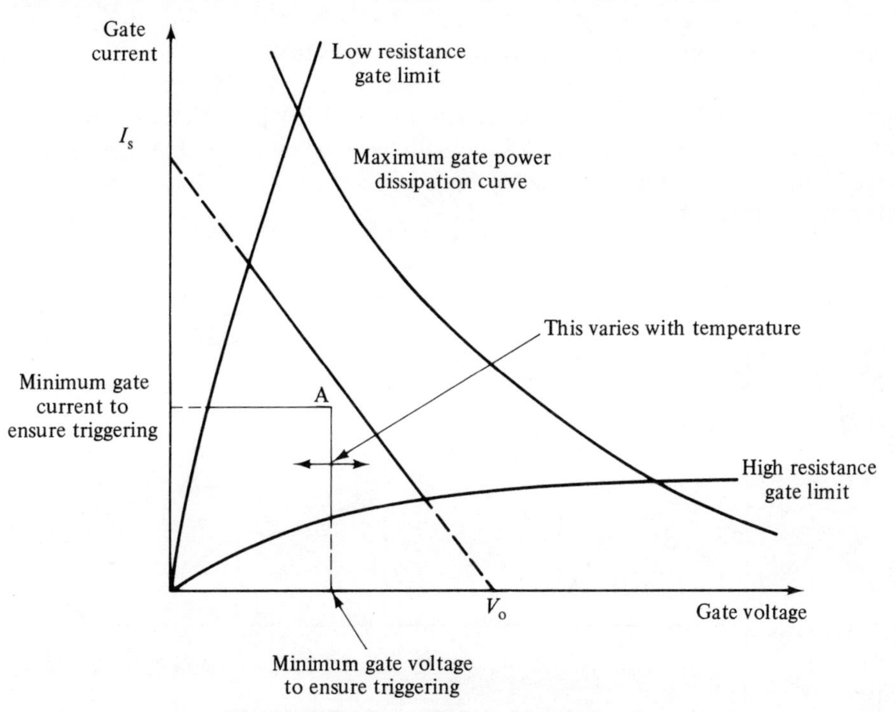

Fig. 12.3 Thyristor triggering requirements

The load line of the gate's external driving circuit must lie outside the point A, but inside the maximum dissipation curve. The load line in Fig. 12.3 is produced by a pulse generator with an open circuit voltage of V_0 volts and a short-circuit current of I_s amps. The duration of a pulse sufficient to trigger a thyristor is normally in the order of 10 μs. For anode circuits containing high inductance, it may be necessary to increase this width.

Typical values of thyristor ratings are given below.

V_f	100–5000 volts
V_r	100–5000 volts
I_A	1–5000 amps
dv/dt	100–300 volts per μs
dI/dt	100–500 amps per μs
gate V_0	10 volts
gate I_g	1 amp

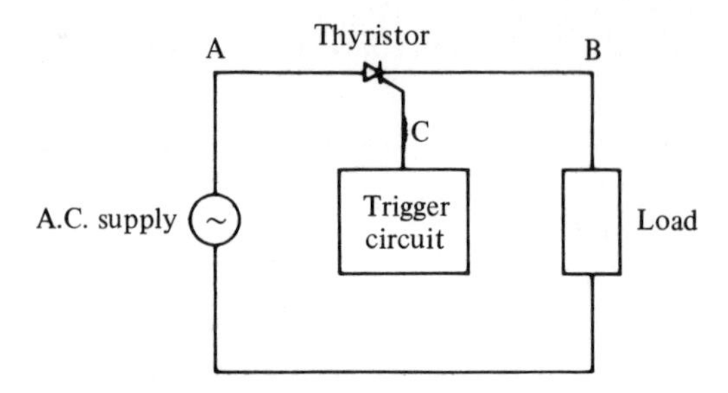

Fig. 12.4 Basic thyristor half-wave a.c. control circuit

In its simplest power control application, phase control, the thyristor rectifies and controls the current flow at the same time. For instance the circuit in Fig. 12.4 is a half-wave single-phase control circuit. The thyristor is triggered on by pulses, and automatically turns off at the end of the half cycle, when the voltage reverses. The waveforms at A–B and the trigger pulse C are shown in their correct phase relationship in Fig. 12.5.

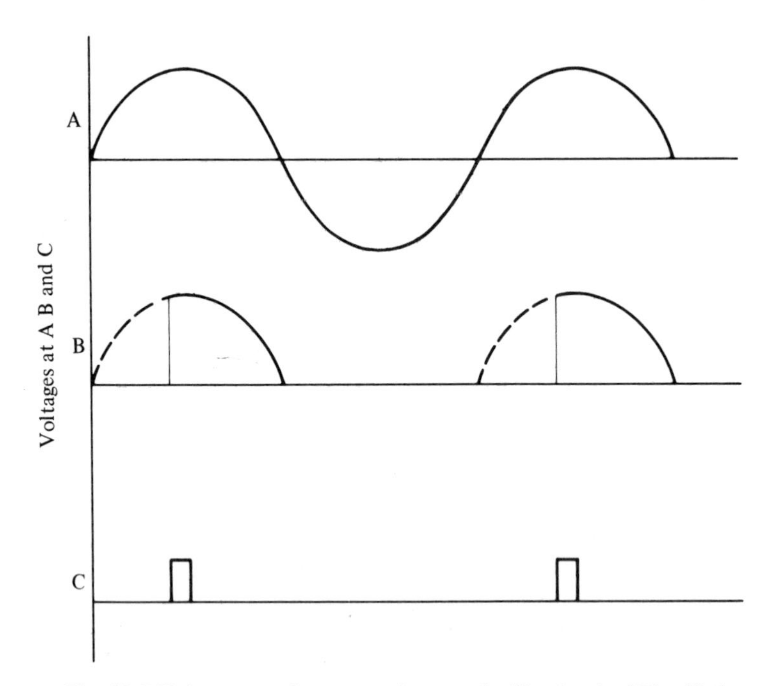

Fig. 12.5 Voltage waveforms at points marked in circuit of Fig. 12.4

From this basic configuration it can be appreciated that more complex circuits are possible—for instance, single and three-phase controlled rectifiers. There are two basic types of thyristor bridge circuits—half controlled bridges and fully controlled bridges. In a half controlled bridge, only the positive or negative half cycles are controlled, but not both. In a fully controlled bridge, both positive and negative half cycles are controlled. Single-phase half and fully controlled bridges are shown in Fig. 12.6. There is no difference in the waveform produced by either circuit, and so there

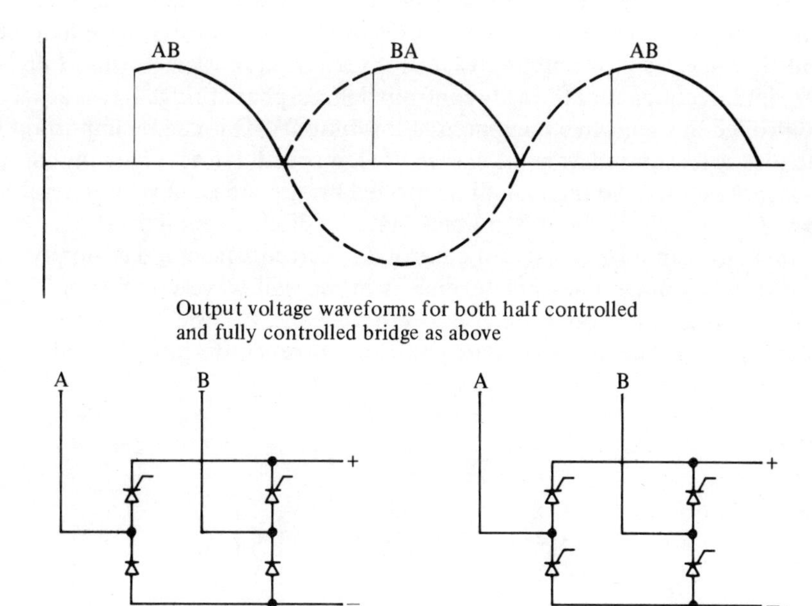

Fig. 12.6 Half and fully controlled single-phase thyristor bridge circuits

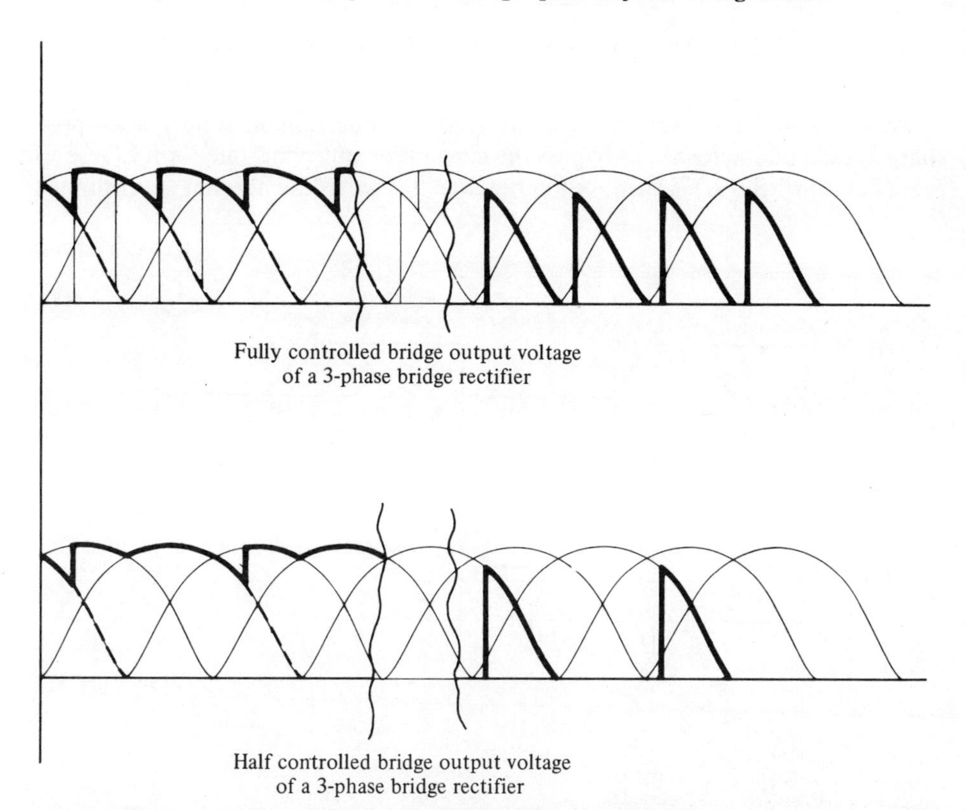

Fig. 12.7 Output voltage waveforms of half and fully controlled three-phase thyristor bridge circuits

is little or no advantage in using a single-phase fully controlled bridge. However, there is a difference between the waveforms produced by three-phase half or fully controlled bridges as shown in Fig. 12.7. It is seen from the waveforms of Fig. 12.7, that the ripple frequency of the fully controlled three-phase bridge is twice that of the half controlled bridge, when compared at low outputs. This can be important when smooth torque is required from a d.c. motor at low speed. However, a fully controlled bridge is more expensive than a half controlled bridge, since, in very general terms, the cost of a thyristor is about 3–6 times that of a diode of similar rating.

The thyristor can also be used to control d.c. currents from a d.c. supply. In this case the supply voltage does not reverse as in the half wave rectifier of Fig. 12.4, hence the thyristor must be turned off by other means. A basic d.c. control circuit is shown in Fig. 12.8. The thyristor is triggered and current flows into the load. The only

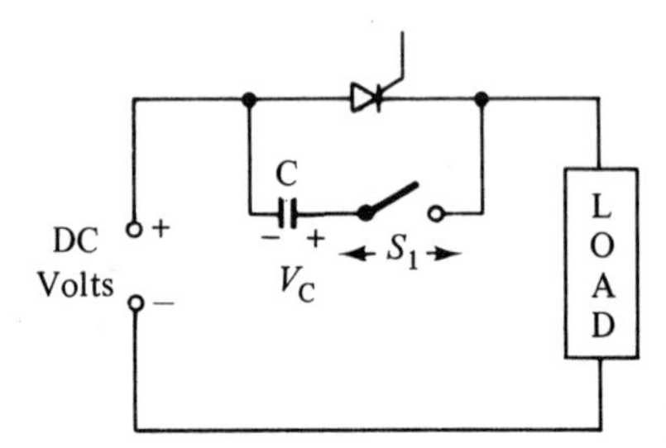

Fig. 12.8 Forced turn-off method for thyristors in d.c. circuits

way now to switch the thyristor off, apart from disconnecting the supply, is to apply a charged capacitor across it, to bypass the current for sufficient time for it to reset. In Fig. 12.8, a switch (S_1) is used, but in practice this would be another thyristor.

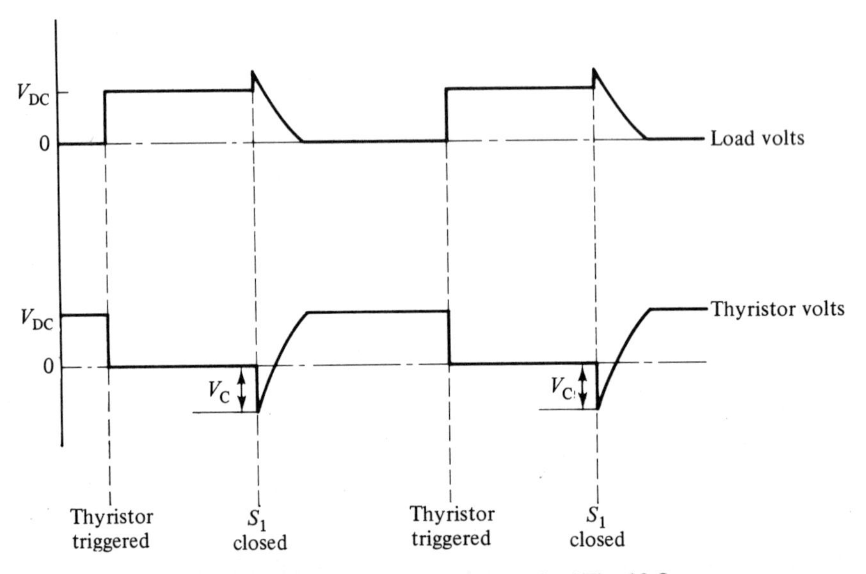

Fig. 12.9 Voltage waveforms in circuit of Fig. 12.8

The voltage waveforms across the load and the thyristor are shown in Fig. 12.9. By varying the ratio of 'on' time to 'off' time, the mean level of output d.c. voltage is varied. If the load is inductive it is a considerable advantage to connect a diode (D) across it, as shown in Fig. 12.10. This has two purposes, one to smooth the current in the load inductance, and the other to prevent large voltages appearing in the circuit, due to interruption of the inductive current.

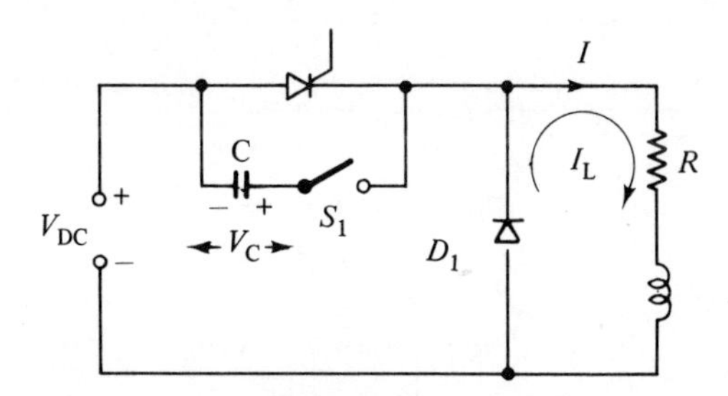

Fig. 12.10 Showing flywheel diode for use with inductive loads

The action of the circuit is as follows. The thyristor is triggered and load current can flow, building up to a value $V_{DC} \div R$. When the switch S_1 is closed, the charged capacitor bypasses the current from the thyristor, causing it to turn off. When the capacitor becomes sufficiently charged to oppose the supply voltage, diode D turns on, and the inductive current (I_L), due to the collapse of flux in the inductor's magnetic circuit, flows through the diode. In this situation the current is not interrupted, but allowed to decay naturally, at a rate determined by the L/R time constant of the inductor/diode circuit. The waveforms of this circuit are shown in Fig. 12.21.

12.2.2 *The triac*

The triac is a device which behaves like two thyristors connected back to back as shown in Fig. 12.11. The characteristics of the triac are very similar to the thyristor,

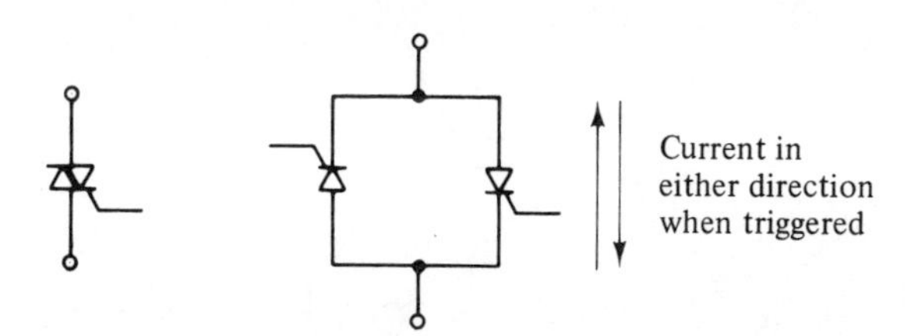

Fig. 12.11 The triac symbol and its equivalent thyristor circuit

as shown in the anode voltage/current curve of Fig. 12.12. Either positive or negative pulses on the gate will trigger the triac in either direction. The triac is designed specifically for a.c. power control, using phase control techniques. A typical circuit is shown in Fig. 12.13, and appropriate waveforms in Fig. 12.14.

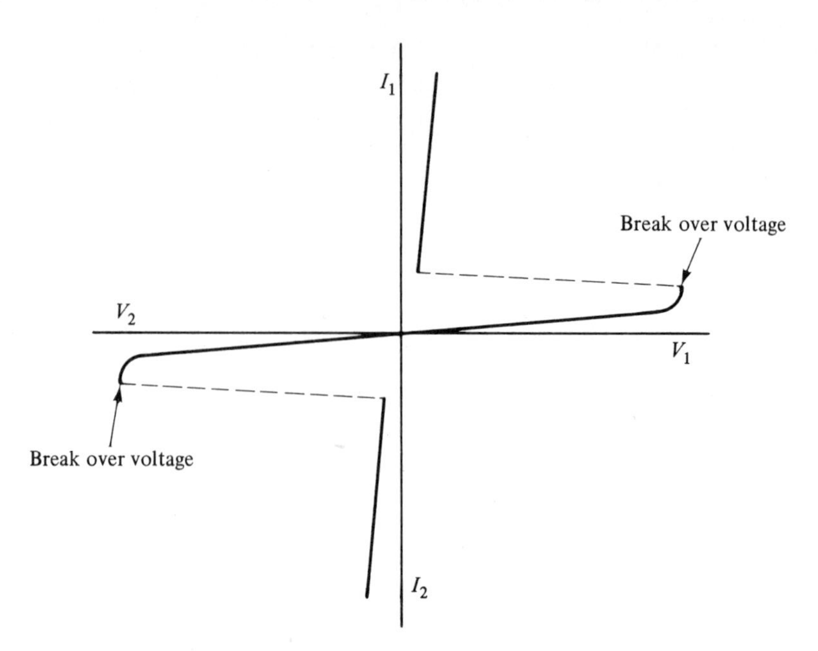

Fig. 12.12 Triac anode–cathode *V/ I* characteristics

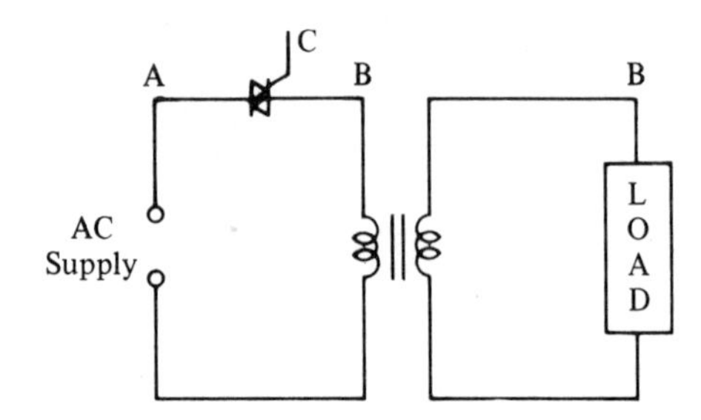

Fig. 12.13 Basic triac circuit for controlling a.c. supplies—with or without transformer

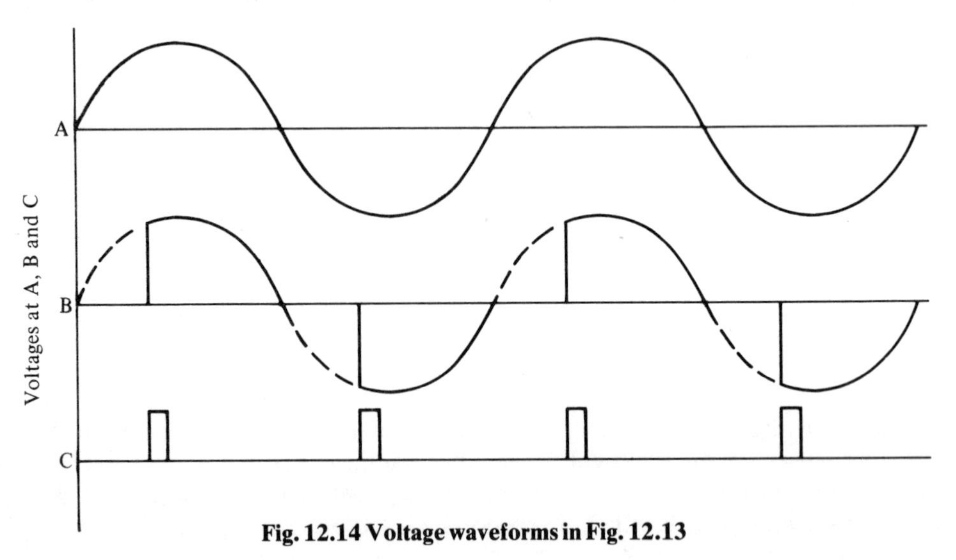

Fig. 12.14 Voltage waveforms in Fig. 12.13

12.2.3 *The transistor*

The transistor is a device which only passes current when a base current is applied, and ceases to do so when the base current is removed. Connections and currents are shown in Fig. 12.15 and the typical collector–emitter characteristics for various

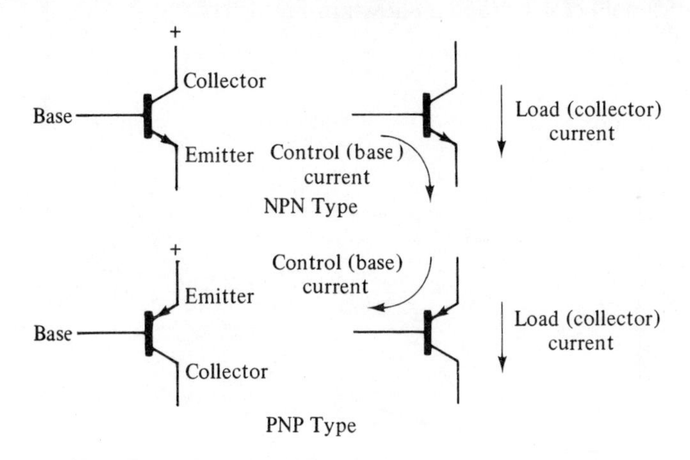

Fig. 12.15 NPN and PNP transistor symbols and currents

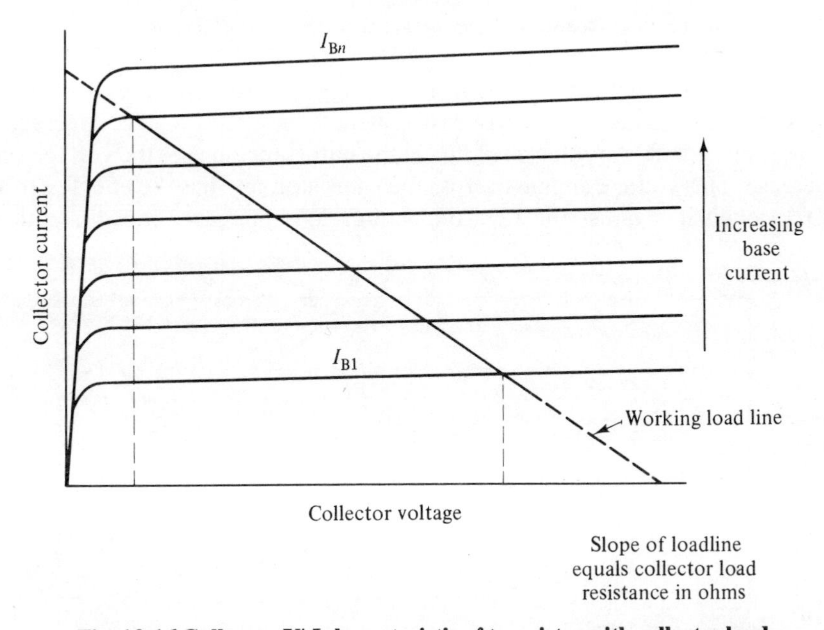

Fig. 12.16 Collector *V*/*I* characteristic of transistor with collector load

values of base current are shown in Fig. 12.16. For most power control work, the transistor is used as a switch. Because of the current gain of the transistor (typically >10) a relatively small base current will control a large collector current. The transistor is inherently more reliable in its switching than the thyristor, since no collector current can flow if there is no base current. The transistor is therefore unlikely to turn on when not required. Although the transistor is a more reliable switch, it is somewhat less rugged than the thyristor. Its main problem is its limitation on peak currents. If the collector current exceeds a figure equal to base current times the current gain (βI_h), the transistor forward volts drop will rise considerably, and hence will drastically increase its power dissipation. A typical curve of collector current against collector voltage is shown in Fig. 12.17.

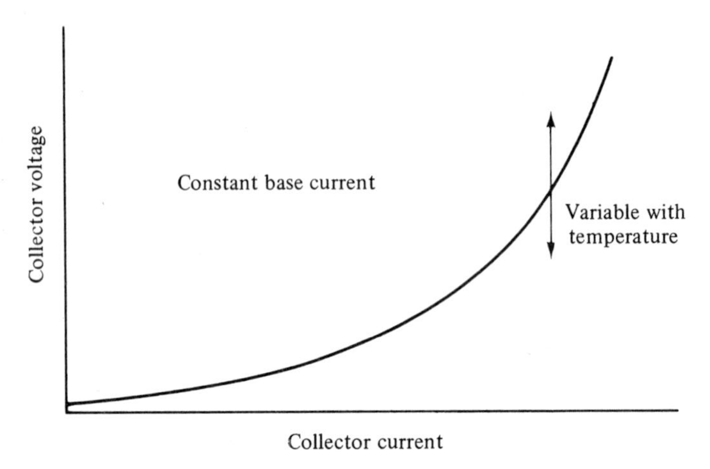

Fig. 12.17 Collector *V/I* characteristic with constant base drive

In one other respect the transistor is not as rugged as the thyristor. This is when switching inductive loads. When a transistor, with an inductive load connected to its collector, is turned off, the collapse of flux in the inductance causes the voltage across it to increase. This voltage appears across the transistor, and has the effect of forcing current through it, against the turn-off action. This increases the dissipation in

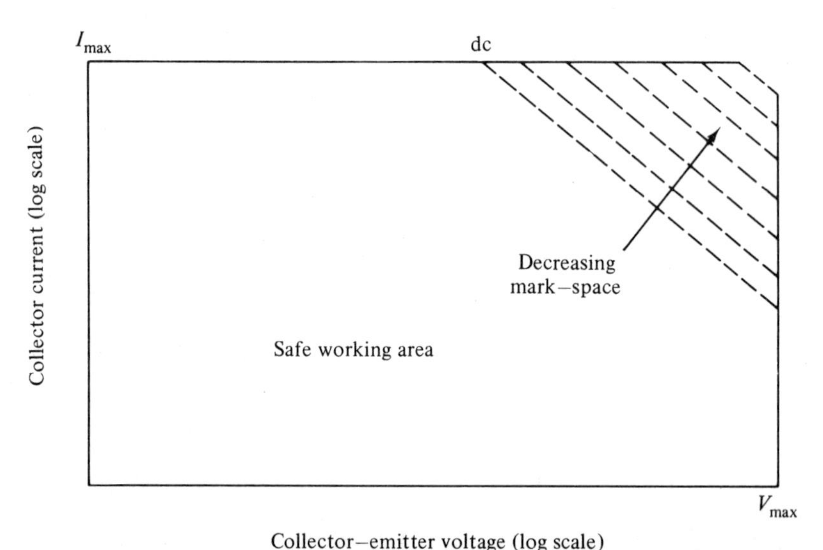

Fig. 12.18 Showing safe working area. Transistor *V/I* switching and static values must remain within safe envelope

relatively small areas in the transistor and causes localized hot spots. If allowed to exceed certain values, this increased dissipation can destroy the transistor. Manufacturers have produced curves of safe working areas for their transistors during switching. This effect is called secondary breakdown. The safe working area is shown in Fig. 12.18. A typical example of a power control application for transistors is now described. The waveforms of Fig. 12.19 indicate that diagonal pairs of transistors are

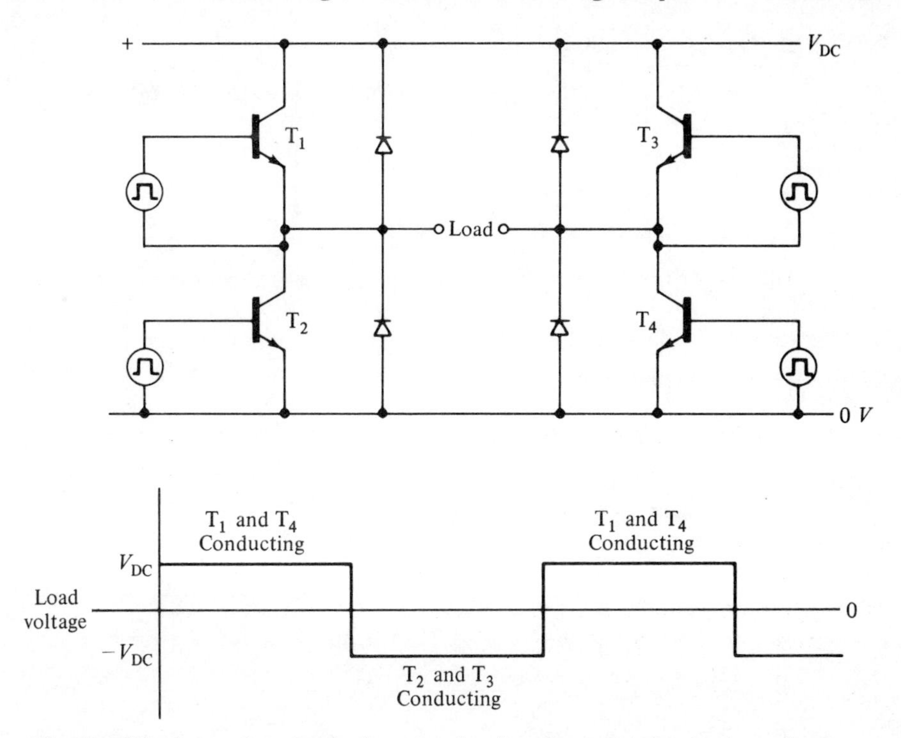

Fig. 12.19 Basic transistor bridge invertor circuit and associated output waveform

switched on together, causing the voltage across the load to reverse each time the pairs are changed. This circuit provides an alternating squarewave output from a d.c. supply, and is called an invertor. By adding another two transistors and switching them all in the right sequence, a three-phase invertor can be easily obtained. There are commercially available a.c. variable-frequency motor control systems that use transistors in this type of circuit.

The diodes across each transistor enable inductive currents to be switched without damaging the transistors. Inductive current can flow through the diodes when the transistors turn off. The diodes also clamp the load voltage to that of the supply.

12.3 Variable Voltage Using Thyristors, Triacs, and Transistors

12.3.1 *Variable voltage from a d.c. supply*

We have seen that solid-state power control devices can be used for d.c. and a.c. voltage control. The d.c. voltage control will be considered first as applied to d.c. motor speed control.

12.3.1.1 Variable d.c. voltage from a fixed d.c. supply. The production of a variable d.c. voltage from a d.c. supply, was shown in section 12.2.1, where a method of chopping the d.c. into pulses, and varying the mark to space ratio, varied the effective mean level of the output d.c. This system, applied to a d.c. motor, is shown in Fig. 12.20. The control device may be either a transistor, or a thyristor with its associated

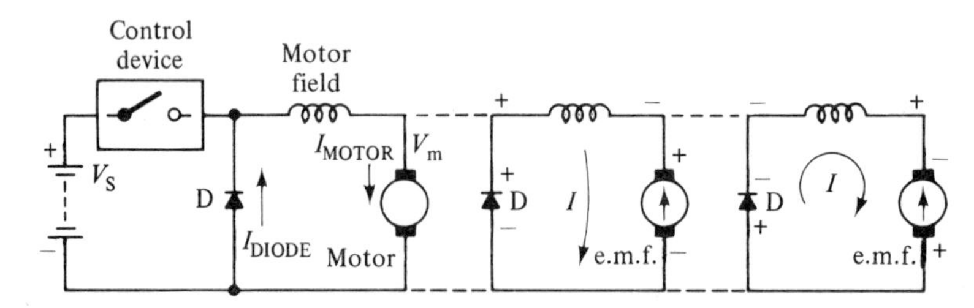

Fig. 12.20 Speed control of a d.c. motor showing flywheel diode action

commutation circuit. The diode is used to provide a flywheel action to the motor current, as described in section 12.2.1. Current flowing through the motor will be forced to continue after the control device is switched off. This forcing action is caused by the collapse of flux in the magnetic circuit of the motor. The collapsing flux

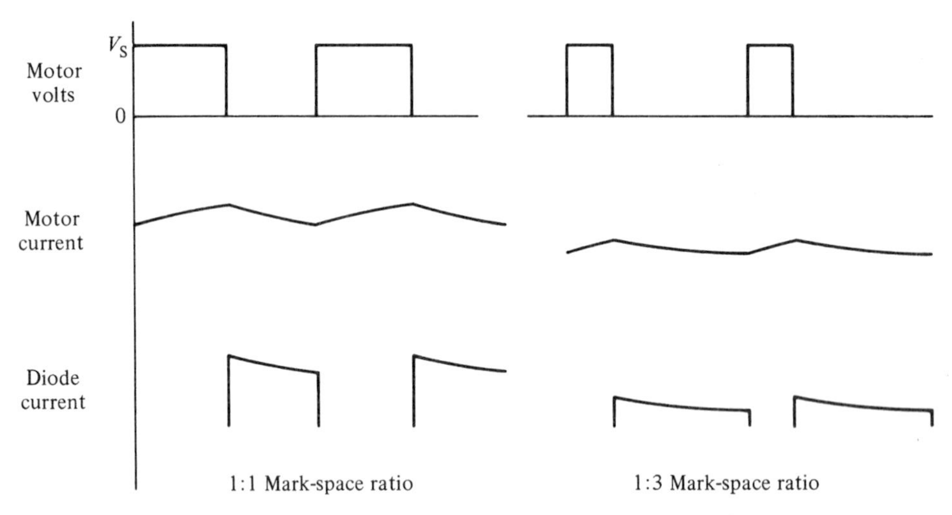

Fig. 12.21 Voltage and current waveforms of Fig. 12.20

causes a reversal of the motor and field winding polarities opposing the motor back e.m.f. as shown, and as the control device is off, the only path for the current is through the diode. Since the diode resistance is low, and inductance time constants are proportional to L/R, the time constant of the motor/diode circuit is normally relatively long compared with the periodic time of the chopping frequency. The waveforms of the various voltages and currents in the circuit are given in Fig. 12.21. The main application for this system, is for the speed control of the traction motor in electric vehicles. However, it can be used anywhere where it is necessary to control the speed of a d.c. motor driven from a d.c. supply.

12.3.2 *Variable a.c. voltage from an a.c. supply*

This technique is most likely to use triacs or thyristors plus diodes. It can be used to control the speed of an induction motor, by variable slip control. A typical basic circuit is as shown in Fig. 12.22, for a three-phase induction motor, driven from a

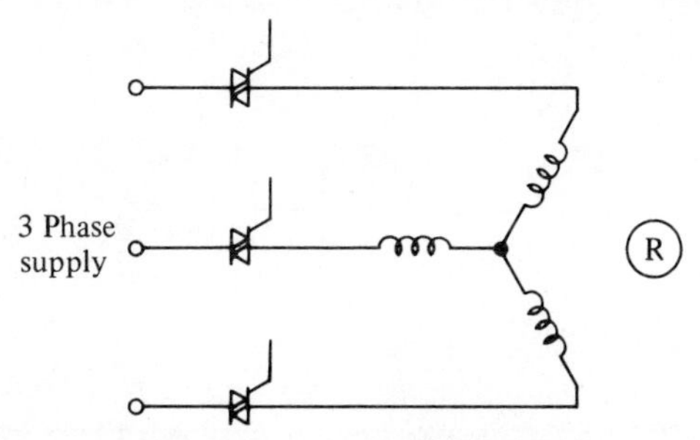

Fig. 12.22 Three-phase triac a.c. controller with induction motor load

three-phase supply. The waveforms associated with this system, are shown in Fig. 12.23. This method of control is only suitable for use on a standard low-resistance-rotor induction motor, when the motor load is of a type that increases with speed. For instance, a fan or centrifugal pump.

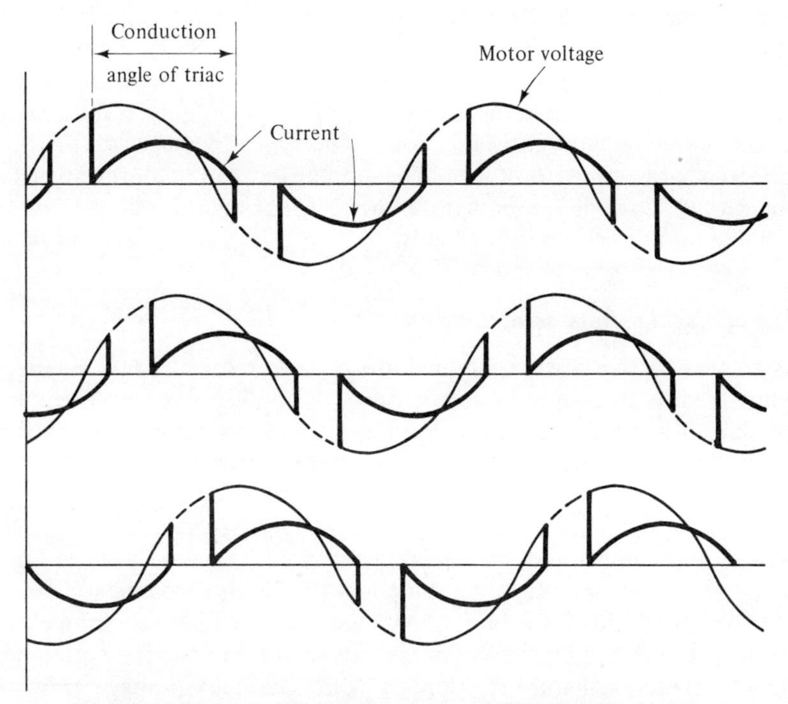

Fig. 12.23 Output voltage and current waveforms of Fig. 12.22

The torque/speed curves for an induction motor with a high-resistance rotor, and one with a low-resistance rotor, together with a fan-type load characteristic, are shown in Fig. 12.24. From these characteristics, it is seen that using either a

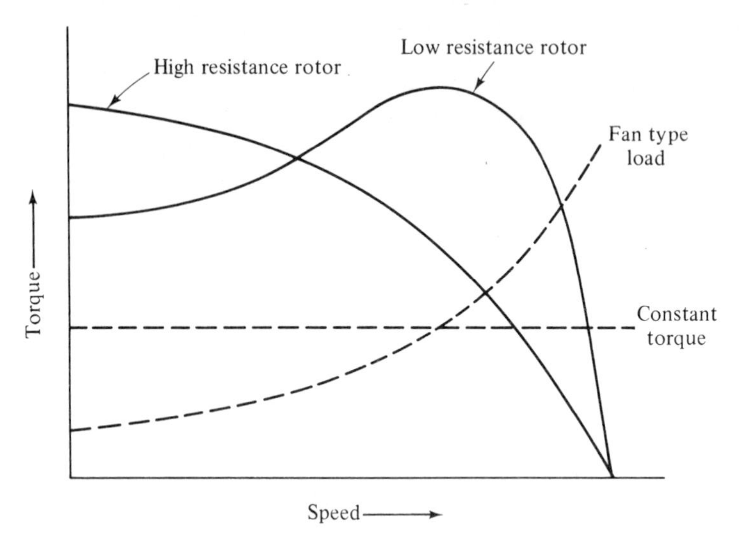

Fig. 12.24 Torque/ speed characteristics of induction motor with high/ low resistance rotor and constant/fan-type load

high-resistance rotor or a low-resistance rotor, the motor torque/load situation is always stable for a fan-type load, but it is only stable when using a high-resistance rotor on a constant torque load. In this situation, stability is defined as a decreasing speed of the motor being compensated by a surplus of motor torque over that required by the load. A high-resistance-rotor induction motor is less efficient than an induction motor with a low-resistance rotor. This type of voltage control, therefore, used with a high-resistance rotor, would be suitable for constant torque type loads for intermittent use, and when used with a low-resistance-rotor induction motor, it is only suitable for fan or square law loads.

12.3.3 *Variable d.c. from an a.c. supply*

This is at present the most common form of motor speed control system using solid-state devices. It is used in a wide range of applications, and there are many commercial units on the market. Good speed control characteristics are easily obtainable, and with tachogenerator or armature voltage feedback, good speed holding down to very low speeds (>1 rev/s) are possible. This system is at present the most popular because of its good performance and relative simplicity. Typical circuit diagrams of basic single- and three-phase control systems are shown in Fig. 12.25.

Both these circuits operate in the same way. In the three-phase system, the field winding is supplied with a constant d.c. voltage from the full-wave bridge consisting of diodes D_2–D_7. Diode D_1 is a flywheel diode with similar action to that described earlier. The motor armature is supplied with a variable voltage from the half controlled bridge, consisting of diodes D_5–D_7 and thyristors TH_1–TH_3. The phase angles of the trigger pulses are varied with respect to the supply voltage, to provide phase control of the rectified supply. Waveforms of the single-phase system and the three-phase system are shown in Fig. 12.26. The rectified supply waveforms and the back e.m.f. of the motor are also shown.

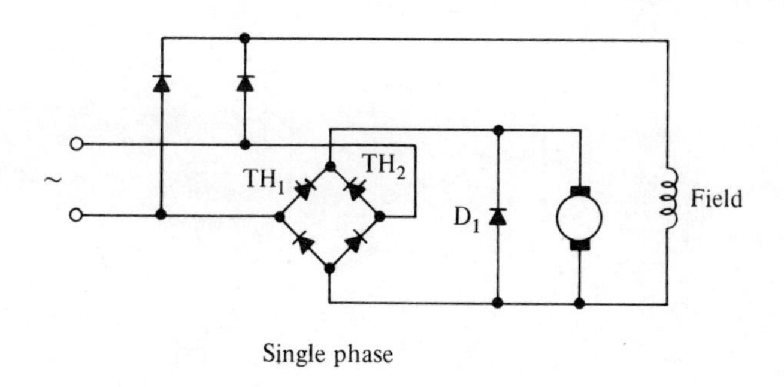

Single phase

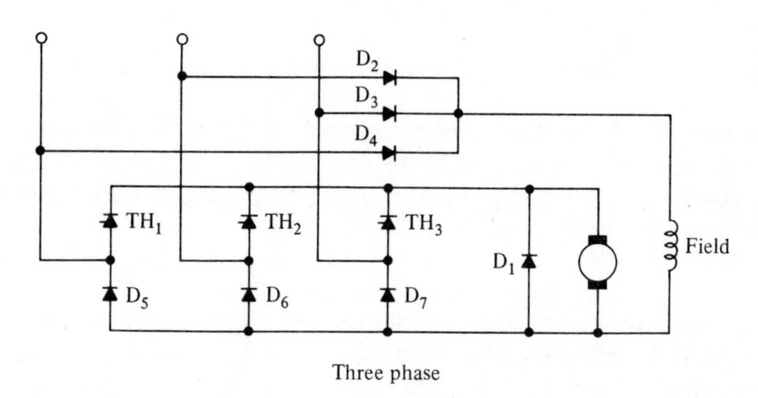

Three phase

Fig. 12.25 Single- and three-phase d.c. motor speed control from a.c. supplies

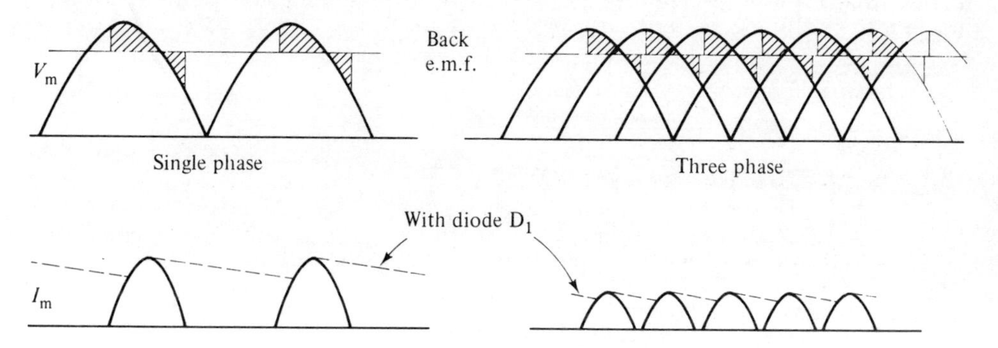

Fig. 12.26 Voltage and current waveforms of circuit in Fig. 12.25

12.4 Variable Frequency from an a.c. Supply

In theory, this is a very simple form of motor control. In practice it is somewhat more complex to implement than d.c. motor speed control (i.e., d.c. from a.c.). The object is to obtain a variable frequency supply to drive an induction motor at variable speed. The circuit to achieve this is shown in block diagram form in Fig. 12.27.

A motor, as with transformers and chokes, requires a certain amount of volt-seconds to energize it. Hence if the frequency changes, the voltage must also be changed to keep the V/f ratio constant. To achieve this in practice the system uses

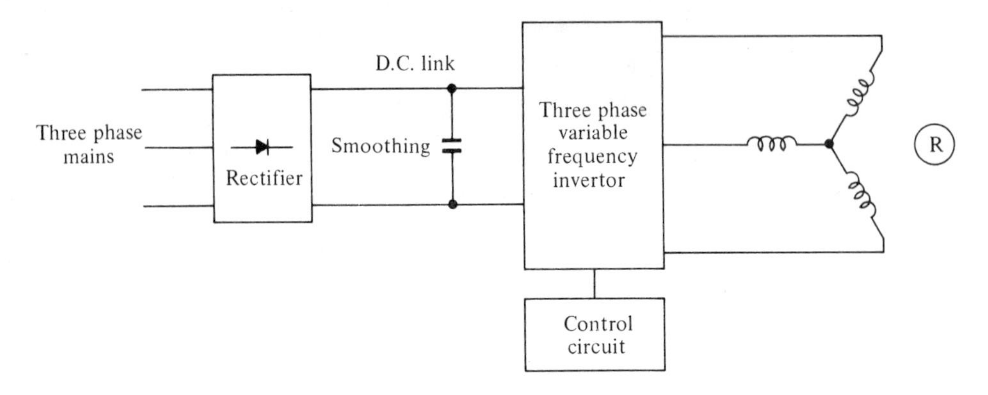

Fig. 12.27 Basic variable frequency a.c. motor speed control circuit

either a variable d.c. link voltage, which is controlled by means of a phase-controlled bridge, and is made to vary with the frequency, or a fixed d.c. link, when the voltage variation is achieved within the invertor using pulse width modulated techniques. Both techniques are used in commercial systems and they both have their uses. The fixed d.c. link system is probably cheaper in the cost of components, but requires a more complex invertor, whereas the variable d.c. link allows a simpler invertor circuit but is overall a more complex system. The performance of the two systems is similar, but differs in minor points. The difference between the two systems, with respect to output voltage waveforms, is given in Fig. 12.28. Harmonic reduction in the output waveform can be achieved in both cases, by careful control of the pulse widths. Graphs of the typical harmonic content of the two pulse waveforms are given in Fig. 12.29. The basic circuit for a fixed d.c. link invertor using thyristors is shown in Fig. 12.30.

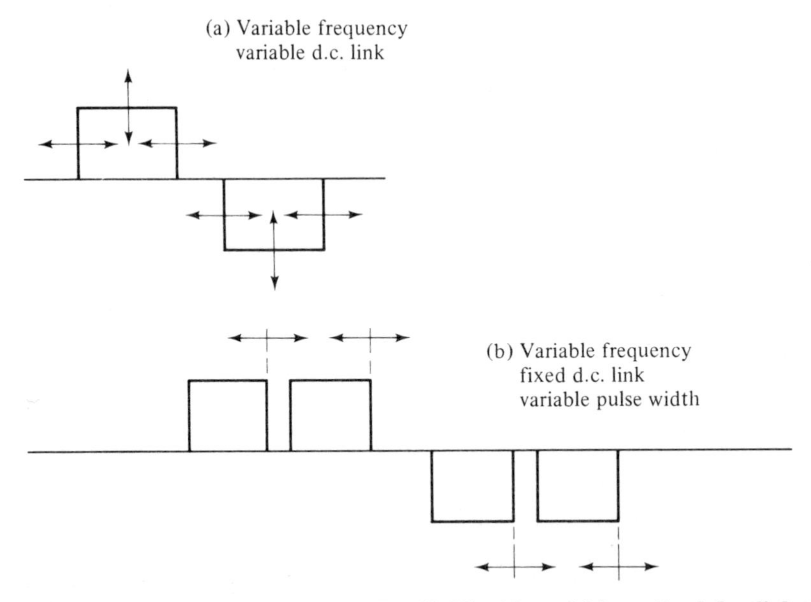

Fig. 12.28 Output voltage waveforms of Fig. 12.27 with variable or fixed d.c. link voltage control

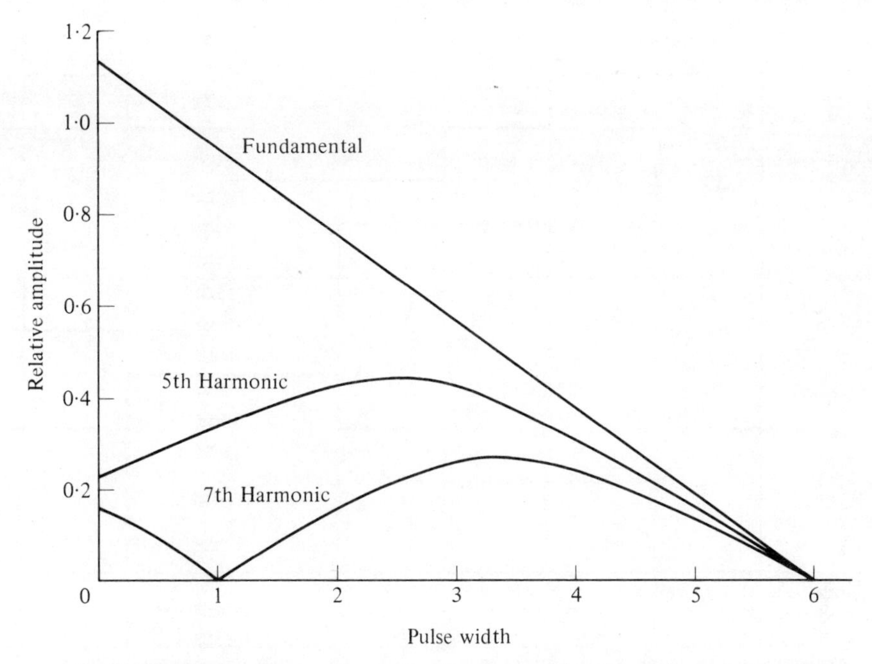

Fig. 12.29 Relative harmonic content for single pulse stepped square waveform

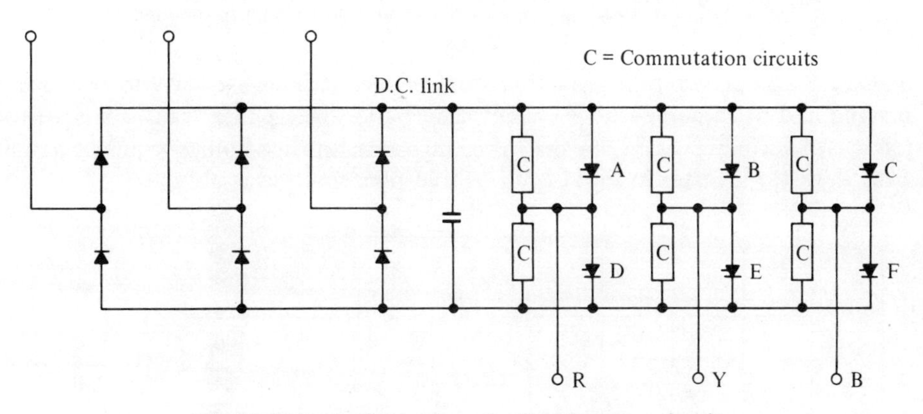

Fig. 12.30 Basic three-phase thyristor invertor circuit

The output waveforms for a basic six-step square-wave invertor, together with the thyristor conduction periods to produce the waveforms, are given in Fig. 12.31, from which it is seen that at any one time there are three thyristors conducting. One-sixth of the time they carry the full motor line current, and two-sixths of the time, half the motor line current.

12.4.1 *Transistor circuits for variable frequency*

Variable-frequency invertor systems using transistors are on the market, but as yet these are not so popular as thyristor invertor circuits, mainly because transistor technology has been slow in producing economic transistors to compete with the

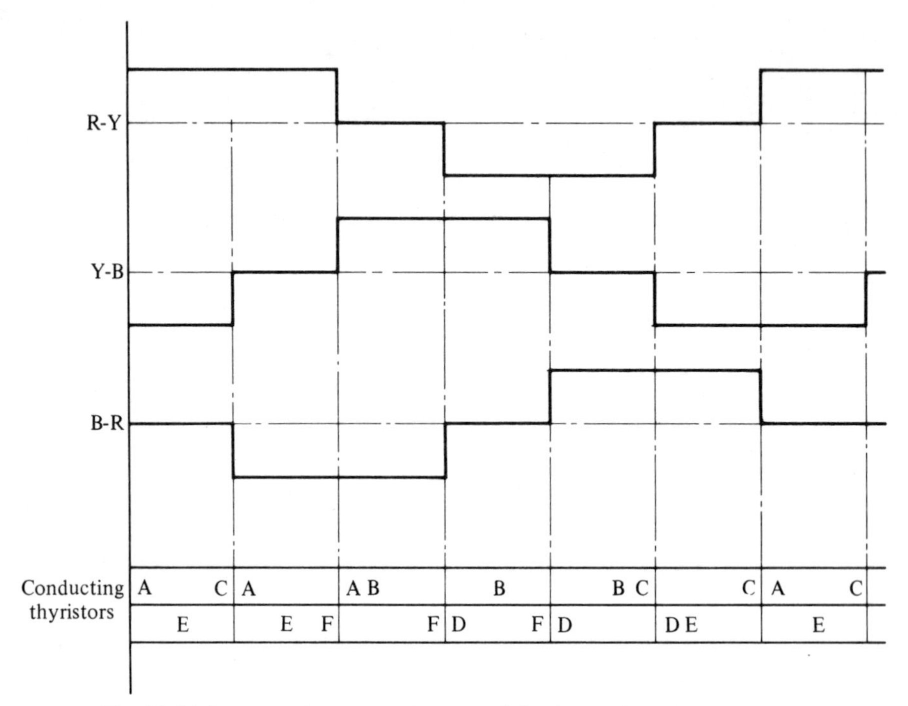

Fig. 12.31 Output voltage waveforms and thyristor triggering sequence

robustness and power ratings of thyristors. Hence it is not so easy to produce a versatile and price competitive system. The basic three-phase transistor invertor system is given in Fig. 12.32. By switching the transistors in a similar sequence to that of the thyristor invertor in Fig. 12.30, a three-phase output is obtained.

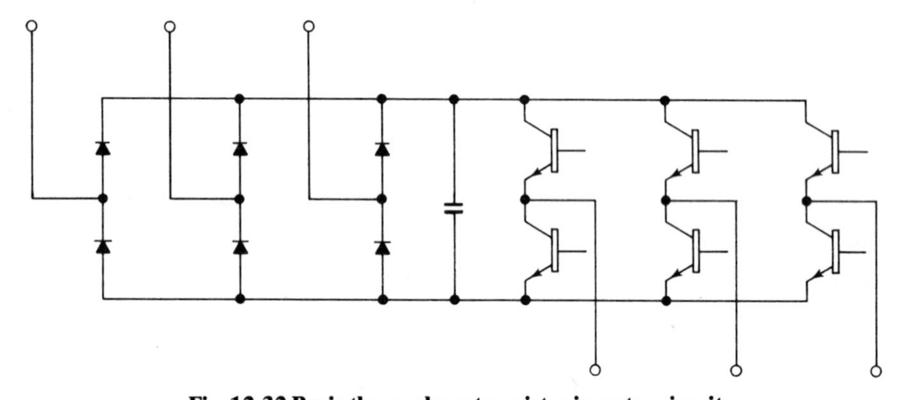

Fig. 12.32 Basic three-phase transistor invertor circuit

Transistors have one main advantage over thyristors, and that is their speed of switching. Consequently, more subtle methods of pulse width modulation can be used for both voltage, frequency, and harmonic reduction. Types of transistor are gradually becoming available, which are an attractive alternative to thyristors, and possibly in the near future they may become the first choice for this application.

12.5 Examples of Systems using Electronic Control

12.5.1 *D.C. motor speed control*

Although there are numerous circuits for d.c. motor speed control using electronic power control methods, they are all very similar in basic form, differing only in detail. A typical practical system is shown in Fig. 12.33. This circuit is similar to that of Fig. 12.25, but with the additional circuit refinements found necessary in practice. Going

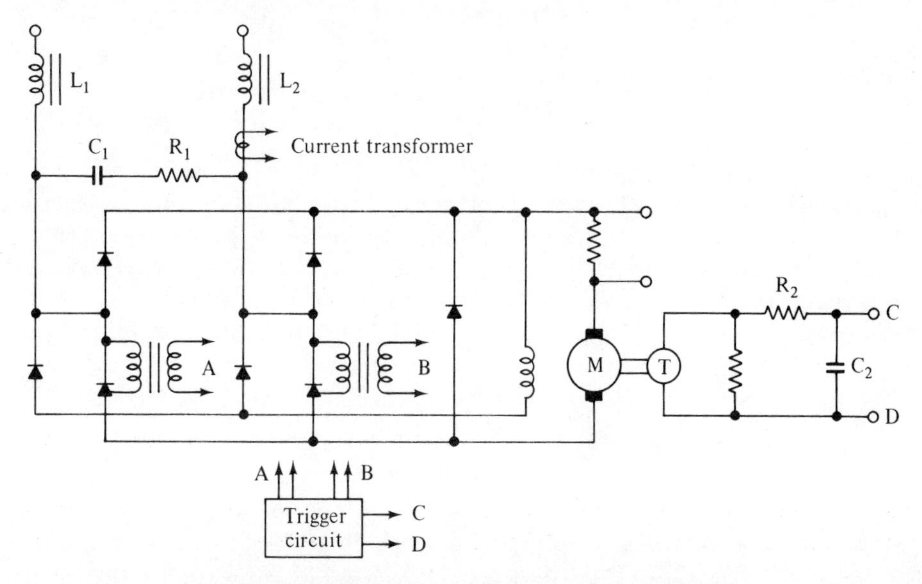

Fig. 12.33 Single-phase d.c. motor speed control circuit

through the circuit in detail and giving reasons for these additions to the basic circuit, we see:

(a) Capacitor C_1, inductors L_1 and L_2, and resistor R_1 prevent transient voltage spikes from reaching the thyristors, and turning them on inadvertently.
(b) The C.T. (current transformer) monitors the supply current (which is also a measure of the motor current) and by overriding other signals to the thyristor phase angle control circuit, imposes a limit to the amount of supply and motor current, hence providing current limit.
(c) The tachogenerator (T) generates a d.c. voltage proportional to motor speed; this voltage is compared with a reference voltage, and if there is any difference, the control and power circuits automatically correct the speed of the motor, to readjust the tachogenerator voltage. In this circuit, two transistors connected in a configuration termed a 'long tailed pair' or an integrated circuit operational amplifier, is used for the comparison.
(d) The smoothing circuit, consisting of R_2 and C_2 across the tachogenerator is necessary to minimize the effect of tachogenerator brush ripple and brush noise. This value of smoothing has to be carefully selected, since too large a capacitor will cause instability, by introducing a lag into the closed loop.
(e) The trigger circuit for the thyristors is described in section 12.2.1.

The relationship between the voltage output from the controlled bridge and the motor speed can be obtained. The motor equation is:

$$\text{motor speed} = \frac{E - IR}{\phi} \cdot C.$$

Where

E = applied voltage
I = armature current
R = armature resistance
ϕ = field flux
C = constant for the machine

As R is very small IR is also small compared with E and so the speed is very nearly proportional to E if the flux remains constant. For most purposes when considering control characteristics, the flux can be assumed constant. The speed can therefore be considered proportional to the applied voltage for most practical purposes. A curve of a thyristor bridge output voltage against phase angle is shown in Fig. 12.34. The speed of the motor is therefore of the same relationship to phase angle, and since

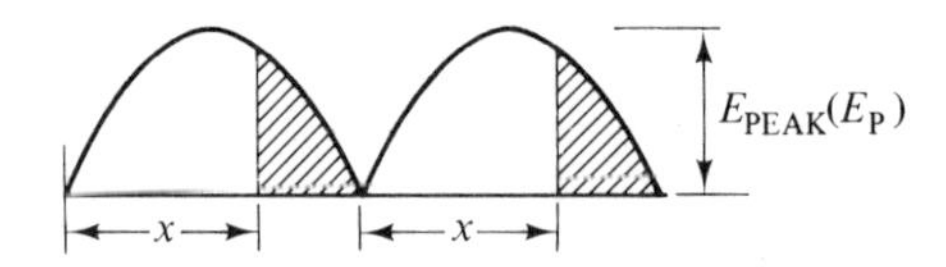

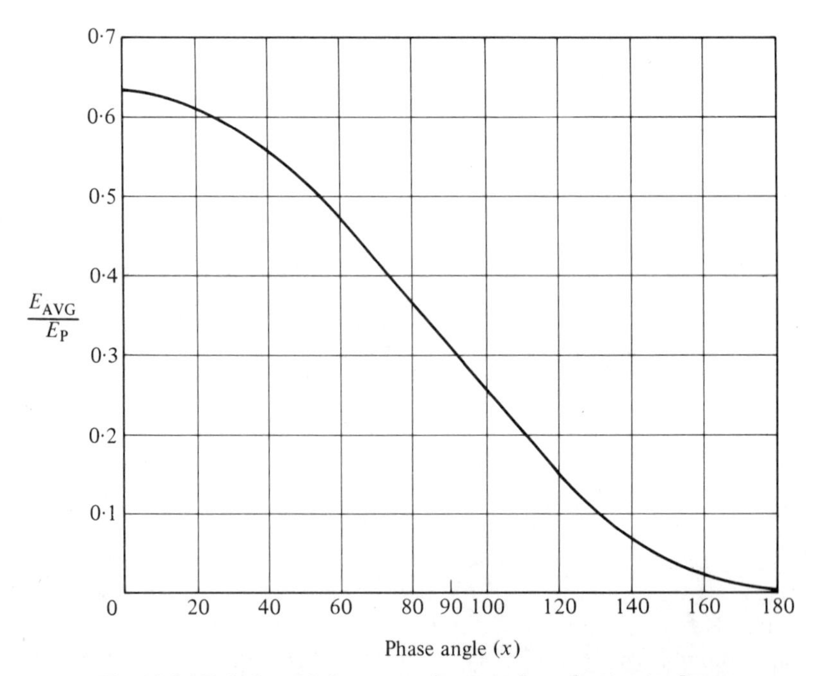

Fig. 12.34 Relationship between phase angle and output voltage

phase angle is usually proportional to the amplitude of a d.c. control signal, the speed is also of the same relationship to the amplitude of the signal. It is possible to build into the phase angle control circuit a sine function that causes the motor speed to have a linear relationship with respect to the d.c. control signal. Further information on this subject can be found in Chapter 14.3.

12.5.2 *A.C. motor speed control*

12.5.2.1 Slip control. This method of control allows the slip of the motor to increase by decreasing the motor volts. Obviously this can only be done with induction motors. In hysteresis and synchronous motors there is no slip until pull-out torque is reached. It has been mentioned earlier that a satisfactory method of motor control for square law or fan-type loads, is an induction motor fed with a variable voltage a.c. supply. The family of curves for such a system is given in Fig. 12.35.

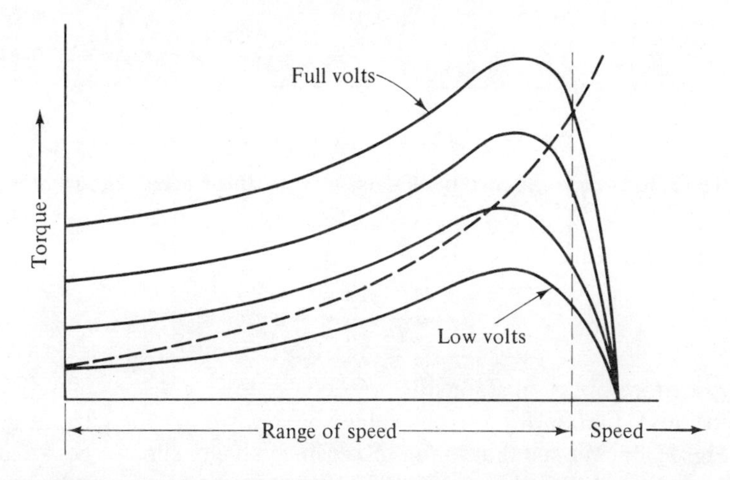

Fig. 12.35 Induction motor speed control with fan-type load by variable voltage

At low motor speeds, the stator flux rotational speed is still at synchronous speed, which is 1500 rev/min for a four-pole motor on a 50 Hz supply, whereas the rotor speed is almost at a standstill. The difference between the stator flux rotational speed, and the rotor speed, is the slip given as $N_s - N$ where N_s = synchronous speed, and N = actual rotor speed. The torque equation for an induction motor is:

$$\text{Torque} = K\frac{N_2}{N_1^2} \times \frac{E_1^2 R_2}{(R_2^2 + S^2 x_{2s}^2)}$$

Where

N_1 = supply frequency
N_2 = rotor voltage frequency
S = slip factor N_2/N_1
E_1 = supply voltage
R_2 = rotor resistance
x_{2s} = rotor reactance at standstill

We see that the torque is proportional to E^2 and so small changes in voltage give much greater changes in torque. A plot of this equation with respect to slip gives the curve of Fig. 12.36.

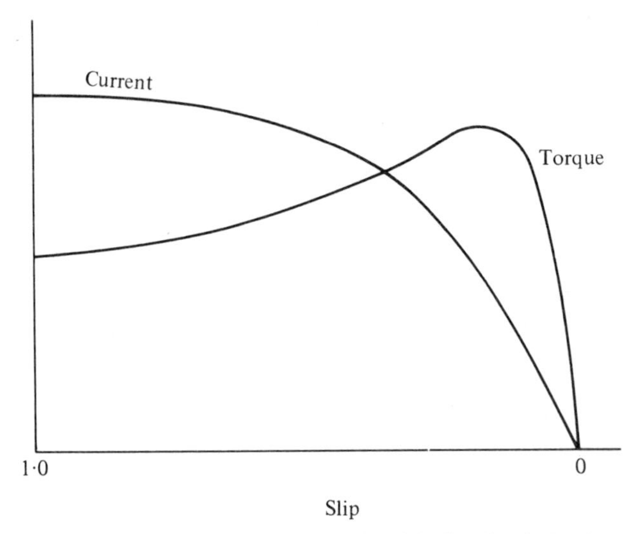

Fig. 12.36 Torque and current relationship with slip of an induction motor

The equation for rotor current is:

$$I_2 = \frac{SE_{2s}}{\sqrt{(S^2 x_{2s}^2 + R_2^2)}}$$

Where E_{2s} = rotor voltage at standstill.

If we plot this equation for various values of slip, we get the current curve also shown in Fig. 12.36. We see that for an increasing value of slip, we get an increasing value of rotor current. It is therefore evident that motors running at large values of slip, are inefficient and would overheat. However, in our fan-load situation, the load torque requirement decreases with speed, and the motor voltage at low speeds is also reduced, and so the system is reasonably efficient. It is not easy or convenient to use this method of speed control on a constant-torque load because the system is inherently unstable, and very inefficient. Consider the graphs of Fig. 12.37. The motor can be controlled by feedback, using a tachogenerator, to remain at point A on the torque/speed characteristic. However, the current is equivalent to that at B and therefore is about five times that of point C where the motor would normally run at that torque; consequently the copper losses are about 25 times (I^2R). There are stability problems with this system on constant-torque loads because the feedback control circuit requires a wide range of loop gain.

The same system can be used to good purpose using an induction motor with a high-resistance rotor. If the rotor resistance is equal to its reactance, the peak torque occurs when the slip is equal to the synchronous speed, i.e., when the rotor is at a standstill. The torque/speed characteristic is then as in Fig. 12.38.

By varying the voltage to the motor, the speed of the motor can be controlled over a wide range. This system is inherently stable, and can be used on open and closed

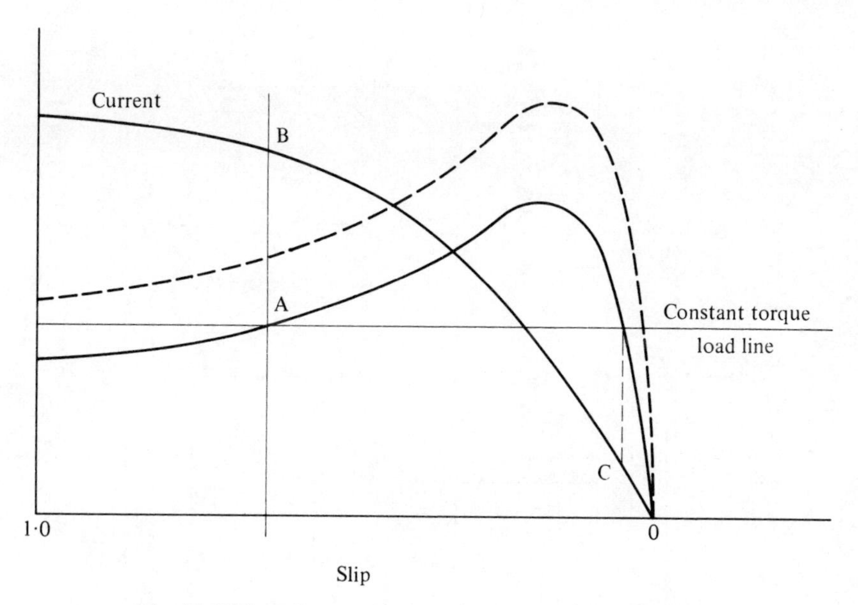

Fig. 12.37 Speed control in unstable region using feedback

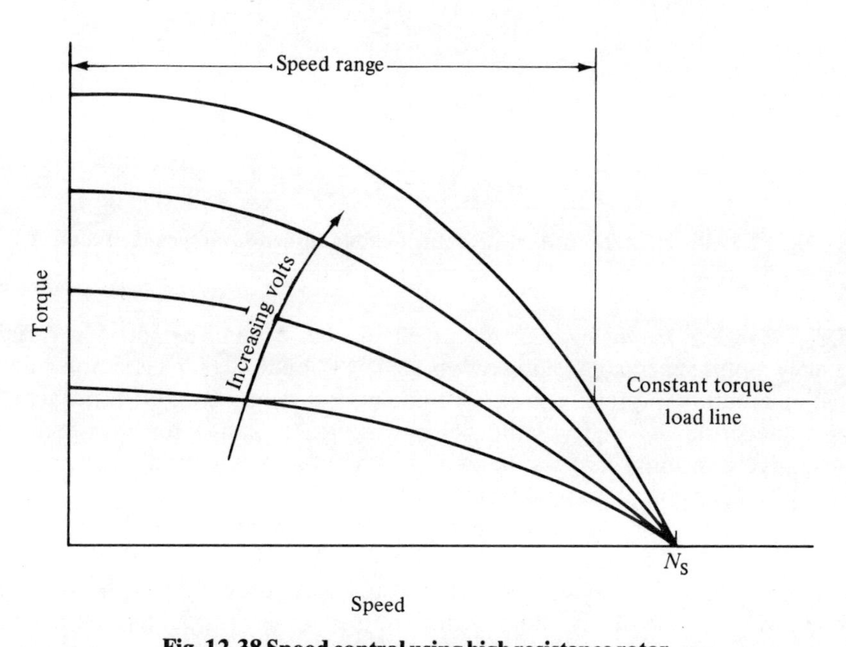

Fig. 12.38 Speed control using high resistance rotor

loop systems. The current is still high compared with a normal system using a low-resistance rotor, and the motor has to be derated by a factor of about three. It is still a useful and relatively simple system, providing the motor is rated accordingly. The basic control circuit for this type of system was shown in section 12.3.2.

A more detailed and practical circuit is shown in Fig. 12.39. This system will satisfactorily drive a low-resistance motor for fan-type loads, or a high-resistance motor on constant-torque loads.

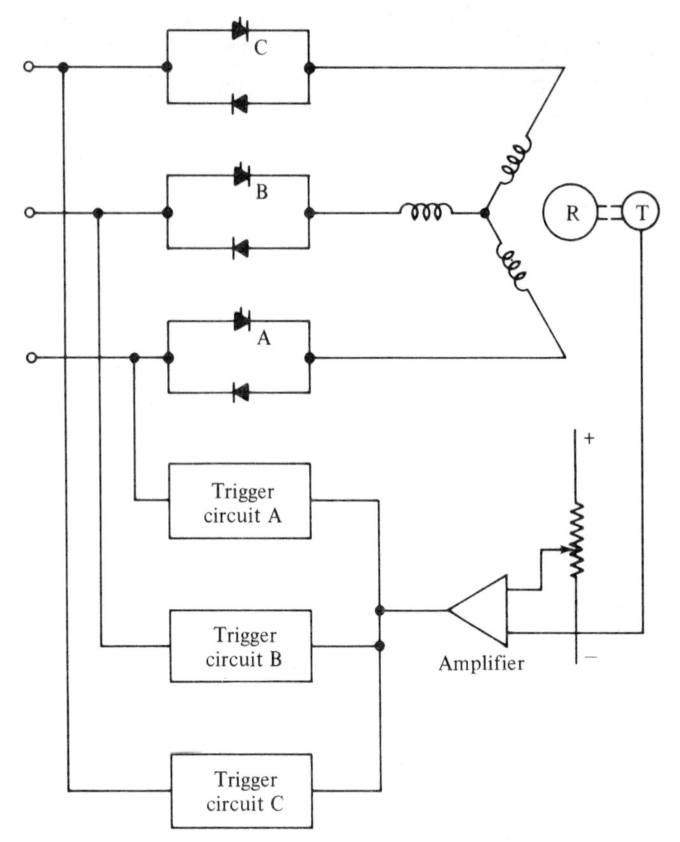

Fig. 12.39 Basic half controlled a.c. controller for induction motor speed control

12.5.2.2 Variable frequency. We discussed the basic requirements for variable-frequency motor speed control in section 12.4. Although the idea is simple enough, its implementation in practice is rather difficult. For a very long time, perhaps fifty years, engineers have toyed with the idea of variable-frequency control systems. It is only relatively recently that the flexibility of solid-state electronic power control devices has made complex static invertors possible.

Since the thyristor came into the life of the engineer, about 1958, work on invertors using thyristors has progressed so that today there are available some very sophisticated variable-frequency invertor systems. As previously stated, the major problem with the use of thyristors in d.c. control circuits is commutation. This, of course, also applies to invertors, since they convert d.c. to a.c. and therefore have to commutate d.c. currents. If a semiconductor device were available that could switch on and off reliably with gate pulses, variable-frequency a.c. motor speed control would be in much greater use than it is now. However, considerable work has been done in developing reliable and efficient commutation circuits, and the use of a.c. motor speed control is gaining impetus.

In practical circuits there are a wide variety of ways to commutate the thyristors. A typical commutation circuit for one limb of a three-phase variable-frequency invertor, is shown in Fig. 12.40. Thyristors A and B are the main thyristors that

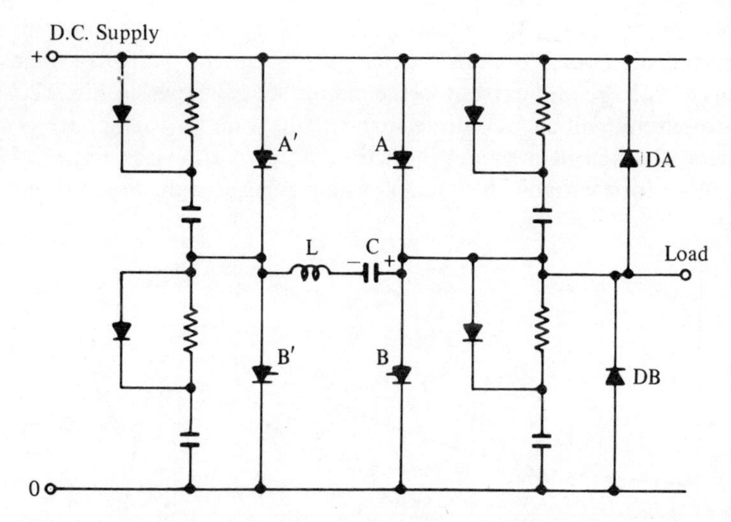

Fig. 12.40 Practical circuit of single-phase thyristor bridge invertor

supply the load current, and A′ and B′ are the commutating thyristors. The circuit action is as follows. Thyristors A and B′ are triggered, this allowing load current to flow through thyristor A and also allows capacitor C to be charged to the polarity shown. When the capacitor is fully charged, thyristor B′ turns off because no further current can flow through it. When it is necessary to commutate thyristor A, thyristor A′ is triggered. The capacitor C then discharges partly through the load, and partly round the path A′, L, DA. The volts drop across DA reverse biases thyristor A, and providing it is reverse biased for a time greater than its specified turn-off time (typically 20 μs) it will turn off (see section 12.2.1). The capacitor has in the meantime become partially charged in the reverse direction, so that part of the commutation energy has been conserved. The inductance L has two functions: one in limiting the peak current during commutation, and the other causing C to be charged in the opposite direction, by means of resonance with C. The next action of the circuit is to trigger thyristors B and A′ so that current can flow out of the load, also allowing C to be topped up to the supply voltage.

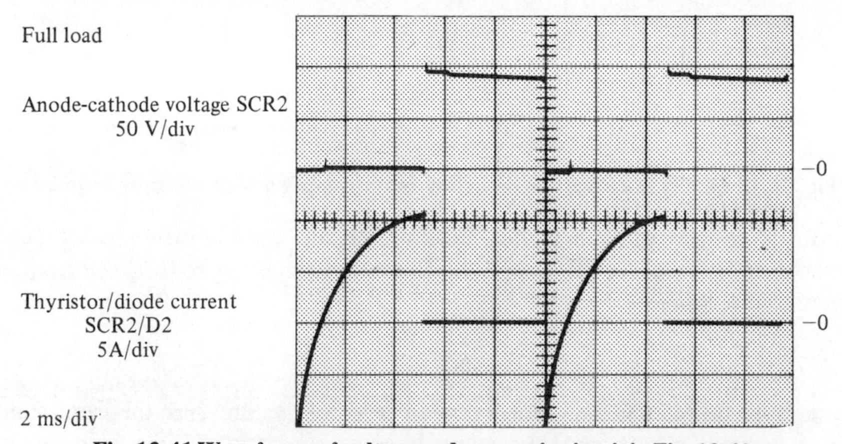

Fig. 12.41 Waveforms of voltage and current in circuit in Fig. 12.40

The diode, resistor, capacitor network across each thyristor, prevents high-voltage transients, and high rates of dv/dt from turning on the thyristors inadvertently. Waveforms of voltage and current in the circuit are as shown in Fig. 12.41 and the entire power circuit will include three such circuits, one for each phase. The normal torque/speed characteristic for an induction motor is shown in Fig. 12.42 and the family of curves for a variable-frequency system using an induction motor is shown in Fig. 12.43.

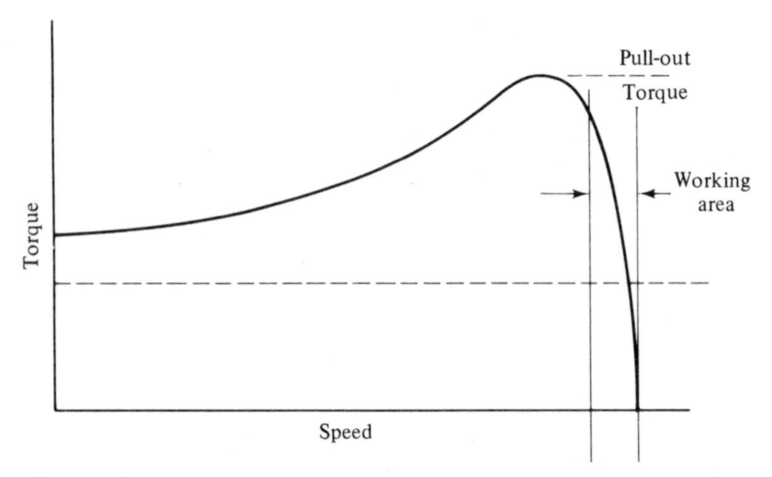

Fig. 12.42 Induction motor torque/ speed characteristic showing working area

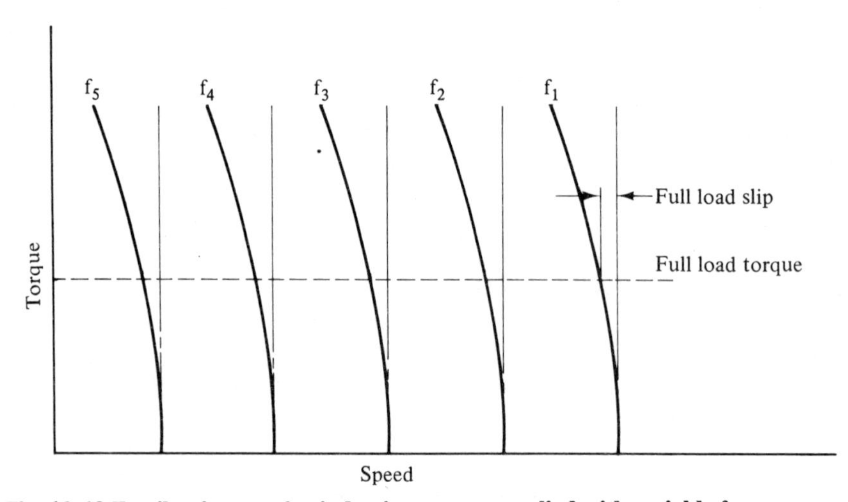

Fig. 12.43 Family of curves for induction motor supplied with variable frequency

Feedback control using a tachogenerator, or slip compensation using current feedback, can be employed to stiffen the characteristic, by increasing the frequency for increased torque.

12.5.3 *Reversing and braking*

The variable-frequency a.c. motor speed control system has inherent control facilities that d.c. motor control systems have to have added. One such facility is reversing. This is achieved in d.c. motor speed control by reversing either the field voltage or armature voltage by contactors or dual bridge systems. In a.c. motor speed control, it is achieved simply by changing the sequence in which the invertor thyristors are triggered and this is done at the low level logic stage. Another facility inherent in a.c. motor speed-control systems is braking. By reducing the frequency of the invertor, the induction motor torque becomes negative, and causes the rotor rapidly to slow down to the new speed. This can be more clearly seen by studying the torque/speed curve of Fig. 12.44.

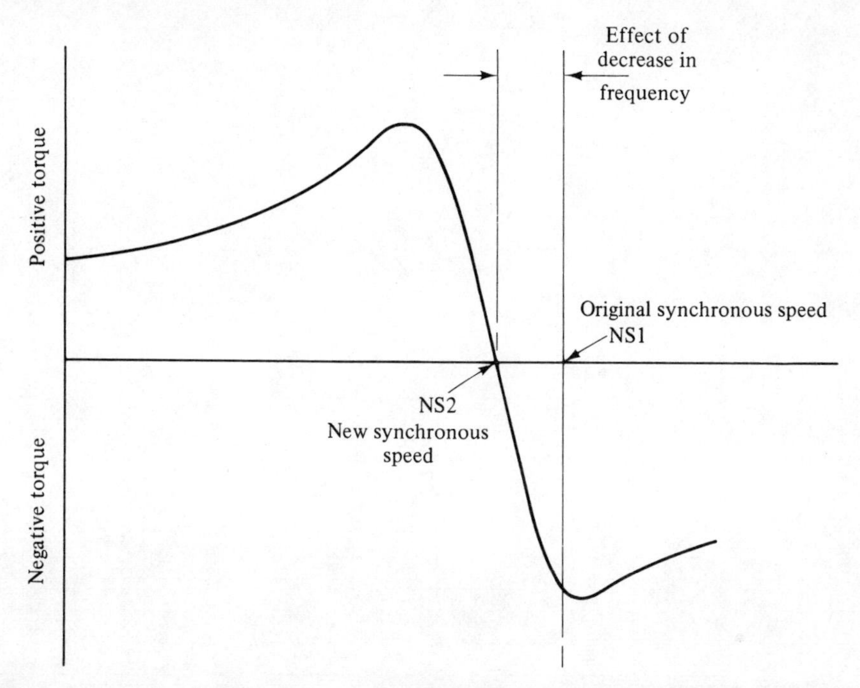

Fig. 12.44 Induction motor torque/ speed characteristic showing braking effect with decrease of frequency

In d.c. machines, braking is achieved by regeneration or dynamic braking, both of which require additional circuitry.

The excess energy fed back in an a.c. system does not return to the mains supply but is absorbed in the invertor and motor, as increased power losses. This is not as bad as it seems, since the deceleration rate of a 4 h.p. induction motor can be as fast as 0·1 second over a 10:1 speed change. Although the basic circuit of an a.c. variable-frequency system is much more complex than that of a basic d.c. system, the a.c. system has more inherent facilities. If the d.c. system facilities are increased to include reversing, braking, and good speed holding (long term drift) then it becomes as complex and as costly as a variable frequency system. Alternating current variable-frequency systems have been built for ratings greater than 500 h.p. for specialized applications, but for more general applications in industry the range is about 2–50 h.p. A photograph of a typical 5 h.p. variable-frequency a.c. motor speed control using a circuit similar to that described is shown in Fig. 12.45 and its operating characteristics are shown in Fig. 12.46.

Fig. 12.45 Industrial a.c. motor speed control unit

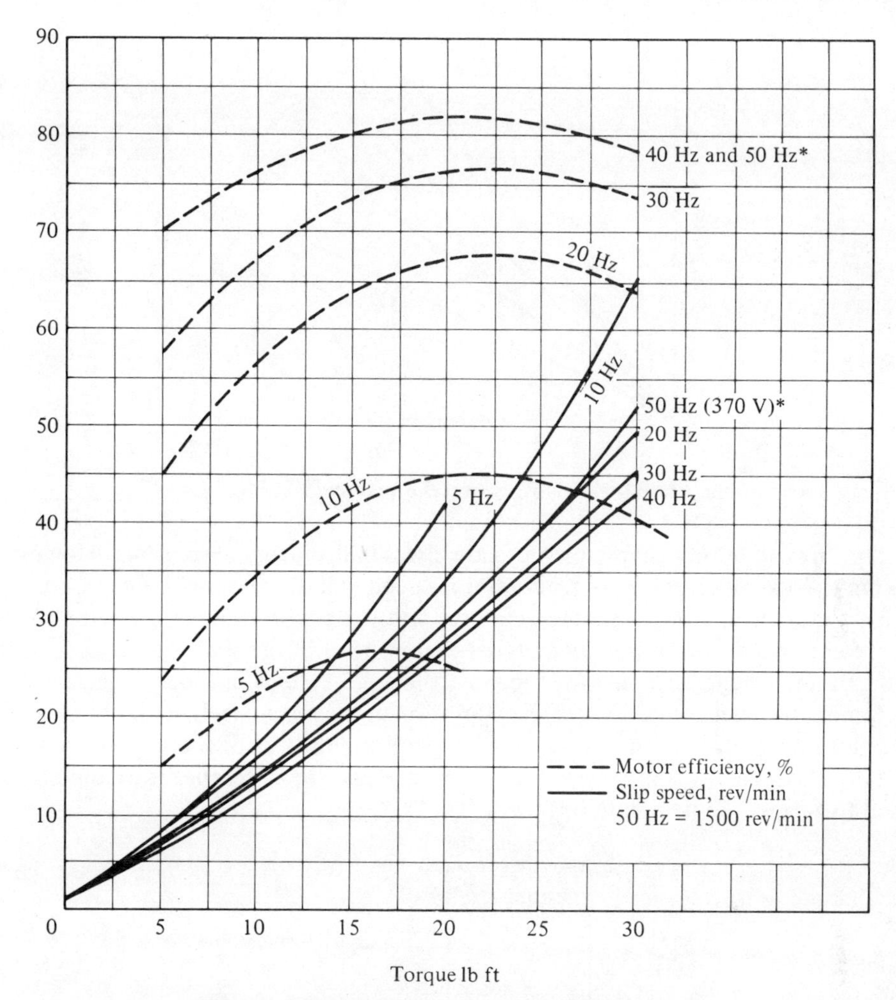

Fig. 12.46 Performance curves for unit in Fig. 12.45

12.6 Practical Problems

The motor and its speed control gear can be considered for most purposes to be a single unit with an electrical control signal input and a mechanical output in the form of a rotating shaft. It can be represented as shown in Fig. 12.47.

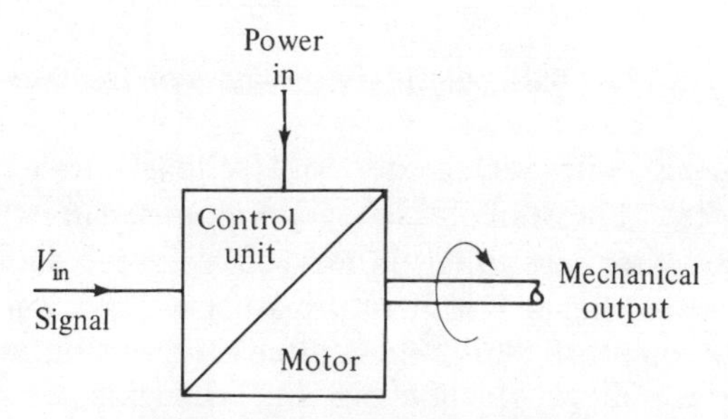

Fig. 12.47 Basic control system

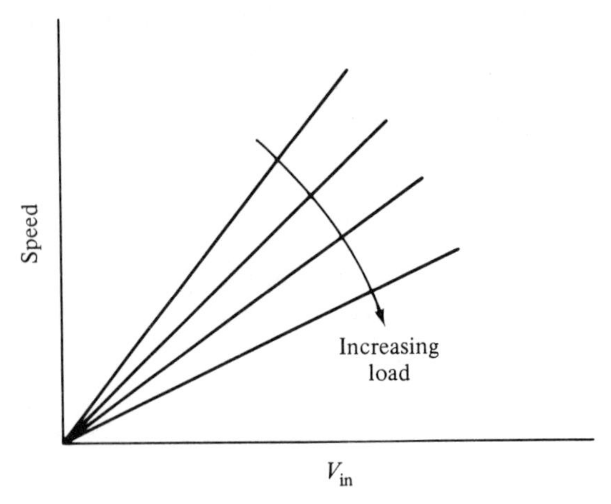

Fig. 12.48 Ideal input/ output characteristics of Fig. 12.47

The input to output characteristic is ideally as that shown in Fig. 12.48. However, owing to loading on the motor, the motor speed will decrease with increased load (unless it is a synchronous motor on a very stiff variable-frequency a.c. supply). To overcome this droop in the characteristic, the speed of the motor is sensed, using either a tachogenerator or other means; then the system in block diagram form becomes as shown in Fig. 12.49. The operation of the system is then that the load on the machine tends to reduce its rotational speed and the tachogenerator signal starts to fall. This signal is compared with a reference and the difference is amplified and used to correct the deviation in speed.

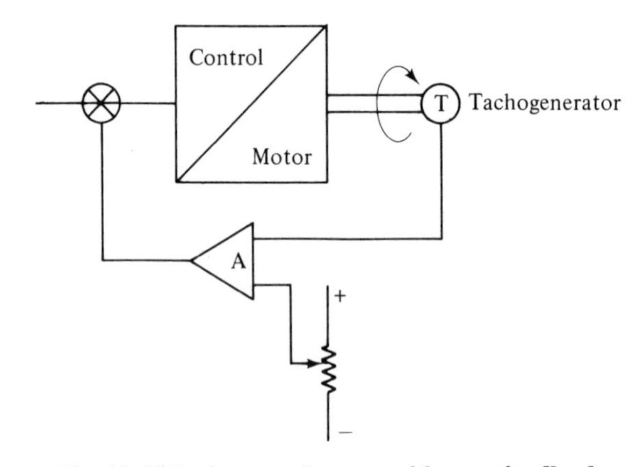

Fig. 12.49 Basic control system with error feedback

The major problem with tachogenerator feedback systems starts with the tachogenerator itself. The most likely tachogenerator used in industrial applications is the d.c. generator type, and this will have brush ripple and brush noise as an inherent part of its output signal. It is therefore usual and necessary to smooth out the ripple and noise by connecting a low-pass filter across its output as shown in Fig. 12.50. This has the effect of delaying the changing output signal of the tachogenerator as shown in Fig. 12.51.

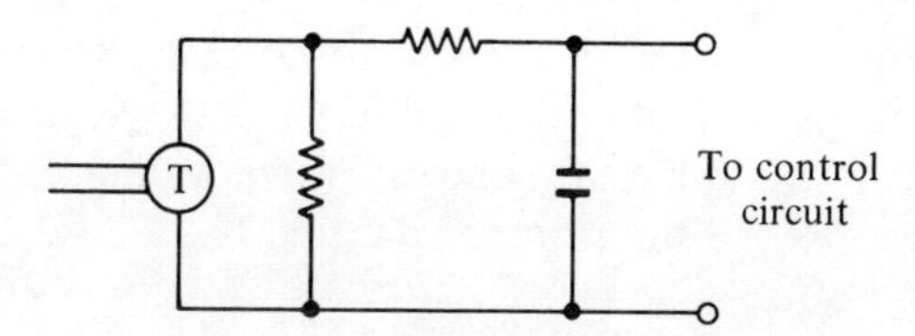

Fig. 12.50 Tachogenerator smoothing circuit

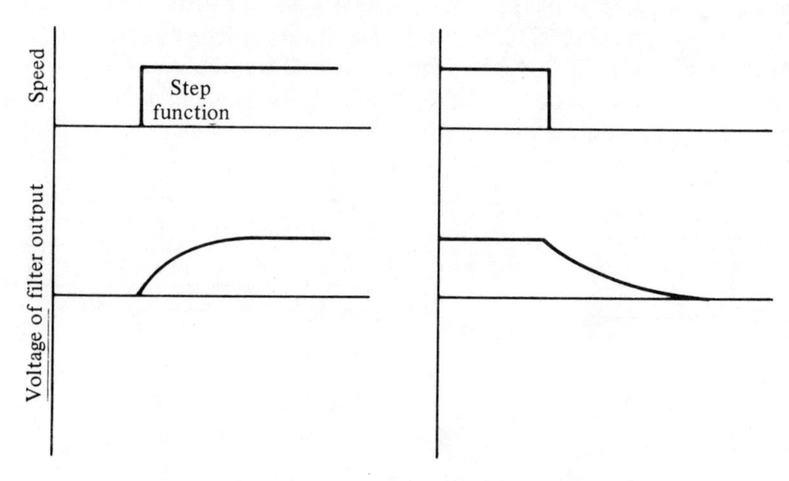

Fig. 12.51 Control waveforms showing lag in response

If the load is taken off a motor with such a filter connected across its tachogenerator the speed will rise. The corrective signal is delayed and so the motor speed is allowed to rise some considerable way before the control circuit regains control and the motor speed overshoots. This will also happen when the motor is initially switched on and its speed will overshoot the preset value as shown in Fig. 12.52. Because the delay (lag) is present in both the increase of speed and decrease of speed conditions there will be several overshoots. If the entire system is stable the

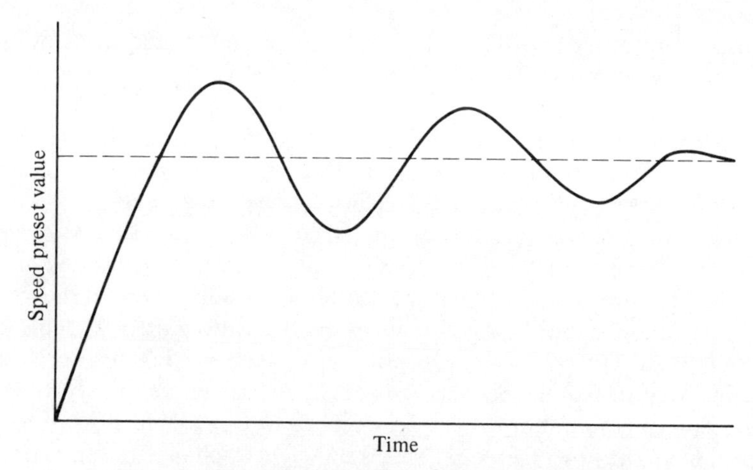

Fig. 12.52 Effect of lag in motor speed control system

overshoots will die down and the motor speed will settle at the preset level. If the system is unstable the overshoots will continue and the system will oscillate. It is therefore important to minimize the amount of smoothing in the tachogenerator while still eliminating most of the ripple and noise.

12.7 Mains Disturbance and Distortion

Phase control using thyristors (section 12.2.1) will cause a disturbance on the mains. This is because the current is, under some conditions, discontinuous and the load is not connected to the mains over the entire cycle. The regulation of the mains will therefore cause the mains voltage to drop when current is being taken. An example is shown in Fig. 12.53 where a half controlled bridge is feeding a resistive load.

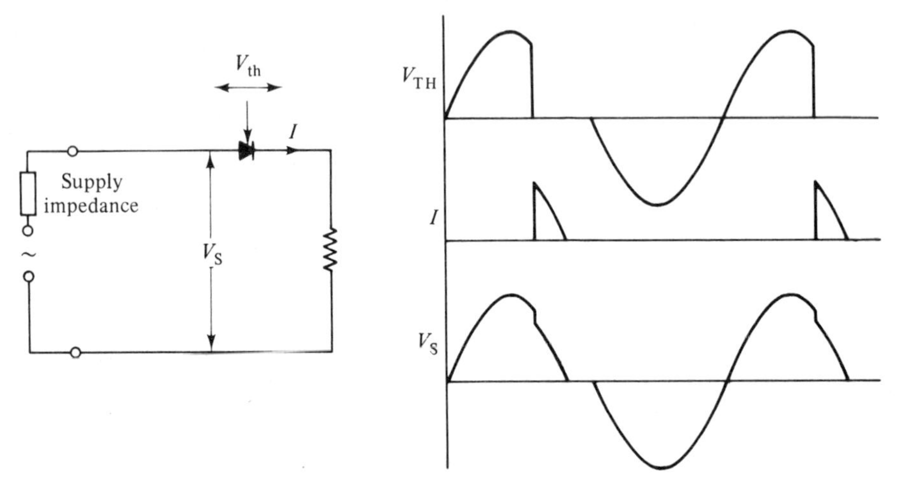

Fig. 12.53 Mains disturbance caused by thyristor phase control

In large installations the disturbance can be considerable and there is a regulation which limits the proportion of phase-controlled power to a given percentage of the overall power demand.

The distortion caused by fixed d.c. link a.c. motor speed control (variable frequency) is somewhat less. This is one of its advantages. The distortion is the same as that for a full-wave diode bridge with a capacitor load. With regard to radio frequency interference (RFI) both the d.c. motor system (phase control) and the a.c. motor system (variable frequency) radiate and transmit by conduction, considerable interference. This is because the thyristors turn on relatively fast (in about 1 μs) and as they can switch high currents at this rate (up to 5000 A) and high voltages (up to 5000 V) the radiated energy can be considerable. Special precautions have to be taken to minimize the effect of this radiation on other equipment in the vicinity.

Special care in the layout of the circuits of the thyristor control systems (d.c. and a.c.) has to be taken to prevent malfunction of the system. For instance, low level logic used in the control of the power thyristors has to be kept relatively free from the interference. Voltages of about 1 V can be picked up in short lengths of wire in the neighbourhood of the power circuits. This is sufficient to change the state of the logic circuits and cause the control circuit to malfunction.

The normal precautions are as follows:

(a) Twisted pairs of leads for thyristor gates.
(b) Built-in threshold voltages in the logic.
(c) Keep control circuits clear of power circuits.
(d) Decouple power supply leads.
(e) If necessary suppress thyristor gates.
(f) Use high-noise-immunity logic (MOS).

12.8 Maintenance

The main object in going to solid-state systems is to improve the overall reliability of the system over previous methods of control. For example, the variable-frequency a.c. motor-control system is theoretically very reliable and free from the need for maintenance. The solid-state control unit is freed from moving parts by using semiconductors, which have a theoretically long life as shown in table 12.1. It has an easily maintained and reliable motor (induction motor) and therefore it should give years of trouble-free service. However, in any system there are likely to be faults which can cause failures in components due to over stress (electrically) and also a large number of components are used. All these things cause periodic failures of the system, but good design will minimize the rate of failure.

Table 12.1 Electronic component reliability

Component	Failure rate per cent per 1000 hours
Capacitors, plastic film	0·01
Capacitors, electrolytic	0·2
Resistors, composition	0·005
Resistors, wire wound	0·01
Transistors, silicon >1 W	0·08
Transistors, silicon <1 W	0·008
Thyristors	0·1
Integrated circuit, silicon digital	0·01
Integrated circuit, silicon linear	0·03
Connections, soldered	0·001
Connections, wrapped	0·0001
Connections, multipin per used pin	0·005
Diodes >1 W	0·05
Valves, diode	1·0
Valves, triode	1·8

Although the solid-state control units are much more versatile and more reliable than earlier systems, faults will occur. It is therefore essential that the maintenance engineer understands the basics of the system and knows how to locate faults and replace faulty parts. Most units have printed circuit boards for the low level control circuits, and complete boards are replaced to get the system working quickly, and are then repaired as a separate exercise.

In order to maintain electronic control gear efficiently the maintenance engineer must have a good idea of how the system works. It is necessary that manufacturers of control gear have the problems of the maintenance engineer in mind when they design their gear. It can be made much easier to maintain control equipment if manufacturers include test points and give complete details of voltage, resistance, and/or waveforms to be expected at these points.

The maintenance engineer will need suitable equipment such as the following to maintain relatively complex equipment satisfactorily.

Oscilloscope plus probes and current probe.
Multimeter: Volts, Ohms, Amps.
High input resistance voltmeter.

It is a good idea to include in the maintenance engineer's stores a complete set of spare printed circuit boards fully tested and working. Most manufacturers of control gear are happy to supply spares. This minimizes breakdown time.

12.9 Specification Considerations

When specifying motor control gear it is necessary to consider the following points.

(a) Matching the drive to the load requirements.
(b) Capital cost versus running cost.
(c) Maintenance capability inhouse.
(d) Environment.

In (a) it is necessary to study the requirements and decide which drive system meets these requirements or most of them. Points to consider are as follows:

horsepower
speed range
torque/speed characteristic
efficiency

These points will obviously vary from application to application, but it is most essential that the specifying engineer takes all these points into consideration. In most instances the only item of data given to the manufacturer is horsepower, and he has to wheedle out of the user the rest of the specification. It is also common for the user to ask for a tighter specification than he really needs.

The second point, capital cost versus running cost, is rarely considered. Mostly the user only considers the capital cost and ignores the efficiency and maintenance prospects of a system, but an inefficient system uses more power and is hence more costly to run. An unreliable system, or one that is difficult to maintain, may mean long and/or frequent shutdown periods that are costly especially in production situations. It is therefore essential to select a system that is both reliable and easy to maintain if prolonged periods of stoppage are undesirable.

Inhouse maintenance capability is also an important point to consider. It is no use installing the latest and most versatile piece of equipment in a key production system if there is no-one to repair it should it develop a fault. Manufacturers cannot be expected to provide an instantaneous repair service. They expect the user to provide the first-line maintenance service. The following indicates the possible division between the manufacturer and the user for various faults that may occur.

Manufacturer	*User*
Repair consistent faults	Fuse replacement
Repair boards	Change boards
Recalibrate	Change power devices
Modify	Clean air filter

Environmental consideration is essential when choosing a motor control gear. Electronic equipment, especially electronic power-control gear, dissipates heat and

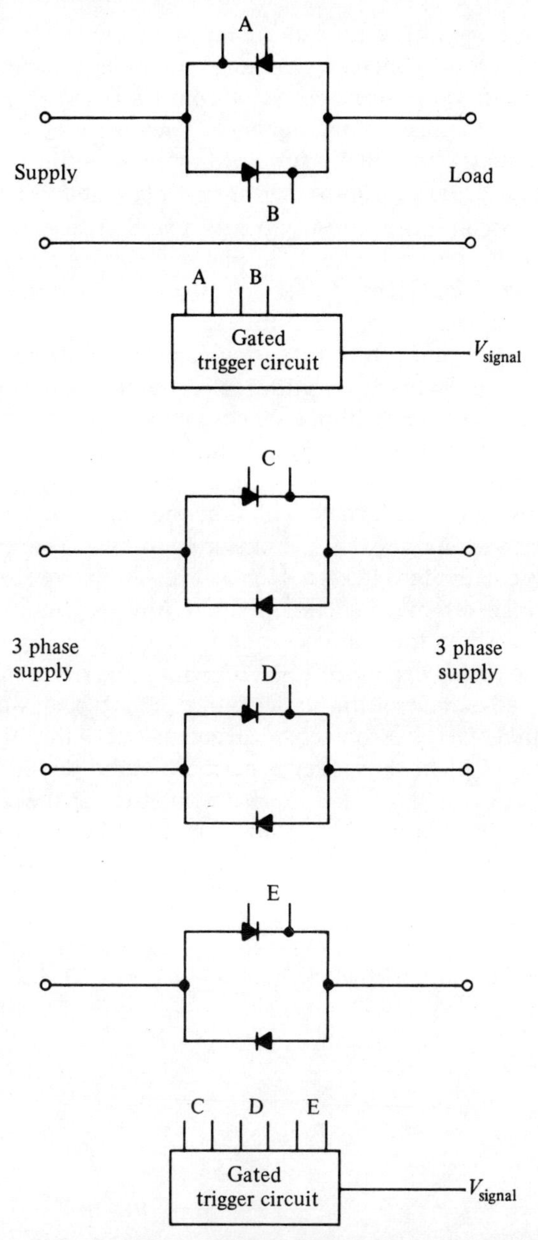

Fig. 12.54 Basic single- and three-phase static contactor circuits

needs to be kept cool. It is usually situated next to the motor being controlled and from an environmental viewpoint the motor is more robust than its control gear. Items such as swarf, lubricants, dust, filings, liquids, etc must not be allowed to infiltrate into the cabinet of the control gear. This can sometimes be difficult and it may be necessary to install the electronic equipment in an adjoining but separate room.

12.10 Static Contactors

At present, electromechanical contactors are normally used to switch the supplies to motors on and off. These contactors require maintenance and are not readily adaptable to electronic control circuits. Static contactors use thyristors or triacs to switch the supply and require no maintenance. They are electronically controlled devices and consequently are ideally suited to integration into electronic control systems. A single-phase and a three-phase contactor circuit are shown in Fig. 12.54. In the single-phase circuit it is necessary to have two thyristors to control both half cycles of the supply. However, in the three-phase circuit each limb need only have one thyristor and one diode, since, if the current going into the motor is controlled, the current coming out must also be controlled.

The thyristor gating circuits can be very simple, just a gated pulse generator. Since we are only interested in the thyristors either fully conducting or completely off, then continuous pulses to the gates without synchronization to the mains is sufficient. Apart from the zero maintenance requirement of static contactors, one of their major advantages is that they switch off under zero supply current conditions. Because the thyristor and triac turn off when their anode currents fall to zero, after their gate pulses have been removed, the device continues to conduct until the next time the a.c. supply current falls to zero. Hence, even inductive currents are allowed to turn off naturally. Electromechanical contactors, however, interrupt the current at any point depending when the coil is de-energized and consequently in inductive circuits considerable arcing can occur together with large transient voltages.

The one main disadvantage of the static contactor is its cost, which may be up to seven to ten times that of its electromechanical equivalent. If the number of switching operations per hour is high, then over a period of time the static contactor will become economical because of the replacement rate of the electromechanical contactor due to burnt contacts, etc. Static contactors also become desirable in hazardous atmospheres, for instance in chemical plants and in coal mines.

13 Materials and Components

13.1 Bearings and Lubrication

13.1.1 *Introduction*

In electric motors the function of the bearings is not only to allow armatures and rotors to rotate freely but to support them in their correct position in the machine.

Since magnetic attraction is proportional to the square of the distance, excessive eccentricity can make the shaft flex just sufficiently to cause physical contact between stator and rotor when a motor is started on line. This will be the case particularly in control motors pulsed with overvoltage to produce starting torques of the order of ten times their continuous rating. Even when motors are running normally, unbalanced air gaps can cause additional forces on the bearings and increase noise. It is equally important that the bearings position the rotor in the correct axial position, and are themselves adequately aligned. Such alignment is often facilitated by the bearing or the housing being made 'self' aligning.

The two main types of bearing are defined by the manner in which the friction between the two bearing elements is reduced. The earliest forms of bearing depended on sliding friction, reduced by using suitable dissimilar metals and a lubricant.

Developments in design, materials, and methods of manufacture have made the often very complex modern version of this type of bearing one of the most reliable and useful elements of large electric motors. In high-speed motors the losses due to the shearing of the oil film are large enough to justify the inclusion of means of cooling the oil before it is recirculated. If the oil is replaced by air or gas, friction practically disappears but the supply of the high-pressure air and precautions against its sudden failure make this a rather costly bearing reserved for special applications. The other main type of bearing involves additional elements such as spheres or cylinders rolling between the two elements. Most bearings of this type constitute complete assembly consisting of an inner and outer ring with the rolling elements between them held in position by a cage. This makes possible bearing replacement without regrinding either shaft or housing.

13.1.2 *Plain and oil film bearings*

For small motors where the loading on the bearings is comparatively small it suffices to maintain the presence of oil or a low-friction solid lubricant. The bearings in their simplest form are cylindrical bushes manufactured by sintering such materials as brass and phosphor bronze, and subsequently impregnating them with a suitable lubricant. This method of manufacture allows the outside of the bearing to have spherical seating surfaces and flanges for self-aligning assembly, and to accommodate end thrust without substantially increasing the cost. To replenish the oil supply the bearing may be surrounded by a thick felt washer to act as an oil reservoir. Where the presence of oil is unacceptable and speeds are low, plastics materials and graphite impregnation preceded the now very common PTFE impregnated sintered bearings. In certain applications carbon bushes have been used to provide a completely inert bearing. These bearings are extremely simple and in nearly all cases fit directly into the end frame of the motor. The shafts, whose diameters must be controlled to very fine limits, are usually made of high grade steel and polished to reduce running-in time.

In the case of the larger machines the aim is to maintain an oil film and bearings are usually lined with white metal and incorporate ducts to ensure that a plentiful supply of oil floods the shaft. In the simplest form, oil is carried to the top of the bearing by oil rings, which nowadays are often made of nylon or other suitable plastics material. Large bearings are often lubricated by special oil pumps, and heat exchangers and filters are fitted in the circuit.

When the shaft is at rest there is usually metal-to-metal contact, and on starting the shaft will tend to climb up the bearing wall and oil will be drawn into the space. As oil begins to cover the shaft it will slide back and provided there is a sufficient supply of oil and the shaft has reached a high enough speed a wedge of oil will be formed. The pressures which have been measured in such a wedge are very high and considerable loads can be supported while at the same time a very low coefficient of friction is achieved. Bearings operating on this principle are known as hydrodynamic bearings, providing thick film lubrication. During running-up there will be some wear during the direct metal contact stage, but if the viscosity of the oil is sufficiently high, mixed friction conditions will rapidly take over. Under mixed friction conditions there are only localized areas of lubricating film, while in the remaining area there is still sliding and rubbing. Such a condition can also be brought about during running when operating conditions become too severe or the oil supply becomes inadequate. Heating will then become severe and unless timely corrective action is taken the bearing will be damaged. In order to detect impending bearing failure temperature transducers are fitted to bearings and these are used to activate alarms or trips which automatically stop the motor.

So far, journal bearings have been considered but even in machines running with their shafts horizontal there may be end thrust which must be taken up without causing excessive friction losses. Since plain faces cannot form wedgelike films the tilting pad bearing, invented at the beginning of the century and now produced with many developments and refinements, is used. Many large motors are required to run with their shafts vertical and be capable of prolonged reliable operation. These are now fitted with such bearings.

Figure 13.1(a) shows a typical thrust-ring assembly which consists of a carrier ring with equally spaced thrust pads. Lubrication is usually by flooding the bearing and allowing the oil to circulate through slots at the back of the carrier ring. At high speeds the generation of heat due to the rapid churning and shearing of the oil makes it desirable to improve the flow of oil. One such method is directed lubrication (Fig. 13.1(b)) which not only provides the more rapid circulation of oil required but by

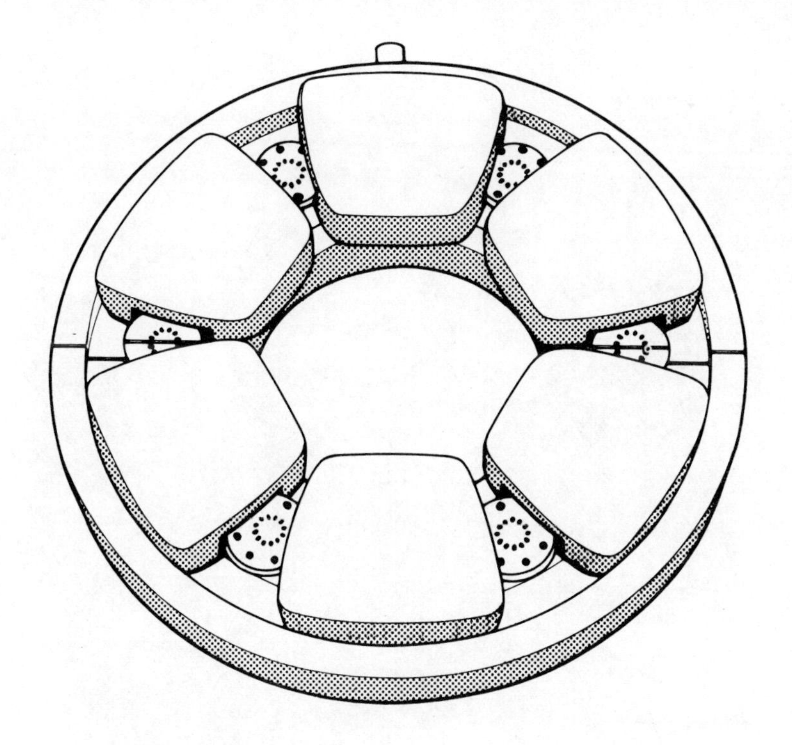

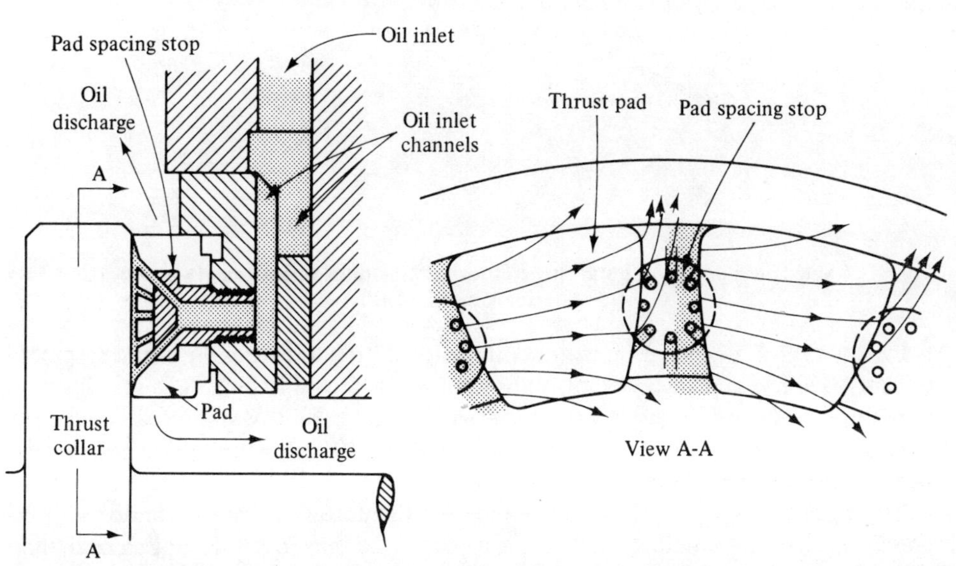

Fig. 13.1 Standard thrust bearing adapted for directed lubrication. (Courtesy Glacier Metal Co Ltd)

directing the oil onto the pads reduces power loss, lowers the pad temperature, and increases the film thickness at high speeds.

Figure 13.2 shows a typical self-contained horizontal thrust and journal bearing assembly with viscosity pump for a 60 mm diameter shaft. Attention is drawn to the cooler below the main pedestal.

Fig. 13.2 Self-contained horizontal thrust/ journal assembly with viscosity pump. (Courtesy Glacier Metal Co Ltd)

Figure 13.3 shows a special water coil cooled double thrust and journal assembly for a 12 000 h.p. nuclear coolant pump motor. Hydrostatic jacking is provided on both upper and lower thrust bearings: the supply piping for this system can be seen as well as the grooving provided in the thrust pads. A 450 mm, 8 pad tilting pad journal bearing with adjustable pivots is also used.

The calculation of oil flow, power losses, diametral clearances under various conditions of combined steady and varying load has been computerized, and manufacturers can rapidly advise on any modifications which may become necessary should operating or load conditions change on an existing installation. It is, of course, necessary to supply properly quantified information of loads, speeds, lubricants, and ambient conditions.

13.1.3 *Rolling bearings*

The most used types are deep grooved single-row ball bearings and roller bearings for standard motors.

Fig. 13.3 Special water coil cooled double thrust and journal assembly for 12 000 h.p. nuclear coolant pump motor. (Courtesy Glacier Metal Co Ltd)

For standard motors with outputs up to 300 kW ball bearings are normally fitted. Some manufacturers can, at little or no extra charge, provide roller bearings at the drive end if the machine has to withstand appreciable radial loading or is required to have a longer life. The permitted radial and axial loadings for one or more average values of bearing life in running hours are given in current catalogues showing standard types. 'Average life' is usually understood to be the expected life of 90 per cent of such bearings under proper working conditions. As the applications and ambient working conditions of motors vary widely, other factors affecting bearing life must also be considered. As a rough guide for example, the life can be considered to be inversely proportional to the cube of the load on the one hand and inversely proportional to speed on the other.

With the lubricants now available and the consistency of manufacturing standards the starting point is usually about 50 000 hours or nearly six years continuous running.

Properly mounted, protected, and lubricated rolling bearings hardly show any wear so that the running clearances which are in any case much smaller than for plain bearings are maintained throughout their useful life. The need for lubrication arises from the fact that due to elastic deformation of the rolling elements definite areas of contact develop between them and their tracks, and some frictional resistance occurs. There is, furthermore, friction between the rolling elements and cages or separators as well as end flanges in roller bearings. Lubricants are also required to perform three further important functions, namely, protect surfaces from corrosion, help to conduct heat, and, finally, to keep out foreign matter. The usual lubricant for bearings in electric motors is grease. Grease is a homogeneous mixture of mineral oil and a soap or base. In small motors bearings are charged with grease during manufacture and protected as well as located by bearing housing covers. Alternatively, sealed bearings may be used, but in either case there may be instances where even after only one year under severe operating conditions the grease has to be renewed. Sealed bearings have to be replaced, but where access to the bearings is possible they may be carefully cleaned and re-lubricated. On large motors grease nipples are fitted and old grease can be replaced while the motor is running. Care must be taken to ensure that the old grease has been forced out through the valve provided. Excessive grease in ball bearings will cause overheating.

The roller bearing is thus not only less demanding on lubrication than plain bearings, but its low friction makes for ease of starting and lower bearing losses.

The type of bearing used at the drive end frequently involves a choice between a ball and a roller bearing. The following parameters will govern the choice:

Magnitude and direction of applied load.
Nature of load (possibly out of balance).
Speeds.
Ambient temperature conditions.
Life expectancy.

It should be noted that parallel roller bearings have no axial load capacity and, therefore, either combined units or taper roller bearings must be used. Where a user intends to change bearings on an existing drive the advice of the motor manufacturer and/or the bearing manufacturer should be sought.

Because there is direct metallic contact, ball and roller bearings have a higher inherent noise level; any change or significant increase in noise indicates that some fault has developed. Many bearing problems can be traced to a fault in the assembly of the motor or the coupling connecting it to the load. If, for example, the load is a pump and a solid coupling is used, only one shaft must be constrained in an axial direction and special care must be taken to ensure alignment. See also chapter 15.

13.2 Commutators, Brush-holders, and Brushes

13.2.1 *Introduction*

The functions of the commutator, as well as the theory and practice of commutation in d.c. motors are dealt with in chapter 4. In a.c. commutator motors there is, in addition to the reactance e.m.f. in the short-circuited coil, rotational e.m.f. due to its movement relative to the air-gap flux, which the brush has to contend with. The considerable commutation difficulties have been solved and industrial three-phase commutator motors can be expected to have brush lives of 2 to 3 years under normal operating conditions. Commutator motors in general will work satisfactorily as supplied and provided the necessary preventive maintenance is carried out there should be little change in commutation. However, changes which occur over a period of time can provide valuable clues for the timing of brush changes and possible improvements in the drive or its control.

A number of theories have been advanced to explain the phenomena occurring when current flows from a rotating commutator to a brush. The following is that given by ASEA:

A brush which is functioning properly has no mechanical contact with the commutator, as it rides on a cushion of water molecules. The current flows through a series of points which move rapidly over the whole face of the brush contact surface. An important condition for a good sliding contact is, therefore, sufficient moisture in the atmosphere surrounding the machine. The minimum requirement quoted in the literature is a minimum of 3 to 4 g/m^3 absolute humidity, or 100 per cent relative humidity at -5°C and 20 per cent at 20°C. If the humidity is too low, which can occur in places with low temperatures or at high altitudes, it may become necessary to use specially impregnated brushes or humidify the air around the motor.

In addition to humidity a certain minimum current density is required for good commutation. In this context current density is synonymous with temperature since the temperature at the interface changes the humidity to vapour, providing what is popularly called 'electrical greasing', and helping to form a skin on the commutator. Thus, there are the following layers: an oxide skin on the commutator surface covered by a graphite skin, with free particles of water vapour forming the interface on which the brush floats.

Tests on brush friction have shown that this is very temperature dependent, which, as mentioned above, implies dependence on current density. Typical values are 0·3 at 40°C (5 A/cm^2) and 0·1 at 80°C (10 A/cm^2) which clearly indicates that at the lower current density with a coefficient of friction three times as large the brush and commutator wear will be greater. It must be emphasized that the above temperatures are the result of heating under the brush where conduction actually takes place, and that no manner of artificial heating will produce the reduction in the coefficient of friction. Motors which run on light loads for appreciable periods may, therefore, be subject to rapid brush wear and the advisability of changing the grade of brush used should be taken up with the manufacturer or one of the leading brush manufacturers.

The brush grade should, however, not be changed until all other possible causes of wear have been eliminated. In d.c. motors working from thyristor or other static converter supplies (see chapter 3) in particular the control circuits should be checked to ensure that they are working correctly before investigating the motor for causes of bad commutation. Excessive sparking should, in such cases, be investigated as soon as possible since prolonged running under fault conditions can cause flats on the commutator and make it necessary to turn the commutator within months of installation. As mentioned above, regular inspection and recording of changes are valuable not only for improvements but for the prevention of unexpected break-downs.

13.2.2 *Commutators*

With few exceptions commutators are made up of special hard-drawn copper bars separated from each other and their clamping rings by micanite, or some other suitable insulating material. Commutators for motors around and below 1 kW, where economy is important, will be moulded. Such commutators are largely used on domestic appliances and portable tools where their heat-resistance, insulating properties, and mechanical strength are more than adequate.

Since accuracy of spacing is important for good commutation and is necessary to facilitate economic manufacture, the copper and insulation segments must be held to close tolerances. Where the intersegment insulation remains, the brush grade used must be sufficiently abrasive to keep it flush with the copper. On large machines the intersegment insulation is removed and the commutator is said to be 'undercut'. Chapter 19 deals with the correct maintenance of undercutting.

13.2.3 *Brush holders*

Brush pressure and the maintenance of contact are provided by the brush holders. The ideal arrangement would provide constant pressure for the whole of the useful brush life and have no friction to prevent the free radial movement of the brushes. As with current density there is an optimum value and insufficient pressure will cause increased wear. Generally minimum brush wear occurs at about 17·5 kN/m^2 (2·5 lbf/in^2), but to cater for friction between brush and the inside of the brush box and any resistance by the flexible connection between the brush and the holder (pig-tail) the pressure finger is adjusted to between 20 and 25 kN/m^2.

The brush holder and commutator are perhaps the most vulnerable parts of commutator machines with respect to dust and other airborne contamination. According to ambient conditions brush holders should be examined at more or less frequent intervals to ensure that the brushes are not sticking in their holders. In the plastics industries dust from thermoplastics which fuse on contact with warm parts, as for instance the motor brush, can have particularly disastrous effects on brush movement. Oil, either from overlubrication of the motor or the driven machinery, will not only be absorbed by the brush but form a thick skin on the commutator. Under such conditions it will be advisable to provide filtered air or use a totally enclosed machine to at least IP 5. standard. It should be pointed out that the d.c. or a.c. commutator motor is here at no great disadvantage since the PVC would eventually have equally disastrous effects when fused to the cage of an induction motor.

During maintenance and when changing brushes the distance between the bottom edge of the holder and the commutator should be verified to have a nominal value of about 2 mm—depending on size and application.

In large machines brushes may be staggered axially—that is, there may be two or more brushes in parallel, usually in separate brush boxes. This often makes it possible to renew brushes without elaborate bedding-in procedure. About a quarter of the total number of brushes or half the number on any brush arm may be changed.

Brush holders and their leading dimensions are becoming more and more standardized so that the replacement of the holders, should they after several years be too worn, presents no problems.

13.2.4 *Carbon brushes*

It is clear that the manufacturer of the motor will have carefully selected the most suitable grade and, if advised of special requirements, even changed the standard to meet them. Great care must be taken to note any such changes in the plant records and avoid the assumption that all d.c. motors in the plant or factory will use the same grade. Once established, great care must be taken to ensure that the same grade is used.

There is an increasing tendency to use electrographitic brushes, mainly because the firing process leaves a tough brush free from impurities and having a low coefficient of friction. It must be pointed out that the adjectives 'hard' and 'soft' can be very misleading, particularly if the belief is that 'hard' brushes cause excessive commutator wear. A 'soft' natural graphite brush running on a flush mica commutator is certain to cause more wear than the 'hard' electrographite brush capable of dealing with such a condition. In defining the properties of carbon brushes abrasiveness and hardness are by no means the same.

The essential constituents of carbon brushes are few—carbon, possibly a metal, and a binder. Carbon may be in the form of natural or electro-graphite, derived from coal, lamp or furnace black, or charcoal. The binder can be tar, pitch or resin and the metal copper or silver. The processing of these materials will now begin to open up many possibilities of grinding, mixing, moulding, and finally heat treatment. There are, furthermore, a considerable number of impregnations such as PTFE and barium fluoride which can be applied to reduce friction and wear at low temperatures. It has been suggested that with ten practical variants at seven definable production stages some ten million grades readily distinguishable to a brush technologist could be produced.

The table 13.1 shows the most commonly used grades and their typical applications.

13.3 Insulating Materials

The development of synthetic resins and plastics materials has made a considerable impact on electric motor size. The effect has been particularly pronounced in small machines where relatively the insulation takes up more space. In particular, slot liners were the subject of considerable development both in the material used and its method of insertion. A method gaining ground and proving successful is epoxy powder coating. This may be applied to the stator either by fluid bed or spray application with stator pre-heat temperatures of between 130 and 220°C. Though up to now approval is only to class B (130°C) ongoing tests indicate that continuous operating temperatures in excess of 145°C will be achieved. Problems which faced manufacturers in the early stages were the tendency of the coatings to fume if the preheat temperature was not exactly controlled and the presence of dust clouds over the normal fluidizing beds. The all-important edge covering is expected to achieve 50 per cent so that with a dielectric strength of 32 kV/mm a very useful saving in space is achievable. The machine winding of stators is now common practice and the tougher and more temperature-resistant coatings combine well with such improved methods of slot insulation.

The life of insulating materials and their resistance to repeated temperature cycling is a complex subject and difficult to assess in the short term. For accelerated life tests must of necessity put additional stresses on the materials and in many cases only prolonged field trials can prove their worth. Breakdowns in large machines are particularly costly and as such machines usually operate at high voltages the stator insulation is generally based on mica. Mica is a complex mineral silicate found almost

Table 13.1 Characteristics of standard grades of brushes

Grade	Specific resistance ohm/in	Specific resistance ohm/cm	Specific gravity	Relative strength lb/in²	Relative strength kg/cm²	Hardness kg/mm²	Hardness lb/in²	Normal load amp/in²	Normal load amp/cm²	Max. speed ft/min	Max. speed m/s	Contact drop	Coeff. of friction	Range of applications
Hard Carbon														
HC	0·0015	0·0038	1·56	3400	238	26	36 981	50	8	3500	18	M	L	FHP motors—d.c. machines
HC10	0·0025	0·0064	1·43	3700	259	38	54 049	40	6	3000	15	H	M	Hand tools—Hair dryers—Vacuum cleaners
HC11	0·0020	0·0051	1·45	3500	245	21	29 869	40	6	3000	15	H	M	FHP motors—Contacts—Traction
HC20	0·0016	0·0041	1·44	2500	175	18	25 602	45	7	5000	25	H	M	Tools—d.c. machines—Mechanical
HC25	0·0027	0·0069	1·40	4750	332	35	49 782	40	6	3000	15	H	M	d.c. machines with difficult commutation—a.c. motors
HC28	0·0017	0·0043	1·50	4500	315	30	42 670	50	8	6000	30	L	L	LV generators—d.c. and a.c. machines—Carbon bearings
HC40	0·0021	0·0053	1·45	3500	245	38	54 049	40	6	3000	15	M	H	FHP Motors—Old machines under arduous conditions
HC51	0·0014	0·0036	1·55	5000	350	46	65 427	50	8	6000	30	L	L	High speed FHP motors
HC60	0·0081	0·021	1·42	2750	193	30	42 670	30	5	3000	15	H	L	FHP motors with difficult commutation
Natural Graphite														
NG13	0·0014	0·0036	1·55	3400	238	20	28 447	50	8	4000	20	M	L	Machines up to 15 h.p. at 250 volts—Contacts
NG14	0·0009	0·0023	1·48	3500	245	15	21 335	65	10	5000	25	M	L	Automobile generators—250 V machines—Contacts
NG15	0·0008	0·0020	1·62	3400	238	15	21 335	65	10	5000	25	L	L	Automobile generators—Low power machines
NG16	0·0008	0·0020	1·57	3750	263	14	19 913	65	10	5000	25	L	L	Automobile generators—Small motors
NG17	0·00025	0·00064	1·75	1700	119	5	7112	65	10	7000	36	L	L	Generators—Motors 25–120 volt
NG21	0·0009	0·0023	1·50	600	42	—	—	65	10	9000	46	L	L	Old d.c. machines
NG22	0·0004	0·0010	1·40	1400	98	—	—	65	10	9000	46	L	L	d.c. machines to 120 h.p.
NG28	0·0004	0·0010	1·45	600	42	—	—	65	10	12 000	61	VL	L	Steel rings—Rotary converters
NG30	0·0015	0·0038	1·46	1100	77	3	4267	65	10	10 000	51	L	L	d.c. machines with difficult commutation
High Resistance Graphite														
HG0	0·0024	0·0061	1·9	2700	189	15	21 335	40	5	7000	36	M	L	High speed FHP and miniature motors
HG7	0·0048	0·012	1·8	4600	322	16	22 757	50	3	8000	41	H	L	Repulsion motors—3 phase variable speed
HG14	0·0080	0·020	1·8	4300	301	17	24 180	50	3	8000	41	H	L	Repulsion motors—3 phase variable speed—Shavers
HG16	0·014	0·036	1·75	5000	350	26	36 981	30	5	7000	36	VH	L	FHP machines with excessive vibration
HG20	0·040	0·102	1·70	4750	333	30	42 670	30	5	7000	36	VH	L	FHP machines with difficult conditions

Electrographite														
VG01	0·0003	0·00076	1·65	700	49	2	2845	65	10	14 000	71	L	L	d.c. steel rings
VG04	0·0004	0·0010	1·30	850	59·5	—	—	65	10	12 000	61	L	L	d.c. high speed rings
VG10	0·0005	0·0013	1·65	3500	245	15	21 335	80	12	10 000	51	L	L	Steel and bronze rings—Automobile generators
VG22	0·0008	0·0020	1·50	2800	196	12	17 068	80	12	6000	30	L	L	Up to 250V d.c.—Small quiet-running machines
VG24	0·001	0·0025	1·60	3000	210	14	19 913	80	12	7000	36	L	L	Small d.c. machines—Rail lighting generators
VG26	0·0013	0·0033	1·55	4000	280	25	35 558	80	12	10 000	51	L	L	a.c. and d.c.—d.c. welding—Rotary converters
VG27	0·0015	0·0038	1·59	4250	298	26	36 981	80	12	10 000	51	M	L	a.c. and d.c.—d.c. welding generators
VG28	0·0019	0·0048	1·60	3600	252	25	35 558	70	11	10 000	51	H	L	a.c. commutators—a.c./d.c. undercut micas
VG31	0·0012	0·0030	1·70	3250	228	25	35 558	80	12	10 000	51	L	L	Tramway motors—Mine traction
VG32	0·0016	0·0041	1·70	3150	221	32	45 515	80	12	10 000	51	M	L	Rail Traction—Lift and crane motors
VG33	0·0014	0·0036	1·60	4250	298	26	36 981	80	12	10 000	51	H	M	48 volt FHP motors
VG150	0·0013	0·0033	1·63	2400	168	18	25 602	80	12	8000	41	VH	L	Mill motors—Will withstand no-load conditions
VG180	0·0016	0·0041	1·60	2200	154	20	28 447	80	12	8000	41	VH	L	Mill motors—Will withstand no-load conditions
VG190	0·0018	0·0046	1·70	2500	175	28	39 825	80	12	8000	41	VH	L	Ward-Leonard—HT machines
VG200	0·0020	0·0051	1·60	1650	116	20	28 447	65	10	9000	46	VH	H	Ward-Leonard—HT machines
Metal Graphite														
CC1	0·00035	0·00088	2·40	1750	123	6	8534	80	12	7000	36	M	L	12 V—120 V d.c.—48 V lighting generators
CC2	0·00045	0·0011	2·50	3100	217	12	17 068	75	11	6000	30	L	L	Heavy duty automobile generators
CC3	0·00035	0·00088	3·00	4250	298	22	31 291	80	12	5000	25	L	L	24 V—60 V d.c.—Contacts
CC4	0·00015	0·00038	3·60	4800	336	23	32 713	90	14	4000	20	L	L	12 V—36 V d.c.—Starters—Contacts
MC8	0·0000032	0·0000081	5·40	8750	613	20	28 447	160	25	—	—	VL	—	Contacts—Toy motors
MC9	0·0000032	0·0000081	5·20	9250	648	18	25 602	150	23	4000	20	VL	L	Starters—Heavy duty rings
MC26	0·000046	0·00012	4·50	8000	560	17	24 180	105	16	6000	30	VL	L	Rotary converters—Sliprings
MC30	0·000011	0·000028	4·20	6000	420	14	19 913	105	16	7000	36	VL	L	d.c. machines—Sensitive rings
MC35	0·00012	0·00030	4·20	4500	315	15	21 335	100	15	8000	41	VL	L	a.c. machines—Sensitive rings
MC40	0·00012	0·00030	3·80	3700	259	14	19 913	90	14	8000	41	VL	L	Heater motors—Sensitive rings
MC70	0·000016	0·000040	3·80	1650	116	10	14 223	100	15	7000	36	L	L	Dynastarters—LV machines
MC76	0·000035	0·000089	3·60	1450	102	7	9956	90	14	8000	41	L	L	Dynastarters—LV machines
MC84	0·000045	0·00011	3·50	2250	158	10	14 223	85	13	8000	41	L	L	Low voltage machines—Shavers
MK31	0·0000030	0·0000077	5·30	8000	560	15	21 335	225	35	4000	20	VL	VL	Earthing—Replacing copper gauze
MK70	0·000028	0·000071	5·20	8500	595	17	24 180	130	20	5000	25	VL	L	Dynamos—12 V starters—Rings—Earthing
MK80	0·0000039	0·0000099	6·30	8450	592	16	22 757	140	22	5000	25	VL	L	Self-exciting 6 V d.c. generators

Symbols:

Volt drop
up to 1·0 V = VL = VERY LOW
1·0–1·8 V = L = LOW
1·8–2·5 V = M = MEDIUM
2·5–3·5 V = H = HIGH
above 3·5 V = VH = VERY HIGH

Coefficient of friction
up to 0·15 = VL = VERY LOW
0·15–0·20 = L = LOW
0·20–0·26 = M = MEDIUM
above 0·26 = H = HIGH

A wide range of Silver Graphite materials from 2 to 99 per cent silver for every class of electrical application is also available. All grades of carbon can be impregnated to produce a variety of differing characteristics such as quiet running, extra density for improved commutation, cool running, etc.

Courtesy: T. M. Harding (Worthing) Ltd.

exclusively in India. It usually occurs in bulk but it can be easily fragmented into thin sheets called mica flakes or splittings. These splittings are themselves excellent insulators and highly impervious to electrical discharges. Generally, stator insulation is composed of mica splittings held together by a resin binder and forming a micanite sheet. As the supplies of mica are dwindling, splittings may be further crushed and then bonded on glass laminate backing tape, the composite being called mica paper. Stator conductors may be sheet wrapped or taped along their slot length while the end windings are usually taped. Over the past decade, there has been a gradual change in high-voltage machine insulation through replacement of traditional copal bitumen or shellac by epoxy resins as bonding agents for mica flakes. With taped or sheet systems there are two basic methods in use. The most common is the 'rich resin' system in which the sheet or tape is fully saturated with epoxy resin. The resulting material is dry and readily malleable. By the use of heat and pressure the material can quickly be moulded onto coils to form electrical insulation.

In general there are four main causes of insulation ageing:

(a) Electrical: due to discharges in insulation voids or cavities the insulation may be eroded and tracks or so-called treeing occur.
(b) Mechanical: some vibration is almost inevitable in rotating machinery, and where there is no strong bond the windings can actually move relative to the core thus causing fretting of the insulation. Differential expansion of the core and conductors can furthermore result in tearing of the insulation. For this reason many of the large machine manufacturers have installed large tanks so that all but the very largest machines can actually be dipped in varnish.
(c) Chemical in air: insulation can crack and age by the oxidation of organic materials so that synthetic varnishes which do not oxidize and are resistant to many more chemicals than are organic substances represent a considerable step forward.
(d) Thermal: excessive heat generation immediately adjacent to or actually within the winding can cause degradation of the mica/resin bond. On traditional insulation systems using thermoplastics this would actually cause delamination. With the newer thermosetting insulations embrittlement may occur which can equally affect the mica/resin bond. Some manufacturers now use insulation systems which can be rated as class 'F' but are only loaded to class 'B', thus giving extra overload capacity and longer operating life.

If started on full voltage the endwindings are subjected to considerable stresses and must therefore be braced.

13.4 Lubricants

Lubrication is a design element which today, mainly on the basis of experience and testing, is considered when designing a motor or gearbox. The recommended lubricants and relubrication times are, therefore, extremely important and should be rigorously observed. The effect of an inferior lubricant will be to cause overloading of the bearings or sliding contact surfaces.

An important quality of grease is its consistency as this determines its effectiveness. Too stiff a grease will be thrown off by centrifugal force and so fail to provide a seal against ingress of dirt and other contaminants. In spite of the wide temperature range over which ball and roller bearings operate, experience has shown that one consistency of grease is suitable for most general applications. Thermal stability or resistance to separation when the motor reaches and maintains its operating temperature for prolonged periods is also important. Lithium bases have been popular for such applications not only because of their wide temperature range but their high degree of water repellency. The consistency of a grease is determined by allowing a cone-shaped weight of standard size to drop into it and measuring the distance it penetrates in hundredths of a centimetre—the penetration value of a grease must naturally be related to a specified temperature.

Most greases are a mixture of petroleum oil and a soap, which is generally used to define the type of grease. Typical greases are based on either calcium, sodium or lithium soaps which must be carefully selected to ensure that they have no corroding or oxidizing properties which might affect the highly polished surfaces of the bearings.

The oil used must be equally carefully selected, and where necessary should have extreme-pressure characteristics. In the search for lubricating oils whose viscosity remains more constant and whose resistance to oxidation is greater than that of the petroleum oils, various synthetic oils have been developed. The most notable of these are the members of the silicone family, in particular dimethyl silicone polymers and methylphenyl silicone polymers. These are available in various viscosities and to date are the most temperature stable of the synthetic lubricants known. In almost all cases their volatilities are exceptionally low when compared with those of other lubricants of equivalent viscosity. Their oxidation resistance characteristics are excellent from low to very high temperatures. The chemical solvency properties are very different from those of most other synthetic lubricants and they do not readily accept additives. One of the drawbacks, however, is that their lubricating properties are relatively poor, and though considerable progress has been made in improving them silicone lubricants are generally only used in very special cases. An improvement in the lubricating properties of silicones has been achieved by adding a chlorinated molecule to the phenyl group. This makes their lubricating properties equivalent to those of petroleum oils to which they are superior in that they have been found capable of providing satisfactory lubrication at some 230°C for 500 hours.

Synthetic diester greases have been primarily developed to provide lubrication under extreme conditions. They are generally required to have low torque characteristics at temperatures of -55°C or lower, yet be sufficiently viscous to 150°C and have high oxidation resistance at 200°C and over. Their cost and the need for frequent replacement has restricted their use mainly to military applications.

Silicone greases are produced by mixing the silicone fluid with lithium soap, silica gel, bentone, and the like. Lithium-based silicone greases have a good all-round performance but for high temperatures phthalocyamines, for example, are showing high promise as thickeners.

13.5 Copper

The electrical as well as thermal conductivity of high purity copper is only exceeded by silver which, because of its lesser presence in the earth's mineral stocks, is too costly to use for the windings of electrical machines. In general all additions of small amounts of other metals, except silver and cadmium, lower the conductivity of copper to a lesser or greater extent. Because it is subjected to secondary refining operations copper can be specified in a variety of conditions, but the essential differences depend on the presence or absence of oxygen. For electrical applications where high conductivity is of paramount importance, and it may have to be brazed or welded, oxygen-free copper is essential.

For incorporation into electrical machines copper is mostly supplied in the form of insulated wires or bars. For very large machines bare sections are often formed into suitable coils and insulated as described in the section 'Insulation'.

British Standards distinguish between coppers for electrical and non-electrical application, and most of the high conductivity grades also appear in the non-electrical standards which merely makes the conductivity test non-mandatory. The basic material specification is BS 1861: 'Oxygen-free high conductivity copper'—specifying a minimum copper content of 99·5 per cent. For finished forms the 'Copper for electrical purpose standard'—BS 1433, Rod and bar, BS 1434 Commutator bars, and BS 4109 Wire for general electrical purposes are those most likely to be needed. In the case of wires several forms of coverings each have their own specification.

Where high conductivity copper is heavily cold worked it should be noted that the conductivity may be reduced by as much as 3 per cent.

In the case of metals which are to be used as conductors, figures for conductivity in what used to be known as reciprocal ohms (mohs), and are now Siemens, are more useful than resistivity.

As far back as 1913 the International Electrotechnical Commission adopted a standard of resistivity for annealed copper of 0·01724 $\mu\Omega$m at 20°C, and defined it as a conductivity of 100 per cent IACS (International Annealed Copper Standard). This was based on the highest purity copper then commercially available and all other materials were expressed as a percentage of the agreed standard and denoted 'per cent IACS'. Today commercial high conductivity copper may have an electrical conductivity as high as 101·5 per cent IACS.

Where the copper is used for the bars of cage rotors, or in heavy duty stator windings, the mechanical strength must be carefully considered and when a machine is checked or overhauled as part of the planned maintenance particular attention should be paid to any signs of the steel banding of rotors and bracings on stator windings becoming loose. At the same time joints between rotor bars and end-rings should be checked and any tell-tale signs of heating carefully investigated.

Like the resistance of all metals that of copper varies significantly with temperature. The exact temperature coefficient at 0°C is 0·004265 per °C, but in practical calculations it is usually sufficient to take 0·004 if referring to a basic temperature of 0°C. The resistance R_t at any other temperature is then obtained from:

$$R_t = R_0(1 + 0{\cdot}0043t) \text{ ohms}$$

For a reference temperature of 20°C the IEC has adopted 0·00393 per °C as the correct value of the constant mass temperature coefficient, so that

$$R_t = R_{20}(1 + 0{\cdot}00393t) \text{ ohms.}$$

The formulae quoted are for high purity annealed copper of at least 100 per cent IACS conductivity between −100°C and +200°C. The resistivity at cryogenic temperatures is extremely sensitive to both the presence of impurities and the degree of cold work applied during manufacture.

The most important properties of copper are:

Melting point · · · 1083°C
Density, standard value · · · 8890 kg/m^3
Linear coefficient of thermal expansion between
−190°C and 16°C · · · $14{\cdot}1 \times 10^{-6}$/°C
between 25°C and 100°C · · · $16{\cdot}8 \times 10^{-6}$/°C.
and 20°C and 200°C · · · $17{\cdot}3 \times 10^{-6}$/°C.

with 17×10^{-6} taken as the usual value in practice.

Specific heat (thermal capacity) at

20°C · · · 0·386 kJ/kg K.

. . .

The thermal conductivity at 20°C, valid around normal operating temperatures of say −20°C to 100°C is 394 W/mk dropping to 381 at 200°C.

13.6 Aluminium

Because of its relatively good conductivity and lightness, aluminium is used as a conductor—as well as a structural material in rotating electrical machines. The electrical conductivity of chemically pure aluminium is 61 per cent of that of the International Copper Standard (IACS). Since, however, the metal is too weak and soft in this pure state, it must be alloyed with small quantities of other metals, such as zinc, which affect its conductivity to a considerable extent. Thus, for example, aluminium conductors made to BS 3242 are specified to have a conductivity of only 52 per cent IACS. To obtain the necessary mechanical strength and achieve homogeneity in die-cast rotors, further additions and the actual casting process may reduce the conductivity of the bars and end-rings to below 40 per cent IACS. In large motors, cages may be fabricated from bars and end-rings of higher conductivity material, but here the brazed joints between the two must be taken into account. The use of aluminium casting in the manufacture of cage rotors for small- and medium-sized induction motors was promoted by reduced costs in mass production rather than any comparison of its relative cost with that of copper as in the case of bus-bars and overhead lines.

The most important properties of aluminium are:

	Resistivity	Conductivity
Pure aluminium	0·02873 $\mu\Omega$-m	34·807 MS-m
Conductors to BS 3242	0·0325 $\mu\Omega$-m	30·769 MS-m
Typical casting	0·045 $\mu\Omega$-m	22·22 MS-m
Temperature coefficient of resistance 0·045 to 0·0040/K		
Density	2700 kg/m^3	
Melting point	660°C	
Specific heat	0·94 kJ/kg	
Thermal conductivity	204 W/mK	

14 Motor Control Gear

14.1 Introduction

The full advantages of an electric motor drive can only be realized when there is complete integration of the three main components: the driven machine, the electric motor, and the control gear. Such complete integration requires a full understanding of the characteristics of motor, control gear, and how different methods of starting, protection, and control can be used to meet the requirements of a particular application. There are many forms of speed control, and braking may either be electrical or mechanical to meet many requirements and particular site conditions.

While the prime function of the starter is to connect the motor to the mains without disturbance it can also play an important part in ensuring its long and trouble-free service. Considerations must, therefore, be given to:

(a) The control of starting and accelerating torque from the standpoint of the driven motor.
(b) Protection of the motor against overcurrent due to overloading and overheating due to this or other causes.
(c) Isolation of earth faults particularly in hazardous areas where fast action to localize the fault is essential.
(d) Protection of the operator and provision of emergency stopping of the motor.
(e) Provision for interlocking with other starter or mechanical devices to prevent incorrect operation.
(f) Speed control.
(g) Reversal of direction of rotation.

The application engineer should also familiarize himself with the relevant British or other applicable standards, which are:

BS 587 Motor Starters and Controllers.
BS 140 Liquid Starters and Controllers.
BS 775 Contactors.
BS 229 Flameproof enclosure of electrical equipment.
BS 2817 Types of enclosure for electrical apparatus.
BS 4070 Performance of a.c. control equipments for use on high prospective fault-current systems.
BS 822 Recommendations for terminal markings for electrical machinery and apparatus.
BS 3939 Graphical symbols for electrical and electronic diagrams.
BS 2771 Electrical equipment of machine tools.

The 'Regulations for the electrical equipment of buildings' refer principally to the installation of the equipment and are, therefore, of especial value to the installation engineer. These regulations which are issued by the Institution of Electrical Engineers, Code of Practice CP 1011, and cover the maintenance of motor control gear, should also be in the possession of every works engineer.

14.2 Starters for d.c. Motors

Many small d.c. motors up to 1·5 kW (2 h.p.) can be started directly on line but, as indicated in chapter 3, page 69, this may cause commutator trouble. This method should, therefore, only be used with the approval of the motor manufacturer.

To prevent burning on the commutator and control the torque applied to the driven machine, rheostatic (resistance) starters are used. The direct relationship between current and torque in d.c. motors makes consideration of starting and acceleration problems relatively simple. This is particularly so for the shunt or permanent magnet motors where the field flux is constant and the torque is directly proportional to the armature current.

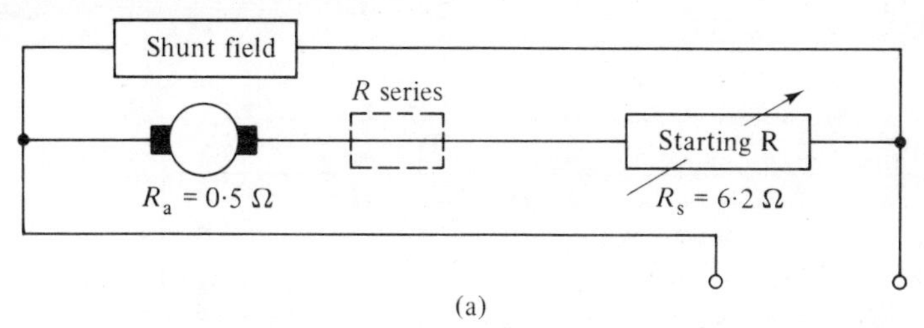

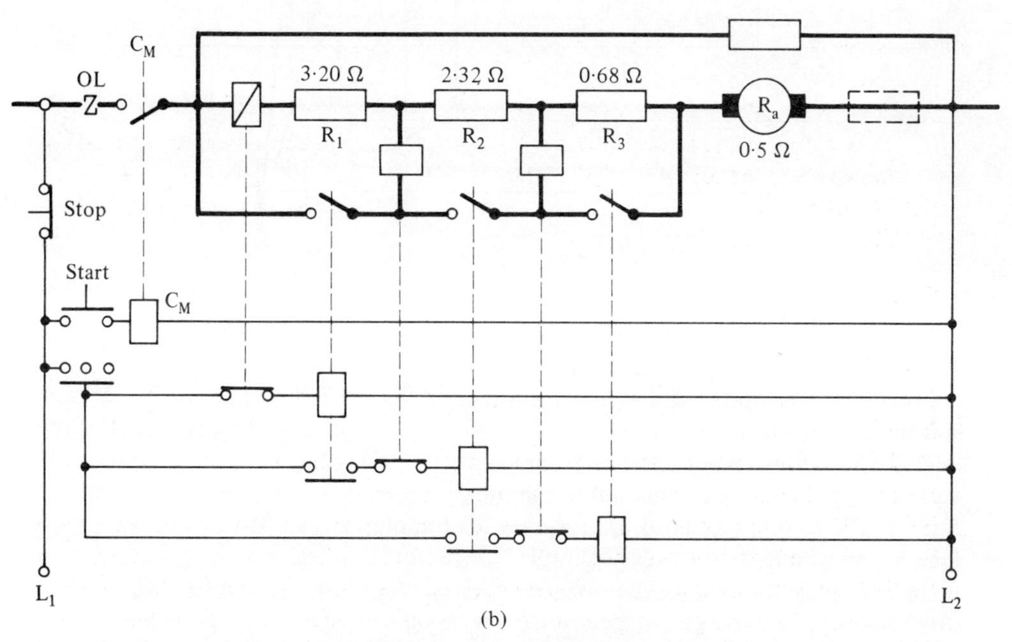

Fig. 14.1 (a) Basic circuit for resistance-start of typical d.c. shunt (or compound) motor. (b) Automatic current element resistance starter

Figure 14.1 shows the circuit diagram of a 19 kW (25 h.p.) shunt motor designed for 460 V with an armature resistance of 0·5 Ω and taking 46 A at full load. Assuming a starting torque of 1·5 times full load is required the current at starting must be 1·5 times full load current, i.e., 69 A. To allow this to flow and ignoring possible voltage drops the total resistance in circuit at starting will be approximately 6·7 Ω. Deducting the armature resistance gives the required starting resistance of 6·2 Ω. The motor will accelerate until the torque developed balances the load resistance, assumed to be 60 per cent of the motor's rated full-load torque. At this equilibrium condition there will be a voltage drop in the starting resistance $46 \times 0{\cdot}6 \times 6{\cdot}2 = 171$ volts, and since the speed is proportional to the voltage across the armature (the field being constant) it will only be 62 per cent. When the series resistance is removed the current surges up to 69 A and provides further accelerating torque to enable the motor to reach the rated speed for 60 per cent full-load torque. In practice current surges would be controlled by cutting out the starting resistance in graded steps at predetermined speeds or current values. This is shown graphically in Fig. 14.2 for a shunt motor required to produce, as in the previous example, 150 per cent starting torque against a constant-load torque of 60 per cent and started with a four-step resistance starter.

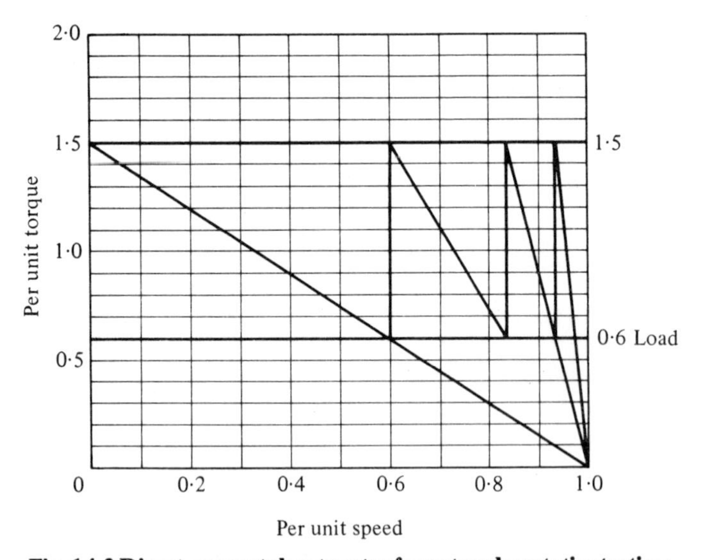

Fig. 14.2 Direct-current shunt motor four step rheostatic starting

In practice the load torque is likely to vary and Fig. 14.3 shows the corresponding starting curves for a centrifugal fan. Mostly the speed/torque characteristics of the load are not known when general-purpose starters are designed so that appropriate values of acceptable maximum and minimum accelerating torques are assumed. A ratio of 150 to 60 per cent is appropriate for the majority of drives, but where the load has special features, as for example considerable inertia breakaway, torques of up to 2·5 times full load torque may be required. As shown in chapter 2, the time a drive takes to reach a given speed or even be able to start at all is dependent on adequacy of starting and accelerating torque. If a motor–starter combination for too low a starting torque has been chosen the motor may only start when the second step

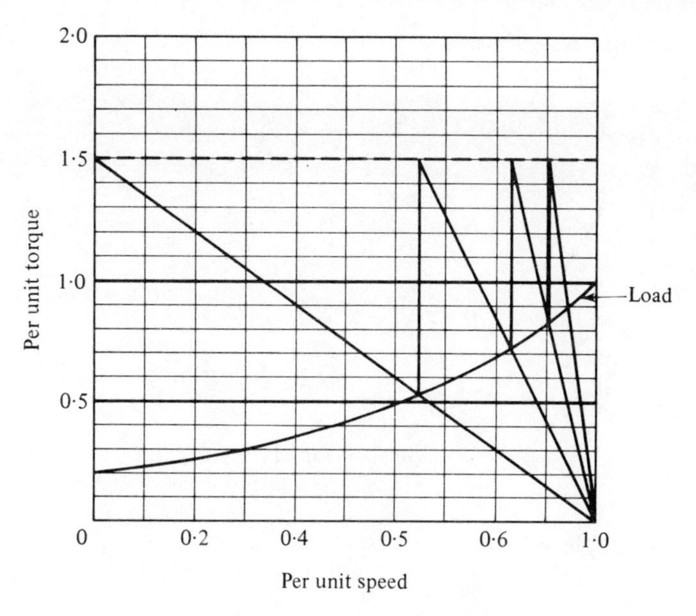

Fig. 14.3 Direct-current shunt motor rheostatic starting on fan load

of the starter has been reached. This will cause excessive current which is liable to damage not only the commutator, but burn the starter's contacts.

The nature of the start and the way in which the load is accelerated to full speed is thus in the hands of the starter designer. A close ratio between maximum and minimum accelerating torques will give a very smooth start but will necessitate a large number of starting steps. A wide ratio on the other hand while needing fewer starting steps results in violent current and torque peaks.

For any given starter the current peaks may be equalized if the total resistance in circuit in successive steps forms a geometric progression. The ratio between maximum and minimum currents, hence torques, may be determined from the formula:

$$K = \sqrt[n-1]{\frac{R_1}{R_A}} \tag{14.1}$$

where

$$K = \frac{\text{maximum accelerating current (torque)}}{\text{minimum accelerating current}}$$

n = number of steps in starter

R_1 = total resistance in circuit on first step.

The actual ratio will be somewhat less than the calculated value due to inductance in the motor. It follows that

$$\text{maximum current} = K \text{ minimum current}$$

The total resistance in successive steps graded according to the above formula will be:

$$R_1 \qquad R_4 = \frac{R_3}{K}$$

$$R_2 = \frac{R_1}{K} \qquad R_n = \frac{R_{n-1}}{K}$$

$$R_3 = \frac{R_2}{K}$$

The number of steps necessary to limit the peak current to a specific value can be determined by extending eq. (4.1)

$$n = \frac{\log R_1 - \log R_A}{\log K} + 1$$

Taking the example considered above

$$K = \sqrt[3]{\frac{6{\cdot}7}{0{\cdot}5}} = 2{\cdot}375$$

$R_1 = 6{\cdot}7$	starter $r_1 = 6{\cdot}2$
$R_2 = 2{\cdot}82$	starter $r_2 = 2{\cdot}32$
$R_3 = 1{\cdot}18$	starter $r_3 = 0{\cdot}68$
$R_4 = 0{\cdot}496$	starter $r_4 = 0{\cdot}00$

Approximate values of K can be obtained from Fig. 14.2 by measuring the distance between the peaks, when it will be found that the ratio between successive measurements is that already calculated.

In d.c. series motors the field flux alters with changes in armature current which modifies the acceleration curves, but the same methods are employed to design the starting resistances.

In the first quarter of this century when d.c. motors were extensively used the faceplate type of starter with no-volt and over-current solenoid was generally adequate for most applications (Fig. 14.4). The rate at which resistances were taken out of circuit depended on the operator, who, in the case of larger power drives may have had an ammeter to show him when he could safely cut out the next resistance, otherwise too fast operation would result in the overload trip being activated requiring the starting sequence to be repeated. The no-volt coil was an important safeguard against the supply being reconnected to a stationary or very much slowed up motor after a supply failure. To remove the human factor and consistently achieve the starting sequence best suited to a particular application the operation was made automatic and dependent on the speed or current of the motor. In the second quarter of the century many automatic starters were designed and manufactured, based on principles which, with the advent of controlled rectifiers, found renewed application in electronically controlled drives. Currently 'solid-state' power electronics and new ranges of specially designed d.c. motors make it not only possible to control the supply to the motor accurately and fully automatically during starting but extend such control over a wide range of operating conditions. Whatever means are used to

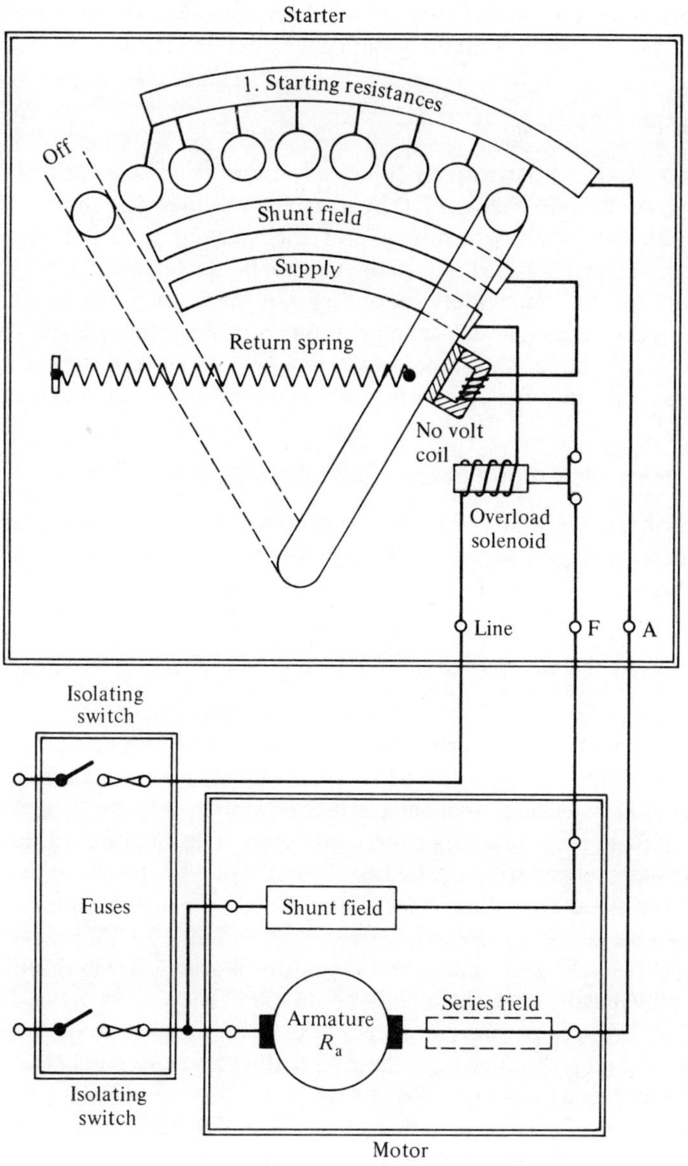

Fig. 14.4 Automatic d.c. shunt motor rheostatic starter

control the input to the motor in reduced voltage starting of d.c. motors, the modes can be classified as follows:

14.2.1 *Time-element starter (definite time acceleration)*

In this the resistances are reduced or power input is increased according to a definite time sequence—in the electromechanical starter this may be done by timing relays or dashpots or in controlled rectifier equipment it could be done by an electromechanical or electronic timer unit. This method does not take account of either load or motor conditions and necessitates separate protection circuits.

14.2.2 *Current element starter (current limit acceleration)*

This type of starting is particularly suitable for motors driving variable loads where the starting conditions also vary. The current at which the motor is allowed to accelerate further is predetermined and the starting time will depend on the magnitude and inertia of the load. In the automatic starter each successive contactor is controlled by that immediately preceding it by means of auxiliary contacts. These are mechanically latched in the open position and connected to a spindle connected to the plunger of a separate solenoid carrying the load current so that the plunger is still held up until the value of current falls to a predetermined value.

14.2.3 *Counter e.m.f. starter (speed responsive acceleration)*

Here the sequential shorting out of the starting resistances is determined by the back e.m.f. value reached, and as with current element starters the time the motor takes to reach this speed will depend on the magnitude of the load.

14.2.4 *Time–current acceleration*

This is a combination of the first two methods and in effect of the third, and consists in the time increments between the closing of the contactors being modified by the motor current so that with increased load on the motor it is given more time to reach a certain speed. This ensures that the current has fallen sufficiently and prevents the next current peak from reaching excessive values. This method will ensure that the motor always starts and will only be interrupted when the overload device, which in correct design is tailored to suit the thermal capacity of the motor, overrides the starting sequence.

Mechanical switching not only has finite limits of practical timing and adjustments but invariably means that there will be contacts subject to wear. Maintenance, overall reliability, and perhaps most of all the disappearance of d.c. supplies caused d.c. motors to be used only where their versatility and special characteristics could not be matched by a.c. motors. Nowadays controlled rectifiers make it possible to vary the input power to the motor by voltage 'chopping' and thus eliminate the power lost in the starting resistances which, even for medium sized motors, can cause space and heat dissipation problems. Electronic control, furthermore, makes smooth control of acceleration and speed relatively easy to integrate with other requirements and variables such as current and temperature limits which must be observed for all load and duty cycle variations. The modular design techniques employed for motor controllers make it possible to meet a wide variety of requirements by suitably combining appropriate logic and control modules.

14.3 Speed Control of d.c. Motors

The speed of a d.c. motor is directly proportional to the back e.m.f. and inversely proportional to the strength of the field. Some aspects of speed control by field weakening are discussed in section 3.28 of chapter 3. Field control, which gives shunt motors a constant power output characteristic, is useful and economical for fine speed adjustments over a limited range. The usually appreciable inductance of the field windings, while useful in some cases to limit armature current surges and maintain stability, is detrimental in others to a necessary rapid response. A typical example of this is the reversal of rolling-mill motors, where, to overcome the field time constant field-forcing—that is the application of over-voltage—is employed. To ensure satisfactory starting the field must be at its full strength and field regulating systems are, therefore, usually interlocked with the starter, which will only operate when the field control is correctly set.

Since the back e.m.f. is given by the difference between the applied voltage and the volt drop across the armature, it follows that a reduction in speed can be obtained by introducing resistance in the armature circuit as for starting, Fig. 14.1(a). The speed/torque curves for various values of series resistance are shown in Fig. 14.5

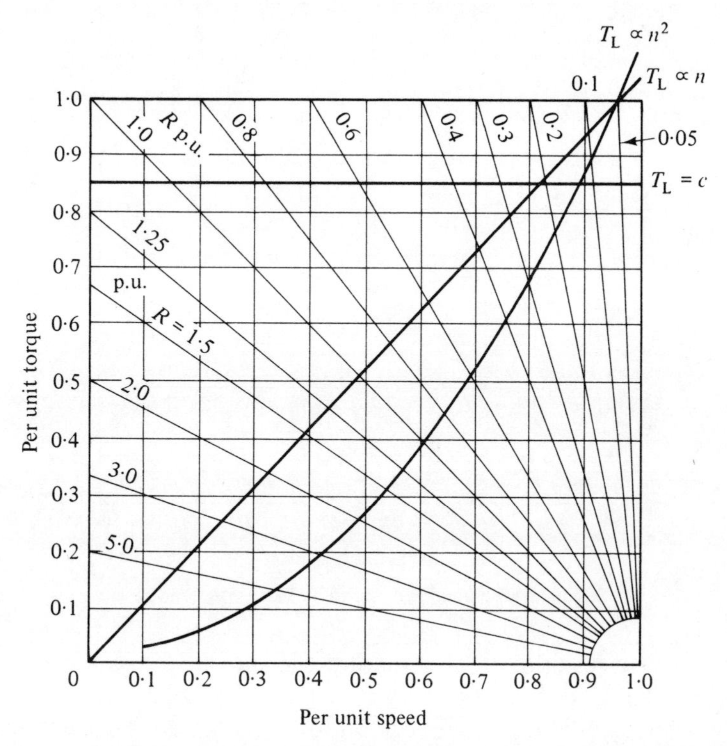

Fig. 14.5 Speed control of d.c. shunt motor by resistance in armature circuit—*F* remains constant

together with characteristic load curves. The graph illustrates that the equilibrium speeds for given values of R are entirely dependent on the load and that for the specification of a suitable controller it is essential to know the load characteristic. The following numerical examples illustrate other features of this form of speed control:

If the speed of the 460 V motor used in the previous examples is to be reduced to 50 per cent the back e.m.f. must be of the order of 230 V. If it is also assumed that the value of the load at the reduced speed is 50 per cent of full load, and the current is

25 A then the resistance required will be 230 : 25 = 9·2 Ω. Deducting from this value the armature resistance of 0·5 Ω gives the actual value of resistance to be placed in series as 8·7 Ω. This resistance is now left in circuit rather longer than in starting so it must be rated accordingly, i.e., it must be capable of dissipating 5 kW for the appropriate rating. It must also be noted that apart from being wasteful in power it will, as Fig. 14.6 shows, give the motor a pronounced series characteristic. A 40 per

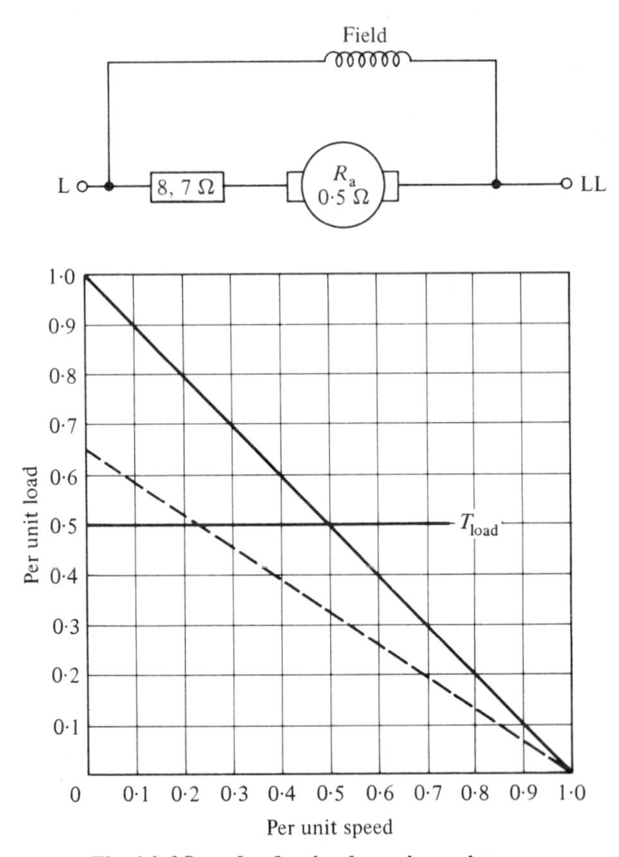

Fig. 14.6 Speed reduction by series resistance

cent load increase resulting in an armature current of 35 A will cause a drop of 9·2 × 35 = 322 volts. The back e.m.f. now becomes 460 − 322 = 138 volts corresponding to a speed of about 30 per cent of full rated speed. This method of speed control is, therefore, only suitable for substantially speed-proportional loads and very occasional short-term reduction. It is not normally acceptable for speeds below half speed where it is more usual to employ an armature diverter control.

Considering the same motor as before and for the sake of simplicity ignoring the armature resistance which has a negligible effect on the result we find that to obtain a 50 per cent reduction in speed and assuming a diverter current of 20 A at 50 per cent speed gives us a diverter resistance of 11·5 Ω and a series of 230/45 = 5·1 Ω. Should the current increase to 35 A the total line current will increase to 55 A resulting in a volt drop across the series resistance of some 280 V and an armature volt drop of 180 V corresponding to a speed of 39 per cent full speed. In other words, with plain series control a 40 per cent increase in load resulted in a 40 per cent decrease in speed, whereas with armature control a similar increase in load reduced the speed by only 21 per cent. The greater the diverter current, the less will the speed be reduced by the increasing load, but this can only be achieved at the expense of an increased energy loss in the resistors.

On many applications, such as cranes, haulages, winches, etc, creeping speeds are often required, where the load can vary over very wide limits and for these drives armature diverter control is normally used. The value of the series and diverter resistances can be varied by the operator manipulating the controller.

For other drives, particularly those calling for a close control of speed with varying load, it is now more usual to employ armature voltage control where the applied voltage is obtained from a separate source.

Increase of speed can be obtained by weakening the field, using a shunt field regulating resistance for shunt or compound wound motors. It is generally the responsibility of the motor maker to specify the value of resistance necessary and if it is desired that the speed increments are equal on each step of the regulator, then the starter designer must have the motor maker's curve showing the relationship of speed plotted against shunt current.

Where relatively small increases in speed are concerned no interlocking is necessary to ensure full field starting but if the speed increase is more than say 2:1 then an interlock is necessary to ensure that the shunt regulator is in the full field position before the motor can be started, otherwise unduly high currents may flow in the armature circuit. On some applications the requirements of the drive are such that the field regulator must be left at a predetermined speed, and when started the motor must always accelerate to this speed. A full field interlocking relay is provided for this purpose. This relay will respond to armature current during the starting period so that an excess of current will operate the relay to cut out the regulating resistance and give full field excitation. As the motor accelerates the back e.m.f. increases, the armature current falls to de-energize the relay and re-introduce the field resistance. This flip-flop action of the relay will continue until the motor attains the speed selected.

Though thyratrons and ignitrons are now being superseded by solid-state devices such as thyristors, some are still likely to be in service for some time. In such systems the field current, which must in any case be switched on before the armature for shunt machines, is provided by independent rectification. This makes it possible to control the armature supply, as shown in the simplified diagram of the AEI Emotrol ignitron motor control scheme, Fig. 14.7. The phase-shift controller determines the instant at which the ignitron connected as three-phase full-wave rectifiers fire. The power drawn from the a.c. supply is limited by series reactors. For better speed control the field is weakened by linking a field rheostat with the main speed control rheostat, but the wide speed range of up to 60:1 is essentially obtained from the phase-shift controller. Since the maintenance of a given voltage across a d.c. motor armature is not sufficient to maintain a constant speed under varying load conditions it is usual to incorporate a tachogenerator (armature back e.m.f. with varying field also being unsatisfactory). The difference between the tacho output and the voltage analogue of the desired speed is amplified and used to control the firing point of the ignitrons in such a manner that it is advanced to increase and retarded to decrease the speed of the driving motor.

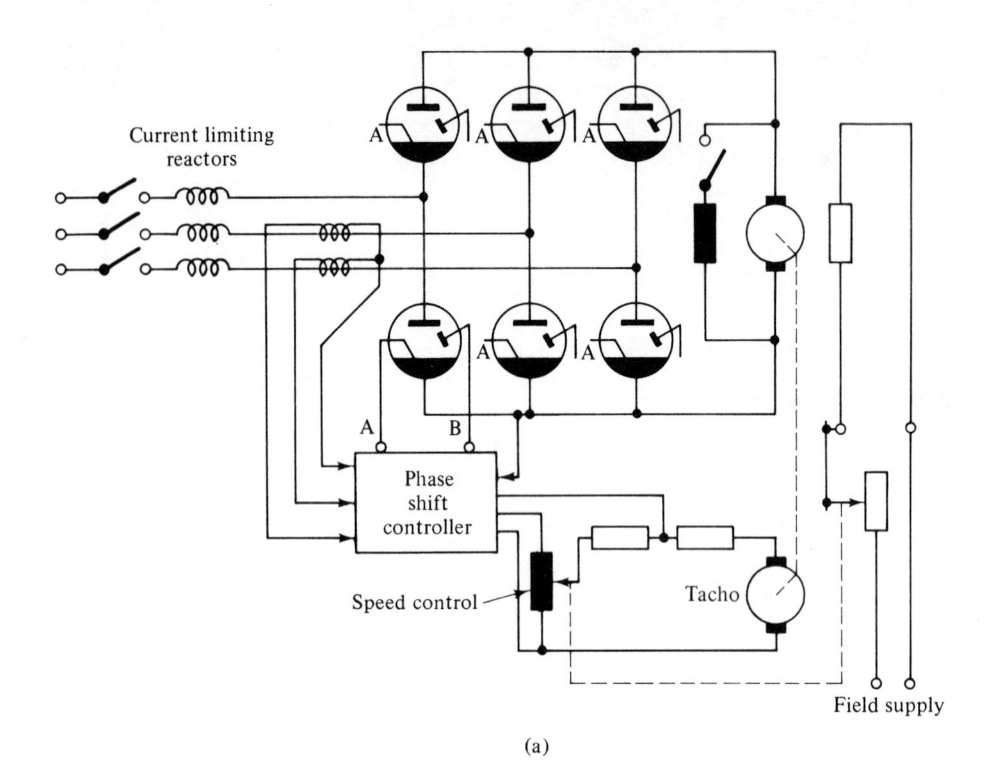

(a)

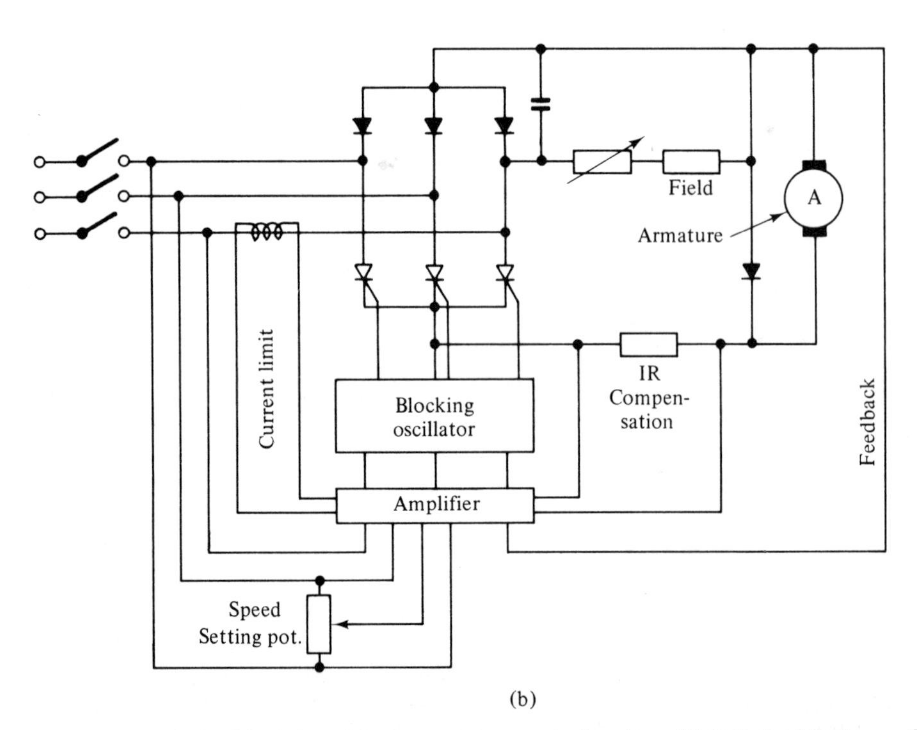

(b)

Fig. 14.7 (a) Principal features of ignitron motor control system. (b) Principal features of thyristor motor control

Regenerative braking is not possible with simple controlled rectifier equipment and where this is required for emergency or rapid stopping the braking methods described in the following sections are used.

Reversal of the motor's direction of rotation can be achieved by changing over the connections to the armature. Since the magnitude of the armature current is limited by the series reactors mentioned above the shunt field may, subject to appropriate precautions, be reversed instead.

The comparative ease with which the speed and torque of a d.c. motor can be controlled still makes it the most suitable motor for large power reversing and variable-speed drives such as mine winders and steel rolling mills. In heavy engineering applications the powers required run into many thousands of kilowatts and special rotary convertors used in systems such as the Ward–Leonard Controls were used. Since most power is generated and distributed as alternating current, grid-controlled mercury arc rectifiers were subsequently developed. Another method was to use saturable reactors (transductors) and semiconductor rectifiers where the supply to the motor is rapidly and efficiently controlled by a special winding on the transductor controlling the armature current. These devices have proved themselves very robust and reliable since they contain no moving parts. They have the disadvantages of not being able to provide regenerative braking facilities and requiring considerably more space and power than the solid-state controlled rectifiers now readily available commercially with a wide range of outputs. Thus powers of 400 kW can now be controlled by six thyristors in a three-phase bridge configuration. Electronic controls are also dealt with in chapter 12.

14.4 Speed Control of d.c. Motors under Overhauling Conditions

According to type and method of control the speed of d.c. motors increases as the load falls to a greater or lesser extent. With no external resistance the variation from no load to full load on a shunt machine is very small, but with external resistance the variation is considerable. In a series machine the speed increases appreciably when the load is reduced and in crane or hoist drives where the load torque on lowering becomes negative and adds to the torque developed by the motor then dangerous over-speeding could occur and special steps must be taken to avoid this. Mechanical brakes can guard against this but the form of electrical braking most commonly used is dynamic or rheostatic braking, as shown in Fig. 14.8.

It will be noted that in the shunt motor case the armature is disconnected from the line and connected across a braking resistor, and the field left energized thus turning the motor into a generator driven by the load. The degree of braking torque will depend upon the value of the braking resistor, R.

In the series motor case both the field and armature are disconnected from the supply, and the field connections reversed to turn the machine into a generator. Again, the braking torque can be controlled by varying the external resistance.

Potentiometer braking is a more sophisticated form used almost exclusively for the control of cranes, and the line diagram is shown in Fig. 14.9. This form of control is particularly suited to cranes and hoists because the change-over from motoring to overhauling conditions is entirely automatic. When hoisting, the motor is connected in the orthodox manner with two or more steps of armature diverter control; but when lowering it will be noted that the machine acts as a shunt generator when the

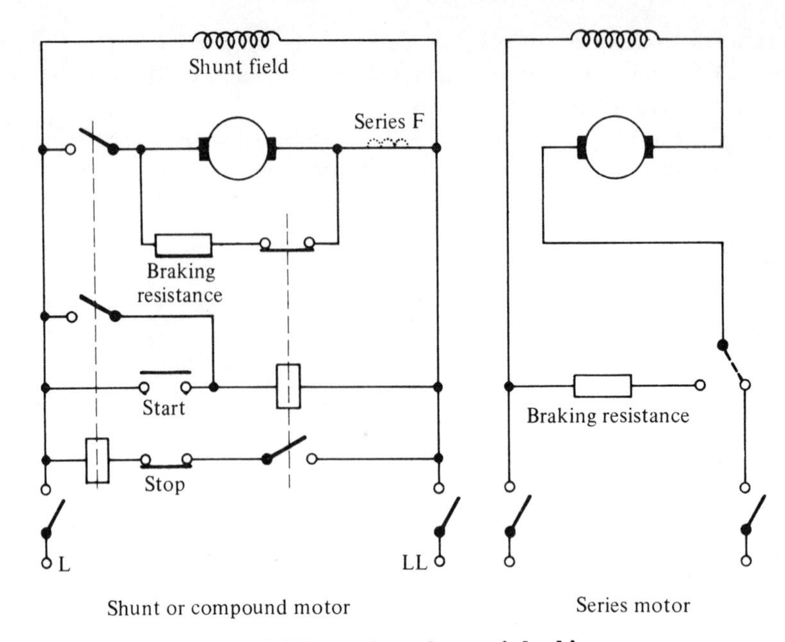

Fig. 14.8 Dynamic or rheostatic braking

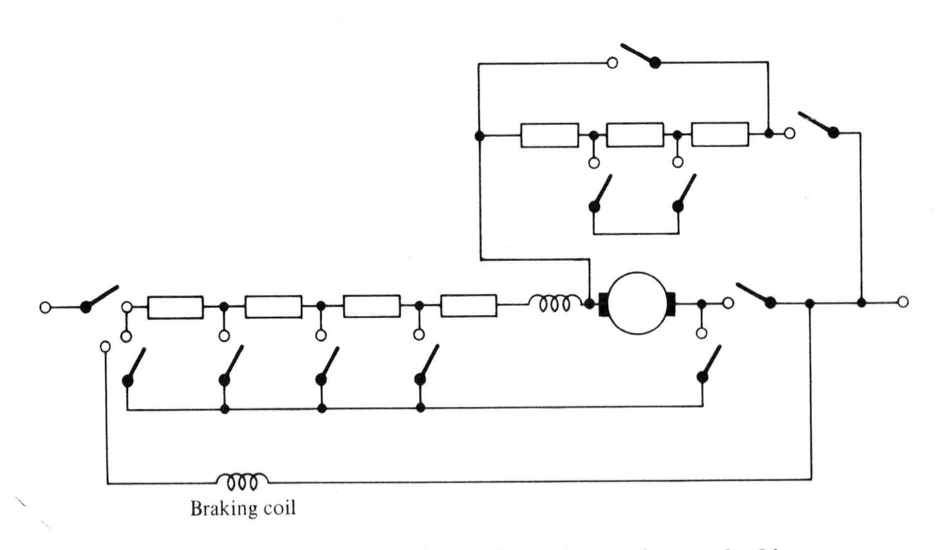

Fig. 14.9 Series motor control—starting and potentiometer braking

load is overhauling the motor, but as a shunt machine when the motoring torque exceeds the load torque.

Figure 14.10 shows typical speed/torque curves for a series motor controlled by a controller provided with plain series control on hoisting including two steps of armature diverter for a slow-speed creep and with potentiometer lowering.

It should be noted that maximum braking torque per armature ampere requires maximum field current so that with appreciable field control (as might obtain in an adjustable speed drive) it may be necessary to strengthen the field during braking,

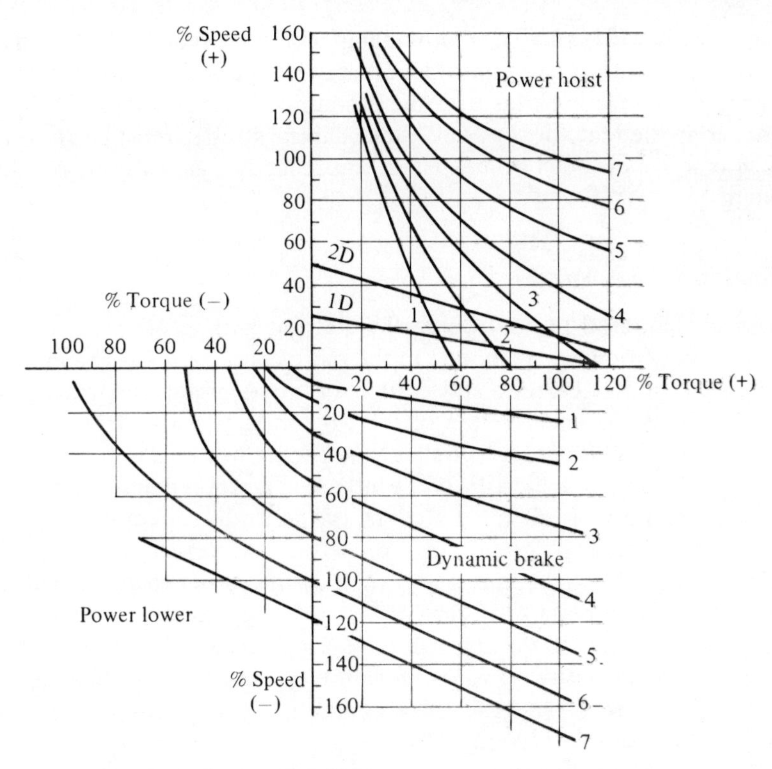

Fig. 14.10 Direct current series motor speed/ torque curves

bearing in mind that field time constants for shunt fields can be very significant (of the order of 1 s).

The magnitude of the braking resistor must be such as to limit peak armature currents to within the machine's commutating ability. Here again there will be a time constant during which the motor cannot develop any significant electromagnetic torque, but this is only likely to be of the order of 0·1 s in thyristor controls. By using switching logic and special circuits involving another rectifier one manufacturer has achieved dead times as low as 20 ms. The ultimate virtually instantaneous reversal using a back-to-back connection with two circulating current chokes and involving the most equipment and stresses on motor and load is justified only where extremely rapid dynamic performance is mandatory.

14.5 Emergency Stopping of d.c. Motors

On the majority of drives it is satisfactory either to allow the machine to coast to rest or, when switching off, allow one of the many forms of brakes to come into operation. The considerable energy stored in high inertia loads can be dissipated partly by dynamic braking in suitable resistances, partly in the mechanical brake which must be applied some time before the motor ceases to generate sufficient voltage for the dynamic braking to contribute significant retarding torque. For really fast stopping, reversal of the motor, or plugging, may be required—this means that at the instant of reversal the supply voltage and the back e.m.f. are additive and very high currents and torque peaks will occur. The motor manufacturer must be consulted on the effect of high currents on the commutator and it may be necessary to limit the current by first having resistance in series with the armature. A careful study of the energies which are converted into heat as well as mechanical stresses must be made to ensure that such emergency stops do not affect the reliability and life expectancy of the equipment.

14.6 Starters for a.c. Motors

Problems encountered in controlling the starting and accelerating of polyphase induction motors are principally due to the more complex relationship and control characteristics of the current and torque of these motors compared with d.c. machines. This is particularly true of the cage induction motor which is used in probably more than 90 per cent of all a.c. motor applications.

The torque/speed characteristic of the induction motor is dealt with in some detail in chapter 4, and a more detailed analysis of the torque equations will show that torque which, as in most other motors, results from the interaction of a magnetic field and current-carrying conductors is governed by the rotor reactance which is maximum at starting and practically zero at no-load. An induction motor designed to have maximum torque at starting will have considerable slip on load, conversely optimum relationship between rotor resistance and reactance for maximum output will result in a very poor starting torque (see graph, Fig. 4.7 of chapter 4). Double and triple cage rotors achieve, at some extra cost, very good compromise and particularly favourable torques per ampere of starting current. The starting characteristics of cage induction motors are defined in BS 4999: Part 41 which is dealt with in chapter 5, section 5.2.2.1. From tables 5.4 and 5.5 it will be seen that for relatively modest starting torques (twice down to half rated torque) the starting currents range from about eight down to six times rated full-load current for standard continuously rated motors. Because of these large starting currents, supply authorities limit the size of motor which may be started directly on line. Where independent supplies are generated, or in power stations where no disturbance can be caused to consumers, quite large motors may be started on line particularly if, like fans, blowers or pumps, they are not required to supply a large starting torque. The disturbance caused in the supply system by the starting of induction motors is often aggravated by loads with considerable inertia which, because of the induction motor's relatively low starting torque, may take a considerable time before the current drops to its rated value.

Various types of reduced voltage starters have been produced to reduce the starting current of induction motors, but as the starting torque is proportional to the square of the supply voltage even a modest reduction to 80 per cent will bring the starting torque down to 64 per cent of its nominal rated value. Figure 14.11 shows a typical induction motor characteristic for 80, 100, and 120 per cent of rated supply voltage.

In the application of induction motors great care must be taken that, when starting under reduced voltage conditions has to be adopted, the starting torque is adequate for all conditions of supply and load. The motor torque must be adequate to accelerate the system to a sufficiently high speed so that when the full voltage is applied the current surge does not cause any of the protective devices to operate or inject a more severe disturbance into the supply network than would have occurred with direct-on-line starting. Table 14.1 shows some approximate values of starting torque and current which obtain with the different starting methods used for cage induction motors.

Table 14.1

Method	Starting torque % FLT	Starting current
Direct-on	100/150	6/9 times FLC
Primary-resistance	25/40	$3/4\frac{1}{2}$ times FLC
Star-Delta	33/50	2/3 times FLC
Series-Parallel	25/40	$1\frac{1}{2}/2$ times FLC
Auto-transformer	16/80 According to tapping	$1/4\frac{3}{4}$ times FLC
Part Wound	45/70	4/5 times FLC

The figures given in table 14.1 apply to standard cage induction motors. Special high starting torque motors giving some 50 per cent more starting torque with possibly even less than average standard starting current have been made and, where the higher cost can be justified, may be considered.

The calculation of load power requirements is considered in chapter 2. Accelerating and average run-up torques are defined and these, after taking into account the reduction of motor torque due to reduced voltage starting, must still be adequate to accelerate the drive to full speed in reasonable time as calculated from the given formulae. The definition of reasonable time will vary widely with size and type of motor as well as frequency of starting. If necessary, losses and the resulting temperature rise can be estimated, but consultations with the manufacturers should take place wherever possible.

14.7 Primary Resistance or Reactance Starters for Induction Motors

In this method of starting a resistor or reactor is connected in series with the motor during starting to reduce the voltage at the motor terminals. When the motor has accelerated to 75–80 per cent of rated full-load or synchronous speed the resistor or reactor is short circuited, and full rated voltage applied to the motor. While the starting current is reduced in direct proportion to the reduction in voltage, the torque is reduced as the square of the voltage. This means, for example, that for a 50 per cent reduction in starting current the motor will produce only 25 per cent of direct-on-line starting torque. When current limitation is the principal consideration then other methods of starting such as star-delta or autotransformer are to be preferred. The resistance required for primary resistance starting is generally more expensive and larger than one required for a slipring motor of equal rating and the reactor type gives rise to a greatly reduced power factor at starting. As both these methods are costly in equipment and operation neither method is extensively used.

Since it has, however, two distinct advantages, as listed below, it has been widely used in the USA and may still today prove suitable for certain applications.

(a) Since the voltage drop across the resistor or reactor falls as the motor accelerates, there is a corresponding increase in motor volts giving a gradually increasing torque during acceleration.

(b) Since the motor is not disconnected from the supply when the change-over from START to RUN occurs there is no transient peak current or torque, resulting in the acceleration to full speed being particularly smooth.

A special type resistance starter is the liquid resistance starter marketed under the name of 'Statormatic' (Fig. 14.11). Its operating principle is based on the decrease in

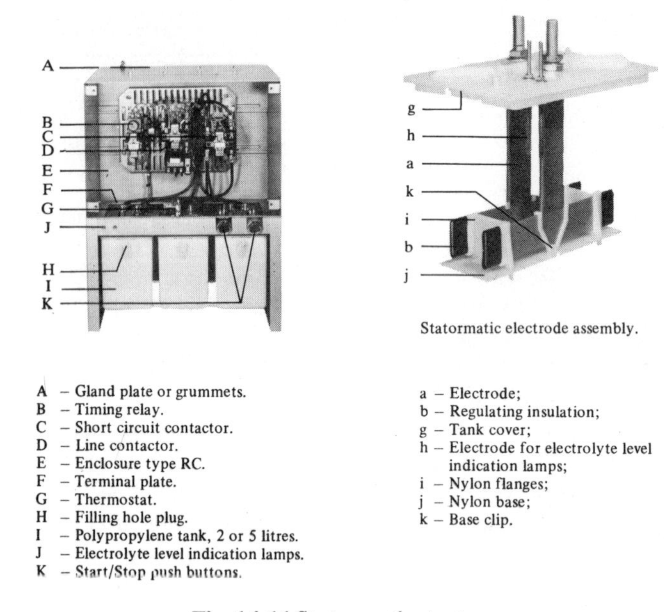

A – Gland plate or grummets.
B – Timing relay.
C – Short circuit contactor.
D – Line contactor.
E – Enclosure type RC.
F – Terminal plate.
G – Thermostat.
H – Filling hole plug.
I – Polypropylene tank, 2 or 5 litres.
J – Electrolyte level indication lamps.
K – Start/Stop push buttons.

a – Electrode;
b – Regulating insulation;
g – Tank cover;
h – Electrode for electrolyte level indication lamps;
i – Nylon flanges;
j – Nylon base;
k – Base clip.

Fig. 14.11 Statormatic starter

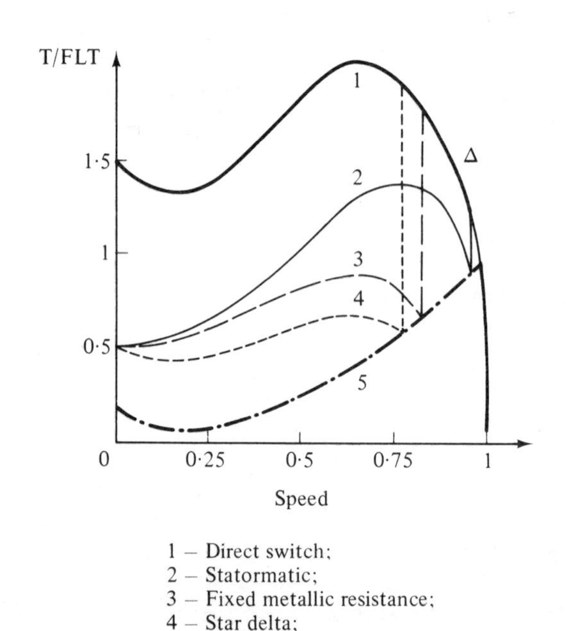

1 – Direct switch;
2 – Statormatic;
3 – Fixed metallic resistance;
4 – Star delta;
5 – Load torque.

Fig. 14.12 Torque/ speed curves for induction motor with various methods of starting

the resistance of an electrolyte when heated inside an electrode chamber. The resistance is connected in series with the motor windings to reduce terminal volts and thereby the current drawn from the supply. As the motor runs up to speed there is an automatic decrease in resistance value with a consequent increase in terminal volts giving progressive acceleration without peaks of torque or current. At the end of the run-up period a timed contactor closes and short circuits the residual resistance.

Figure 14.12 shows a typical speed/torque curve for an induction motor started by a 'Statormatic' automatic starter (curve 2) and compares it with direct-on-line starting (curve 1), fixed metallic resistance starting (curve 3), and star-delta starting (curve 4). A hypothetical load/torque curve (5) allows a comparison of the net acceleration torque to be made for the various methods of starting.

Starting times for motors up to 30 kW in typical industrial drives range from 3 to 30 seconds. Larger motors in the range of 30 to 150 kW may require up to 60 seconds to reach their rated speed. In both cases the maximum values apply to loads with considerable inertia such as fly-wheel presses and dust extractor fans. Exceptionally long starting times as high as 300 seconds may be encountered in centrifuges.

14.8 Star–Delta Starting of Cage Induction Motors

For this method of starting both ends of each phase winding must be brought out to a terminal box, which will have to contain at least six terminals. The reduced voltage for starting is obtained by connecting the windings in 'Star'. When the motor has reached 75 to 80 per cent of rated speed, full rated voltage is applied to the windings by reconnecting them in 'Delta' (Fig. 14.13).

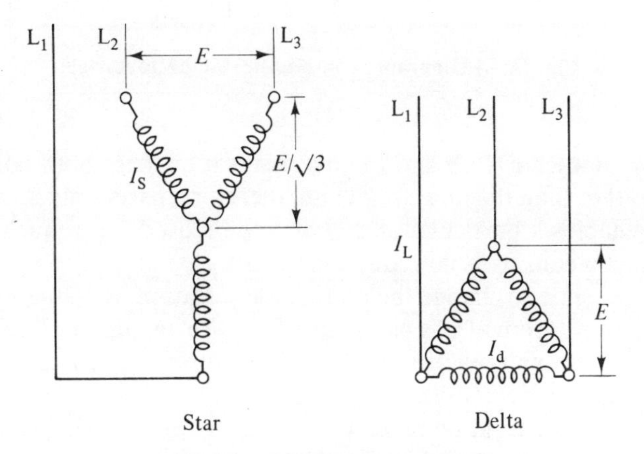

Fig. 14.13 Star-delta starting

If the line current in delta is I_L then the phase current in delta is I_L divided by $\sqrt{3}$, or 58 per cent I_L. By reconnecting the windings of a three-phase motor in star when their normal connection on rated voltage is delta, the voltage across the phases and the line currents are reduced to 58 per cent of their rated value. The starting torque is reduced to $\frac{1}{3}$ of normal value and if this is sufficient for the application under consideration this method of starting should, because of its simplicity and therefore low cost, be given serious consideration.

The diagram of a typical contactor-type star-delta starter is shown in Fig. 14.14 and two points are worthy of note.

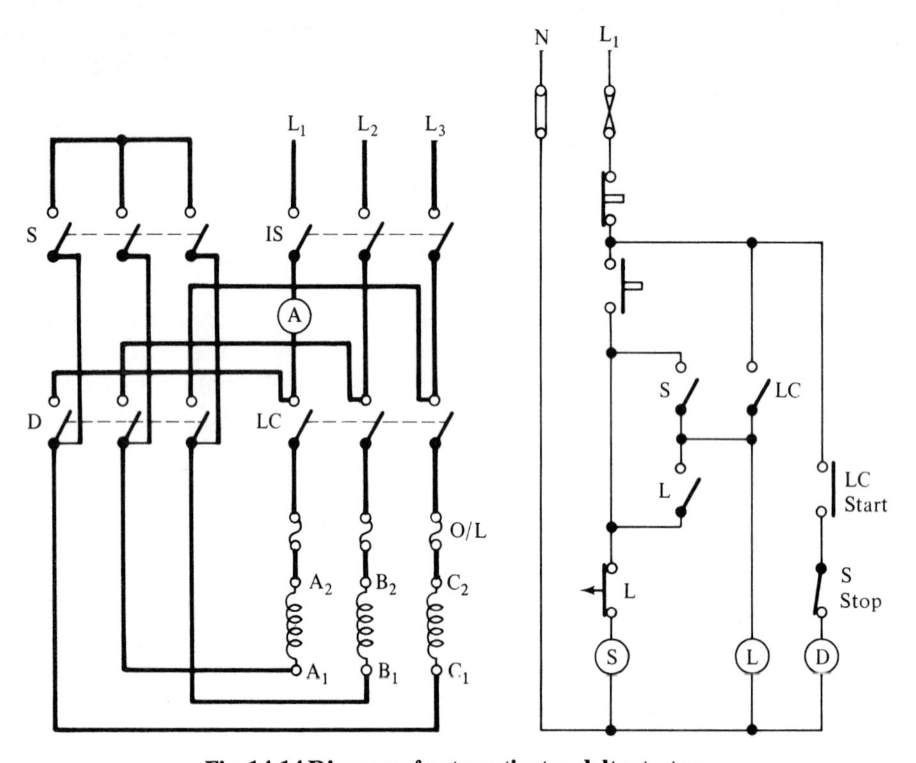

Fig. 14.14 Diagram of automatic star-delta starter

(a) The line contactor (LC) and the overloads (O/L) are both connected in the phase circuit rather than the line circuit and, therefore, carry only 58 per cent of line current. This enables a particular starter size to be used for a much higher horsepower and effect a considerable reduction in cost.

(b) It is not possible to change the direction of rotation by changing over any two phases in the motor terminal box as can be done with any three-terminal motor and any attempt to do so will result in blowing the distribution fuses when the starter changes from star to delta. When a change of direction is required two of the incoming phase lines should be reversed. The effect of reversing two phases at the motor terminal box is illustrated in Fig. 14.15.

The starting procedure and automatic operation of the control gear is as follows (Fig. 14.14):

(a) Close main isolator IS.
(b) Close main contactor LC by push button LC START.
(c) Start contactor closes automatically and connects motor windings to supply.
(d) Timing device which is usually adjusted to the required value when the drive is commissioned comes into operation when the motor is nearly up to full speed, opens the star contactor and closes the delta contactor.

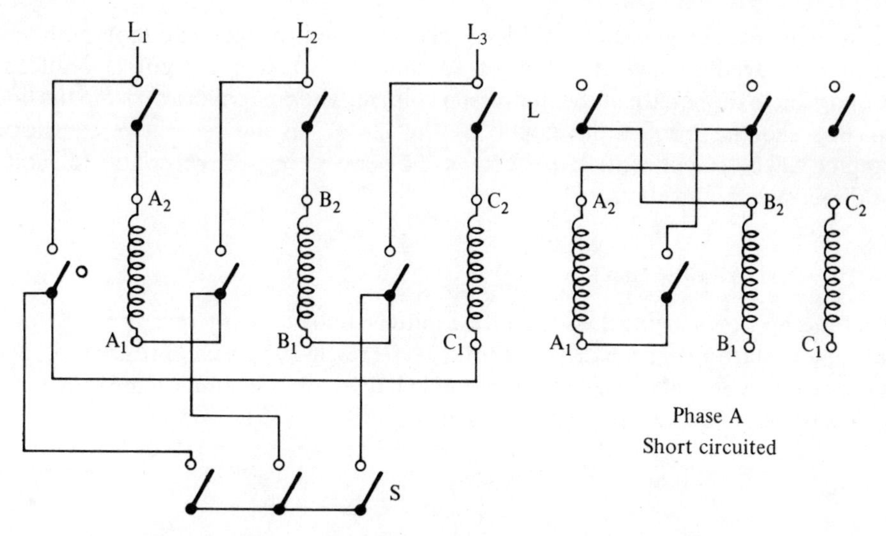

Fig. 14.15 Star-delta—effect of reversing two motor phases

The transient peaks which occur when this changeover occurs are common to other types of starters and are considered in section 14.11. A type of star-delta starter which considerably reduces the current and voltage peaks is described in section 14.12.

14.9 Auto-transformer Starting of Induction Motors

Figure 14.16 shows the connection for an auto-transformer starter from which it can be seen that three distinct starting performances are possible to meet the requirements of different drives. This is made possible by providing three voltage tappings

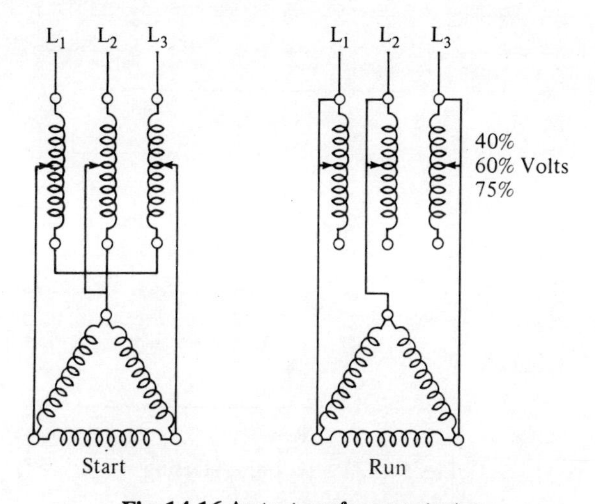

Fig. 14.16 Auto-transformer starter

on the transformer to give open circuit voltages of say 40 per cent; 60 per cent; or 75 per cent of line volts at the motor terminals. The corresponding starting torques will be 16 per cent; 36 per cent; or 56 per cent of rated torque. The motor current is further reduced by the transformer so that the starting current is reduced in proportion to the square of the impressed voltage. The current drawn from the line at starting and the torques developed by the motor on the respective transformer tapping will be 16 per cent; 36 per cent; or 56 per cent respectively of the full voltage or direct switching values.

14.10 Part Winding Starters

This method of starting has found a limited application in recent years for the infrequent starting of such drives as rotary type refrigeration compressors. A special motor winding in which the stator is either in halves, or in sections in the ratio of $\frac{2}{3}$–$\frac{1}{3}$ is necessary. Connections for both types are as shown in Fig. 14.17.

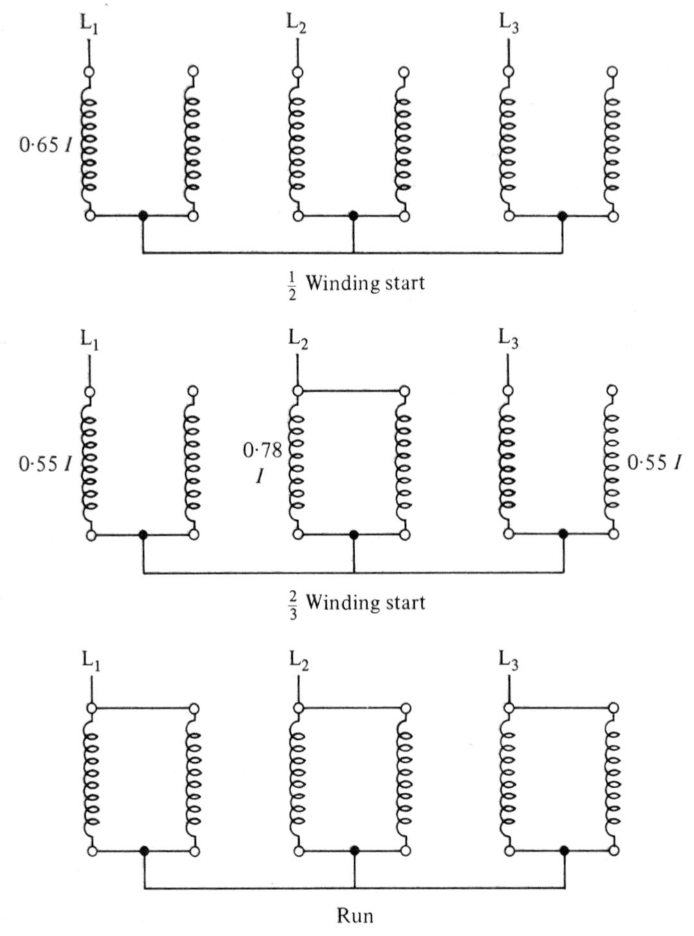

Fig. 14.17 Part winding starting

Typical speed/torque and speed/current curves are shown on Fig. 14.18. It will be noted that this method of starting produces a pronounced dip in torque at about half speed which, under certain circumstances, could prevent the motor accelerating

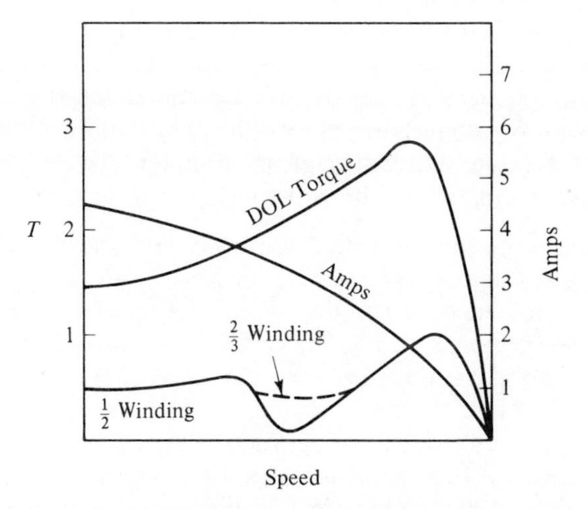

Fig. 14.18 Part winding starting—characteristic curves

beyond this speed and result in a high peak current when switching in the second part of the winding. The dip in torque is particularly noticeable in the centre-tapped winding and although it is much smaller in the two-thirds winding, the improvement is obtained at the expense of higher current peaks at starting.

The starting torque obtained in both methods is approximately 45 per cent of that obtained from normal single winding machines, and while the starting current of 65 per cent of normal locked rotor current is indicated as being the same on both methods of switching, this is not strictly accurate, since, with the two-thirds connection, this is the average of the current drawn from the three phase lines. The starting currents in the three phases are actually unbalanced; L_1 and L_3 taking some 55 per cent and L_2 80 per cent of the direct-on-line starting current. The high current in L_2 could, in certain circumstances, be unacceptable and to overcome this a single step of primary resistance has also been used in some applications. This increases the cost of the equipment and reduces the starting torque still further, suggesting that it may be more advantageous to consider the use of a standard motor and other methods of starting. Apart from these considerations, part wound motors tend to be noisy during starting due to inbalance and cannot, therefore, be recommended for duties where frequent starting is called for.

14.11 Transient Peaks

It was noted that with the primary resistance or reactance method of starting a very smooth start was possible due not only to the rising torque developed by the motor, but also to the absence of a break in positive motor torque during the changeover from 'Start' to 'Run'.

If when using star–delta or auto-transformer starting the accelerating torque of the motor has been sufficient to bring the drive up to a sufficiently high speed (75 to 80 per cent) in the start position the secondary current and torque peaks should be moderate in value. The value of current when the windings are reconnected to the full supply voltage does not normally correspond to the current the motor would draw at that speed with the appropriate connection. In the case of star-delta changeover the theoretical current increase is a trebling of the star current at the instant of changeover. In practice it is impossible to achieve the instantaneous changeover with simple starting equipment which completely disconnects the motor from the supply and re-connects it after a short, but in terms of mains frequency, random time (a 50 Hz supply goes through one complete cycle in 20 ms). The factors which can influence the nature of the switching current surge are as follows:

(a) Unless the energy stored in the rotating system and the load is capable of maintaining the speed against the resistive load the motor will slow up during the changeover from one connection to the other. The lower the speed of the motor when re-connected the higher will be the current and torque peaks.

(b) When the motor is running the back e.m.f. opposes the applied voltage. When the supply is disconnected the field created by the back e.m.f. does not collapse instantaneously and is still likely to be appreciable when the supply is reconnected. At the instant of reconnection the phase angle between the applied voltage and the back e.m.f. is indeterminate. The worst instant is when they are in phase and therefore give rise to current peaks proportional to their sum. Values as high as twenty times the motor's normal full-load current have been measured.

Current peaks of order of magnitude two to three times the current for direct-on-line starting are not only unacceptable to the supply, but can give rise to torque surges likely to cause damage to motor and driven machine.

For motors of higher power rating which are finding increased use on such drives as pumps and compressors these problems are particularly serious and for such applications starters in which the motor is not disconnected from the supply during changeover from start to run are now available.

14.12 Continuous Torque Closed Transition (Wauschope) Star-delta Starters

The line diagram of a typical closed transition star–delta starter is illustrated in Fig. 14.19.

It will be noted that a resistance is connected in parallel with each phase winding. This resistance has a separate star point so that when the star point of the phase windings is opened the resistance in each phase is left in series with the windings in delta formation. The resistance is finally short circuited. A continuous torque is thus maintained during the changeover, but the need for a fourth contactor and resistances to withstand starting currents makes the starter much more expensive than the conventional star-delta starter.

14.13 Auto-transformer (Korndorfer) Starter

The connections for this method of starting are shown on Fig. 14.20.

From the diagram it can be seen that in the changeover from start to run a section of the transformer winding is left in series with the motor acting as a choke, which in the run position is finally shorted out. A continuous motor torque during changeover is thus maintained and so long as the accelerating torque produced at starting is adequate, then moderate secondary peaks can be expected.

It is worth noting that in this method of starting high voltage peaks can occur at the star point of the transformer, because the open circuiting of the star point of the

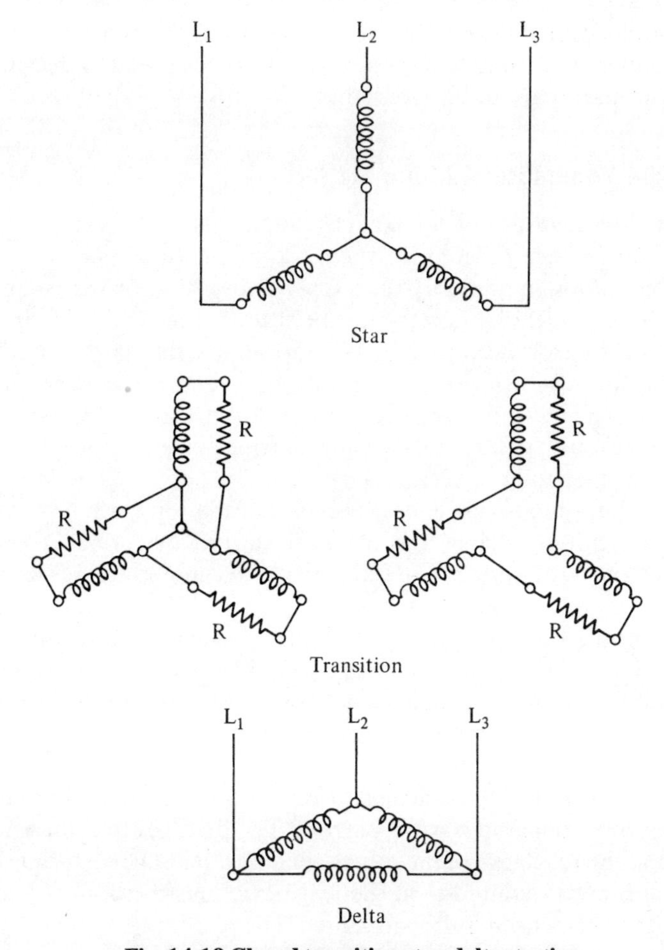

Fig. 14.19 Closed transition star-delta starting

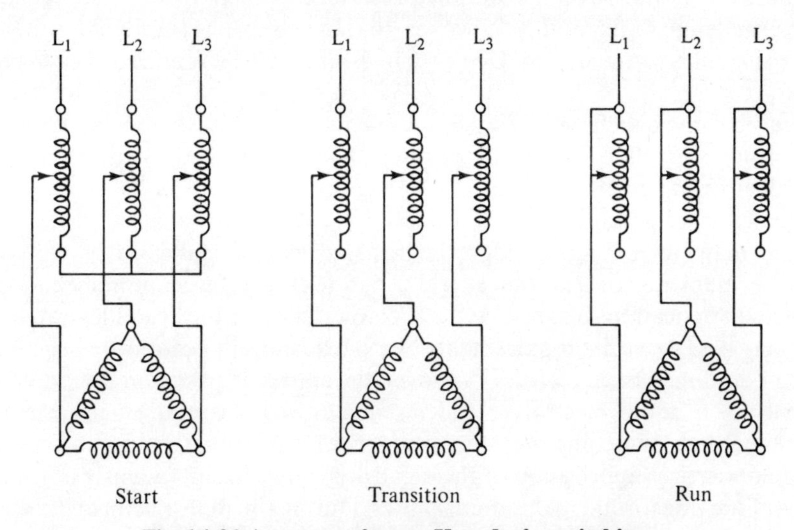

Fig. 14.20 Auto-transformer Korndorfer switching

transformer produces conditions similar to those occurring in an open circuited current transformer. The transformer must, therefore, be specially designed if 'flash overs' within the starter are to be avoided.

14.14 Polyphase Wound Rotor Motor Starters

Despite the predominant use of squirrel cage machines and the improved starting methods which have been devised for them there are still applications where the current surge limitation imposed by the supply authorities, or the torque requirements of the driven machine, necessitate the use of slipring motors in which the control of both starting torque and current is more akin to the control of a d.c. motor.

The speed/torque characteristic of an induction motor is dependent upon the relationship of resistance and reactance in the rotor circuit, and it has been shown how the starting torque can be increased and starting current reduced by increasing the rotor resistance (chapter 4, section 4.6).

In the case of the wound-rotor motor the ends of the rotor windings are brought out to sliprings, and it is possible to control the starting performance and design a starter to meet the requirements of the driven machine and special conditions stipulated by the supply authority exactly.

The speed/torque characteristics of a slipring machine with different values of external motor resistance in circuit are the same as those shown for a cage motor with rotors of various resistances (chapter 4, Fig. 4.7). The resistance may be incorporated in the starter, or in the larger sizes or for special applications mounted separately.

With external resistance the starting torque produced by a slipring motor is approximately proportional to starting current. The divergence from strict proportionality is due to the reactance of the motor windings. If a starting torque of 150 per cent full-load torque is required then the resistance must be designed to limit the starting current to 175 per cent full-load torque. The resistance is gradually cut out of circuit in steps so that the successive values form a geometric series exactly, as was the case in the d.c. motor starting resistance described in section 14.2. The calculations will be exactly the same, noting, however, that the voltage is rotor voltage per phase and the current is rotor current. One such resistance will be required in each phase of the rotor.

Balanced rotor switching (Fig. 14.21) is normal practice on contactor-type starters but on drum controllers and hand-operated starters balanced switching is generally impractical owing to the number of contacts required and unbalanced switching arrangements are employed.

In the majority of applications the resistances are made up of wire wound resistance elements or resistance grids, but for very large powers and certain specialized applications a special type of rotor starter of the type illustrated in Fig. 14.22 may best meet the requirements of a particular application.

The operating principle of the 'Vapormatic' starter is based on the difference in resistivity of a liquid electrolyte and its vapour when contained in an electrode chamber. When switching on, the passage of the initial rotor current causes immediate partial vaporization of the electrolyte and instantaneously adjusts resistance and starting torque to optimum values. During the motor's run up to speed the thermal interchanges which take place give rise to a liquid/vapour mixture whose

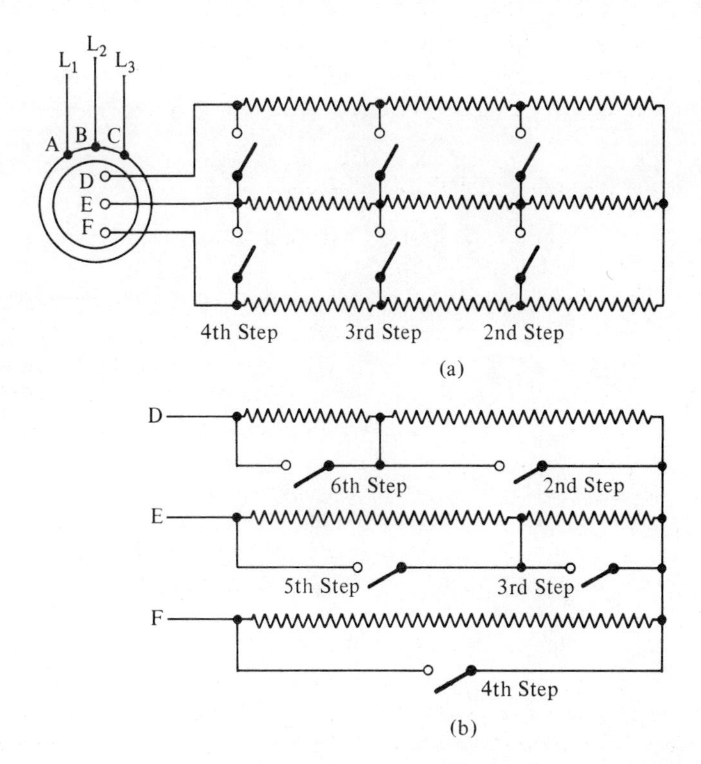

Fig. 14.21 Balanced (a), and unbalanced (b), rotor resistance switching

resistance decreases progressively. A timed contactor finally shorts out the low residual resistance.

Figures 14.23(a) and (b) show a comparison between the change in external rotor resistance and the speed/torque curve when the 'Vapormatic' starter and a conventional three-step metallic resistance starter is used.

Automatic slipring motor starters generally use a three-pole contactor in the stator circuit and, depending on the power of the motor, one or more two-pole contactors in the rotor circuit. These rotor or accelerating contactors have their operation controlled by preset automatic timing devices. In certain cases it may be more economical to use a three-pole contactor in the final stage since this, by also connecting the rotor circuit in delta, reduces the current per contact to 0·577 of the value with the two-pole contactor and makes it possible to use a smaller model.

14.15 Single-phase Motors

A wide variety of different types of single-phase motors, many mainly distinguished by their method of starting, exist and are described in chapter 7. In industries such as agriculture, cottage-type industries where hitherto it has been found impractical or uneconomic to provide a three-phase supply, the use of electricity has increased to such an extent that the difficulties are being overcome, and the more robust and economical three-phase induction motor is replacing single-phase machines. On the other hand, the use of very small fractionals (less than 1 kW), miniature and micro-motors in homes, offices, and other commercial premises has increased

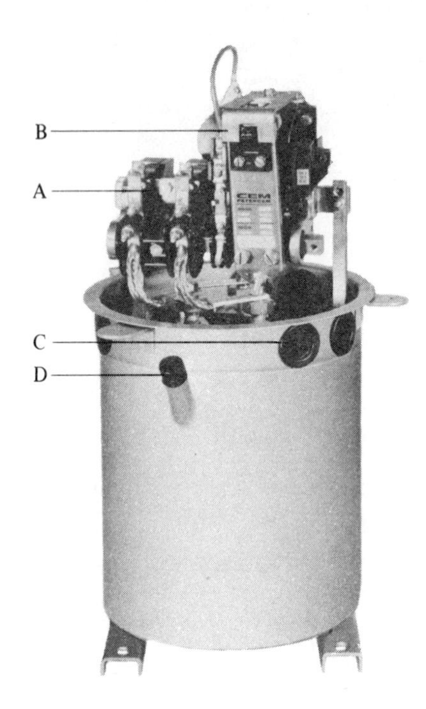

A – Short circuit contactor.
B – Timing relay.
C – Rubber grummets (or cable compression glands).
D – Filling hole and dipstick.

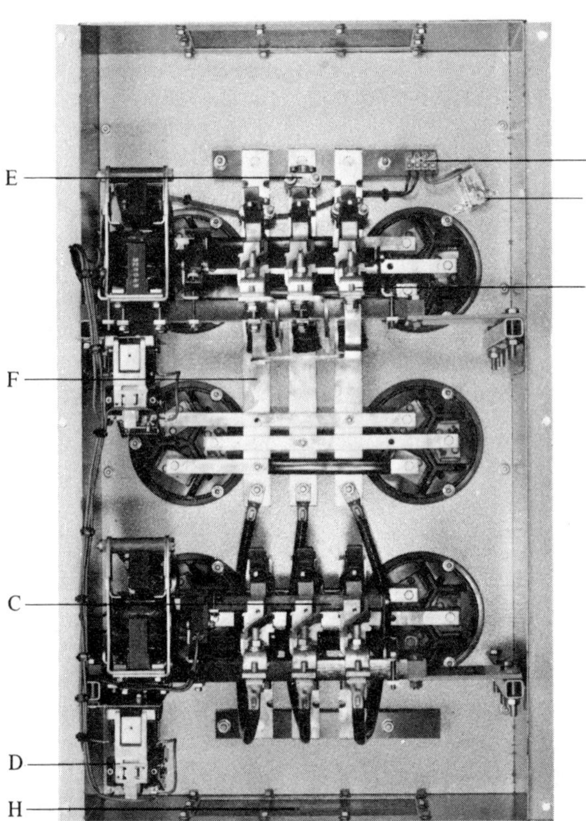

A – Short circuit contactor
B – Thermostat
C – Insertion contactor.
D – Timing relay.
E – Terminal connexion.
F – Connecting bars.
G – Connecting strap.
H – Terminal plate.

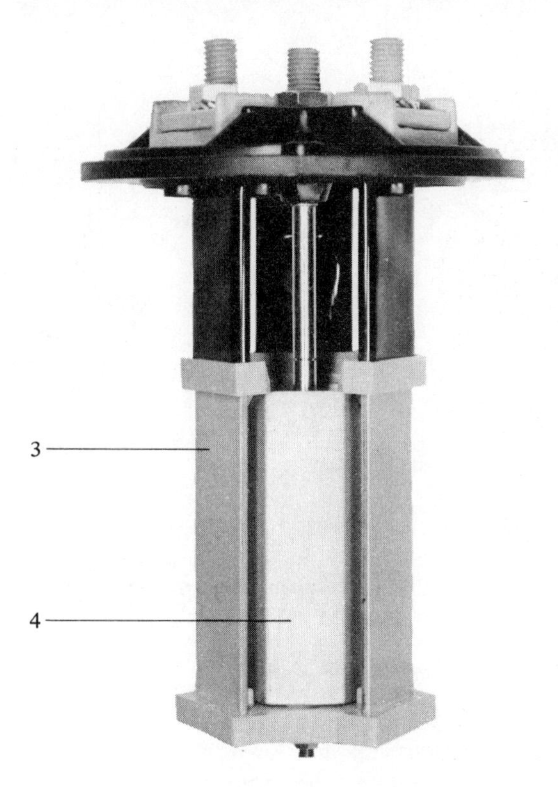

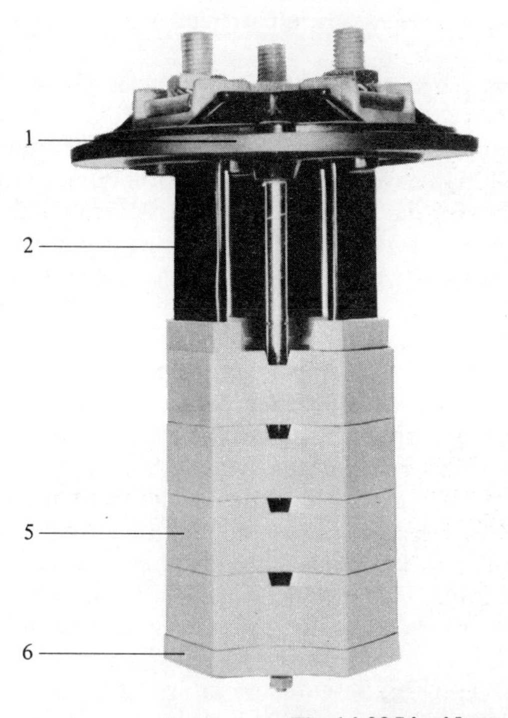

1 – Terminal plate.
2 – Insulation sheath.
3 – Electrode.
4 – Prism (if required by regulation).
5 – Polypropylene ring (0, 1, 2, 3 or 4 as required).
6 – Polypropylene cover.

Fig. 14.22 Liquid rotor starter

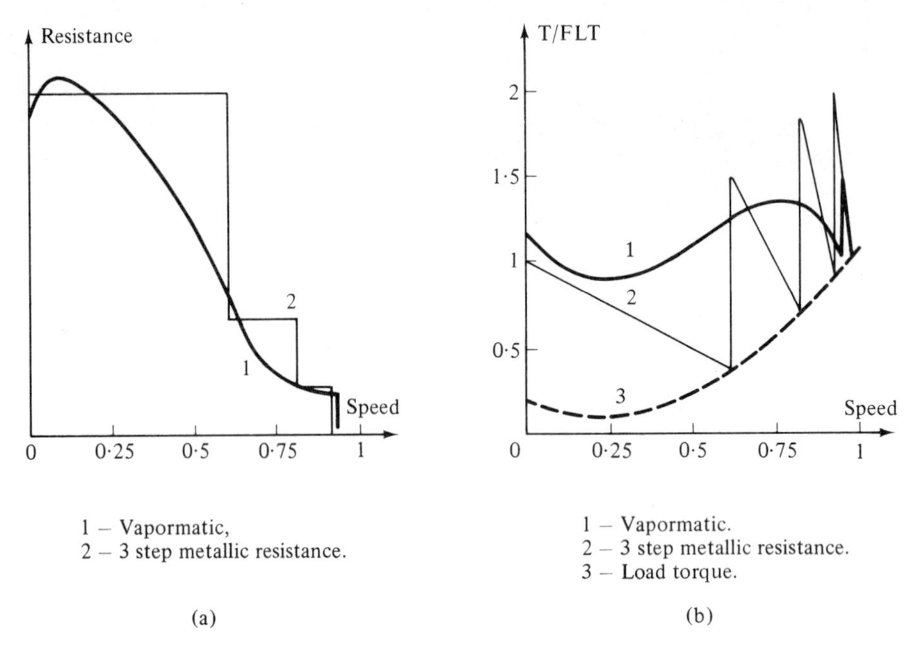

Fig. 14.23 Vapormatic starter

manyfold. Portable appliances invariably fitted with their own on/off switch have mostly either series (universal) or shaded pole motors, both of which are self-starting. Circulating pumps, refrigerator, freezer, and similar motors may use capacitors to give them sufficient starting torque, but either these are left in circuit or automatically cut out by a built-in centrifugal switch.

For stationary single-phase motors requiring no more than an on/off switch, miniature circuit breakers with thermal overload protection and fuse back-up are generally the most suitable. Small motors where current and operating conditions make external protection impractical may have a thermostat incorporated in the windings which will disconnect the motor from the supply when it is overloaded.

In small workshops or the like where several single-phase motors are used an artificial three-phase system can be created by using the Ferraris–Arno system.

This consists of using one of the three-phase motors as a pilot motor, starting and running it without significant load on the single-phase supply. The voltage induced in the third phase of the pilot motor combined with the single-phase supply produces an artificial three-phase system to which other three-phase motors can then be connected. The pilot motor must be capable of providing enough power in the third or phantom phase to start the largest of the other motors which, though standard three-phase types, will draw most of their power from the single-phase supply. Standard three-phase starters may be used, but since the stator currents are unbalanced special attention may have to be given to the overload devices. In some cases the overload protection in the middle phase may have to be omitted. A typical arrangement for a Ferraris–Arno artificial three-phase system is shown in Fig. 14.24.

An alternative scheme for producing an artificial three-phase system is shown in Fig. 14.25. This system is particularly suited to the smaller installation.

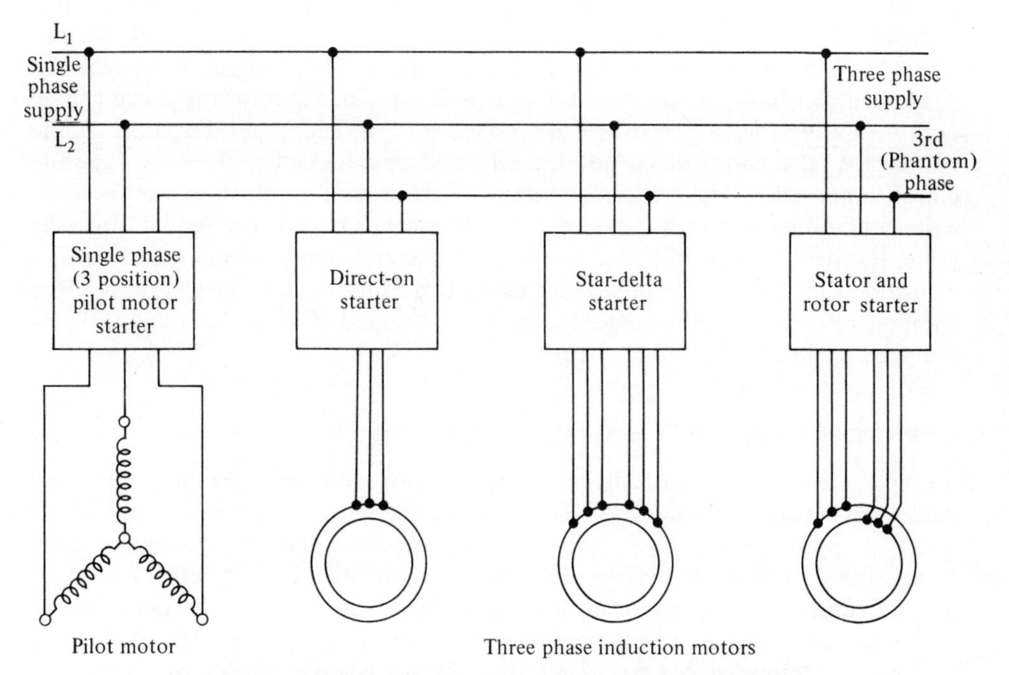

Fig. 14.24 Artificial three-phase supply by Ferraris-Arno System

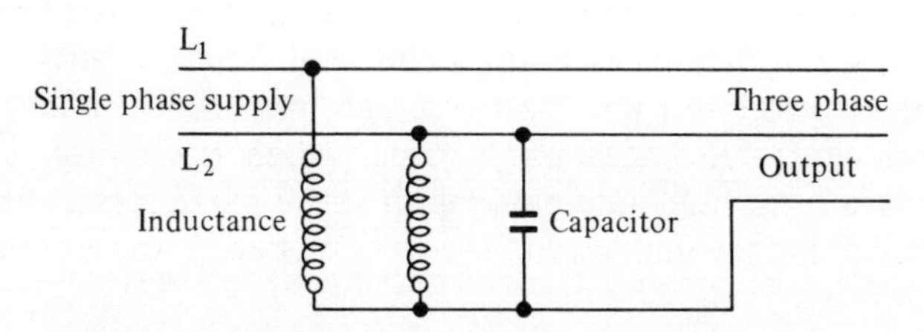

Fig. 14.25 Westinghouse artificial three-phase system

14.16 Speed Control—a.c. Induction Motors

The cage induction motor is inherently a constant-speed machine in which the speed varies very little from no load to full load when it runs at about 2–5 per cent below synchronous speed. The speed is a function of the supply frequency and the number of pairs of poles and motors which are able to operate at several speeds by pole changing are available. These are dealt with in sections 4.9 and 4.10 of chapter 4. The PAM wound induction motor perfected by Professor Rawcliffe represented a major breakthrough in multi-speed motors which today are produced with a wide choice of speed ratios.

The switching arrangements are made within the starter, and while the smaller motors can be switched directly from high speed to low speed it is advisable on the higher horsepowers to provide a time delay between speed changes to avoid the transient current peaks. These can occur if the field created by the back e.m.f. at the previous speed has not collapsed before the motor is again connected to the supply.

Some speed variation can also be obtained by voltage control and since the advent of the thyristor, which provides a simple method of voltage control, many schemes have been developed. Using a motor with a high-resistance rotor the speed can be reduced appreciably by decreasing the applied voltage. The torque developed by the motor also falls, and if this is not accompanied by a reduction of load torque the motor temperature may become excessive. This method must, therefore, be used with care and only after the motor manufacturer has been consulted. It is usually limited to drives where the load torque varies as the square of the speed, as in a centrifugal fan. The method is used extensively for the control of refrigeration compressor condenser fans where the speed of the fan is controlled automatically by the pressure in the refrigerant circuit.

14.17 Speed Control of Wound-rotor Induction Motors

Slipring motors are also basically constant-speed machines in which the slip, i.e., the difference between synchronous speed and running speed, is relatively small (in the order of approximately 5 per cent). This slip is proportional to the rotor copper losses, and since these losses can be varied by the addition of an external resistance via the sliprings the speed of such motors can be adjusted with a permanently connected controller. Since the rotor current is proportional to the torque developed by the motor it follows that the rotor losses will also vary with the load and therefore affect the speed. This method of speed control therefore exhibits a pronounced series characteristic and the speed variation due to a change in load is greater the higher the resistance in the rotor circuit. For this reason speed control by adding to the rotor resistance is only satisfactory as long as the load remains fairly constant at the reduced speed. Speeds lower than half speed are not usually practicable. The very substantial iron and I^2R losses which occur when the motor is made to run appreciably below its normal induction motor speed make it essential to check that the temperature rise of the motor and the cost of wasted power are within acceptable limits.

Three-phase rotor regulators are similar in construction to stator rotor starters but have rotor resistances which are rated to remain in circuit continuously. The regulators may be either operated manually or by a small motor in a closed loop control system. In either case the switching of resistances is complex and precludes fine adjustment of speed so that for larger motors it is better to use liquid resistances. A typical automatic speed control by contactors is shown in Fig. 14.26 which shows that it is possible to obtain up to seven speed settings with only three contactors.

Where a wider speed range, better efficiency, and accurate adjustment of speed over a wide range is called for it is necessary to consider other motors and forms of speed control. The Schrage motor which is still popular for certain types of variable speed drives is described in some detail in chapter 4, section 4.12.3. Electronic speed controls for a.c. motors are dealt with in chapter 12.

The series characteristic of the slipring motor with rotor resistance makes it particularly suitable for drives which are fitted with a flywheel and are subject to periodic peak loads of short duration. A typical example of such a drive is the power press. If a small section of the external resistance is permanently connected in the rotor circuit it will have little effect on the speed of the motor at light loads, but when the load peak occurs it will increase the slip and allow the flywheel to contribute some

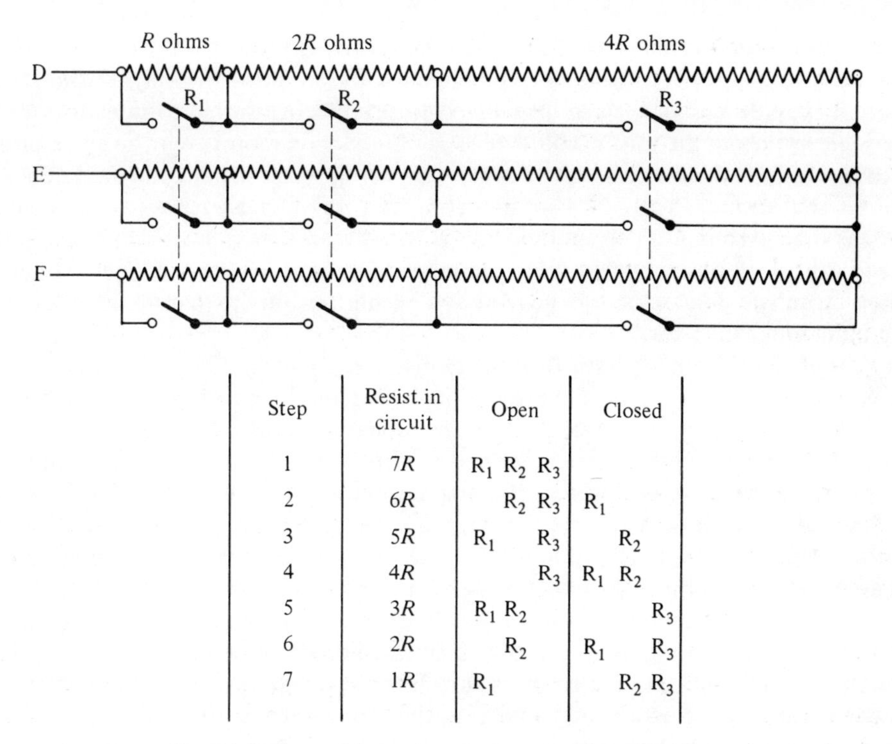

Step	Resist. in circuit	Open	Closed
1	$7R$	R_1 R_2 R_3	
2	$6R$	R_2 R_3	R_1
3	$5R$	R_1 R_3	R_2
4	$4R$	R_3	R_1 R_2
5	$3R$	R_1 R_2	R_3
6	$2R$	R_2	R_1 R_3
7	$1R$	R_1	R_2 R_3

Fig. 14.26 Stages of speed control using three contactors

of its stored kinetic energy. When the load peak has passed the high rotor characteristic causes the motor to develop appreciable torque and rapidly accelerate the flywheel to its original speed thus restoring its kinetic energy. By using induction motors with moderately high rotor resistance characteristics it is usually possible to satisfy the power requirements of drives with pulsating loads with a cage motor whose power output rating is somewhat below the maximum intermittently demanded peak power, and thus reduce the capital cost.

14.18 Speed Control of Single-phase Motors

Since the torque of most fans is approximately proportional to the square of their speed, simple resistance regulators are adequate for most applications employing shaded-pole motors whose normal power consumption seldom exceeds 100 W. For larger motors and universal motors thyristor or triac energy regulators can control the motors from a crawling speed up to the normal running speed for that load. Speed regulators used with portable drills can extend their usefulness, enabling them to perform essentially low-speed operations such as tapping and coil winding.

Repulsion motors described in chapter 7 can provide variable speed if they have movable brush-gear, but because of their cost are no longer in common use.

14.19 Speed Control of a.c. Motors under Overhauling Conditions

The tendency to use a.c. motors for hoists, winches, cranes, and conveyors made it necessary to provide suitable speed controls and safeguards for operating conditions where gravitation enables the load to provide a driving torque. If the motor is of the wound-rotor type and the stator is controlled by an auto-transformer the stator phases may be unbalanced and reversed to produce a reverse torque. In fact by suitable switching the torque characteristic of induction motors may be given either positive or negative slope at low speeds. Another dynamic braking method depends on the self-excitation capability of induction motors when capacitors are connected across their terminals. This method can be used with cage motors which when the need arises are disconnected from the mains and have resistors connected across their terminals. The capacitors provide self excitation and the motor will now act as an induction generator.

Though such methods have proved themselves, their relative inflexibility, complexity, and cost has caused them to be overtaken by the now more generally used and more efficient injection of d.c. into the stator windings. If the motor is of the wound-rotor type with rotor resistance control very fine adjustment of braking torque is possible. A typical d.c. braking system is shown in Fig. 14.27. The low voltage d.c. injection current is obtained from a bridge rectifier fed from an auto-transformer in series with a saturable reactor. The control winding of this reactor is fed from the rotor circuit via another rectifier. In this way the d.c. braking current is made to depend on the current in the rotor circuit and the braking torque made to depend on the load. The speed of the motor can be further controlled by varying the resistance in the rotor circuit. In hoist or lift applications the lowering speed control is, therefore, very similar to that obtained with the rheostatic braking of a d.c. series-wound motor.

Since only stator I^2R losses have to be supplied in d.c. braking of induction motors the power required for a given torque is appreciably less than if the motor's normal

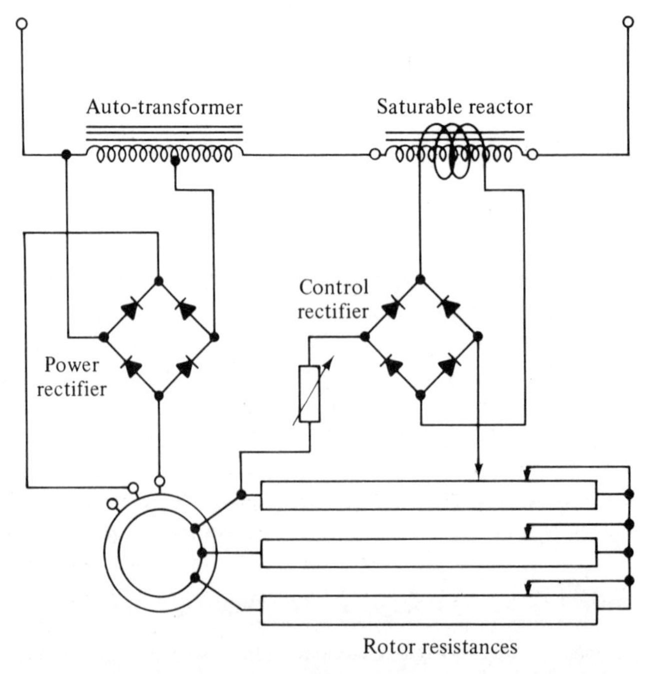

Fig. 14.27 Elementary diagram—dynabraking control

a.c. supply is used. Since in the a.c. case considerably more power has to be switched and windings may have to be re-arranged (e.g., delta to star) or transformers interposed to keep currents and torques within acceptable limits, d.c. braking may also be found to have economic advantages. The speed/torque curve with d.c. excitation has a much sharper peak, which also occurs at a very much lower speed than on a.c. In the case of slipring machines it is, therefore, usual to insert a fairly large rotor resistance in order to obtain satisfactory braking torque over a wide range of speeds. The peak braking torque obtainable is also limited by magnetic saturation, and in certain cases it may be only of the order of full-load torque. Direct-current excitation does not produce torque at zero speed.

14.20 Emergency Stopping of a.c. Motors

A very quick stop can be obtained by injecting d.c. into the stator as shown in Fig. 14.28. Subject to limitations mentioned above, the time the rotating system takes to come nearly to rest will be in inverse proportion to the value of the d.c. current in the windings.

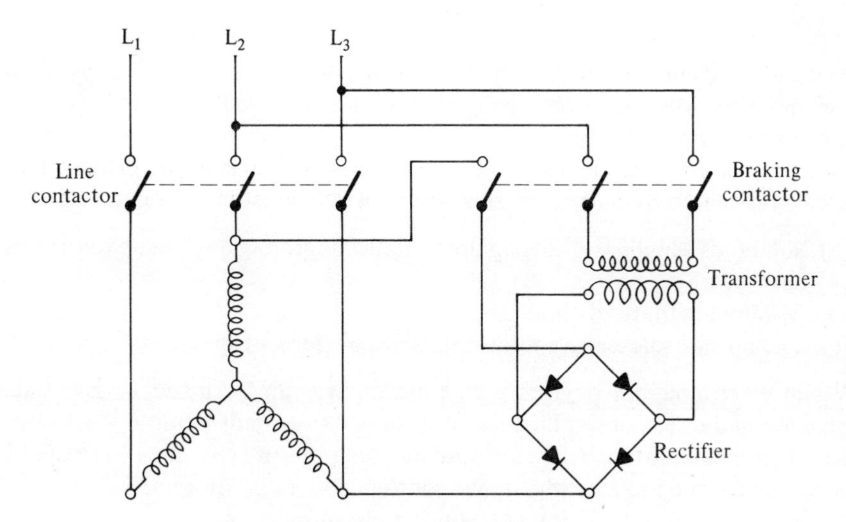

Fig. 14.28 Quick stopping by d.c. injection

A very rapid deceleration can be obtained by reversing the motor, usually referred to as plugging. This method has several disadvantages which must be considered:

(a) Very high currents are drawn from the supply when the motor is re-connected after reversal of connection.
(b) Zero speed detectors consisting of centrifugal switches, frequency sensitive relays, timed relays or the like must be fitted to ensure that the motor does not accelerate in the reverse direction, and is disconnected the instant it stops.
(c) The mechanical stresses on the shaft, bearings, and transmission elements, as well as windings are considerable and may significantly affect the reliability and life of the motor. There will also be considerable heating effect and thermally this may be equivalent to three starts.

14.21 Protective Devices

14.21.1 *Undervoltage protection*

To prevent power being connected to the motor without going through its proper starting sequence after a power failure most starters are fitted with a 'no volt' solenoid. The plunger of the solenoid is connected to a latch which either retains the starting handle in the run position or keeps the main contactor closed. This solenoid is energized as soon as the motor starting sequence is commenced and latches the main contactor in the closed position as long as the supply to it is maintained. Most modern contactors are designed to hold in with a temporary fall or even a complete supply failure lasting only a few cycles. For this purpose a capacitor/resistance slug is frequently used to delay the drop-out of the contactor. In some industrial plants there may be system disturbances lasting several seconds which would cause the normal no-volt release to operate. In such cases motor control gear having an automatic reclosing feature, which operates if the voltage recovers within a preset time interval, may be installed.

14.21.2 *Overload protection*

Motor overloads normally manifest themselves by the current drawn by the motor rising above its rated value, though a number of exceptional circumstances can arise where normal specified operating conditions (as for example ventilation, supply conditions) are no longer met and the motor becomes 'thermally' overloaded without this being detected by one of the following current-sensitive devices:

(a) Solenoids usually working against a time integrating device such as an oil-filled dash-pot.
(b) Bi-metallic thermal elements.
(c) The now obsolescent 'solder pot' thermal elements.

While electromagnet devices with time delays can be made to give excellent protection in the case of significant and sustained overloads to motors operated well below their maximum capacity of output and temperature rise, they were found to be inadequate for the modern maximum continuously rated motors.

In the case of thermal elements, which were fitted to the majority of small- and medium-power motor starters, a coil carrying the motor current surrounds a bi-metal strip, the thermal characteristic of which should, as near as possible, match that of the motor it is to protect. The operating current of such a device can usually be adjusted by altering the distance its operating point has to travel before tripping the starter.

A well-designed thermal relay will operate fast enough to prevent the motor burning out should it be stalled or very severely overloaded. Three-phase models should have an element in each phase and have these arranged for differential tripping to give single-phasing protection. Those which also respond well to moderate unbalance tend to be more costly, but even a partially burnt-out motor will cost considerably more than the saving which has been achieved by using low-cost thermal relays.

To overcome the disadvantage of thermal relays motors are being fitted with temperature detectors on or actually in their windings, and often also on bearings.

The two main types of devices which can be used and shown to have advantages for particular applications are the thermostat and the thermistor. The thermostat consists of a bi-metallic element, which when heated above a certain temperature will deflect and break a contact. If this element is placed in series with the no-volt coil of the starter it will stop the motor when the critical temperature is reached. In a three-phase motor reasonably complete protection requires a thermostat in each phase. As these thermostats cannot have any adjustable time delay they are not particularly useful in modern motors whose winding insulation is well able to withstand short occasional periods at temperatures above their rating. If the normal sluggishness of the thermostat is increased it will very likely fail to protect the winding against a heavy overload, particularly a stall condition, and when the motor is cold. It is, therefore, essential that the starter should still be fitted with a current-dependent cut-out to afford reasonable overall motor protection. Since thermostats are made as small as possible to give them a reasonably short reset time their contacts can at best handle the currents of very small motors (of the order of 100 W output maximum). Thermostats are generally of the normally closed type and placed in series with the off pushbutton and the no-volt coil. Any break in the leads from the thermostat will therefore cause a 'fail safe' condition.

As has been indicated heavy overloads near stall and actual blocked rotor conditions are very satisfactorily dealt with by the overcurrent relay which, if there is additional protection, can be set above the normal over-current limit. Before considering the thermistor as an alternative to the thermostat (which, because of its re-setting time, which may be several minutes, cannot be used) some of the conditions which are either impossible or difficult to cover by the methods of protection discussed will be outlined:

(a) As mentioned above, ventilation or heat radiation may deteriorate and become totally inadequate.
(b) The frequency of starting or duty rating of the motor may be exceeded.
(c) The ambient temperature may rise above specification.
(d) A star-delta starter may remain in the star position.
(e) A slipring motor has too much resistance left in its rotor circuit.

The thermistor is a small semiconductor resistance with a very pronounced temperature coefficient. The resistance is minimum at the low (ambient) temperature (cold conducting) and increases by several powers of 10 when hot. Its behaviour can be likened to a contactless normally closed contact device, and therefore again be safeguarded against open circuits in the leads to the control equipment. Such transistors are also known as 'positive coefficient' thermistors. The changeover temperature is generally referred to as the Curie temperature and the material used is mostly a doped polycrystalline titanium ceramic.

As thermistors only handle comparatively minute currents they require back-up electronics. The proven reliability and increased application of solid-state electronics in industry appears to herald an era of 'electronic' motor protection and thermistor relays are now readily available. To afford effective protection the thermistor must be located in the windings and, as previously mentioned, one in each phase so that introduction of this form of protection is generally limited to new motors and rewinds.

Though the thermistor has a considerably shorter time constant than the thermostat it is unlikely to afford adequate protection if the motor fails to start. For direct-on-line starting small circuit breakers now afford adequate protection against this eventuality but make it generally impractical to also fit a thermistor relay. Those concerned with specifying motor and control gear must, therefore, weigh the risks and consequences of a possible breakdown against the cost of a one time insurance premium in the form of more adequate and therefore costly protection.

It should also be noted that thermostats and thermistors are self-resetting devices, and if the starter is on two-wire control the motor will restart automatically. On some types of process plant this may be desirable, but in such cases the control engineer should be advised by the monitoring or data-logging equipment usually associated with such plants. Control gear which is under three-wire control, that is, which requires pressing of a start button after the tripping of a protective device, should be used wherever possible, since it is also highly desirable that the cause be established and possible damage prevented.

The importance of protection against the failure of a phase in a three-phase system has already been stressed. A substantial number of motor breakdowns (reference 2 quotes 15 per cent) are due to this and it is generally accepted that only a specially designed low-current relay can provide adequate 'single-phasing' protection. Such relays make use of the fact that even on no load, induction motors draw some 40 per cent full load current, so that if the current in any phase drops to some 20 per cent of

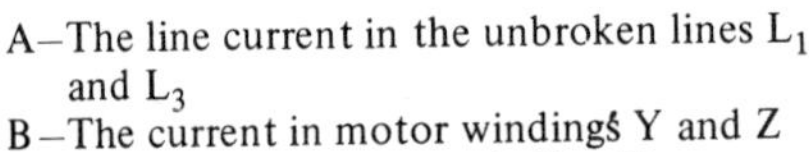

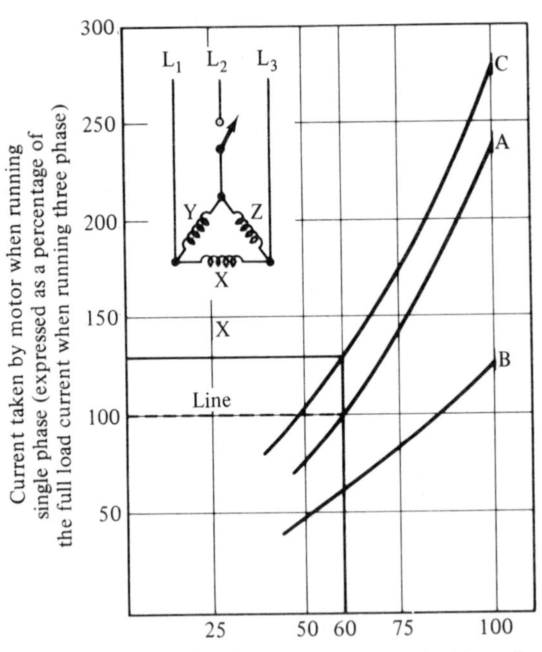

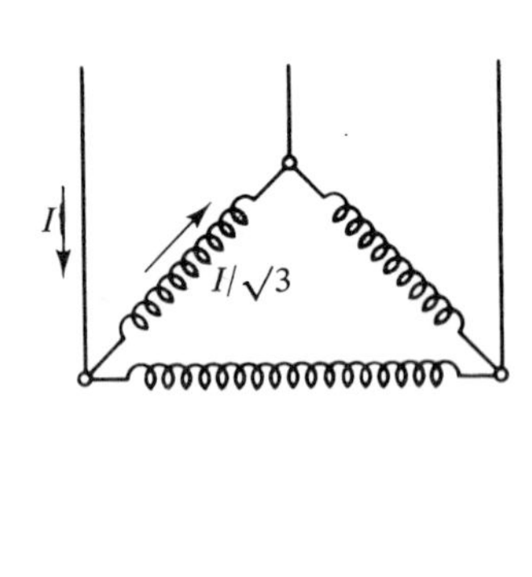

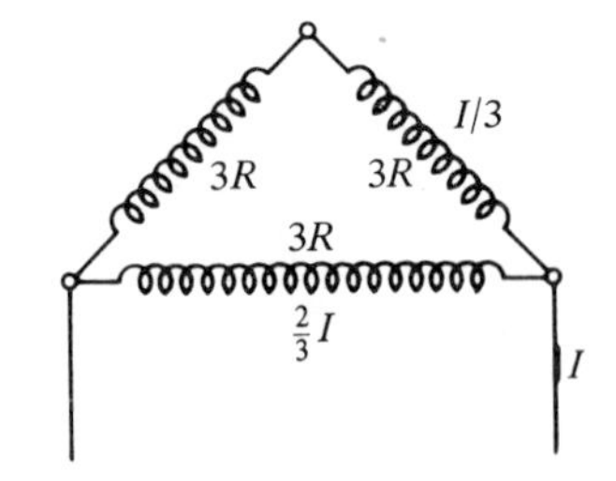

Fig. 14.29 Effect of phase failure on motor current

normal rated current it indicates a phase failure and the main contactor must be opened.

It is well known that a three-phase motor will continue to run if one phase fails. If the motor is operating well below its rated load the currents in the remaining phases will not rise sufficiently to operate the overload protection. As the load is increased it is possible to have conditions where the line current-sensitive protection does not operate and if there should be a load cycle with permitted overload conditions, thermal protection may be too late to save the winding.

Figure 14.29 illustrates the redistribution of currents which occurs when one of the supply leads to a delta-connected three-phase motor becomes open circuited. If the failure occurs when the motor is running at 60 per cent load then the current in the

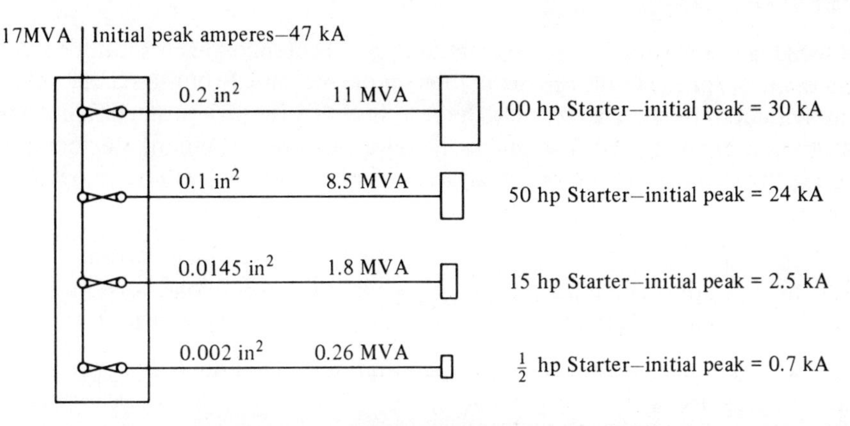

(a) Separate starters–showing prospective fault level at starter terminals

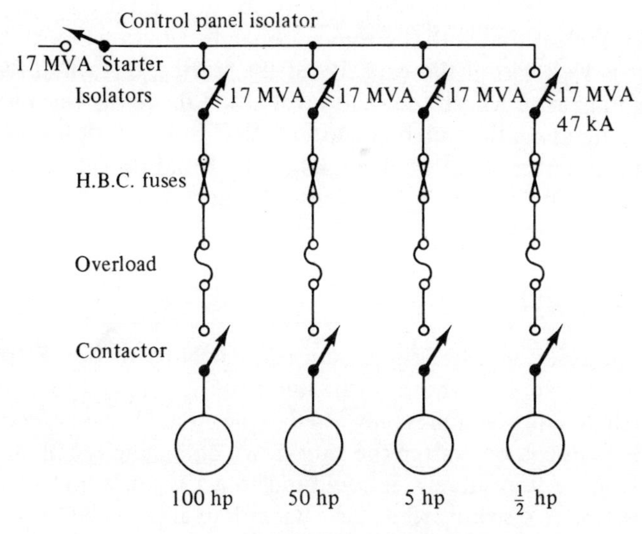

(b) Cubicle type control panel–showing prospective fault level at starter terminals

Fig. 14.30 Comparison of fault levels with separate starters and with bus-bar chamber

heavily loaded or 'single' (rated voltage energized) winding rises to 130 per cent of rated full load current, whereas the other two phases only carry some 60 per cent. Since the two remaining live lines only carry 100 per cent rated current the protective devices will have no cause to operate and since BS 587 allows in addition a 10 per cent tolerance on overload devices the danger of one phase being burnt out becomes even greater.

A form of single-phasing protection which may be found on some starters with bi-metal thermal overloads is a mechanism which can trip the starter by differential action as well as by the uniform and adjustable deflection of all three strips when an overload occurs.

14.21.3 *Fault protection*

Standard motor starters are not expected to give protection against fault conditions, such as phase/phase faults or earth faults on the motor side of the starter, and so far as individual starters are concerned the responsibility for such protection lies with the distribution engineer. BS 587 makes it quite clear that standard starters are only intended to protect the motor and system from excess current due to overloading, stipulating maximum ratings as follows:

d.c. motors, all sizes	four times full load current
a.c. motors up to 100 h.p.	eight times full load current
a.c. motors above 100 h.p.	six times full load current

If the stalled motor current is higher than that indicated above then a larger starter may be necessary.

Since it is now common practice to feed a bank of meter starters from low-impedance bus-bars the question of protection against fault currents becomes the responsibility of the starter designer. Figure 14.30 compares the fault levels at the starter terminals when fed from a distribution box and when fed from a bus-bar chamber. From this it is clear that with the conventional distribution board the fault level at each starter is well within the capacity of the starter, but when fed from bus-bars a situation may arise where the fault level at the incoming terminals of a $\frac{1}{2}$ h.p. starter is the same as on the 100 h.p. starter. HRC fuses must, therefore, be incorporated within the starter, ideally before the starter isolating switch, but if this cannot be done the individual starter isolaters must be capable of handling the fault level at the incoming terminals.

14.21.4 *Earth fault protection*

This is also, in the majority of installations, the responsibility of the distribution engineer, but there are exceptions where earth fault protection must be provided in the starter. If the earth fault protection is only fitted to the circuit breaker feeding the starters then a minor fault could result in the shut-down of a complete plant, which could well be disastrous. Alternatively, it could fail to act when a high resistance earth fault occurs and lead to sparking which, in a hazardous area, could have equally disastrous results.

The simplest form of earth leakage device—used extensively on distribution circuits—is the voltage operated earth leakage unit, designed to trip the circuit breaker when the leakage voltage exceeds a predetermined value. The type normally used on motor control gear comprises a relay operated by the out-of-balance currents created by an earth fault.

14.21.5 *Control circuit protection*

Under 5 h.p. the main fuses can be relied upon to protect the starter coils and wiring, but above 10 h.p. it is advisable to provide control circuit fuses in the starter. This is particularly true where remote control devices such as pushbuttons are used, since an undetected high resistance earth fault could impose a dangerously high potential on the remote device and endanger the operator.

On three-phase three-wire systems some national standards and many large users call for a step-down transformer for control circuit supplies. When these are provided the secondary winding should be earthed at one end and not at the mid-point.

References

1. R. W. Jones, *Electric Control Systems*, Chapman & Hall Ltd, London (1953).
2. K. Goodchild and P. Rayner, 'Nearer The Brink (Trends in overload protection for a.c. motors)', *Electronics & Power* (February 1976).

Note: References, certain additions and Section 14.21.2 have been added by the Editor and contain statements and opinions of those authors.

15 Installation and Mounting

15.1 Preliminaries

Many electrical machines are dispatched as completely assembled units, sometimes with minor adjustments to minimize risk of damage during transportation and storage, such as partial removal of brushes, axial locking of rotor and shaft, protective coverings of bare shaft extensions, etc. Others are dismantled before dispatch and have to be reassembled before or during erection on site.

It is important to check the condition of any machine and its protective packing immediately it is received to ensure that neither has suffered any damage or to report any deterioration to the supplier. If the machine cannot be erected immediately the manufacturer's recommendations for storage, etc, must be complied with; for example, machines having high-voltage windings may require to have their anti-condensation heaters energized.

Before erection is started the following checks should be made:

(a) The machine rating details, etc, are correct.
(b) The manufacturer's erection/maintenance instructions and drawings are available,
(c) The indicated direction of rotation is correct.
(d) The machine is complete, including items such as shims, holding-down bolts, terminations, shaft coupling, etc.
(e) The foundations are adequate as regards seatings, dimensions, cabling access, drainage, etc.
(f) The necessary handling equipment is available, including cranes, jacks, slings, lifting beams, dummy shaft ends, etc, as required.
(g) The plant to which the electrical machine is to be coupled is correctly erected and fitted with the appropriate coupling.

15.2 Forces in Electrically Driven Shaft Systems

There are many possible variations of mechanical arrangements of electrical machines and the equipment to which they are coupled and the design used is selected for average conditions unless all specific conditions existing on site are known. It is important, therefore, to ensure that when erected on site the electrical machine will operate under the mechanical conditions for which it was designed. The most significant factors can be determined by an examination of a very generalized form of machine shown in Fig. 15.1, the main components of which are:

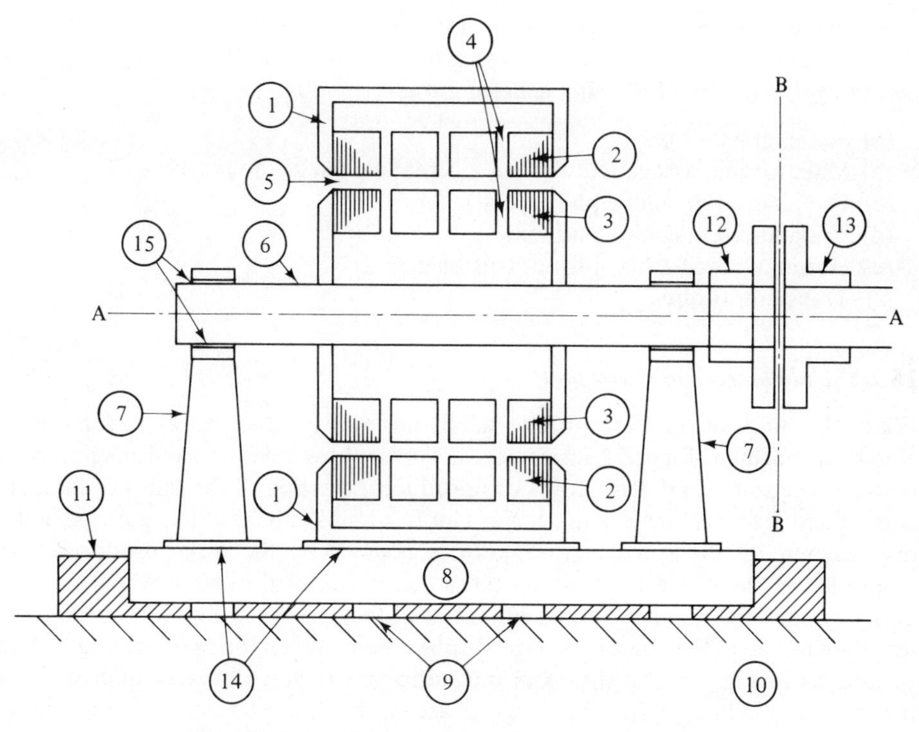

Fig. 15.1 General arrangement of an electrical machine

(1) Stator frame which supports the magnetic circuit;
(2) Stator core which carries the stator winding;
(3) Rotor core which carries the rotor winding;
(4) Ventilating ducts in cores;
(5) Radial airgap;
(6) Shaft;
(7) Pedestals for bearings;
(8) Baseplate;
(9) Wedges (for levelling baseplate);
(10) Foundation;
(11) Grouting;
(12) Motor half-coupling;
(13) Driven half-coupling;
(14) Shims under stator frame and pedestals;
(15) Journal bearings.

The most significant difference between an electrical machine and a purely mechanical rotating device is the magnetic flux which crosses the airgap and links the stator and rotor windings. This flux, in conjunction with currents flowing in the stator or rotor windings produces torque and is essential to the operation of the machine. It also has secondary effects, however, which can affect the operation of the machine by producing forces other than a uniform tangential one. These may be unidirectional, oscillatory, or transient, and may be periodic with various frequencies, and may have components in the axial, radial, or tangential directions.

Although they are not of significance in all electrical machines, it is important that these forces be considered when designing the associated shaft system, and they can be segregated into the following general groups:

(a) Static, gravity loads.
(b) Main torque, associated with the normal output of the machine.
(c) Unbalanced magnetic pull (radial).
(d) Unbalanced magnetic pull (axial).
(e) Harmonic oscillatory torques (sustained).
(f) Transient torques.

15.2.1 *Unbalanced magnetic pull*

When the 'lines of flux' are purely radial they will produce no axial force in an electrical machine. Figure 15.2 shows, however, how relative axial displacement between the stator and rotor cores can produce distortion of the 'lines of flux' and consequent reaction between the cores. This flux tends to behave like a spring in that displacement of the rotor in the stator is resisted by the force produced. The amplitude of this force depends on the amount of flux distortion which, in turn, depends upon the number of discontinuities of the cores and the ratio of the radial length of airgap to the amount of axial displacement. When the two cores have ducts in line, as in Fig. 15.2(a) the axial magnetic pull is much greater than with the constructions in Fig. 15.2(b) and (c) which are typical of most synchronous machines.

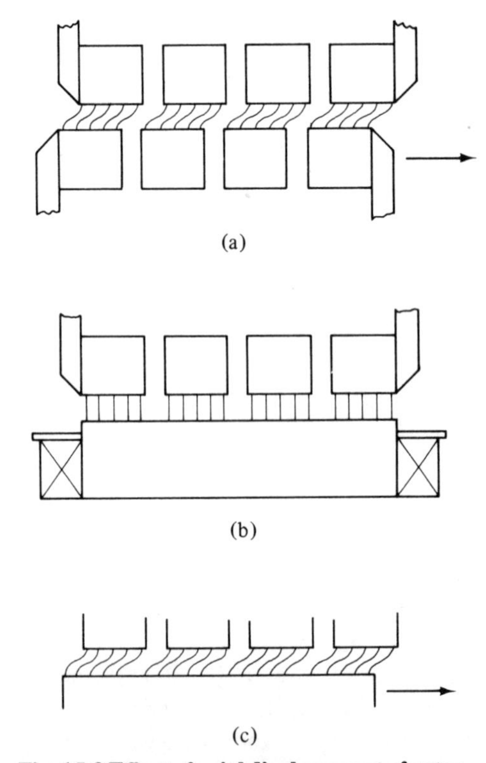

Fig. 15.2 Effect of axial displacement of rotor

The radial component of flux produces a radial magnetic force but if the flux is symmetrical round the periphery of the rotor then there is no resultant unidirectional force between the stator and rotor. If, however, the airgap is not uniform the flux will not remain uniform and there will be a resultant unidirectional pull between the stator and the rotor. This phenomenon is described as 'unbalanced magnetic pull'.

When the amount of airgap asymmetry is known then the amount of pull can be determined for any value of flux density. Unfortunately this force usually produces a further distortion of the airgap. As the pull is greatest where the airgap is smallest this subsequent distortion then increases the amount of the pull and the process is cumulative. The final distortion can be calculated in the form of a convergent series but, of course, the stator may touch the rotor before this final condition is reached

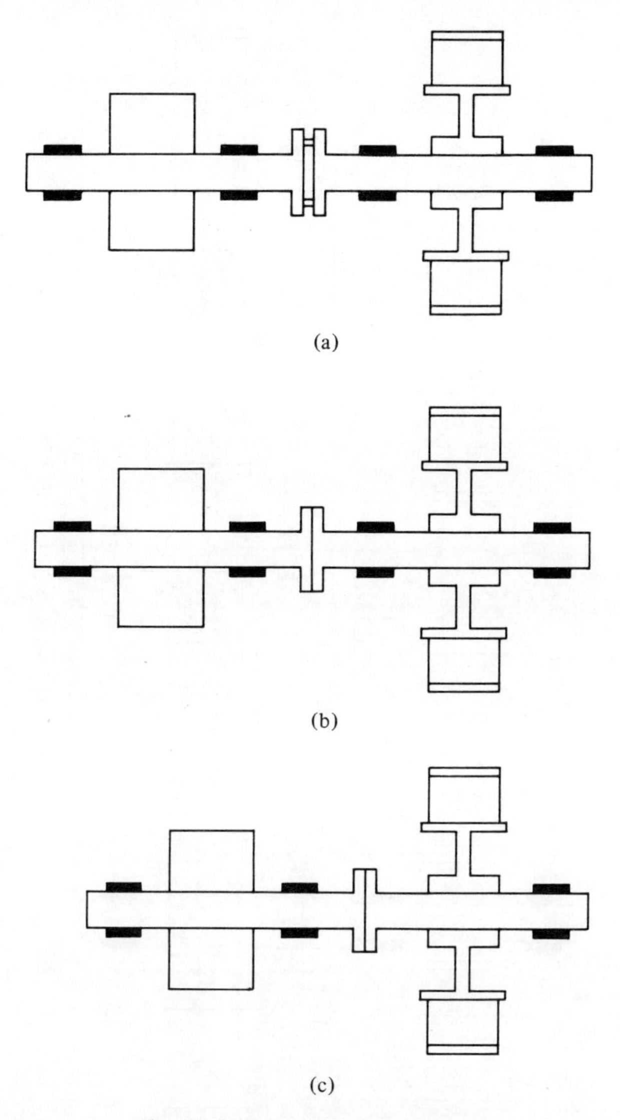

Fig. 15.3 Types of shaft systems

and the machine is then said to have 'pulled-over'. This phenomenon is normally only significant in induction motors which have very small radial air-gaps, usually less than one-tenth of an inch. It is consequently much more important to maintain uniform airgaps in induction motors than on other types of electrical machines which have longer radial airgaps.

It is essential therefore to erect any machine in such a manner as to keep the rotor concentric in the stator. In theory this is simple but in practice many factors have to be considered. Where the machine is mounted on an integral baseplate which is rigid the rotor can easily be manipulated by adjusting the position of the pedestal bearings and adjusting the shims under them as necessary. However, none of the components are completely rigid and items do distort and move when holding-down bolts are tightened. This must be allowed for in lining-out, and also the fact that the sequence

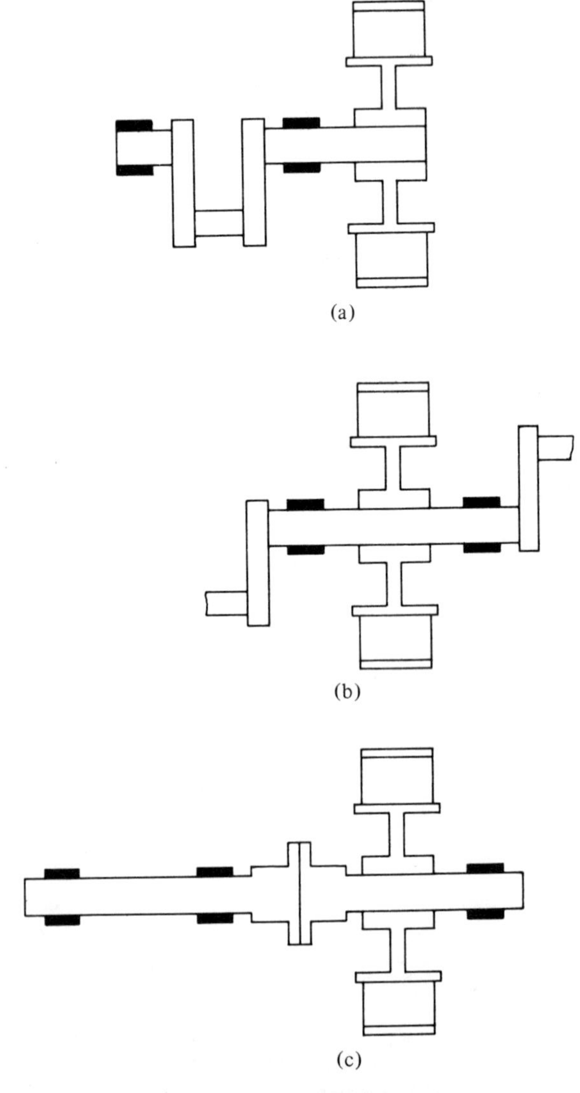

Fig. 15.4 Types of shaft systems

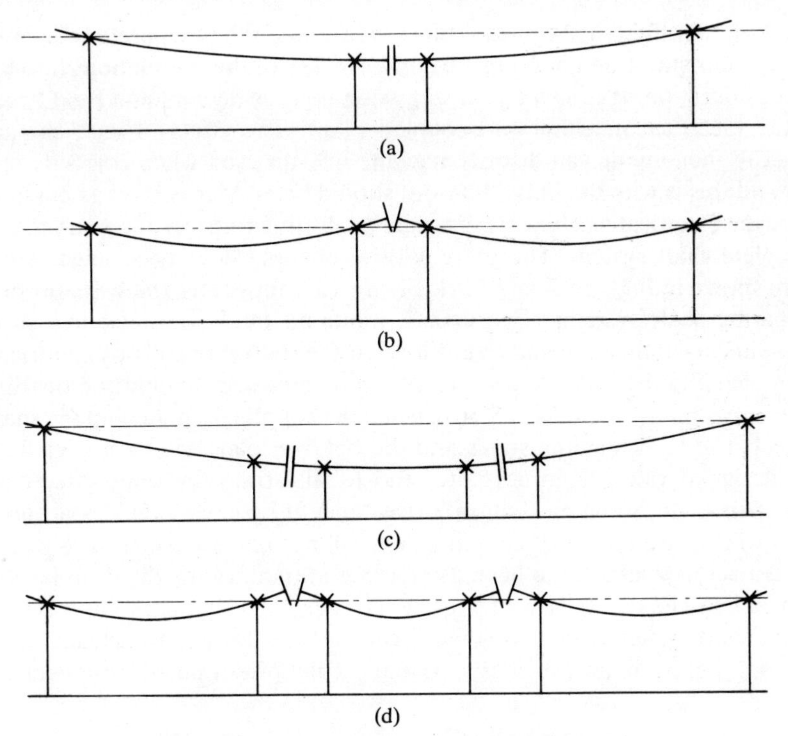

Fig. 15.5 Arrangements of multi-bearing sets

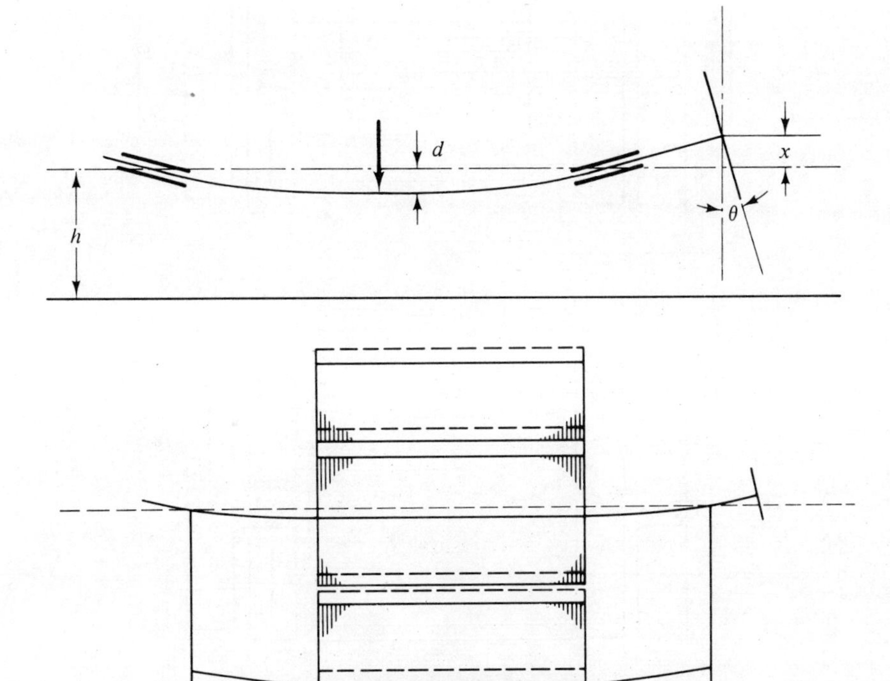

Fig. 15.6 Effects of shaft and foundation deflections

in which bolts are tightened can also affect the results. When separate soleplates are provided under stator and bearings then the rigidity of the foundations becomes of prime consideration. It is also significant when there is no common rigid baseplate under the electrical machine and coupled plant. The effect of age, climate, or geophysical phenomena can disturb machine line-up even when correctly erected initially and this is why the initial line-out should be as near perfect as possible.

Before the electrical machine itself can be finally lined-up it is necessary to line-up the complete shaft system. There are several possible bearing arrangements and some are shown in Figs. 15.3 and 15.4. Figure 15.5 illustrates the basic problem of multi-bearing shaft systems. Every shaft mounted freely between two supports exhibits a normal deflection and to ensure that the shafts are truly concentric at the common coupling the outer bearings have to be elevated. Should the bearings be lined-up horizontally, as in Fig. 15.5(b), when the coupling is bolted up the shaft will be subjected to high bending stress and the bearing loads will be higher than the normal designed value. Figure 15.5(c) and (d) illustrate the same effects with a three-section shaft. Some mechanical drives benefit by having their bearings horizontal in which case their half coupling face will not be vertical. In such a case the electrical machine will have to have its bearings at significantly different levels. It is sometimes assumed that the use of a resilient or flexible coupling avoids this need for lining-out shaft systems, but this is incorrect as these devices are usually provided because of other problems, such as torsional oscillations or possible movement of a

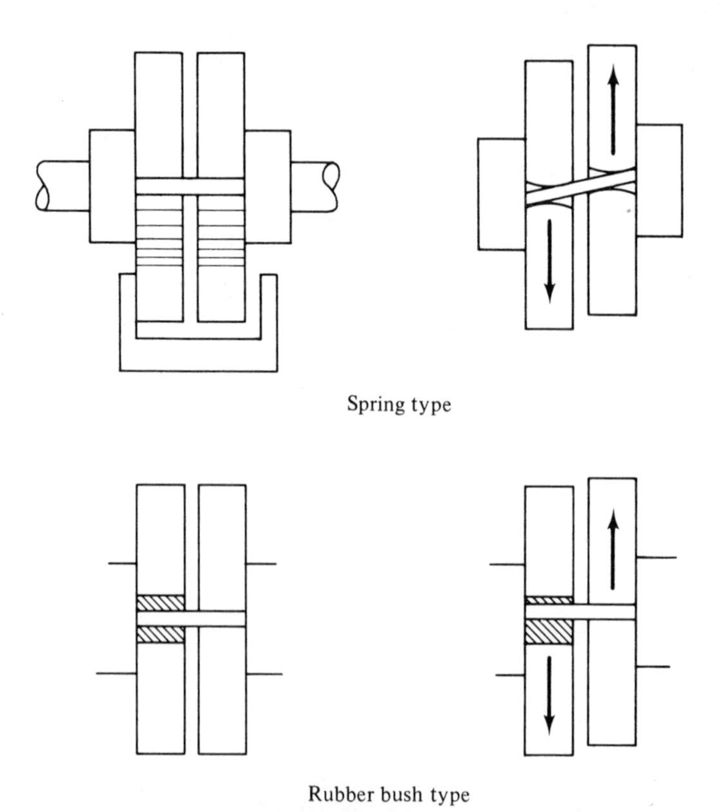

Fig. 15.7 Some flexible couplings—simplified

random nature of predictable magnitude. If incorrectly aligned such couplings can rapidly fail.

In addition to the problem of shaft-line-out, the axial location of a shaft system is important. Figure 15.8 indicates a typical journal bearing and shows that there is a

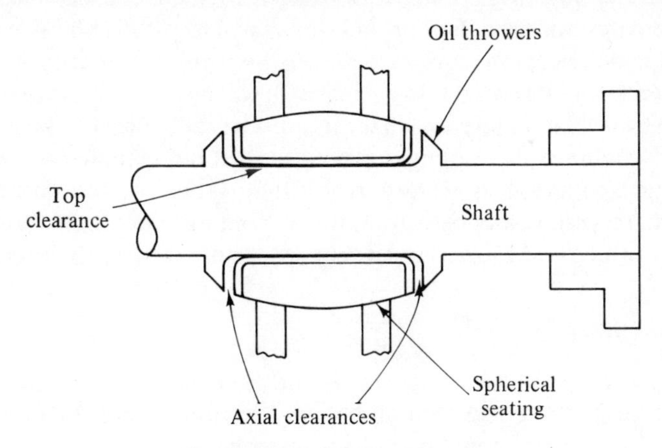

Fig. 15.8 Bearing arrangement

finite limit to the axial movement which the shaft can be permitted before interference occurs. Anti-friction bearings have also limits, some being virtually zero, that is acting as location bearings. Shafts can move axially due to impressed mechanical forces, e.g., some compressors and pumps cause reaction displacement; due to temperature effects, e.g., coal pulverizers in boiler rooms; or due to change in axial dimensions of devices such as resilient or flexible couplings which can occur due to changes in torque loading or other effects. Any shaft system in which two locating bearings are provided is susceptible to trouble through overloading one of them by thrust loading if axial expansion or contraction occurs. Manufacturers give recommendations regarding axial bearing clearances to be used on their machines and advise where restrictions have to be applied. On complicated shaft systems an axial line-out sketch is usually provided and should be strictly adhered to during erection.

When all the necessary information listed above has been checked and is available the following sequence should be followed as relevant to the particular installation.

15.3 Foundations

These should be of high grade concrete, the upper surface being left 25 to 50 mm lower than required when finished to allow for the baseplate or soleplates to be supported on steel packings before being finally grouted in. The quality of the concrete should be checked and also that it has dried out properly before any load is applied to it. Should a pit be provided in the concrete then it must be provided with an adequate drain hole. Any holes or ducts for drainage, cables, etc, should not be located directly below a load-bearing point. Where holding-down bolts are to be set in the foundation it should be ensured that the holes for them have adequate clearance for adjustment and that the bolts can be located in their correct dimensional position.

The electrical machine baseplate or soleplates should be set in their approximate positions with holding-down bolts, etc, loosely located. Steel packing or double taper wedges should be fitted below the plates at the points indicated by the manufacturer, to ensure correct loading on the plates. The level of the top pads of the plates should be checked in relation to the plant to be driven by the electrical machine. Where separate soleplates are provided, or on large machines, it may be necessary to use water-level or other accurate devices to ensure the correct level of each point. Where the manufacturer has indicated that the machine will not be operated with a perfectly level base, the specified differences in levels must be set at the specified points. After a provisional levelling, the holding-down bolts must be tightened correctly and the levelling rechecked and, if necessary, readjusted. When found to be correct when bolted down, the recommended grouting of holding-down bolts and baseplates should be performed and allowed to set properly before further erection is started.

15.4 Machine Erection

When the base or soleplates are correctly set the machine components should be provisionally located to check that all holding-down bolts are suitably located and that the electrical machine coupling will line up with that on the coupled plant. At this stage all auxiliary components and devices should be checked to ensure that they can be adequately located and fitted. Cable access, conduit, piping, etc, to the machine should also be checked at this stage.

After ensuring that the machine to which the electrical machine is being coupled is correctly erected and the shaft is in the correct position both vertically and horizontally, offer up the coupling faces to each other and start to line up the electrical machine. The coupling faces should be parallel and truly co-axial but there is always a possibility that the coupling periphery or face may not be true with respect to the machine shafts and the method of checking coupling line-up should eliminate such errors.

To check parallel faces the axial clearance measured top, bottom, and both sides must be equal. To prevent errors due to untrue coupling faces these values should be measured with the two coupling faces in different relative positions, i.e., one rotor should be rotated relative to the other. Preferably values should be measured at the four positions with the rotor of one machine rotated 90 degrees, then 180 degrees, then 270 degrees, and then 360 degrees which should give the same values as the initial position. If checks have shown that both couplings are not true then it is necessary to check the values at four positions for each, four positions of one rotor being rotated at each of four positions of the second rotor, that is 4 values for 16 relative shaft positions. From these can be calculated the errors in both coupling faces and hence the actual gaps which will correspond to the two shafts being axially parallel for any particular relative position of the two rotors.

Having established parallel shafts it is then necessary to establish true co-axial relationships. This is normally done by lining up the peripheries of the two half couplings. If these are true with relation to the respective shafts then a simple straight edge applied at the four locations will indicate the required condition. In practice this procedure is adopted for the first rough line-up so that pedestals or bearings can be adjusted in their height quickly. Trial checks are done by clamping a micrometer device on one half coupling periphery and indicating the relative radial position of the adjacent half coupling periphery. If both peripheries are not true then values should be taken with the two rotors in the 16 relative rotor positions used for parallelism checks. From these readings the errors can be calculated and hence the actual values which will correspond to the two shafts being co-axial for any particular relative position of the two rotors.

Having established co-axial shafts it is necessary to ensure correct axial relative positions of the shaft systems and bearings. This can be done simply if a shaft axial-location drawing is available for the set. If not, the machine and coupling drawings will indicate the axial clearances which should be provided when the machines are solidly coupled up ready for running. It is also necessary to ensure that the correct axial clearances are obtained simultaneously in the complete shaft system.

Finally adjust the rotors and bearings to obtain correct axial clearances, and shafts co-axial simultaneously after the bearings have been finally bolted down to the plants or foundation, and final checks should always be made in this condition.

When the rotors and shaft system are located it may be necessary to adjust the stator of the electrical machine axially and radially to obtain concentricity between the rotor and stator. Arrangements provided for this operation will be described on the machine drawings or instruction books. Again the air gaps must be checked after the stator holding-down bolts have been finally tightened.

Some electrical machines are supplied as integral, pre-assembled, units and the procedure for lining up consists of adjusting the whole machine until the coupling line up, described above, has been completed. When the machine is finally bolted down and coupled up it is desirable to check the air gaps for equality, ensuring correct stator/rotor concentricity, to confirm that the machine has been correctly assembled.

Machines with special mechanical constructions, such as rotatable stators, axial-racking stators, etc, should have these features checked over the required range of operation and should then have the above-mentioned checks for correct line-up repeated to ensure that the machine components return to the correct position for any required running conditions. Such a recheck should also be made on machines driving through belt and chain drives for the complete range of tensions likely to be used in service.

When line-up has been completed and finally rechecked, any dowelling should be completed as recommended by the manufacturer.

15.5 Auxiliaries and Interfaces

When lined up, the machine should have all necessary services connected and checked. Bearing lubrication should be checked and the condition, type, and quantity of lubricant confirmed. When a separate oil lubrication system is provided the flow to each bearing should be checked, and, if necessary, adjusted by reducing valves, or similar means. Filters, drains, heat exchangers or coolers incorporated in the circuit should be checked and finally all joints, flanges, etc, checked to ensure that no leaks exist. Operation of motor driven or other pumps, pressure, flow, and level switches, and all automatic control systems should be proved under service conditions.

Cooling water for heat exchangers should be checked for suitable standard of cleanliness, flow, and pressure, and all pipe work and joints checked for leaks. Any air locks in the circuit should be bled off and correct flow established. Anti-condensation heaters or other devices requiring power supply should be insulation tested and then connected to the correct supply and tested for correct operation.

Instrumentation circuits such as thermocouples, ETDs flow, and pressure switches should be connected up and checked and, if necessary, re-calibrated.

Excitation circuits should be completed and electrically tested.

The main power cables should be made off and terminated correctly.

Care should be taken before terminating all electrical cables to ensure that the cable is of a suitable type for the duty and rating and is adequately insulated, for both normal and abnormal service conditions. It should also be compatible with the form of machine termination provided as regards socket sizes, glands, sealing box, armour glands, etc. Terminations designed for copper cable are usually unsuitable and sometimes even impossible for aluminium, and where aluminium cables are to be used special precautions have to be taken to ensure good termination.

In general compression sockets produce the best cable termination provided the correct socket is used and the correct tool (preferably power operated) is used. When terminated, adequate clearances should exist between all live conductors and earth and all insulation barriers should be fitted as specified by the machine supplier. Glands used should be of the correct size and type for the cable used, and earthing and armour clamping, also, should be as recommended. Cables should all be adequately supported and clamped to resist short-circuit forces and to take unnecessary weight off the machine terminations.

Electrical machines connected to power systems with high fault levels should be provided with terminal boxes which can withstand the effect of internal faults. Such boxes must be used exactly according to the manufacturer's instructions if they are to function properly and safely.

All necessary earth connections should be made to ensure personnel safety. The electrical machine frame, base, etc, should be solidly connected by a conductor of adequate capacity to some point of virtually zero potential. Where an insulated bearing has been provided to prevent damage due to circulation of shaft currents, care must be taken that no deliberate or accidental electrical circuit shall be made which might short-circuit such insulation and hence permit circulating currents to flow through the machine bearings. For this reason oil or water pipes, or electrical cables connecting the bearing to the machine base should have an insulation barrier.

15.6 Testing

If the machine has been stored for some time and could have absorbed moisture into its insulation, tests should be taken before energizing it. If the test value is below that recommended by the manufacturers then it should be dried out either by circulating warm dry air through it, or passing a low value of current through the windings or preferably by a combination of these methods. High voltage machines should be dried out slowly or the moisture being removed can cause physical damage and manufacturers recommend suitable rates of dry-out. When the insulation level of the machine windings and connecting cables is satisfactory it is safe to energize it electrically. Before doing this, however, a final mechanical check should be made to ensure that on the complete set every component is correctly fitted, that all auxiliary systems such as water, oil, etc, are operating correctly and that the rotors are free to rotate. If possible the shaft system should be rotated by hand to ensure that no obstructions are present.

Finally check the protection setting levels on the switch or control gear connected to the electrical machine and set them to a low value initially.

Electrical generators should be run to speed in the de-energized condition and given a full mechanical check-over after conditions have settled down. Checks should include vibration, lubrication, etc. When running satisfactorily it should be excited and voltage control checked on open circuit, and then synchronized with the rest of the system. It should then be checked over its full range of load and for any other specified loading conditions.

Electric motors should have the stator switch closed for a few seconds and then tripped. During this period it will be possible to check the direction of rotation, bearing lubrication, etc. The motor can then be restarted and run up to full speed if vibration, lubrication, etc, are satisfactory. Shutdown should be immediate if any dangerous abnormality is observed. When up to full speed the motor should be left running for a settling down period under close observation. During this period the driven plant should be unloaded as far as possible to make running-in easier and load should not be applied until the motor has run satisfactorily for several hours.

When machines have been proved mechanically and have been on load for a reasonable proving period the protection should be set to the design values recommended by the manufacturer and the machines can then be put on load for formal acceptance.

16 Geared Motors

16.1 Low-speed Drives

The design of electric motor drives has developed to a stage which now makes it possible to match the motor with its control gear precisely to the load requirements. The continuing development of geared motors plays an important part in making such design objectives feasible.

Linear velocities up to about 3 m/s (10 ft/s) seem typical in machinery used in a wide range of manufacturing industries. If one now assumes a mean diameter of some 200 mm (say 8 in) for the driving pulley this will require to be driven at 300 rev/min since

$$n_2 = \frac{60v}{\pi d}$$

where

n_2 = rotational speed of the drive, rev/min
v = linear velocity speed m/s (ft/s)
d = pulley diameter of drive in m (ft).

16.2 Limitations of Speed obtainable from Three-phase Motors

As pointed out in chapter 4 the cage-rotor induction motor is, because of its low cost and robustness, the most commonly used type of electric motor. The internationally standardized range of electric motor frame sizes covers motors up to 130 kW output. Connected to the European standard mains frequency of 50 Hz such motors normally provide a choice of the following speeds:

Number of poles	2	4	6	8
Synchronous speed, rev/min	3000	1500	1000	750

Apart from the economic limitations discussed in the next section, motors with a larger number of poles, because of the large number of coils and slots required, often present undue manufacturing problems.

16.3 Economic Limits to the Speed of Three-phase Induction Motors

It has also been shown in chapter 4 that the magnitude of torque obtainable from an induction motor is a function of the product of the rotating field Φ and the rotor current I_2 and, as the latter is induced by the field, the torque produced by an electric motor can be said to be proportional to Φ^2. Since the power output of the motor is the product of torque and speed, the low-speed multi-pole motor must produce more torque and thus requires a greater flux. Increase in flux in a motor is limited by magnetic saturation and, therefore, an increase of active material becomes necessary. Thus, for a given power the frame size and, therefore, the cost of an electric motor must increase with a decrease in its rated speed and the proportional increase in torque. Figure 16.1 illustrates this with the standard sizes of 22 kW (approximately

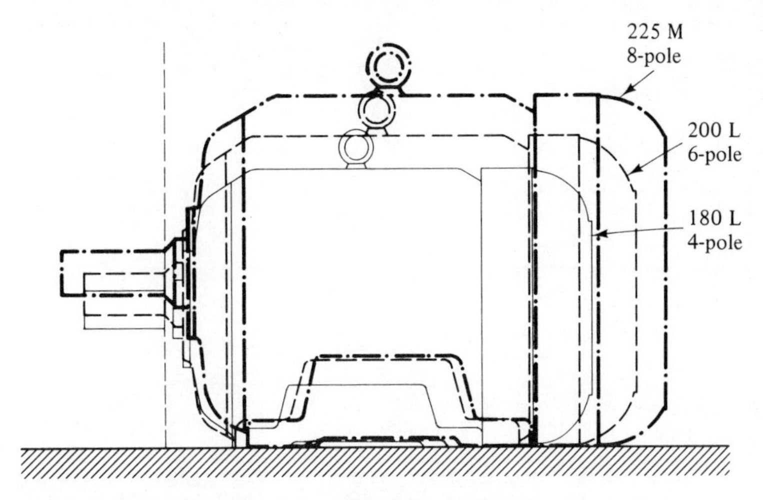

Fig. 16.1 Comparison of frame sizes for 22 kW (30 h.p.) motors, 4 pole (1500 rev/min), 6 pole (1000 rev/ min), and 8 pole (750 rev/ min)

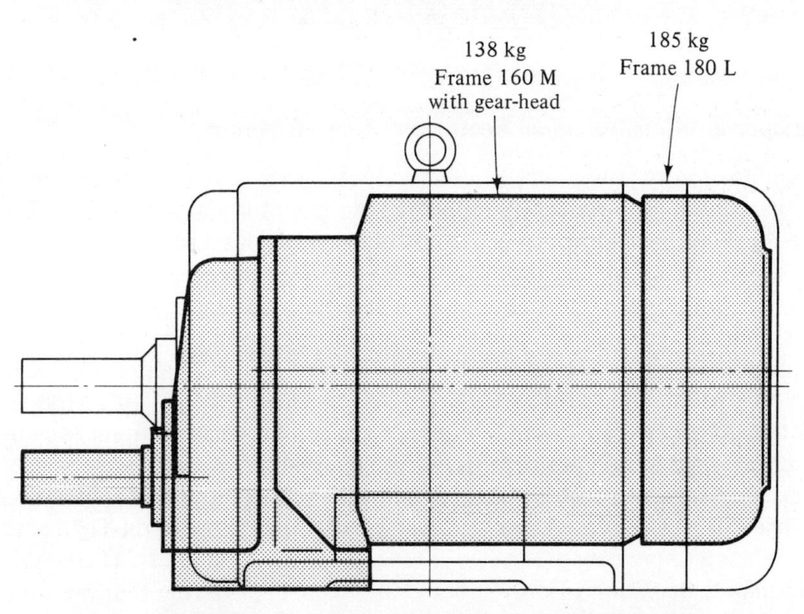

Fig. 16.2 Comparison between IEC standard motor and geared unit with nominal output of 11 kW (≈15 h.p.) and nominal speed of 700 rev/min

30 h.p.) motors with synchronous speed ratings of 1500, 1000, and 750 rev/min. Figure 16.2 shows that already from 700 rev/min downwards the high-speed four-pole motor with a single-stage reduction unit requires less space than an eight-pole motor without gearing. This pre-supposes, however, that the compact design described in section 16.4 is used. Though there may be cases where for motors of other power ratings such comparisons are less clear cut, the price comparison graphs in Fig. 16.3 show that the lower the required output shaft speed the greater is the economic advantage of the geared motor.

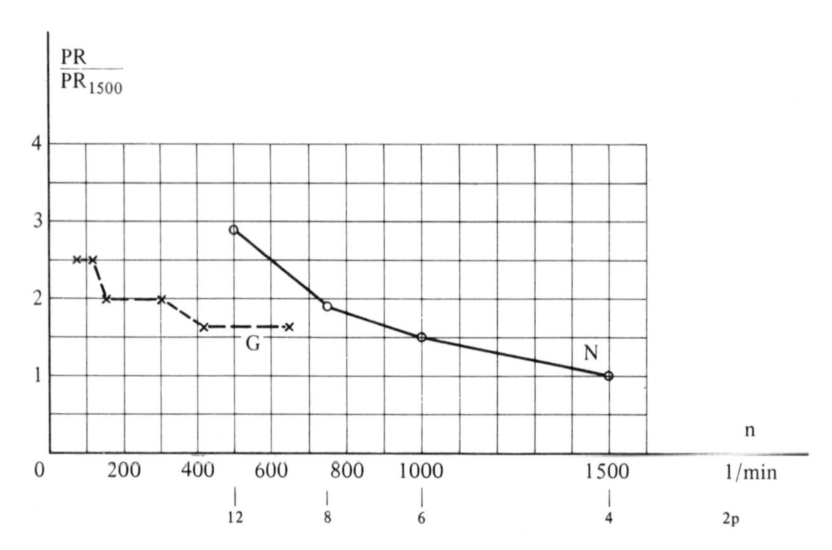

Fig. 16.3 Price comparison between standard motor (N) and geared (G) for nominal output of 55 kW (≈57 h.p.) taking as a reference PR_{1500}, the price of a four-pole motor

16.4 Important Constructional Features of Geared Motors

The combination of a high-speed electric motor with a reduction gearbox into a single geared unit requires careful matching of the motor and gearing to achieve the best design. Figure 16.4 illustrates the advantages of the integral form more clearly.

The determination of the individual steps between speeds in a range of geared motors is a compromise between the often very special demands of the machine- or process-plant design engineer, and the economic utilization of the standard matched pairs of gearwheels produced by the gear manufacturers. Figure 16.5 shows an example of one manufacturer's standard range of geared 0·55 kW (0·75 h.p.) motors where the steps between successive speeds are smaller than those in the standard R 10 series which has 25 per cent steps.

In the range of 15 to 150 rev/min, within which most industrial applications fall, the individual steps between ratios are even smaller than those of the R 20 grading in which there is a 12 per cent difference between successive numbers. This requires the user to allow a deviation of only 6 per cent which, considering that the difference between no-load and full-load speed of an induction motor is of that order, means that most practical speed requirements can be met.

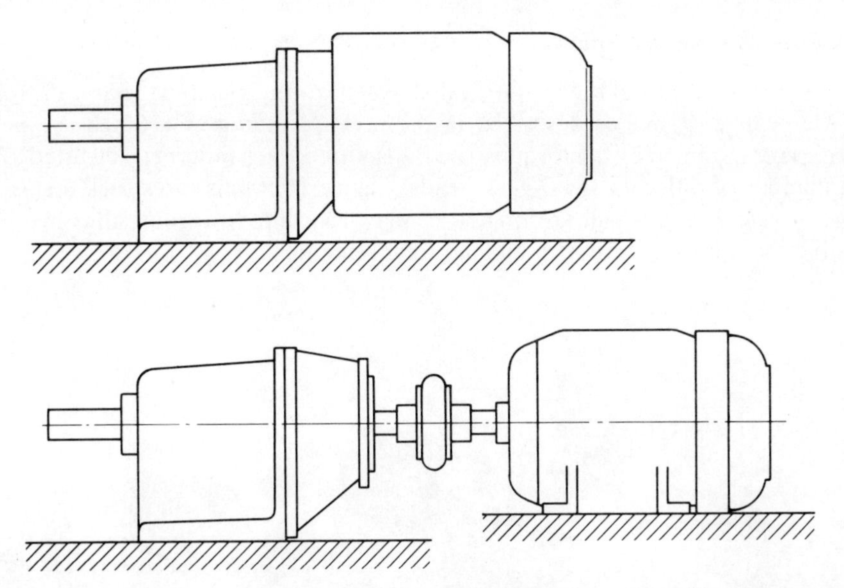

Fig. 16.4 Comparison of space and mounting requirements for a geared unit and separate motor and gear box

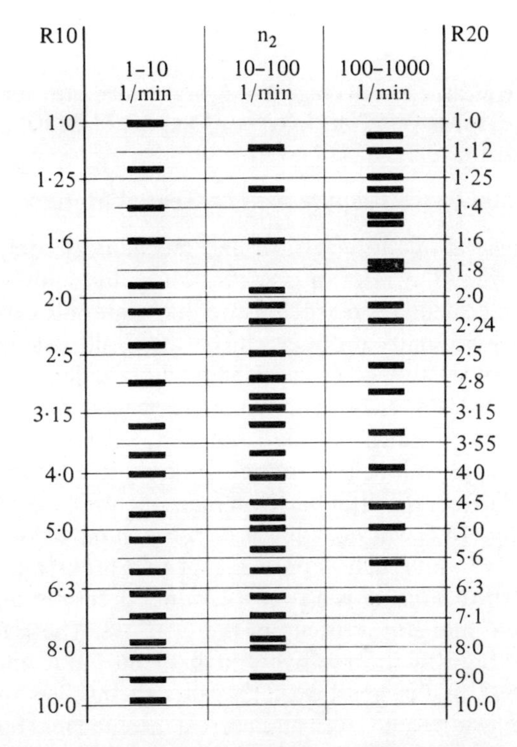

Fig. 16.5 Comparison of manufacturer's range of 0·55 kW (0·75 h.p.) geared motors with standard R 10 and R 20 series

16.5 Sizes of Geared Motors

For a given motor output the torque available at the output shaft will naturally follow the wide range of speeds described in the previous paragraph. Since the size of gearing is determined by the torque to be transmitted, each motor can be fitted to one of a number of different sized gear-heads. Figure 16.6 illustrates such a standard range of gear-heads which are graded in size according to torque, thus making it possible economically to match all load requirements.

Fig. 16.6 Example of typical gearhead range. The figures beside each unit represent its torque capacity in Nm (1 Nm = 0·1 kgf = 0·74 lbf ft)

16.6 Construction and Space Requirement of Geared Motors

Some of the dimensions of the electrical and mechanical components of geared motors will be determined by relevant specifications and regulations. Other dimensions the designer will calculate so as to ensure that materials and processes used in manufacture produce a quality product which, under all reasonable conditions of usage consistently and reliably gives its specified life and performance. For certain applications, such as tools or airborne equipment, special materials and ratings are used to construct motors which, at increased cost, save space and weight. Units designed for similar applications and ratings will tend to a limiting size, and where standard motors are used vary little between manufacturers. A range of gear heads such as shown in Fig. 16.6 will not only accept the standard motors described in chapter 5, but many variants such as pole-change and brake motors.

Typically the gearbox housing is a robust casting, precision machined to ensure accuracy of alignment and concentricity of the bearings. These features are of the utmost importance, both at the relatively high-speed input and the high-torque low-speed output end of the gearbox. The gears themselves must be accurately finished and surface heat treated to ensure correct meshing and hard, wear-resistant yet resilient teeth. A considerable contribution to wear resistance and, therefore, long service life is made by the now almost universally used high-pressure lubricants,

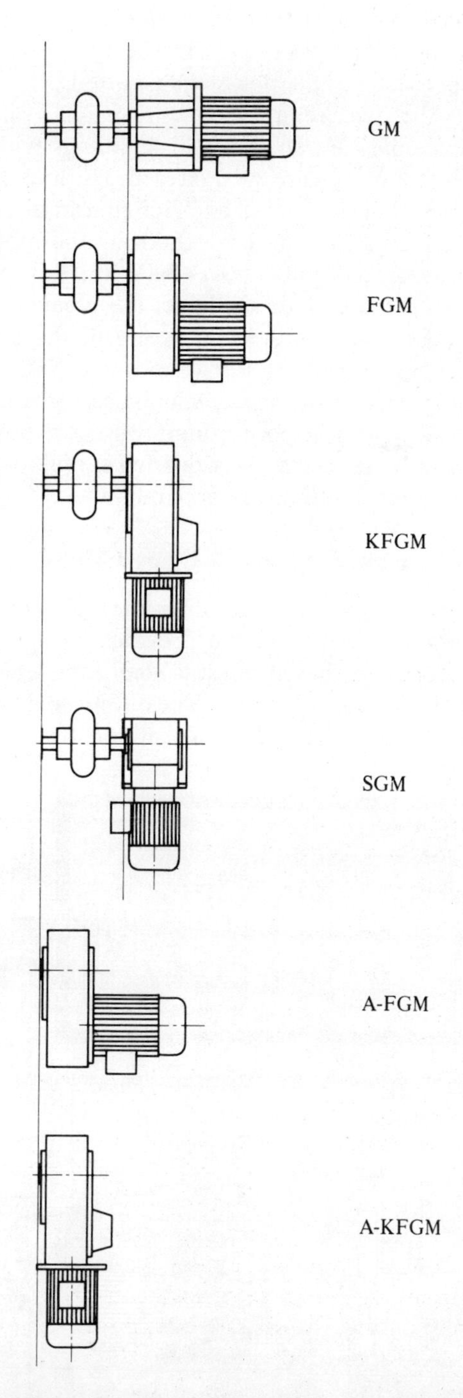

Fig. 16.7 Space required by various motor drives: GM—coaxial shaft helical geared motor; FGM—parallel shaft helical geared motor; KFGM—right-angle shaft bevel geared motor; SGM—right-angled shaft worm geared motor; A-FGM—parallel shaft mounted helical geared motor; A-KFGM—right-angle shaft mounted bevel geared motor. (No coupling is required with shaft mounted units)

whose presence must be assured at all times. Gearing losses are almost entirely due to tooth friction which, in the case of spur or helical gearing, may cause losses only of the order of 1–2 per cent per stage of the power transmitted.

Figure 16.7 compares the space required by some typical geared motor configurations. The first four examples are shown with couplings which are essential for applications where high shock loads are encountered. The most important advantage of the right-angle drives is that the overall axial length can be substantially reduced and the designer can actually save space by choosing the most convenient motor position. Multi-mounting geared units up to 45 kW (60 h.p.) with options such as back stops and brakes are commercially available. The great variety of types of gears and motors and many possible permutations are outside the scope of this work and the designer or plant engineer seeking a drive motor is best advised to list all his requirements, preferably on one of the questionnaires provided by most reliable manufacturers, before accepting a poor compromise or, possibly, an out-dated solution. Great care must be taken when evaluating suppliers' proposals to ensure that their specifications agree on all essential parameters.

16.7 The Drum Motor as a Special Form of Geared Motor

The method of operation of a typical drum motor as well as its similarity to conventional geared units described in the previous paragraph is shown in Fig. 16.8(a). The geared motor is mounted inside a steel tube which is supported by a bearing on the non-driving end of the motor. The output shaft of the gearbox passes through another bearing and is clamped in the support frame. This means that the

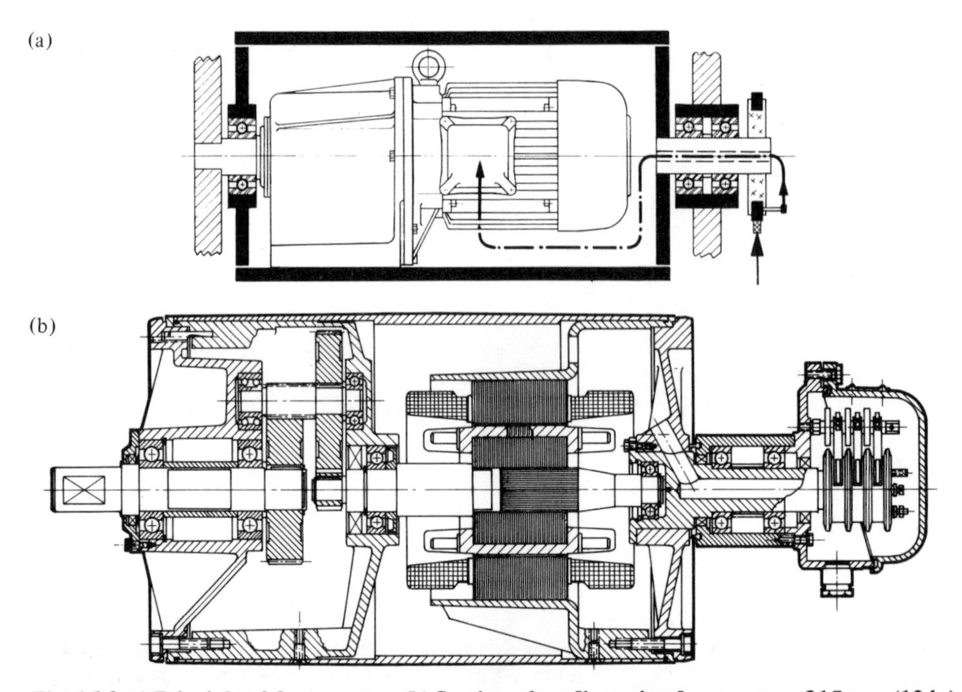

Fig. 16.8 (a) Principle of drum motor. (b) Section of medium-size drum motor. 315 mm (12 in) diameter; 50 mm (21 in) facewidth; rated at 2·2 kW (3 h.p.); 1·31 m/s (4 ft/s) peripheral speed

geared unit itself will rotate at the speed determined by the motor speed and the gear reduction, and with it the drum. To allow the motor to rotate continuously it must be supplied over sliprings, which, in turn, are connected to the stator by means of leads passing through the hollow supporting shaft.

Figure 16.8(b) shows one way in which this principle is executed in practice. The stator housing is specially designed to fit snugly inside the steel shell and forms one end of the drum. The other end is formed by the gearbox, which becomes a completely self-contained unit, usually packed with enough grease for some 10 000 running hours. Where the diameter of the drum remains the same and only the length requires to be changed, all components except the pulley shell and the rotor shaft remain identical. The intimate contact between the stator housing and the drum transfers the heat losses of the motor by conduction to the pulley, which because of its large area and contact with the conveyor belt is usually capable of dissipating the heat generated by the motor windings.

A slightly different construction in which the motor winding remains stationary is shown in Fig. 16.9. The separation of the motor from the outer shell means that the motor must be cooled by transferring the heat it generates to the outer shell either by an integral fan or by partly filling the drum with oil. In the latter case the motor and gearbox communicate through ports and the lubricating oil used for the gears also serves to keep the motor cool. In such motors care must be taken when selecting the oil for refilling to provide the extreme pressure additives required for the gear teeth and, at the same time ensure that the lubricant does not attack the motor insulation chemically.

Generally, however, both types of drum motor incorporate a separate grease-filled gearbox, which, assembled with precision-made gears, results in a very quiet, well-protected, and compact driving unit good for some 8000 to 10 000 running hours between grease or oil changes. For operation in particularly hostile surroundings such as cement works, or similar dusty environments and also in wet ambient conditions drum motors can be sealed even more effectively by fitting labyrinth seals at both ends of the drum. Reliability and running hours between regreasing are very much a function of correct selection and operating conditions as in any other motor application. When costed realistically the difficulties of mounting, aligning, and protecting a separate drive motor will not only reflect in high capital investment, but significant operating costs due to the inevitably higher failure rate probability. For most exacting applications the robust and simple-to-mount drum motor will, therefore, be found to be the most cost-effective alternative.

To meet practical requirements one manufacturer offers the following range:

Pulley diameter	*mm*	160	215	315	400	510	630
	in	6	8	12	16	20	24
Output up to	*kW*	0·75	2·2	3·7	7·5	11	18·5
	h.p.	1	3	5	10	15	25
Belt speeds	*m/s*						
	from	0·04	0·016	0·35	0·14	0·35	0·28
	to	1·68	2·9	2·1	2·2	2·9	3·0
	ft/min						
	from	8·5	3·2	69	27·5	69	55
	to	331	571	413	433	571	591

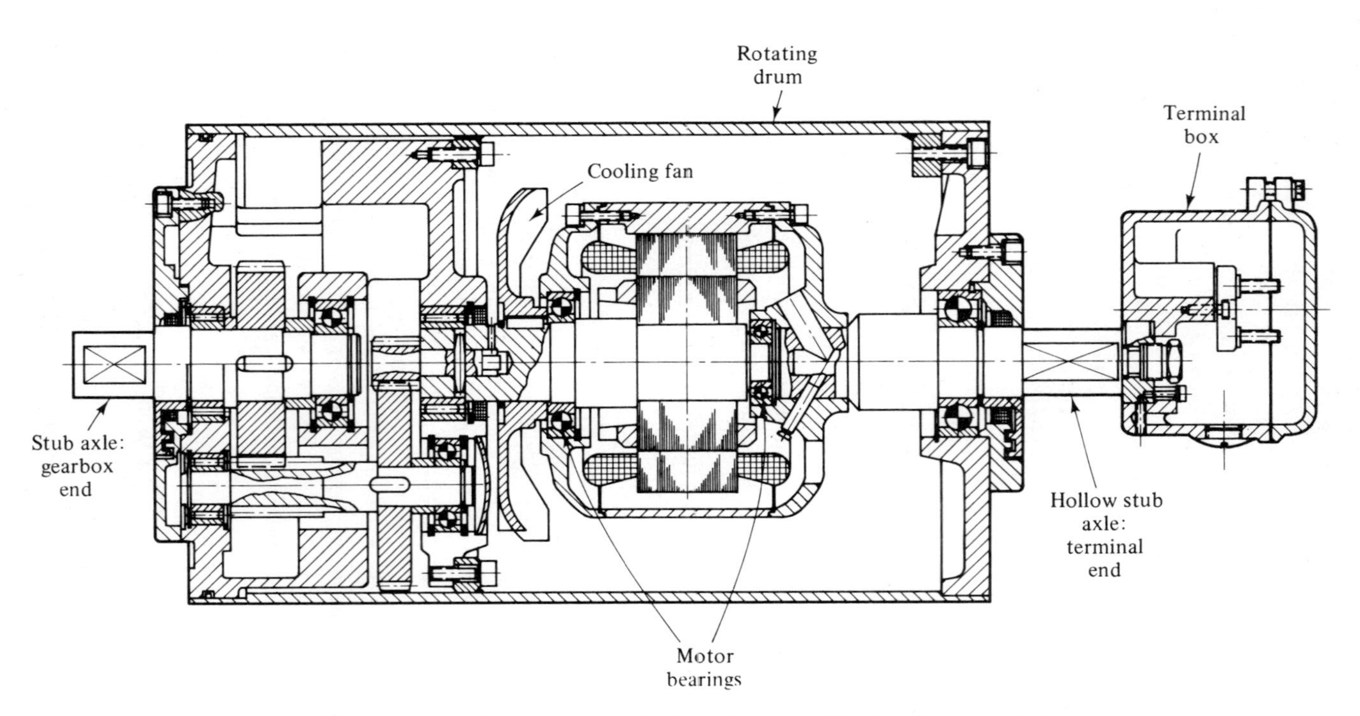

Fig. 16.9 Drum motor with stationary winding. (Courtesy Eberhard Bauer, W. Germany)

The drum motor is ideal for most conveyors but cannot be used where very hot materials such as billets or extruded sections have to be transported. The heat transferred to the motor is equivalent to a high ambient temperature whose value is limited more by degradation of the high pressure lubricant in the gears than deterioration of the insulating materials used in the windings. In roller table or similar applications where hot materials are transported, externally mounted geared units driving individual rollers through special couplings have been found more reliable and easier to maintain than group drives involving mechanical linkages which are also vulnerable in the hostile environment.

16.8 Nominal Torque

In the case of drives with low rotary or linear speeds the nominal torque rather than the power gives a better and more easily visualized picture of the drive which will be required. Thus, for example, a drum motor rated at only 0·075 kW ($\frac{1}{10}$ h.p.) having a pulley diameter of 215 mm (8 in), and developing a nominal torque of 215 Nm (158 lbf ft), produces a belt tension of 2000 N (450 lbf). In many cases the linear force required can simply be determined by means of a spring balance.

If the force in newtons (1 lbf = 4·448 N) is F and the pulley radius in metres is r then the torque in newton metres is Fr and motor power in kilowatts = (torque × rev/min) ÷ 9550 as shown in more detail in chapter 2.

Since the driving torque required by most machines is mainly composed of a load and a frictional component, and remains substantially constant for all speeds, the power required from the motor is directly proportional to the conveyor speed.

For completeness the formula for obtaining the rotational speed n in rev/min from linear speed is also repeated here:

$$n = \frac{v_s}{\pi D} = \frac{v_{min}}{\pi D}$$

where

v_s is the conveying speed in metres or feet per second.
v_{min} is the conveying speed in metres or feet per minute.
D = diameter in compatible units, i.e., metres or feet of pulley or drum.

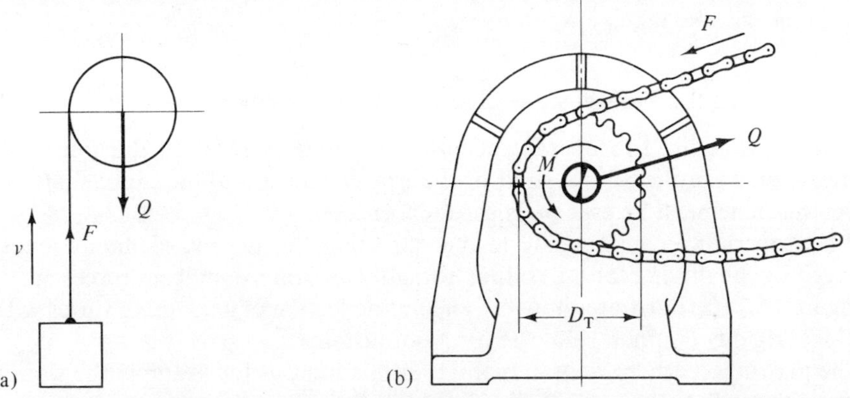

Fig. 16.10 (a) Loading of output shaft bearings—Overhung load. Q = loading on shaft. (b) Loading of output shaft bearings—Chain or sprocket drive. M = driving torque; Q = torque reaction on shaft

16.9 Loading of the Output Shaft Bearings

Where geared motors having relatively low speeds are used, correspondingly large torques are as a rule required. Depending on the arrangement of the drive, the driving forces also act on the bearings of the output shaft of the gearbox.

The driving forces encountered are illustrated by two typical arrangements in Fig. 16.10(a) and (b) for which the forces acting are:

	SI Units	*FPS Units*
Hoisting a load (translation force)	$F=\frac{1000P}{V}$	$F=\frac{550P}{V}$
Chain or sprocket drive (torque load)	$F=\frac{19\,100p}{n\times D}$	$F=\frac{12\,600p}{n\times D}$
	F—Force (N) P—Power (kW) v—Linear velocity (m/s) n—Rotational speed (rev/min) D—Pitch circle diameter (m)	F—Force (lbf) P—Power (h.p.) v—Linear velocity (ft/s) n—Rotational speed (rev/min) D—Pitch circle diameter (in)

These two examples indicate the magnitude of torque reaction acting on the output shaft of the driving element, to which, in the case of a geared motor, must be added the torque reaction on the internal driving teeth. According to the direction of the external torque reaction the effect on the bearing can be the vector sum or difference of the two forces. Each geared unit has a limit of radial loading and the user is advised to consult the manufacturer with the details of his proposed drive, comprising, in particular, the axial distance of Q from the bearing face and the pitch circle diameter to ensure maximum possible bearing life.

When choosing the type of transmission and the correct dimensions for a geared unit, in particular those of its output shaft bearings, additional loadings due to particularly arduous operating conditions such as shock loads, inching or reversing duties must also be taken into account.

16.10 Selecting the Transmission Elements for Reversing Drives

If a drive is required to operate in both directions of rotation the transmission elements between the output shaft of the geared motor and the input shaft of the driven machine must be especially carefully chosen.

When operating continuously in one direction the driving elements are kept engaged by the driving torque so that normally no other significant forces act.

Figure 16.11(a) illustrates how the angular deflection of a coupling varies with its torsional rigidity (stiffness) and the torque transmitted.

The play inherently present in many transmissions, as for example the slack in a chain or the radial play of a coupling, is particularly detrimental in reversing drives. The torque characteristic in Fig. 16.11(b) shows the radial play φ_s which is taken up every time the motor is reversed and represents the angular displacement during

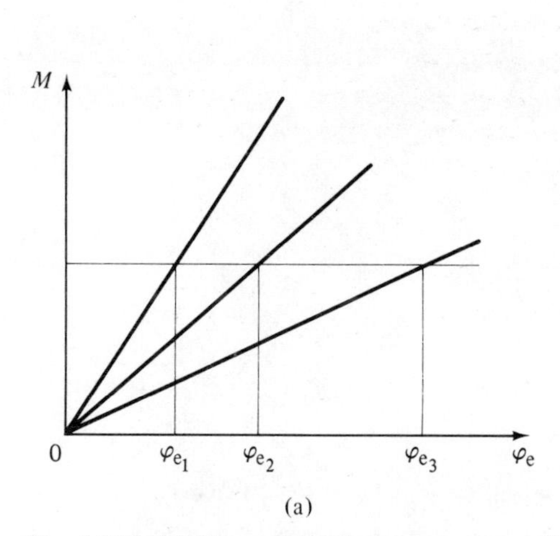

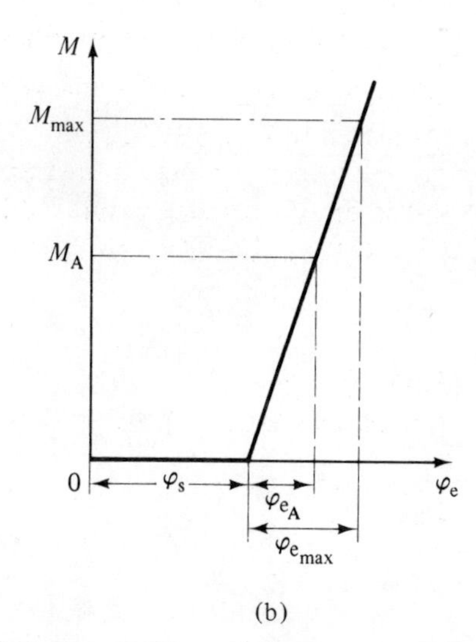

Fig. 16.11 (a) Torque/ displacement curves for backlash-free elastic couplings or varying stiffness. (b) Torque/ displacement curve for elastic coupling with backlash

which the motor is not driving the load. The elastic angular displacement φ depends on the torque M.

It must also be noted that relatively small play at the output shaft of the gearbox appears magnified by the gear ratio i at the rotor shaft.

Should the radial play φ_s, that is to say the 'free travel' of the output shaft, be of the order of magnitude of the no-load running-up displacement, the motor can reach full speed before the load has begun to move. The shock loadings which occur when after 'free travel' the load is suddenly engaged are far in excess of the torque normally developed by the drive, because they depend only on the elasticity of the transmitting elements. Thus the aspects for choosing the transmitting elements for low- or high-speed shafts are entirely different.

The following numerical examples will illustrate this point: A four-pole three-phase motor with a rated output of 7·5 kW (10 h.p.) is used in a reversing drive with a coupling which has radial play of $\varphi_s = 5°$; the motor inertia is 0·045 kgm^2. At no load there is no resisting torque so that the rotor would require an angular displacement of 250° to reach its full rated speed of nearly 1500 rev/min.

Within the 'play' of $\varphi_s = 5°$ the rotor would only have been accelerated to a speed of about 210 rev/min before the driving half of the coupling engaged with the still stationary half on the load side. At that instant the rotor contains energy of rotation given by:

$$W = \frac{Jn^2}{182{\cdot}5} = \frac{0{\cdot}045 \times 210^2}{182{\cdot}5} = 11\text{J}$$

W = work (energy) in Nm = Ws = J

J = inertia in kgm^2

n = speed in rev/min.

If a similar coupling is now connected to the output shaft of a geared unit having a gear reduction to give about 20 rev/min ($i = 75$), then the free travel at the rotor

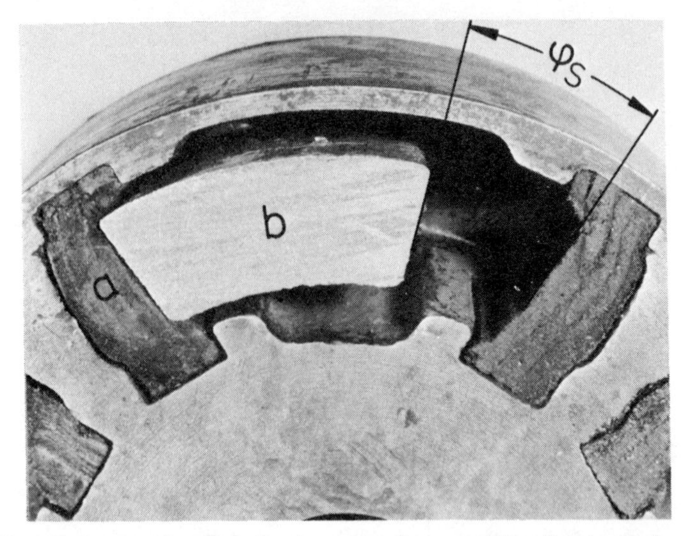

Fig. 16.12 Greatly increased radial play in a claw type coupling due to the leather packing pieces: (a) Having been compressed by claws and (b) due to reversing duty

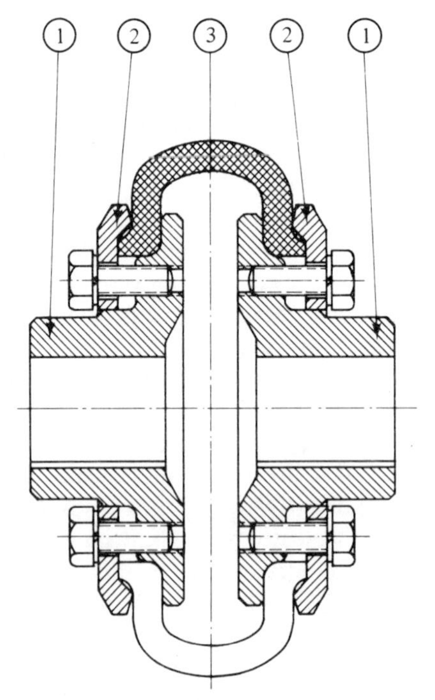

Fig. 16.13 Highly elastic backlash-free coupling consisting of boss (1), Pressure plate (2) and fabric reinforced rubber tyre (3)

shaft is now $\varphi_s = 5 \times 75 = 375°$. This angle of rotation is sufficient to allow the rotor to reach its rated speed of nearly 1500 rev/min without engaging the load. As the driving half of the coupling engages the stationary load the total amount of kinetic energy of $W = 560$ J will be redistributed over the combined masses of the geared motor, the load, and the energy lost due to plastic deformation. Elastic deformation

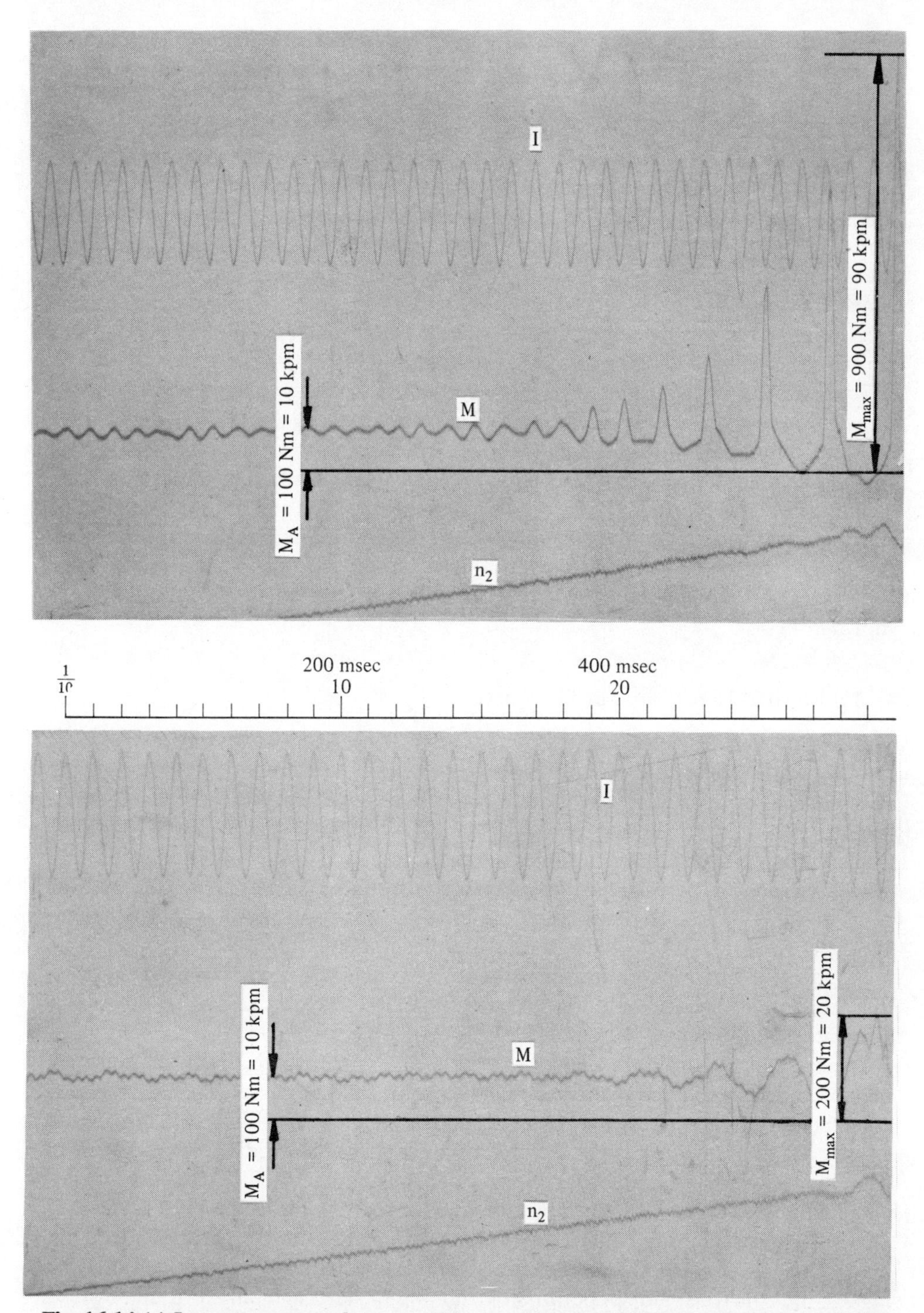

Fig. 16.14 (a) Instantaneous peak torque (M_{max}) produced by kinetic energy compared with starting torque (M_A) developed by the motor when reversing with a claw-type coupling as shown in Fig. 16.12. (b) Instantaneous peak torque (M_{max}) produced by kinetic energy compared with the starting torque (M_A) developed by the motor when reversing a highly elastic backlash free flexible coupling of the type shown in Fig. 16.13

will return stored energy to the rotating masses but must also be strictly limited since it can give rise to torsional oscillations. The complexity of the motor, gearhead, and load combination makes a calculation of the individual energies extremely difficult. In unfavourable conditions very large forces are developed, as in the above calculation where the energy available from the geared unit is 50 times larger than it would be for a direct drive. The example, furthermore, shows that couplings with backlash which are quite satisfactory for driving continuously in one direction can give rise to very large forces when the motor is reversed and should be replaced by one of the special backlash-free couplings on the market. Claw couplings with elastic inserts, which are very satisfactory for continuously rated, unidirectional drives, can develop considerable backlash when reversed or started frequently, mainly due to the permanent hardening and deformation of the inserts (Fig. 16.12).

For severe reversing duties special types of couplings are available and Fig. 16.13 shows a section through typical design. The coupling is similar to the well-known bellows coupling used in instrument work. In selecting this type of coupling care should be taken not to apply too large a safety factor since this would make the coupling excessively stiff and prevent it from absorbing shock loads caused by on-line starting, braking, reversing or plugging. The oscillograms, Fig. 16.14(a) and (b), show the difference between the large torque peaks due to backlash in normal claw coupling with the very much smaller transients in the special resilient couplings as shown in Fig. 16.13.

16.11 Electrical Protection of the Motor

The cost of a motor breakdown in a process plant is appreciably higher than the installation of appropriate protective devices as, for example, those prescribed in the IEC publication No. 204. The general guidelines for the various forms of motor protection are given in chapter 14. Where built-in thermostats are used it should be remembered that these permit the windings to reach the limiting temperature irrespective of ambient conditions. These contrast with line-current (I) operated devices whose operation depends on the total power supplied to the motor. Figure 16.15 shows how a motor with thermostat protection can be severely overloaded under low ambient temperature conditions. In all cases where low ambient temperature conditions are encountered protective devices which can take these into account, such as current-dependent relays, should be used.

Special consideration must also be given to the protection of frequently started, reversed or plugged geared units. Induction motor starting currents tend to be large and, though of short duration, the cumulative effect of frequent switching could easily trip a thermal overload device before the permissible winding temperature has been reached. The thermal capacity and time constants of such devices vary widely and the number of starts allowed by manufacturers are in the relatively low range of 25 to 60 starts per hour. The actual limit in a particular application depends, however, on motor size, type of duty, and loading so that it may actually be somewhat higher; thus the true value of permissible number of starts can only be found by tests under actual working conditions.

Where motors are operated according to one of the intermittent periodic duty cycles S4, S5, S7, and S8 (as defined in section 5.2.1.1 of chapter 5) it is, therefore, advisable to use protective devices which respond to the actual winding temperature.

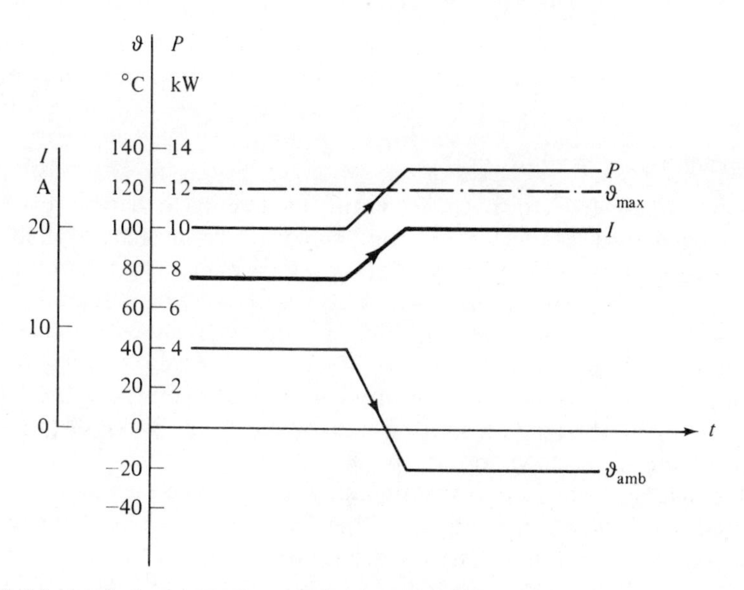

Fig. 16.15 Mechanical overloading of thermally (TMS) protected geared motor at low ambient temperatures

Fig. 16.16 Sprocket wheel with mechanical overload protection in the form of slip coupling on hub

Thus, the thermostats already mentioned or thermistors will fully protect the motor for the actual number of starts and/or stops a particular configuration permits.

16.12 Mechanical Protection

The torques available from the output shafts of geared units tend to be very large. As regulations stipulate that the whole of the rated power must be available at the output shaft of the gearbox, the inclusion of some form of mechanical protection such as a slip coupling or shearing pin in the transmission may be essential to prevent catastrophic damage in the driven machine. Figure 16.16 shows a clutch incorporated in the hub of the sprocket.

A further form of protection against torque peaks due to reversing the drive or its suddenly picking up the load on starting is the resilient but backlash-free coupling already mentioned and shown in Fig. 16.13.

From the diagrammatic representation in Fig. 16.17 it is clear that the motor can be protected better than the gearbox, so that for essential drives there should not merely be a *spare motor* but a complete geared unit.

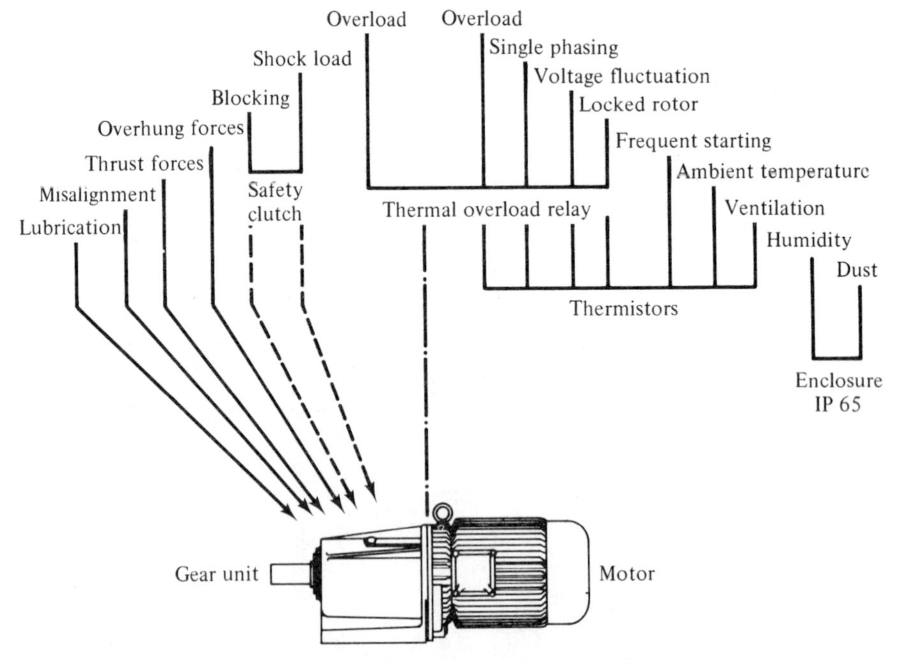

Fig. 16.17 Limits to which a geared motor can be protected

16.13 Special Geared Units

In some cases it is essential to prevent the load from driving the motor once power has been switched off. For applications where the motor will always run in the same direction, for example in the inclined conveyor, most manufacturers can fit back-stops to their motors. A typical back-stop is shown in Fig. 16.18. For applications where the motor operates in both directions brakes with electromagnetic release can be used.

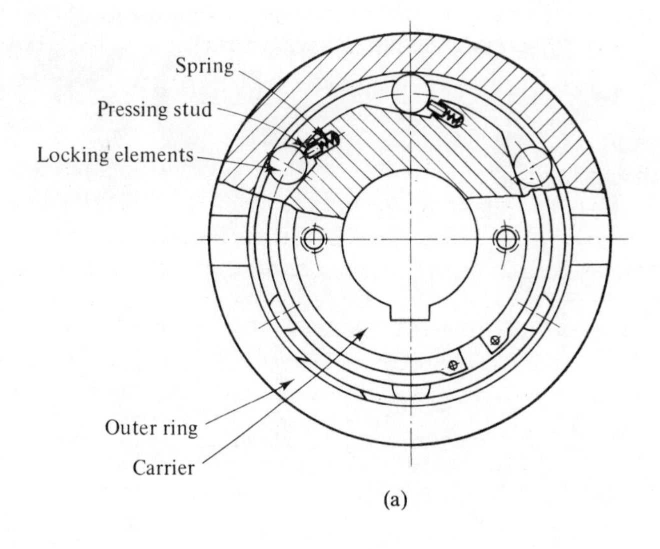

(a)

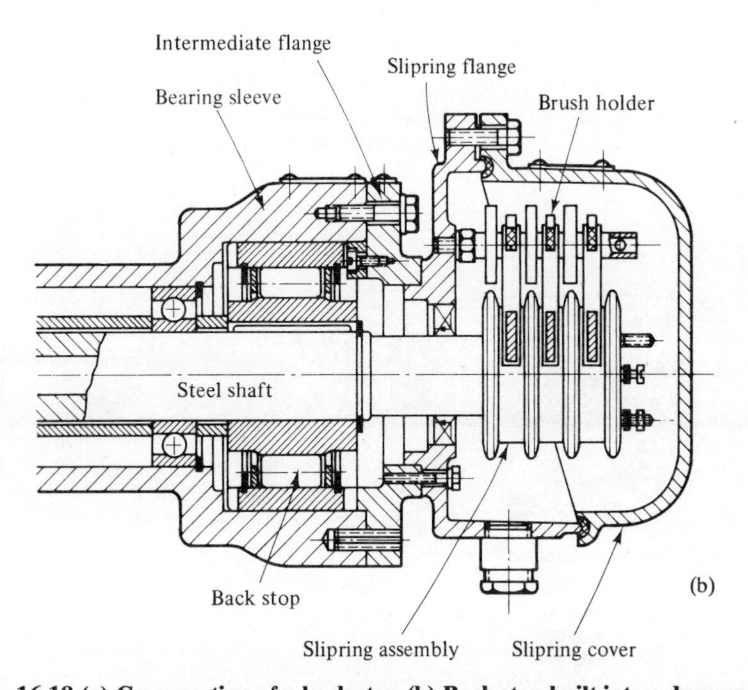

(b)

Fig. 16.18 (a) Cross section of a back stop (b) Back stop built into a drum motor

The motor shown in Fig. 16.19 makes use of the magnetic pull on a conically shaped rotor to release a conical brake. The double motor arrangement has been devised for applications such as machine tools where precise positioning is required at the end of, say, a fast slewing run. It can be seen that when the small motor drives for fine adjustment the brake of the large motor acts as a coupling.

Another useful arrangement is the mechanically variable speed geared unit in Fig. 16.20. In this the speed can be finely and steplessly adjusted over a range of 1 : 7 by

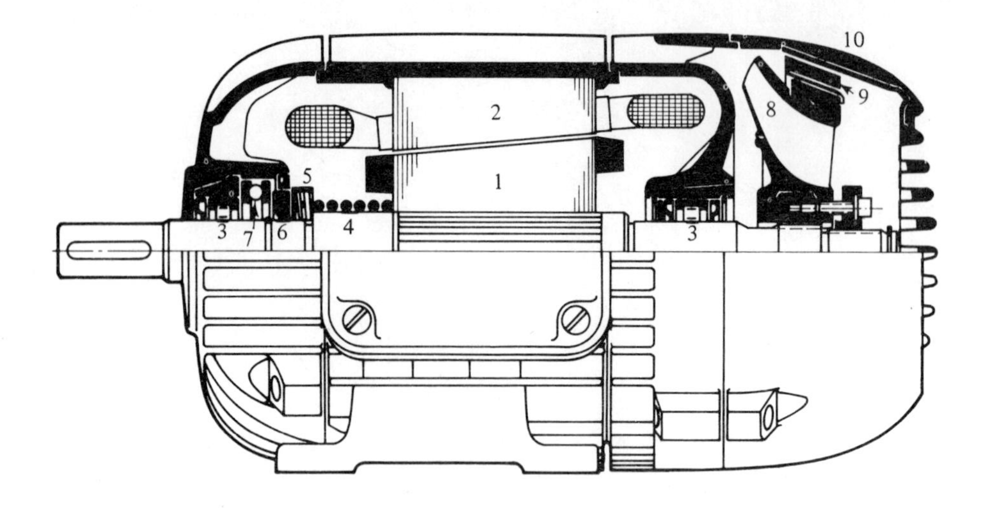

1 Conical rotor
2 Stator with corresponding conical bore
3 Roller bearing permitting rotor to move axially
4 Against spring 4 when motor energized
5 Disc springs limiting rotor travel
6 Pressure ring bearing against
7 Thrust bearing
8 Combined fan and brake disc
9 Conical brake pad supported by
10 Brake pad housing

(a)

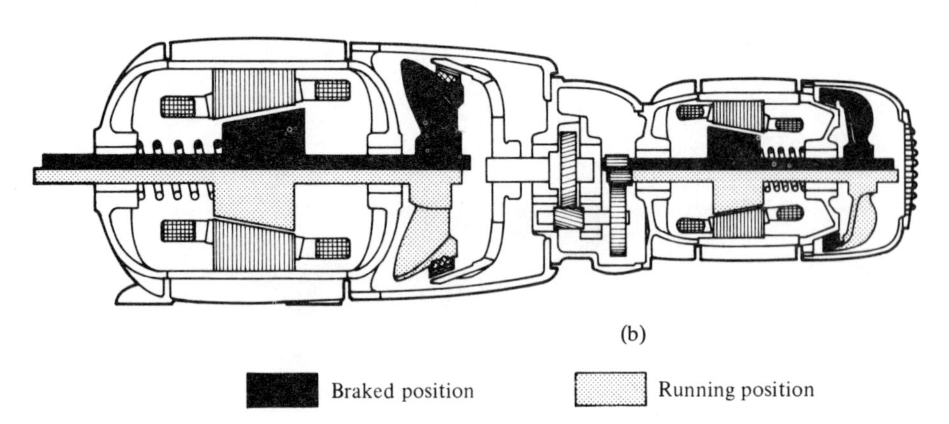

Fig. 16.19 (a) Brake motor utilizing magnetic pull of rotor (b) Microspeed unit (DEMAG) for indexing and accurate positioning

the hand-wheel shown. To overcome the appreciable forces acting on the special operating linkage a lead-screw is used and, if required, the hand-wheel can be replaced by a servomotor for remote control. In this type of adjustable speed gearing the motor drives the gearbox through a special broad sectioned belt which runs between two pairs of conical sheaves. This belt is made up of a large number of high-tensile fibres embedded in special oil and heat-resistant rubber, which also

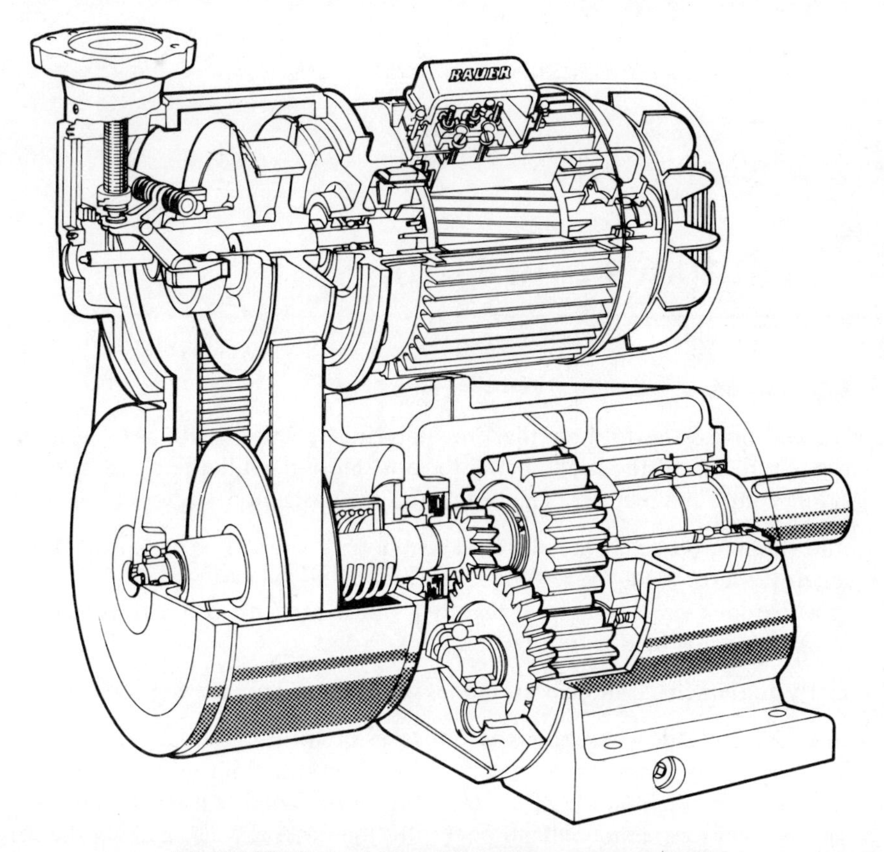

Fig. 16.20 Mechanically adjustable speed geared unit

forms the driving flanks. The profile of the belt is specially designed to ensure that the load is evenly distributed over the reinforcing fibres.

Geared units can usually be fitted with any of the many other types of motors including a.c. commutator motors and d.c. motors, which, for speed adjustment, may be supplied by one of the many forms of electronically controlled rectifying units (see chapter 12).

17 Clutches and Brakes

17.1 Introduction

Clutches and brakes perform similar but opposite functions within a mechanical system, the function of the brake being to resist motion while the clutch transmits motion in a controlled manner. The brake has two basic modes of operation:

(a) the slipping mode where the brake imposes load torque to absorb kinetic energy, thereby slowing and ultimately stopping the system.
(b) positive lock on a rotational member which prevents motion, engagement normally being made after motion has stopped.

Similarly, the clutch has three basic modes which equate to the two brake modes:

(a) the take-up mode where slip is used to transmit torque gradually from one part of a system to another thereby providing a smooth build-up of speed.
(b) the disengagement mode where the clutch is designed to transmit a predetermined load torque and will slip over this load thereby disengaging the drive from the driven member.
(c) the non-slip mode where there is a positive physical connection between the drive and driven member, engagement normally being made when both members are stationary, or rotating at the same speed.

Whatever type of equipment is used, if there is slip, kinetic energy is absorbed and heat generated. Well-designed equipment from a reputable manufacturer will perform satisfactorily within the heat environment it has itself produced; however, this might not always be true of adjacent equipment which is required to function in the same environment. While this is obvious, it is not always given sufficient consideration and an otherwise satisfactory system may fail owing to heat failure of a component in the proximity of heat-dissipating equipment.

17.2 Definition of Terms

Given below is a short list of terms commonly used in association with brakes and clutches; it also indicates the characteristic which requires consideration when selecting a brake/clutch for a specific application.

Duty cycle:	The number of times the device may be operated in a given period.
Response time:	The time required for the device to develop 63 per cent of its maximum torque in response to a step input.

Life (Duty):	The duty cycle hours or number of operations under the specified load and environmental conditions, that the equipment can be expected to perform to specification before overhaul is necessary. This should not be confused with MTBF (mean time between failures) as brakes and clutches often form part of the safety equipment and a failure could not be tolerated.
Life (total):	The product of the duty life and number of overhauls that can reasonably be performed before the equipment becomes unsafe and/or uneconomic to repair.
Drag torque:	The torque exerted by the clutch/brake in the free or non-energized state; also known as Zero excitation torque.
Magnetic torque:	A form of drag torque usually caused by non-useful eddy currents or residual magnetism.
Accelerating torque:	Torque over the steady-state load torque of a clutch, required to accelerate the load up to speed.
Break-away torque:	The torque imposed on the prime mover by the clutch/brake that is caused by bearing friction or restoring spring pressure.
Slip torque:	The retarding torque exerted on the system when there is differential motion between the clutch/brake and the system.
Torque to inertia ratio:	The output torque of the clutch/brake divided by the combined inertia of the input and output rotating members.
Crown tooth face:	The engaging face of the clutch/brake which has gear-like teeth that, on engagement, provides a slip-free contact between drive and driven members.
Backlash:	The rotational angular error caused by play in shaft couplings on engagement of clutches and brakes; of specific importance when there is a change of rotational direction.
Chatter:	Vibration or bounce of engaging faces caused by the restoring spring or variations in input power.
Thermal capacity (or Rating):	A measure of the clutch/brake's ability to dissipate heat.
Power Gain:	The ratio between mechanical output power and electrical input power.
Armature:	The moving element of magnetic material which produces engagement of the torque transmitting faces.

17.3 Types of Clutches and Brakes

While there is a great variety of styles and sizes of equipment from which the designer may select, these generally fall into a few defined types of which a selection is given in table 17.1. Friction clutches and brakes may also be operated pneumatically or hydraulically and there are a number of purely hydraulic devices. In some motors the

Table 17.1 Clutch and brake types

Type	Type of control power	Operational Mode		Duty cycling capability
		Brake	Clutch	
Magnetic solenoid	Electric	Zero speed torque	No slip on–off	Up to several hundred cycles/min
Hysteresis	Electric	Zero speed torque	No slip continuous control	Limited to long time cycles
Eddy current (Air/water cooled)	Electric	No zero speed torque	Slips continuous control	Limited to long time cycles
Magnetic (Dry Particle/fluid)	Electric	Zero speed torque	Limited slip continuous control	Limited by heat dissipation capability and low inertia loads
Multi-disc friction	Electric or spring	Zero speed torque	No slip on–off	12 stops/min maximum
Spring	Spring	Zero speed torque	Limited slip continuous control	Medium speeds and cycles/min
Air	Compressed air	Zero speed torque	No slip on–off	Many cycles/min
Shoe	Electric or spring	Zero speed torque	No slip on–off	3 stop/min maximum

armature or rotor is made to move axially against a spring force, which sets the brake when the motor is disconnected from the supply. In other motors the clutch or brake, though functionally independent, is already built into the motor to facilitate installation. In such cases the control equipment is particularly important and must conform to the specification for the type of brake or clutch fitted to the motor.

A form of braking which uses the inherent characteristics of the prime mover can also be useful. In general terms this involves switching the machine which normally functions as a motor into the generating mode to absorb some of the kinetic energy of the rotating masses. Under certain conditions this energy can even be fed back into the supply and such regenerative braking is generally useful as an addition, rather than a substitute for the more conventional forms of braking.

Important application factors for clutches are their ability to deliver high, intermittent torques with modest control power to allow the driving motor to reach its operating speed before connecting it to the load. When the stationary portion of a friction clutch is first contacted by the driving member, there will be more or less slip until both parts are in synchronism. The amount of slip and slipping time is determined by the magnitude of the load torque and the inertia of the load which must be subtracted from the clutch slipping torque to obtain the effective accelerating torque.

17.3.1 *Solenoid operated clutches and brakes*

Electromagnetically operated brakes and clutches provide a simple and adaptable means of mechanical torque control in low and medium power systems. The construction of the friction types may vary from simple shoe or radial types to single or multiple disc types for dry or wet operation. Toothed clutches or brakes are used where fast switching has to be carried out on relatively large torques and no drag torque is allowable in the disengaged position. Figures 17.1.a to 17.1.f show schematically the basic forms of solenoid clutches and brakes with typical combinations in the last two. In general the d.c. magnetic field, which is built up when the winding is energized, causes the central member to move axially to make strong physical contact with a mating surface and thus develop the required locking action up to its specified torque limit. When the excitation current is removed, the spring forces the two members apart until complete disengagement is achieved. When used on d.c. motors the solenoid may be connected either in series or in parallel with the motor winding. In the latter case it is desirable to separate the brake and armature circuits to avoid the possibility of regeneration current flowing through the brake coil and thus preventing the brake from setting. When selecting a brake or clutch the compatibility of the time constant of the operating solenoid with load and system requirements must be considered. While sluggish operation may give smooth acceleration or deceleration it becomes unacceptable when rapid and accurate stopping is demanded. The role of the solenoid and the spring can be reversed so that the spring provides the engaging force and the d.c. supply provides the power for disconnecting the device.

A typical single plate spring-operated brake is shown in Fig. 17.2. The housing (1) contains the magnetizing coil (2), which, to minimize hot spots, protect it from the ingress of moisture or fumes, and retain it in position, is usually thoroughly impregnated and cast in or encapsulated with a suitable resin. The housing, generally

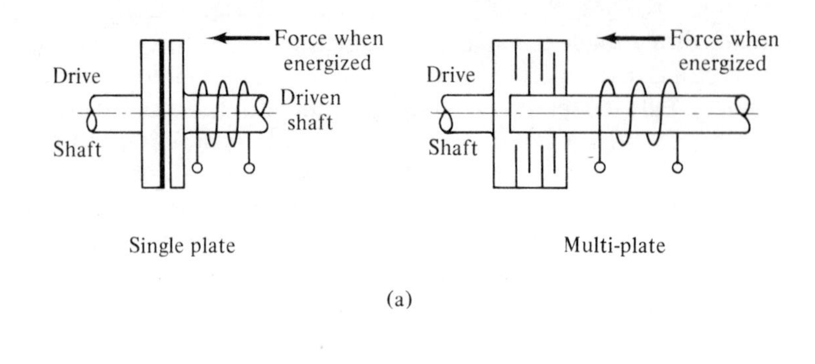

(a)

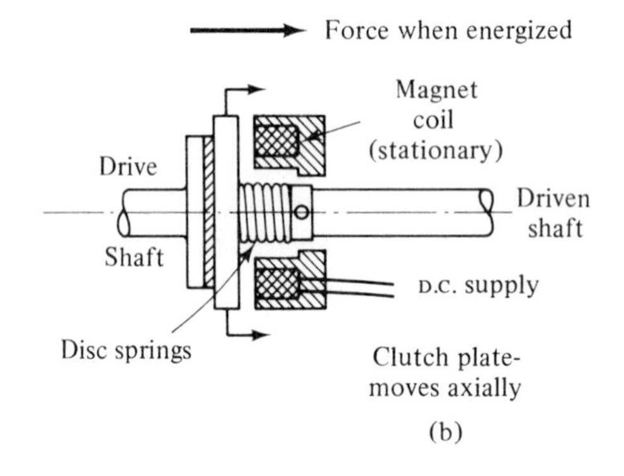

(b)

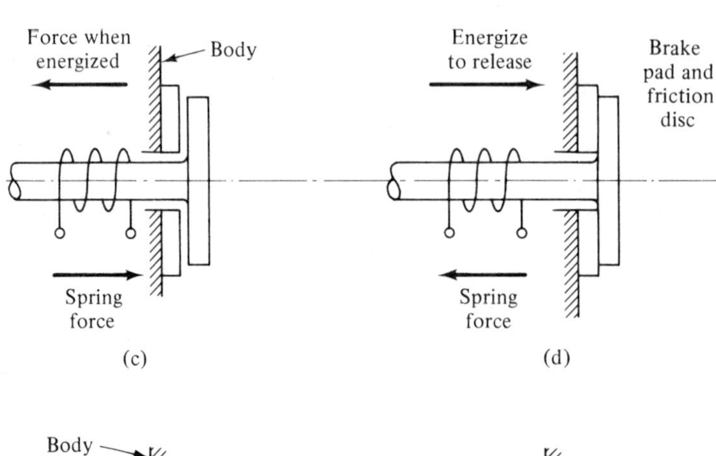

(c) (d)

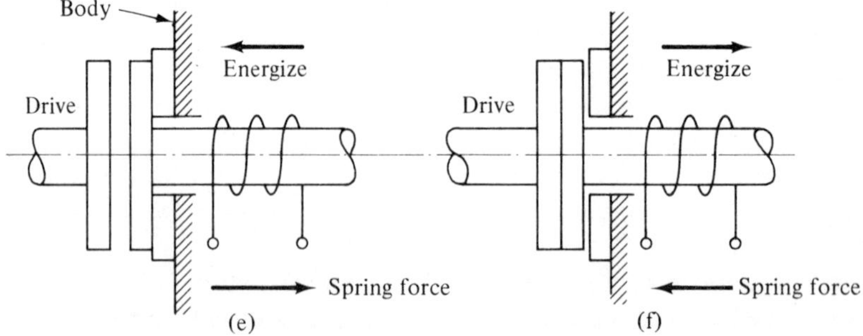

(e) (f)

Fig. 17.1 Schematic representation of various forms of electro-magnetic brakes and clutches

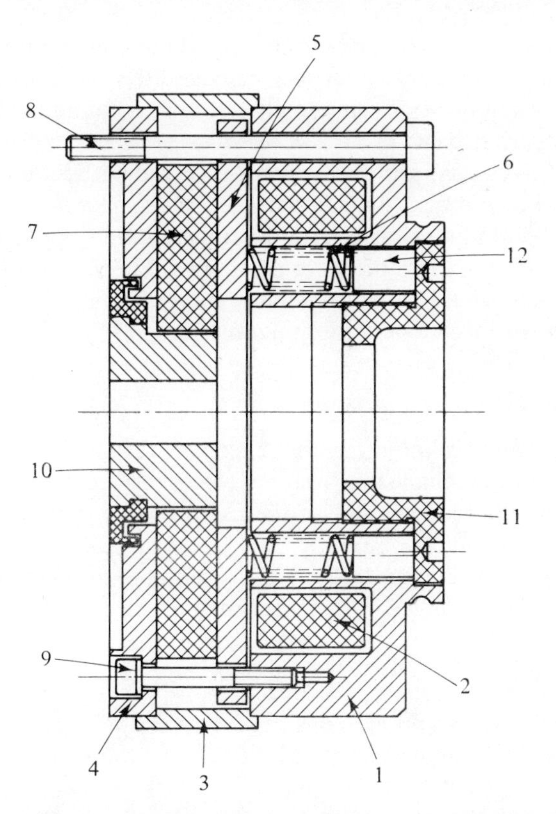

Fig. 17.2 Single-plate spring-set brake with d.c. electromagnetic release

made of soft iron with good magnetic properties, is attached to a flange (4) from which it is spaced by the ring (3) to provide the space for the armature (5) and the radial brake disc (7). The disc brake is forced against the flange by the springs (6) located in a number of holes drilled in the housing and supported radially by the bolts (8) and (9). When the coil is energized the armature is attracted towards the coil housing and, overcoming the spring force releases the brake. The friction disc (7) is free to move axially on the carrier sleeve (10) which is mounted on the non-driving shaft extension of the motor. Since there are no provisions for wear adjustment or compensation, the magnet is made sufficiently powerful to ensure reliable release of the brake up to the maximum permissible wear level. The magnitude of the braking torque may be adjusted by the threaded bush (11) which controls the spring pressure transmitted by the thrust blocks (12).

Many different types of electromagnetic brakes and clutches exist and the requirements of any particular application must be carefully defined so that meaningful competitive quotes can be compared. Reliability must also be given its proper rank in the selection criteria and must be linked with inspection and maintenance as recommended by the manufacturer of the device.

Due attention to safety is also necessary and a 'fail safe' mode should be chosen wherever possible. Normally it is best to arrange operation so that the solenoid function is less than that of the spring since this will save power and reduce the heat dissipated by the coil. The stiffness of the spring usually determines the operating time so that, for example, stiffer springs in a multiple plate clutch will increase the engaging time, but produce faster release. Provided the user adequately defines his requirements the manufacturer can usually not only recommend the most suitable type, but from experience can predict performance.

Clutches and brakes used in servo-systems received particular attention between about 1935 and 1945 and typical delay times for small solenoids were of the order of 5 milli-seconds. To obtain faster operation and smoother application some of the other types mentioned in the table were developed. For industrial applications cost becomes an important characteristic but many automatic machine tools achieve their fast cycle times by using clutches and brake motors.

17.3.2 *Hysteresis clutches and brakes*

Hysteresis clutches and brakes utilize the magnetic lag in a permanent magnet rotor to develop output torque. Since this hysteresis torque is independent of speed, these devices permit not only proportional torque control, but either synchronous drive or subject to adequate cooling continuous slip with little torque variation. Since the power gain is low—of about 5 to 20—their use is limited to small instrument devices where they are usually supplied from transistorized, electronic power packs.

17.3.3 *Solenoid operated clutch-brake combinations*

Magnetic solenoid brakes and clutches provide a simple and adaptable means of mechanical control in low-power systems. Working on the same principle as a simple solenoid with a spring return to normal, this clutch acts as an on–off switch for controlling power flow.

Figure 17.1 shows various configurations and applications of this type of clutch and brake. Typically, with the application of a d.c. magnetic field, the central member moves axially to make a strong physical contact with the other cylinder thus developing the required surface locking action up to the specified torque limit. When the d.c. current is removed, the spring forces the contact surfaces apart, disengaging the central member from the other cylinder. Obviously, the roles of the spring and solenoid can be reversed with the spring providing the engagement force and the d.c. current being used for disengagement. In practice it is best to arrange that the solenoid function is less than the spring function; this will save power and reduce life-shortening heat produced when the coil is energized.

Proper selection of magnetic materials, adjustment of spring constants and air gaps are important design considerations. However, with this type of compact clutch/brake it is possible to predict operating characteristics at the design stage; it is also possible to design the unit as an integral part of a servo activator and other similar devices. The small units used in control systems can have response time of better than 5 ms while in the larger sizes it is possible to obtain a power amplification of several thousand so that a conventional low-power circuit can control loads in the kilowatt range.

17.3.4 *Hysteresis clutch-brake combination*

The hysteresis clutch or brake uses the magnetic lag in a permanent magnet rotor to develop output torque. It is a proportional torque control device and permits synchronous driving or continuous slip with little torque variation at any slip differential, provided that the generated heat can be removed. Its control-power requirement is small enough to use a transistorized electronic power pack.

Figure 17.3 shows the components of a typical hysteresis clutch. The rotor rotates in the field of a group of magnetic poles, the flux passing through and magnetizing the rotor ring. When the rotor cuts the field of the input assembly, the resultant discrete poles generated in the rotor ring by the input field lead the poles of the input assembly field. The lead angle depends upon the control excitation of the stator and the hysteresis-loss in the permanent magnet material of the rotor. Attractive forces between the induced poles of the rotor ring and those in the input assembly result in a torque opposing rotation. Efficient rotors use high cobalt content iron alloys.

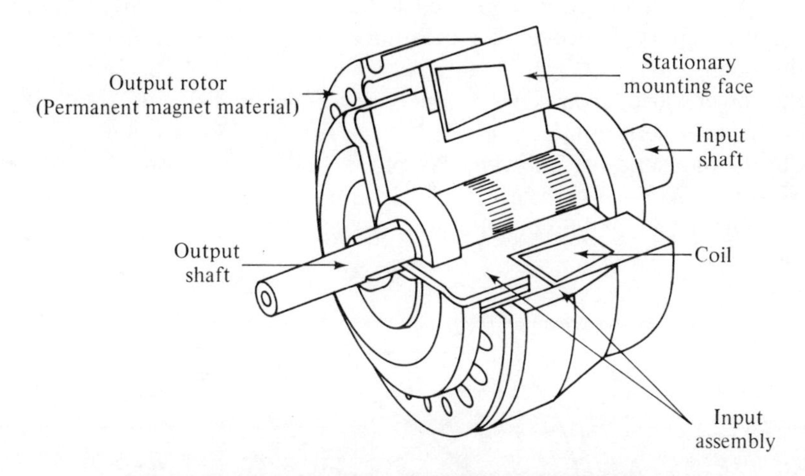

Fig. 17.3 Components of a typical hysteresis clutch

The strength of the input field depends on the field-coil current, where a few watts of power produces magnetic saturation and maximum torque. Typical characteristics of both hysteresis clutches and brakes include:

Output torque independent of speed, except for a slight torque increase (3 to 10 per cent) at the higher speeds because of eddy currents.
Smooth operation due to rotor symmetry, but some noticeable cogging at very low slip speeds due to the discrete poles of the input assembly.
Prolonged life owing to the absence of friction surfaces to generate wear.
Low residual, zero-excitation, drag.
Low power gain (5 to 20).
Indefinitely maintained initial performance.
Millisecond response time.
Synchronous input and output rotation at full load unless unit is overloaded.

Ideal applications for the hysteresis clutch include torque limiting and shock-loads cushioning. Its uniform torque characteristics also permit its use in tension-control devices. The hysteresis brake is excellent for space-borne applications where power is limited and wear unacceptable.

Notable misapplications include duty cycles with rapid reversals that require damping of backlash oscillations. During reversals a deadband period occurs before opposing torques are developed.

17.3.5 *Eddy current clutches and brakes*

Speed-sensitive eddy current clutches and brakes require relative motion between the rotor and stator to develop torque by dissipating eddy currents through the electrical resistance of the rotor ring. Generated torque increases in proportion to the square of the rotor field strength and almost linearly with the speed difference between the input and output members. Because there is no relative motion between the coupling members, the eddy current brake has no holding torque at stand-still. However, eddy-current coupling is well suited to speed control applications and as oscillation dampers. However, for maximum clutch efficiency the amount of slip must be small.

Field power varies from several per cent of output power in small units, to a fraction of a per cent in large units. Eddy current units come in ratings from a few watts up to thousands of kilowatts. Moreover, one can obtain low-slip units and high-slip units. While less efficient, high-slip units can develop relatively high starting torques (Fig. 17.4). Modifying the pole design of the d.c. field supplies various characteristics.

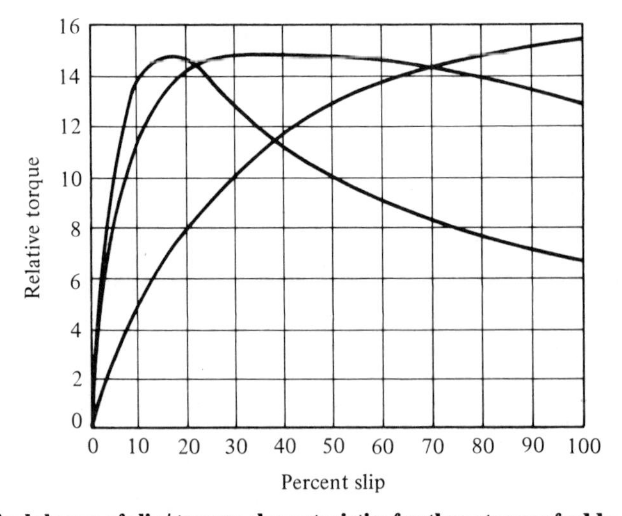

Fig. 17.4 Typical shapes of slip/ torque characteristics for three types of eddy-current clutch

17.3.6 *Magnetic fluid and magnetic powder units*

Magnetic fluid and magnetic powder clutches and brakes consist of two plates separated by a finely divided dry magnetic powder or powder-oil slurry. Normally one plate is part of the input shaft while the other connects to the output. When energized, a field coil surrounding the annular magnetic particle chamber produces a magnetic field that causes the powder to become a rigid mass, oriented along the lines of flux and capable of transmitting torque. The finely divided iron particles cohere with a high shear strength and the extent of the coupling varies directly with the excitation current. Magnetic particle brakes work in the same manner, except that braking action occurs between rotating input shaft and stationary housing. Slipping a clutch generates heat in the clutching space owing to the friction between the magnetic particles. The amount of heat generated depends on the product of speed, torque, and the percentage slip. If clutches are to be slipped, they may have to be derated, taking into consideration the duty cycle, speed, required transmitted torque, heat transfer coefficient of the mounting and ambient temperature. Because many of these factors vary from application to application, it is difficult to establish a general derating curve for any particular unit. Although slipping in the overloaded fluid-type clutch generates heat, it does not cause excessive wear since the sliding magnetic particles and all interior surfaces are lubricated, usually with talc and molybdenum based lubricants.

Because of its highly uniform magnetic particle distribution, fluid type clutches and brakes operate more smoothly than their dry counterparts. However, the fluid makes the unit more prone to leakage through the seals, one of the most critical parts of the clutch. The seals prevent humidity from invading the clutching space, and prevent the magnetic particles from migrating into the bearings, but they also add to the de-energized torque.

Applications for magnetic dry particle clutches and brakes, essentially analogue devices, principally include spooling, tensioning, and positioning devices that require a continuous change of speed. A clutch may be used as a continuously variable speed control in other applications. A second important application is in devices which require very short engagement and disengagement times. Inherently, magnetic dry particle clutches and brakes have fast response times, because of the low inertia of the output member and the lack of any axial or radial movement as the clutch engages or disengages.

Magnetic dry particle clutches are small and deliver high torque. Since these clutches may slip continuously, they find use in bi-directional actuators. In this application, one clutch continuously slips, overpowered by the other clutch. This overpowering action determines the direction of rotation of the output shaft, even though the prime mover maintains unidirectional rotation. Suitable biasing can provide approximately linear control.

Magnetic dry-particle clutches and brakes also find important applications in systems that require a constant torque for a given input, over the life of the clutch. One can use the brakes for maintaining a constant drag regardless of the direction of shaft rotation. Clutches may act as protective torque limiters, energized to transmit a predetermined value, slipping upon a sudden excess demand.

Typical torque–excitation current characteristic for these types of clutches and brakes is smooth and extremely linear in the operating region, as shown in torque vs excitation current plot (Fig. 17.5(a)). Response characteristics of Lear Siegler clutch-brake are shown in Fig. 17.5(b).

17.3.7 *Overrunning spring clutches*

Overrunning spring clutches use the wrapping and unwrapping ability of a coil spring when subjected to torsion, to exert friction on the coupling shafts (see Fig. 17.6). As the outer member rotates in the drive direction the rollers are caused to create an

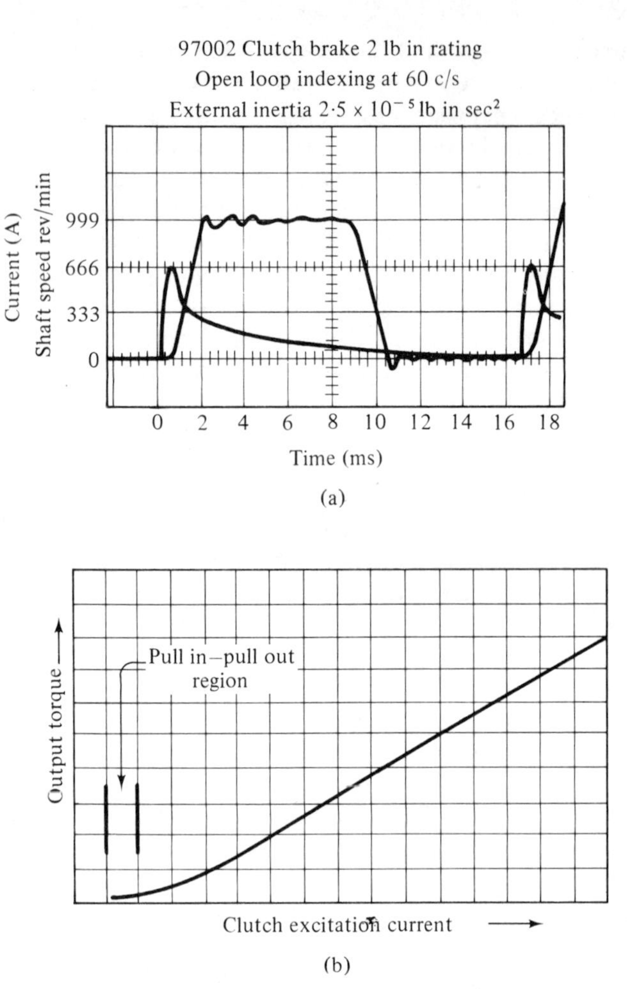

Fig. 17.5 (a) Torque-excitation current characteristic for Vibrac dry particle clutch. (b) Response characteristics of Lear Sieger clutch/brake

interference fit on the driven shaft thereby transmitting motion. When the direction of rotation is reversed the springs retract the rollers thereby freeing the driven member.

This form of clutch can drive heavy masses. The overrunning feature prevents the momentum of the driven mass from driving the driver when the driving mechanism is stopped. It provides accurate increment or indexing motion with a mechanism consisting of a continuous rotary motion crank that oscillates an arm attached to the clutch. The resulting uniform intermittent forward motion corresponds to the indexing that a pawl and ratchet with an infinite number of teeth could achieve. It is particularly adaptable to feeding presses and printing machines.

On–off spring clutches allow selective intermittent output motion from a continuously rotating input. In this application, the driven output coasts to a stop after the input has been disengaged by, say, a solenoid-operated actuator.

Single or partial revolution spring clutches are similar to the on–off type, except that their output stops at approximately the same relative angular position on every

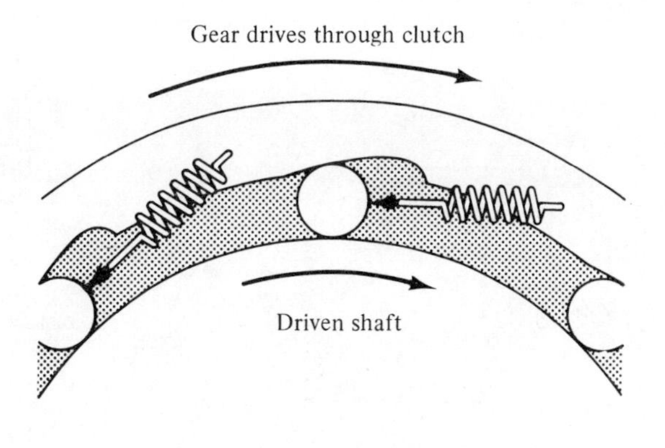

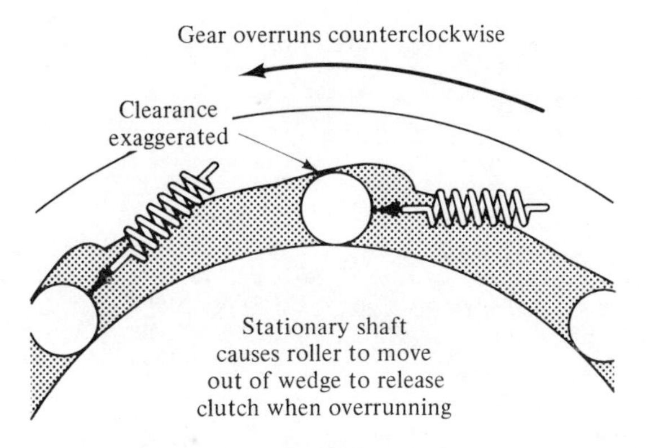

Fig. 17.6 Overrunning roller clutch

cycle with a non-cumulative error. Output accuracy can be maintained to within plus or minus half a degree. Like the on–off type, the single revolution clutch needs an actuator. If two clutches are mounted in series, a bi-direction no-back-motion device can be constructed. By attaching this combination clutch to a stationary member, the input may drive in either direction of rotation, while prohibiting the output from backdriving the input. No outside actuating force is required with a combined unit.

17.3.8 *Air clutch or brake*

This form of device is ideal for power transmission systems; air actuated clutches put machine elements into motion or brake them to a stop rapidly, and do this without shock or vibration while maintaining a precise torque. To some extent the air assists the dissipation of heat with constant slippage. Output torque proportional to applied pressure can be as high as five times that obtained with a conventional magnetic clutch of comparable size.

The device self compensates for wear and requires no adjustment or lubrication, except for the bearings (see Fig. 17.7). Since it generates no heat during idling, it requires very little cooling. Further, since no limiting temperature considerations exist, such as those imposed by windings in a magnetic clutch, a properly designed clutch of this type can operate in environments of 500°C. The air that actuates the clutch can also serve for bearing support if hydrostatic air bearings are used;

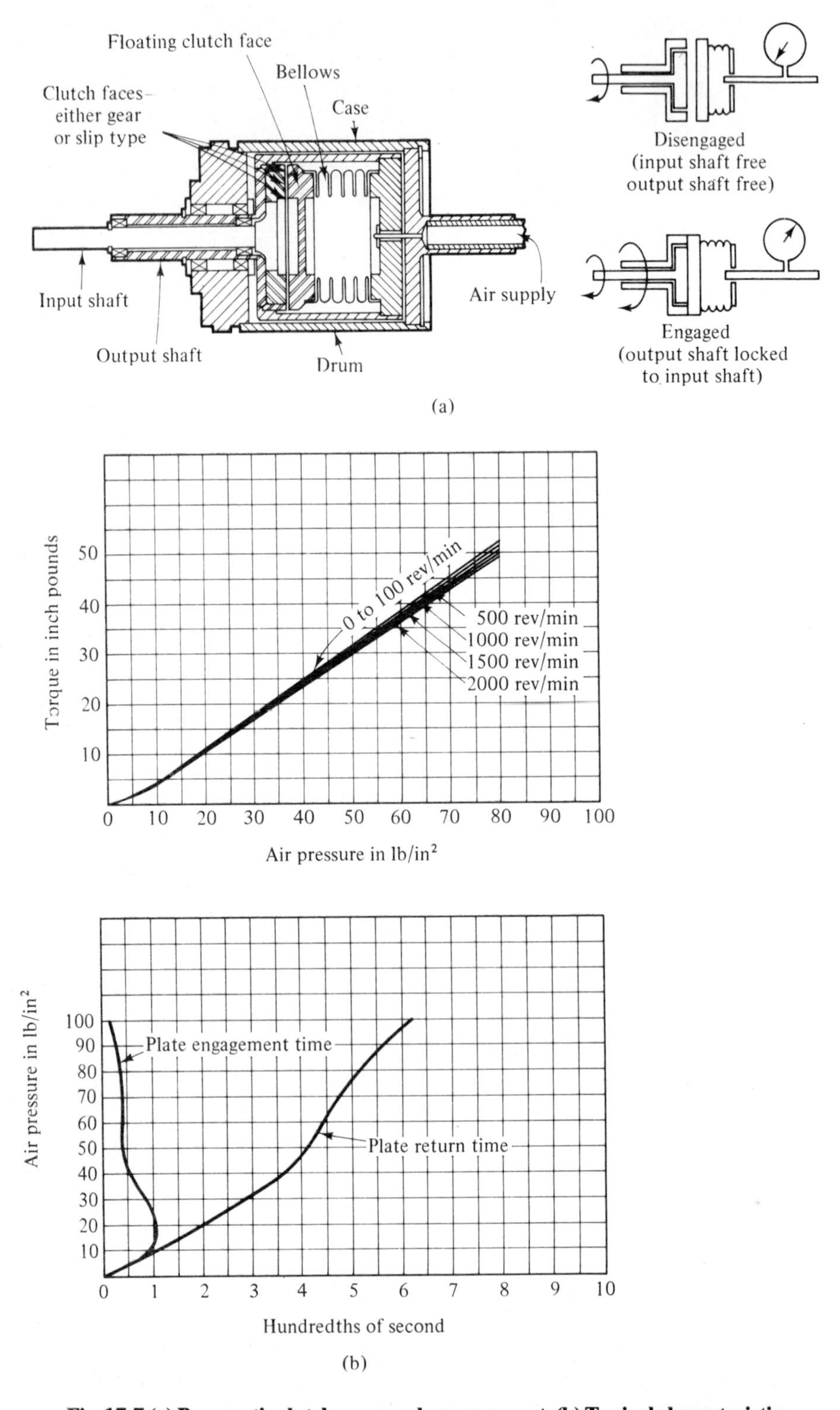

Fig. 17.7 (a) Pneumatic clutch—general arrangement. (b) Typical characteristics

consequently the material characteristics set the limit to high temperature applications and the unit is suitable for working at very high speeds.

Air pressure force exerted on the floating clutch face does not transmit to shaft bearings; it is self contained within the rotating drum. Therefore, magnitude of the applied air pressure does not influence bearing friction. Consumption of air is primarily dependent on the number and frequency of engagements of the clutch. During operation (shaft rotation) there is near zero (minute leakage) air consumption. Except during initial engagement, there is no angular error and therefore zero backlash.

Crown tooth faces on air-actuated clutches can be used where high torque is required and a small engaging error can be tolerated; slip type friction faces serve where there is no engaging error and torque levels can be maintained.

17.3.9 *Spring set frictional braking*

Spring set brakes/clutches and hold loads or transmit torque as a result of spring pressure that clamps two or more frictional faces together. Release may come from mechanical, pneumatic, hydraulic or electrical forces. Usually, these devices take the form of shoe brakes or disc brakes and clutches. The shoe brake usually consists of two hinged cylindrical segments of friction material while the disc units, as their name implies, consist of discs which are forced into contact with the braking member. In both cases the spring may be used to engage or disengage the braking members, the choice being dependent upon application.

Electrically actuated brakes come in a variety of configurations and use spring pressure opposing mechanisms such as clapper magnets, thrusters, and solenoids. They can operate on alternating or direct current and sometimes they are combined with a clutch to form an integral clutch–brake combination.

Spring/solenoid disc brake/clutch packages come in a wide variety of enclosures and devices are available for use in most environments including such as exploding atmospheres. It is probably one of the least expensive types of unit and rugged in construction. In most applications it requires no additional wiring, controls or auxiliary electrical equipment. It is simple to maintain since the expendable items (the friction discs) can be inexpensively replaced with little effort. Many brakes are equipped with features such as simple wear adjustment to provide optimum lining life, wear indicators, torque adjustment, and manual releases. In some cases the units are contrived to automatically adjust for friction disc wear which is particularly advantageous when the unit is located in a relatively inaccessible position.

Electric disc clutches can have single or multiple discs operating with stationary or rotating field coil design. For a given torque level, the multiple disc clutch is generally smaller in diameter but longer axially than the single disc unit. Stationary field coil designs eliminate the need for brushes and slip rings that transmit the electrical energy in the case of a rotating coil design.

Figure 17.8 shows a multiplate disc brake with a three-phase release magnet. The advantage of a.c. electromagnets is their shorter electrical time constant and therefore their faster response. They also obviate the need for rectification of the a.c. supply or the provision of a suitable d.c. supply. The multiple disc clutch or brake provides not only a more powerful and consistent braking torque, but enables better heat dissipation. The a.c. magnet is less efficient and the disadvantages of having to

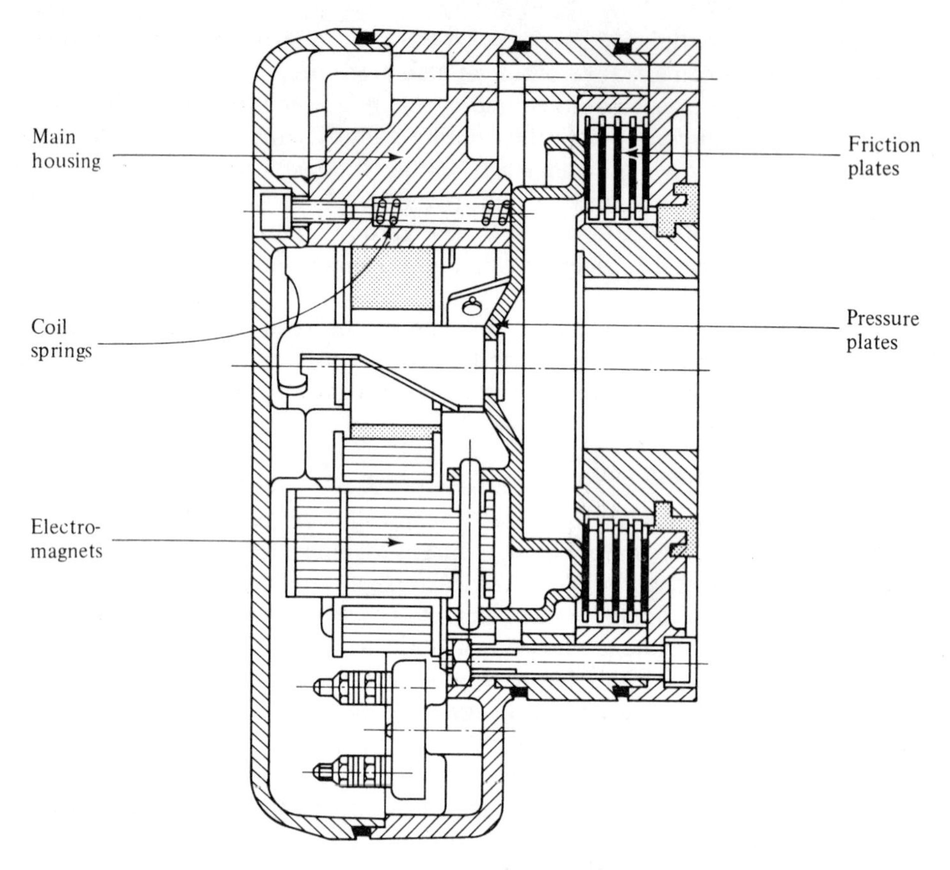

Fig. 17.8 Multiplate spring-set brake with a.c. electromagnetic release

provide substantially improved cooling must be weighed against the absence of rectifiers, special switch-gear and generally a slightly faster operation when the supply is disconnected or manually controlled.

17.4 Friction Materials

The reliability of contact represents the biggest single application difficulty in friction disc clutches and clutch/brake devices. Contact surfaces may be of metal-to-metal varieties or non-metallic materials such as cork or silicone rubber. It is very important to specify precisely the actual application and environmental conditions in which a friction unit is to operate in order to ensure that the correct mating materials are used. While continuing research is producing better and more reliable friction materials, none of these are universally applicable.

It is frequently necessary to 'run in' friction devices in order to guarantee a uniform torque characteristic. This may consist of slipping the engaged surfaces a specified number of times until a levelling off occurs in the torque characteristic. Traces of contamination, such as oil on friction faces, result in torque reduction.

Crown and jaw tooth faces, offering a positive lock, more approximate to couplings than true clutches. Some models develop torques 400 per cent higher than comparable smooth faced types. Their one disadvantage is the need to ensure that the meshing teeth align perfectly on engagement; consequently, during engagement, a certain amount of indexing is unavoidable. Actuation at high speed can severely damage the engaging faces and in the extreme case completely strip the teeth. Manufacturers do not recommend even the more expensive models, especially hardened to withstand abuse, for speeds exceeding 300 rev/min.

17.5 Brake Motors

The brake motor is a very useful composite device in which the brake is integral with the motor and requires no separate excitation from the motor. The motor can be of virtually any type and has a tapered armature which can move axially. The brake element is mounted on the armature shaft and, when the machine is stationary, is held in contact with the stationary brake member by the action of a spring. When the motor is excited, the reaction between the stationary field windings and the tapered armature draws the armature into the winding cavity against the action of the spring; this has the effect of releasing the brake and allowing the motor to rotate. When excitation is removed from the motor the spring pressure forces the braking surfaces into contact thereby bringing the motor to rest (see Fig. 17.9).

In this kind of device, as the brake is integral with the motor, application and suitability is a function of the motor rather than the brake. It logically follows that if the motor is suitable for the application in terms of power, control, and speed, the

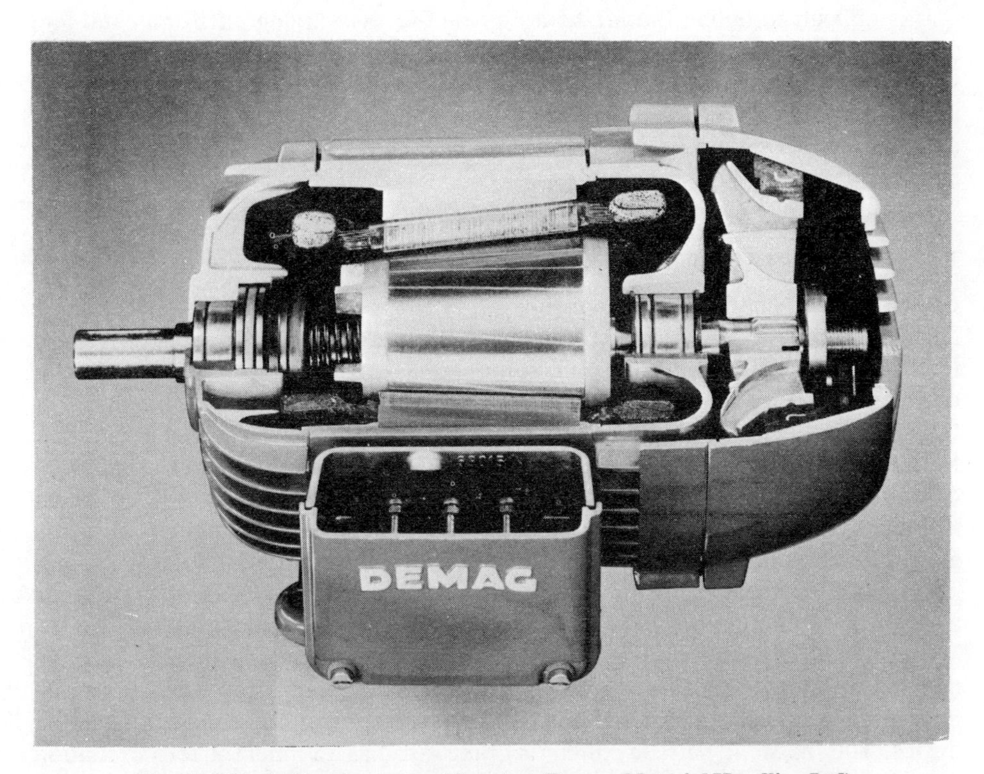

Fig. 17.9 Typical brake motor. (Courtesy Demag Material Handling Ltd)

integral brake will have been designed to operate under the same conditions. Consequently, with this type of device, over and above the basic motor characteristics one need only consider the time to release, time to stop, and the total heat dissipated by the unit to determine whether the unit is suitable for a given system application.

17.6 Life and Reliability

Bearing failures under severe environments limit clutch reliability; contamination in spite of bearing shielding becomes almost unavoidable. Moisture condensation under humidity conditions can cause some small fall-off in torque, especially on brake surfaces of clutch/brake combinations. Good internal shielding, rugged plating, and non-oxidizing brake surfaces contribute to resist severe environments.

Hysteresis, eddy current, and magnetic particle clutches and brakes have longer life expectancy than clutches of other types and they maintain a given torque for a given input voltage over the entire life of the clutch. However, like all mechanical devices, hysteresis and magnetic particle clutches and brakes eventually fail. Failure occurs almost invariably due to bearing or seal failure. Consequently, while clutches and brakes of this type already have a long life expectancy, reliability can be significantly improved by specifying high grade bearings and seals. Additionally, shaft misalignment, excessive radial loading, and high temperature operation should be avoided as this will materially affect the seal and bearing life.

It is difficult to define the life of a unit due to the varying environmental and operating conditions. When a manufacturer quotes a life expectancy for a device he will either quote or assume certain operating parameters; deviation from these parameters can significantly affect the life expectancy and where there is the slightest doubt reference should be made to the manufacturer for confirmation of device suitability.

17.7 Disengaged Friction and Drag

Clutch friction generally exceeds that of the bearings alone. In the case of magnetic clutches, the additional friction comes from the heavy magnetic forces developed internally in order to operate the unit. Some clutch designs impose heavy thrust loads on the bearings which sharply increases friction and reduces life at high speed. For this reason, one may specify 'no load friction' (i.e., with no externally applied shaft torque) separately from the frictional load of the unit in the energized and de-energized state.

Some manufacturers minimize bearing loads by having only radially directed magnetic forces between the moving members and the frame. These radial forces balance out to avoid thrust on the bearings; however, the inertia of these units exceeds those of other designs.

Because of the nature of hysteresis clutches or brakes, unavoidable residual magnetism in the output rotor results in de-energized drag. However, this can be minimized by applying a reversed polarity input after each normal application of the device. The magnetic particle clutch or brake also has an inherent zero excitation drag caused by the friction of the magnetic particles and the seals.

17.8 Excitation and Switching

It is possible to vary the applied voltage rating for magnetic devices over a wide range by specifying a suitable excitation coil winding. As long as the total copper volume of the coil remains constant, the power rating remains unchanged. However, the rated applied voltage is proportional to the number of turns, which in turn depends upon the wire size. The principal limitation, the fineness of wire winding, sets the upper voltage limit in miniature components where available wire room is limited.

Since magnetically actuated devices depend upon ampere-turns, response diminishes at high temperatures where increase in winding resistance reduces the current and the power in accordance with Ohm's law.

Magnetic devices may be energized from an a.c. supply either directly or through a suitable rectifier circuit. The major difficulty encountered in high-speed low-line-frequency applications is dither resulting from the pulsating character of the supply or rectified current; although the coil inductance provides some measure of smoothing, in certain applications this may not be sufficient. Unfortunately, filtering

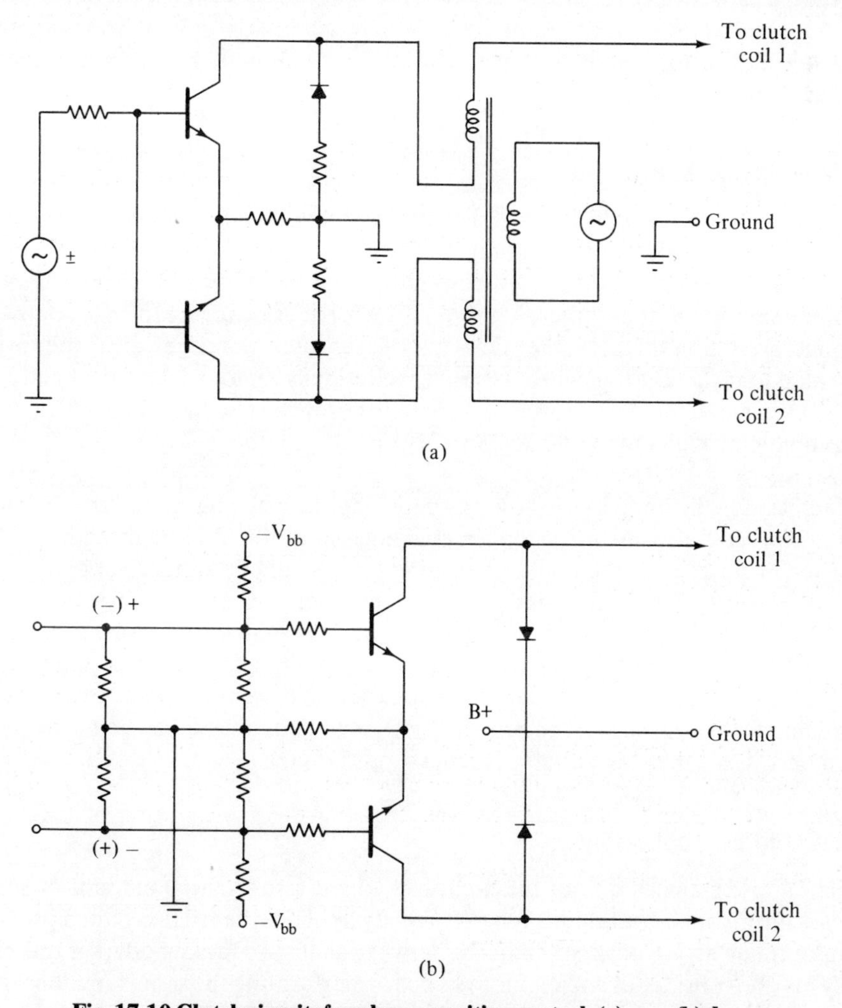

Fig. 17.10 Clutch circuits for phase-sensitive control. (a) a.c.; (b) d.c.

circuits introduce time delays that add to those already occurring within the unit due to the inductance of the coil.

Voltage ratings for direct-acting clutches or brakes range from 200 V d.c. or higher to 6 V d.c. or less. The supply for the higher voltage rated devices is usually obtained through a simple half wave rectifier while the low voltage devices are suitable for transistor control. To facilitate inclusion in mobile systems and aircraft applications, a large number of devices are designed to operate from 12 V d.c. and 28 V d.c. Figure 17.10 shows basic examples. Sensing circuits using a.c. or d.c. can activate either of these two clutch circuits for phase-sensitive control. The addition of feedback, stabilizing means, and a preamplifier enables the resulting system to control a variety of functions. If input is low level, then you must amplify input signal; if input source has very high impedance, you must buffer by an emitter-follower amplifier. Transistors driving clutches act as switches.

Clutch excitation signals must be d.c. to achieve high forces at a high efficiency. In ordinary single-ended control circuits, a full wave rectifier bridge can supply direct current; a capacitor resistance filter normally smoothes the clutch chatter caused by the a.c. excitation current component. Where excitation comes from a polarity-sensitive a.c. error detector, one must use a combination of a phase-sensitive detector and a d.c. amplifier.

17.9 Response Time

In many high speed-applications and control systems, clutch or brake response speed becomes a key design consideration. Major factors affecting response time include the current build-up lag in inductive windings and the delay in response of moving components due to their inherent inertia. The force of restoring springs opposes the applied excitation force, further delaying the response time; however, the spring is essential to fast operation when external excitation is removed. Unfortunately, as in all electro-magnetic devices, the requirements of high speed and high efficiency are incompatible. Current build-up lag corresponds to the conventional L/R time constant.

Increasing efficiency by closing magnetic gaps or packing more copper into the energizing coil also means a longer coil time constant. Although adding a large resistor in series with the coil and exciting from a higher voltage supply increases speed, power is lost in the series resistor. A parallel R–C network in series with the coil speeds response more efficiently than a series resistor by passing a high initial current through the uncharged capacitor. One can also adopt other techniques from magnetic amplifier theory to adapt response time. Where a constant current source is available for energization, the principal lag occurs due to the inertia of the moving members. Careful design minimizes inertia and total travel.

17.10 Clutch Applications

Friction disc clutches do not function well where a substantial amount of slipping occurs during their useful life. When correctly applied, the friction clutch provides a firm coupling action between shafts or between shaft and brake body; for this reason they may be considered analogous to switches, performing the same function within a mechanical system as a switch does in an electrical circuit. However, it should be noted that a great variety of types exist from which one can make a suitable choice for a given application.

The location of a clutch in a system is generally dictated by the system itself but one may have some choice as to whether it should be installed on the high or low side of a speed reducer. Positioning a clutch on the high-speed side reduces the torque requirements, permitting the use of a smaller device; however, a larger, cheaper clutch on the low-speed side dissipates heat more efficiently. Where there is a step down in speed from motor to load, the various elements between the motor and the clutch do not have to be accelerated at the same rate and are not subjected to shock of starting and stopping.

17.11 Clutch/Brake Selection

While specifications vary from application to application it is useful to tabulate performance parameters in order that a fair comparison can be made between competing units. To facilitate this, a list of the 22 main performance parameters is given below which generally covers most applications.

1. Type
2. Diameter
3. Length
4. Clutch torque
5. Brake torque
6. No-load torque (unenergized)
7. No-load torque (energized)
8. Backlash
9. Inertia (input)
10. Inertia (output)
11. Power input
12. Response time
13. Thermal capacity
14. Mounting preparations
15. Wear adjustment
16. Life for:
    - Load inertia
    - Load friction
    - Load slip considerations
    - Overload period
17. Duty cycle
18. Speed
19. Current source
20. Voltage
21. Environmental conditions
22. Applicable military specs

Once the most suitable unit has been selected, the following precautions will ensure optimum performance:

(a) Make certain that the supply lines to the device are of adequate size to handle the inrush of current on starting without excessive voltage drop.

(b) Provide good mechanical alignment between shafts.
(c) Ensure that all mounting bolts and screws are securely tightened.
(d) Periodically inspect wear indicators or friction surfaces and adjust as necessary. This is not required on devices equipped with automatic wear compensation.
(e) Maintain a relatively constant voltage to the device; usually a variation of 10 per cent is permissible.
(f) Do not increase cycling speed beyond the capacity of the device.

18 Control Motors

18.1 General Introduction

The motors considered in this chapter have, in general, ratings ranging from a few watts to 3 to 4 kilowatts. Their continuously rated output is mostly less significant than their ability to produce large peak torques with low-inertia rotors or armatures in response to the voltages applied at their terminals. Their function is to control by responding to and being themselves controlled by signals which represent the difference between an actual and a desired condition. According to the *Oxford English Dictionary* the term 'servo-motor' dates back to 1899, and was then described as an auxiliary motor, e.g., one used for directing the rudders of a Whitehead torpedo or the reversing gear of a marine engine. Today the servo-motor can be defined as any motor suitable for remote control and in which the response characteristics are sensibly identical in both directions of rotation. The range of types, constructional, and performance variants is almost as great as the manifold applications.

The order of weighting of the various performance and constructional characteristics specified for motors used in control is largely a function of application. Thus, motors used in the aerospace industry in general and navigational instruments in particular must give maximum performance for minimum weight in a well-defined environment and with assured skilled maintenance. Industrial instrumentation and instruments demand motors which are less costly yet are more robust, better protected against hostile environment and free from maintenance for long continuous periods of operation. This can usually be achieved by increasing the size of motor and thus reducing the difficulties of manufacture of the servo-motor.

Servo-motors are used in the numerical control of machine tools, through paper or magnetic tape, which is itself transported by a special motor able to start and stop rapidly and then run at constant speed. Military applications and later space research stimulated very thorough studies of control systems and the development of special components. To achieve excellence of performance and maximum resistance to adverse ambient conditions, materials were specially produced and manufacturing tolerances reduced to a minimum. Such devices were not only too costly for industry but proved unreliable because they generally lacked endurance and could not be given the necessary skilled maintenance. The control motors used in industry today are, therefore, designed to give the required performance at economical prices, and run continuously without maintenance for many thousands of hours. As such motors with special features are usually manufactured in smaller numbers they are inevitably more costly than absolutely standard motors of the same output rating.

The selection of the most suitable motor for a given application cannot be made without considering its method of control and coupling to the load. Whether the system is required to control position, speed or constitute a phase-locked loop, that is speed and position for absolute synchronism, the required performance will usually be obtainable with various different motors. Incremental applications are not necessarily best solved with stepping motors, particularly when for fast traverse the stepping rate begins to introduce stability problems. Similarly in applications requiring constant speeds synchronous motors may not always offer the best solution.

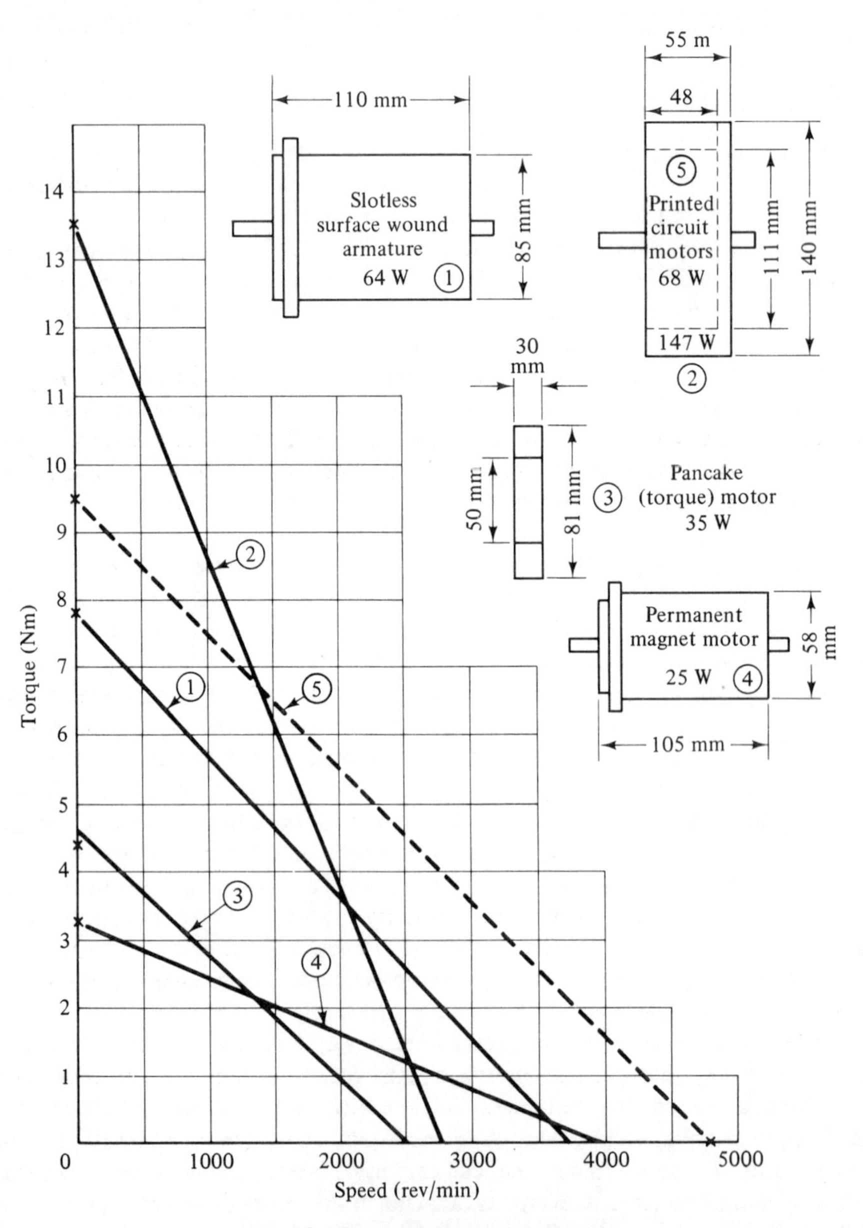

Fig. 18.1 Permanent magnet motors

In control motors especially, their geometry can play a significant part in the design of the equipment. The printed circuit motor's short axial length is a feature which may strongly influence its choice for certain applications, but an almost identical performance is obtainable from a motor with a low inertia cylindrical armature. The graphs illustrating the versatility of the permanent magnet motor in Fig. 18.1 have been adapted from catalogue data and, therefore, may not be the best match in performance obtainable. While the system designer will obviously strive to obtain the best performance, commercial aspects such as price and assured availability must receive due consideration in commercial designs.

In printed circuit motors and d.c. motors with low-inductance armatures brush life is largely determined by speed. The precise construction and careful installation requirements of pancake motors limit their application mainly to navigational and scientific instruments such as telescopes where speeds are so low that brush life is usually no problem.

It should be noted that the motors chosen to illustrate the versatility of permanent magnet d.c. motors do not necessarily represent the most commonly used characteristics or output ratings. The printed circuit and surface wound armature motors have particularly low armature inductances which allow them to be pulsed with currents several times their normal starting current. Thus the 68 W motor is capable of peak torques up to 24 Nm and represents a motor with a higher than usual no-load speed. If the field is made stronger the maximum torque will be increased at the expense of speed. The ratios of starting torque to maximum peak torque for the printed circuit motors are 13·6 and 22, respectively. These torque pulses are valuable for fast response systems, but the production of the current pulses naturally requires a specially designed control amplifier.

18.2. Alternating-current Servo-motors

18.2.1 *Introduction*

Most a.c. servo-motors are two-phase induction motors, with high-resistance, low-inertia rotors in the fractional to sub-fractional output range. The production of a rotating field by a symmetrical two-phase winding as well as the principle of starting single-phase induction motors is dealt with in chapter 4. Servo-motors which were developed from such special electromagnetic devices as Magslips and 'M' motors for servo mechanisms and automatic control systems usually have one phase—the 'reference' winding–permanently energized by the supply and the other phase, or control winding, connected to a power amplifier. Motors with diameters of 15 to 60 mm and stall torques of between 0·7 and 70 mNm were developed extensively in the 1940s for military instrument applications which necessitated the use of 400 Hz as well as 50/60 Hz supplies.

Such motors subsequently found many other applications of which potentiometric recorders, plotting tables, navigational instruments, remotely controlled cameras, machine tools, and computer peripherals are typical examples. In the machine tools and process control systems as used in nuclear reactors, automatic acceleration controllers, and valve positioners larger powers and torques are required, and motors with stall torques of 1400 mNm (1·4 Nm) are now being manufactured.

The interdependence of the characteristics of individual elements in systems and the overall effect on component selection is illustrated by the effect of the developments in, for example, power electronics and permanent magnet materials. Thus, in some fields two-phase servo-motors are now being replaced by stepping motors or permanent magnet d.c. motors which do not have the same problem of the heat losses in a reference phase when the system is balanced.

18.2.2 *Characteristics*

The effect of rotor resistance on the speed/torque characteristic of an induction motor is dealt with in chapter 4, section 4.4. Figure 18.2 shows typical curves and attention is drawn to the effect amplifier characteristics may have on curves obtained with a perfect, low source impedance two-phase supply. Many practical factors such as manufacturing tolerances, effects of ambient temperatures as well as supply

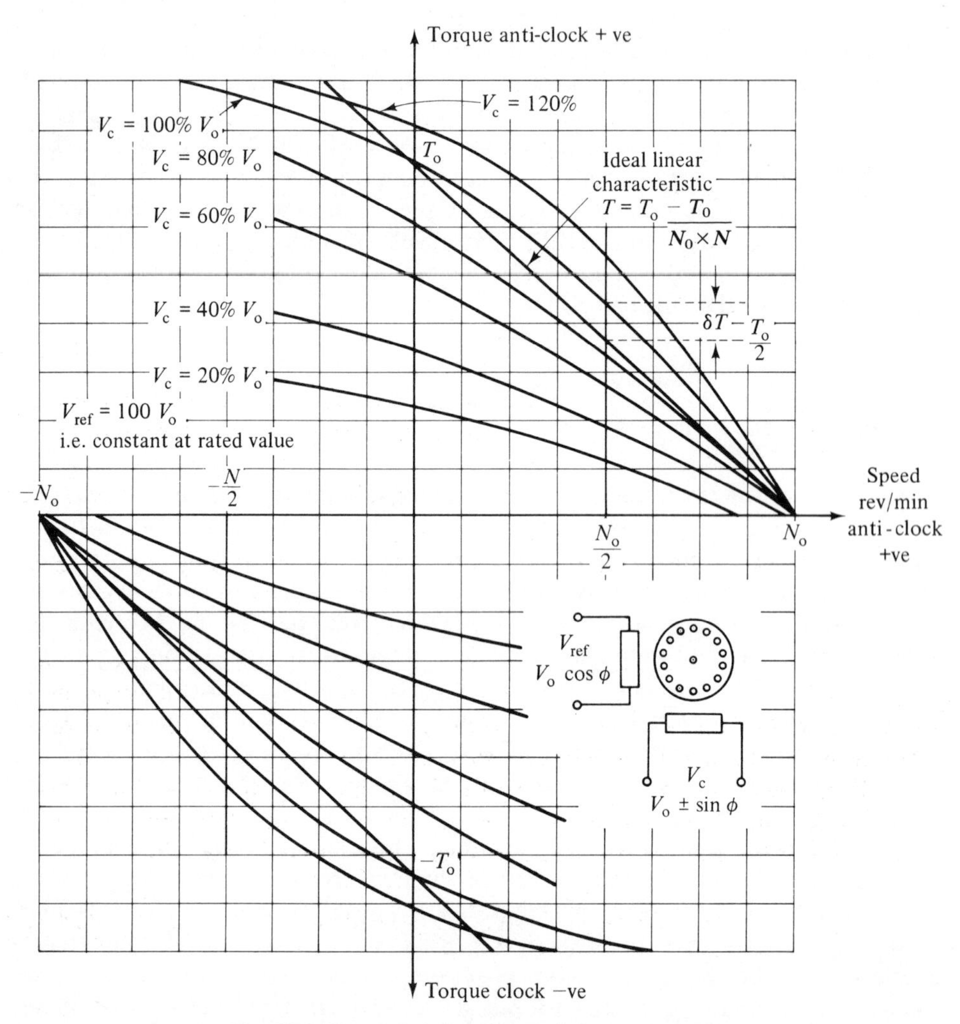

Fig. 18.2 Characteristics of two-phase servo-motor

variations must be allowed for in design calculations, and these are dealt with in detail in the more specialized literature.

The following is a summary of some more important formulae and definitions of the parameters:

18.2.2.1 Speed/Torque. At balanced two phase operation, i.e., when $V_C = 100$ per cent $= V_R$.

In absolute terms

N = speed in rev/min δT = deviation at $N_0/2$ (see Fig. 18.2)

T = torque Nm.

$$T = T_0 - \frac{T_0 - 4\delta T}{N_0} N - \frac{4\delta T}{N_0^2} N^2 \text{ Nm} \tag{18.1}$$

This equation assumes that the actual speed/torque curve can be represented by a quadratic equation of the form:

$$T = T_0 + DN + MN^2$$

If this curve deviates from the actual curve plotted from test results then further terms must be added to the equation.

In dimensionless per unit terms
where

$t = T/T_0$ δ = curvature at half per unit speed

$n = N/N_0$

$$t = 1 + (4\delta - 1)n - 4\delta n^2 \tag{18.2}$$

Considering now the family of speed/torque curves produced by varying the voltage applied to the control winding the equation for the per unit torque is:

$$t = 2v_c + \frac{d}{2}(1 + v_c^2)n + 2v_c mn^2 \tag{18.3}$$

where v_c = per unit control voltage and d and m determined again from the balanced condition per unit speed/torque curve.

18.2.2.2 Damping factor. The initial slope of the balanced input torque curve or coefficient of n is a measure of the motor's inherent damping factor, which expressed as a function of the control voltage v_c is given by $d/2(1 + v_c^2)$. If the damping factor is plotted against v_c for $n = 0$, it will be seen that its value at very low control voltages is only half of the balanced conditions value. For comparing motors and as a first approximation the damping factor can be calculated from the stall torque and the no-load speed at rated control voltage 100 per cent v_c from the equation

$$D = k\frac{T_s}{N_0} \text{ Nm seconds per radian} \tag{18.4}$$

If T_s is in:

Nm then $k = 9{\cdot}55$

If T_s is in gf cm	then	$k = 9{\cdot}36 \times 10^{-3}$
ozf in	then	$k = 6{\cdot}74 \times 10^{-2}$
dyne cm	then	$k = 9{\cdot}55 \times 10^{-7}$

The inherent damping factor which is analogous to viscous friction may not be adequate in high gain systems and a tachogenerator must be used. If a servo-motor with its reference phase energized is driven it will act as a generator and a voltage which is proportional to the speed will be induced in the control phase. To reduce inertia and harmonics tacho rotors are usually in the form of a copper cup attached to the end of the motor shaft. A stationary stack of laminations which in some forms may carry the output winding completes the magnetic circuit. Another method of stabilizing high-speed high-acceleration systems is to allow a highly conductive cup to rotate in the field of a powerful permanent ring-magnet which is free to rotate. This method, known as inertial damping, is generally only applicable to high precision instrument systems where the freedom of rotation of the magnet can be maintained at its proper value.

18.2.2.3 Electrical breakaway torque (*starting voltage*). This is the value of voltage which has to be applied to the control phase of an unloaded motor to cause the shaft to commence and continue to rotate from any initial angular position. For the test the reference phase should be energized at rated voltage and frequency.

18.2.2.4 Maximum power output. The maximum power output of a servo-motor is said to occur when, with both phases energized at their rated values, the shaft speed is reduced to one-half of its no-load value by the application of a load. Its numerical value may be calculated from:

$$P_0 = kT_mN \text{ watts} \tag{18.5}$$

T_m = value of torque developed at half no-load speed rev/min	value of k
Nm . . .	$0{\cdot}1047$
gf cm . . .	$1{\cdot}027 \times 10^{-5}$
ozf in . . .	$7{\cdot}396 \times 10^{-4}$
dyne cm . . .	$1{\cdot}048 \times 10^{-8}$

18.2.2.5 Time constant. The time constant is the time required for the shaft to reach 63·2 per cent of the no-load speed after the control winding has been energized by the rated voltage; the reference winding already being energized at the rated voltage. It is equal to:

$$TC = J_r/D \text{ seconds}$$

where J_r is the rotor moment-of-inertia in kilogramme-metres^2, D is the damping factor.

18.2.2.6 Reversing time. The reversing time is the time required for the shaft to reach 63·2 per cent of the no-load speed upon phase reversal of the control winding voltage when the shaft has initially been rotating at no-load speed in the opposite direction. It is approximately equal to:

$$RT = TC \times 1{\cdot}69 \text{ seconds, where } TC \text{ is the motor time-constant.}$$

18.2.2.7 Theoretical synchronous speed. As for all induction motors the theoretical synchronous speed is equivalent to:

$$N_s = \frac{\text{supply frequency (Hz)} \times 60}{P}$$

where P is the number of pole pairs of the motor, N_s is the theoretical synchronous speed in revolutions/minute.

18.2.2.8 Velocity constant. In order to assess the ability of the motor to follow a uniformly changing command signal the relation between speed and control volts has to be known.

$$\text{velocity constant} = \frac{\text{no load speed in radians/s}}{\text{control volts}}$$

where the velocity constant is, therefore, expressed in radians per second/volts, no load speed is expressed in radians per second.

18.2.2.9 Theoretical acceleration at stall. The theoretical acceleration at stall is an indication of the ability of the servo-motor to accelerate from the stall conditions when energized with the rated voltage on the reference and control windings. It is equal to:

$$A = T_s/J_r \text{ radians per second}^2$$

where

T_s is the stall torque in newton-metres,
J_r is the rotor moment of inertia in kilogramme-metres^2

18.2.3 *Construction*

Alternating current servo-motors are in many cases of very similar construction to ordinary fractional motors. Because they are required to have special characteristics their construction is more precise and materials and components rigidly controlled. Thus, to supply and withstand high starting and accelerating torques, shafts, which are normally splined, are made from hardened corrosion-resistant steels. The rotors they carry may be made up from special laminations into which a pure aluminium rotor is pressure die-cast. The whole assembly is carefully ground to ensure concentricity and precision bearings are selected to ensure that their torque loading does not increase when the motors are fully assembled. The influence of military applications resulted in the creation of a standard range of servo-motors now widely known as international or NATO frame sizes. These are common with other rotary components such as synchros, resolvers, tacho-generators, and d.c. servo-motors. The standardization of spigot mountings and associated clamping hardware, shafts, splines, and principal performance characteristics has made some contribution to reduction of cost and increased availability. Typical sizes in this range are 08, 11, 15, 18, 23, 31, and 37 with respective diameters ranging from 20·3 to 94 mm. The designation of the frame-size is the maximum body diameter to the next highest tenth of an inch.

The housings of the international frame size motors are of stainless steel with accurately machined spigots at the drive end. The preferred mounting method makes use of special clamps which fit into a groove on the driving end on the smaller motors and a clamping ring on the larger frame sizes.

The stator and rotor laminations are made from corrosion resistant nickel irons and special assembly techniques are used to avoid impairing the uniformity and quality of the iron. For the stator windings copper wire with suitable covering such as polyester is used and the stator assembly is then impregnated with a high temperature varnish. Normally materials are selected to achieve class F (155°C) but class H (180°C) motors employing special insulating materials are available for special applications at extra cost.

In the smaller frame sizes terminals are difficult to accommodate and such motors are frequently supplied with 'fly leads' which are brought out directly from the stator.

18.2.4 *Specifications and standards*

Servo-motors are not covered by the British Standards Institution or international standards dealt with in chapter 5, but because of their applications the international frame sizes mentioned are the subject of specifications issued by the Defence Departments of the United Kingdom and the United States.

Within NATO, member countries have been keenly aware of the need to use standard instruments, and in recent years a series of documents NETRs (NATO Electrical Technical Recommendations) have been produced covering the whole range of servo components. NETR documents are not generally available but are used as a basis of the specifications which are made available through the usual channels to manufacturers of motors and equipments incorporating them.

Current documents covering the international frame sizes are:

UK Ministry of Defence

DEF STAN 61-14 (Part 1): Precision Instrument, Rotating, Servo-Components Part 1—General Requirements and Common Tests.

DEF STAN 61-14 (Part 2): Precision Instrument, Rotating Servo-Components Part 2—Motor, Control a.c. (Servo-motor—Two Phase Induction).
Precision Instrument, Rotating Servo-Components Part 3—Servo-motor Tachometer-Generators a.c. (Two Phase Induction).

US Military Specifications

MIL-S-22432: Servo-motors General Specification.

MIL-S-22820A: Servo-motor Tachometer Generator a.c. General Specification.

18.3 Direct-current Servo-motors

18.3.1 *Introduction*

The theory and principal types of d.c. motors are dealt with in chapter 3. Because of the ease with which they can be controlled over a wide range of speeds and their inherent high torques they have been used in closed loop control systems in sizes covering outputs from a few watts to hundreds of kilowatts. The following summary of d.c. motors will only deal with motors with maximum outputs of a few kilowatts. Among the advantages of d.c. motors are:

(a) higher output than from a 50/60 Hz motor of the same size;
(b) can be controlled from a smaller amplifier than an a.c. motor with the same output;
(c) easier to stabilize;
(d) more efficient, particularly in variable speed applications.

Many of the disadvantages of d.c. motors can now be overcome by the very compact semiconductor control amplifiers and advanced designs of motors.

In the mid 1960s several ranges of high performance d.c. servo-motors appeared on the market. Designers had, with the help of new improved materials and sophisticated production techniques, been able to extend the principles of motor design and produce motors with the required special characteristics. Small diameter surface-wound armatures, ironless cylindrical and disc type (printed circuit) armatures had low inertia as well as low inductance, and could be placed in powerful fields provided by new improved permanent magnets. Such configurations produced significantly greater torque to inertia ratios and faster build-up of maximum torque. To the servo-system designer this meant torque constants (K_t) and voltage constants (K_v) which combined to give mechanical time constants of a few milliseconds. The figures of merit obtained were comparable with those of hydraulic motors, but before a final selection can be made the system must be considered as a whole since in both cases special supplies are necessary to provide the energy.

18.3.2 *Shunt motors with split-field control*

In the early days of servo-system development, power-amplifiers with outputs of more than a few watts were not readily available so that a range of field-controlled motors with relatively large field stacks was specially developed. The fields which were (as shown in Fig. 18.3(a)) connected in push-pull carried quiescent currents of 40 mA while the armature was connected to either a 24 V or 220 V d.c. supply through a current-limiting resistance. The substantial losses in the motor and in the external resistance were only acceptable as long as no better alternative was readily available. Typical speed/torque curves are shown in Fig. 18.3(b).

18.3.3 *Split series motors*

This type of motor was used with either relay or thyraton control amplifiers and gave similar characteristics to the above-described field controlled motor. This system at least avoided the large losses or costly current limiter in the split field motor which

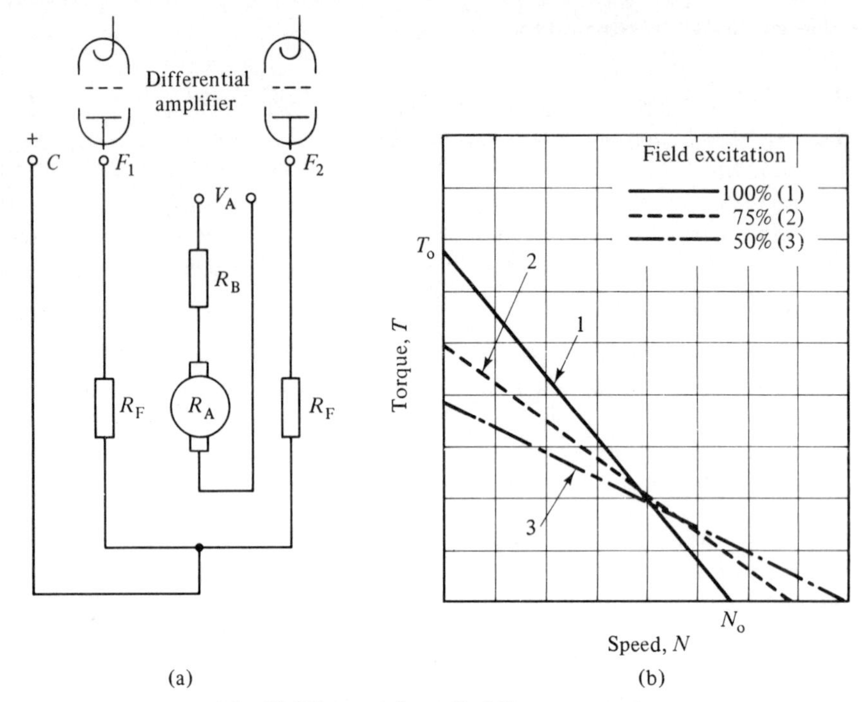

Fig. 18.3 Separately excited d.c. servo-motor

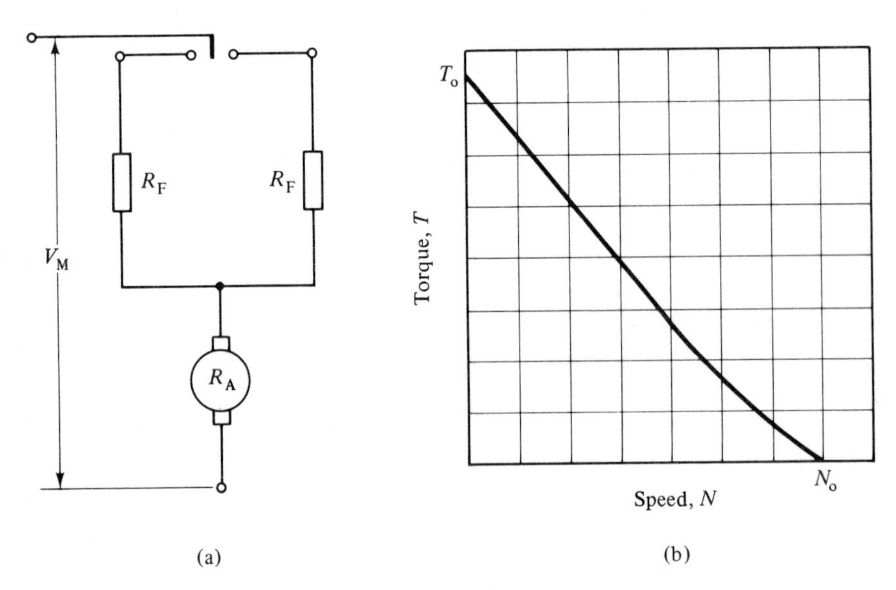

Fig. 18.4 Split series d.c. servo-motor

kept the armature current at balance to a safe value. The internal losses of the motor are still high and larger motors are cooled by a separate blower mounted either axially in line or on the outside with its shaft parallel to the armature shaft of the servo-motor, Fig. 18.4.

18.3.4 *Special d.c. motors for electronic control*

In many automatic control systems which either involve position or speed control synchronization the power for the motor is nowadays derived from the mains supply and fed to the motor through controlled rectifiers, (s.c.r's, thyristors, etc) so that the motor has to cope with a d.c. supply which contains considerable ripple. This and the rapid rise of field and/or armature currents causes additional heating and stresses on the windings. Before ordering controls and motors from separate sources the buyer should satisfy himself that the two are in fact compatible and will produce the required performance without detriment to either motor or controller.

These motors as well as the complete systems have been studied in detail and their performance can be evaluated and compared by means of the formulae and analytical methods developed. The return of the d.c. motor as one of the most widely used prime movers in control systems is not only due to the many applications which stimulated further developments but, in sizes up to several kilowatts, attributable to considerable advances in permanent magnet performance.

18.3.5 *Permanent magnet d.c. motors*

The types dealt with in this section are those with a cylindrical wound armature, but again attention is drawn to printed circuit, moving coil, and torque motors all of which develop torque by the interaction of magnetic fields and current carrying conductors specially arranged to produce motors with characteristics making them particularly suitable for certain applications. The saving in copper losses is considerable and the absence of field windings will improve robustness and reliability and at the same time reduce costs.

Typical performance curves are shown in Fig. 18.5, and from these it can be seen that for intermittent ratings the considerable starting torque of the motors can be exploited. Under such conditions it is, however, advisable to monitor the armature temperature, and at a given temperature prevent the control equipment from further overloading the motor.

One of the limiting factors in the performance of permanent magnet d.c. motors, namely the demagnetizing effect of large armature current transients, has been almost eliminated with the introduction of ceramic magnets. Though such magnets can only match the energy product of the conventional Alnico magnet their resistance to demagnetization is such that they can be used without the usual partial demagnetization to ensure stability.

The recently introduced rare-earth magnets (samarium cobalt, $SmCo_5$) may not only produce very significant improvements in d.c. motor performance, but by making it possible to place the armature windings on the stationary part of the motor may actually revolutionize permanent magnet servo-motor design. Recently published figures indicate that the energy product of the samarium cobalt can be as much as three times that of the best Alnico. It remains to be seen, however, if the material can be made available at economically justifiable prices, and if the potential benefits of such high flux densities can be exploited in suitable motor designs.

18.3.6 *Permanent magnet motor constants*

The characteristics typical of a permanent magnetic motor, (Fig. 18.5(a)) show that for torques below the limit line the motor can be operated over a wide range of

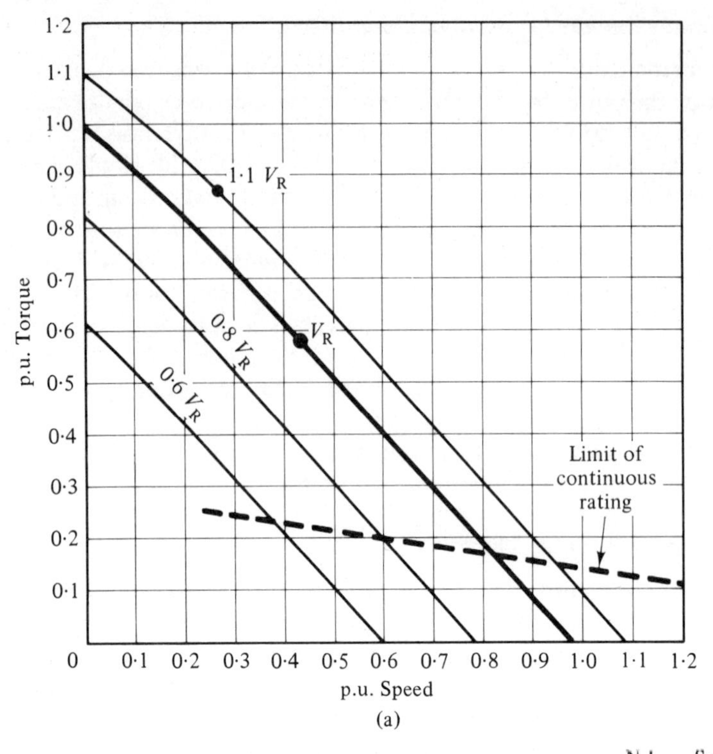

(a)

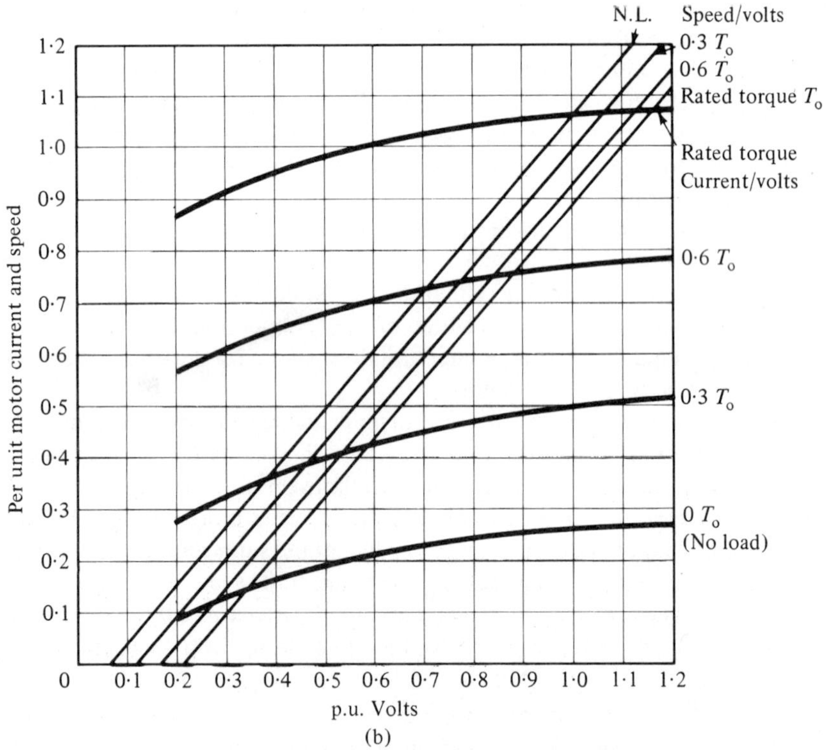

(b)

Fig. 18.5 Typical permanent magnet characteristics

speeds by controlling the applied voltage. The designation of one particular speed/torque curve corresponding to V_R is to some extent arbitrary, and will depend on whether the higher power output is necessary.

Since fast response is one of the most desirable features of control motors the systems designer will make every effort to utilize the very high starting torque capability of permanent magnet motors. The extent to which this is possible will depend on the duty cycle and the thermal capacity of the motor. The high coercivity of modern magnet materials permits control motors not only to be started and reversed on full voltage, but where required given short (millisecond) current pulses in excess of normal starting current. For such operation it is not only essential to have an armature with very low inductance, but ensure that it is mechanically strong enough to withstand the high accelerating forces.

In the temperature rise calculations the possibility of local hot spots should not be overlooked and new application should be the subject of proper consultation between manufacturer and user.

The substantially linear behaviour of permanent magnet motors allows their characteristic constants to be easily defined:

$$\text{EMF constant } K_E = \frac{\text{no-load speed}}{\text{applied volts}}$$

$$\text{Torque constant } K_T = \frac{\text{stall torque}}{\text{stall current}}$$

$$\text{Damping constant} = \frac{\text{stall torque}}{\text{no-load speed}}$$

$$\text{Mechanical time constant} = \frac{\text{motor inertia}}{\text{damping constant}} = \frac{Jr}{fr + K_E K_T}$$

$$\text{Electrical time constant} = \frac{L}{r}$$

J = moment of inertia of rotating parts
f = friction torque per unit speed

$$\text{General motor circuit equation } V = K_E N + Ir + L\frac{di}{dt}$$

18.4 Moving Coil Motors*

Small d.c. motors are used in very large numbers in playback equipment, toys, measuring recorders, etc. These motors owe their wide applications to a number of features which can be made use of singly or in combination, depending on the particular requirements. For example, the efficiency is high, even for very small motors; the delivered torque is uniform, and there is ample scope for selection and control of the motor speed. Further features which may be of importance are the rapid start, the ease with which the direction of rotation can be reversed and the possibility of using a battery as power supply. The rotor of such motors often consists

* (Courtesy of Philips, Ref. 11)

of a laminated ferromagnetic material, with the rotor windings wound round the stack of laminations. Figure 18.6 shows by way of example a simple version with three rotor lobes. The stator generally consists of a permanent magnet in the form of a hollow cylinder, which is sometimes divided into a number of segments.

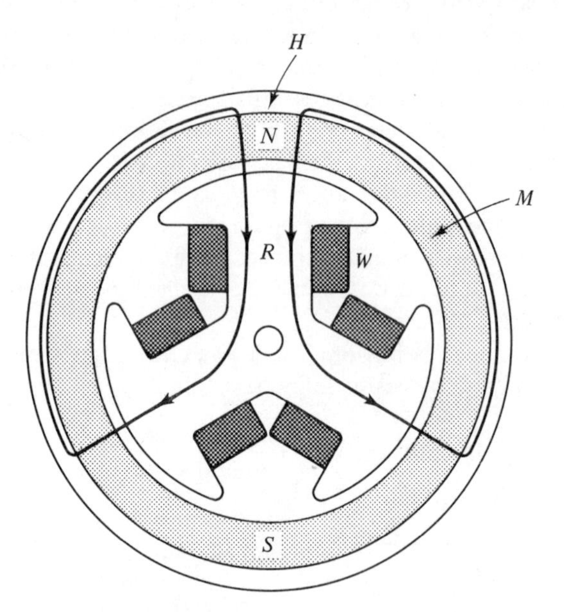

Fig. 18.6 Example of d.c. motor with iron rotor

Despite the utility of this type of motor for many applications, it has certain disadvantages when a very uniform torque is required, e.g., in tape recorders. Since there are slots in the rotor, it has a number of preferred positions, sometimes called 'reluctance positions'. The corresponding magnetic reluctance torques can be felt when the shaft of the motor is turned by hand. It will be clear that these reluctance torques give fluctuations in the torque delivered by the motor; this has an adverse effect on the quality of audio recording and playback. The reluctance torques can also cause motor noise, since there is always a little play in the bearings. The reluctance torques can be reduced by increasing the number of rotor lobes, but for given motor dimensions this can only be done at the expense of the efficiency, since the space factor of the winding is then decreased. Another solution is to minimize the width of the rotor slots, but this also can only be done within certain limits since the slot has to be wide enough for the winding process.

The impossibility of eliminating the reluctance torques from motors with an iron rotor while continuing to satisfy all the other requirements involved led to the design of a d.c. motor with no iron in the rotor. Such motors, which are called moving coil motors, offer the following advantages:

(a) Reluctance torques cannot arise since there is no iron.
(b) Since there are no discrete rotor lobes, the rotor can be wound uniformly, which makes it possible to divide the winding into a large number of coils. This also helps to reduce the fluctuations in the delivered motor torque, since a larger number of coils with corresponding commutator segments reduced the

torque pulses resulting from commutation. This feature makes this type of motor very suitable if a low and uniform speed of rotation is required.

(c) The moment of inertia of the rotor can be made smaller than that of a rotor with an iron core in a comparable motor (same power consumption, same torque-speed characteristic).

(d) Iron losses (hysteresis losses and eddy-current losses) cannot occur in the rotor. This is especially important in small motors, where the delivered motor torque can be of the same order of magnitude as the loss torque caused by an iron core (a situation which leads to a very low efficiency).

(e) The inductance of the rotor is low, which improves the commutation; there is less electrical interference, and the life of the commutator (and hence of the motor) is increased.

Consider now the basic principle of the d.c. moving coil motor, with reference to Fig. 18.7. The stator consists of a permanent magnet M, which produces a magnetic

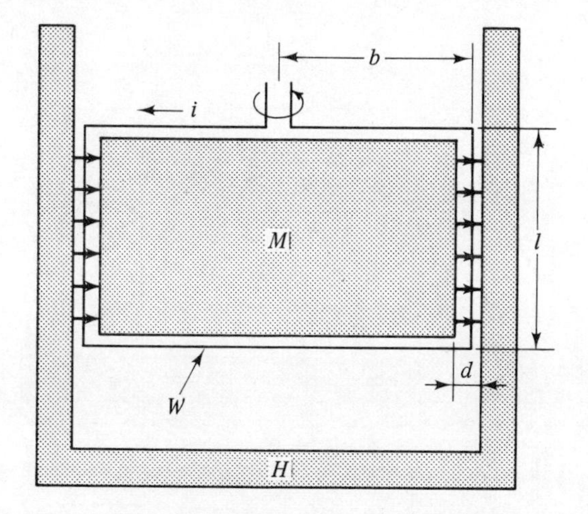

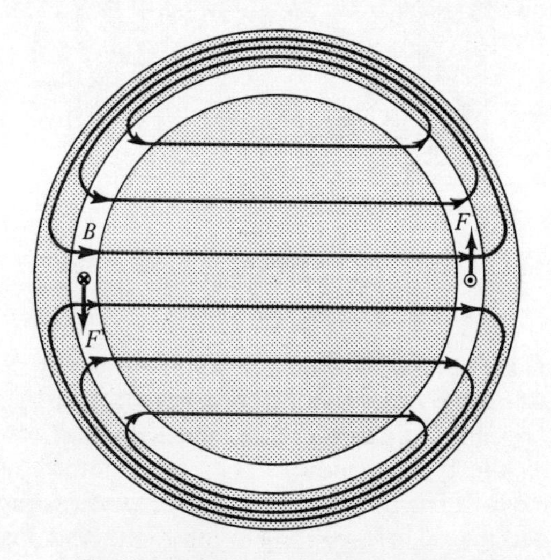

Fig. 18.7 Principles of moving coil motor

flux density B in the air gap d between the magnet and the housing H (which is made of a ferromagnetic material). The rotor winding W rotates round a vertical axis in this air gap (only one turn of the winding is shown). If the flux density B and the current i in the turn have the directions shown in the figure, the rotor will rotate anticlockwise under the influence of a Lorentz force F applied to one side of the turn; the magnitude of this force is liB, where l is the length of the side of the turn in question. The torque delivered per turn, Te, is equal to $2Fb$ (b is the radius of the winding).

18.4.1 *Construction*

Figure 18.8 shows a cross-sectional view of a typical moving coil motor. The stator consists of a permanent magnet M mounted on the steel motor housing H by means of a support 7. This steel housing is also a part of the magnetic circuit; the magnet has two diametrically opposite poles.

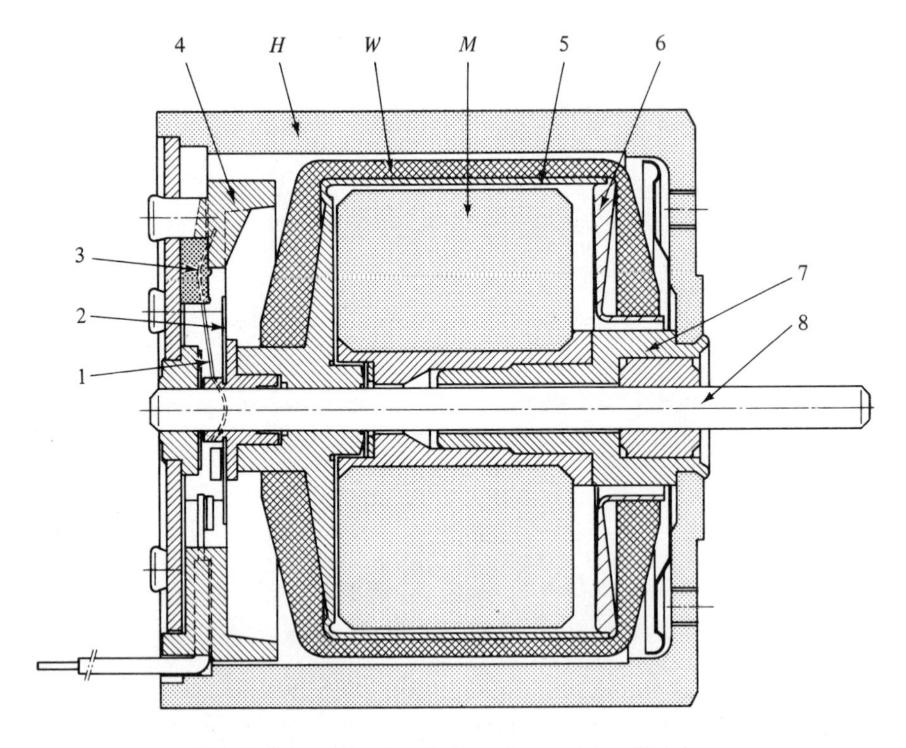

Fig. 18.8 Cross-sectional view of a moving coil motor

The rotor is formed by a winding W wound on a coil former 5 with cover 6. The coil former is fixed on a shaft 8. The winding consists of a large number of turns distributed uniformly round the circumference, and is divided into several coils, the number depending on the type of motor. The commutator 2 is built up of flat segments, one per rotor coil. The brush unit consists of two brush springs 1, mounted in an injection-moulded plastic bridge 4. Damping compound 3 is applied to damp the vibrations of the brush springs.

18.4.2 *Winding and magnet*

It can be seen from Fig. 18.8 that the coil former has central projections at each end: these are necessary for the insertion of the magnet support and for attaching the coil former to the motor shaft. This arrangement does not permit purely diametral winding, and the winding method shown in Fig. 18.9 has to be used, in which the plane of the winding is slightly skewed with respect to the centre-line of the coil former.

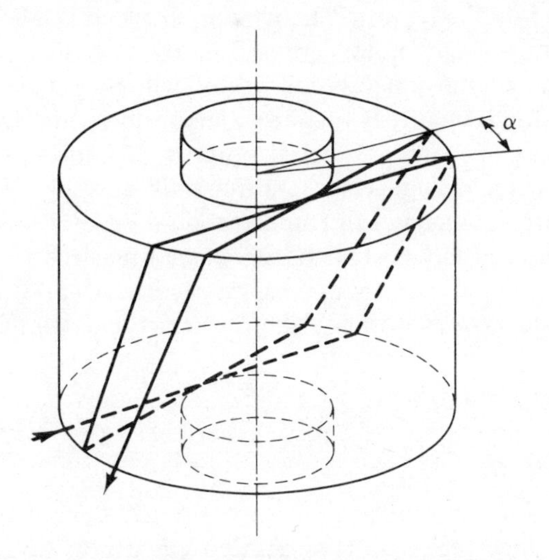

Fig. 18.9 Principle of the 'ball winding' method

A regular distribution of the turns around the circumference is obtained by rotating the coil former through a small angle α after each turn of the coil has been wound; the value of α is chosen so that the coil former will have rotated through exactly 360° after the desired number of turns. Because of the skewed position of the turns, successive turns will cross previously wound ones during the winding process. The coil obtained in this way looks rather like a ball of string and this winding method is therefore called 'ball (or basket) winding'.

The air gap between the magnet and the housing is considerably wider than the thickness of the winding, since room must be left for the coil former and for clearance on each side. Since the space can only be provided at the expense of the magnetic flux, the moving coil motor is at a disadvantage here compared with the iron-rotor motor (see Fig. 18.6). Moreover, in the motor with iron rotor the magnetic flux can be concentrated at a small cross-section through the winding. As a result, the total length of the winding in a motor of a given size can be made less than in a moving-coil motor. Since the motor losses are mainly the sum of copper-resistance losses and iron losses, the higher dissipation in the coil of the moving coil motor could partially cancel out the advantage of the absence of iron losses.

This difficulty can be avoided by making the permanent magnet of material with very good magnetic properties. This makes it possible to obtain a motor with good characteristics (in particular, high efficiency and short starting time). If a material combining a reasonably high remanance with a high coercivity is used, (such as

'Ticonal 550'*) the high magnetic flux produced in the air gap as a result of the use of this material, and which is not of course limited in this type of motor, by magnetic saturation of the core, compensates for the above-mentioned disadvantages. A further advantage is that with the geometry of coil and magnet used here, the ends of the coil are situated in the stray field of the magnet, so that the winding wire there does not merely increase the resistance of the coil, it also makes an appreciable contribution to the motor torque.

Magnet materials such as 'Ticonal 550' are expensive compared with the materials conventionally used in d.c. motors. The winding method used for the coil, and the complete construction of the motor, are also more expensive. It is therefore likely that the application of the moving coil motor will be restricted to cases where high-quality operation is required—in particular, uniform rotation, a low loss torque and a small moment of inertia. The last point is of importance for rapid motor starting, and rapid speed changing; the 'starting time constant' of the motor, which is a measure of its performance in this respect, should be kept low.

Typical applications of moving coil motors of varying sizes are in cameras, pocket dictating machines, digital and audio cassette recorders, tape recorders and other professional equipment such as measuring recorders and computors.

18.5 Printed Circuit Motors

18.5.1 *Introduction*

An in-house servo-motor requirement in a French Company[19] resulted in the application of the new technology of printed circuits to the axial airgap motor concept known, but found to be impractical to produce for industrial use, for more than a century. Since the first patents were filed in the late 1950s the printed circuit technique has been superseded by a stamped armature disc forming 100 or more conductors which are assembled and bonded in stacks of two to six to form the disc armature. The absence of iron in the armature makes it not only very light, but results in very small values of inductance. This together with a large number of commutation periods (the brushes actually bear on the 117 to 196 outer armature conductors) eliminates cogging and ensures excellent commutation (Fig. 18.10). In high performance servo-motors four, six, or in large motors even ten, permanent magnets are mounted on either side of the armature to provide the constant field (Fig. 18.11). In 'general purpose' motors with disc armatures only one side carries the magnets, while the front end-plate of the motor, which must also now have the necessary magnetic properties, forms the return path for the magnets. The servo-motor version which is the subject of this section will, because of the greater number of more powerful magnets and higher grade bearings, be the more expensive, but when compared with a servo-motor of similar performance show a similar manufacturing cost.

One significant difference between the disc armature or pancake motor and one with a long thin cylindrical armature, and therefore considerably limited number of commutator segments, is the former's significant ability to run evenly at speeds as low as 1 rev/min. By virtue of the evenness of the torque for any position of the armature the disc motor can reasonably be considered for incremental motion applications in competition with the stepping motor.

* 'Ticonal' is a registered trade mark of N.V. Philips' Gloeilampenfabrieken.

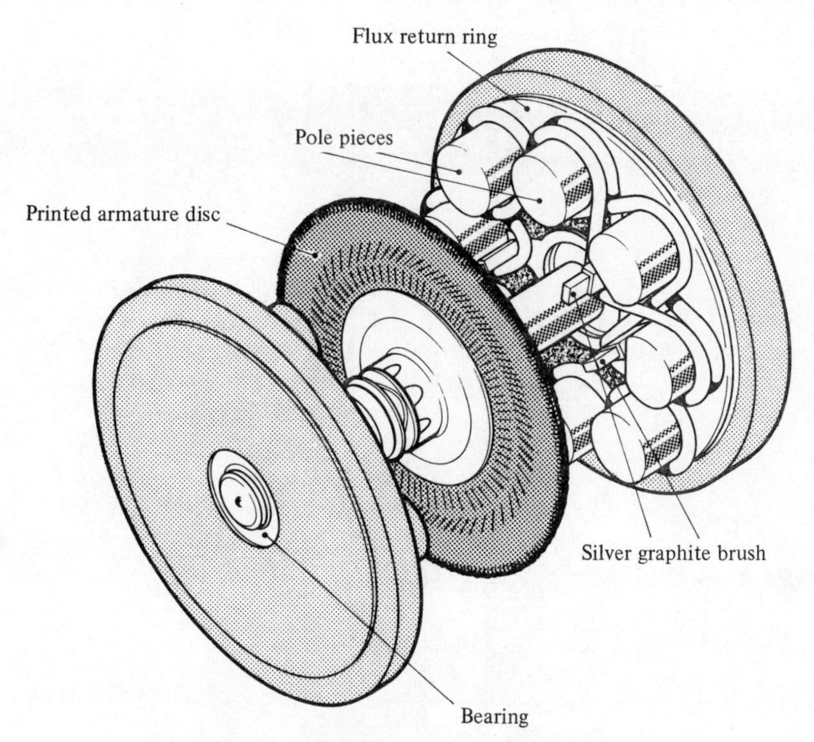

Fig. 18.10 Exploded view of a typical printed motor

18.5.2 *Construction*

In the smaller sizes comprising motors with outside diameters of 60, 90, 120, 160, 170, 230, and 260 mm the housings are aluminium tubes, while the kilowatt sizes have finned cast aluminium housings to improve heat dissipation. The shafts, typically 12, 7, 19, 27, and 31 mm diameter, are generally of stainless steel and run in precision ball bearings. The motors can be supplied with integral tacho and/or gearbox, but even in the maximum configuration the motor unit will be considerably shorter than a similar unit with cylindrical armature. This feature is particularly useful in such applications as digital tape decks, valve actuators, and machine-tools, where their wide speed range, fast response, high power to weight ratio, and convenient geometry combine to make them superior to other forms of servo-motors. Great care must be taken in the design of mountings, couplings, and gear train to ensure that the large peak torques do not cause instability due to mechanical compliance in the short term or failure due to fatigue in the long term. The manufacturer will also specify the effect of the mounting as a heat sink as well as the increase in performance obtainable by providing a specified amount of cooling air.

18.5.3 *Performance*

The absence of iron in the armature and therefore freedom from saturation effects results in printed motors having linear torque/speed characteristics. The characteristics shown in Fig. 18.12 for a motor with a rated output of 244 W are typical. In practice there will be some deviation from this linearity due to temperature effects,

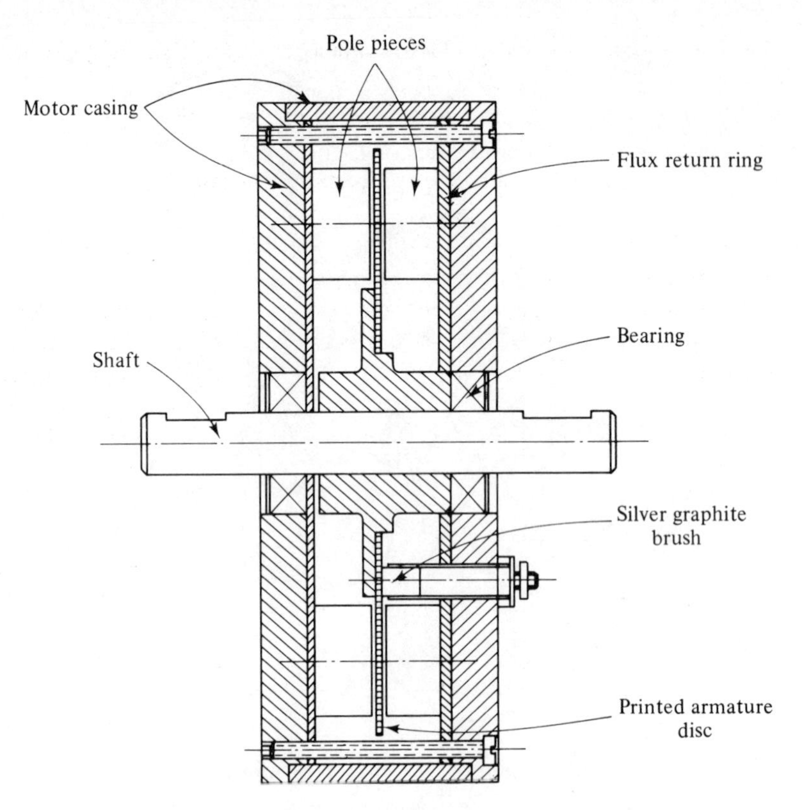

Fig. 18.11 Cross section of a typical printed motor

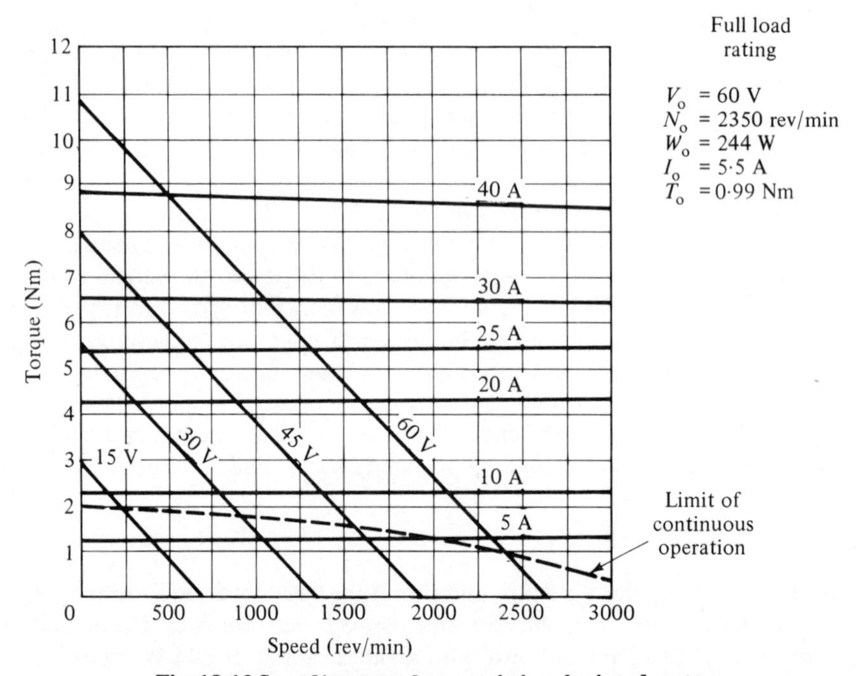

Fig. 18.12 Speed/ torque characteristics of printed motor

but if all parameters are related to normal full-load operating temperatures sufficiently accurate results will be obtained.

The fundamental motor constant is that relating the applied voltage to speed, which is variously quoted as millivolts per rev/min or volts per 1000 rev/min. These numerically equal values are designated the voltage constant K_E.

The motor torque constant K_T defines the gross torque per ampere, is directly related to K_E and expresses the conversion of electrical to mechanical energy for a particular motor. For any given current I the torque available at the motor shaft will, therefore, be $K_T I$ less the torques necessary to overcome coulomb friction and viscous damping. The viscous damping losses are produced by eddy currents in the armature conductors and friction losses are due to brushes and bearings. The constant K_D is used to combine these two torque losses and define them as a function of speed. The units used depend on the system used and would be Nm/1000 rev/min in the SI system.

The following motor equations are given without dimensions of some of the constants since their validity is not affected as long as consistency in the choice of units is observed.

$$V = E + Ir \tag{18.6}$$

If the gross torque is $T = K_T I$, and $E = NK_E$

$$V = K_E N + r\frac{T}{K_T} \tag{18.7}$$

If T_0 is the output/torque and the friction and damping torque are T_F and $T_D = K_D N$, respectively

$$T = T_D + T_F + T_0 \tag{18.8a}$$

or

$$T = K_D N + T_F + T_0 \tag{18.8b}$$

then V becomes

$$K_E N + \frac{r}{K_T}(K_D N + T_F + T_0)$$

which can be solved for the net torque T_0 available at the shaft

$$T_0 = \left[\frac{VK_T}{r} - T_F\right] - n\left[\frac{K_E K_T + K_D r}{r}\right] \tag{18.9}$$

This expression defines the torque/speed curve for a constant applied voltage, with the reservation about temperature made above.

The motor regulation R_M, which is the reduction in speed per unit change in load torque is the negative slope of the torque/speed characteristic

$$R_M = \frac{r}{K_T K_E + K_D r} = \frac{1}{(K_T K_E / r) + K_D} \tag{18.10}$$

From this expression the effective terminal resistance is

$$r = \frac{K_T K_E / 1}{R_M - K_D}$$

which gives another expression for net torque

$$T_0 = \frac{VK_T}{r} - T_f - \frac{N}{R_M} \tag{18.11}$$

The no-load speed is

$$N_{NL} = R_M\left(\frac{VK_T}{r} - T_F\right) \tag{18.12}$$

and the stalled torque

$$T_{ST} = \frac{VK_T}{r} - T_F \tag{18.13}$$

A further useful expression can be derived from (18.8(a)) for as $T = K_T I$

$$I_{run} = \frac{K_D N + T_F + T_0}{K_T} \tag{18.14}$$

18.5.4 *Motor rating*

Most servo-motors are used intermittently, though in some applications such as computer peripherals there may be periods of running approaching continuous operation. The continuous rated output, as well as the permissible pulse current, is usually quoted by the manufacturer and when the two are combined a careful calculation of the temperature balance and resulting increases in temperature must be made. The maximum operational temperature of the armature is dependent on the type of insulation used, but a maximum of 135°C is suggested for motors operating in ambient temperatures of up to 40°C. To achieve this and allow for hot spots in the vicinity of the brushes the armature temperature should not exceed 75°C.

These recommendations do not preclude temperatures of up to 150°C being permitted under exceptional circumstances, but the mechanical stability of the armature, the bearing lubrication and the insulation will be subject to accelerated ageing. If incorporation of a larger motor presents economic problems the monitoring of the armature temperature and the incorporation of cutouts or selective currents limiting in the motor control unit should be considered.

For a given load torque and corresponding speed the following calculation can be used to check the power dissipated by the motor:

$$\begin{aligned} \text{Power output } W_0 &= \frac{T_0(\text{Nm}) \cdot N(\text{rev/min})}{9{\cdot}55}\text{Watts} \\ &= \frac{T_0(\text{kgf cm}) \cdot N(\text{rev/min})}{97{\cdot}4}\text{Watts} \\ &= \frac{T_0(\text{oz in}) \cdot N(\text{rev/min})}{1352}\text{Watts} \end{aligned} \tag{18.15}$$

The current I_0 drawn by the motor can be calculated from Eq. (18.14) and the applied voltage from:

$$V_0 = K_E N + I_0 r \tag{18.16}$$

The losses are then calculated from:

$$W_L = V_0 I_0 - W_0 \tag{18.17}$$

Usually the manufacturer will have calculated and verified by tests the maximum dissipation for various operating conditions. Thus, as shown on the graph, a limit line defining the maximum continuous load condition for the full range of operation can be drawn.

Figure 18.13 shows the limits of current and output as a function of speed which defines the limit of continuous operation. As there is no significant cooling at stall and very low speeds, and as the graph shows the current tends to be greater than rated current, operation around that region may make some reduction in current necessary.

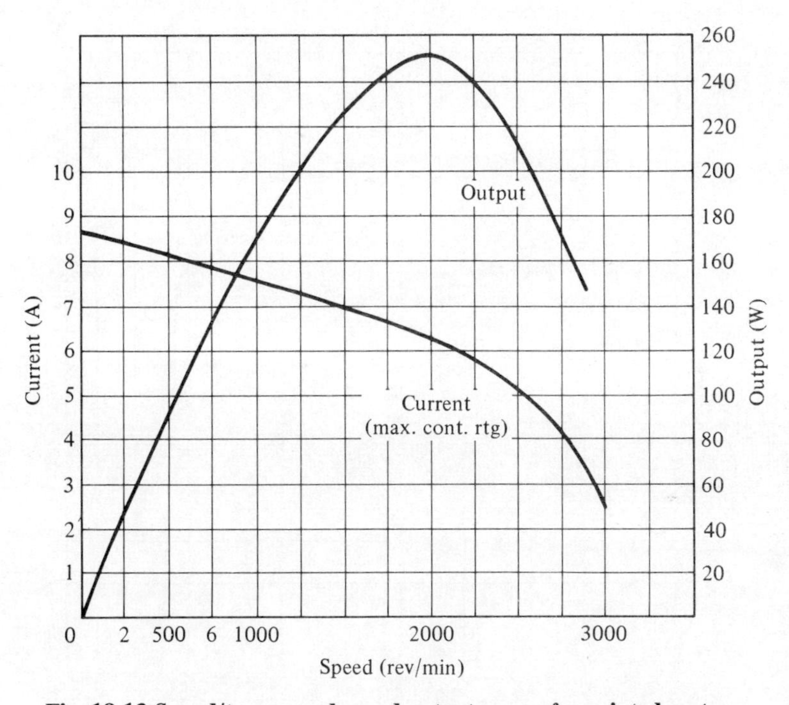

Fig. 18.13 Speed/torque and speed output curves for printed motor

Very substantial increases in output can be obtained by forced air cooling, but this also increases the danger of a catastrophic failure if, through some malfunctioning of the auxiliary equipment, the air flow ceases.

18.5.5 *Pulsed operation*

Pulsed operation is a useful means of obtaining a fast response in a control system, and today's printed circuit motors are capable of delivering torque peaks in excess of 10 times their rated torque. The fact that the current peak is now 10 times rated value is still more due to the effect of increased electric loading than any magnetic saturation. In normal practice currents 5 times the rated value will be the more usual limit, the only practical limits being more the design of suitable drive circuits. This is best illustrated by the typical motor which, at a current of 5·5 A, delivers a torque of nearly 1 Nm. The absolute peak torque quoted is 19 Nm at a current of 87 A when the armature I^2r will have risen from a normal 36 W to 9·8 kW, and it is clear that the motor power-supply would have to be very large and costly. The rating must, furthermore, be carefully studied and the duration and magnitude of the current pulse rigidly controlled. One manufacturer quotes an impulse rating of 50 milliseconds with a 1 per cent duty cycle, which means that each torque pulse must be followed by a period of at least 4·95 s with no current applied to the motor. Figure 18.14 shows more comprehensively the permissible ratings up to 45 A for which the motor will deliver a torque of nearly 10 Nm.

The peak temperature reached by the motor after a number of current pulses will be the averaging out of the relatively short time constant of the armature and very

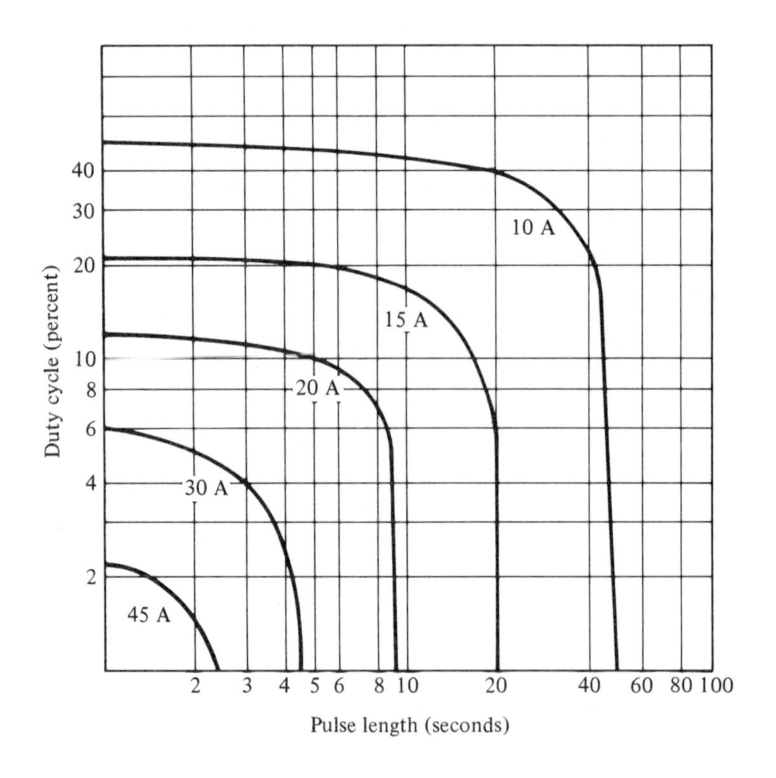

Limiting conditions

Maximum continuous torque 0 to 1000 rev/min	1·695 Nm
Maximum continuous stall current	8 A
Peak torque (87 A T_{max})	19·1 Nm

Basic motor constants

e.m.f. constant $K_E = \dfrac{60}{2{\cdot}61}$	23 V/1000 rev/min
Torque constant K_T	0·22 Nm/A
Damping constant K_D	0·061 Nm/1000 rev/min
Regulation constant	235 rev/min/Nm

Fig. 18.14 Pulse operation limits

much longer one of the motor where the pulse duration will influence the former by rapid heating cycles whereas the other will show proportionality to the duty cycle.

18.5.6 *Power rate*

The relationship between the time constant of the motor and its inertia has already been dealt with in a previous section.

When a motor is coupled to a load, the time constant of the system becomes

$$\tau_s = \tau\left(\frac{J_M + J_L}{J_M}\right) \text{seconds} \tag{18.18}$$

where

τ_s is the time constant of the system

τ is the motor time constant

J_M is the motor inertia

J_L is the load inertia

From these expressions it can be seen that although the lowest time constant (highest acceleration) is obtained without a load inertia, the optimum acceleration can only be obtained with a load inertia similar to that of the motor. When the load inertia is significantly larger than that of the motor, the optimum conditions can be restored by coupling it to the load by gearing having a ratio N such that

$$J_L = N^2 J_M \tag{18.19}$$

Under these optimum conditions the acceleration dn/dt at the load imparted by a motor developing a peak torque of T_p is

$$\left(\frac{dn}{dt}\right)^2 = \frac{T_p^2}{4J_L J_M} = \frac{T_p^2}{J_M} \tag{18.20}$$

This parameter is designated as the power rate. As it is defined entirely in terms of the motor constants, it is useful as an indication of the motor's peak capability.

18.6 Stepping Motors

18.6.1 *Introduction*

In stepping motors rotation is produced by sequentially switching suitably connected windings to produce discrete angular steps of essentially uniform magnitude.

Their first use was as receiving elements of remote indication systems such as the 'M' type transmission in ships. The transmitter was a multi-segment switch which commutated a d.c. supply to a three-phase type winding. Today the stepping motor competes with other types of control motors thanks to the development of electronic devices which can be made to switch substantial currents rapidly and reliably. The wide application of digital computers has also played a significant part in their increasing use on machine-tool-, valve-, and shutter-positioners, plotting tables, and computer peripherals such as disc stores and printers.

18.6.2 *Classification of stepping motors*

There are three basic types of stepping motors:

18.6.2.1 Electromechanical types. These can be divided into two groups, namely, ratchet and pawl drives and rotary solenoids. The mechanism of a typical ratchet and pawl type consists of a ratchet wheel rotated by a solenoid, and a pawl, locked between pulses by a spring-loaded detent. The rotary solenoid type employs an inclined cam and multiple rollers or ball bearings to translate an axial pull into shaft rotation. Two solenoids are required for bi-directional rotation.

In this chapter only electromagnetic and electrohydraulic stepping motors will be dealt with in detail, particularly as the use of electromechanical types is mainly limited to specialized applications such as rotary switches.

18.6.2.2 Electromagnetic types. These are normally brush-less electrical machines which rotate in fixed angular steps when fed with d.c. pulses. When supplied with an alternating current, the motors can be made to rotate continuously due to their similarity with induction motors.

The classification of motors commonly used is given in table 18.1.

Table 18.1 Classification of electromagnetic stepping motors

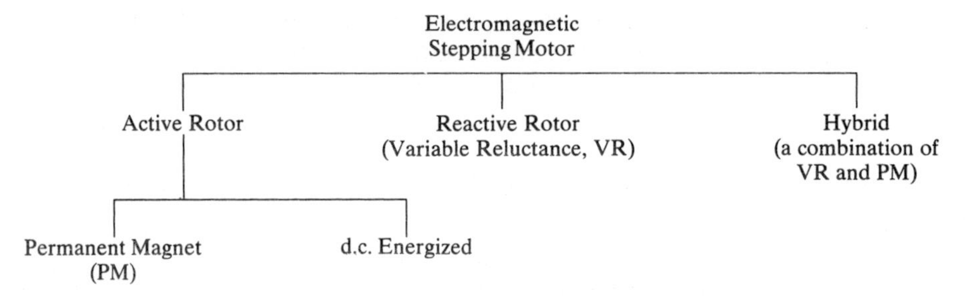

18.6.2.3 Electrohydraulic types. These consist of an electromagnetic stepping motor and a hydraulic torque amplifier, i.e., a hydraulic motor controlled by a servo valve.

18.6.3 *Constructional features of electromagnetic stepping motors*

The most popular types of stepping motors are:

(a) Permanent Magnet (PM)—Active Rotor.
(b) Variable Reluctance (VR)—Reactive Rotor.
(c) Hybrid Type—A Combination of PM and VR.

(*a*) *Permanent magnet type.* This type of stepping motor has a permanent magnet rotor similar to that of a conventional two-phase or three-phase induction motor. Figure 18.15 shows the basic permanent magnet (PM) configuration.

(*b*) *Variable reluctance type.* This type of stepping motor uses a magnetically soft iron rotor having teeth and slots similar to the rotor of an inductor alternator or salient-pole type rotor. The stator is similar to that of a permanent magnet stepping motor. A stepping motor with a salient-pole type rotor is shown in Fig. 18.16.

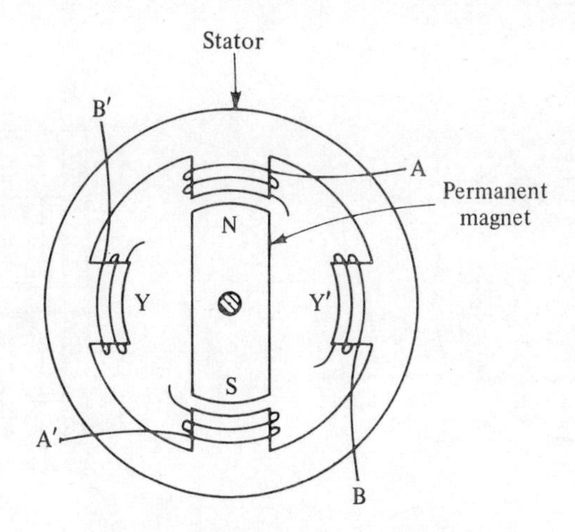

Fig. 18.15 Active type stepping motor

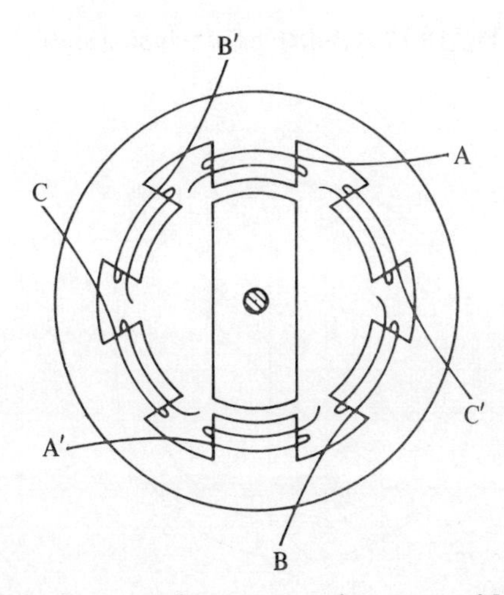

Fig. 18.16 A three-phase reactive type stepping motor with salient type rotor

(*c*) *Hybrid type*. This type of motor uses a combination of permanent magnet and variable reluctance rotor structure and is similar to an inductor motor in construction. Figure 18.17 shows a motor of this type. In this figure, the stator winding consists of eight coils, one around each tooth. These coils are connected as in a conventional two-phase winding. Rotor laminations, with teeth, are mounted in two stacks on the shaft separated by a permanent magnet which is axially magnetized.

18.6.4 *Basic principles of electromagnetic stepping motors*

18.6.4.1 Excitation torque. If the stator is excited by a direct current, as shown in Fig. 18.18, a simple bi-polar field is produced. If the rotor is a bipolar magnet, it will turn so that unlike poles on the rotor and stator magnetic axes align. The m.m.f.s will act in the same direction round the magnetic circuit and the flux as well as the stored energy

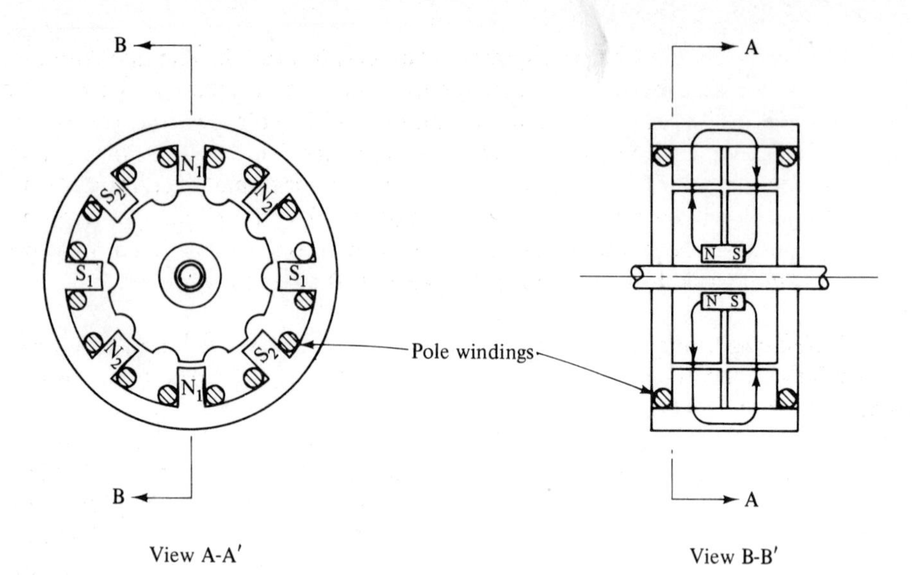

Fig. 18.17 Hybrid type of stepping motor

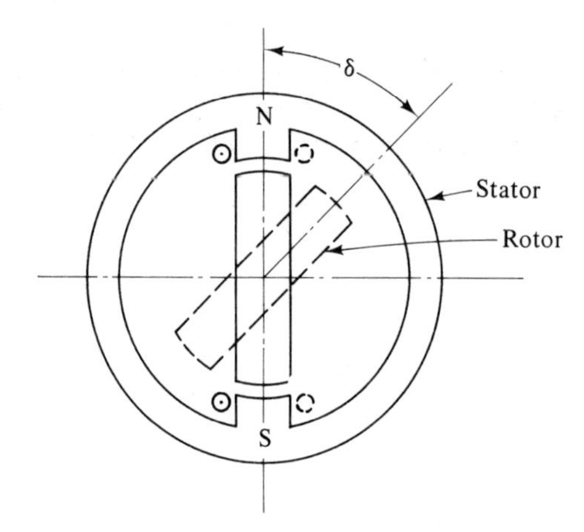

Fig. 18.18 Excitation torque

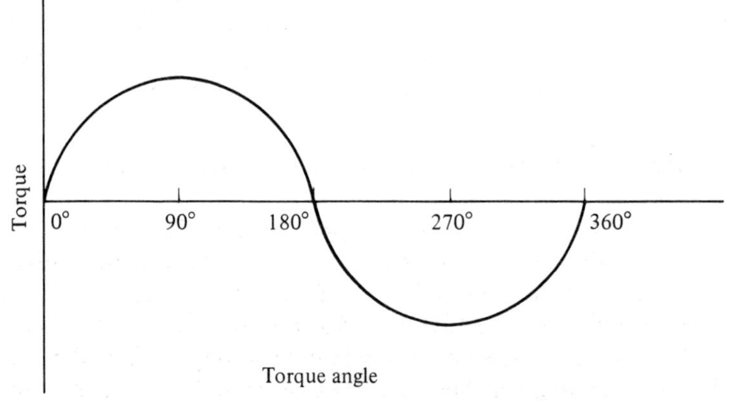

Fig. 18.19 Variation of excitation torque with angle

will be a maximum. If the rotor is displaced from this position, the alignment torque will increase with the angle, reach a maximum, and then decrease. If the rotor is turned through 180° from the original position so that like poles face each other the m.m.f.s will be in opposition and the flux and stored energy will be a minimum. The torque will be zero but the position will be one of unstable equilibrium. The maximum torque will evidently occur at an angle of 90°. The torque, which is the result of excitation of both rotor and stator, is referred to as the excitation torque and its variation with rotor position or torque angle δ, is shown in Fig. 18.19.

18.6.4.2 Reluctance torque. If the stator is excited by direct current but the rotor is a salient soft iron type, the rotor will align with the stator magnetic field. If the rotor is turned from the position of alignment with the stator, a torque tending to realign it with the stator field will be developed. Since the rotor polarity is induced by the stator there will now be as many stable positions of alignment as there are numbers of poles.

For a given m.m.f. the flux and the stored energy will be greatest in the position of alignment. As the torque developed is due entirely to there being a position of minimum reluctance, it is called a reluctance torque. The flux density produced in the air-gap will not therefore be uniform, but will be a maximum on the vertical centre line and will decrease on each side. The torque will be zero both at $\theta = 0$ and if rotor and stator are at right angles ($\theta = 90$), a position of unstable equilibrium. Maximum torque will occur at an angle approximately midway between these positions. The variation of reluctance torque, T_r, with the angle between the axis of symmetry is shown for a complete revolution of the rotor in Fig. 18.20.

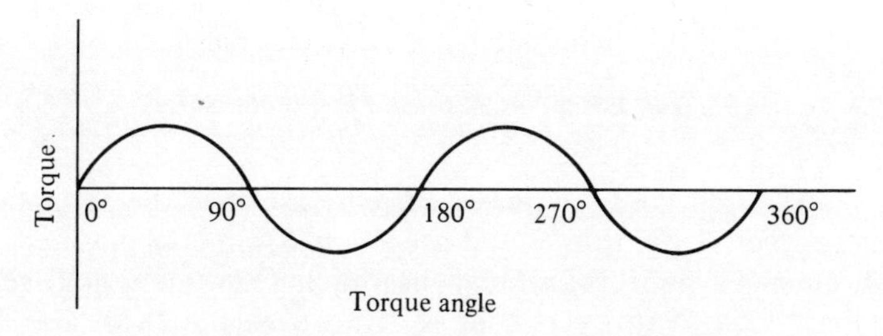

Fig. 18.20 Variation of reluctance torque with angle

18.6.4.3 Combined action. When the stator is excited and the rotor is a PM type and has a saliency, the excitation torque will be superimposed on the reluctance torque. The resulting torque/angle characteristics are given in Fig. 18.21.

18.6.5 *Basic theory of operation of electromagnetic stepping motors*

18.6.5.1 Operation of permanent magnet stepping motors. The motor shown in Fig. 18.22 is a typical two-phase stepping motor with a two-pole permanent magnet rotor which will be used to illustrate the principle of stepping motors.

When the stator is not energized the rotor is held in one of the four positions, i.e., at X, Y, X′ or Y′ where the air-gap is a minimum, by a torque which depends on the air-gap length and the magnetic field. This torque is the detent torque, and each of these positions acts as a detent.

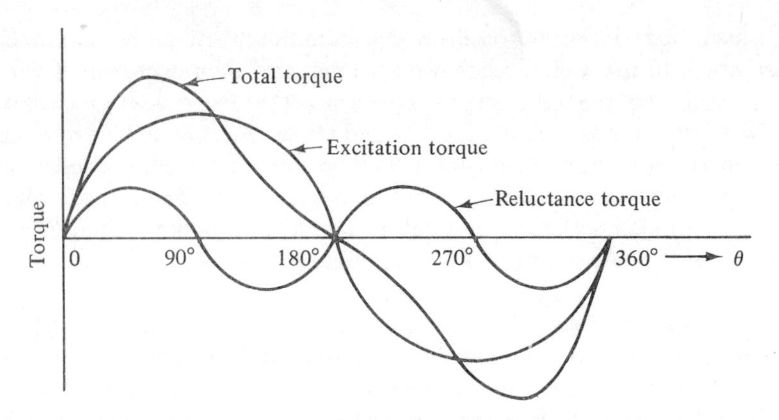

Fig. 18.21 Component and total torque of doubly excited salient device as applicable to PM stepper motors

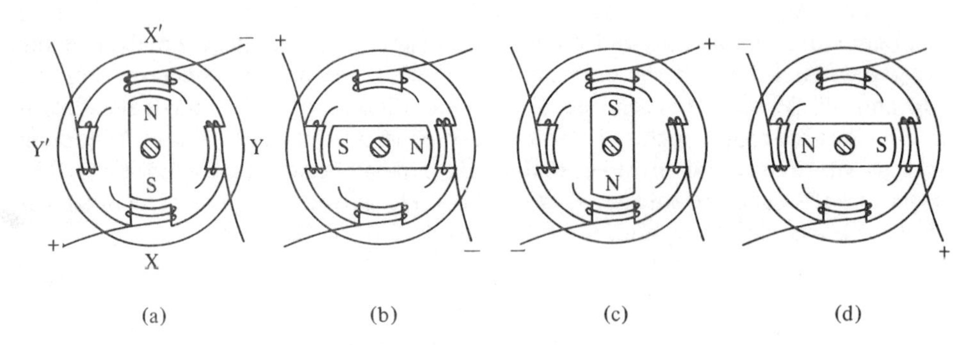

Fig. 18.22 Stepping action of PM stepper motor

Exciting the stator winding by direct current results in the formation of two magnetic stator poles, one North (N) and one South (S) either in the line XX′ or YY′. On the application of a mode, the rotor aligns with these magnetic fields sequentially as shown in Fig. 18.22(a) to (d), where the polarities of excitation are also shown. It is seen that the rotor changes position with an angular movement of 90° for a step. This angular movement is called a step position. The change of rotor position with time from step position 1 to step position 2 follows the curve in Fig. 18.23. If the switching rate is increased the period τ becomes shorter.

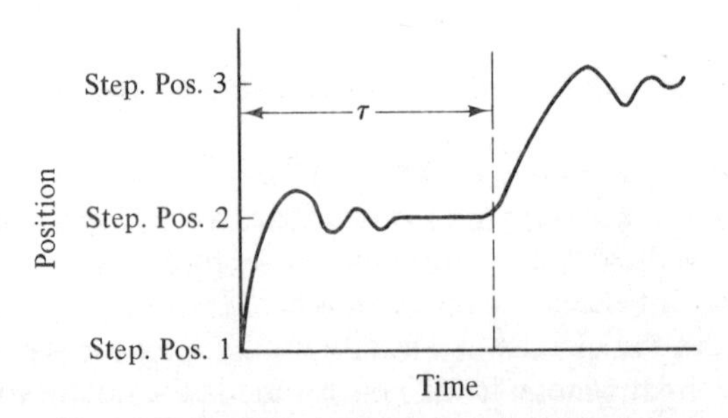

Fig. 18.23 Variation of step position with time

As the rotor follows each 90° shift of the stator field, this particular configuration has four steps per revolution. The speed with which the rotor moves is determined by:

$$T = Ia \times T_{max} \sin \theta \tag{18.21}$$

where

$$T = \text{torque}$$
$$T_{max} = \text{maximum torque}$$
$$a = \text{angular acceleration}$$
$$\theta = \text{step angle}$$
$$I = \text{moment of inertia}$$

This results in an oscillating type of motion before settling to the second step position.

Considering Fig. 18.23, the torque produced in the alignment position is zero. If a resisting torque is applied, the rotor will be displaced through $\theta°$, that is until this torque is balanced by that developed by the motor. If a uniformly distributed field is assumed, the torque–displacement characteristic, Fig. 18.24, is a sine wave. At $\theta = 0°$ or $\pm 180°$ (the rotor and stator field alignment positions) the torque exerted is zero.

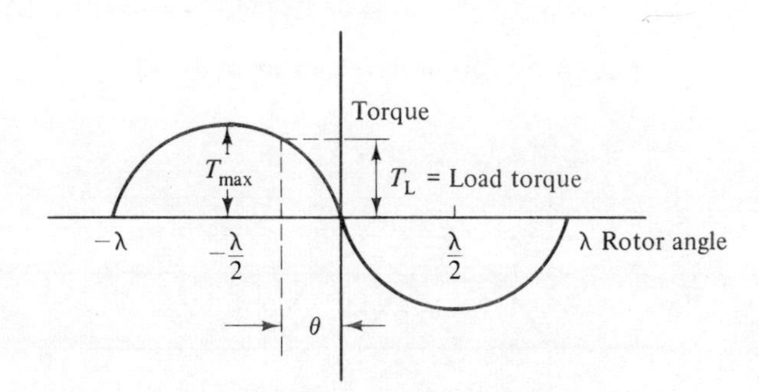

Fig. 18.24 Torque-displacement curve

The maximum holding or static torque is developed at $\theta = 90°$, i.e., when the magnetic axes of the rotor and stator are in quadrature by 90° and 270°. If an increasing torque is applied to the rotor until $\theta = 90°$ and if then the second winding is excited, an angle of 180° will be created between the rotor and the new resultant stator field. The torque developed at this point is given by:

$$T = T_{max} \sin 180° = \theta \tag{18.22}$$

This situation is shown schematically in Fig. 18.25. Thus the holding torque is not usable at the switching rate, i.e., under dynamic conditions. The maximum torque usable under dynamic conditions occurs when θ does not exceed 45°. This can be seen from Fig. 18.26. When field 1 is excited, the torque given by $T = T_{max} \sin 45° = 0{\cdot}707 T_{max}$ at a rotor position of 45° from the stator field. If at this instant field 2 is excited and field 1 is switched off, as shown in Fig. 18.27, a resultant field 135° from

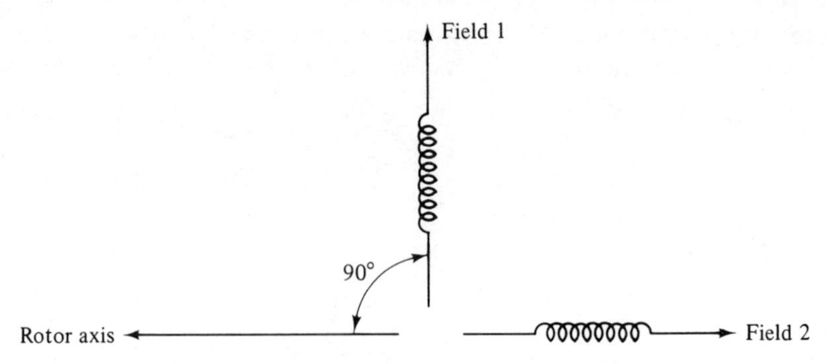

Fig. 18.25 Torque development at $\theta = 90°$

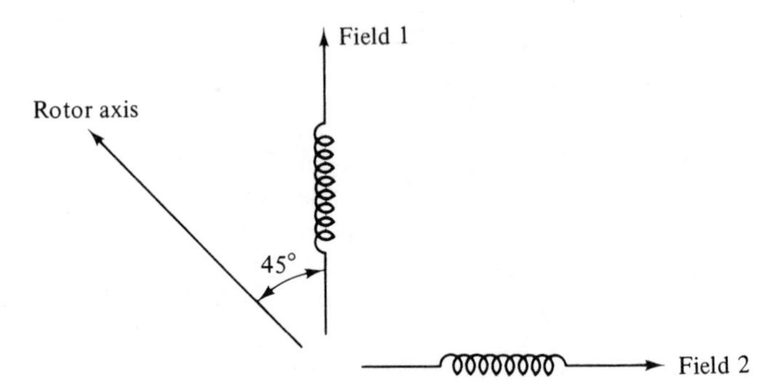

Fig. 18.26 Torque development at $\theta = 45°$

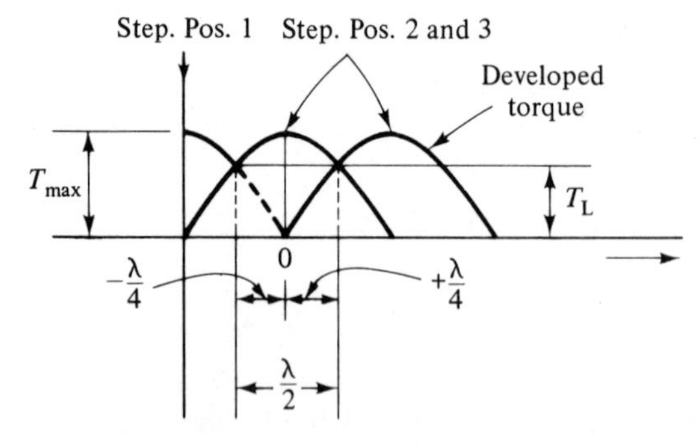

Fig. 18.27 Torque development with an acceleration torque

the rotor is created and the torque is:

$$T = T_{max} \sin 135° = 0{\cdot}707\, T_{max}$$

In the interval between step positions 1 and 2 no acceleration torque is available as the torque developed equals that of the load. It is, therefore, necessary to reduce the load applied until the phase displacement is less than 90°. Figure 18.28 shows the case for a phase displacement of 30°. In this situation, acceleration torque exists. Therefore, theoretically it is necessary that the torque angle should be limited to 45°. In practice this angle should be such that it gives sufficient acceleration torque.

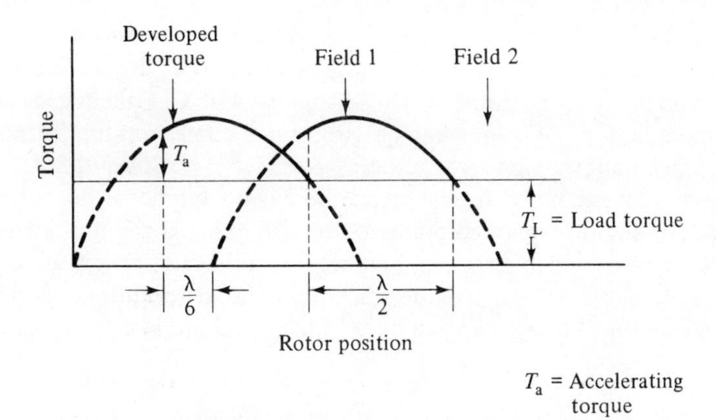

Fig. 18.28 Torque variation with acceleration torque

18.6.5.2 Operation of variable reluctance stepping motors. The operation of these motors is based on the action of an electro-magnetic field, produced by the stator excitation, on a ferromagnetic rotor. The operation follows the alignment principle of reluctance torque already discussed. When winding A of Fig. 18.29 is energized by direct current, the rotor assumes the position of least magnetic reluctance in the stator field, i.e., axis 1–2 aligns with A. In this position the torque on the rotor is practically zero. If winding A is de-energized and winding B supplied with current, the rotor will rotate through an angle of 30° so as to take up the position in which

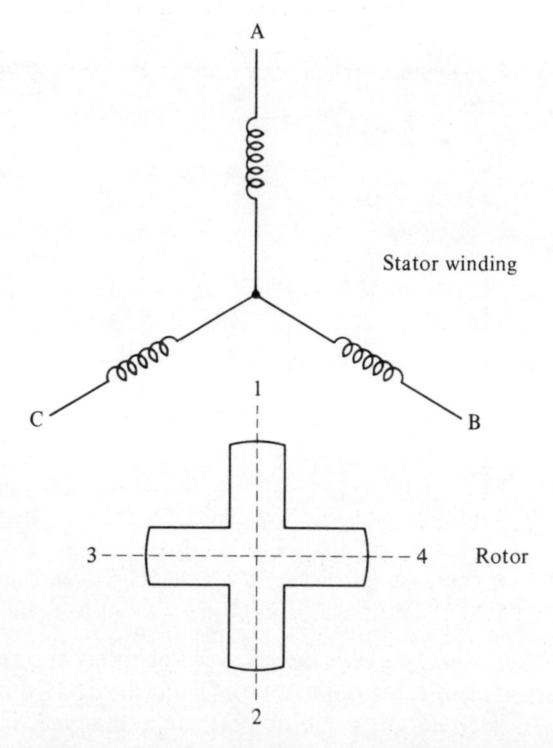

Fig. 18.29 Representations of a three-phase VR stepping motor

there is the minimum reluctance to the new stator field. This position is that in which 3–4 is aligned with B. When the excitation is switched to C, the rotor turns through another 30° and takes up a position with 1–2 aligned with C. For each rotation of the stator magnetic field of 360° mechanical, there is a corresponding rotation of the rotor by 90° mechanical. This arrangement gives a step position of 30°.

When there is no excitation on the stator there is no detent torque corresponding to that of a permanent magnet stepping motor and hence the rotor can assume any position. As the rotor will move to a minimum reluctance position when the motor is energized, the characteristic provides flexibility in generating various stepping angles. Two ways of obtaining 15° steps are illustrated in Figs. 18.30 and 18.31.

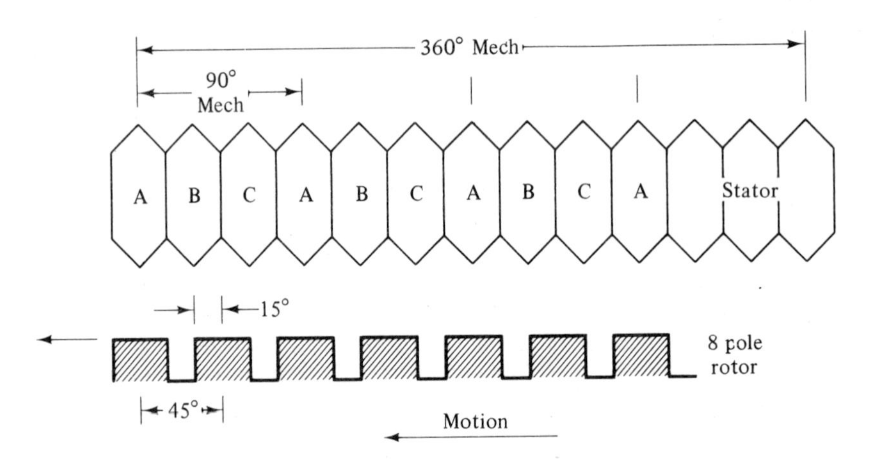

Fig. 18.30 Stepping action at a VR motor: four-pole stator

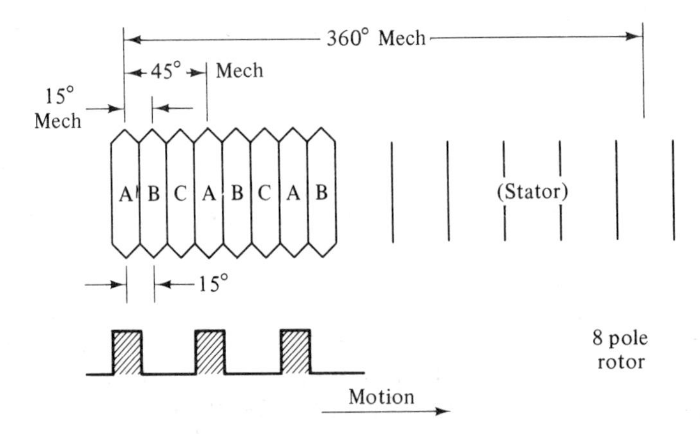

Fig. 18.31 Stepping action at a VR motor: eight-pole stator

18.6.5.3 Operation of hybrid stepping motors. Several different configurations of PM stepping motors are available, the popular types being derived from a.c. synchronous inductor motors. As such motors use both permanent magnets and soft iron in the rotor, they are commonly known as hybrid stepping motors.

Figure 18.32 shows a simplified cross-section and longitudinal view of a two-phase synchronous inductor motor with four salient poles on the stator and a toothed soft iron rotor with cylindrical permanent magnets magnetized axially.

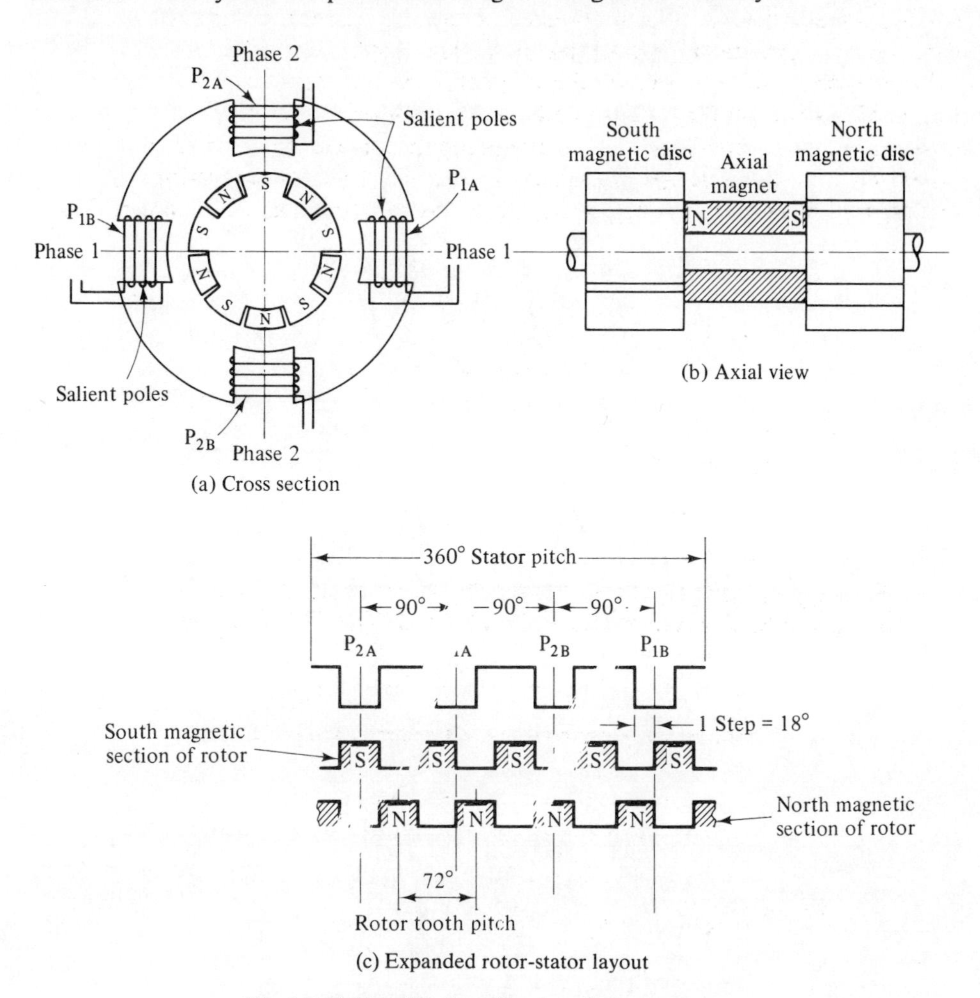

Fig. 18.32 Simplified synchronous inductor motor

The rotor soft iron has five teeth, and hence the rotor tooth pitch is 72°. The stator salient poles occur every 90°. Hence, one step corresponds to $\frac{1}{4}$ of a rotor tooth pitch or a rotor movement of 18°; one complete revolution of the rotor corresponds to 20 steps. Figure 18.33 also gives a graphical representation of the developed torque as the rotor moves relative to the stator salient poles. Stepping the rotor is accomplished by reversing the direction of the current in one phase while holding the other phase constant. For this motor, the salient poles can form four possible magnetic pole combinations.

Figure 18.33(a) shows the developed torque as a function of rotor position for combination 1. The resultant curve is the summation of the torque curves due to the individual phases and is assumed to be sinusoidal. The equilibrium points of the resultant curve lie midway between the equilibrium points of the individual phases.

▲ Stable equilibrium points ● Unstable equilibrium points

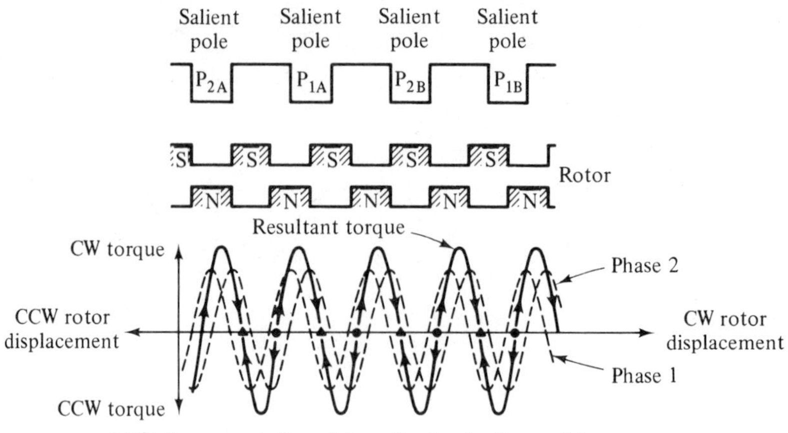

(a) Stator currents I_1 and I_2: salient poles P_{1A} and P_{2A} are north magnetic poles; P_{1B} and P_{2B} are south magnetic poles

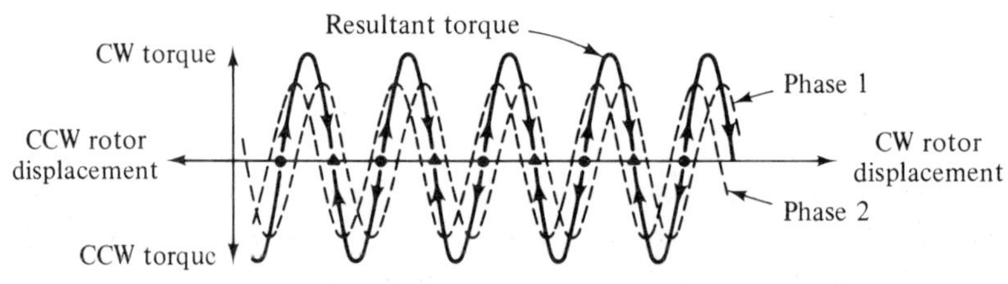

(b) Stator currents $-I_1$ and I_2: salient poles P_{1B} and P_{2A} are north magnetic poles; P_{1A} and P_{2B} are south magnetic poles

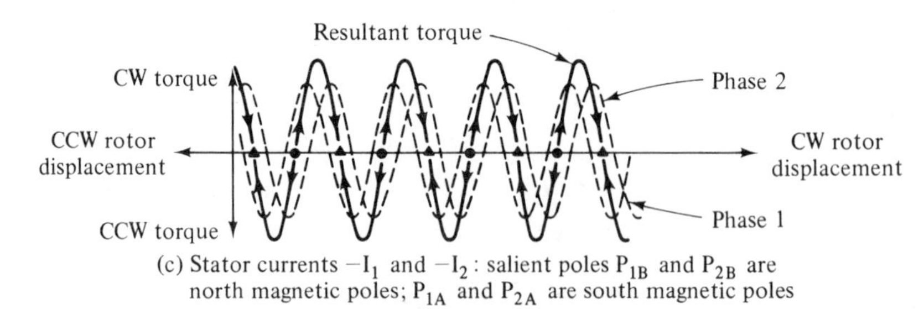

(c) Stator currents $-I_1$ and $-I_2$: salient poles P_{1B} and P_{2B} are north magnetic poles; P_{1A} and P_{2A} are south magnetic poles

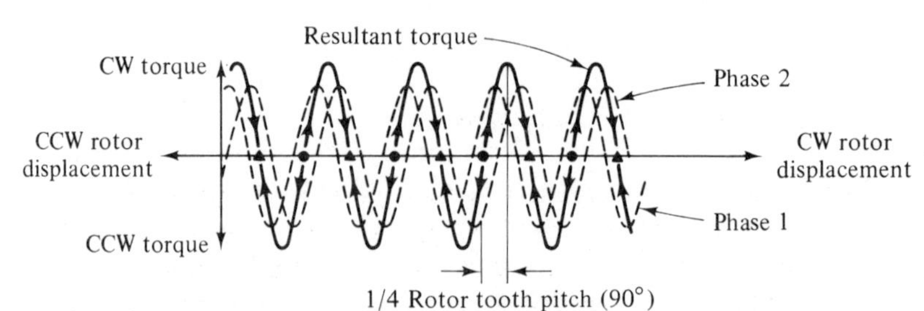

(d) Stator currents I_1 and $-I_2$: salient poles P_{1A} and P_{2B} are north magnetic poles; P_{1B} and P_{2A} are south magnetic poles

Fig. 18.33 Torque as function of rotor position—both phases energized. Arrows indicate direction of developed torque about equilibrium points

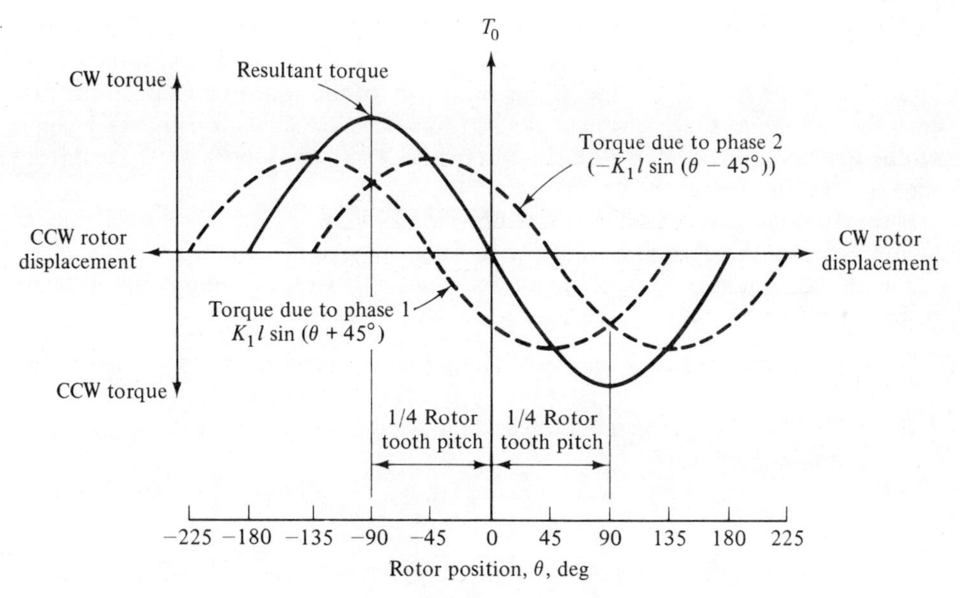

Fig. 18.34 Torque as function of rotor position for generalized permanent magnet stepping motor—both phases energized

Figure 18.34 shows the torque as a function of rotor position for a generalized case. Assuming the magnitude of the stator currents are equal, the developed torque when both the phases are energized can be shown to be:

$$T_D = \sqrt{(2)}K_T I \sin\theta \tag{18.23}$$

where

T_D = developed torque, Nm

k_T = motor torque constant, Nm/A

I = constant value of stator input current

$\theta = N_{RT}\phi$ deg.

N_{RT} = number of rotor teeth

ϕ = step angle, deg.

18.6.6 *Stepping motor dynamic characteristics*

18.6.6.1 Torque/speed. Most practical systems in which stepping motors are employed involve frequent increases to fast stepping rates for quick return (slewing) in addition to the rapid and accurate initial response demanded of a servo-motor.

Thus the first and most obvious observation is that a motor must deliver sufficient torque to operate a servo system; at the same time it must be capable of performing within a defined stepping rate range. But in spite of modern advancement in the designs of both stepping motors and drive units, the general speed of most conventional systems is faster than that at which most available stepping motors can run without error, or, if the motor does run fast enough it may not have enough torque at the desired speed. These speed limitations may arise from two sources, the design limitation of the motor, and the design limitation of the control circuit. In the first case, the motor may reach a form of mechanical resonance which constitutes the limit on the maximum speed response. In other cases, the rotor inertia prevents it from starting and stopping at the desired rate.

A third limitation is possible with stator windings with long time constants, since at high pulse rates the current cannot rise to fully energize the winding.

The problem of speed limitation may rest not only with the motor, but with the respective drive circuit. Possible reasons for this would be:

(a) Drive circuit not capable of switching the necessary currents at the desired rate.
(b) The width of stepping pulse fed into the windings is too great, causing overlap at high pulse rates.
(c) General overheating of components causing degradation of switching characteristics.

The power available from a stepping motor is in many ways the key to its ability to stop or start on load without error. In spite of the normally low efficiency (measured in terms of mechanical power output) one of the main features of a stepping motor is its relatively high torque output at low speeds.

Figure 18.35 outlines the general shape of the torque/speed relationship for stepping motors while Fig. 18.36 compares similar characteristics for servo-motors, normal induction motors, and synchronous motors. In practice, the curve for stepping motors is rarely observed because of the resonance effects noted below.

The holding torque quoted by manufacturers can be used as a figure of merit for the torque output of individual machines. However, in practice, the motor cannot drive a load requiring this amount of torque since this is a maximum torque level achieved at only one relative displacement position. A more meaningful indication of the loading capacity is the pull-out or pull-in torque. The former is limited by

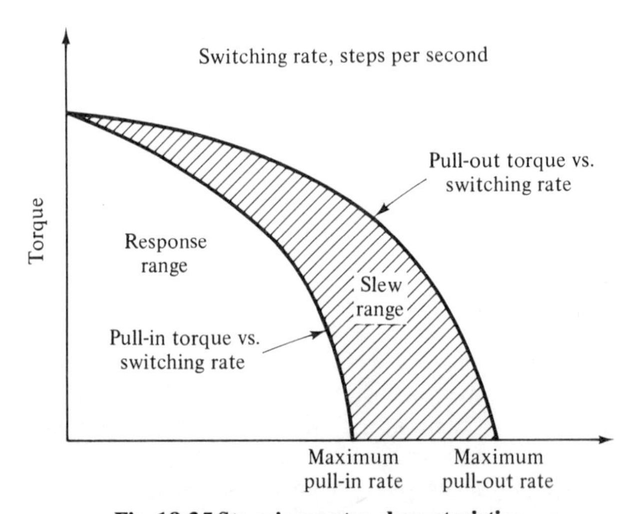

Fig. 18.35 Stepping motor characteristics

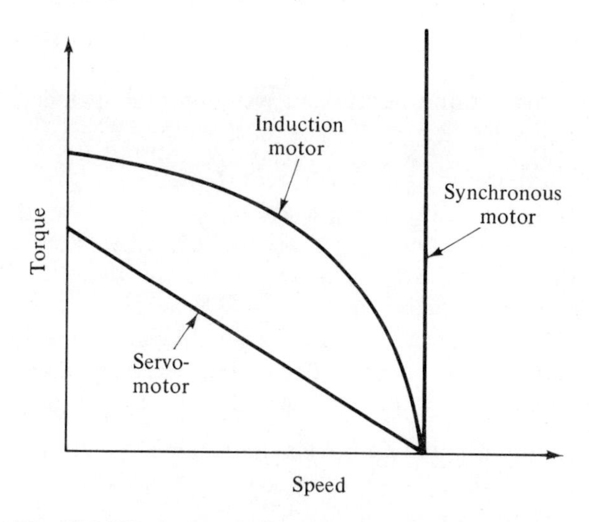

Fig. 18.36 Typical torque/ speed curves for other motors

inherent damping in the motor and load inertia has little or no effect. The latter depends on the total inertia and is a measure of the switching rate at which the motor can start without losing steps.

The pull-in curve is the upper boundary of the response range or start range. A motor can start, stop, and reverse on command in the response range; in this range, therefore, a stepping motor can be successfully used as an open-loop positioning device. The area between the pull-in and pull-out curves is the slew range. The slew range must be entered while the motor is running and within it the motor cannot start, stop, or reverse on command. However, the slew range is useful since the motor can run unidirectionally and follow the switching rate within a certain acceleration without step losses while it develops enough torque to overcome the load torque.

Within a particular speed range, a stepping motor may behave erratically. Stepping occurs but with irregular step angles. Pull-out torque varies considerably and at certain switching rates can actually be lower than at higher rates.

This unstable operation is caused by the combined inertia of load and rotor causing overshoot and, subsequently, an oscillatory motion at the end of a step. Friction, eddy currents, and hysteresis damp these step-end oscillations. Stable operation is achieved through proper selection of the physical and operational parameters of the system, such as load inertia, damping, and switching rate.

The greatest torque demand on the motor is made when starting the system because, in addition to the torque required to drive the load (including friction, eddy currents, and hysteresis effects), inertial effects must be overcome. The total torque requirement may be expressed as:

$$T_{\text{total}} + T_{\text{load}} = T_{\text{inertia}} \tag{18.24}$$

Remembering that the starting and stopping stage may involve only one step of the motor, the torque demand for stopping is dependent on the relative values of inertia and load torques. The load torque aids the slowdown process and, if it is greater than inertia-torque effects, it is entirely possible to stop the motor, in synchronism, without providing a counter-acting torque via an electrical input to the motor.

However, as noted above, there are times when not enough torque is produced at the desired speed.

Methods of matching motor output and system torque include:

(a) Improving drive circuitry.
(b) Reduction in the winding time constant (L/R).
(c) Use of gearing (*at the expense of speed*).
(d) Another technique often more desirable, decreases system torque requirements by permitting the motor to accelerate over the first few starting pulses and decelerate over the last few.
(e) A different scheme for reducing the torque necessary for error-free starting of inertial loads uses a 'compliant' coupling between motor and load; this solution is appropriate where increased stopping oscillation can be tolerated.

However, there is a limit to the improvements attainable even by a combination of these methods.

18.6.7 *Resonance and damping of stepping motors*

The rotor of a stepping motor oscillates after each step before settling to its final position as shown in Fig. 18.37. This oscillation depends on the moment of inertia of the rotor and the connected load, the frictional torque of the load, and the type of electrical control. The rotor oscillates until its kinetic energy is dissipated.

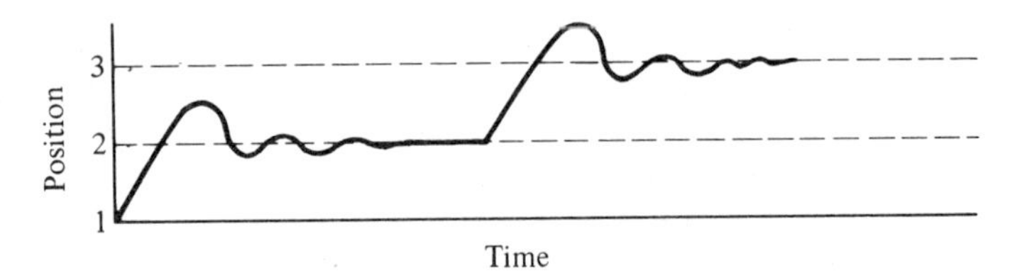

Fig. 18.37 Stepping motor oscillations

In PM stepping motors, these oscillations are damped by the rotor's interaction with the stator field, and by eddy current and hysteresis losses. In VR stepping motors, eddy current and hysteresis losses provide little damping effect.

Resonance or the inability of the rotor to follow the step input command results when the torque angle increases beyond the attractive capability of the advancing magnetic field. Under resonant conditions, the rotor will oscillate at random and lose

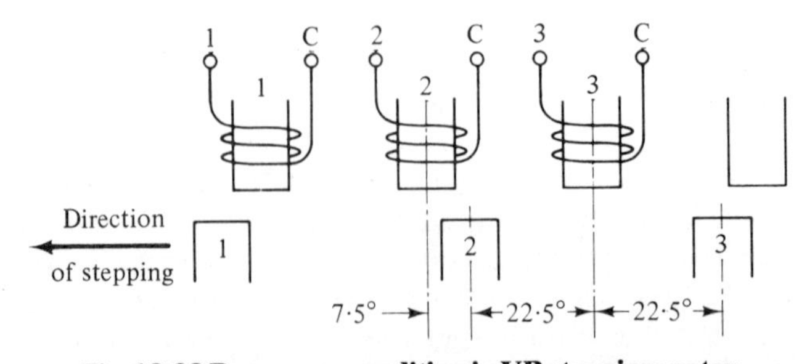

Fig. 18.38 Resonance condition in VR stepping motor

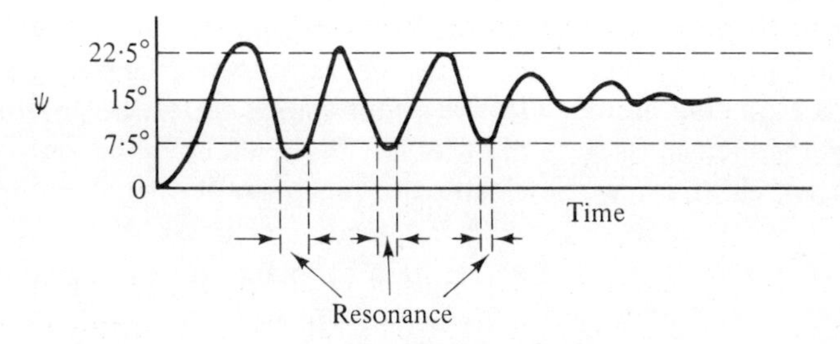

Fig. 18.39 Excessive oscillation

step integrity. Figures 18.38 and 18.39 show this situation in a 15° VR stepping motor.

In Fig. 18.38 rotor pole 2 is 7·5° away from the step position, with the other stator winding, 3, 22·5° apart on either side. If at this precise moment current is switched from stator winding 2 to winding 3, it is possible for the rotor to step in either direction regardless of the programmed command. In a 15° VR stepping motor, resonance may typically occur at several points as shown in Fig. 18.39. To eliminate resonance in VR stepping motors the amplitude of oscillation must be kept within 7·5° deflection from each 15° step position.

Resonant conditions distort the torque/speed curve causing a drop in torque, and are very disturbing phenomena in stepping motors. A speed/torque characteristic with a resonant condition is shown in Fig. 18.40. It is highly desirable to remove this

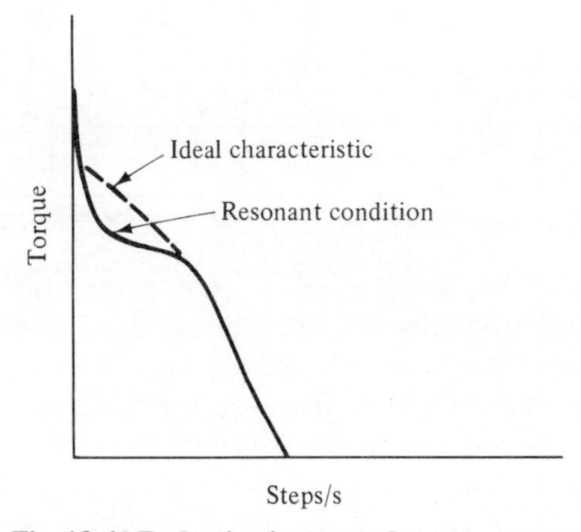

Fig. 18.40 Reduction in torque due to resonance

phenomenon from the motor by improving damping. This may be achieved by several methods. However, the performance characteristics of a stepping motor are so interrelated that to achieve better damping, another characteristic has to be sacrificed. For example, reducing the air-gap length tends to increase the torque and rotor damping but also has the effect of reducing the pull-in and pull-out rates. Different types of damping methods are available and these have been described by some writers and manufacturers.

Mechanical damping by introducing friction is very effective but this also reduces the pull-in and pull-out rates. One such device consists of an inert mass connected to the motor shaft by a frictional coupling. The inert mass is not able to follow rapid rotational oscillations of the motor shaft. Therefore the frictional coupling has a damping effect on the motor. On the other hand, when the motor is running continuously, the inert mass runs idly with it without affecting the motor. In order to achieve high switching rates, the electrical time constant of the winding should be kept small. A resistance in series with the winding achieves this. It is theoretically possible to achieve a very small time constant by increasing the series resistance but there are limits as this increase in resistance reduces the electrical damping of the motor.

Other suggestions include the use of a shock absorber in the form of a short-circuited cage, and pneumatic or hydraulic mechanisms to reduce the oscillation. Oscillations can be damped by applying a brake to the shaft and Fig. 18.41 shows the

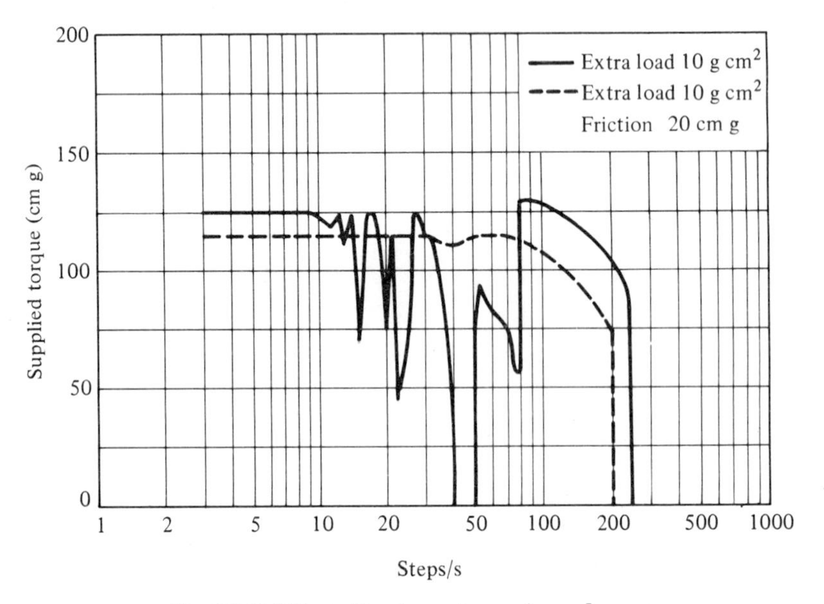

Fig. 18.41 Effect of brake on torque/ speed curve

torque/speed characteristic with and without such a brake. It is seen that the torque is reduced by the brake. Other methods of damping include:

(a) *Slip-clutch damping.* This mechanical damper utilizes a heavy inertia wheel sliding between two collars, Fig. 18.42. As the rotor moves, the inertia wheel resists movement and adds load to the system. This has the effect of reducing the rotor speed and consequently decreasing the oscillation. The amount of friction is controlled by spring pressure, using PTFE discs to separate the steel members. The damping will vary as the system wears.

(b) *Resistive damping.* External resistors across the stator windings, Fig. 18.43 allow rated current through one phase while limiting the current through the remaining two phases. As a result, a slight retarding torque is applied to the rotor by the two windings not carrying rated current. This prevents the rotor from accelerating as quickly and limits overshoots. Resistive damping increases power consumption by an average of 20 per cent for most inertia loads.

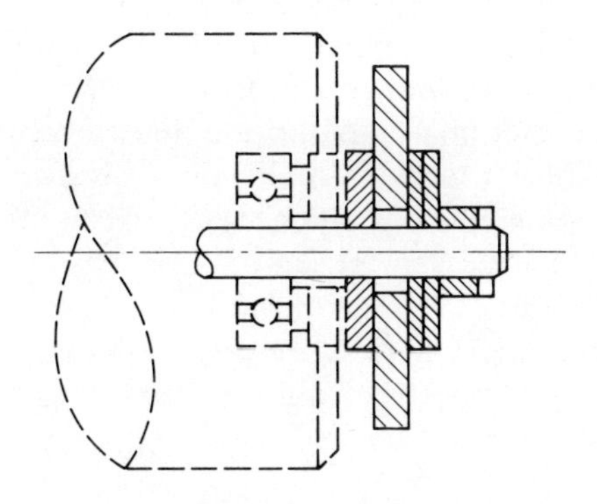

Fig. 18.42 Slip-clutch damping

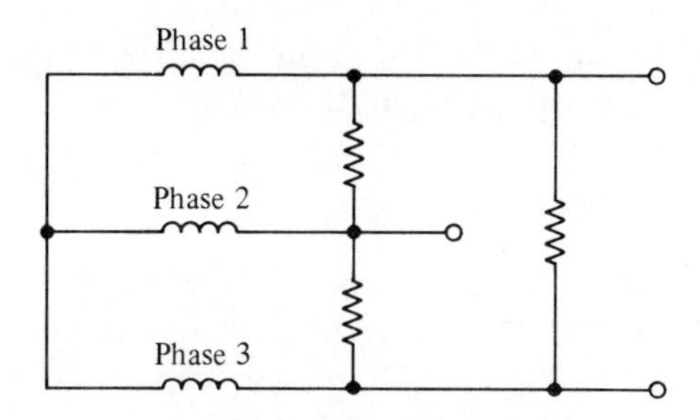

Fig. 18.43 Resistive damping

(c) *Capacitive damping.* In place of resistors, capacitors can be used to provide retarding torque, Fig. 18.44. At the moment phase one is de-energized and field two energized, the capacitor on phase one slowly discharges. The discharging current applies retarding torque to the rotor. This action repeats as the remaining phases are shifted. This method offers the advantage of lower power consumption as compared to resistive damping.

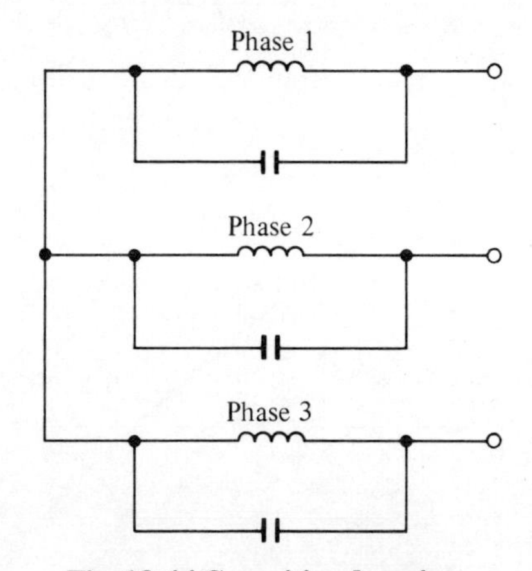

Fig. 18.44 Capacitive damping

(d) *Two-phase damping.* In this method, two stator windings are energized simultaneously, causing the rotor to index to a minimum reluctance position, midway between the stator poles. The step angle is not changed from single phase energization, but the final rotor position will be different from that of single phase energization. While advancing to this final step position, the two adjacent rotor teeth exert equal and opposite torque, which in combination with the stator magnetic field, produce twice the damping effect of single-phase excitation. The disadvantage of two-phase damping is that power consumption is approximately double that of single-phase excitation, while speed and torque remain essentially unchanged.

(e) *Retarding torque damping.* In this method the motor is excited in a single-phase mode and is damped by supplying a pulse of power to the stator winding last energized. Drive circuit complexity and increased expense may be factors for consideration.

Conflict arises when the stepping motor system is used as a high-speed positioning device where angular accuracy is critical. Here, both start-stop speed and damping characteristics must be considered. Unfortunately, drives that yield high start-stop rates tend to be poor in settling characteristics.

18.6.8 *Inertial aspects of the stepping motor*

The dynamic characteristics of a motor are modified by the effects of the driven load. *Inertial* load has an enormous effect on the ability to start and stop a stepping motor without error and therefore has a direct influence on its use in open-loop systems. The rotor moment of inertia gives an indication as to the limitations of load inertia which may be carried without affecting motor performance. With a load inertia, or referred load inertia when gearing is used, greater than the rotor moment of inertia, the maximum synchronizing rate is reduced, response time is increased, and overshoot may be caused. With friction load there is no problem.

Figure 18.45 illustrates the effect of inertia upon error-free starting characteristics. In system selection it is therefore necessary to consider the motor's own rotor inertia as a guide to inertia-accelerating abilities. However, inertias above this level can be tolerated depending upon how much compliance exists in the system couplings, etc. These are empirical factors in the consideration of motor life.

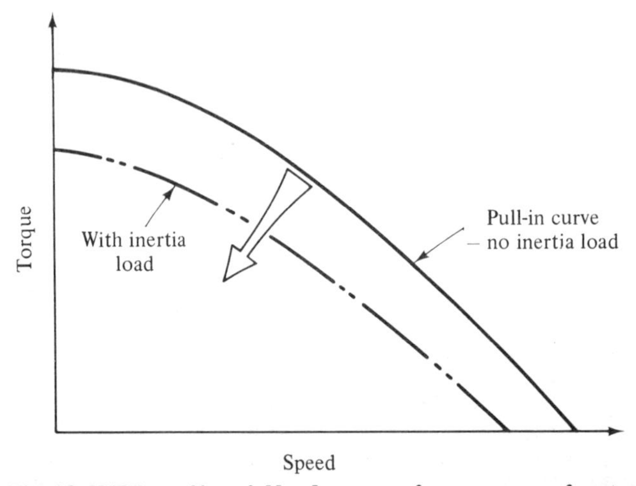

Fig. 18.45 Effect of inertial load on error-free response of motor

In all practical systems an explicit estimate of the moment of inertia of the moving system should be made before selecting a motor for a particular application.

Inertia becomes the critical parameter for motors operating at high stepping rates. Most motors achieve high synchronizing rates utilizing rotors with very low internal inertia (such as thin wall tubular rotors). This means that apart from systems involving very little inertia (such as positioning via small shafts, hollow cylinders or gears through which the reflected load is negligible) the motor may not be able to supply sufficient pull-in torque for many applications because load inertia will be greater than rotor inertia.

Further, operation at the high stepping rates accentuates the problem that the stepper motor is not a constant torque mechanism. In fact, for the first part of a given step, the motor is producing more than twice rated torque and then for the last part of the step, torque is falling off. For these reasons, in the case of high-speed motors, the inertia–speed relationship overrides the usual torque/speed characteristics in determining practical performance. During selection it is therefore essential to consult the stepping rate/inertia curve for each suitable motor. Figure 18.46 illustrates such a

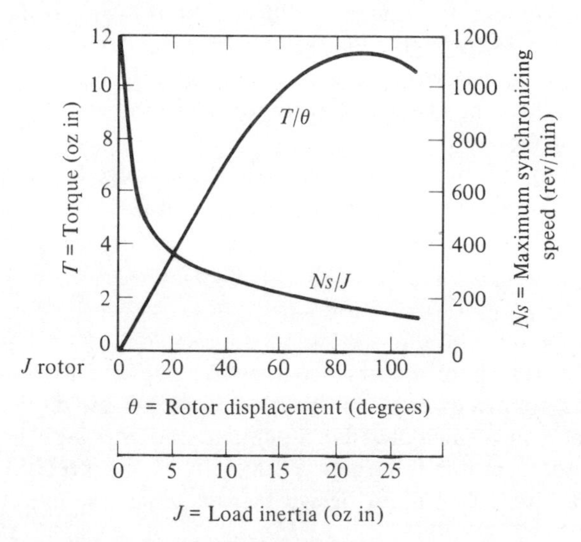

Fig. 18.46 Torque/ displacement and speed/ inertia curves for size 23 permanent magnet-stepping motor

curve for a typical size 23 permanent magnet motor indicating, under the rated conditions, the inertia that can be started and stopped at the speeds shown without losing steps. Obviously, in every application, it is necessary to check that sufficient torque will be produced at the required speeds but there is some disagreement between manufacturers as to the relative significance of this relationship. Although it is a most important constraining parameter, inertia is not generally a determining factor in most systems except those at high stepping rates—as noted.

Between them friction and inertia, combined with the required load, speed, and acceleration, determine the basic requirements of the system, since friction and load determine the power output and inertia and speed define the amount of kinetic energy that must be fed into the system on starting and removed when stopping.

18.6.9 *Step angle (resolution)*

The motor step angle affects the resolution or output positioning of the system. If all other things are equal, the smaller the step angle the more closely the output can be positioned to any predetermined point. At the same time step angle determines, along with output speeds, the stepping rate of the system. Stepping rate fixes the angular acceleration, α, which must be known in practical applications for calculating the inertia torque required.

Obviously, step angle is thus a very important parameter in all stepping motor applications. The ideal stepping motor must therefore have not only the correct torque/speed relationship for a particular requirement but also the ideal step angle. However, it follows that the smaller the step angle, the longer the output mechanism will take to move from one point to another. Therefore, a choice of step angle must be a compromise between resolution and speed.

By virtue of their design, the larger type reluctance and hybrid stepping motors tend to have the smaller step sizes, while the permanent magnet types have the larger ones.

If a particular application requires a different step angle to those offered, it is possible to achieve this through gearing. Unfortunately this increases complexity, possibility of failure, inertia, expense, and system error. If the step angle is decreased by gearing, output shaft speed is also decreased; and if step size is increased, then so is the inertia seen by the motor.

A second method of reducing the step angle is to change the driving logic. Some motors can have the number of steps per revolution doubled by using a different mode of energization to the windings.

18.6.10 *Accuracy*

In the application to stepper motors, accuracy indicates the tolerance limits within which the motor stops at a given step. This can be expressed in degrees or percentage (e.g., a motor has a step angle of $1{\cdot}8° \pm 5$ per cent or $1{\cdot}8° \pm 0{\cdot}09°$).

The accuracy of a motor is largely a function of the precision of its air-gap machining and a number of other design factors which are extremely difficult to overcome.

Accuracy is not often a problem in the application of stepping motors because many have good accuracies and in many systems involving mechanical couplings, etc, the stepping motor is the most accurate system component. However, in other applications where accurate and highly repeatable positioning are of the utmost importance, motor accuracy could become a limiting consideration. An obvious solution in these cases is to use gearing to increase system resolution and hence absolute (but not relative) accuracy.

18.6.11 *Mechanical endurance*

There is a minimum of wear and maintenance with stepping motors and as they do not employ brakes or commutators, the units are remarkably trouble-free. An operational life in the order of twenty years has been quoted for some types, but in the case of all units this will depend upon usage and environment. However, it is fairly safe to say that the mechanical life of a stepping motor is a direct function of the life of its bearings under conditions of normal use.

18.6.12 *Open or closed loop operation*

One of the great advantages of stepping motors, is that they can be used in open loop control systems. When properly applied, such systems provide for a fast response control without introducing the instability or hunting problems encountered with closed loop systems. Careful fabrication can reduce zeroing error below a small known quantity. Proper component selection can also save money as a suitable stepping motor can replace amplifier, servo-motor, tachometer, gear train, transducer, and stabilizing networks.

Complex system synthesis becomes more straightforward. Performance of elements and sub-systems are easier to predict and interaction between system loops is easier to control.

Load variation can distort the synchronized relation between pulse input and motor stepping, causing the motor to miss one or more steps. Without positional feedback, open loop operation requires that the motor chosen must provide sufficient torque and stepping speed within its *response range*. It is only within this range that the motor can start, stop or reverse on command instantaneously. In practice, it is advisable to check the alignment of the system at a particular point in its operating cycle so as to avoid possible accumulative errors.

One of the major limitations of the stepping motor is that it is only self-aligning within the symmetry of the rotor and any misalignment between motor and input pulses which may occur when the motor is de-energized, will remain undetected. Therefore the load position should always be checked when the motor d.c. supply is restored.

Recent work on the control of stepping motors has shown that the open loop performance can be increased considerably. In fact, a performance similar to a d.c. servo-motor is obtained by closing a minor loop around the motor and its controller. In this instance, the feedback signal is of the velocity type which is used to control the pulse rate into the motor controller.

In one such system a pulse generator is mounted on the motor shaft which produces a feedback pulse for each motor step. That in turn initiates the next driving pulse. A different technique is used in another system which uses the back e.m.f. from the motor to control the frequency of the oscillator producing the driving pulses.

However, in many applications open-loop operation may not be satisfactory. Generally this occurs under the following conditions.

(a) System alignment cannot be checked.
(b) Speed/load requirements approach the motor performance limitation.
(c) High accuracy is required with high speed.
(d) Load is entirely inertial and no external damping can be used.
(e) Motor may operate in the slew range.
(f) Fine resolution is required, hence a variable reluctance motor must be used and its ambiguity cannot be tolerated.
(g) If a direct readout of physical position is required.
(h) If position transducers already exist in the system for other reasons.

(i) If a programming system is used which is likely to accumulate unacceptable errors over a relatively long period.

Closed loop operation overcomes these problems.

As an example, with open loop a particular stepping motor can be pulsed at a maximum of 200 steps/s, but may drop out of synchronism when a minor load disturbance occurs. With closed loop operation, the same stepping motor can operate smoothly at 3000 steps/s and can be completely stalled without losing control.

18.6.13 *Environmental conditions*

Consideration must also be given to several environmental factors possibly affecting the choice of motor type or even particular model. Ambient temperature obviously affects motor performance, winding resistance, and bearing life. Vibration may be a problem in causing loss of angular position. Here a motor can be controlled by the detent torque of permanent magnet motors. Similarly when a stepping motor is used in an external a.c. magnetic field, some rotor demagnetization may occur in permanent magnet motors and hence a variable reluctance motor would be preferable.

18.6.14 *Stepping motor selection*

18.6.14.1 Permanent magnet. The permanent magnet motor is in general capable of higher rotational speeds than the variable reluctance type; this arises because permanent magnet motors usually rotate at higher step angles, i.e., 90° and 45°, than for the variable reluctance type; this offsets the lower pulse rate response of this motor.

Because of its inherent damping when driven from low impedance drive circuits, the pulse response rate of the permanent magnet motor is lower than for the variable reluctance type.

The rotors of permanent magnet motors are held in a detent position (due to the permanent magnetic field formed by a magnetic lock) when the excitation is switched off.

The torque output of a permanent magnet motor is lower than for a variable reluctance motor of corresponding size. While a reduced air-gap between rotor and stator would increase the holding torque, it would also reduce maximum pulse rate response since the rotor would be overdamped. Because of the low pulsing rates employed, permanent magnet motors do not exhibit marked 'resonance' effects, and in two-pole types the rotor positioning is unique.

The permanent magnet motor can be driven by simple drive logic, or electromechanical switching devices.

18.6.14.2 Variable reluctance. A feature of variable reluctance motors is the high output over a wide pulse rate range.

The small incremental steps usually associated with variable reluctance motors provide for a high order of resolution. In some applications it eliminates the need for output gearing. It is also a feature of particular importance when the stepping motor is used to control a distribution valve in an electrohydraulic control system.

Variable reluctance motors of the multi-stack type, in addition to providing a small incremental step angle, also have a high response characteristic in that the maximum pull-in rate is in the order of twice that of a permanent magnet motor. Resonance is more marked than for permanent magnet motors because of lack of inherent damping.

18.6.14.3 Hybrid. Hybrid types of stepping motors are used when the requirement is for an output torque higher than that of variable reluctance types together with a small step angle and a high stepping rate.

Another important feature of the hybrid stepping motor is the very fast stopping characteristic. This is achieved by simply de-energizing the motor, no dynamic braking being necessary. The simplicity of control, plus the fast starting, stopping and reversing characteristics make it a suitable 'ON–OFF' servo-motor.

The torque/step-rate curves in Fig. 18.47 show an interesting comparison between the performance of a number of different types of stepper motors all of the same size, i.e., size 11.

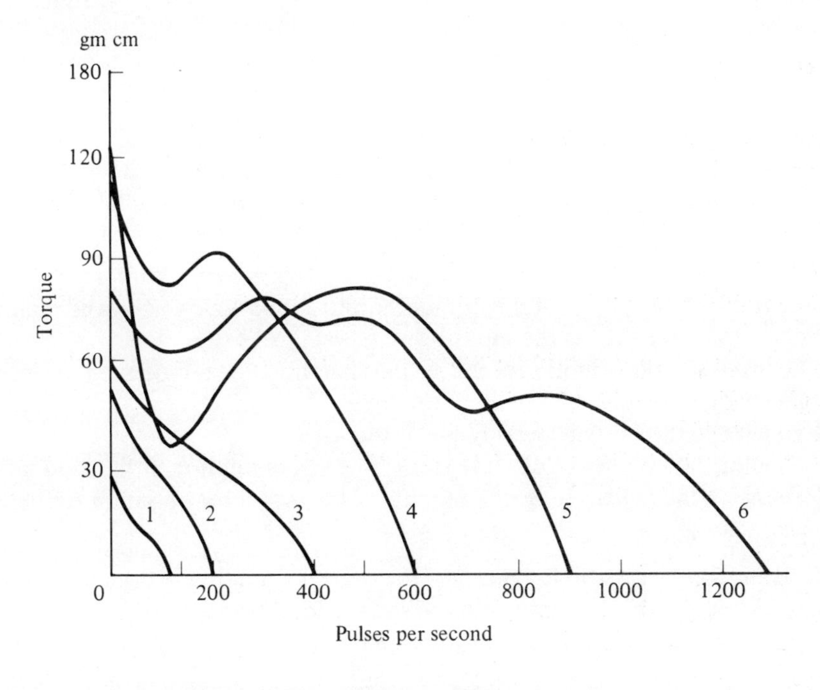

Curve No. (1) Permanent magnet 4 Phase 13 Watts Input
Curve No. (2) Permanent magnet 4 Phase 20 Watts Input
Curve No. (3) Permanent magnet 3 Phase 10 Watts Input 30° Step angle
Curve No. (4) Variable reluctance 3 Phase 10 Watts Input 15° Step angle
Curve No. (5) Variable reluctance 3 Phase 29 Watts Input 15° Step angle
Curve No. (6) Variable reluctance 3 Phase (Multi-stack) 5° Step angle

Fig. 18.47 Torque/ step-rate curves for six different types of size 11 stepper motors

When the curves of Fig. 18.47 are drawn on the basis of torque versus speed in rev/min (see Fig. 18.48) the performance of permanent magnet motors is presented more favourably.

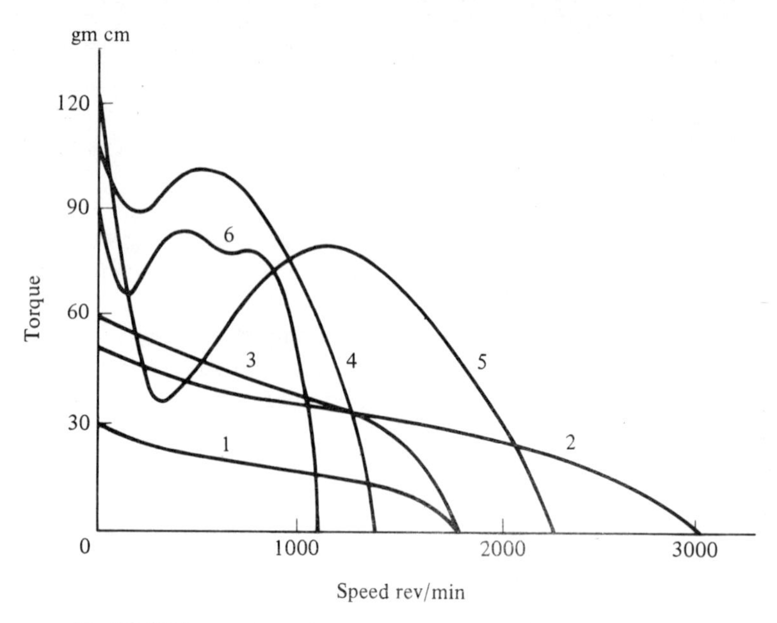

Fig. 18.48 Torque/ speed curves for permanent magnet motors

18.6.15 *Gearing*

There are several instances when gearing might be used with stepping motors. These are:

(a) To transmit the power of the output shaft to a load which cannot conveniently be coupled directly to the motor.
(b) To translate rotary motion of the output shaft into linear motion of the load.
(c) To change the torque applied to the load.
(d) To change the output speed range at the load.
(e) To alter the number of discrete steps for each revolution of the load motor.
(f) To match the inertia of the load to that of the motor for optimum acceleration characteristics.

The third and fourth reasons for gearing are often encountered in the adaptation of a standard motor to a specific application where the torque and/or speed ranges of the motor do not adequately fulfil the requirements.

18.6.16 *Number of phases*

In conventional a.c. machines, the number of phases of a motor being selected for a particular application will often be influenced by the power supply that is most convenient and available. With stepper motors, this is not relevant as the drive circuitry, which is itself supplied from a d.c. source, will be designed to meet the particular requirements of the motor.

However, in addition to any power output considerations, the main criterion for determining the number of phases on the motor to be selected will be the step angle required. The correct angle will be obtained from a combination of the number of phases; the mode of operation (as described elsewhere) and, in the case of permanent magnet types, the number of poles on the rotor.

18.6.17 *Driving circuits and methods of excitation*

The purpose of stepping motor drive circuitry is, in addition to producing the usual motor excitation source, to monitor the successive input pulses applied to the circuit in order to control the sequence of energization of the motor stator winding, thus producing the required stepped angular motion of the rotor.

Drive circuits for stepping motors can be designed to cater for the need of a particular application or system, considering economy and ease of operations as well as obtaining the greatest advantage from the technical possibilities (i.e., performance, step angle, and speed) of the motor itself. It is therefore important, from the point of view of a user, that the operation of both stepping motors and their drive circuits is fully understood to obtain maximum benefit from improvement in design.

There are many types of switching units available at much reduced size and cost to suit every need and application. These circuits utilize modern discrete components and integrated circuits, and may be obtained in modular packages or on edge connector type printed circuit boards for incorporating in 'plug in' type racking systems, etc.

Generally, drive circuitry consists of logically controlled power output stages, designed to accept information in the form of an external pulse train (of any shape) and convert it into separate appropriately timed pulses which are amplified to drive three, four, five or more phases of a motor. The timing of these triggering pulses is arranged in various combinations or modes which control the sequence by which the windings are energized and the rate at which the rotor steps. Typical stepping modes are shown in tables 18.2 and 18.3. Suitable logic circuits are used to feed these pulses, in the sequence required, to the various phases and the rotor advances by a unit step per pulse. The sequence of rotation is controlled by means of a 'switch' used for feeding binary levels pertaining to two states, clockwise and counter-clockwise. For controlling the sequence in accordance with a change in signal voltage level the switch takes the form of a suitable Schmitt Trigger.

18.6.18 *Choice of driving circuit.* Stepping motors may operate from a variety of logic circuits giving single or double d.c. pulses, or commutated d.c. square waves. The amplitude and length of pulses must be considered along with the availability of standby power. As will be appreciated from the different types of control circuits discussed, the selection of a particular motor is greatly influenced by the drive circuit and frequently this is *considered to be the major factor in motor selection.*

In most cases, the individual suppliers will recommend a particular control circuit for use with each motor to give optimum motor performance. Such motors will either be described in the product literature, or will certainly be available on contact with a particular supplier.

The majority of larger suppliers provide their own control circuits and therefore tend to recommend these units. However, there are a number of independent suppliers of control systems specifically for stepping motors. Because of the great influence of the control unit on stepper motor performance, it is always worth while investigating control circuit design in detail when selecting a stepping motor. In fact,

Table 18.2 Typical stepping motor modes

Mode	Figure	Rotor position	Step A	Step B	Step C	Step D	Step position	Step/rev
1	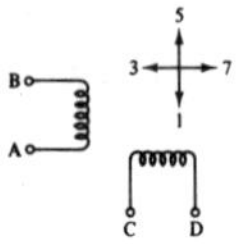	1	+	−	0	0	90°	4
		3	0	0	+	−		
		5	−	+	0	0		
		7	0	0	−	+		
2	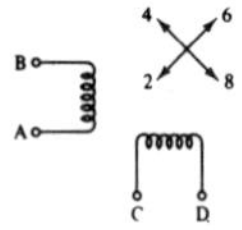	2	+	−	+	−	90°	4
		4	−	+	+	−		
		6	−	+	−	+		
		8	+	−	−	+		
3	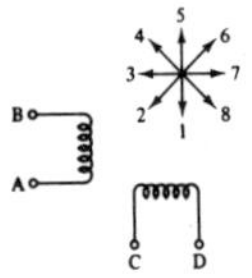	1	+	−	0	0	45°	8
		2	+	−	+	−		
		3	0	0	+	−		
		4	−	+	+	−		
		5	−	+	0	0		
		6	−	+	−	+		
		7	0	0	−	+		
		8	+	−	−	+		
Mode 3 is a combination of modes 1 and 2								
4	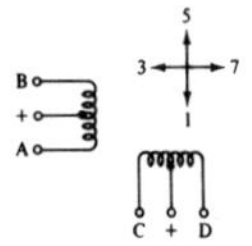	1	0	−	0	0	90°	4
		3	0	0	0	−		
		5	−	0	0	0		
		7	0	0	−	0		
5	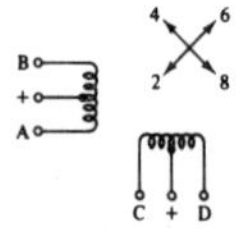	2	0	−	0	−	90°	4
		4	−	0	0	−		
		6	−	0	−	0		
		8	0	−	−	0		
6	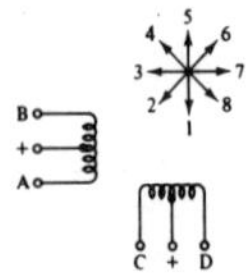	1	0	−	0	0	45°	8
		2	0	−	0	−		
		3	0	0	0	−		
		4	−	0	0	−		
		5	−	0	0	0		
		6	−	0	−	0		
		7	0	0	−	0		
		8	0	−	−	0		
Mode 6 is a combination of modes 4 and 5								

Table 18.3 Modes for three-phase motor

Mode	Figure	Rotor position	Step A	Step B	Step C	Step position	Step /rev
1		1	+	−	−		
		2	−	+	−	120°	3
		3	−	−	+		
2		1	+	+	−		
		2	−	+	+	120°	3
		3	+	−	+		
3		1	+	−	−		
		2	+	+	−		
		3	−	+	−	60°	6
		4	−	+	+		
		5	−	−	+		
		6	+	−	+		
4		1	−	+	+		
		2	−	0	+		
		3	−	−	+		
		4	0	−	+		
		5	+	−	+		
		6	+	−	0	30°	12
		7	+	−	−		
		8	+	0	−		
		9	+	+	−		
		10	0	+	−		
		11	−	+	−		
		12	−	+	0		

Similarly it can be shown that by suitable modes with five-phase excitation, basic step angles of 36° and 72° can be obtained while six phases give basic angles of 30° and 60°. These have increased torque compared with that available with a single-phase mode.

in many applications, users prefer to design their own control circuits in common with the other electronics involved in their system, rather than incorporate a special 'control unit'. For machine tool applications, stepping motors are generally supplied in a complete system, comprising motor and controls. Figure 18.49 illustrates the influence of particular control systems on the output performance of a particular motor.

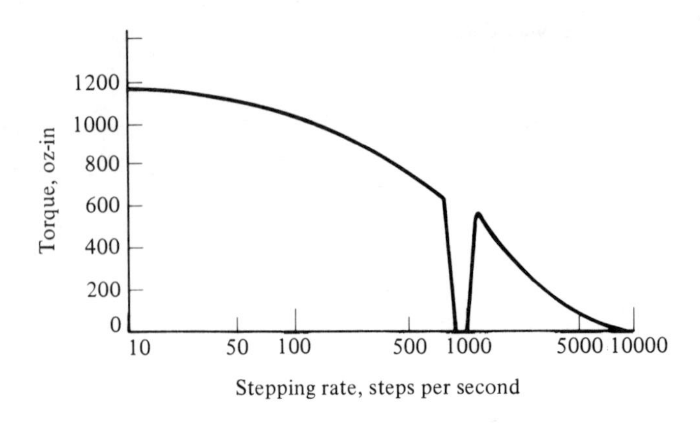

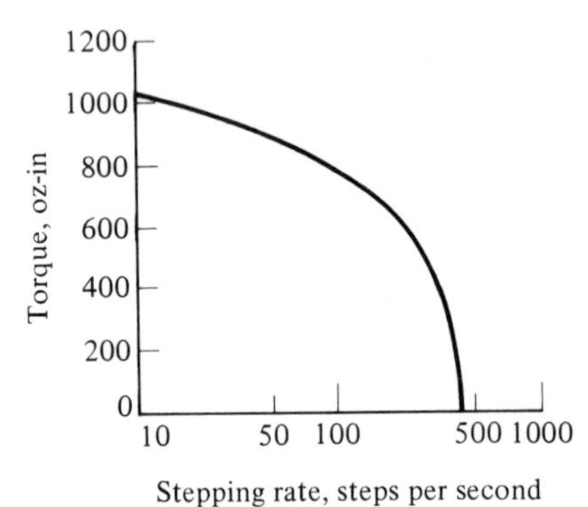

Fig. 18.49 Motor performance using differing drive circuits

18.7 Direct-current Torque Motors

18.7.1 *Introduction*

It could be argued that any motor which produces a torque could be labelled a 'Torque Motor' and indeed in the past there have been many types and varieties produced which have been so called. These have included a completely housed type resembling a conventional d.c. permanent magnet motor but having a limited angle of shaft movement and with the internal circuitry analogous to that of a polarized magnetic bridge. The direction and angle of rotation depend upon the sign and magnitude of the current flowing in the windings.

However, nowadays the term 'torque motor' has become synonymous with the precision units developed for computer peripherals, business machines, space applications, and other such systems. For the purpose of this book these are the types which will be considered.

18.7.2 *Conventional d.c. torque motors*

In general, torque motors are designed essentially for high torque 'standstill' operation in positioning systems, and for high torque at low speeds in speed control systems. They are manufactured in sizes ranging from torques of several millinewton metres to several kilonewton metres.

Most torque motors are frameless and require very little space, offering the greatest flexibility and adaptability in application. The 'Pancake' versions are, as the name suggests, thin compared to their diameter and have a relatively large axial hole through the rotor to enable the motor to be directly attached to the load, therefore no gearbox is required, with the result that backlash is eliminated and a high torque-to-inertia ratio exists at the load shaft. A high coupling stiffness and hence a high mechanical resonance frequency, results in a high servo stiffness. With conventional servo-gearing, backlash contributes towards reducing positional accuracy, but with a direct-drive torque motor positional accuracy is limited only by the error transducer system.

Because torque motors have low values of self-inductance, armature currents have a very fast rise time and hence torque is rapidly produced; as a result the motor responds rapidly at all operational speeds. Special design features such as high level magnetic saturation, together with the use of a large number of poles, reduce the armature self-inductance still further to very low values, thereby producing a very fast time response to changes in voltage. The torque increases directly with input current, independent of speed or angular position and is linear through zero excitation, assuring no dead band due to torque non-linearities.

18.7.3 *Construction*

In construction, the torque motor consists of an anisotropic magnet, either cast in ring form and annularly grain orientated, or in block form mounted in a low reluctance steel ring to enable a larger number of poles to be formed.

The armature laminations are made from high permeability silicon steels, containing a large number of slots into which the winding is machine wound and insulated from the iron with fluidized bedded resins. Polymide coverings are normally employed, allowing high temperature rises.

The windings, in general, are wave wound, having the connections made at the non-brushgear end of the printed circuit copper commutator which is laid in the slots on top of the winding. The whole armature assembly is then encapsulated with an epoxy resin to produce a very robust construction with good heat dissipation.

The brushgear essentially consists of a moulded glass, nylon, or plastics ring containing small beryllium copper brush arms onto which the carbon brushes are secured. The complete assembly is finally screwed to the magnet ring.

18.7.4 *Brushless d.c. torque motor*

As with the conventional d.c. torque motor the need to produce sufficient torque to actuate servo-systems with stringent performance characteristics led to the development of brushless d.c. torque motors capable of direct application to a load. The interaction of a permanent magnet and induced electromagnetic fields gives sufficient torque to actuate servo-systems in nulling, positioning, and actuator applications, without the need for gear trains.

However, in conventional d.c. torque motors, preformed windings inserted into slots create a variable reluctance path for the field flux as a function of motor position. Slotting and commutation effects produce ripple torques approaching peak-to-peak values of 15 per cent of the peak torque. The elimination of ripple torques could increase control system resolution, sensitivity, and linearity.

The shortcomings on conventional d.c. torque motors are reduced with a brushless d.c. torque motor originally designed as a precision zero-friction actuator for gyro-stabilized aerial camera platforms. The motor employs a constant reluctance magnetic circuit and a precision toroidally distributed coil winding. No undesirable slot ripple or varying torque effects due to motor rotation are therefore experienced. Used over its proper angular range, the motor is capable of providing extremely high linearity, resolution, efficiency, and reliability.

Mechanical isolation between the driving and driven members, a result of brush elimination, ensures an operational life at full-rated power not limited by wear considerations. Another advantage is that power is not dissipated in a fixed field phase (no delivered torque), so the power supply is utilized at optimum efficiency. Depending on the duty cycle and power available, peak torque capacity may be considerably higher than the rated continuous duty torque, without demagnetization.

The d.c. torquer (Fig. 18.50) requires a current direction in each internal winding capable of exerting a torque on the polarized rotor, such that the torque contributed

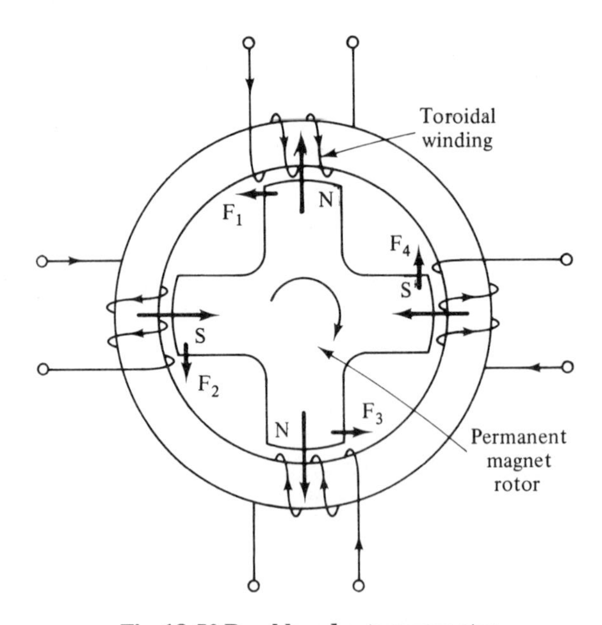

Fig. 18.50 Brushless d.c. torque motor

by each section is accumulative. Typical angular position versus torque curves (Fig. 18.51) show the ripple-free operating characteristics.

The permanent magnet rotors are cast and machined solid magnets. For designs in which pole pairs become numerous and casting impractical the rotors are fabricated assemblies of magnetic pole pieces. The annular stator core completes the magnetic circuit.

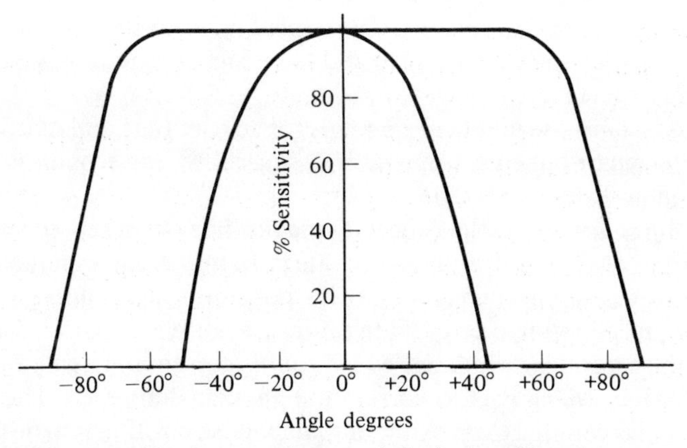

Fig. 18.51 Motor position/ torque characteristics for brushless d.c. torque motor

The selection of the number of polar pairs is determined by the following considerations:

(a) In very small units, more than two poles are physically impractical.
(b) As the number of pole pairs increase, the useful rotation angle decreases.
(c) Increasing the pole pairs increases the peak torque available for the same power input.
(d) Increasing the pole pairs increases the rotor inertia.

Although the two-pole magnet suggests 180° rotation capability without commutation, and the four-pole device suggests 90°, the width of the poles results in a practical motion of approximately 120° and 60° for the two- and four-pole constructions respectively.

The reliability of a brushless torque motor is a function of the insulation system and its operational environment. The distributed winding technique eliminates local hot spots, and the lack of stator slots avoids mechanical wire stresses.

Direct-current torque motors have successfully passed many of the stringent military environmental tests such as vibration, shock, humidity, temperature, and high altitude as well as having operated successfully both in space and when submerged in a variety of liquids from salt water to flammable hydrocarbons. The smooth, linear, constant reluctance winding technique also suggests application as a sensitive directly driven tachometer generator, particularly useful at low velocities.

18.8 Selsyns

18.8.1 *Introduction*

Selsyns are electromechanical devices which receive and transmit rotational movement. They may be used to operate equipment requiring mechanical imputs, such as dial indicators, valves, and dampers. The hardware is inexpensive and rugged, and therefore provides practical means of remote positioning and indication in rough-service industrial environments.

Selsyn units are interconnected electrically. To provide an equivalent interconnection with mechanical linkages increases costs and limits the relative locations of the transmitter and receiver. Electronic systems analogous to selsyns generally involve components which are more expensive and have higher maintenance requirements with lower life expectancy.

Most selsyn systems use identical elements for transmitters and receivers. Primary windings are brought out to a common a.c. source, while secondary windings are interconnected in series opposition.

The line voltage applied to the primaries makes the system self-synchronous. The primary winding of each unit induces a voltage between corresponding secondary leads. If the rotors of the two units are in position synchronism, the secondary voltages are equal and sum to zero. If the rotors are not synchronized, a net voltage is produced in the secondary windings, which causes current to circulate. This current produces a torque, which acts to correct the angular difference. The rotor of the transmitting unit is coupled to an external shaft, whose position is to be monitored. If allowed to run free, the receiving unit will track this position.

If a receiver is to drive a light load, a flywheel may be attached to the shaft through a slip clutch. This will limit acceleration and decrease overshoot after rapid speed changes. Flywheels may also be used on transmitters.

Operation principle of the system of synchronous transmission with three-phase or single-phase selsyns is based on the variation of the phase and the amplitude of the secondary e.m.f.s of the transmitting selsyn with regard to the secondary e.m.f.s of the receiving selsyn. Selsyns in these systems operate in two duties: one is the duty of rotor turn and the other the duty of continuous rotation with definite speed.

In the duty of rotor turn, the rotors of transmitting and receiving selsyns are only turned by a definite angle and then they are stopped. The angle of displacement between rotor axes is called the 'static error' of the system.

In the duty of continuous rotation the shafts of both selsyns are synchronously revolved. Such duty is called dynamic duty and the angle of displacement of rotor axes is called the 'dynamic error' of the system. The static error of the system is always less than the dynamic one.

The angle of displacement in the system arises from loss of alignment of selsyn rotors in space because of bearing friction, brush friction on rings, and load on the receiving selsyn shaft. This angle in indicating and measuring systems must be as small as possible. In systems of high accuracy it must not be more than half a degree.

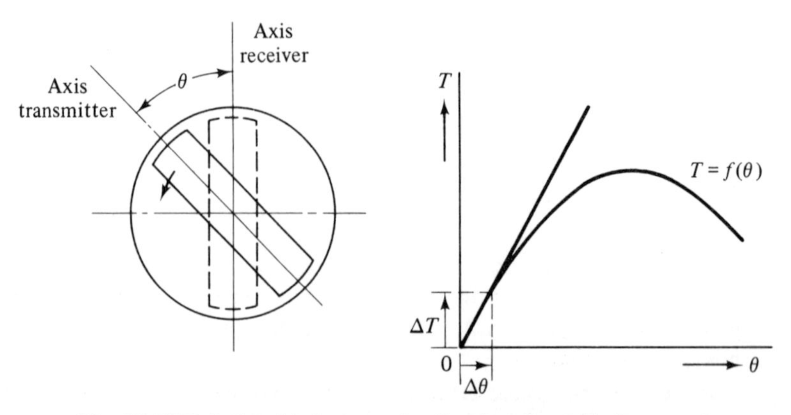

Fig. 18.52 Relationship between load and angle of displacement

For a load on the receiving selsyn shaft the angle of displacement can reach several degrees or even tens of degrees (Fig. 18.52).

18.8.2 *Single-phase selsyns*

Single-phase selsyns are induction machines with single-phase primary and three-phase secondary windings. The secondary winding is called the synchronization winding. The single-phase winding is usually put on the rotor of the selsyn for the purpose of reducing the number of slip-rings, and the three-phase winding is placed on the stator. It is possible to reverse the order of windings in the selsyn, i.e., the primary winding can be placed on the stator and the secondary winding on the rotor.

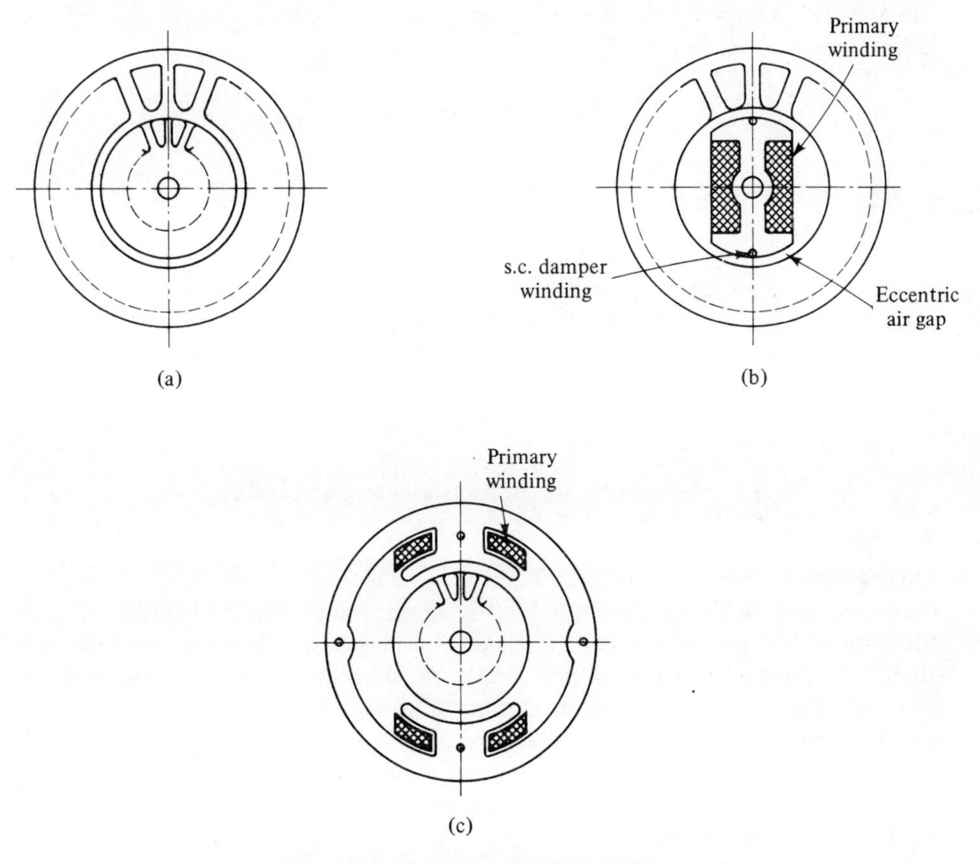

Fig. 18.53 Types of single-phase selsyns

However, this configuration is now normally found only in obsolete types of selsyns. In practice there are the following varieties of single-phase selsyns (Fig. 18.53):

(a) With the primary winding distributed in the slots of rotor and with a uniform air-gap (Fig. 18.53(a)).
(b) With the primary winding distributed in the slots of rotor and with short-circuited damper winding in them, mutually displaced along the periphery of rotor by one half of a pole pitch.

(c) With the primary winding on the salient pole rotor (Fig. 18.53(b)).
(d) With the primary winding on the salient pole stator and with three-phase windings distributed in the slots of rotor with slip-rings (Fig. 18.53(c)).

High accuracy of synchronous transmission is secured by high specific synchronizing torque and small friction. For the reduction of mechanical friction in selsyns, ball-bearings and light brushes on slip-rings are used.

The most usual construction of selsyns is with salient pole rotor and with three-phase winding of synchronization placed on the stator (Fig. 18.53(b)). The air-gap in salient pole selsyns is made eccentric for obtaining a sinusoidal distribution of the field in air-gap. For the reduction of higher harmonics associated with the stator or rotor teeth, the slots of one are skewed.

Figure 18.54 represents curves of synchronizing torques as functions of the displacement angle θ for different constructions of selsyns.

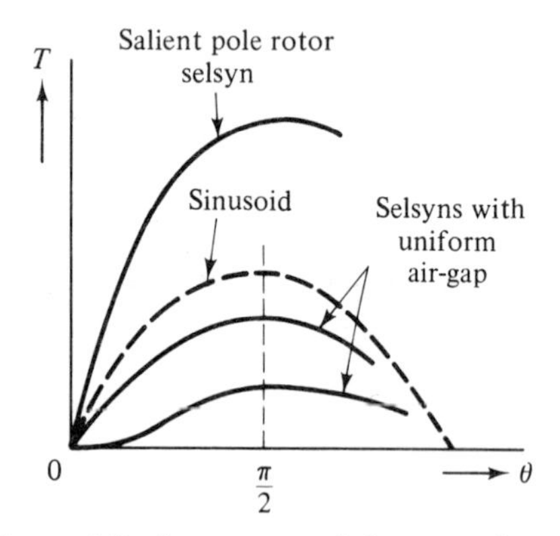

Fig. 18.54 Torque/ displacement angle for several types of selsyns

Selsyns with non-salient pole stator and rotor have the least specific synchronizing torque, m_s, and they are usually used in transformer duty (Fig. 18.53(a)).

Selsyns with non-salient pole stator and rotor and with a short-circuited damper winding on the rotor have a specific torque, m_s, greater than the previous type of selsyn and can be used in indicating systems (Fig. 18.53(b)).

Selsyns with a salient pole rotor have the highest specific torque, m_s, of them all and, therefore can be used in any system; such as in indicating systems, and in control systems with the load on the receiver selsyn shaft. These selsyns are the most popular of all single-phase selsyns.

18.8.3 *Three-phase selsyns*

Three-phase selsyns are small three-phase induction machines with wound rotors and slip-rings, the construction of which is the same as that of ordinary induction motors. Primary windings of the selsyn stators normally have a star connection to the supply and the rotor windings are used as the windings of synchronization, being connected differentially through slip-rings and observing proper phase sequence.

Figure 18.55 represents the connection diagram of a system utilizing three-phase selsyns. Principle of operation of this system is based on the variation of the phase of

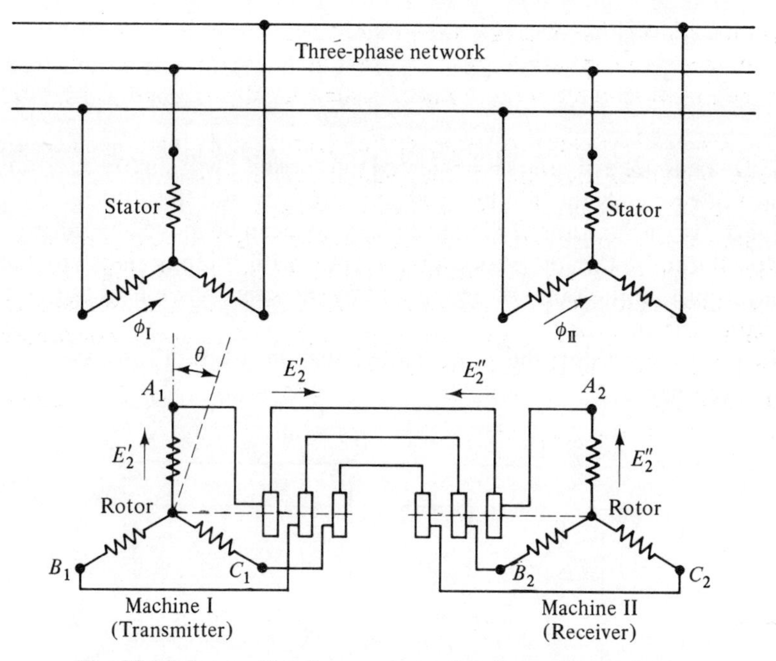

Fig. 18.55 Connections for a system using three-phase selsyns

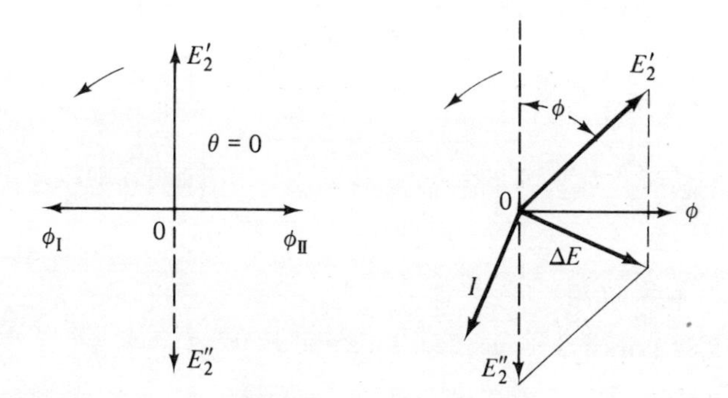

Fig. 18.56 Variation of phase of secondary e.m.f. of transmitting selsyn related to secondary e.m.f. of receiving selsyn

the secondary e.m.f.s of the transmitting selsyn with regard to the secondary e.m.f.s of the receiving selsyn (Fig. 18.56).

If the rotor axes of transmitting and receiving selsyns coincide, i.e., if similar rotor phase windings are in similar positions with regard to corresponding windings of the stators, equal and opposite e.m.f.s are induced in them. In this case current circulating in the rotor circuit is equal to zero, similar to the normal parallel connection of two identical three-phase synchronous machines. A phasor diagram of e.m.f.s for this case is represented in Fig. 18.56.

If the rotor of the transmitting selsyn is turned with regard to the rotor of receiving selsyn by some angle θ in the direction of rotation of the field, the change in phase of e.m.f. of the transmitting selsyn E'_2 with regard to e.m.f. of receiving selsyn E''_2 causes the appearance of resultant e.m.f. ΔE in the rotor circuit. On account of this a circulating current appears in this circuit given by:

$$I=\frac{\Delta E}{2Z_2}\approx\frac{\Delta E}{2x_2}$$

If the active resistance of rotors is neglected the current I will lag ΔE by an angle $\pi/2$. The rotor of one machine will be in generating duty and the rotor of the other in motoring duty. As the rotor of the receiving selsyn can be moved by the influence of torque produced by the interaction of current I with rotating magnetic field Φ'', it will tend to turn in the same direction and by the same angle θ as the rotor of the transmitting selsyn.

Figure 18.57 represents the principal connection diagram for a system utilizing single-phase selsyns.

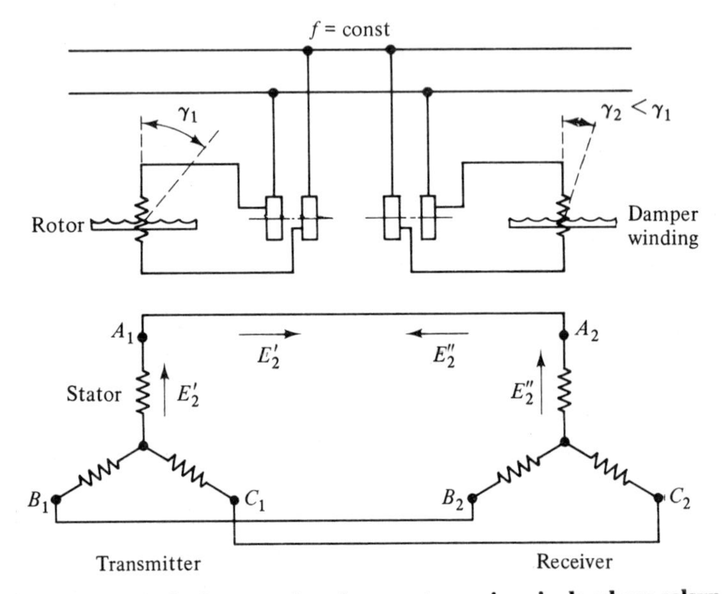

Fig. 18.57 Principal connections for a system using single-phase selsyns

The principle of operation of these systems is based on the variation of the value of the secondary e.m.f.s of the transmitting selsyn with regard to the secondary e.m.f.s of the receiving selsyn. Single-phase alternating current, supplied to the primary windings of the selsyns, produces the pulsating magnetic flux in each selsyn. This flux induces e.m.f.s in each phase of the secondary three-phase windings, coinciding in phase but differing in value. This difference depends on space position of phase windings with regard to the axis of primary winding. At the coincidence of the position of both selsyn rotors the secondary e.m.f.s in the circuit of the synchronization windings will be equal and opposite. Therefore, the current in this circuit will be zero. When the transmitting selsyn rotor is turned to some angle its e.m.f.s are decreased and torques will be produced on the selsyn shafts which will tend to move their rotors to the synchronous position. The rotor of the transmitting selsyn is coupled to the feed mechanism and the rotor of the receiving selsyn is free, and is turned by the synchronizing torque to the synchronous position with the rotor of the transmitter.

18.8.4 *Performance*

Selsyns are available in a range of torque, speed, and no-load accuracy ratings. The latter specification refers to the transmitter displacement required to cause receiver motion, with no external loading.

Speeds are usually limited by the lubricant used in the bearings rather than inherent electrical characteristics. Low-speed units are typically supplied with low-viscosity instrument oil, to minimize friction and maximize accuracy. Faster models often use grease-packed bearings, at some sacrifice in accuracy. A typical torque/displacement characteristic for two similar units in a simple system has a parabolic shape (Fig. 18.58).

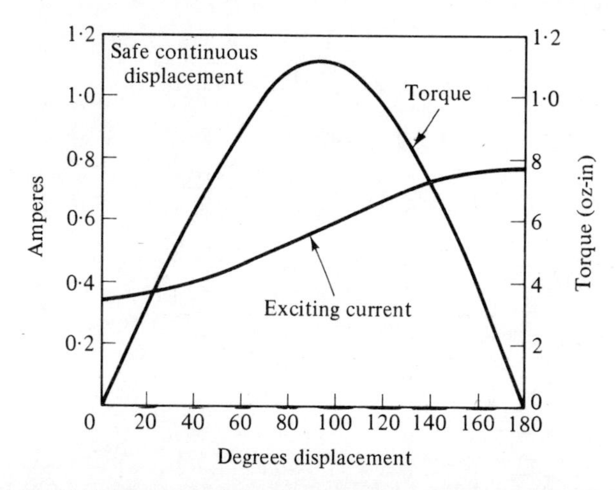

Fig. 18.58 Typical torque/ displacement for two similar units

References

1. G. Weiss, 'Optimum Design of Induction Torque Motors and Servomotors', *Trans. A.IEE* (Oct. 1955).
2. S. A. Davies, 'Converting Ideal to Working Data for Application of Two-Phase Servomotors', *Electrical Manufacturing* (September 1956).
3. J. C. West, *Servo Mechanisms*, The English Universities Press (1957).
4. S. A. Davies and B. K. Ledgerwood, *Electromechanical Components for Servomechanism* McGraw Hill (1961).
5. N. P. Yermolin, *Small Electrical Machines*, Rajkamal Ltd. (1963).
6. A. R. Upson and J. H. Batchelor, '*Syncho Engineering Handbook*', Muirhead & Co. Ltd. (1965).
7. E. H. Werninck, 'Servomotors, Actuators and Stepping Motors', *Proceedings of IEE Conference on Servocomponents*, Conference Publication No. 37, (November 1967).
8. M. Marrero and L. Reynolds, *Electro Mechanical Prime Movers—Mechanical Engineering Evaluations*, Macmillan Press Ltd. (1971).
9. Anon, Optimizing Servo and Geared Motor Performance Reduces Your Design Time, *Electromechanical Design* (August 1972).
10. B. Longden and S. E. Spriggs, 'a.c. & d.c. Servomotor and Tachogenerator Features and Design', *Control and Instrumentation* (November 1972).
11. J. H. M. Hofmeester and J. P. Koutstaal: Moving-coil motors, *Philips Tech. Rev.* **33**, 244–248, 1973 (No. 8/9).
12. S. K. Pal, *A Critical Review of Stepping Motors* ERA Report No. 73-5103 (1973).

13. Hy Natkin, 'Guide to Selecting Servomotors', *Instrument and Control Systems* (June 1973).
14. *Machine Design*, Electric Motor Reference Issue (April 1974).
15. B. A. Kennerly, Advantages of Permanent Magnet d.c. Motors', *Electrical Times* (17 October 1974).
16. D. E. Knights, 'Prospects for the Printed Motor', *Engineering* (March 1975).
17. S. Jenkins, 'Servomotors: Matching the Equipment to the Applications', *Control and Instrumentation* (July/August 1975).
18. J. M. Brown, 'Rare Earth Magnets—A New Trend in d.c. Motors', *Control Engineering* (October 1975).
19. J. Henry-Baudot (Société d'Electronique et d'Automatisme). Un nouveau moteur dans les servomécanismes. *Automatism* Tome IV, No. 3, 1959, pp. 107–113.
20. G. I. Biscoe, A. S. Mills, and D. Crooks, 'A Bibliography of Published Works on Stepping Motors', ERA Publication No. 77–1023, March 1977.

19 Maintenance

PART 1. ECONOMICS

19.1 Introduction

Maintenance policies can vary over a wide range of philosophies and are affected by many factors, among them:

Cost of repair.
Cost of avoidance of breakdown.
Consequence of breakdown.
Cost of alternative or stand-by plant.

It is essential at the outset to establish clearly the maintenance policy for each company or establishment, or in some cases even different policies for each department or group. This policy must be integrated with the company's general industrial activity and policy. For example, a major breakdown in a beet-sugar factory during the 'campaign' season could have serious consequences. Whereas a motor breakdown in a processing plant could create a hazard at any time, it would do so only occasionally, albeit more vitally, in applications like flood control pumping.

No other examples are quoted because each executive responsible for maintenance must identify his own priorities which will be determined by service needs, ranging from motors in non-essential applications such as entertainment to the vital need for continuity in such as nuclear plants.

Where there are many identical machines it may be preferable to await a breakdown and to have a quick replacement drill.

19.2 Clarification and Ratification of Policy

Each company or group should appoint a person responsible for the 'thinking' on its own overall policy, and when this is agreed by the management it should be put on record as Standard Company Procedure.

In large organizations estimates of production and therefore revenue losses when plant has broken down or is being maintained can, with advantage, be coordinated by the divisional accountant, who is in the best position to point out the fallacies of overstatement to the departmental staff estimating the actual loss in output. In some processes a breakdown can mean not only the loss of a batch but costly plant cleaning and start-up. On the other hand the significant cost of providing stand-by plant or preventive maintenance may be justified in consideration of the possible loss of immediate and future business.

Where probable loss of life is a factor, this must be recognized and recorded for future reference, particularly when at a later date cost reductions have to be considered.

19.3 Planning of Maintenance

When planning maintenance it is appropriate first to identify some of the alternatives and comment on their relative merits.

19.3.1 *No plan*

The simplest scheme is to have no plan at all but to await breakdown and accept the consequences. In a few cases where the continuity of the process is not important, or some non-essential, luxury or domestic equipment is affected this is satisfactory. If, after consideration of all known factors, it is decided to adopt this policy, it should be supplemented by a quick replacement drill. This may vary from case to case and should be put on record for quick reference when required.

The economics of this method are rarely analysed, but the success of several 'round the clock' repair and rewind companies may be partly the result of its fairly wide adoption. Since it is impossible to assess the value of work which such repair companies would lose by proper planned maintenance their contribution to better overall national cost effectiveness cannot be determined. In addition, the service which these quick-repair companies render in case of accidents or urgent modifications to plant must also be borne in mind.

While this practice of awaiting a breakdown (little or no maintenance) can hardly be taken as an ideal solution, there are installations containing large numbers of identical or similar motors where routine maintenance confined to lubrication only is found to be satisfactory.

19.3.2 *Fixed period maintenance*

This consists of maintenance performed on a calendar basis, each machine or item of plant being maintained at scheduled times. The main advantage is that all concerned know in advance when a machine is to be taken out of service and that experience shows what operations are likely to be needed.

When such a scheme is operating properly it is easy to plan the movements and functions of the maintenance staff and facilities.

The disadvantage is that machines may be taken out of service which do not need attention, and indeed may not have been used between 'call in' dates.

19.3.3 *Fixed running time maintenance*

In this it is necessary to have records or estimates of the running times of the motors. It gives a more accurate method of control of the 'need for maintenance' but creates the possibility of peak and idle periods for the maintenance personnel and other resources over which there is little control. It also takes no account of the factor which should be recognized as 'deterioration during idle time'—covering such phenomena as 'brinelling' of bearings and humidity effects.

19.3.4 *Monitoring*

Since the monitoring of cost and effectiveness of maintenance is essential there is need for some degree of record keeping. This need not be involved or costly—more detail is given later—however, it is necessary to know not only the cost of avoidance of breakdown but its consequences when it occurs.

When required other important details can be deduced from these records by analysis, such as:

The durability of different manufacturers' versions of similar machines.
The effectiveness of different maintenance teams or methods.
The effects on similar machines of different types of use or production methods.
The performance of different types of machines or methods of control in similar application.

In some cases maintenance monitoring can be dispersed and delegated to the operator or user. For example: it is rare that the failure of a portable electric tool would be serious, but if the earth continuity conductor becomes severed it could be dangerous. Frequent checking of this continuity by maintenance electricians is one solution, but there are devices available which allow this to be done by the operator himself.

19.3.5 *Preventive maintenance*

This term is often used in the context of arranging maintenance to prevent the occurrence of a breakdown. The decisions to be taken again must take account of the cost of the prevention which, in extreme cases, could exceed the cost of the breakdown. With the operation of a clearly defined maintenance policy and appropriate monitoring it will be possible to establish its effectiveness and that of subsequent changes in it. Since some of the effects are short term and some only become evident after a considerable period of time, a clear policy of breakdown analysis is necessary to identify their causes.

Though aircraft are very different from electrical machines their safe operation required such extensive study of maintenance procedures and disciplines as well as the determination of MTBF's (mean time between failures) that the experience gained can serve as a useful guide for machine and equipment care in industry. The following points are taken from an article[1] which shows that scientific management of maintenance resulted, not only in direct economies, but reduced capital investment by increasing the number of hours each aircraft was operational.

(a) The chief of maintenance is the responsible manager of all maintenance resources comprising personnel, equipment, and facilities.
(b) Standardized and formalized procedures are required to enable such management to operate effectively and ensure that all safety aspects are covered.
(c) Chronological records of individual service actions are required for security and analysis:

What was done and replaced
By whom
When
Where
How long did it take.

By critically reviewing planned overhauls or replacements, extending time intervals, eliminating the examination of certain components, and progressively extending operating hours in the light of recorded trouble-free running hours the cost of maintenance and its interference with production is minimized.

Electric motors in large processing plants may be almost as vital as those on an aircraft and where the provision of stand-by equipment would prove too difficult or costly it may be essential to service to a very strict schedule to keep well within the manufacturer's guarantees for life under specified operating conditions before complete overhaul or replacement.

The foregoing underlines the need to identify the individual requirements and draw up standard procedures. These must be referred and adhered to, and technical administrative improvements must be made in an organized and controlled manner.

19.3.6 *Disturbances caused by examination or test*

It is accepted that the operating temperature of an electric motor affects the life of its insulation, lubrication, and perhaps even its bearings. There are formulae and empirical rules which attempt to relate these factors. Thus, after a period of say 20 000 hours at normal operating temperatures it might be considered that a rewind is advisable. However, so long as the insulation is not disturbed a further useful period of 10 000 hours or more may be possible. A decision is needed whether or not to disturb and check, or to renew, or to allow the machines to run on undisturbed. Account must be taken of the consequences of a possible breakdown, and of the costs and benefits involved in the rewind. Today built-in thermal protective devices are readily available in standard motors, but there may be isolated cases in plants where it may be necessary to continue running a motor under fault conditions. In these cases fault alarms can take the place of cut-out devices. Though a motor may not have actually been 'burnt out' or even show any outward signs of damage, its complete overhaul, or at the very least, careful examination becomes essential after such an incident.

19.4 Standardized Maintenance Management

Mention has already been made of some of the benefits of standardized maintenance management and further advantages will become apparent from the following basic rules:

(a) All concerned with maintenance procedures should, as far as possible, have agreed them or at least have clearly understood them.
(b) The decisions on policy must, as outlined in section 19.2, also be put on record and circulated to those concerned.
(c) It must be recognized that many departments are inevitably involved in either direct administration or executive action. These include:

Accounts
Plant purchase (and even suppliers)
Production
Operators.

(d) To ensure that the procedures can be understood, implemented, and enforced they must be written clearly and concisely, and be readily available for reference at any time.

(e) Procedures should be improved and updated in the light of experience gained.

19.5 Plant Utilization

It will be very difficult to develop a logical maintenance routine without some real knowledge of plant utilization and the ability to identify how much of any down-time is due to maintenance.

Basically, all plant is in existence for 168 hours per week. The amount of actual use varies with the industry and application, but an ordinary single-shift production plant is only active for some 40 hours per week. Even during this time some plant is idle because of setting up, maintenance, breakdown or other unplanned reason.

As really accurate records of machine utilization require dedicated study, which, in most cases, is not considered to be economically justified, statistical methods have to be used. Devices recording the actual number of running hours or working strokes give some guidance when re-lubrication, bearing changes or rewinds are due, but give no information on the loading, abnormal supply or ambient conditions. The records must, to be of use to maintenance planning, also give information on idle times due to breakdown, possible causes, and most important of all, the time required for repair. When percentage utilization factors are computed the calculation should always be based on the maximum possible weekly running hours of 168, since 'standard' working weeks or shift hours are too variable.

Though elaborate systems of plant utilization recording exist, it is mostly up to the maintenance staff to collect and evaluate such information, particularly where specific elements of plant or machinery, namely the electric motor, are concerned. Information so collected may be recorded on a suitably pre-printed motor history card and/or in a plant record book, which may also become of considerable use to other departments such as production, plant-purchase, plant-design, and general factory management. In larger factories a case can often be made for such records to be maintained by administrative personnel specially employed for this specific function.

One acceptable method of determining running time is activity sampling. This is basically the procedure of evaluating plant activity on a statistical basis from observations made at random times.

The frequency of such observations will depend on the working cycle and for the determination of the times at which they should be made, random number tables 0–24 and 0–60, to give hours and minutes, may be used. Clearly the number of observations made must be sufficient to be representative and where seasonal conditions play a part it may take a year before any valid conclusions can be drawn. Where such tables include numbers outside the required range, i.e., above 24 and 60 respectively, these may be omitted without affecting the randomness of those used. The number of observations should be sufficient to give a 95 per cent confidence level of ± 5 per cent accuracy. This is given by the formula:

$$N = \frac{4P(100 - PO)}{L_2}$$

N = number of observations

P = approximate percentage of time of the activity observed.

$I = L$ Limits of accuracy.

In the smaller factory card index and schedules can be drawn up and maintained manually and supervised by those responsible. In larger plants where the number of electric motors may run into thousands the sorting and indexing can be done by computer. Plant utilization plays an important part in good company management and constructive collaboration between those responsible for plant, production, and maintenance must be maintained to ensure its continued economic viability. The danger of misleading figures being produced by using any other basis than the full 24-hour day is a real one and must be avoided at all costs. Equally, records must be looked upon as measurements on whose accuracy the effectiveness of cost control depends.

19.6 Planning Recurrent Activities

The nature of the plant and its utilization provides the information which is necessary to plan routine maintenance tasks whose frequency and nature will vary considerably. It will be necessary to compile 'job cards' detailing man-hours, materials, and, where applicable, (e.g., lift motors) any tests and adjustments required by law. From these it should be possible to assess the total number of man-hours which must be made available, and in conjunction with a site plan to rationalize the work as far as possible by grouping the motors in particular plants, or in the same section of the building.

19.7 Aggregated Total Cost

When considering the economics of maintenance, it is important to use the aggregated total cost of plant. This is the total expenditure on a machine during its lifetime, and includes original cost, maintenance, repair, and overhaul. Aggregated total cost figures of the various choices or options should be used at all times since they enable more rational decisions to be made, even if such considerations are likely to be more concerned with the penalties of temporary unserviceability. During its life a machine will not only need maintenance, repair, and overhaul but will consume power. Whether or not the inclusion of the cost of fuel or power is a valid or significant factor is a matter for decision in each case; but if it is decided to exclude it, it should be a deliberate decision—not an oversight.

In the first case let us take the case of a variable speed motor with its control-gear; costing say £20 000 in 1965. Some production managers might say that 10 years is an appropriate useful life. In fact, few people in Britain accept this figure and most machines are retained in use much longer than this. Although one sometimes finds accountants and others who assume depreciation at 10 per cent per annum, the cost or value of a machine is nevertheless still a factor used in cost calculations, long after the machine is more than 10 years old and written off. This is a valid point for clarification for those pursuing marginal details in economic analyses, and the need to assess as accurately as possible the real aggregated total cost of plant at any point in its life. The curves of these costs will be of the nature shown in Fig. 19.1.

The aggregated cost rises continuously with time, and in practice, due to the need for more frequent attention also with increasing age, the curve often rises more steeply as time goes on.

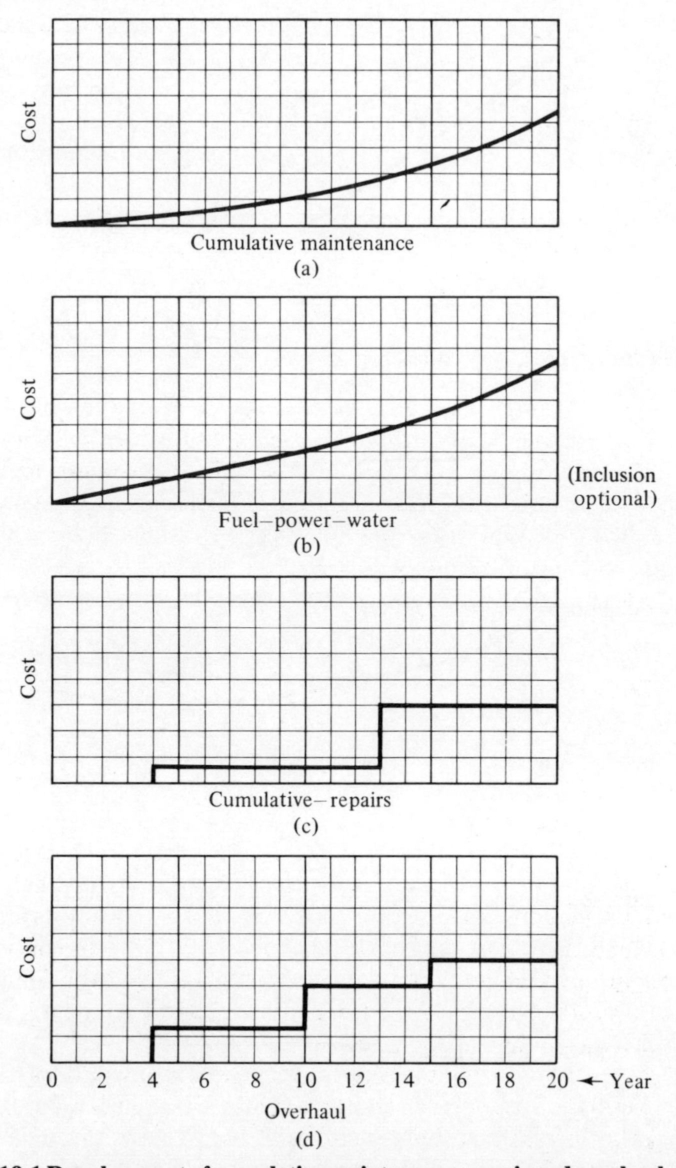

Fig. 19.1 Development of cumulative maintenance, repair and overhaul costs

Factual figures for cost comparisons can be obtained from the already mentioned record or history cards. Their value becomes ever clearer when the information which is necessary to arrive at a balanced decision on maintenance, repair or purchase of new and more productive plant has to be made available (see Fig. 19.2).

Original cost (Capital—installation—running).
Depreciation.
Present market value (a) as scrap, (b) in working order.
Aggregated total maintenance and repair costs.

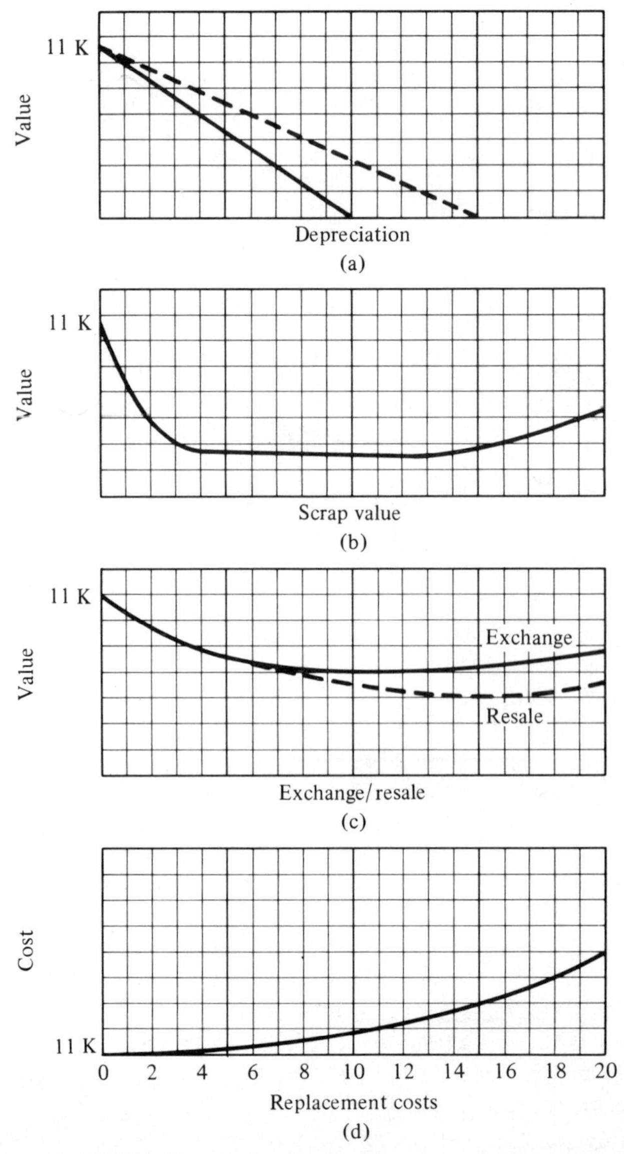

Fig. 19.2 Curves showing trends of value using record card figures

If such costs are up-dated annually and presented graphically, trends can be seen and the time at which it becomes more economical to purchase a new machine more accurately predicted. One of the methods of taking inflation into account is to vary the slope of the base line as shown in Fig. 19.3.

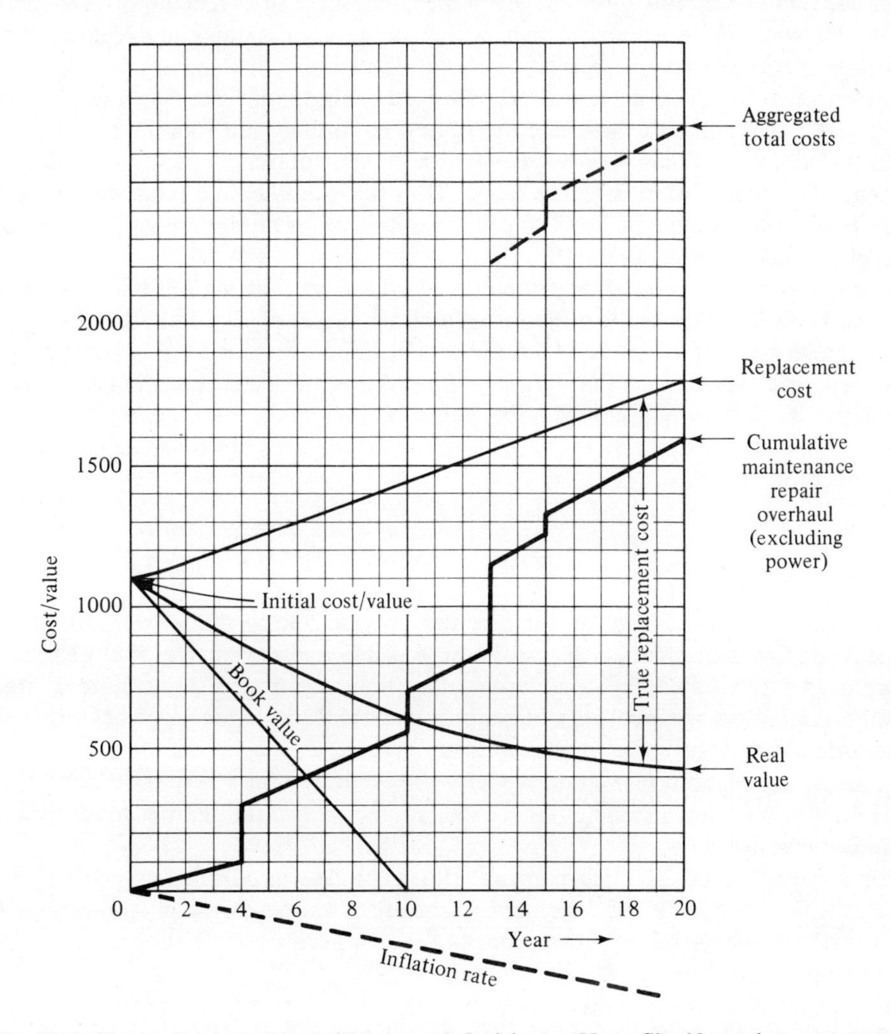

Fig. 19.3 Composite cost curves for renewal decisions. (*Note*: Significant improvement in motor efficiencies and very much higher power charges can materially advance replacement)

The optimum effective economic life of a machine will almost certainly vary inversely with its complexity. Thus, a simple motor such as a cage rotor polyphase induction motor will require minimum maintenance and rarely require overhaul. Sophisticated machines such as commutator and variable speed motors are not only likely to show a higher rise of aggregated costs, but advances in technology are likely to make them uneconomical to operate. Examples are Ward–Leonard controls and some of the more complex motors.

19.8 Standard Motors

One of the most significant benefits of standardization is the ease of obtaining and fitting replacements. Nevertheless, the time taken to obtain even stock motors from the manufacturer may be significant and the maintenance engineer may have to keep a limited number of motors on site. Even then, well-kept stores records and efficient communication and transport systems will be necessary if a quick replacement is to be made in a large factory complex.

When older types of machines are involved replacement can become a major operation and involve the manufacture of new mountings and couplings.

In the setting up of large or even some small plants it may be worth considering limiting the number of ratings in a range. Thus for example, in a range of standard motors covering outputs of 1 to 100 kW the optimum for minimum cost, warehousing, etc, could be shown to be 10.

While it is often assumed, and sometimes even true, that the optimum range of standard motors contains the minimum practical number of sizes it must be borne in mind that this causes a number of machines to be fitted with motors of larger capacity than required. This means that larger and heavier motors incorporating costly raw materials are also running considerably below optimum efficiency and power factor. This, in a world of rising power and material costs, requires careful consideration of all aspects of each case.

19.9 Over Capacity

The fitting of an 'oversized' motor may not only be due to the above-mentioned endeavour to standardize on the minimum possible number of sizes, but generous margins of safety having been added at more than one design stage. Where large numbers of motors are involved this not only leads to a poor system power factor but a considerably higher power consumption.

In many applications maximum power is only required for a very short period, as for example in drilling machines where reduction of motor rating by as much as 3 : 1 has been possible.

Since the maintenance department should, in any case, as outlined in sections 19.4 and 19.8, be monitoring and documenting the use of motors, company policy should also make it responsible for the reporting, and where possible, improvement of motor utilization. The penalties of poor power factor and high consumption could make the introduction of an incentive scheme a very attractive proposition.

Repairs, overhauls, plant modifications or extensions provide the opportunity for fitting smaller motors or having existing motors uprated by rewinding them with class 'F' insulation. The rating of motors, particularly of the CMR (continuous maximum rating) and insulation classes are contained in relevant British Standards, which are dealt with in chapter 5.

Control and protective gear (see chapter 14) plays an important part in the safe and reliable operation of electric motors, and where these are modified or replaced by lower power motors appropriate adjustments to the external protective devices may have to be made. Motors are now being fitted with temperature-sensitive devices such as thermostats or thermistors, and in such cases the compatibility of the control gear must also be ensured.

PART 2. PROCEDURES

19.10 Introduction

Most machinery, provided that it is cleaned regularly, lubricated correctly, and maintained properly will give trouble-free service for the greater part of its useful life. The electric motor is no exception and, provided that its particular needs for cleanliness and routine maintenance are met, it will frequently outlast the machinery it is driving. These requirements are simple electrically: insulation resistance must be kept up and contact resistances kept down. Mechanically, running clearances and

Fig. 19.4 A large motor stator severely damaged by overheating. (Courtesy CEGB)

bearings must be maintained. Treated thus, with an occasional complete overhaul, it will give trouble-free service for many years.

Motor failures can be very expensive, particularly because the plant they drive ceases to be available. Often it is impossible, for economic or logistic reasons, to provide a stand-by machine, so that when a major failure occurs a repair specialist must be called in. If all materials are available the motor may be back in service in 48 hours, or even less, but a good maintenance programme, conscientiously carried out, will often locate faults before they develop into breakdowns or cause serious damage to the machine (Fig. 19.4).

19.11 Cleaning

The entry of foreign matter such as grit, dust, moisture or oil is the cause of many motor failures. Cleanliness is therefore the first essential in maintenance.

Insulation failures of motor windings are frequently caused by moisture or oil; a high proportion of bearing failures are caused by grit and dirt. Bearings will fail also through lack of, or improper, lubrication. A regular cleaning routine should be established to ensure that any deposit of dirt or dust is removed immediately it is found, and oil or grease leaking from bearings should be cleaned away and the bearing seals attended to as soon as possible. The entry of moisture into a motor must always be prevented, unless the motor has been designed specifically to work in a moist atmosphere.

The frequency of cleaning necessary will depend mainly on the local conditions and the type of motor involved. In some large industrial plants the motors are blown out once a week by means of a supply of dry compressed air, or electric blowers.

Warning. If either of these means is used, care must be taken to ensure that dust and dirt is blown out of the machine and not further into it. If hoses are used, the air delivered must be absolutely dry, and the pressure should not exceed 1·75 bar because higher pressures will tend to blow the dust and dirt into the insulation. When one machine is being blown out, other machines in the vicinity should be covered to protect them from flying dust. A suction cleaner, with a dust container, should be used in preference to compressed air if one is available—especially in confined spaces.

It is a straightforward job to blow out a ventilated motor, for it is necessary only to remove covers or doors to gain free access to the interior of the motor and then direct the air onto the end windings of the armature or rotor and the field or stator windings until dust ceases to issue from the motor and all visible parts appear to be clean. If the motor incorporates brushgear, commutator or slip rings, these should be wiped clean and the brushes checked as described in section 19.12.4.5.

A totally enclosed motor need not be opened up provided that the covers or doors are fitted correctly and the local dust is not abnormally fine, unless the motor incorporates brushgear, commutator or slip rings, which must be serviced. But its external surfaces must be cleaned regularly. The cleaning of the external surfaces is very important because it is through these that heat generated inside the motor by its internal losses is dissipated. A thick layer of dust can retard heat dissipation and may result in excessively high temperatures, particularly in continuous-running motors. This condition, if allowed to continue, will reduce considerably the life of the winding insulation.

For types of motors other than those mentioned already, cleaning should depend on whether or not contaminated or dirty air comes into contact with the windings. Summary of cleaning routines:

(a) Ventilated machines: remove all loose dirt and dust by extraction or blowing out.
(b) Totally enclosed machines: clean external surfaces.
(c) Remove any oil or grease which has escaped from bearings and lubricate according to manufacturer's instructions.
(d) Inspect and clean brushgear, slip rings, and commutators.
(e) Clean dust extractors, end shields, and filters; or renew them if necessary.

In addition, a maintenance engineer should keep a close watch and an attentive ear for any undue vibration or noise from any motor, drive or driven equipment. By doing so, he may be able to detect a fault developing and by taking immediate action avoid a more serious breakdown.

19.12 Overhaul

The annual holiday shut-down periods which are spreading across industry are ideal opportunities for plant overhaul.

Whenever possible, motors requiring overhaul should be removed to the maintenance workshop or working area, where facilities for the work should be satisfactory. If the motor is too large, or local conditions prevent removal, the area around the motor must be cleared of obstructions and cleaned.

19.12.1 *Safety*

Before work is started on any motor its identity must be proved and a check made to ensure that it has been isolated from all points of supply, particularly if there are several identical motors in the same area.

19.12.2 *Disconnect and remove*

Having identified the motor and proved that it is safe to start work, first remove the terminal box cover and disconnect the supply cables, taking care to mark each connection clearly to avoid any confusion when the time comes to reconnect the motor. With high-voltage machines, protect the cable tails to avoid damage while the motor is away.

Disconnect the coupling bolts, driving belts or whatever other means is used to transmit the drive. Remove the holding-down bolts or nuts. When lifting the motor ensure that any packing shims under the motor feet are tied to the appropriate feet or fixed to the holding-down studs. If the motor cannot be removed completely, it may be necessary to lift it and turn it on its bedplate to facilitate dismantling.

19.12.3 *Dismantling*

When any motor is dismantled it is essential that all parts and positions are marked clearly, particularly terminals and connections, brushgear, and end shield positions, to avoid problems and to save time when the machine is reassembled.

All bolts, nuts, screws, and other small parts should be put into a suitable container where they will not get lost or mixed with parts from another machine. Where several motors are being dismantled at the same time the boxes containing the parts should be marked clearly. It should not be necessary to use excessive force when dismantling, because modern machines are made to jigs and the fitting of every part is assured before the manufacturers release the machines.

When the motor has been dismantled completely each part must be cleaned thoroughly, particular attention being given to parts which cannot be cleaned at any other time. Some parts of the motor may require soaking in cleaning fluid, or even scrubbing, especially end shields and stator housings, and a bath of fluid may be required. There are many proprietary brands of cleaning fluid on the market and for safety one which has a high flash point and is non-toxic should be used. When all parts have been cleaned thoroughly they must be inspected in detail.

19.12.4 *Inspection*

19.12.4.1 Stator. The stator windings or field assembly must be inspected for damage to the insulation, signs of overheating and movement. The insulation must be examined carefully for mechanical damage, which can be caused by carelessness when replacing the rotor in the stator, and when removing and replacing end shields, particularly on large motors. Discolouration of the insulation may be an indication that the windings have been overheating, and in such cases the cause must be found. Some of the reasons for the overheating of windings are discussed under 'breakdowns'.

The slot wedges must be checked for tightness as must the windings and turn bracings, to ensure that there is no movement of the windings. If movement is found it must be eliminated, because any movement will cause fretting of the insulation, and if allowed to continue will result eventually in failure of the insulation. The winding tails, terminal connections, and terminal insulation should be examined. If all appears to be in order, the windings should be tested with an insulation resistance tester to obtain an indication of the quality of the winding insulation between the conductors and the motor frame. If the reading is low (say 10 000 Ω or less) then the previous readings on the motor's history sheet should be checked. These will show if there has been a sudden deterioration in the insulation or it has been deteriorating over a long period. The former usually denotes recent damage to the insulation, or the entry of moisture. The latter usually denotes a general ageing of the insulation, which can sometimes be controlled by the application of a coat of high quality varnish to the windings.

The stator core should be examined carefully for signs of damage, particularly if the motor has suffered a bearing failure and the rotor has rubbed the stator core. If the core plates are damaged it may be necessary to dismantle them in order to clean the plates and restore the insulation between the laminations. Failure to do this will result in a hot spot occurring where the plates are damaged, and if the motor is allowed to run in that condition will soon result in winding insulation failure due to the excessive heat generated in the iron circuit at the point of damage. In d.c. motor field assemblies it will be necessary to check also the inter-coil connections to ensure that they are in good condition.

If all appears to be in order, the insulation resistance of the windings must be measured and recorded and the windings given a coat of high quality insulating varnish.

19.12.4.2 Rotor and armatures. The rotor must be examined carefully for signs of damage. The end rings of a cage-type motor should be inspected for cracks, and each rotor bar inspected particularly where it is brazed to the end rings and where it enters and leaves the slot in the iron core. Each bar should be checked for movement in the slot, and the slot wedges checked to ensure that they are tight. The core laminations should be examined for signs of damage and overheating, and it is essential to prove that all the ventilating ducts in the core are free from obstructions. If an internal fan is fitted on the rotor shaft it is necessary to ensure that all the blades are in good order. Failure to maintain a good air circulation will result in the motor overheating and consequently a reduction in its useful life. If the motor has a wound rotor, the winding insulation must be examined and the end turn bindings checked for tightness. The winding connections to the sliprings or commutator should be checked to ensure that they are in good order. The insulation resistance of the windings must be measured and recorded, and if all is in good order a coat of high quality insulating varnish should be applied to the windings.

19.12.4.3 Sliprings. These must be examined for signs of overheating and undue wear. The insulation between the rings, and between the rings and the shaft, must be inspected for damage and signs of tracking. The winding connections to the sliprings should be checked for tightness. If all appears to be satisfactory, then the insulation resistance should be measured and recorded. Should the insulation resistance be found to be low it will be necessary to disconnect the windings from the sliprings to prove if the cause of the low reading is in the rotor windings or the sliprings. Dirty slipring insulation may well be the fault. The insulation should be cleaned thoroughly.

If short circuiting gear is fitted it must be inspected for signs of wear and the operation of the mechanism must be checked. The shorting switch contacts should be examined and cleaned, and if necessary adjusted to ensure proper contact is made when the shorting gear is operated.

If the sliprings have become pitted, or show signs of uneven wear, they should be trued up and polished, preferably in a lathe; but in general sliprings give very little trouble in service.

19.12.4.4 Commutators. These require careful examination of the connections between the windings and the risers, the segments, and the slot insulation.

Loss of solder from the winding connections on the risers indicates usually either winding troubles developing, or that the commutator has been getting too hot.

The segments should be inspected for signs of overheating and burning, and checked for flat spots and for ridging. If the motor has been running correctly the commutator will have formed a 'skin', and it should have a dark and smooth appearance; in such cases the surface should only be wiped clean. If the commutator has become ridged badly, or shows signs of the development of flat spots, it may be necessary to have it turned in a lathe to restore the surface to the correct concentric shape.

With very large machines it may be preferable to skim the commutator while turning it in its own bearings. This will require a lathe saddle and traverse gear fixed firmly to the bedplate or foundations, and carefully lined up with the commutator, which must be turned at a controlled speed.

The slot insulation must be examined to ensure that the mica insulation has not become proud of the segments. In most motors the commutator segment insulation is recessed or undercut below the level of the copper segments to a depth of approximately 0·8 mm, except for some special cases such as silent-running motors. So it is important that the slots are cleaned, for if they become filled with carbon dust and dirt, shorting between the segments can occur.

When a commutator is turned or skimmed the mica insulation must be undercut. For this operation the armature is set up on vee blocks at a convenient height. (Undercutting can be a very tedious job, particularly on commutators having many segments, and so it is important to study the needs and comfort of the operator if a first-class finish is to be achieved.) The mica insulation must be removed to the correct depth using an undercutting machine fitted with the correct type and size of cutting blade or file as recommended by the tool manufacturers for the particular commutator. After undercutting it will be necessary to slightly bevel the edges of the segments to remove any sharp corners, and this can be done by drawing a shaped piece of hard steel along the grooves.

Finally, the armature should be blown out to remove the mica and copper dust which may have dropped between the windings and into the iron circuit ventilating ducts.

19.12.4.5 Brushgear. Brushes and brushgear require regular inspection and cleaning—in continuously running motors brushgear may well need inspection once a week.

At motor overhaul the brushgear should be cleaned and examined carefully. The brushbox arms must be cleaned and the insulation inspected for signs of tracking and to ensure that there are no cracks which can become filled with carbon dust, thus making a conducting path across the surface of the insulation. The brush-boxes must be cleaned thoroughly and the springs and pivot pins checked.

When a motor is overhauled it is advisable to fit a new set of brushes, but if this is not done all brushes must be checked for signs of overheating and the tails must be checked to ensure that they are firmly fixed into the brushes. If new brushes are fitted it is essential to ensure that the correct type and grade are used and that they are bedded-in correctly; i.e., the brush faces must assume the contours of their commutators or sliprings. This can be done by placing a strip of glasspaper under the brush (rough side upwards) when it is fitted into its box and the spring pressure is applied. The glasspaper is then moved back and forth until the brush is the correct shape, the whole of its bearing surface making good contact with the face of the commutator or slipring. Having fitted the brushes it will be necessary to check carefully the spring pressure on each brush, and this can be done by means of a spring balance. The spring pressures must be set to the motor manufacturer's recommended values.

19.12.4.6 Bearings. Some motors are fitted with two-piece white-metal bearings which can be either integral with the end shields or fitted to external pedestals. This type of bearing is usually fitted to large, low-speed motors (up to about 1000 rev/min). It requires regular maintenance and must always be well lubricated. Where oil pick-up rings are fitted it is most important that these rings are free to rotate with the shaft. Failure of an oil ring, if not attended to quickly, will result in damage to the bearing. Great care must be taken to avoid contamination of the bearings, particularly when replenishing or renewing the lubricant, for a high proportion of failures of this type of bearing can be attributed to a lack of lubrication or contamination of the lubricant.

The same care and attention is necessary for motors fitted with sleeve bearings except, perhaps, for some smaller types of motors fitted with oil-impregnated sleeve bearings. These are impregnated with oil during manufacture and are designed so that the oil content is sufficient for the working life of the bearing under normal operating conditions.

The majority of motors manufactured today are fitted with ball bearings and/or roller bearings. Because of the precision of their manufacture these are robust and reliable, and give very little trouble provided that they are fitted correctly, kept absolutely clean, and always lubricated with the correct quantity of the correct grade of grease.

There is a strong body of opinion that if a bearing appears to be performing reasonably well then it should not be disturbed, but an engineer must be sure that the bearing will last until the next major overhaul, and if any doubt exists it is advisable to change a bearing rather than risk its failure within the next working period. There are, of course, instruments available today designed to detect deterioration in bearing performance, but the sure way of detecting impending troubles is by the senses of an experienced engineer. Before opening a bearing housing to clean and inspect the bearing it is most important that the outside surfaces are cleaned thoroughly to ensure that no dirt will enter. When the housing has been opened, both bearing and housing should be cleaned with a non-toxic cleaning fluid, and when all the old lubricant has been washed out the bearing should be oiled lightly. Although it is generally possible to clean a ball or roller bearing while it is in place on the shaft, the most effective method is to remove the bearing from the shaft so that it can be cleaned thoroughly and inspected for signs of wear. The removal of a bearing from its shaft requires the use of wedges and pullers, which are designed to facilitate the removal of bearings without damage to either the bearing or the shaft. Provided that the correct sizes of wedges and pullers are used (Fig. 19.5), the removal of even the largest bearing should not be difficult. With the bearing removed, a thorough inspection can be made: if any traces of metal dust, or filings, such as brass are found this will indicate cage wear and the bearing must be replaced by a new one. If no dust or filings are found, the bearing should be washed thoroughly in cleaning fluid and when it is perfectly clean the tracks and the balls or rollers must be examined carefully for signs of wear or damage. If no flaws are seen, the outer race should be revolved slowly with one hand while the inner race is held stationary. If any noticeable increase in force is required at any point, or there is a tendency to stick, the bearing should be rewashed and the races revolved in the fluid. Should the checking or sticking persist the bearing must be rejected and a new one substituted. Sometimes bearings can be damaged through lack of use: i.e., if a motor has remained stationary for a considerable time in a building subject to vibration. Under these conditions the hammering effect between the balls or rollers and the tracks will cause damage where

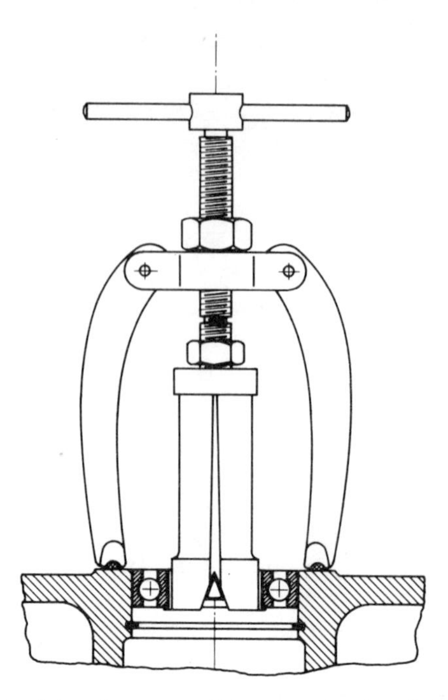

Fig. 19.5 Bearing puller. (Courtesy Eberhard Bauer)

the weight is resting. In these circumstances it is advisable to turn the motor shaft frequently.

The bearings should also be examined for signs of creeping, i.e., movement on the shaft and/or movement of the outer race in its housing. Wear of the shaft and the housing is not uncommon, and a method of overcoming this defect is to build up the worn surfaces by metal spraying, and then turn the surface to its original size. Having established that either the bearings are in good condition and fit for further service, or that new ones are required, the shaft must be cleaned thoroughly, a thin film of light oil applied, and the bearing back plate replaced. The bearing must be started square on the shaft and pressure must be applied evenly to the inner race as close to the shaft as possible. Special tubular drifts can be obtained for this purpose. It is important to ensure that the bearing is pushed right home against the shaft shoulder.

With large bearings, shrink fitting with pre-oiling is recommended if the facilities are available. To do this, immerse the bearing in a tank of oil heated to an average temperature of 77°C (170°F) for 10–15 minutes. Lift it out of the hot oil and allow it to drain lightly, then grasp it with a non-fluffy cloth and thread it onto its correct seating position. Be sure that it is pressed hard against the shaft shoulder. The operation must be completed within two minutes of the removal of the bearing from the oil.

When the bearings have been replaced, the locking rings (if applicable) must be refitted, and the bearings lubricated with the correct type and quantity of grease as recommended by the manufacturer.

19.12.4.7 Terminal connections. The insulation of the terminal box must be examined for cracks and signs of tracking, the windings connections must be examined, and the cable lugs checked for signs of overheating and to ensure that the cable is firmly secured in the lug. Where soldered lugs are used, the conductors should be examined for signs of fracture adjacent to the solder.

19.12.5 *Reassembly*

When all the parts have been inspected thoroughly, and defective parts repaired or replaced by new, the motor can be reassembled and, provided that everything was marked correctly when the motor was dismantled, no trouble should be experienced.

When replacing the rotor or armature in its stator or field assembly care must be taken to ensure that no damage is done to the stator or field winding insulation. The rotor or armature must be positioned centrally within the stator or field assembly. This is most important, particularly with large machines and especially those fitted with pedestal bearings. The air gap must be checked if possible, by inserting feeler gauges between the stator and the rotor core. The measurement must be taken from iron to iron and not to slot wedges, and should be taken at four positions 90° apart, the rotor then turned 90° and the gaps measured again. Ideally, the air gap should be the same size at all positions.

If the gaps are uneven, adjustments will have to be made to the bearings and/or endshields until the best values possible are obtained. Some motor endshields are fitted with inspection covers to allow feeler gauges to be inserted.

When replacing endshields, care must be taken to ensure that they are tightened up evenly, and this can be checked by rotating the shaft to prove that it is free to move during the final stages of tightening the endshield bolts.

When reassembly has been completed, and free rotation proved, the motor can be remounted in its working position and aligned to its load. It is most important to prove that the alignment is correct, because misalignment can cause excessive wear and overheating of bearings, and can result also in the rotor or armature not being central in the stator, and thus an uneven air gap.

When the motor has been aligned and bolted down it can be coupled to the load, and the supply leads connected to their correct terminals. Before replacing the terminal box cover it is necessary to check the insulation resistance between the terminals and earth. Care must be taken while refitting the cover to make sure that it fits correctly, so that the entry of moisture and foreign matter is prevented as far as possible. It is also recommended that the terminals of the higher voltage motors are taped individually as an extra precaution against a flashover.

Before the motor is returned to service, it should be given a test run to prove that all is in order.

19.13 Breakdowns

Provided that electric motors are maintained properly, failures in service will be the exception rather than the rule. However, it is inevitable that some breakdowns will occur, and then it is essential that the fault be identified and the necessary repairs or replacements put in hand as quickly as possible.

Some faults will be obvious, but the majority will require a detailed investigation to find the cause and the extent of the damage. When tracing motor faults it is important to remember that in many cases it will be only by a process of elimination that the actual cause of the trouble will be located. Many of the faults likely to occur in motors, such as breakdown of insulation, loss of power supply, and bearing failure, are common to all types of motors.

An insulation failure can develop in several ways: e.g., mechanical damage, which is usually confined to the end turns of the windings, caused by movement of loose windings in their slots or slackness of the end turn bracing. This condition will cause rubbing and fretting of the insulation where the movement occurs and will eventually cause failure. Another way is by ageing of the insulating materials. The ageing of the insulation will be accelerated considerably if the motor is subjected to excessive temperature rises, because these materials tend to dry out and shrink, thus developing cracks through which moisture and dust can penetrate and seriously reduce the effective insulation values. The windings may be subjected to voltage surges, which will stress the insulation and under certain conditions may cause electrical breakdown. The quality of the winding insulation is usually measured by taking readings of the insulation resistance (IR) with an insulation resistance tester (a Megger), or in

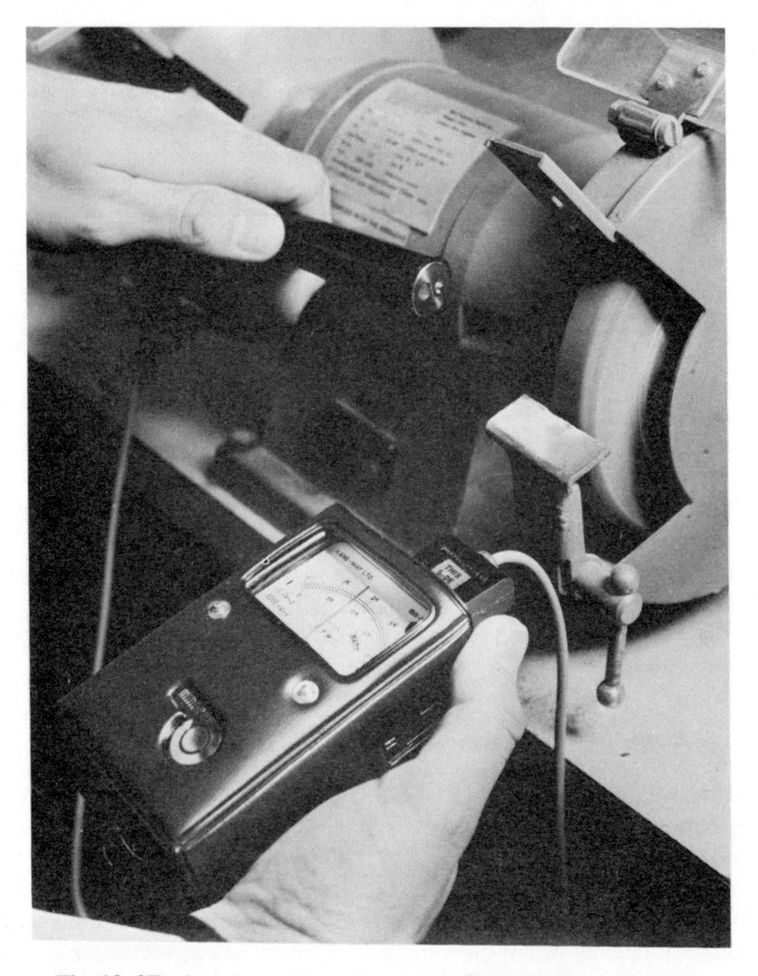

Fig. 19.6 Typical electronic tachometer. (Courtesy Kane May Ltd)

high-voltage machines by using an a.c. bridge to measure the loss tangent and capacitance of the insulation.

Breakdowns due to insulation failure usually result in an earth fault or short-circuited turns, and in some cases where three-phase windings are involved, a phase-to-phase fault may develop. A number of very good instruments are available which have been designed specifically to locate fault conditions in armatures, field coils, stators, and rotors—faults such as earth faults, shorted turns, and open-circuit coils. Most winding faults involving insulation breakdown will necessitate a complete rewind; except, perhaps, if the failure was due to mechanical damage to the end turn insulation, in which case a repair may be successful.

Another cause of electrical failure, particularly in high-voltage machines, is a flashover in the terminal box. In these cases no detection is required as the damage will be obvious. Special care must be taken when closing high-voltage terminal boxes

Fig. 19.7 Typical thermometer and probe for measurement of surface temperature (Courtesy Kane May Ltd)

to ensure that they are moisture proof, and as an extra precaution the terminals should be taped with a good quality insulating tape. Most motors are protected against supply failure with 'no-volt' protection, temperature alarms, and, in some cases, single-phase protection. But three-phase motors, which may have overload protection only, can be damaged very quickly if one phase is lost while the motor is running. In such a situation it is important that the motor is switched off immediately, or the windings will overheat and burn out very quickly.

Some motor failures are caused by defective or worn bearings. It is essential, therefore, that bearing performance is monitored regularly. Badly worn sleeve bearings will result in an uneven air gap which, in turn, will result in overheating and, in severe cases, can result in the rotor or armature rubbing the stator or field assembly iron core.

Instruments are available for detecting vibration, roughness and wear, and for measuring and recording speeds and temperatures. Figure 19.6 shows a typical modern electronic tachometer which makes it possible to measure speeds without danger to the operator. The lightweight indicator and separate sensor makes speed checks of motors in awkward positions practical. Once a strip of reflective tape has been affixed speed can be measured in a few seconds.

The reasonably accurate measurement of temperature can present problems if the simple thermometers involving expansion of liquids or gases are used. Figure 19.7 shows an electronic thermometer being used to measure surface temperature. Special probes for surface temperature measurements are available and despite its small size the instrument displays the temperature on a clear linear scale. Where many measurements have to be made under difficult conditions the higher cost of the digital version is likely to justify itself.

A bearing failure can result in extensive damage to the motor and a costly repair or replacement, so no chances should be taken. Some motor breakdowns may be the direct results of troubles in the driven loads, and this possibility must be checked during the investigation into the cause of a breakdown.

There are motors known to be working satisfactorily after more than fifty years service, and this is due mainly to the constant care and attention they have received throughout the years. Regular cleaning and good maintenance in general will ensure a high level of availability of a motor, and should therefore be considered a worthwhile investment.

20 Measurement and Rating of Machine Noise

20.1 Introduction

Noise, which is one of the many annoying environmental hazards to which man is increasingly exposed, has recently received more and more attention. While some extraneous noises, particularly those due to traffic on land and in the air are still, within limits, being accepted as inevitably increasing, those connected with working conditions are more successfully controlled. In Britain there is now the 'Code of Practice for reducing the exposure of Employed Persons to Noise' (HMSO 1972) soon to become enforceable by law, and similar measures exist in other countries. Employers and trade unions alike have become increasingly conscious of the long term effects of noise and after studying problems peculiar to their industries have issued recommendations. These laws, regulations, and recommendations affect not only original equipment manufacturers (OEM's) and plant engineers, but the manufacturers of the prime movers.

The trend in electric motors has been to significantly reduce their size by increasing the electrical and magnetic loading and take advantage of insulating materials allowing higher temperature rises. To keep size to a minimum and maximize efficiency, large-power high-speed motors are used wherever possible. Such motors are inevitably a source of considerable magnetic, windage, and mechanical noise, particularly if they are then also coupled to a gearbox. These factors are taken into account by motor manufacturers, who must try to minimize noise without prohibitively increasing the price. In some cases the user will, in any case, accommodate the motor outside and, if necessary enclose it not only for its own protection, but protection of the environment. Normally electric motors, particularly in the smaller sizes (below about 150 kW) do not present a problem but it must be recognized that the most elaborate and costly noise suppression on the motor will be of no avail if the driven plant or machinery is not similarly screened and structurally transmitted noise avoided.

20.2 Fundamental Aspects of Noise

In general, noise is unwanted or unpleasant sound and, as will be seen is extremely difficult to quantify or qualify in a universally acceptable manner. The first difficulty is inherent in the definition of sound as a sensation produced in the ear due to wave motion of the surrounding air. This means that it is not only dependent on the sensitivity and physical conditions of the ear, but the state of the brain which receives the generated nervous impulses. Other problems arise from the definition of loudness of sounds which vary in frequency (or pitch), intensity, duration, and the effect surroundings have on the binaural detection.

Sound originates from mechanical vibrations of solids, fluids or in gases by relatively fast moving objects, such as fan blades in air. The laminations of a motor will be induced to vibrate by the electromagnetic action of alternating current, and motor endshields will act as radiators of any bearing noises.

The classical source of a pure sound or tone is the tuning fork, which, when vibrating, will cause longitudinal sound waves to be radiated. These waves cause variations of pressure, which are sinusoidally varying according to frequency and travelling through the air at an approximate propagation speed of 330 m/s. The propagation speed of sound is a function of the medium's density so that in fresh water, for example, the sound would travel at about 1435 m/s. More important than the propagation speed is the frequency which for the acoustic range lies between 16 Hz and 20 000 Hz. The intensity or loudness, determined by the energy radiated, can be represented by the amplitude of the sound wave. Most sounds, particularly noises, are rich in harmonics, and because they are radiated from complex physical shapes into spaces surrounded by acoustically reflective surfaces can only be statistically analysed.

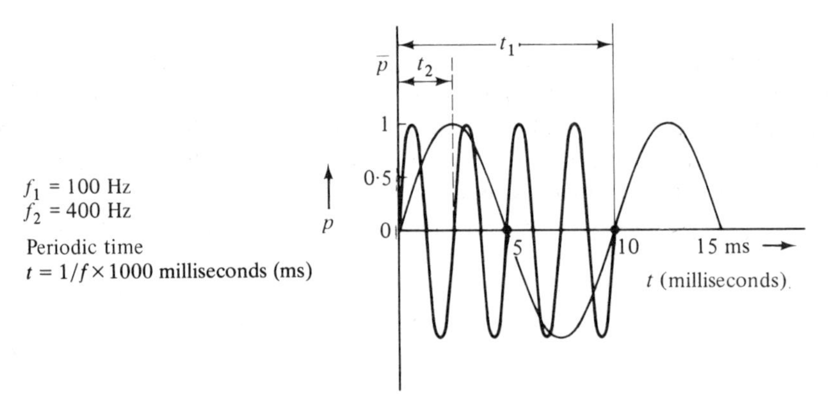

Fig. 20.1 Equal pressures for two pure tones two octaves apart

Figure 20.1 shows equal pressure, which in the SI system of units is in N/m^2, for two pure tones two octaves apart (the doubling of the frequency represents an octave). The pure tone can now be fully defined by its frequency and pressure and attempts made to relate this to subjective loudness as judged by a human observer. The intensity will be judged from the threshold of hearing to the level of pain and will be found to cover a range of intensity of one to one tera (10^{12}) units of intensity or 2×10^{-5} N/m^2 (or 20 μN/m^2) to 20 N/m^2. This is not only outside the scope of normal instruments but makes it difficult to express meaningfully the intensity of the resulting response to the human ear. Sound-energy in watts is the energy radiated by the source as sound in 1 second and is also unsuitable for everyday use. Telephone

Note: 1 N/m^2 = (1 Pascal) = 10 μbar or dyne/cm^2

engineers faced with the same problem more than half a century ago defined a logarithmic unit to express power ratios and named it after Alexander Graham Bell. The actual ratio used is 1/10 of the bel, that is the decibel (dB), with the reference level the threshold of hearing given above. Each increase of twenty decibels is equivalent to an increase of sound energy of a hundred times, or to an increase of sound pressure of ten times. This results in numbers ranging from 0 to 120 and table 20.1 is intended to give a guide with typical examples.

Table 20.1

Decibels	W/m^2	N/m^2	
0	10^{-12}	20×10^{-6}	threshold of hearing
20	10^{-10}	20×10^{-5}	rustling sound, whisper
			(20–30 quiet suburban home)
40	10^{-8}	20×10^{-4}	50 motor car
60	10^{-6}	20×10^{-3}	50–60 normal conversation
80	10^{-4}	20×10^{-2}	70–80 heavy street traffic
100	10^{-2}	2	riveting gun
120	10^{-1}	20	thunder

20.3 *Loudness*

If the sensitivity of the human ear is examined over the audio range it will be found that for constant power (N/m^2) the ear is most sensitive around 4000 Hz and generally cuts off below 20 and above 8000 Hz. Usually 1000 Hz is taken as reference frequency and with a pressure intensity of 60 dB equal loudness for 50 Hz is at 82 dB, for 300 Hz at 56 dB and for 8000 Hz at 68 dB. These comparisons are, however, only valid for pure tones since for noises with random frequency composition, but the same total pressure the divergence is even greater.

In addition to the complex frequency response the psychological factor plays a considerable part. A machine operator may be quite accustomed to the regular though loud noise of his machine, but be quite annoyed by his son's motorcycle noise. Needless to say the roar of the engine is sheer music in the proud owner's ears.

To overcome the variable frequency response of the human ear sound meters are fitted with weighting networks which endeavour to follow the equal loudness curves (researched by Fletcher and Munson as well as Robinson and Dadson). It should however, not be found too surprising if subjective tests and instruments do not agree on the quietest motor, particularly if the character of the noise can be described as a hissing or howling which does not have to be loud to be objectionable. In addition it has been found that not only does the sensitivity of the human ear change with age but that there is a significant difference at the higher frequencies between that of men and women. Not to be neglected either in the population of industrial nations is the temporary threshold shift detectable after exposure to loud noises.

The subjectively determined loudness is expressed in 'phons' which endeavours to overcome the disadvantage of the decibel in expressing loudness over a range of frequencies. The loudness of a sound in 'phons' is numerically equal to the decibel value of sound intensity for a pure tone of 1000 Hz. For pure tones or noises covering only a narrow frequency band the equal loudness curves must be used. The accuracy of subjective noise evaluation of a representative group of, say, ten observers is considered to be of the order of ±2 phon.

In practice loudness measurements are not used for the determination of noise levels; instead a number of objective measurements are made to obtain a noise rating. Among the most widely accepted recommendations are those of the International Standards Organization (ISO) R 1996. An ISO noise rating NR 70 means that no peak of the frequency spectrum exceeds the limit of 70 dB at 1000 Hz. The NR 70 curve, as Fig. 20.2 shows, takes a very approximate account of the sensitivity of the human ear and by taking a frequency spectrum and analysing it a general noise level can be determined.

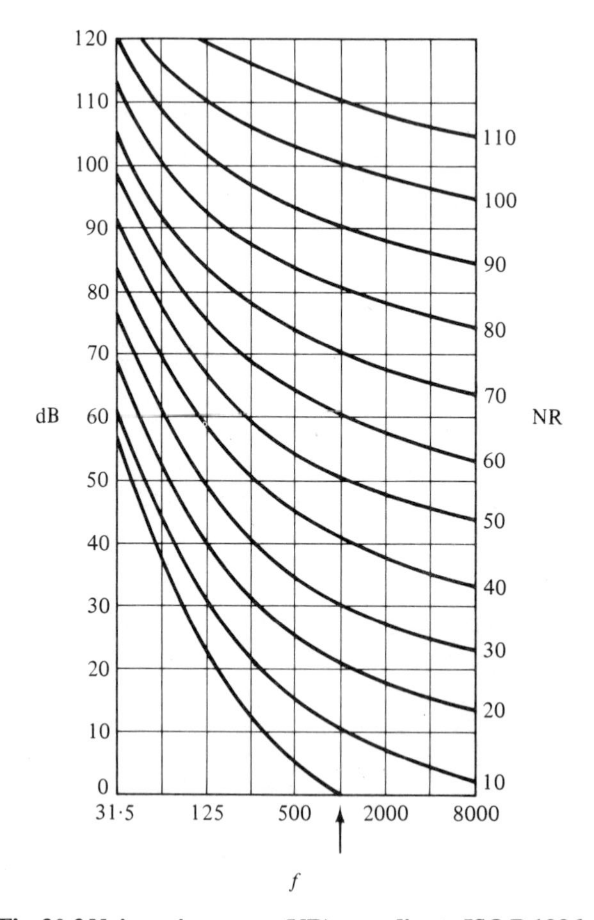

Fig. 20.2 Noise rating curves (NR) according to ISO R 1996

Noise level measurements were made with three frequency weighted networks labelled A, B, and C derived from 30, 70, and 100 dB equal loudness contours to represent three levels: A below 55 dB, B for 55–85 dB, and C above 85 dB. It is now accepted practice to make measurements on the dB(A) scale for a noise spectrum analysis. The details of this can be found in the IEC Publication 179.

20.4 Noise Power Levels

While measurement of the pressure of the sound waves allows a relatively simple comparison of similar noises it still does not evaluate the nuisance value. The frequency spectrum allows the experienced sound engineer to give a much better indication but even this, though clearly objective if carried out in an anechoic test chamber, does not represent actual conditions.

It must be repeated that the motor and the machine it drives represent a very complex radiator of sound energy and that wherever the machine is situated there will be sound reflecting and sound absorbing surfaces which considerably modify the noise level.

To take account of this, free field measurements are approached as nearly as possible. Free field conditions are said to exist if at any point surrounding the sound source no appreciable reflected sound energy is present. This is verified by measuring the sound-pressure level which for every doubling of the distance from the source should reduce by 6 dB.

The room constant is defined by the following equation:

$$R = \frac{A\alpha}{1-\alpha}$$

where A is the total internal area of the room and α the absorption coefficient, typical values of which are given in NEMA MG 3-1974 is shown in Table 20.2.

Table 20.2 Absorption coefficients

Material	Coefficient α
Raw concrete (breeze) blocks	0·36
Raw concrete when painted	0·07
Plasterboard	0·07
Glassfibre boards 25 mm thick	0·75
Windowglass	0·16
Large plate glass	0.04
Wood	0·09
Parquet floor on concrete	0·06
Lino, asphalt, rubber, cork on concrete	0·03
Metal plates	0·03
Acoustic panels	0·42
Heavy curtains	0·65
Painted brickwork	0·02

Source: NEMA MG 3-1974

Motors tested under free field conditions should have measurements taken on a hypothetical envelope at points which represent equivalent surface areas. In the case of small machines this may be a hemisphere but for large machines an envelope 1 to 3 m distant from the surface with many more measuring points will be necessary.

Theoretically the room in which noise tests are carried out should be large enough to take measurements beyond the immediate radiation field of the source and at least $0{\cdot}25\lambda$ from the wall, where λ is the wavelength of lowest frequency of interest ($\lambda = c/f$ where c is the speed of sound and f the frequency in hertz and average value for air is 340 m/s). Since in practice frequencies below 100 Hz are not considered, a distance of 0·85 m suffices. For sound spectrum analysis measurements an anechoic room is best but where there are reverberant or semi-reverberant conditions these can be measured and subtracted.

20.5 Methods of Measurement and Interpretation of Results

The frequency spectrum is best measured in one-third octave bands. Each octave is defined as an interval of frequency the upper limit of which is double that of the lower one. The mid-frequencies of the now preferred bands in hertz are 63, 125, 250, 500, 1000, 2000, 4000, and 8000.

Where only some six readings are taken the positions most commonly used are one or two directly above the motor and four in a horizontal plane at shaft level, one of them in line with the shaft. Since the load mostly contributes to noise, tests are usually made on no-load or in the case of geared motors, with just the inherent loading of the gearbox.

For the measurement of the sound-power level L_n the meter is set to the position at which dB(A) is measured and the arithmetic mean over a time period of at least 5 s may be taken if the excursions of the meter are within 5 dB(A). Ensure that the measurements are representative of the machine itself and that any cowlings, baseplates, pipes, busbars or other rigidly connected devices do not materially contribute to the airborne noise.

Where extraneous noise must be corrected for this is first measured with the machine under test standing still and the corrected value determined as follows:

If L_G is the total power measured and L_E the value of extraneous power, then the noise power level of the machine alone is, because of the logarithmic scale used $L_M = L_G - K_1$. The table of K_1 is as follows:

$L_G - L_F$ in dB(A)	3	4	5	6	7	8	9
K_1 in dB(A)	3	2	2	1	1	1	1

Where the difference is less than 3 dB(A) the measured value cannot be corrected and it can only be stated that the noise to be measured will not be greater than the measured value.

The distance of the points of measurement is of great importance and whereas certain standards (e.g., DIN 45 635) specify 1 m, others may specify 3 m. For purposes of comparison it is possible to determine a correction factor K_2 with the aid of an auxiliary test sound source. The following are guide values from the now obsolescent DIN 45 632

Test distance a in metres	0·25	0·5	1	2	4
Correction factor K_2 dB(A)	−8	−4	0	+4	+8

The value for comparison then becomes

$$L_1 = L_a + K_2 = \text{(sound power level referred to 1 m)}$$

$$L_a = \text{actually measured sound power level at distance } a.$$

It is important that when such values are quoted, their origin is also specified, e.g., $L_A = 70$ dB(A) calculated from an L_{25} (measurement at 25 cm) value.

To calculate the total sound power level, again because of the logarithmic units used, simple addition is not possible. To illustrate this it will be found that two equal sound-power levels add up to a level 3 dB higher, e.g., 60 + 60 dB = 63 dB. In general the following will apply:

$$L_1 = 10 \log I_1/I_0$$

where

L_1 = sound level at a certain pressure p_1

I_1 = sound intensity at a certain pressure p_1

I_0 = sound intensity at pressure p_0 of threshold

for n sound sources of equal intensity

$$L_T = L_1 + 10 \log \frac{nI_1}{I_0} + 10 \log n = L_1 + 10 \log n$$

thus for two sources (equal) $n = 2$ and $L = L_1 + 10 \times 0{\cdot}3 = L_1 + 3$. This addition is furthermore only admissible if the two sound sources are close to each other. For widely spaced n machines the mean sound power level is approximately:

$$L \approx L_1 + 5 \log n \text{ in dB}$$

The addition laws make it clear that if two machines have an almost equal noise level there is little point in trying to reduce the noise level of only one of them.

Where it is necessary to average n sound power levels this must either be done with the aid of tables or by using the following equation to convert to power levels, the n sound pressure level readings.

$$L_p = (L_p) \text{ av.} = 10 \log \frac{1}{n}\left(\text{antilog} \frac{L_{pl}}{10} + \cdots + a \log \frac{L_{pn}}{10}\right)$$

The most important aspect of any statement on noise is that the values given must be related to clearly defined test methods and conditions. Thus if the British Standard 4196:1967 entitled 'Guide to the selection of methods of measuring noise emitted by machinery' or ISO R 1680 or any other national specification has been used this must be stated and appropriate references given where alternatives exist.

Before any noise suppression methods can be considered it is essential not only to know the magnitude of the sound power but its frequency spectrum.

The foregoing has attempted to give an overview of noise measurement and interpretation and is now followed by a designer's view of the problems and their solution.

20.6 Induction Motor Noise

In the past the development of the standard electric motor has taken little account of noise produced. In motors of up to approximately 150 kW (200 b.h.p.), lowest cost per unit output has been the principal design criterion and the present output/dimensions relationship for squirrel cage induction motors laid down in BS 3979 is based on the smallest practical size for a given output, based on the most commonly used insulation classes (classes E and B: winding temperature rises of 75°C and 80°C respectively, measured by resistance). To accommodate the prescribed output within these sizes on large high-speed motors, and to keep temperature rise within permitted limits, imposes ventilation arrangements which make a motor fairly noisy if a simple form of construction is used.

The cage motor which is used for the majority of industrial drives because of its comparative simplicity (hence inherently low cost and reliability) has two basic forms of ventilation. On the drip-proof motor air is normally drawn through the motor by an internal fan so that the outside air directly cools the windings. On the totally enclosed fan cooled (TEFC) motor an external centrifugal fan mounted on the shaft extension at the opposite end to the drive shaft is covered by a cowl which directs cooling air over the external surface of a ribbed frame. By their nature drip-proof motors are not such a problem as TEFC motors of a similar size. As the TEFC motor is the most popular type of motor due to its suitability for use in dusty or outdoor conditions with appropriate treatment, more attention has been paid to reducing noise of this type of motor.

Using a 50 Hz supply the synchronous speeds for two, four, and six pole motors are 3000, 1500, and 1000 rev/min. A standard motor must be capable of either direction of rotation which restricts the fan design to straight radial blades. It can be imagined that a radial blade fan rotating at 3000 rev/min and large enough to cool the surface of a large induction motor is a major source of noise due to its tip speed. Noise problems progressively decrease with four- and six-pole motors of a given output but note that large four-pole motors are still comparatively noisy. There are three principal sources of noise: electromagnetically produced noise, bearing noise, and fan noise. The former is a function of the electromagnetic design, and the latter two sources are a function of speed. If we took a typical standard two-pole TEFC motor of 75 kW (100 h.p.) output we might expect a sound pressure level of approximately 95 dBA at a distance of 1 metre from the surface. This would be principally fan noise and to the human ear the other sources of noise would not be heard. Removing the fan would reduce the noise level by at least 20 dBA and bearing noise would then be the predominant source of noise. Whether electromagnetic noise is significant would be found by switching off the motor supply while listening to or taking noise measurements on the motor (a large motor unloaded takes a long time to slow down). By these techniques the relative importance of the various noise sources can be estimated and this will depend on the size and speed of the motor.

Various national standards give values of maximum noise levels for motors of specific outputs and speeds. Unfortunately compliance with these standards does not necessarily mean that a motor has an acceptable noise level in an industrial environment as these standards are usually written in consultation with motor manufacturers. Major motor users including the petrochemical industries have noise standards which are far more rigorous to satisfy the codes of practice which apply to industrial sites to protect the hearing of workers and to minimize disturbance where industrial sites are adjacent to residential areas.

The Code of Practice for Reducing the Exposure of Employed Persons to Noise (published HMSO 1972) states that a person should not be exposed constantly to a sound pressure level exceeding 90 dBA. The Health and Safety at Work Act 1974 implies that the employer has a statutory duty to comply with this code. It is considered that exposure to sound pressure levels above 90 dBA over a considerable period of time may cause impairment to hearing.

It is worth explaining a little about noise levels. Noise can be specified as a sound power level in decibels which is a measure of the total noise power radiated from the source and is independent of distance from the source. Alternatively and perhaps more usefully noise can be specified as a sound pressure level in decibels at a specific distance from the source. The decibel scale is a logarithmic scale relative to threshold values of power and sound pressure level and an increase of 3 dB represents a doubling of the intensity. Sound pressure levels may be specified in dBA and the 'A' means that relative sound pressure levels of the component frequencies which make up the sound have been modified to match the sensitivity of the human ear to different frequencies. Sound pressure levels may also be specified as an octave band analysis in dB so that the actual character of the noise (high and low frequencies) is specified. If we assume the motor to be a point source of noise with no reflecting surfaces in the vicinity then the noise intensity is inversely proportional to the square of the distance from the source. This means that under these ideal conditions the sound pressure level decreases by 6 dB each time the distance from the point source is doubled. If we put a second identical motor close to the first the total sound pressure level would increase by approximately 3 dB.

Note that in practice the sound pressure level at a given distance from a motor depends on the character of reflecting surfaces such as walls and floor and so the motor manufacturer publishes noise figures measured under specific conditions. The user must understand that these figures may be modified by conditions on site and the way the motor may be fixed to reverberant structures.

Workers' hearing can be protected by:

(a) providing ear defenders to people who must work in dangerous noise conditions;
(b) keeping workers at a suitable distance from the noise source;
(c) using machinery and motors silenced to safe limits;
(d) whenever possible using low-speed motors.

Motors below 11 kW are not normally a problem. On TEFC motors up to 150 kW it has been found that attention to fan noise is all that is required to bring the motors within safe limits. Using a carefully designed unidirectional fan can make some reduction in noise level. However, if the motors are to have the facility of rotation in either direction it is necessary to reduce the diameter of the fan or provide an acoustic hood over the fan or both. The acoustic hood can replace the normal fan cowl and would usually be fabricated in sheet steel with a lining of at least 25 mm thick sound absorbing material on the inside. A plate facing the fan inlet would have a similar sound absorbing lining. This type of construction can reduce noise level by 10 to 13 dB. These techniques will increase motor temperature rise and so if a user requires a standard frame size for the output it may be necessary to use a higher temperature class of insulation (class F). If the user insists on the normal insulation class it may be necessary to go to a larger motor. It must be emphasized that reducing motor noise levels cannot be done cheaply; keeping within acceptable noise limits will increase the cost of industrial plant.

Sometimes the user prefers to enclose the motor and the driven equipment in an acoustic enclosure. If this is done sufficient attention must be paid to the cooling air flow of the motor. A standard motor is suitable for use in an ambient temperature of 40°C maximum and the air temperature in the vicinity of the motor must be checked to see that this is not exceeded under any condition. If necessary the manufacturer can supply a motor for use in a higher ambient temperature.

Index